Operator Theory: Advances and
Applications
Vol. 124

Editor:
I. Gohberg

Recent Advances in Operator Theory

The Israel Gohberg Anniversary Volume

International Workshop in Groningen, June 1998

A. Dijksma
M.A. Kaashoek
A.C.M. Ran
Editors

Springer Basel AG

Editors:

A. Dijksma
Department of Mathematics
University of Groningen
P.O. Box 800
9700 AV Groningen
The Netherlands
e-mail: A.Dijksma@math.rug.nl

A.C.M. Ran
Division of Mathematics and Computer Science
Faculty of Sciences
Vrije Universiteit
De Boelelaan 1081a
1081 HV Amsterdam
The Netherlands
e-mail: ran@cs.vu.nl

M.A. Kaashoek
Division of Mathematics and Computer Science
Faculty of Sciences
Vrije Universiteit
De Boelelaan 1081a
1081 HV Amsterdam
The Netherlands
e-mail: kaash@cs.vu.nl

2000 Mathematics Subject Classification 47-06; 46, 45, 30, 34, 35, 93

A CIP catalogue record for this book is available from the
Library of Congress, Washington D.C., USA

Deutsche Bibliothek Cataloging-in-Publication Data
Recent advances in operator theory : the Israel Gohberg anniversary volume ;
international workshop in Groningen, June 1998 / A. Dijksma ... ed.. - Basel ; Boston ;
Berlin : Birkhäuser, 2001
  (Operator theory ; Vol. 124)
  ISBN 978-3-0348-9516-3   ISBN 978-3-0348-8323-8 (eBook)
  DOI 10.1007/978-3-0348-8323-8

© 2001 Springer Basel AG
Originally published by Birkhäuser Verlag, Basel - Boston - Berlin in 2001
Softcover reprint of the hardcover 1st edition 2001

Printed on acid-free paper produced from chlorine-free pulp. TCF ∞
Cover design: Heinz Hiltbrunner, Basel

ISBN 978-3-0348-9516-3

9 8 7 6 5 4 3 2 1

www.birkhasuer-science.com

Israel Gohberg

Contents

Picture of Israel Gohberg v

Editorial preface xi

Speeches given at the conference dinner in celebration
of Israel Gohberg's 70-th anniversary xiii

Curriculum vitae of Israel Gohberg xxiii

Rien Kaashoek
A review of the mathematical work of Israel Gohberg xxvii

List of publications of Israel Gohberg xxxiii

Vadim Adamyan, Reinhard Mennicken and Vjacheslav Pivovarchik
On the spectral theory of degenerate quadratic operator pencils 1

Daniel Alpay, Vladimir Bolotnikov and Leiba Rodman
**Two-sided tangential interpolation for Hilbert-Schmidt operator
functions on polydisks** 21

D. Alpay, A. Dijksma and Y. Peretz
**Nonstationary analogs of the Herglotz representation theorem:
realizations centered at an arbitrary point** 63

Yury Arlinskii
Characteristic functions of maximal sectorial operators 89

T.Ya. Azizov, A.I. Barsukov, A. Dijksma and P. Jonas
**Similarity between Krein space bicontractions
and Hilbert space contractions** 109

Hari Bercovici and Thomas Smotzer
Classification of cyclic invariant subspaces of Jordan operators 131

Paul Binding and Rostyslav Hryniv
**Elliptic spectral problems of higher order with eigenparameter
dependent boundary conditions** 145

V.B. Bogevolnov, A.B. Mikhallova, B.S. Pavlov and A.M. Yafyasov
About scattering on the ring 155

Ramon Bruzual and Marisela Domínguez
Dilation of generalized Toeplitz kernels on lexicographic $\Gamma \times R$ 175

Harry Dym
On Riccati equations and reproducing kernel spaces 189

Torsten Ehrhardt
**A status report on the asymptotic behavior of Toeplitz
determinants with Fisher–Hartwig singularities** 217

Volker Hardt and Reinhard Mennicken
**On the spectrum of unbounded off-diagonal 2×2 operator
matrices in Banach spaces** 243

J. Janas and S. Naboko
**Multithreshold spectral phase transitions for a class
of Jacobi matrices** 267

M.A. Kaashoek and H.J. Woerdeman
Positive extensions and diagonally connected patterns 287

Eizaburo Kamei
Parametrized Furuta inequality and its applications 307

I. Karelin L. Lerer and A.C.M. Ran
J-symmetric factorizations and algebraic Riccati equations 319

A.G. Kostyuchenko and A.A. Shkalikov
Scattering of waves by periodic gratings and factorization problems .. 361

Heinz Langer, Alexander Markus and Christiane Tretter
Corner of numerical ranges 385

M.M. Malamud
**On some classes of extensions of sectorial operators
and dual pairs of contractions** 401

S.A.M. Marcantognini and M.D. Morán
Liftings of intertwining operators 451

Alexander Markus and Vladimir Matsaev
**Some estimates for the resolvent and for the lengths
of Jordan chains of an analytic operator function** 473

David Ogle, Nicholas Young
The Parrott problem for singular values **481**

B. de Pagter, F.A. Sukochev and H. Witvliet
Unconditional decompositions and Schur-type multipliers **505**

Vyacheslav Pivovarchik
Scattering in a loop-shaped waveguide **527**

Steffen Roch
Pseudospectra of operator polynomials **545**

Editorial Preface

This volume contains a selection of papers in modern operator theory and its applications. Most of them are directly related to lectures presented at the International Workshop on Operator Theory and its Applications held at the University of Groningen (IWOTA-98) in Groningen, the Netherlands, from June 30-July 3, 1998. The workshop was attended by 97 mathematicians–of which 12 were PhD or postdoctoral students–from 19 countries. The program consisted of 19 plenary lectures of 40 minutes and 72 lectures of 30 minutes in 4 parallell sessions.

The present volume reflects the wide range and rich variety of topics presented and discussed at the workshop. The papers deal with operator polynomials and analytic operator functions, with spectral problems of (partial) differential operators and related operator matrices, with interpolation, completion and extension problems, with commutant lifting and dilation, with Ricatti equations and realization problems, with scattering theory, with problems from harmonic analysis, and with topics in the theory of reproducing kernel spaces and of spaces with an indefinite metric. All papers underwent the usual refereeing process.

This volume is dedicated to Israel Gohberg, one of the initiators of the series of IWOTA workshops and president for life of its steering committee, on the occasion of his 70-th birthday . The workshop at Groningen, which is the tenth in the series, turned out to be a particularly appropriate occasion for a pre-celebration. The texts of the speeches delivered at the workshop diner, honoring Israel Gohberg and reflecting on his outstanding achievements and wonderful personality, are included in this volume. We have also added to this volume his CV and complete list of publications (listing 21 books and 401 articles), and a short description of his mathematical work.

The Editors.

Speeches Given at the Conference Dinner in Celebration of Israel Gohberg's 70-th Birthday

At the conference dinner a number of speeches were given, as a pre-celebration of Israel Gohberg's 70-th birthday later that year. The texts here below have been prepared by Harm Bart, Harry Dym, Bill Helton, Heinz Langer, Alek Markus, Cora Sadosky, and Hugo Woerdeman. The order of the texts corresponds to the way the speeches were presented at the dinner.

1 To You, Izea!, by Alek Markus

I have known Israel Gohberg most of my life – since 1951. And for most of my life I have been proud of being his first disciple. I don't mean the best, but the first I was.

In 1959 we began to work together, for almost 15 years he was my boss and a very good boss, I have to say. Unfortunately I was not a very good subordinate, as I understand now. Whenever he asked me to do something, I immediately asked him if it was worth doing at all. He never lost patience with me and very soon I realized that he was almost always right and I was nearly always wrong. He realized that too, because he always told me: "You should never argue with me, because I am old and wise." He did seem old to me then – he was already 31 in 1959. But about 10 years before the time described, Israel was not quite so old and, I suspect, not quite so wise.

To illustrate the last point I would like to tell you a story. There were two characters in the story – one Israel Gohberg, the name of the other one will come up later. The events in the story are connected with elections to the Soviet Parliament. Those who lived in the former Soviet Union know that the Soviet Parliament was nothing like a parliament and the elections were nothing like elections. There was always one candidate for every vacant position and the result of the election was always 99.99% "for".

On the morning of some election day Israel came to his polling station and took a ballot paper with the name of the only candidate ... Joseph Stalin! In accordance with his previous firm decision Israel went into the booth, which itself was a very unusual and risky action, and crossed out the hated name. He then carefully folded the ballot paper and put the paper into the ballot box. Next morning he opened the newspaper. He was eager to know how many people were as brave as him. What he read was: "The result of the voting for Joseph Vissarionovich Stalin was one hundred percent." To make sure that the figures were not rounded up the newspaper said: "All the voters as one person voted comrade Stalin their deputy."

It seems to me it was the last time Israel acted naively. But he also acted bravely and that was not the last time in his life. Many years later Israel was one of the first to decide to leave the Soviet Union. He needed all his courage for this step. And this step proved to be very wise.

To you, Izea! To your courage! To your wisdom! To your health!

2 Dear Israel, by Hugo Woerdeman

First of all I would like to wish you a happy 70th birthday. One thing that has always amazed me, and still does, is your endless energy. Traveling all year round to do mathematics in all corners of the world. As I said ten years ago: you seem so young! At the time I was the youngest participant at the conference in Calgary honoring your 60th birthday, and while I am no longer the youngest participant you are still as young as ever. During your travels you inspire many mathematicians, including myself, to work on challenging problems. You build bridges between different projects and make connections with other disciplines, most notably engineering. I have profited from your energy and am still profiting from it. Under the guidance of both Rien and you I have been provided with an excellent foundation on which I am still building. Sure, you were demanding at times, though you would probably argue you were not demanding enough! You have taught me to pursue problems that are versatile and applicable. Further, I admire your true artistry in explaining hard problems and complicated notions in a way that makes them seem easy and transparent without compromising the mathematics. Israel, I am grateful for the education and inspiration you have given me and look forward to future interaction with you, as I know that you will remain active for a long time to come.

Thank you!

3 Dear Israel, by Heinz Langer

While Hugo Woerdeman, who spoke before me, is one of the youngest professors among your students here, I belong to those who have known you for a very long time. I do not remember if it was our first meeting, but in December 1961, you were at the age of 33 and I was just 26, you gave a lecture at Krein's seminar at the Odessa Civil Engineering Institute. I am still keeping my notes of this lecture. It was about your abstract factorization theorem in rings. You considered two elements such that one is a one-sided inverse of the other, and then you proved the existence of a factorization for certain functions on the circle. After your lecture, as usually, Krein made some remarks and said "You made something out of nothing", meaning that you have built up a beautiful theory starting from only very few assumptions.

I think, in some sense, these words of Mark Grigor'evitsch are also true for your activities in general. Wherever you have worked, you started something new, you created a school and influenced people very much by your knowledge and enthusiasm, under whatsoever conditions. First in Kishinev, then in Tel Aviv and other places in Israel, in Amsterdam, Maryland, Calgary etc. You started to issue new journals, first one in Kishinev, and now the internationally leading journal "Integral Equations and Operator Theory", which is *your* journal, and you started a new series of monographs and books in which until now more than 100 volumes have appeared. Your activities and your mathematics, which evolved in almost 400 articles and about 20 books, had a tremendous influence on the development of operator theory in the whole world. And last but not least you also initiated the IWOTA conferences, of which we enjoy right now already the 10th very much. And almost at the age of 70, your enthusiasm hardly seems to diminish. For all this we are very grateful to you!

Let me also mention something more personal. Some people consider family like relations between mathematicians, the teacher is the father (or mother) and so on. So did Krein sometimes, and he proved that Euler was our grandgrand ... grandfather. From this point of view we are brothers. Of course, you are the older brother, or the big brother, whom I always really admired and who was always watching me. I do not mean this in the ironic or dark sense of Orwell. I mean that you always took great personal interest in me and in general in other people, in your colleagues and friends. In particular, and this is also one of my personal experiences with you, after you went to the West, you tried to support those who were left in the East in all possible ways, including really practical help. Maybe it is hard to imagine for somebody who never lived in the East what it meant to have such a trustworthy friend.

So, in conclusion, I would like to thank you very much for everything you did for me personally and for us as mathematicians and friends. I wish you, Bella and your family all the best! I hope you will be inspiring and watching us for many years to come!

4 Wishing Professor Gohberg a Happy 70th Birthday and Many More to Come!, by Cora Sadosky

We are together to honor our friend Israel Gohberg, who has taught us so much, not only through his mathematics but through his way of being. Those who spoke before already mentioned some of the special qualities of our honoree: great mathematical talent, a formidable capacity for collaboration, enduring generosity and loyalty towards friends, extraordinary organizational skills as well as an enormous energy to put them into practice. I want to add a few words on a most remarkable trait of Israel Gohberg: his inclusiveness.

The conjunction of all of Gohberg's talents have yielded many realizations. Among them, the writing of a collection of seminal books on operator theory, the editorial leadership of a main international journal and of an outstanding series of books, the gathering of groups around the world to work in interdisciplinary problems, and the organization of many international meetings, quite especially, the IWOTAs.

The IWOTAs, as well as **his** journal, Integral Equations and Operator Theory, bear Israel's trademark: inclusiveness. These meetings running every two years in conjunction with the MTNS meetings of engineers and mathematicians are a prime example of the Gohberg vision. A place to discuss mathematics with people coming from different strands of operator theory, in the broadest sense, from applied linear algebra, functional and harmonic analysis, complex function theory, differential equations, mathematical physics, and their applications to systems theory and H^∞ control. At IWOTA we learned what was going on far beyond our sometimes initially narrow range of specialization, and we had the opportunity to interact with senior mathematicians as if they were our peers. Beyond that, Israel always succeeds in bringing together all sorts of people: not just from different mathematical backgrounds, different schools, and different levels of maturity, but from many countries, including those in the peripheries. And to do so he had to work hard, since enacting inclusiveness requires resources and cooperation.

Thanks to Israel's vocation for inclusiveness, the IWOTA meetings are a true intellectual melting pot. He helps mathematicians from all over to become acquainted with the tumultuous, exhilarating world of operator theory as he sees it: inclusive. It was Israel who made this international community of mathematical interests possible, and this feat is the result of his clear and deep appreciation of the universe of mathematics, that is, as himself, open, unrestricted, non-sectarian, and multifacetic.

Thank you, Israel, and mazel tov on your young seventy years!

5 Rehovot Days Redux, by Harry Dym

Israel the mathematician, arrived in Israel the State towards the end of July 1974. Shortly thereafter he was offered a position in the Department of Pure Mathematics (as it was then called) of the Weizmann Institute of Science, as well as at a number of Israeli Universities. Israel accepted a full time position at Tel Aviv University. The Institute then offered him a half time position in the hope that he would eventually leave Tel Aviv and come to the Weizmann on a full time basis. He never did. However, he continued on a half time basis at the Institute for almost ten years. This period came to an end when the Institute went through a financial crisis and cut back on all part time positions.

Israel's presence at the Institute was like a breath of fresh air for me and my students and the enlightened few who were interested in operator theory and its applications. Even before Israel arrived I was an ardent admirer of the two wonderful Gohberg-Krein books. The first one was a marvelous source of information for some problems on trace formulas connected to a generalization of the Szegö formula that I was working on. The second was useful for a problem I had worked on with Naftali Kravitsky (Z"L), my first doctoral student. The problem was to recover a string from its principal spectral function when that spectral function is a small perturbation of the principal spectral function of a known string. Gohberg-Krein factorization along a chain of orthoprojectors enters in an essential way in the resolution of this problem. It was mind boggling to have one of the principal architects of this theory show up in the flesh and on top of it to be so friendly and personable.

As it happened, Naftali was also an immigrant from the Soviet Union. He arrived in Israel in 1971 and, although he had already written a dissertation for the degree of "Kandidat" under the supervision of V.B. Lidskii at Moscow University in 1968, he was advised by a well meaning official who did not understand the Soviet system to get a "real" PhD. Presumably, the unfamiliar term "Kandidat" contributed to the confusion. Naftali was a wonderful student and a wonderful lecturer. He went on afterwards to make important contributions to the theory of commuting nonselfadjoint operators. Unfortunately, he passed away prematurely in August 1998 and never obtained a full measure of the recognition that he so richly deserved.

During Israel's Rehovot period he would generally come to the Institute twice a week: a full day on Sunday and a longish afternoon later in the week. On Sundays Israel would usually lecture from 11am to 1pm. Then, after lunch, we would work together for a couple of hours, take a tea break and then continue on until 6:30 or so. The pattern on the second afternoon was more or less the same. During breaks between work our conversation would drift into other topics. I was always keen to hear about M.G. Krein and, on a different tac, about how one coped with life under communism. Humor helped. Two of my favorite jokes stem from this period:

A high government official comes to a communal farm to see how the poultry division is being run. He approaches one of the local workers and asks him what he feeds his chickens. "Wheat," was the reply. "What, you feed your chickens wheat when we do not have enough wheat to bake bread for the citizens of our great country. Off to Siberia!" The official then approached a second worker, Ivan, with the same question. "Chocolate," Ivan replied, "I feed my chickens chocolate." "What, you feed your chickens chocolate when we do not have sweets to give to our children. Off to Siberia!" The pattern continued this way until finally the official asked a little old Jewish man, Yaakov, the same question. "Feed them," replied Yaakov. "I don't feed them! I give them a ruble and tell them to buy what they want."

A second story from that period concerns a Bakery that is selling bread for two rubles a loaf. A customer purchases a loaf and, after he has it safely under his arm, says to the baker. "Tell me, why do you charge two rubles a loaf when the bakery down the street only charges one ruble?" "Ah," says the baker, "thats because they don't have any bread. If I didn't have any bread, then I would also charge only one ruble."

As our friendship grew, Israel shared some of his plans with me and in fact incorporated me into them. Thus, I became involved with the journal Integral Equations and Operator Theory and a conference organizer for the second (of what was to become) IWOTA (International Workshop on Operator Theory and its Applications). It is hard to be a passive bystander when Israel is involved. His enthusiasm and optimism is infectious. I do, however, remember a luncheon meeting at Tel Aviv University with Bill Helton, Tom Kailath, Rien Kaashoek (and possibly others) in which we managed to dissuade him from starting a second journal, dedicated to engineering applications of operator theory. I believe that although initially disappointed at our lack of enthusiasm for this project, Israel was ultimately grateful not to have this extra responsibility to contend with.

One of the dominant aspects of Israel's personality is optimism. It manifests itself in a number of different ways: enthusiasm, vitality, generosity of spirit. I never heard Israel deprecate other mathematicians.

By and large Mathematics is not a particularly sociable activity. There are collaborations and a limited contacts with the few others who are working on the same set of problems. This changed when Israel arrived on the scene. In short order he established a network of collaborators and colleagues that extended over large parts of Europe, the United States and Canada, with branch offices in Amsterdam, Calgary, College Park, Rehovot, Tel Aviv, to mention a few of the longer lasting ones. In addition to the Journal, he began the OT book series (which at last count had more than 107 volumes), initiated periodic workshops (which evolved into the IWOTA), started the biannual Toeplitz Lecture Series at Tel Aviv University and was a principal organizer of a number of Oberwolfach meetings.

The Oberwolfach meeting on Operator Theory in the Fall of 1976 was akin to a family reunion of distant relatives who had been heard of but not seen (at least by many of us). It was Israel's first visit to Oberwolfach and also his first visit to a Michelin three star restaurant in nearby Strasbourg. The participants that I remember included Harm Bart, Kevin Clancey, Lewis Coburn, Chandler Davis, Ron Douglas, Paul Fuhrmann, Bernhard Gramsch, Bill Helton, Rien Kaashoek, Leonid Lerer, Erhard Meister, Reinhard Mennicken, Joel Pincus, plus, of course, many others.

The first of the Toeplitz Lecture series featured Peter Lax and Ciprian Foias as the main speakers. There was a party in their honor at the Gohberg home in Raanana catered by three generations of Gohberg women: Israel's mother Clara, his wife Bella, his sister Fanny, and his two daughters Zvia and Yanina. Bella and Fanny, both of whom were practicing physicians, took three days off from work to help cook and bake. The apartment was wall to wall people and food. It was an amazing event, especially considering the fact that Israel was a relatively recent immigrant from the Soviet Union (that was none trivial in those days; you can read about it in OT40) and that Ciprian had even more recently made a dramatic escape from Romania via Finland. (In fact, Peter also had an "interesting" history, having made the passage from Hungary to New York in December 1941.)

I would like to close with two stories about Israel that I reported on earlier in the first of the two volumes dedicated to Israel's sixtieth birthday and will undoubtedly wish to repeat for his eightieth because, though short in length, they are rich in content.

When Israel first met my youngest son Michael, he asked him "How many children are there in your class?" "Forty one," was the reply. "Wonderful," said Israel, "so many friends."

One Sunday morning Israel came to the Institute much later than usual. It turned out that he had had a traffic accident enroute. Another car had hit his car from the rear. As accidents go, it was relatively minor, but still the expense and inconvenience was far from negligible. After settling in, he called his wife to tell her what happened. "But why did he do this to you?" was her immediate wifely response. "Belochka," said Israel, ever so gently, "this question you must put to him, not to me."

6 Wishing Israel Gohberg a Very Happy 70th Birthday, by Bill Helton

Rien has been asking everyone to be quiet, so the speakers can be heard. This is not actually necessary while I am speaking, because I have brought a large collection of slides which I did not get time to finish at my talk this afternoon, and if the noise becomes great I can just show them to you.

Toasting Israel Gohberg is a great pleasure. At a personal level because of his warmth and friendship. At a scientific level because of the many wonderful things he has done which effected me.

I first met Professor Gohberg at a conference sponsored by Sz-Nagy in Hungary in 1970. Several of us here met for the first time at that conference. I had just gotten out of graduate school and this was the first conference where I was invited to give a talk. I was very very nervous and it must have showed, because after the talk the famous Professor Gohberg patted me on the back and gestured that the talk had gone well. This made me feel much better, but soon I came to realize that Professor Gohberg's reassurance was firmly based on fact that he did not speak any english. Throughout the conference though we did communicate some in german and Israel spent most of his time trying to phrase jokes in german so simple that I could understand them. At the time Israel worked in pure mathematics and he had done many important things with a number of collaborators. It was a great thrill as a beginning mathematician to meet him.

When Israel came to the west he took up engineering mathematics with great energy and had a remarkable impact on the subject. At that time there were just a few of us in the West who worked on the boundary between operator theory and linear systems theory. On one side was Paul Fuhrmann and I, and on the other side (always) is Patrick De Wilde, who is sitting over there in the corner. Soon the landscape changed. Israel and collaborators like Rien trained a whole generation of very strong young mathematicians who extended and greatly united engineering systems theory. The operator side of the field became lively and stimulating with many people to talk to. In less than a decade Israel built on his strong base in pure mathematics to become proficient in applied mathematics[1].

Israel has created much more than many theorems as his personal achievement. He has created a whole culture. Please let us toast a remarkable person, with important achievements, whose great warmth makes him a friend who is cherished by us all.

7 Total Quality Management, by Harm Bart

Ladies and gentlemen, dear friends, it is a pleasure to say a few words in anticipation of Israel Gohberg's seventieth birthday later this summer. In 1988 Israel's sixtieth birthday was celebrated with an impressive conference in Calgary. In the ten years that since then have passed, my professional life changed considerably. The change is – to use a metaphor – that from a mathematical butterfly living in a world of scientific beauty to an administrative caterpillar running a faculty in financial distress.

Why to bring this up in a short speech that should focus on Israel Gohberg and not on myself? There is a simple reason, lying in the fact that the deanship that I hold is not in a Mathematics Department. It is in a School of Economics, including a large section of Business Economics, a Dutch variant of what Americans call Business Administration. This gives new insights, some of which can be used to enliven this dinner.

One colleague from Business Administration recently told me his view on the difference between the Mathematical Sciences and the Behavioral Sciences, including Business Administration. Mathematics, he said, is like mining. You dig coal or precious stones and bring them to the surface. There they can lie an indeterminate time and still keep their value and usefulness. Business Administration, on the other hand, is like fishing. What you catch is valuable also, but you should not let it lie for too long. Otherwise it will spoil and become a nuisance.

Our friend Israel Gohberg certainly is a miner. He brought up diamonds where others would only expect coal. Precious stones that gladdened the hearts of many mathematicians and will continue to do so in the future.

There is a part of Business Administration where one studies developments and trends. Sometimes a certain development is not so much a trend but a fad. Recently, in my faculty a thesis was produced with the title "Beyond Fads". Such an expression certainly applies to Israel Gohberg. He is a trendsetter, leading his followers into mines full of precious stones. In this, he is usually gentle. Like Einstein, who wrote in 1943 to junior high school student Barbara Wilson: "Do

---

[1] He also became proficient in English.

not worry about your difficulties in mathematics; I can assure you that mine are still greater".

But Israel can also be demanding! This brings me to the third issue from my present background: "Total Quality Management", often abbreviated as TQM. Indeed, Israel practiced TQM long before the notion was invented in modern management theories. Those that have worked with him know his relentless drive for perfection. They will appreciate the following small anecdote.

A singer gives a concert. After performing a beautiful aria, the audience applauds and shouts: "again, again". She is very pleased and repeats the aria. After she finishes, the same thing happens. Elated she sings the piece for the third time, thinking that this will be enough, and that she can look back at a very satisfactory performance. But once more, the audience shouts: "again, again". This makes her uneasy and hesitatingly she asks: "Again?" Then a heavy voice booms out: "Yes! Again, again and again! Until you do it right".

Back to real life. For me it was a privilege to work with Israel. I am grateful for the many things he taught me, especially for his lessons in "mathematical tast". But also for the warm personal contacts in which our families were involved too. Where of course- I also wish to mention emphatically Israel's wife, Bella. She played an important role in all of this!

Thank you very much.

Curriculum Vitae of Israel Gohberg

Date and place
  of birth: August 23, 1928, Tarutino, USSR
Marital status: Married, two children

Education:

> Two years studies in the State Pedagogical Institute,
> Frunze, Soviet Republic of Kirgizia, USSR,
> Date of award: July 1948
>
> M.Sc. Mathematics
> Kishinev University, Kishinev,
> Moldavian Soviet Republic, USSR
> Date of award: July 1951
>
> Kandidat of Sciences, Ph.D. Mathematics
> Leningrad Pedagogical Institute, USSR
> Date of award: April 1954
>
> Doctor of Sciences, Mathematics
> Moscow State University, USSR
> Date of award: February 1964

Academic and professional experience:

1951–1953   Assistant Professor, Soroki Teacher's Institute,
                Moldavian Soviet Republic, USSR

1953–1959   Assistant Professor, Associate Professor,
                Chairman of the Department of Mathematics,
                Beltsky Pedagogical Institute,
                Moldavian Soviet Republic, USSR

1959–1974   Senior Researcher and Head of the Department of
                Functional Analysis,
                Institute of Mathematics of the Academy of
                Science, Moldavian Soviet Republic, USSR

1966–1973	Professor at Kishinev University, USSR (part-time)
1974–1996	Professor, Tel Aviv University, Israel
1996–	Professor Emeritus, Tel Aviv University, Israel
1975–1983	Professor, Weizmann Institute of Science, Rehovot, Israel (part-time)
1981–1998	Incumbent of the Nathan and Lily Silver Chair in Mathematical Analysis and Operator Theory, Tel Aviv University
1983–1998	Professor, Vrije Universiteit, Amsterdam, The Netherlands (part-time)
1974–2000	Visiting and adjunct professor for various extended periods at State University of New York at Stony Brook, New York, U.S.A. University of Calgary, Alberta, Canada University of Georgia, Athens, GA, U.S.A. University of Maryland, College Park, MD. U.S.A. Vrije Universiteit, Amsterdam, The Netherlands. Virginia Polytech Institute and State University, Blacksburg, Virginia, U.S.A. University of Connecticut, Storrs, CT, USA University of Regensburg, Regensburg, Germany University of Mainz, Mainz, Germany Institut fur Angewandte Analysis und Stochastik, Berlin

Academic awards:

1970	Elected corresponding member of the Academy of Sciences of MSSR, USSR
1974	Removed from the list of members of the Academy of Sciences of Moldova SSR
1996	Reinstated as a corresponding member of the Academy of Sciences of Moldova
1985	Elected Foreign Member of the Royal Netherlands Academy of Arts and Sciences

1976	Awarded Landau Prize in Mathematics (Israel)
1986	Awarded Rothschild Prize in Mathematics (Rothschild Foundation, Israel)
1992	Awarded A. Humboldt Research Prize (Humboldt Foundation, Germany)
1994	Awarded Hans Schneider Prize in Linear Algebra (International Linear Algebra Society)
1997	Awarded Honorary Doctoral Degree, Technische Hochschule Darmstadt (Technische Hochsule, Darmstadt, Germany)

Editorial work:

1978–   Founder and editor of the international journal
"Integral Equations and Operator Theory"
published by Birkhauser Verlag, Basel

Founder and editor of the book series
"Operator Theory: Advances and Applications"
published by Birkhauser Verlag, Basel

Member of editorial board of the journals:
"Asymptotic Analysis" (North-Holland);
"Applied Mathematics Letters" (Pergamon Press).
"Calcolo" (Springer Verlag)

Supervised more than 40 doctoral students

Israel Gohberg
January, 2000
Tel-aviv University
Raymond and Beverly Sackler
Faculty of Exact Sciences
School of Mathematical Sciences
Department of Pure Mathematics

A Review of the Mathematical Work of Israel Gohberg

M.A. Kaashoek[1]

A short review is given of the mathematical work of Israel Gohberg, on the occasion of his seventieth birthday.[2]

The mathematical work of Israel Gohberg is extensive and influential. He is author or co-author of more than 400 publications and of more than 20 books. Except for two attractive little books on elementary geometry and combinatorial problems, which he wrote together with Boltyansky (1965, 1971), his contributions belong to the fields of Analysis, Operator Theory, Linear Algebra, and Numerical Linear Algebra. He is a leading authority in the following research areas:

- Integral equations.
- The theory of nonselfadjoint operators.
- Spectral theory and factorization of matrix and operator functions.
- Inversion problems for structured matrices.

His papers and books, which mainly aim at the development of mathematics rather than its applications, are widely known and often quoted outside mathematical circles, in particular by (astro)physicists and engineers (from electrical engineering, control theory and system theory).

Gohberg's first major research topic was the theory of Fredholm operators and its different generalizations. He was among the first to start a systematic investigation of these operators. His results in this area were published in 1951 in "Doklady" while he was still a student. The main achievements were general theorems about Fredholm operators (perturbation, index and representations), a theorem about the spectra of analytic Fredholm operator valued functions, and the necessary and sufficient condition for a singular integral operator with continuous coefficients to be Fredholm. These results were in his master thesis. His Ph.D. thesis (which was ready some time in 1952 and defended in 1954) was an extension of his master thesis, and dealt with unbounded Fredholm operators, finite and infinite Jordan

---

[1] Less formal, preliminary versions of this review have been used by the author and his colleagues earlier for different purposes.

[2] For biographical data see also the sections "Mathematical Tales" and "Gohberg Miscellanea" in: *The Gohberg Anniversary Collection, I.* OT 40, Birkhäuser Verlag, Basel, 1989.

chains for Fredholm operators, and Fredholm properties of Toeplitz matrices and Wiener-Hopf operators. In his thesis, and this is one of its most important features, he made an effective connection between the theory of commutative algebras and the theory of singular integral operators by showing that, in today's terminology, the symbol of the operator is the Gelfand transform in an appropriate Calkin algebra. This connection marks the beginning of a new direction in the modern theory of operator algebras. All these results are of interest even today. But at that time, they were truly pioneer achievements.

In 1950 he met Professor M.G. Krein who played an important role in his life and his mathematical career. Their collaboration led to many important developments in integral equations, notably the fundamental treatises in "Uspehi Mat. Nauk" on Fredholm theory (1957) and on systems of Wiener-Hopf equations (1958). The latter contained, among other things, the solution of the factorization problem for matrix functions in Wiener algebras. Both articles have been translated into English and are now classical; they are often quoted in pure and applied literature, up to the present day. Next came the joint work on nonselfadjoint operators which culminated in two world famous books: "Introduction to the theory of linear non-selfadjoint operators" (1965) and "Theory and applications of Volterra operators" (1967). Some of the original and fundamental contributions were: the factorization theory for operators (which is now used in many different areas, such as system theory, probability theory, integral equations, etc.), the theory of singular values for bounded operators (which turned out to be of great importance in interpolation theory of infinite Hankel matrices), the foundation of the theory of operator ideals, completeness of eigenvectors and generalized eigenvectors including its applications to boundary value problems for differential equations, and the theory of characteristic operator functions for operators close to unitary (which provides one of the main tools for the classification of operators up to unitary equivalence).

In 1959, in Kishinev, Gohberg started to build his own group, with A. Markus as one of his first Ph.D. students. He attracted many strong doctoral candidates from various republics of the USSR and even from the DDR. Officially or informally they became his students and later co-workers. In this period Gohberg did fundamental work on the factorization of operator functions which was subsequently (and jointly with J. Leiterer) expanded into a complete theory, that now has important applications to astrophysics and linear transport theory. He also found a complete description of those algebras of matrix-valued functions of which the elements admit canonical factorization (the so-called decomposing algebras), and he developed a new approach to the theory of Wiener-Hopf operators by using functions of maximal symmetric operators. These results, which were published in an Uspehi paper, formed the contents of his second doctoral thesis (defended in 1964). In the late 60's and the beginning of the 70's he returned to singular integral operators, now for the case when the coefficients have jump discontinuities. Jointly with N. Krupnik, he discovered the appropriate description of the symbol of such operators, that surprisingly enough turned out to be in terms of matrix functions even for scalar equations. In this period he also developed numerical methods to solve Wiener-Hopf equations on the half line or on a finite interval. The outcome was the book with I. Feldman which was translated into both English and German. The famous Gohberg-Semencul and Gohberg-Heinig formulas for inversion of finite Toeplitz matrices and their continuous analogues, which are often used in the electrical engineering literature, were obtained during this time. He also returned to his earlier work on analytic operator functions, and developed together with E.I. Sigal a new approach to deal with meromorphic Fredholm operator valued functions. During his Kishinev period Gohberg was the main force behind the journal Matematicheskie Issledovaniya.

A new period in Gohberg's life began in 1974 with his departure from the Soviet Union, when he started a number of major new research projects. This happened first of all in Israel at Tel-Aviv University (where he headed a group in analysis and operator theory) and at the Weizmann Institute of Science (where he held a half time position from 1975 to 1983 and did joint work with H. Dym), and somewhat later also in Amsterdam (collaboration with M.A. Kaashoek), in Calgary (collaboration with P. Lancaster), and College Park (collaboration with S. Goldberg, R. Ellis and D.C. Lay). Typical for this new period (and truly remarkable) is a fruitful interaction with control and electrical engineers, among others via the biennial international conferences on Mathematical Theory of Networks and Systems (MTNS). Inspired by these connections he developed new directions in operator theory, which resulted in new constructive operator theory methods with substantial mathematical contributions to control theory. Three topics are involved.

The first is the state space method. Since the end of the seventies Gohberg and his co-workers have developed systematically a new system theory oriented method to treat problems in analysis and operator theory. This method has its roots in the mathematical system theory of the sixties and is related to the Kalman approach to control problems. It is based on the discovery that analytic functions which are in a natural way related to operators, such as the determinant function of a Fredholm integral operator, the symbol of a singular integral operator or the characteristic functions for operators close to a unitary or a self-adjoint operator, may be viewed as transfer functions of input-output systems, and hence such a function may be analyzed in terms of three or four operators (the coefficients of the corresponding system) which are often much simpler to deal with than the original operator the function came from. For example, for the rational case the symbol of a system of Wiener-Hopf integral equations can be represented in the form

$$W(\lambda) = I + C(\lambda - A)^{-1}B,$$

where A, B, and C are finite matrices, and in this way problems about the original Wiener-Hopf operator can be treated by using linear algebra and matrix theory. This approach has proved to be quite powerful and is nowadays referred to as the *state space method*. A first main result, which he obtained together with Bart, Kaashoek and Van Dooren, was a geometrical principle of factorization for matrix and operator functions which, for example, for the rational case reduces the problem of canonical Wiener-Hopf factorization to a simple matching of finite dimensional spectral subspaces. The state space method proved also to be very effective in the analysis of the zero and pole structure of matrix polynomials (which extended his earlier work with Sigal) and rational matrix functions, and associate inverse problems. In this way in a series of articles Gohberg and his associates developed a new method for solving explicitly the rational matrix-valued versions of the classical interpolation problems of Lagrange-Sylvester, Nevanlinna-Pick, Carathéodory-Toeplitz, Schur-Takagi, boundary Nevanlinna-Pick, and Nehari, and also the more recent bitangential and Nudelman problems. As a by-product, but not less significant, modern control problems, like the sensitivity minimization, model reduction, and robust stabilization, were solved explicitly with the solutions being described in state space form. Gohberg's book with Bart and Kaashoek on "Minimal factorization of matrix and operator functions" (1979), his book on "Matrix polynomials" (1982) with Lancaster and Rodman, and his book on "Interpolation of rational matrix functions" (1990) with Ball and Rodman, which all three employ the state space method systematically, are highlights of this development. The state space method continues to be of great importance, and is now also used in other branches of analysis (for instance, to derive Szegö-Kac-Achiezer limit formulas or to solve direct and inverse problems for canonical differential systems with rational spectral functions).

Gohberg's interaction with control and electrical engineers, also led to new completion and extension problems for partially given matrices or operators. In the eighties he developed, first with Dym (1982/1983) and later with Kaashoek and Woerdeman (1988–1992) and Ellis and Lay (1993), the so-called band method, an abstract scheme which allows one to deal with positive and contractive completion problems from one point of view, and which presents a natural strategy to solve such problems by reduction to linear equations. It led to beautiful and easy to handle explicit formulas for the solutions of various completion and interpolation problems. For example, by using this scheme Gohberg and his co-workers solved the operator-valued versions of the Carathéodory problem, the Nehari problem and its four block generalization, and also the time-variant (nonstationary) analogues of these problems. The connection in the abstract scheme between the central completion and maximum entropy solution is one of the noteworthy aspects of this development; it turned out to be of interest in the commutant lifting setting too. In parallel, with Kaashoek and Van Schagen (1980–1996), he developed the invariants of various partially given matrices and operators, with remarkable applications to eigenvalue completion problems, to stabilization problems in mathematical system theory and to problems of Wiener-Hopf factorization. It led to a new book which appeared in 1995. In another book, number 100 in the OT-series, written jointly with Foias, Frazho and Kaashoek, the state space method and the commutant lifting approach are combined to solve H^∞ control and metric constrained interpolation problems for operator valued functions with interpolation at operator points. This book contains the three chains completion theorem, a nonstationary version of the commutant lifting theorem, which turns out be an effective tool in dealing with time-varying interpolation problems and continues to inspire further research.

A third major topic, related to the previous two, concerns his original contributions to numerical analysis and numerical linear algebra. They include parallel algorithms (developed jointly with Koltracht and Lancaster 1987/1992) for semi-separable integral operators (which are viewed as input-output operators of systems), and finite section and projection methods for convolution operators. For structured matrices and for their block versions Gohberg developed (1987–1995) fast algorithms, first in joint work with Koltracht, and later also with Kailath and Olshevsky. His work in this area, which has been tested on numerical experiments, is first class and of interest to engineers. Here also his recent work with Y. Eidelman should be mentioned.

In the last twenty years Gohberg has made significant contributions to the theory of spaces with an indefinite metric (his 1983 book "Matrices and indefinite scalar products" written jointly with Lancaster and Rodman, is one of the first linear algebra books in which the theory of matrices in indefinite metric spaces is developed systematically), and to the theory of orthogonal polynomials in a number of articles with Alpay (1988), Ellis and Lay (1988/1992–1995), and Lerer (1988). His more recent work includes the development of a time-variant analogue of inertia theorems, of the theory of orthogonal polynomials (both jointly with Ben-Artzi), and of the state space method (jointly with Ben-Artzi and Kaashoek); this orginal work may very well lead to a deep and far reaching nonstationary analogue of the theory of analytic functions on the disc. In the recent years Gohberg also developed further the theory of singular integral equations and wrote together with Krupnik two excellent monographs on the one-dimensional theory.

During these years Gohberg returned a number of times to his earlier research interests, with new ideas and original contributions. His work with Krupnik on singular integral equations has already been mentioned. The book on "Factorization of matrix functions and singular integral operators" written jointly with K. Clancey (1981) and his recent book with S. Goldberg and N. Krupnik on traces and determinants are other examples.

In addition to all this, Gohberg managed

1. to write three textbooks "Basic operator theory" (1981, jointly with S. Goldberg) and "Classes of Linear Operators I and II" (1990/1993, jointly with S. Gohberg and M.A. Kaashoek),

2. to educate and direct more than 40 Ph.D. students (the number includes the students from the USSR period),

3. to found and edit the influential journal "Integral Equations and Operator Theory",

4. to found and edit the series monographs "Operator Theory: Advances and Applications", in which more than 115 volumes have appeared,

5. to create (jointly with J.W. Helton) and direct the IWOTA meetings, a series of international workshops on operator theory and its applications, and

6. to organize the biennial Toeplitz Conferences at Tel-Aviv.

Gohberg's mathematical influence is profound and far reaching. His style and fine mathematical taste has attracted and inspired his colleagues and fellow mathematicians. He is a true leader of a mathematical school.

M.A. Kaashoek
Division of Mathematics and Computer Science
Faculty of Sciences, Vrije Universiteit
De Boelelaan 1081a, 1081 HV Amsterdam
The Netherlands
e-mail: kaash@few.vu.nl

List of Publications of Israel Gohberg

Books

[1]   I. Gohberg, M. Krein, Vvedenie v Teoriju Linejnyh Nesamosopriazhennyh Opera-
        torov v Gilbertovom Prostranstve. Nauka, Moscow, 448 pages (Russian), 1965.

[1a]   I. Gohberg, M. Krein, Introduction to the theory of linear nonselfadjoint operators.
        American Mathematical Society, Providence, 378 pages (translated from Russian)
        1969; second printing 1978, third printing 1983, fourth printing 1988.

[1b]   I. Gohberg, M. Krein, Introduction a la Theorie des Operateurs Lineaires Non
        Autoadjoints dans un Espace Hilbertien. Dunod, Paris, 372 pages (French, translated
        from Russian), 1971.

[2]   I. Gohberg, M. Krein, Teorija Volterovhy Operatorov v Gilbertovom Prostranstve i
        Ejo Prilozhenia. Nauka, Moscow, 508 pages (Russian), 1967.

[2a]   I. Gohberg, M. Krein, Theory and Applications of Volterra Operators in Hilbert
        Space. American Mathematical Society, Providence, 430 pages (Translated from
        Russian), 1970.

[3]   I. Feldman, I. Gohberg, Proektionnye Metody Reshenia Uravnenij Wiener-Hopfa.
        Akademija Nauk MSSR, Kishinev, 164 pages (Russian), 1967.

[4]   I. Feldman, I. Gohberg, Uravnenija v Sveortkah i Proektionnye Metody ih Reshenia.
        Nauka, Moscow, 352 pages (Russian), 1971.

[4a]   I. Feldman, I. Gohberg, Faltungsgleichungen und Projektionsverfahren zu Ihrer
        Losung. Akademie-Verlag, Berlin, 276 pages (German, translated from Russian),
        1974.

[4b]   I. Feldman, I. Gohberg, Faltungsgleichungen und Projektionsvesfahren zu ihrer
        Losung. Mathematische Reihe 49. Birkhauser Verlag, Basel, 275 pages (Translated
        from Russian), 1974.

[4c]   I. Feldman, I. Gohberg, Convolution Equations and Projection Methods for their
        Solution. American Mathematical Society, Providence, 262 pages (Translated from
        Russian), 1974.

[5]   I. Gohberg, N. Krupnik, Vvedenije v Teoriju Odnomernyh Singuliarnyh Integralnyh
        Operatorov. Shtiinta, Kishinev, 428 pages (Russian), 1973.

[5a]   I. Gohberg, N. Krupnik, Einfuhrung in die Theorie der Eindimensionalen Singularen
        Integraloperatoren. Birkhauser Verlag, Basel, 379 pages (German, translated from
        Russian), 1979.

[6]   V.G. Boltyanskii, I. Gohberg, Teoremy i Zadachi Kombinatornoj Geometrii. Nauka,
        Moscow, 108 pages (Russian), 1965.

[6a]   V.G. Boltyanskii, I. Gohberg, Satze und Probleme der Kombinatorishcen Geometrie.
        Deutsche Verlag der Wissenschaftern, Berlin, 128 pages (German, translated from
        Russian), 1972.

[6b]   V.G. Boltyanskii, I. Gohberg, Tetelek es Faladtok A Kombinatorikus Geometriabol.
        Tankoyvkiado, Budapest, 112 pages (Hungarian, translated from Russian), 1970.

[6c] V.G. Boltyanskii, I. Gohberg, Results and Problems in Combinatorial Geometry. Cambridge University Press, 108 pages (Translated from Russian), 1985.

[7] V.G. Boltyanskii, I. Gohberg, Razbienie Figur na Menshie Chasti. Nauka, Moscow, 88 pages (Russian), 1971.

[7a] V.G. Boltyanskii, I. Gohberg, Division de Figuras en Partes Menores. Mir, Moscow, 106 pages, (Spanish, translated from Russian), 1973.

[7b] V.G. Boltyanskii, I. Gohberg, The Decomposition of Figures into Smaller Parts, 75 pages, University of Chicago Press, 75 pages (Translated from Russian), 1980.

[7c] V.G. Boltyanskii, I. Gohberg, Alakzatok Felbontasa Kisebb Reszekre. Tankonyvkiad, Budapest, 93 pages, (Hungarian, translated from Russian) 1976.

[8] H. Bart, I. Gohberg, M.A. Kaashoek, Minimal Factorization of Matrix and Operator Functions. Operator Theory: Advances and Applications, vol. 1. Birkhauser Verlag, Basel, 236 pages, 1979.

[9] I. Gohberg, S. Goldberg, Basic Operator Theory Birkhauser Verlag, Basel, 285 pages, 1981.

[10] K. Clancey, I. Gohberg, Factorization of Matrix Functions and Singular Integral Operators. Operator Theory: Advances and Applications, vol. 3. Birkhauser Verlag, Basel, 234 pages, 1981.

[11] I. Gohberg, P. Lancaster, L. Rodman, Matrix Polynomials. Academic Press, 409 pages, 1982.

[12] I. Gohberg, P. Lancaster, L. Rodman, Matrices and Indefinite Scalar Products. Operator Theory: Advances and Applications, vol. 8. Birkhauser Verlag, Basel, 374 pages, 1983.

[13] I. Gohberg, P. Lancaster, L. Rodman, Invariant Subspaces of Matrices with Applications. Canadian Math. Soc. Series of Monographs and Advanced Texts, John Wiley & Sons, 629 pages, 1986.

[14] J.A. Ball, I. Gohberg, L. Rodman, Interpolation of Rational Matrix Functions. Operator Theory: Advances and Applications, vol. 45, Birkhauser Verlag, Basel, 605 pages, 1990.

[15] I. Gohberg, S. Goldberg, M.A. Kaashoek, Classes of Linear Operators vol. 1. Operator Theory: Advances and Applications, vol. 49. Birkhauser Verlag, Basel, 468 pages, 1990.

[16] I. Gohberg, N. Krupnik, One-Dimensional Linear Singular Integral Equations vol. 1. Introduction. Operator Theory: Advances and Applications, vol. 53. Birkhauser Verlag, Basel, 263 pages, 1992.

[17] I. Gohberg, N. Krupnik, One-Dimensional Linear Singular Integral Equations vol. 2. General Theory. Operator Theory: Advances and Applications, vol. 54, Birkhauser Verlag, Basel, 232 pages, 1992.

[18] I. Gohberg, S. Goldberg, M.A. Kaashoek, Classes of Linear Operators. 2. Operator Theory: Advances and Applications, vol. 63, Birkhauser Verlag, Basel, 562 pages, 1993.

[19] I. Gohberg, M.A. Kaashoek, F. van Schagen, Partially Specified Matrices and Operators: Classification, Completion, Applications. Operator Theory: Advances and Applications, vol. 79, Birkhauser Verlag, Basel, 333 pages, 1995.

[20] C. Foias, A.E. Frazho, I. Gohberg, M.A. Kaashoek, Metric Constrained Interpolation, Commutant Lifting and Systems. Operator Theory: Advances and Applications, vol. 100, Birkhauser Verlag, Basel, 587 pages, 1998.

[21]   I. Gohberg, S. Goldberg, N. Krupnik, Traces and Determinants of Linear Operators. Operator Theory: Advances and Applications, vol. 116, Birkhauser Verlag, Basel, 270 pages, 2000.

Articles

[1]   I. Gohberg, On linear equations in Hilbert space. Dokl. Acad. Nauk SSSR, 76, no. 4, 9–12 (Russian) (1951), MR 13, 46 (1952).

[2]   2 I. Gohberg, On linear equations in normed spaces. Dokl. Akad. Nauk SSSR, 76, no. 4, 447–480 (Russian) (1951). MR 13, 46 (1952).

[3]   I. Gohberg, On Linear operators depending analytically upon a parameter. Dokl. Akad. Nauk SSSR, 78, no. 4, 629–632 (Russian) (1951); MR 13, 46 (1952).

[4]   V.A. Andrunakievich, I. Gohberg, On linear equations in infinite-dimensional spaces. Uch. zap. Kishinev. Univ., vol. V, 63–67 (Russian), (1952).

[5]   I. Gohberg, On an application of the theory of normed rings to singular integral equations. Uspehi Mat. Nauk 7, 149–156 (Russian) (1952); MR 14, 54 (1953).

[6]   I. Gohberg, On the index of an unbounded operator. Mat. Sb 33(75), I, 193–198 (Russian) (1953). MR 15, 233 (1954).

[7]   I. Gohberg, On systems of singular integral equations. Uc. Zap Kisinevsk. Univ. 11, 55–60 (Russian) (1954); MR 17, 163 (1956); MR 17, 75 (1950).

[8]   I. Gohberg, On zeros and zero elements of unbounded operators. Dokl. Akad. Nauk SSSR 101, 9–12 (Russian) (1955); MR 17, 284 (1956).

[9]   I. Gohberg, Some properties of normally solvable operators. Dokl. Akad. Nauk SSSR 104, 9–11 (Russian) (1955); MR 17, 647 (1956).

[10]   I. Gohberg, A.S. Markus, On a characteristic property of the kernel of a linear operator. Dokl. Akad. Nauk. SSSR 101, 893–896 (Russian) (1955); MR 17, 769 (1956).

[11]   I. Gohberg, The boundaries of applications of the theorems of F. Noether. Uch. zap Kishinev Univ, vol. 17, 35–43 (Russian) (1955).

[12]   I. Gohberg, A.S. Marcus, On stability of certain properties of normally solvable operators. Mat. Sb. 40 (82), 453–466 (Russian) (1956); MR 19, 45 (1958).

[13]   I. Gohberg, M.G. Krein, On the basic propositions of the theory of systems of integral equations on a half-line with kernels depending on the difference of arguments. Proc. III Math. Congr. SSSR, vol. 2, (Russian) 1956.

[14]   I. Gohberg, M.G. Krein, The application of the normed rings theory to the proof of the theorems of solvability of systems of integral equations. Proc. III Math. Cong. SSSR, vol. 2, 1956.

[15]   I. Gohberg, On the index, null elements and elements of the kernel of an unbounded operator. Uspehi Mat. Nauk (N.S) 12, no.1 (73), 177–179 (Russian) (1957); MR 19, 45 (1958); English Transl. Amer. Math. Soc. Transl. 2(16) 391–392, 1960; MR 22 #8374.

[16]   I. Gohberg, L.S. Goldenstein, A.S. Markus, Investigations of some properties of linear bounded operators with connection to their q-norm. Uch. zap. Kishinev Univ. vol. 29, (Russian) 1957.

[17]   I. Gohberg, M.G. Krein, The basic propositions on defect numbers root numbers and indices of linear operators. Uspehi Mat. Nauk 12, no. 2 (74), 43–118 (1957);

English transl. Amer. Math. Soc. Transl. (2) 13, 185–264 (1960); MR 20 #3459; MR 22 # 3984.

[18]   I. Gohberg, M.G. Krein, Systems of integral equations on a half-line with kernels depending on the difference of arguments. Uspehi Mat. Nauk 13, no. 2 (80), 3–72 (1958); English transl. Amer. Math. Soc. Transl. 2 (14), 217–287 (1960); MR 21 #1506; MR 22 #3954.

[19]   I. Gohberg, M.G. Krein, On the stability of a system of partial indices of the Hilbert problem for several unknown functions. Dokl. Akad. Nauk SSSR 119, 854–857 (1958); MR 21 #3547.

[20]   I. Gohberg, On the number of solutions of a homogeneous singular integral equation with continuous coefficients. Dokl. Akad. Nauk SSSR 122, 327–330 (Russian) (1958); MR 20 #4748 (1959).

[21]   I. Gohberg, Two remarks on index of a linear bounded operator. Uch. zap. Belz Pedagog. Inst. no. 1, 13–18 (Russian) (1959).

[22]   I. Gohberg, M.G. Krein, On a dual integral equation and its transpose I. Teoret. Prikl. Mat. no. 1, 58–81, Lvov (Russian) (1958); MR 35 #5877.

[23]   I. Gohberg, On bounds of indexes of matrix-functions. Uspehi Nat. Nauk 14, no. 4 (88), 159–163 (Russian) (1959); English Transl. Operator Theory: Advances and Applications, vol. 7, Birkhauser Verlag, Basel, 243–274 (1983) MR 22 #3993.

[24]   I. Gohberg, A.S. Markus, Two theorems on the gap between subspaces of a Banach space. Uspehi Mat. Nauk 14, no. 5 (89) 135–140 (Russian) (1959); MR 22 #5880.

[25]   I. Gohberg, M.G. Krein, On completely continuous operators with spectrum concentrated at zero. Dokl. Acad. Nauk SSSR, 128, no. 2, 227–230 (Russian) (1959); MR 24 #A1022.

[26]   I. Gohberg, A.S. Markus, Characteristic properties of the pole of a linear closed operator. Uch. zap Belz. Pedagog. Inst. no. 5, 71–75 (Russian) (1960).

[27]   I. Gohberg, A remark of standard factorization of matrix-functions. Uch. zap. Belz. Pedegog Inst. no. 5, 65–69 (Russian) (1960).

[28]   I. Gohberg, A.S. Markus, Characteristic properties of certain points of spectrum of bounded linear operators. Izv. Vyso. Ucel Zaved. Matematik no. 2 (15) 74–87 (Russian) (1960); MR 24 #A1626.

[29]   I. Gohberg, L.S. Goldenstein, On a multidimensional integral equation on a half-space whose kernel is a function of the difference of the arguments, and on a discrete analogue of this equation. Dokl. Acad. Nauk SSSR 131, no. 1, 9–12 (Russian) (1960); Soviet Math. Dokl. 1, 173–176 (1960); MR 22 #8298.

[30]   I. Gohberg, On the theory of multidimensional singular integral equations. Dokl. Acad Nauk SSSR 133, no. 6, 1279–1282 (Russian) (1960); Soviet Math. Dokl. 1, 960–963 (1961); MR 23 #A2015.

[31]   I. Gohberg, Some topics of the theory of multidimensional singular integral equations. Izv. Mold. Akad. Nauk no. 10 (76), 39–50 (Russian) (1960).

[32]   I.A. Feldman, I. Gohberg, A.S. Markus, On normally solvable operators and ideals associated with them. Bul. Akad. Stiince RSS Moldoven. no. 10 (76), 51–70 (1960); English transl., Amer. Math. Soc. Transl. (2) 61, 63–84 (1967); MR 36 #2004.

[33]   I. Gohberg, A.S. Markus, One problem on covering of convex figures by similar figures. Izv. Mold. Acad. Nauk 10 (76), 87–90 (Russian) (1960).

[34]   I. Gohberg, A.S. Markus, Some remarks about topologically equivalent norms. Izv. Mold. fil. Acad. Nauk SSSR 10 (76), 91–95 (Russian) (1960).

[35]   I. Gohberg, M.G. Krein, On the theory of triangular representations of non-selfadjoint operators. Dokl. Acad Nauk SSSR 137, no. 5, 1034–1037 (1961); Soviet Math. Dokl. 2, 392–395 (1961); MR 25 #3370.

[36]   I. Gohberg, M.G. Krein, On Volterra operators with imaginary component in one class or another. Dokl. Acad. Nauk SSSR 139, no. 4, 779–782 (1961); Soviet Math. Dokl. 2, 983–986 (1961); MR 25 #3372.

[37]   I. Gohberg, M.G. Krein, The effect of some transformations of kernels of integral equations upon the equations' spectra. Ukrain. Mat. Z 13, no. 3, 12–28 (1961); English transl., Amer. Math. Soc. Transl. 2(35) 263–295 (1964); MR 27 #1788.

[38]   I. Gohberg, A.S. Markus, On the stability of bases in Banach and Hilbert spaces. Izv. Moldavsk. Fil. Akad. Nauk SSSR, no. 5, 17–35 (Russian) (1962); MR 37 # 1955.

[39]   I. Gohberg, A.S. Markus, On some inequalities between eigenvalues and matrix elements of linear operators. Izv. Moldavsk. Fil. Akad. Nauk SSSR, no. 5, 103–108 (Russian) (1962).

[40]   I. Gohberg, Tests for one-sided invertibility of elements in normed rings and their applications. Dokl. Akad. Nauk SSSR 145, no. 5, 971–974 (1962); Soviet Math. Dokl. 3, 1119–1123 (1962); MR 25 #6147.

[41]   I. Gohberg, A general theorem concerning the factorization of matrices-functions in normed rings, and its applications. Dokl. Akad. Nauk SSSR 146, no. 2, 284–287 (1962); Soviet Math. Dokl. 3, 1281–1284 (1962); MR 25 #4376.

[42]   I. Gohberg, M.G. Krein, On the problem of factorization of operators in Hilbert space. Dokl. Akad. Nauk SSSR 147, no. 2, 279–282 (1962); Soviet Math. Dokl. 3, 1578–1582 (1962); MR 26 #6777.

[43]   I. Gohberg, On factorization of operator-functions Uspehi Mat. Nauk 18, no. 2, 180–182 (Russian) (1963).

[44]   I. Gohberg, On relations between the spectra of the Hermitian components of nilpotent matrices and on the integral of triangular truncation. Bul. Akad. Stiiuce RSS Moldoven, no. 1, 27–37 (Russian) (1963); MR 35 #2168.

[45]   I. Gohberg, On normal resolvability and the index of functions of an operator. Izv. Acad Nauk Mold. SSR, no. 11, 11–24 (Russian) (1963); MR 36 #6965.

[46]   M.S. Brodskii, I. Gohberg, M.G. Krein, V. Matsaev, On some new investigations on the theory of non-self-adjoint operators Proc. IV All-union Math. Congress, vol. 2, 261–271 (Russian) (1964); MR 36 #3153.

[47]   I. Gohberg, A.S. Markus, Some relations between eigenvalues and matrix elements of linear operators. Mat. Sb. 64 (106), no. 4, 481–496 (1964); English transl., Amer. Math. Soc. transl. (2), 52, 201–216 (1966); MR 30 #457.

[48]   I. Gohberg, M.G. Krein, Criteria for completeness of the system of root vectors of a contraction. Ukrain. Mat. Z. 16, no. 1, 78–82 (1964); English transl., Amer. Math Soc. Transl., (2) 54, 119–124 (1966); MR 29 #2651.

[49]   I. Gohberg, M.G. Krein, On factorization of operators in Hilbert space. Acta Sci. Math., Szeged, 25, no. 1–2, 90–123 (1964); English transl., Amer. Math. Soc. Transl. (2) 51, 155–188 (1966); MR 29 #6313.

[50]   I. Gohberg, A factorization problem in normed rings, functions of isometric and symmetric operators and singular integral equations. Uspehi Mat. Nauk 19, no. 1 (115), 71–124; Russian Math. Survey 19, no. 1, 63–144 (1964); MR 29 #487.

[51]   I. Gohberg, The factorization problem for operator functions. Izv. Akad. Nauk SSSR, Ser. Mat. 28, no. 5, 1055–1082 (Russian) (1964) MR 30 #5182.

[52] V.G. Ceban, I. Gohberg, On a reduction method for discrete analogues of equations of Wiener-Hopf type. Ukrain Mat. Z 26, no. 6, 822–829 (1964); English transl., Amer. Math. Soc. transl. (2) 65, 41–49 (1967). MR 30 #2244.

[53] M.S. Budjanu, I. Gohberg, A general theorem about factorization of matrix-functions. Studies in Algebra and Math. Anal Izd. "Kartja Mold".Kishinev, 116–121, 1965. MR 36 #726.

[54] I. Gohberg, M.K. Zambitskii, On normally solvable operators in spaces with two norms. Bull. Akad. Nauk Mold. SSR, no. 6, 80–84 (Russian) (1964). MR 36 #3143.

[55] I.A. Feldman, I. Gohberg, On approximative solutions of some classes of linear equations. Dokl. Akad. Nauk SSSR 160, no. 4, 750–753 (1965); Soviet Math. Dokl. 6, 174–177 (1965). MR 34 #6572.

[56] I. Gohberg, M.G. Krein, On the multiplicative representation of the characteristic functions of operators closed to unitary ones. Dokl. Acad. Nauk SSSR 164, no. 4, 732–735, (1965); Soviet Math. Dokl. 6, 1279–1283, (1965); MR 33 #571.

[57] I.A. Feldman, I. Gohberg, On reduction method for systems of Wiener-Hopf type. Dokl. Akad. Nauk SSSR 165, no. 2, 268–271, (1965); Soviet Math. Dokl. 6, 1433–1436, (1965); MR 32 #8085.

[58] I. Gohberg, M.K. Zambitskii, On the theory of linear operators in spaces with two norms. Ukrain. Mat. Z. 18, no. 1, 11–23, (1966); MR 33 #4676.

[59] I. Gohberg, A generalization of theorems of M.G. Krein of the type of the Wiener-Levi theorems. Mat. Issled. 1, no. 1, 110–130, (Russian) (1966); MR 34 #3366.

[60] M.A. Barkar, I. Gohberg, On factorization of operators relative to a discrete chain of projections in Banach space. Mat. Issled. I, no. 1, 32–54, (1966); English transl., Amer. Math. Soc. Transl. 2 (90), 81–103, (1970); MR 34 #6539.

[61] M.A. Barkar, I. Gohberg, On factorization of operators in a Banach space. Mat. Issl. 1, no. 2, 98–129, (Russian) (1966); MR 35 #780.

[62] I.A. Feldman, I. Gohberg, On truncated Wiener-Hopf equations. Abstracts of Short Scientific Reports, Intern. Congress of Math. (Moscow), Section 5, 44–45, (Russian) (1966).

[63] I. Gohberg, M.G. Krein, On triangular representations of linear operators and on multiplicative representations of their characteristic functions. Dokl. Acad. Nauk SSSR, 175, no. 2, 272–275, (1967); MR 35 #7157.

[64] I.A. Feldman, I. Gohberg, On indices of multiple extensions of matrix functions. Bull. Akad. Nauk Mold. SSR, no. 6, 76–80, (Russian) (1967); MR 37 #4658.

[65] I. Gohberg, M.G. Krein, On a description of contraction operators similar to unitary ones. Funkcional Anal. i Priloz, 1, no. 1, 38–68, (1967); Functional Anal. and Appl., vol. 1, 1, 38–60 (1967); MR 35 #4763.

[66] I. Gohberg, On Toeplitz matrices composed of the Fourier coefficients of piecewise continuous functions. Funkcional Anal. i Priloz., 1, no. 2, 91–92, (1967); Function Anal. Appl. 1, 166–167, (1967); MR 35 #4763.

[67] M.S. Budjanu, I. Gohberg, On factorization problem in abstract Banach algebras I. Splitting algebras. Mat. Issled 2, no. 2, 25–61, (1967); English transl., Amer. Math Soc. Transl. (2); MR #5697.

[68] M.S. Budjanu, I. Gohberg, On factorization problem in abstract Banach algebras II. Irreducible algebras. Mat. Issled 2, no. 3, 3–19, (1967); English transl., Amer. Math. Soc. Transl. (2); MR 37 #5698.

[69] I. Gohberg, O.I. Soibelman, Some remarks on similarity of operators. Mat. Issled. 2, no. 3, 166–170, (Russian) (1967); MR 37 #3387.

[70] M.S. Budjanu, I. Gohberg, On multiplicative operators in Banach algebras. I. General propositions. Mat. Issled 2, no. 4, 14–30, (Russian) (1967); MR 379 #1972. (English transl. Amer. Mat. Soc. Trans. (2), vol. 90, 211–223, 1970.

[71] I. Gohberg, N.Ia. Krupnik, On the norm of the Hilbert transfrom in Lp spaces. Funkcional Anal. i Priloz, 2, no. 2, 91–92, (1968); Functional Anal. Appl. 2, 180–181, (1968).

[72] I. Gohberg, N.Ia. Krupnik, On the spectrum of one-dimensional singular integral operators with piece-wise continuous coefficients. Mat. Issled. 3, no. 1 (7), 16–30, (1968); English transl., Amer. Math. Soc. Transl. (2) 103, 181–193, (1973); MR 41 #2469.

[73] M.S. Budjanu, I. Gohberg, General theorems on the factorization of matrices-functions I. The fundamental theorem. Mat. Issled 3, no. 2 (8), 87–103 (1968); English transl., Amer. Math. Soc. Transl. (2) 102, 1–14 (1973); MR 41 #4246a.

[74] M.S. Budjanu, I. Gohberg, General theorems on the factorization of matrices-functions II. Some tests and their consequences. Mat. Issled 3, no. 3 (9), 3–18 (1968); English transl., Amer. Math. Soc. Transl. (2) 102, 15–26 (1973); MR 41 #4246b.

[75] I.A. Feldman, I. Gohberg, On Wiener-Hopf integral difference equations. Dokl. Akad, Nauk SSSR 183, no. 1, 25–28 (1968); Soviet Math. Dokl. 9, 1312–1316 (1968); MR 44 #3096.

[76] I. Gohberg, N.Ia. Krupnik, On the spectrum of singular integral operators in Lp spaces. Studia Math. 31, 347–362 (Russian) (1968); MR 38 #5068.

[77] N.N. Bogolyubov, I. Gohberg, G.E. Shilov, Mark Grigorevich Krein (on his sixtieth birthday). Uspehi Mat. Nauk, 23, no. 3, 197–214 (1968); Russian Math. Surveys, 23, no. 3, 177–192, (May-June 1968); MR 37 #5077.

[78] I. Gohberg, N.Ia. Krupnik, On the spectrum of singular integral operators in Lp spaces with weight. Dokl. Acad. Nauk. SSSR 185, no. 4, 745–748 (1969); Soviet Math. Dokl. 10, 406–410 (1969); MR 40 #1817.

[79] I. Gohberg, N.Ia. Krupnik, On an algebra generated by Toeplitz matrices. Funkcional Anal. i Priloz, 3, no. 2, 46–59 (1969); Functional Anal. Appl. 3, 119–127 (1969); MR 40 #3323.

[80] I.A. Feldman, I. Gohberg, Integro-difference Wiener-Hopf equations. Acta Sci. Math., Szeged, 30, no. 3–4, 199–224 (Russian) (1969); MR 40 #7880.

[81] I. Gohberg, N.Ia. Krupnik, Systems of singular integral equations in Lp spaces with a weight. Dokl. Akad. Nauk SSSR 186, no. 5, 998–1001 (1969); Soviet Math. Dokl. 10, 688–651 (1969); MR 40 #1818.

[82] I. Gohberg, N.Ia. Krupnik, On quotient norm of singular integral operators. Mat. Issled 4, no. 3, 136–139 (Russian) (1968); MR 41 4306. English transl., Amer. Math. Soc. Transl. (2) vol. III, 117–119 (1978).

[83] I. Gohberg, N.Ia. Krupnik, On the algebra generated by Toeplitz matrices in hp spaces. Mat. Issled 4, no. 3, 54–62 (Russian) (1969).

[84] V.M. Brodskii, I. Gohberg, M.G. Krein, General theorems on trianglular representations of linear operators and multiplicative representations of their characteristic functions. Funk. Anal. i Priloz, 3, no. 4, 1–27 (1969); MR 40 #4794.

[85]   I. Gohberg, N.Ia. Krupnik, On composite linear singular integral equations. Mat. Issled. 4, no. 4, 20–32 (1969); English transl., Amer. Math. Soc. Transl. (2) vol. III, 121–131 (1978); MR 43 #996.

[86]   I. Gohberg, N.Ia. Krupnik, The symbols of one-dimensional singular integral operators on an open contour. Dokl. Akad Nauk SSSR 191, 12–15 (1970); Soviet Math. Dokl. 11, 299–303 (1970); MR 41 #9060.

[87]   V.M. Brodskii, I. Gohberg, M.G. Krein, The definition and basic properties of the characteristic function of a knot. Funkt. Anal. i Proloz, no. 4, 1, 88–90 (1970).

[88]   I. Gohberg, N.Ia. Krupnik, On the algebra generated by the one-dimensional singular integral operators with piecewise continuous coefficients. Funkcional Anal. i Priloz 4, no. 3, 26–36 (1970); Functional Anal. App. 4, 193–201 (1970); MR 42 #5057.

[89]   I. Gohberg, M.G. Krein, New inequalities for the eigenvalues of integral equations with smooth kernels. Mat. Issled 5, no. 1 (15), 22–39 (Russian) (1970); English transl. Operator Theory: Advances and Applications, vol. 7, Birkhauser Verlag, Basel, 275–293 (1983); MR 44 #818.

[90]   I. Gohberg, N.Ia. Krupnik, Singular integral equations with continuous coefficients on a composite contour. Mat. Issled no. 5, 1, 22–39 (Russian) (1970).

[91]   I. Gohberg, N.Ia. Krupnik, On singular integral equations with unbounded coefficients. Mat. Issled 5, no. 3 (17), 46–57 (Russian) (1970); MR 45 #980.

[92]   I. Gohberg, A.A. Sementsul, Toeplitz matrices composed of the Fourier coefficients of functions with discontinuities of almost periodic type. Mat. Issled. 5, no. 4, 63–83 (Russian) (1970); MR 44 #7379.

[93]   I. Gohberg, N.Ia. Krupnik, Banach algebras generated by singular integral operators. Colloquia Math. Soc. Janos Bolyai 5. Hilbert space operators, Tihany (Hungary), 239–267 (Russian) (1970).

[94]   I. Gohberg, E.M. Spigel, A projection method for the solution of singular integral equations. Dokl. Akad Nauk SSSR 196, no. 5, 1002–1005 (1971); Soviet Math. Dokl. 12, 289–293 (1971); MR 43 #3755.

[95]   V.M. Brodskii, I. Gohberg, M.G. Krein, On characteristic functions of an invertible operator. Acta Sci. Math. Szeged, no. 32, 1–2, 141–164 (Russian) (1971).

[96]   I. Gohberg, N.Ia. Krupnik, Singular integral operators on a composite contour. Proc. Georgian Akad Nauk 64, 21–24 (Russian, Georgian and English summaries) (1971); MR 45 #4223.

[97]   I. Gohberg, On some questions of spectral theory of finite-meromorphic operator-functions. Izv. Arm. Akad. Nauk, no. 6, 2–3, 160–181 (Russian) (1971).

[98]   I. Gohberg, The correction to the paper "On some questions of spectral theory of finite-meromorphic operator-functions". Izv. Arm. Acad. Nauk, no. 7, 2, 152 (Russian) (1972).

[99]   I. Gohberg, N.Ia. Krupnik, Singular integral operators with piecewise continuous coefficients and their symbols. Izv. Akad. Nauk SSSR. Ser. Mat. 35, 940–964 (Russian) (1971); MR 45 #581.

[100]   I. Gohberg, V.I. Levchenko, Projection method for the solution of degenerate Wiener-hopf equations. Funkcional Anal. in Priloz., 5, no. 4, 69–70 (Russian) (1971); MR 44 #7317.

[101]   I. Gohberg, E.I. Sigal, An operator generalization of the logarithmic residue theorem and Rouche's theorem. Mat. Sb. (N.S.) 84 (126), 607–629 (1971); English transl. Math. USSR, sb. 13, 603–625 (1971); MR 47 #2409.

[102] I. Gohberg, E.I. Sigal, Global factorization of a meromorphic operator-function and some of its applications. Mat. Issled. 6, no. 1 (19), 63–82 (Russian) (1971); MR 47 #2410.

[103] I. Gohberg, E.I. Sigal, The root multiplicity of the product of meromorphic operator functions. Mat. Issled. 6, no. 2 (20), 30–50, 158 (Russian) (1971); MR 46 #2461.

[104] I. Gohberg, E.M. Spigel, On the projection method of solution of singular integral equations with polynomial coefficients. Mat. Issled. 6, no. 3, 45–61 (Russian) (1971); MR 44 #7380.

[105] I. Gohberg, V.I. Levchenko, On the convergence of a projection method of solution of a degenerated Wiener-Hopf equation. Mat. Issled. 6, no. 4, 20–36 (Russian) (1971); MR 45 # 918.

[106] I. Gohberg, J. Leiterer, The canonical factorization of continuous operator functions with respect to the circle. Funk Anal. i Priloz 6, no. 1, 73–74 (1972); Functional Anal. Appl. 6, 65–66 (1972); MR 45 #2519.

[107] I. Gohberg, J. Leiterer, On factorization of continuous operator functions with respect to a contour in Banach algebras. Dokl. Akad. Nauk SSSR, 206, 273–276 (1972); English transl. Soviet Math. Dokl. 13, 1195–1199 (1972).

[108] I. Gohberg, J. Leiterer, Factorization of operator functions with respect to a contour I. Finitely meromorphic operator functions. Math. Nachrichten 52, 259–282 (Russian) (1972).

[109] I. Gohberg, N.Ia. Krupnik, A formula for the inversion of finite Toeplitz matrices. Mat. Issled 7, no. 2, 272–283 (Russian) (1972).

[110] I. Gohberg, A.A. Sementsul, On the inversion of finite Toeplitz matrices and their continuous analogues. Mat. Issled 7, no. 2, 201–223 (Russian) (1972).

[111] I. Gohberg, V.I. Levchenco, On a projection method for a degenerate Wiener-Hopf equation. Mat. Issled. 7, no. 3, 238–253 (Russian) (1972).

[112] I. Gohberg, J. Leiterer, General theorems on a canonic factorization of operator-functions with respect to a contour. Mat. Issled. no. 7, 3, 87–134 (Russian) (1972).

[113] I. Gohberg, J. Leiterer, On holomorphic vector-functions of one variable. Mat. Issled. no. 7, 4, 60–84 (Russian) (1972).

[114] I. Gohberg, J. Leiterer, The factorization of operator-functions with respect to a contour II. Canonic factorization of operator-functions closed to unit ones. Math. Nachrichten, no. 54, 1–6, 41–74 (Russian) (1972).

[115] I. Gohberg, J. Leiterer, The factorizaion of operator-functions with respect to a contour III Factorization in algebras. Math. Nachrichten, no. 55, 1–6, 33–61 (Russian) (1973).

[116] I. Gohberg, J. Leiterer, On co-cycles, operator-functions and families of subspaces. Mat. Issled. no. 8, 2 (1973).

[117] I. Gohberg, J. Leiterer, On holomorphic functions of one variable II. Functions in a domain. Mat. Issled. no. 8, 1, 37–58 (Russian) (1973).

[118] I. Gohberg, N.Ia. Krupnik, On algebras of singular integral operators with a shift. Mat. Issled no. 8, 2 (Russian) (1973).

[119] I. Gohberg, N.Ia. Krupnik, On one-dimensional singular integral operators with a shift. Izv. Arm. Acad. Nauk, no. 1, 3–12 (Russian) (1973).

[120] I. Gohberg, J. Leiterer, Criterion of the possibility of the fatorization of operator-function with respect to a contour. Dokl. Acad. Nauk SSSR, 209, 3, 529–532 (Russian) (1973).

[121]   I. Gohberg, J. Leiterer, General theorems on the factorization of operator-functions with respect to a closed contour. I. Holomorphic functions. Acta. Sci. Math., Szeged, 30, 103–120 (Russian) (1973).

[122]   I. Gohberg, J. Leiterer, General theorems on the factorization of operator-functions II. Generalizations. Acta. Sci. Math. Szeged (Russian) (1973).

[123]   I. Gohberg, J. Leiterer, On a local principle in the problem of the factorization of operator-functions. Funk. Anal. i Priloz. no. 7, 3 (Russian) (1973).

[124]   I. Gohberg, J. Leiterer, The local principle in the problem of the factorization of continuous operator-functions. Revue Rounainie de Math. Pures et Appl., XIX, 10 (Russian) (1973).

[125]   I. Gohberg, N.Ia. Krupnik, On a symbol of singular integral operators on a composite contour. Proc. Tbilisi Simp. Mech. Sploshnich Sred, Tbilisi, (Russian) (1973).

[126]   I. Gohberg, N.Ia. Krupnik, On the local principle and on algebras generated by Toeplitz's matrices. Annalele stiintifice ale Univ. "Al. I. Cuza", Iasi, section I a) Matematica, XIX, F. I, 43–72 (Russian) (1973).

[127]   I. Gohberg, J. Leiterer, Families of holomorphic subspaces with removable singularities. Math. Nachrichten 61, 157–173 (Russian) (1974).

[128]   I. Gohberg, G. Heinig, On the inversion of finite Toeplitz matrices. Math. Issled. no. 8, 3, 151–155 (Russian) (1973).

[129]   I. Gohberg, G. Heinig, Inversion of finite Toeplitz matrices composed from elements of a non-commutative algebra. Rev. Roum. Math. Pures et Appl. 20, 5, 55–73 (Russian) (1974).

[130]   I. Gohberg, G. Heinig, On matrix-valued integral operators on a finite interval with matrix kernels which depend on the difference of arguments. Re. Roumaine Math. Pures Appl., 20, 1, 55–73 (1975).

[131]   I. Gohberg, G. Heinig, Matrix resultant and its generalizations, I. The resultant operator for matrix valued polynomials. Acta Sci. Math. (Szeged), T. 37, 1–2, 1975, 41–61.

[132]   I. Gohberg, S. Prössdorf, Ein Projektionsverfahren zue Losung entarteter Systeme von diskreten Wiener-Hopf-Gleichungen. Math. Nachr., Band 65, 19–45 (1975).

[133]   I. Gohberg, G. Heinig, Matrix resultant and its generalizations, II. Continual analog of the resultant operator. Acta Math. Acad. Sci. Hungar. T. 28 (3–4), 189–209 (1976).

[134]   I. Gohberg, J. Leiterer, Uber algebren steitiger operator functionen. Studia Math., T. 57, 1–26 (1976).

[135]   I. Gohberg, M.A. Kaashoek, D.C. Lay, Spectral classification of operators and operator functions. Bull. Amer. Math. Soc. 82, 587–589 (1976).

[136]   I. Gohberg, L.E. Lerer, Resultant of matrix polynomials. Bull. Amer. Math. Soc. 82, 4 (1976).

[137]   K. Clancey, I. Gohberg, Local and global factorizations of matrix-valued functions. Trans. Amer. Math. Soc., vol. 232, 155–167 (1977).

[138]   I. Gohberg, M.A. Kaashoek, D.C. Lay, Equivalence, linearization and decomposition of holomorphic operator functions. J. Funct. Anal., vol. 28, no. 1, 102–144 (1978).

[139]   I. Gohberg, P. Lancaster, L. Rodman, Spectral analysis of matrix polynomials, I. Canonical forms and divisors. Linear Algebra and its Appl., 20, 1–44 (1978).

[140]   I. Gohberg, P. Lancaster, L. Rodman, Spectral analysis of matrix polynomials, II. The resolvent form and spectral divisors. Linear Algebra and its Appl. 21, 65–88 (1978).

[141]   I. Gohberg, L.E. Lerer, Resultant operators of a pair of analytic functions. Proceedings Amer. Math. Soc., vol. 72, no. 1, 65–73 (1978).

[142]   I. Gohberg, L.E. Lerer, Singular integral operators as a generalization of the resultant matrix. Applicable Anal., vol. 7, 191–205 (1978).

[143]   I. Gohberg, M.A. Kaashoek, F. van Schagen, Common multiples of operator polynomials with analytic coefficients. Manuscripta Math., 25, 279–314 (1978).

[144]   I. Gohberg, M.A. Kaashoek, L. Rodman, Spectral analysis of families of operator polynomials and a generalized Vandermonde matrix, I. The finite-dimensional case. Topics in Functional Analysis. Adv. in Math., Supplementary Studies, vol. 3, 91–128 (1978).

[145]   I. Gohberg, M.A. Kaashoek, L. Rodman, Spectral analysis of families of operator polynomials and a generalized Vandermonde matrix, II. The infinite dimensional case. J. Funct. Anal., vol. 30, no. 3, 358–389 (1978).

[146]   I. Gohberg, P. Lancaster, L. Rodman, Representations and divisibility of operator polynomials. Canad. J. Math., vol. XXX, no. 5, 1045–1069 (1978).

[147]   I. Gohberg, L. Rodman, On spectral analysis of non-monic matrix and operator polynomials, I. Reduction to monic polynomials. Israel J. Math., vol. 30, Nos. 1–2, 133–151 (1978).

[148]   I. Gohberg, L. Rodman, On spectral analysis of non-monic matrix and operator polynomials, II. Dependence on the finite spectral data. Israel J. Math., vol. 30, no. 4, 321–334 (1978).

[149]   I. Gohberg, L. Lerer, L. Rodman, Factorization indices for matrix polynomials. Bull. Amer. Math. Soc., vol. 84, no. 2, 275–277 (1978).

[150]   I. Gohberg, L. Lerer, L. Rodman, On canonical factorization of operator polynomials, spectral divisors and Toeplitz matrices. Integral Equations Operator Theory, 1, 176–214 (1978).

[151]   H. Bart, I. Gohberg, M.A. Kaashoek, Operator polynomials as inverses of characteristic functions. Integral Equations Operator Theory, 1, 1–18 (1978).

[152]   H. Bart, I. Gohberg, M.A. Kaashoek, Stable factorizations on monic matrix polynomials and stable invariant subspaces. Integral Equations Operator Theory, 1, 496–517 (1978).

[153]   I. Gohberg, S. Levin, Asymptotic properties of Toeplitz matrix factorization. Integral Equations Operator Theory, 1, 518–538 (1978).

[154]   I. Gohberg, M.A. Kaashoek, Unsolved problems in matrix and operator theory, I. Partial multiplicities and additive perturbations. Integral Equations Operator Theory, 1, 278–283 (1978).

[155]   K. Clancey, I. Gohberg, Localization of singular integral operators. Math. Z., 169, 105–117 (1979).

[156]   H. Dym, I. Gohberg, Extensions of matrix valued functions with rational polynomial inverses. Integral Equations Operator Theory, 2, 503–528 (1979).

[157]   I. Gohberg, P. Lancaster, L. Rodman, Perturbation theory for divisors of operator polynomials. SIAM J. Math. Anal., vol. 10, no. 6, 1161–1183 (1979).

[158]   I. Gohberg, L. Rodman, On the spectral structure of monic matrix polynomials and the extension problem. Linear Algebra Appl., 24, 157–172 (1979).

[159]   I. Gohberg, P. Lancaster, L. Rodman, On selfadjoint matrix polynomials. Integral
         Equations Operator Theory, 2, 434–439 (1979).

[160]   I. Gohberg, L. Lerer, Factorization indices and Kronecker indices of matrix poly-
         nomials. Integral Equations Operator Theory, 2, 199–243 (1979).

[161]   I. Gohberg, M.A. Kaashoek, Unsolved problems in matrix and operator theory, II.
         Partial multiplicities for products. Integral Equations Operator Theory, 2, 116–120
         (1979).

[162]   I. Gohberg, S. Levin, On an open problem for block Toeplitz matrices. Integral
         Equations Operator Theory, 2, 121–129 (1979).

[163]   E. Azoff, K. Clancey, I. Gohberg, On the spectra of finite-dimensional perturbations
         of matrix multiplication operators. Manuscripta Math. 30, 351–360 (1980).

[164]   I. Gohberg, L. Lerer, L. Rodman, Stable factorizations of operator polynomials,
         I. Spectral divisors simply behaved at infinity. J. Math. Anal. Appl. 74, 401–431
         (1980).

[165]   I. Gohberg, L. Lerer, L. Rodman, Stable factorizations of operator polynomials, II.
         Main results and applications to Toeplitz operators. J. Math. Anal. Appl. 75, 1–40
         (1980).

[166]   I. Gohberg, M.A. Kaashoek, F. van Schagen, Similarity of operator blocks and
         canonical forms, I. General results, feedback equivalence and Kronecker indices.
         Integral Equations Operator Theory, 3, 350–396 (1980).

[167]   I. Gohberg, P. Lancaster, L. Rodman, Spectral analysis of selfadjoint matrix poly-
         nomials. Annals of Mathematics, 112, 33–71 (1980).

[168]   H. Dym, I. Gohberg, On an extension problem, generalized Fourier analysis, and
         an entropy formula. Integral Equations Operator Theory, 3, 143–215 (1980).

[169]   H. Bart, I. Gohberg, M.A. Kaashoek, P. van Dooren, Factorizations of transfer
         functions. SIAM J. Control and Optimization, 18, no. 6, 675–696 (1980).

[170]   I. Gohberg, M.A. Kaashoek, F. van Schagen, Similarity of operator blocks and
         canonical forms, II. Infinite dimensional case and Wiener-Hopf factorization.
         Operator Theory: Advances and Applications, vol. 2, Birkhauser Verlag, Basel,
         121–170 (1981).

[171]   I. Gohberg, M.A. Kaashoek, L. Lerer, L. Rodman, Common multiples and common
         divisors of matrix polynomials, I. Spectral method. Indiana University Mathe-
         matics Journal, 30 no. 3, 321–356 (1981).

[172]   H. Dym, I. Gohberg, Extensions of band matrices with band inverses. Linear
         Algebra and its Applications, 36, 1–14 (1981).

[173]   I. Gohberg, L. Rodman, Analytic matrix functions with prescribed local data.
         Journal d'Analyse Mathematique, 40, 90–128 (1981).

[174]   H. Bart, I. Gohberg, M.A. Kaashoek, Wiener-Hopf integral equations, Toeplitz
         matrices and linear systems. Toeplitz Centennial, Operator Theory: Advances and
         Applications, vol. 4, Birkhauser Verlag, Basel, 85–135 (1982).

[175]   I. Gohberg, P. Lancaster, L. Rodman, Factorization of selfadjoint matrix polynomi-
         als with constant signature. Linear and Multilinear Algebra, 11, 209–224 (1982).

[176]   E. Azoff, K. Clancey, I. Gohberg, Singular points of families of Fredholm inte-
         gral operators. Toeplitz Centennial, Operator Theory: Advances and Applications,
         vol. 4, Birkhauser Verlag, Basel, 57–65 (1982).

[177]   E. Azoff, K. Clancey, I. Gohberg, On line integrals of rational functions of two com-
         plex variables. Proceedings American Mathematical Society, 88, 229–235 (1982).

[178]   I. Gohberg, L. Lerer, On non-square sections of Wiener-Hopf operators, Integral Equations Operator Theory, 5, 4, 518–532 (1982).

[179]   H. Bart, I. Gohberg, M.A. Kaashoek, Convolution equations and linear systems. Integral Equations and Operator Theory, 5, 3, 283–340 (1982).

[180]   H. Dym, I. Gohberg, Extensions of triangular operators and matrix functions. Indiana University Mathematics Journal, vol. 31, no. 4, 579–606 (1982).

[181]   H. Dym, I. Gohberg, Extensions of matrix valued functions and block matrices. Indiana University Mathematics Journal, vol. 31, no. 5, 733–765 (1982).

[182]   I. Gohberg, M.A. Kaashoek, F. van Schagen, Rational matrix and operator functions with prescribed singularities. Integral Equations Operator Theory, 5, 5, 673–717 (1982).

[183]   I. Gohberg, P. Lancaster, L. Rodman, Perturbations of H-selfadjoint matrices, with applications to differential equations. Integral Equations and Operator Theory, 5, 5, 718–757 (1982).

[184]   I. Gohberg, M.A. Kaashoek, L. Lerer, L. Rodman, Common multiples and common divisors of matrix polynomials, II. Vandermonde and resultant matrices. Linear and Multilinear Algebra, 159–203 (1982).

[185]   I. Gohberg, S. Goldberg, Finite Dimensional Wiener-Hopf equations and factorizations of matrices. Linear Algebra and its Applications, 48, 219–236 (1982).

[186]   I. Gohberg, L. Rodman, Analytic operator valued functions with prescribed local data. Acta Sci. Math., Szeged, 45, 189–199 (1983).

[187]   I. Gohberg, L. Lerer, L. Rodman, Wiener-Hopf factorization of piecewise matrix polynomials. Linear Algebra and its Applications, vol. 52, 53, 315–350 (1983).

[188]   H. Dym, I. Gohberg, On unitary interpolants and Fredholm infinite block Toeplitz matrices. Integral Equations and Operator Theory, 6, 863–878 (1983).

[189]   H. Dym, I. Gohberg, Extensions of kernels of Fredholm operators. Journal d'Analyse Mathematique, 42, 51–97 (1982/83).

[190]   H. Dym, I. Gohberg, Unitary interpolants, factorization indices and infinite Hankel block matrices. Journal of Functional Analysis, vol. 54, no. 3, 229–289 (1983).

[191]   H. Dym, I. Gohberg, Hankel integral operators and isometric interpolants on the line. Journal of Functional Analysis, vol. 54, no. 3, 290–307 (1983).

[192]   I. Gohberg, P. Lancaster, L. Rodman, A sign characteristic for selfadjoint meromorphic matrix functions. Applicable Analysis, 16, 165–185 (1983).

[193]   H. Bart, I. Gohberg, M.A. Kaashoek, The coupling method for solving integral equations. Operator Theory: Advances and Applictions, vol. 12, Birkhauser Verlag, Basel, 39–73 (1984).

[194]   I. Gohberg, M.A. Kaashoek, L. Lerer, L. Rodman, Minimal divisors of rational matrix functions with prescribed zero and pole structure. Operator Theory: Advances and Applications, vol. 12, Birkhauser Verlag, Basel, 241–275 (1984).

[195]   I. Gohberg, M.A. Kaashoek, Time varying linear systems with boundary conditions and integral operators, I. The transfer operator and its properties. Integral Equations and Operator Theory, 7, 325–391 (1984).

[196]   I. Gohberg, S. Goldberg, Extensions of triangular Hilbert-Schmidt operators. Integral Equations and Operator Theory, 7, 743–790 (1984).

[197]   I. Gohberg, M.A. Kaashoek, F van Schagen, Non-compact integral operators with semiseparable kernels and their discrete analogues: Inversion and Fredholm properties. Integral Equations and Operator Theory, 7, 642–703 (1984).

[198] H. Bart, I. Gohberg, M.A. Kaashoek, Wiener-Hopf factorization and realization. Lecture Notes in Control and Information Sciences, Mathematical Theory of Networks and Systems, Springer-Verlag, 58, 42–62 (1984).

[199] I. Gohberg, P. Lancaster, L. Rodman, A sign characteristic for selfadjoint rational matrix functions. Lecture Notes in Control and Information Sciences, Mathematical Theory of Networks and Systems, Springer-Verlag, 58, 263–269 (1984).

[200] J.A. Ball, I. Gohberg, A commutant lifting theorem for triangular matrices with diverse applications. Integral Equations and Operator Theory, 8, 205–267 (1985).

[201] H. Bart, I. Gohberg, M.A. Kaashoek, Fredholm theory of Wiener-Hopf equations in terms of realization of their symbols. Integral Equations and Operator Theory, 8, 590–613 (1985).

[202] I. Gohberg, T. Kailath, I. Koltracht, Linear complexity algorithms for semiseparable matrices. Integral Equations and Operator Theory, 8, 780–804 (1985).

[203] I. Gohberg, P. Lancaster, L. Rodman, Perturbation of analytic hermitian matrix functions. Applicable Analysis, 20, 23–48 (1985).

[204] R.L. Ellis, I. Gohberg, D. Lay, Factorization of Block Matrices. Linear Algebra and its Applications 69, 71–93 (1985).

[205] I. Gohberg, I. Koltracht, Numerical solution of integral equations, fast algorithms and Krein-Sobolev equation. Numer. Math. 47, 237–288 (1985).

[206] J.A. Ball, I. Gohberg, Shift invariant subspaces, factorization, and interpolation for matrices, I. The canonical case. Linear Algebra and its Applications, 1–64 (1985).

[207] H. Dym, I. Gohberg, A maximum entropy principle for contractive interpolants. Journal of Functional Analysis 65, 83–125 (1986).

[208] I. Gohberg, L. Rodman, Interpolation and local data for meromorphic matrix and operator functions. Integral Equations and Operator Theory, 9, 60–94 (1986).

[209] I. Gohberg, M.A. Kaashoek, On minimality and stable minimality of time-varying linear systems with well-posed boundary conditions. Int. J. Control 43, 5, 1401–1411 (1986).

[210] I. Gohberg, L. Rodman, On distance between lattices of invariant subspaces of matrices. Linear Algebra Appl. 76, 85–120 (1986).

[211] I. Gohberg, T. Kailath, I. Koltracht, Efficient solution of linear systems of equations with recursive structure. Linear Algebra and its Applications 80, 81–113 (1986).

[212] H. Bart, I. Gohberg, M.A. Kaashoek, Wiener-Hopf factorization, inverse Fourier transforms and exponentially dichotomous operators. Journal of Functional Analysis 68, no. 1, 1–42 (1986).

[213] I. Gohberg, M.A. Kaashoek, Similarity and reduction of time varying linear systems with well-posed boundary conditions. SIAM J. Control and Optimization 24, no. 5, 961–978 (1986).

[214] I. Gohberg, S. Rubinstein, Stability of minimal fractional decompositions of rational matrix functions. Operator Theory: Advances and Applications, vol. 18, Birkhauser Verlag, Basel, 249–270 (1986).

[215] J.A. Ball, I. Gohberg, Classification of shift invariant subspaces of matrices with Hermitian form and completion of matrices. Operator Theory: Advances and Applications, vol. 19, Birkhauser Verlag, Basel, 23–85 (1986).

[216] R.L. Ellis, I. Gohberg, D.C. Lay, The maximum distance problem in Hilbert space. Operator Theory: Advances and Applications, vol. 19, Birkhauser Verlag, Basel, 195–206 (1986).

[217]   H. Bart, I. Gohberg, M.A. Kaashoek, Wiener-Hopf equations with symbols analytic in a strip. Operator Theory: Advances and Applications, vol. 21, Birkhauser Verlag, Basel, 39–74 (1986).

[218]   I. Gohberg, M.A. Kaashoek, L. Lerer, L. Rodman, On Toeplitz and Wiener-Hopf operators with contourwise rational matrix and operator symbols. Operator Theory: Advances and Applications, vol. 21, Birkhauser Verlag, Basel, 75–127 (1986).

[219]   I. Gohberg, M.A. Kaashoek, Minimal factorization of integral operators and cascade decompositions of systems. Operator Theory: Advances and Applications, vol. 21, Birkhauser Verlag, Basel, 157–230 (1986).

[220]   H. Bart, I. Gohberg, M.A. Kaashoek, Explicit Wiener-Hopf factorization and realization. Operator Theory: Advances and Applications, vol. 21, Birkhauser Verlag, Basel, 235–316 (1986).

[221]   H. Bart, I. Gohberg, M.A. Kaashoek, Invariants for Wiener-Hopf equivalence of analytic operator functions. Operator Theory: Advances and Applications, vol. 21, Birkhauser Verlag, Basel, 317–355 (1986).

[222]   H. Bart, I. Gohberg, M.A. Kaashoek, Multiplication by diagonals and reduction to canonical factorization. Operator Theory: Advances and Applications, vol. 21, Birkhauser Verlag, Basel, 357–372 (1986).

[223]   I. Gohberg, P. Lancaster, L. Rodman, Quadratic matrix polynomials with a parameter. Advances in Applied Mathematics 7, 3, 253–281 (1986).

[224]   R.L. Ellis, I. Gohberg, D.C. Lay, Band extensions, maximum entropy and permanence principle. In Maximum Entropy and Bayesian Methods in Applied Statistics (J. Justice, Ed.), Cambridge University Press (1986).

[225]   I. Gohberg, P. Lancaster, L. Rodman, On Hermitian solutions of the symmetric algebraic Riccati equation. SIAM Journal of Control and Optimization 24, 6, 1323–1334 (1986).

[226]   I. Gohberg, M.A. Kaashoek, L. Lerer, Minimality and irreducibility of time-invariant boundary-value systems. nt. J. Control 44, 2, 363–379 (1986).

[227]   J.A. Ball, I. Gohberg, Pairs of shift invariant subspaces of matrices and nonconical factorization. Linear and Multilinear Algebra 20, 27–61 (1986).

[228]   A. Ben-Artzi, R.L. Ellis, I. Gohberg, D.C. Lay, The maximum distance problem and band sequences. Linear Algebra and its Applications 87, 93–112 (1987).

[229]   J.A. Ball, I. Gohberg, L. Rodman, Minimal factorization of meromorphic matrix functions in terms of local data. Integral Equations and Operator Theory, 10, 3, 309–348 (1987).

[230]   I. Gohberg, M.A. Kaashoek, An inverse spectral problem for rational matrix functions and minimal divisibility. Integral Equations and Operator Theory, 10, 437–465 (1987).

[231]   H. Bart, I. Gohberg, M.A. Kaashoek, The state space method in problems of analysis. Proceedings of the First International Conference on Industrial and Applied Mathematics (ICIAM 87), 1–16 (1987).

[232]   I. Gohberg, S. Rubinstein, Cascade decompositions of rational matrix functions and their stability. Int. J. Control 46 2, 603–629 (1987).

[233]   I. Gohberg, M.A. Kaashoek, L. Lerer, On minimality in the partial realization problem. System & Control Letters 9, 97–104 (1987).

[234]   I. Gohberg, I. Koltracht, P. Lancaster, Second order parallel algorithms for Fredholm integral equations with continuous displacement kernels. Integral Equations and Operator Theory, 10, 577–594 (1987).

[235]   I. Gohberg, T. Kailath, I. Koltracht, P. Lancaster, Linear complexity parallel algorithms for linear systems of equations with recursive structure. Linear Algebra and its Applications 88/89, 271–315 (1987).

[236]   R.L. Ellis, I. Gohberg, D.C. Lay, Invertible selfadjoint extensions of band matrices and their entropy. SIAM J. Alg. Disc. Meth. 8, 3, 483–500 (1987).

[237]   I. Gohberg, S. Goldberg, Semi-separable operators along chains of projections and systems. Journal of Mathematical Analysis and Applications 125, 1, 124–140 (1987).

[238]   I. Gohberg, M.A. Kaashoek, Minimal representations of semiseparable kernels and systems with separable boundary conditions. J. Math. Anal. Appl. 124, 2, 436–458 (1987).

[239]   I. Gohberg, M.A. Kaashoek, F. van Schagen, Szego-Kac-Achiezer formulas in terms of realizations of the symbol. J. Functional Analysis 74, 1, 24–51 (1987).

[240]   I. Gohberg, T. Kailath, I. Koltracht, A note on diagonal innovation matrices. IEEE Transactions on Acoustics, Speech, and Signal Processing, vol. ASSP-35, no. 7, 1068–1069 (1987).

[241]   A. Ben-Artzi, I. Gohberg, Nonstationary Szego theorem, band sequences and maximum entropy. Integral Equations and Operator Theory, 11, 10–27 (1988).

[242]   R.L. Ellis, I. Gohberg, D.C. Lay, On two theorems of M.G. Krein concerning polynomials orthogonal on the unit circle. Integral Equations and Operator Theory, 11, 87–104 (1988).

[243]   I. Gohberg, M.A. Kaashoek, F. van Schagen, Rational contractive and unitary interpolants in realized form. Integral Equations and Operator Theory, 11, 105–127 (1988).

[244]   H. Dym, I. Gohberg, A new class of contractive interpolants and maximum entropy principles. Operator Theory: Advances and Applications, vol. 29, Birkhauser Verlag, Basel, 117–150 (1988).

[245]   I. Gohberg, M.A. Kaashoek, L. Lerer, Nodes and realization of rational matrix functions: Minimality theory and applications. Operator Theory: Advances and Applications, vol. 29, Birkhauser Verlag, Basel, 181–232 (1988).

[246]   A. Ben-Artzi, I. Gohberg, Fredholm properties of band matrices and dichotomy. Operatory Theory: Advances and Applications, vol. 32, Birkhauser Verlag, Basel, 37–52 (1988).

[247]   J.A. Ball, I. Gohberg, L. Rodman, Realization and interpolation of rational matrix functions. Operator Theory: Advances and Applications, vol. 33, Birkhauser Verlag, Basel, 1–72 (1988).

[248]   I. Gohberg, M.A. Kaashoek, A.C.M. Ran, Interpolation problems for rational matrix functions with incomplete data and Wiener-Hopf factorization. Operator Theory: Advances and Applications, vol. 33, Birkhauser Verlag, Basel, 73–108 (1988).

[249]   I. Gohberg, M.A. Kaashoek, Regular rational matrix functions with prescribed pole and zero structure. Operator Theory: Advances and Applications, vol. 33, Birkhauser Verlag, Basel, 109–122 (1988).

[250]   D. Alpay, I. Gohberg, Unitary rational matrix functions. Operator Theory: Advances and Applications, vol. 33, Birkhauser Verlag, Basel, 175–222 (1988).

[251]   I. Gohberg, S. Rubinstein, Proper contractions and their unitary minimal completions. Operator Theory: Advances and Applications, vol. 33, Birkhauser Verlag, Basel, 223–247 (1988).

[252]   A. Ben-Artzi, I. Gohberg, Lower upper factorizations of operators with middle terms. J. Functional Analysis 77, 2, 309–325 (1988).

[253]   I. Gohberg, S. Goldberg, Factorizations of semi-separable operators along continuous chains of projections. Journal of Mathematical Analysis and Applications, 133, 1, 27–43 (1988).

[254]   D. Alpay, I. Gohberg, On orthogonal matrix polynomials. Operator Theory: Advances and Applications, vol. 34, Birkhauser Verlag, Basel, 25–46 (1988).

[255]   A. Ben-Artzi, I. Gohberg, Extension of a theorem of M.G. Krein on orthogonal polynomials for the nonstationary case. Operator Theory: Advances and Applications, vol. 34, Birkhauser Verlag, Basel, 65–78 (1988).

[256]   I. Gohberg, L. Lerer, Matrix generalizations of M.G. Krein theorems on orthogonal polynomials. Operator Theory: Advances and Applications, vol. 34, Birkhauser Verlag, Basel, 137–202 (1988).

[257]   I. Gohberg, M.A. Kaashoek, Block Toeplitz operators with rational symbols. Operator Theory: Advances and Applications, vol. 35, Birkhauser Verlag, Basel, 385–440 (1988).

[258]   R.L. Ellis, I. Gohberg, D.C. Lay, On negative eigenvalues of selfadjoint extensions of band matrices. Linear and Multilinear Algebra, 24, 15–25 (1988).

[259]   I. Gohberg, M.A. Kaashoek, P. Lancaster, General theory of regular matrix polynomials and band Toeplitz operators. Integral Equations and Operator Theory 11, 776–882 (1988).

[260]   I. Gohberg, I. Koltracht, Efficient Algorithm for Toeplitz plus Hankel matrices. Integral Equations and Operator Theory, 12, 136–142 (1989).

[261]   I. Gohberg, M.A. Kaashoek, H.J. Woerdeman, The band method for positive and strictly contractive extension problems: An alternative version and new applications. Integral Equations and Operator Theory, 12, 343–382 (1989).

[262]   I. Gohberg, I. Koltracht, P. Lancaster, On the numerical solution of integral equations with piecewise continuous displacement kernels. Integral Equations and Operator Theory, 12, 511–538 (1989).

[263]   I. Gohberg, T. Shalom, On inversion of square matrices partitioned into non-square blocks. Integral Equations and Operator Theory, 12, 539–566 (1989).

[264]   A. Ben-Artzi, I. Gohberg, Inertia theorems for nonstationary discrete systems and dichotomy. Linear Algebra and its Applications 120, 95–138 (1989).

[265]   I. Gohberg, M.A. Kaashoek, H.J. Woerdeman, The band method for positive and contractive extension problems. Journal of Operator Theory 22, 109–155 (1989).

[266]   J.A. Ball, I. Gohberg, Cascade decompositions of linear systems in terms of realizations. Proceedings of the 28th IEEE Conference on Decision and Control. IEEE Control Systems Society, vol. 1, 2–10 (1989).

[267]   I. Gohberg, Mathematical Tales. Operator Theory: Advances and Applications, vol. 40, 17–56, Birkhauser Verlag, Basel (1989).

[268]   I. Gohberg, I. Koltracht, On the inversion of Cauchy Matrices. Signal Processing, Scattering and Operator Theory, and Numerical Methods. Proceedings of the International Symposium MTNS-89, Amsterdam, vol. III, Birkhauser Boston, 381–392 (1990).

[269]   I. Gohberg, T. Shalom, On Bezoutians of nonsquare matrix polynomials and inversion of matrices with nonsquare blocks. Linear Algebra and its Applications, vol. 137/138, 249–323 (1990).

[270] I. Gohberg, M.A. Kaashoek, A.C.M. Ran, Regular rational matrix functions with prescribed local zero-pole structure. Linear Algebra and its Applications, vol. 137/138, 387–412 (1990).

[271] J.A. Ball, I. Gohberg, L. Rodman, Common minimal multiples and divisors for rational matrix functions. Linear Algebra and its Applications, vol. 137/138, 621–662 (1990).

[272] I. Gohberg, B. Reichstein, On classification of normal matrices in an indefinite scalar product. Integral Equations and Operator Theory, 13, 364–394 (1990).

[273] J.A. Ball, I. Gohberg, L. Rodman, T. Shalom, On the eigenvalues of matrices with given upper triangular part. Integral Equations and Operator Theory, 13, 488–497 (1990).

[274] J.A. Ball, I. Gohberg, L. Rodman, Simultaneous residue interpolation problems for rational matrix functions. Integral Equations and Operator Theory, 13, 611–637 (1990).

[275] J.A. Ball, I. Gohberg, L. Rodman, Tangential interpolation problems for rational matrix functions. AMS Proceedings of Symposia in Applied Mathematics 40, 59–86 (1990).

[276] D. Alpay, J.A. Ball, I. Gohberg, L. Rodman, Realization and factorization for rational matrix functions with symmetries. Operator Theory: Advances and Applications, vol. 47, Birkhauser Verlag, Basel, 1–60 (1990).

[277] A. Ben-Artzi, I. Gohberg, M.A. Kaashoek, Invertibility and dichotomy of singular difference equations. Operator Theory: Advances and Applications, vol. 48, Birkhauser Verlag, Basel, 157–184 (1990).

[278] I. Gohberg, M.A. Kaashoek, H.J. Woerdeman, The band method for extension problems and maximum entropy. IMA Volumes in Mathematics and its Applications, vol. 22, 75–94, Springer-Verlag, New York (1990).

[279] J.A. Ball, I. Gohberg, L. Rodman, Sensitivity minimization and bitangential Nevanlinna-pick interpolation in contour integral form. IMA Volumes in Mathematics and its Applications, vol. 23, 3–36, Springer-Verlag, New York (1990).

[280] J.A. Ball, I.C. Gohberg, L. Rodman, Two-sided Lagrange-Sylvester interpolation problems for rational matrix functions. AMS Proceedings of Symposia in Pure Mathematics, vol. 51, 17–84, American Mathematical Society (1990).

[281] A. Ben-Artzi, I. Gohberg, Nonstationary inertia theorems, dichotomy, and applications. AMS Proceedings of Symposia in Pure Mathematics, vol. 51, 85–96, American Mathematical Society (1990).

[282] J. Ball, I. Gohberg, L. Rodman, Two-sided Nudelman interpolation problem for rational matrix functions. Lecture Notes in Pure and Applied Mathematics. Analysis and Partial Differential Equations, vol. 122, 371–416, Marcel Dekker, Inc., New York (1990).

[283] I. Gohberg, S. Rubinstein, Minimal symplectic orbits of rational contractive matrix functions. Integral Equations and Operator Theory, 13, 795–835 (1990).

[284] I. Gohberg, S. Goldberg, Counting negative eigenvalues of a Hilbert-Schmidt operator via sign changes of a determinant. Integral Equations and Operator Theory, 14, 92–104 (1991).

[285] A. Ben-Artzi, I. Gohberg, Band matrices and dichotomy. Operator Theory: Advances and Applications, vol. 50, Birkhauser Verlag, Basel, 137–170 (1991).

[286]   I. Gohberg, M.A. Kaashoek, A.C.M. Ran, Matrix polynomials with prescribed zero structure in the finite complex plane. Operator Theory: Advances and Applications, 50, 241–266, Birkhauser Verlag, Basel, 241–266 (1991).

[287]   I. Gohberg, M.A. Kaashoek, The Wiener-Hopf method for the transport equation: a finite dimensional version. Operator Theory: Advances and Applications, vol. 51, Birkhauser Verlag, Basel, 20–33 (1991).

[288]   I. Gohberg, M.A. Kaashoek, H.J. Woerdeman, A note on extensions of band matrices with maximal and submaximal invertible blocks. Linear Algebra and its Applications, vol. 150, 157–166 (1991).

[289]   I. Gohberg, S. Rubinstein, A classification of upper equivalent matrices. The generic case. Integral Equations and Operator Theory, 14, 533–544 (1991).

[290]   J.A. Ball, I. Gohberg, L. Rodman, Boundary Nevanlinna-Pick interpolation for rational matrix functions. Journal of Mathematical Systems, Estimation and Control, vol. 1, no. 2, 131–164 (1991).

[291]   A. Ben-Artzi, I. Gohberg, Dichotomy, discrete Bohl exponents, and spectrum of block weighted shifts. Integral Equations and Operator Theory, 14, 613–677 (1991).

[292]   J.A. Ball, I. Gohberg, L. Rodman, The state space method in the study of interpolation of rational matrix functions. Mathematical System Theory, Springer-Verlag, 503–508 (1991).

[293]   I. Gohberg, M.A. Kaashoek, The State space method for solving singular integral equations. Mathematical System Theory, Springer-Verlag, 509–523 (1991).

[294]   J.A. Ball, I. Gohberg, L. Rodman, Nehari interpolation problem for rational matrix functions: The Generic case. Lecture Notes in Mathematics, Springer-Verlag, 277–308 (1991).

[295]   I. Gohberg, M.A. Kaashoek, H.J. Woerdeman, Time variant extension problems of Nehari type and the band method. Lecture Notes in Mathematics, Springer-Verlag, 309–323 (1991).

[296]   I. Gohberg, M.A. Kaashoek, L. Lerer, A directional partial realization problem. Systems & Control Letters, 17, 305–314 (1991).

[297]   I. Gohberg, M.A. Kaashoek, H.J. Woerdeman, The band method for several positive extension problems of non-band type. Journal of Operator Theory, 26, 191–218 (1991).

[298]   I. Gohberg, M.A. Kaashoek, H.J. Woerdeman, A maximum entrophy principle in the general framework of the band method. Journal of Functional Analysis, 95, 2, 231–254 (1991).

[299]   I. Gohberg, B. Reichstein, On H-unitary and block-Toeplitz H-normal operators. Linear and Multilinear Algebra, 30, 17–48 (1991).

[300]   R.L. Ellis, I. Gohberg, Orthogonal systems related to infinite Hankel matrices. Journal of Functional Analysis, vol. 109, no. 1, 155–198 (1992).

[301]   I. Gohberg, I. Koltracht, P. Lancaster, Second order parallel algorithms for piecewise smooth displacement kernels. Integral Equations and Operator Theory, 15, 16–29 (1992).

[302]   I. Gohberg, M.A. Kaashoek, A.C.M. Ran, Factorizations of and extensions to J-unitary rational matrix functions on the unit circle. Integral Equations and Operator Theory, 15, 262–300 (1992).

[303] D. Alpay, J. Ball, I. Gohberg, L. Rodman, State space theory of automorphisms of rational matrix functions. Integral Equations and Operator Theory, 15, 349–377 (1992).

[304] I. Gohberg, V. Olshevsky, Circulants, displacements and decompositions of matrices. Integral Equations and Operator Theory, 15, 730–743 (1992).

[305] I. Gohberg, N. Krupnik, Extension theorems for invertibility symbols in Banach algebras. Integral Equations and Operator Theory, 15, 991–1010 (1992).

[306] J.A. Ball, I. Gohberg, M.A. Kaashoek, Nevanlinna-Pick interpolation for time-varying input-output maps: the discrete case. Operator Theory: Advances and Applications, vol. 56, Birkhauser Verlag, Basel, 1–51 (1992).

[307] J.A. Ball, I. Gohberg, M.A. Kaashoek, Nevanlinna-Pick interpolation for time-varying input-output maps: the continuous time case. Operator Theory: Advances and Applications, vol. 56, Birkhauser Verlag, Basel, 52–89 (1992).

[308] A. Ben-Artzi, I. Gohberg, Dichotomy of systems and invertibility of linear ordinary differential operators. Operator Theory: Advances and Applications, vol. 56, Birkhauser Verlag, Basel, 90–119 (1992).

[309] A. Ben-Artzi, I. Gohberg, Inertia theorems for block weighted shifts and applications. Operator Theory: Advances and Applications, vol. 56, Birkhauser Verlag, Basel, 120–152 (1992).

[310] I. Gohberg, M.A. Kaashoek, L. Lerer, Minimality and realization of discrete time-varying systems. Operator Theory: Advances and Applications, vol. 56, Birkhauser Verlag, Basel, 261–296 (1992).

[311] R.L. Ellis, I. Gohberg, D.C. Lay, Distribution of zeros of matrix-valued continuous analogues of orthogonal polynomials. Operator Theory: Advances and Applications, vol. 58, Birkhauser Verlag, Basel, 26–70 (1992).

[312] I. Gohberg, M.A. Kaashoek, The band extension on the real line as a limit of discrete band extensions, II. The entropy principle. Operator Theory: Advances and Applications, vol. 58, Birkhauser Verlag, Basel, 71–92 (1992).

[313] J.A. Ball, I. Gohberg, M.A. Kaashoek, Reduction of the abstract four block problem to a Nehari problem. Operator Theory: Advances and Applications, vol. 58, Birkhauser Verlag, Basel, 121–142 (1992).

[314] I. Gohberg, M.A. Kaashoek, The band extension on the real line as a limit of discrete band extensions, I. The main limit theorem. Operator Theory: Advances and Applications, vol. 59, Birkhauser Verlag, Basel, 191–220 (1992).

[315] I. Gohberg, M.A. Kaashoek, Asymptotic formulas of Szego-Kac-Achiezer type. Asymptotic Analysis 5, 187–220 (1992).

[316] J.A. Ball, I. Gohberg, M.A. Kaashoek, Time-varying systems: Nevanlinna-Pick interpolation and sensitivity minimization. Recent Advances in Mathematical Theory of Systems, Control, Networks and Signal Processing I; Proceedings MTNS-1991, Mitra Press, Tokyo, 1992.

[317] I. Gohberg, A.C.M. Ran, On pseudo-canonical factorization of rational matrix functions. Indagationes Mathematica, N.S., 4(1), 51–63 (1993).

[318] A. Ben-Artzi, I. Gohberg, M.A. Kaashoek, Invertibility and dichotomy of differential operators on a half-line. Journal of Dynamics and Differential Equations, vol. 5, no. 1, 1–36 (1993).

[319] A. Ben-Artzi, I. Gohberg, Dichotomies of perturbed time varying systems and the power method. Indiana Univ. Math. J. 42, 699–720 (1993).

[320]   I. Gohberg, M.A. Kaashoek, H.J. Woerdemann, The band method for bordered algebras. Operator Theory: Advances and Applications, vol. 62, Birkhauser Verlag, Basel, 85–98 (1993).

[321]   J.A. Ball, I. Gohberg, M.A. Kaashoek, Bitangential interpolation for input-output operators of time varying systems: the discrete time case. Operator Theory: Advances and Applications, vol. 64, Birkhauser Verlag, Basel, 33–72 (1993).

[322]   J.A. Ball, I. Gohberg, L. Rodman, Two-sided tangential interpolation of real rational matrix functions. Operator Theory: Advances and Applications, vol. 64, Birkhauser Verlag, Basel, 73–102 (1993).

[323]   I. Gohberg, C. Gu, On a completion problem for matrices. Operator Theory: Advances and Applications, vol. 64, Birkhauser Verlag, Basel, 203–217 (1993).

[324]   R.L. Ellis, I. Gohberg, D.C. Lay, Extensions with positive real part. A new version of the abstract band method with applications. Integral Equations and Operator Theory, 16, 360–384 (1993).

[325]   I. Gohberg, N. Krupnik, Extension theorems for Fredholm and invertibility symbols. Integral Equations and Operator Theory, 16, 514–529 (1993).

[326]   I. Gohberg, N. Krupnik, I. Spitkovsky, Banach algebras of singular integral operators with piecewise continuous coefficients. General contour and weight. Integral Equations and Operator Theory, 17, 322–327 (1993).

[327]   A. Ben-Artzi, I. Gohberg, M.A. Kaashoek, A time-varying generalization of the canonical factorization theorem for Toeplitz operators. Indagationes Mathematica, N.S., 4(4), 385–405 (1993).

[328]   I. Gohberg, I. Koltracht, Condition numbers for functions of matrices. Applied Numerical Mathematics 12, 107–117 (1993).

[329]   I. Gohberg, B. Reichstein, Classification of Block-Toeplitz H-normal operators. Linear and Multilinear Algebra, vol. 34, 213–245 (1993).

[330]   I. Gohberg, I. Koltracht, Mixed, componentwise, and structured condition numbers. SIAM J. Matrix Anal. Appl., vol. 14, 688–709. July (1993).

[331]   J.A. Ball, I. Gohberg, L. Rodman, The structure of flat gain rational matrices that satisfy two-sided interpolation requirements. Systems & Control Letters, 20, 401–412 (1993).

[332]   I. Gohberg, L. Rodman, T. Shalom, H.J. Woerdeman, Bounds for eigenvalues and singular values of matrix completions. Linear and Multilinear Algebra, 33, 233–249 (1993).

[333]   A. Ben-Artzi, I. Gohberg, M.A. Kaashoek, Exponentially dominated infinite block matrices of finite Kronecker rank. Integral Equations and Operator Theory, 18, 30–77 (1994).

[334]   C. Foias, A. Frazho, I. Gohberg, Central intertwining lifting, maximum entropy and their permanence. Integral Equations and Operator Theory, 18, 166–201 (1994).

[335]   I. Feldman, I. Gohberg, N. Krupnik, A method of explicit factorization of matrix functions and applications. Integral Equations and Operator Theory, 18, 277–302 (1994).

[336]   I. Gohberg, V. Olshevsky, Complexity of multiplication with vectors for structured matrices. Linear Algebra and its Applications, 202, 163–192 (1994).

[337]   D. Alpay, J.A. Ball, I. Gohberg, L. Rodman, J-Unitary preserving automorphisms of rational matrix functions: State space theory, interpolation, and factorization. Linear Algebra and its Applications, 197, 198: 531–566 (1994).

[338]   I. Gohberg, V. Olshevsky, Fast inversion of Chebyshev-Vandermonde matrices. Numer. Math. 67, 71–92 (1994).

[339]   I. Gohberg, M. Hanke, I. Koltracht, Fast preconditioned conjugate gradient algorithms for Wiener-Hopf integral equations. SIAM. J. Numer. Anal., vol. 31, no. 2, 429–443 (1994).

[340]   I. Gohberg, N. Krupnik, Szego-Widom-type limit theorems. Operator Theory: Advances and Applications, vol. 71, Birkhauser Verlag, Basel, 105–121 (1994).

[341]   D. Alpay, J.A. Ball, I. Gohberg, L. Rodman, The two-sided residue interpolation in the Stieltjes class for matrix functions. Linear Algebra and its Applications, 208/209: 485–521 (1994).

[342]   I. Gohberg, M.A. Kaashoek, Projection method for block Toeplitz operators with operator-valued symbols. Operator Theory: Advances and Applications, vol. 71, Birkhauser Verlag, Basel, 79–104 (1994).

[343]   A. Ben-Artzi, I. Gohberg, Orthogonal polynomials over Hilbert modules. Operator Theory: Advances and Applications, vol. 73, Birkhauser Verlag, Basel, 96–126 (1994).

[344]   I. Gohberg, Y. Zucker, Left and right factorizations of rational matrix functions. Integral Equations and Operator Theory, 19, 216–239 (1994).

[345]   J.A. Ball, I. Gohberg, M.A. Kaashoek, Bitangential interpolation for input-output maps of time-varying systems: The continuous time case. Integral Equations and Operator Theory, 20, 1–43 (1994).

[346]   I. Gohberg, V. Olshevsky, Fast state space algorithms for matrix Nehari and Nehari-Takagi interpolation problems. Integral Equations and Operator Theory, 20, 44–83 (1994).

[347]   D. Alpay, I. Gohberg, Inverse spectral problems for difference operators with rational scattering matrix function. Integral Equations and Operator Theory, 20, 125–170 (1994).

[348]   J.A. Ball, I. Gohberg, M. Rakowski, Reconstruction of a rational nonsquare matrix function from local data. Integral Equations and Operator Theory, 20, 249–305 (1994).

[349]   I. Gohberg, H.J. Landau, Prediction and the inverse of Toeplitz matrices. International Series of Numerical Mathematics, vol. 119, 219–229 (1994).

[350]   J.A. Ball, I. Gohberg, M.A. Kaashoek, H_∞-control and interpolation for time-varying systems. Akademie Verlag, Mathematical Research, vol. 77, 33–48 (1994).

[351]   I. Gohberg, V. Olshevsky, Fast algorithm for matrix Nehari problem. Akademie Verlag, Mathematical Research, vol. 79, 687–690 (1994).

[352]   I. Gohberg, V. Olshevsky, Fast algorithms with preprocessing for matrix-vector multiplication problems. Journal of Complexity, 10 (1994).

[353]   I. Gohberg, Odessa Reminiscences. Operator Theory: Advances and Applications, vol. 72, xix–xx, Birkhauser Verlag, Basel (1994).

[354]   J.A. Ball, I. Gohberg, M.A. Kaashoek, Input-output operators of J-unitary time-varying continuous time systems. Operator Theory: Advances and Applications, vol. 75, Birkhauser Verlag, Basel, 57–94 (1995).

[355]   A. Ben-Artzi, I. Gohberg, M.A. Kaashoek, Discrete nonstationary bounded real lemma in indefinite metrics, the strict contractive case. Operator Theory: Advances and Applications, vol. 80, Birkhauser Verlag, Basel, 49–78 (1995).

[356]   D. Alpay, I. Gohberg, Inverse spectral problem for differential operators with rational scattering matrix functions. Journal of Differential Equations, vol. 118, no. 1, 1–19 (1995).

[357]   I. Gohberg, I. Koltracht, Structured condition numbers for linear matrix structures. IMA Volumes in Mathematics and its Applications, vol. 69, 17–26 (1995).

[358]   I. Gohberg, I. Koltracht, A fast realization of preconditioned conjugate gradients for Wiener-Hopf integral equations. Applied Mathematics Letters, vol. 8, no. 6, 65–72 (1995).

[359]   J.A. Ball, I. Gohberg, M.A. Kaashoek, Two-sided Nudelman for input-output operators of discrete time-varying systems. Integral Equations and Operator Theory, vol. 21, 174–211 (1995).

[360]   A. Ben-Artzi, I. Gohberg, Inertia theorems for operator pencils and applications. Integral Equations and Operator Theory, vol. 21, 270–318 (1995).

[361]   I. Feldman, I. Gohberg, N. Krupnik, On explicit factorization and applications. Integral Equations and Operator Theory, vol. 21, 430–459 (1995).

[362]   R.L. Ellis, I. Gohberg, D.C. Lay, Infinite analogues of block Toeplitz matrices and related orthogonal functions. Integral Equations and Operator Theory, 22, 375–419 (1995).

[363]   H. Dym, I. Gohberg, On maximum entropy interpolants and maximum determinant completions of associated Pick matrices. Integral Equations and Operator Theory, 23, 61–88 (1995).

[364]   V. Boltyanski, I. Gohberg, Stories about covering and illuminating of convex bodies. Nieuw Archief voor Wiskunde, 13, no. 1, 1–26 (1995).

[365]   I. Gohberg, T. Kailath, V. Olshevsky, Fast Gaussian elimination with partial pivoting for matrices with displacement structure. Math. of Computation, 64, 1557–1576 (1995).

[366]   J.A. Ball, I. Gohberg, M.A. Kaashsoek, A frequency response function for linear, time-varying systems, Math. Control Signal 8, 4, 334–351 (1995).

[367]   I. Gohberg, M.A. Kaashoek, J. Kos, The asymptotic behaviour of the singular values of matrix powers and applications. Linear Algebra and its Applications, 245, 55–76 (1996).

[368]   A. Ben-Artzi, I. Gohberg, On contractions in spaces with an indefinite metric: G-norms and spectral radii. Integral Equations and Operator Theory, 24, 422–469 (1996).

[369]   I. Gohberg, Y. Zucker, On canonical factorization of rational matrix functions. Integral Equations and Operator Theory, 25, 73–93 (1996).

[370]   I. Gohberg, M.A. Kaashoek, J. Kos, Classification of linear time-varying difference equations under kinematic similarity. Integral Equations and Operator Theory, 25, 445–480 (1996).

[371]   I. Gohberg, I. Koltracht, Triangular factors of Cauchy and Vandermonde matrices. Integral Equations and Operator Theory, 26, 46–59 (1996).

[372]   I. Gohberg, S. Goldberg, M. Krupnik, Traces and determinants of linear operators. Integral Equations and Operator Theory, 26, 136–187 (1996).

[373]   D. Alpay, I. Gohberg, A relationship between the Nehari and the Caratheodory-Toeplitz extension problems. Integral Equations and Operator Theory, 26, 249–272 (1996).

[374]   C. Foias, A.E. Frazho, I. Gohberg, M.A. Kaashoek, Discrete time-variant interpolation as classical interpolation with an operator argument. Integral Equations and Operator Theory, 26, 371–403 (1996).

[375]   A. Ben-Artzi, I. Gohberg, Monotone power method in indefinite metric and inertia theorem for matrices. Linear Algebra Appl., 243: 225–245 (1996).

[376]   I. Gohberg, S. Goldberg, A simple proof of the Jordan decomposition theorem for matrices. Am. Math. Mon. 103: (2) 157–159 (1996).

[377]   I. Gohberg, M.A. Kaashoek, L. Lerer, Factorization of banded lower triangular infinite matrices. Linear Algebra Appl., 247: 347–357 (1996).

[378]   R.L. Ellis, I. Gohberg, D.C. Lay, On a class of block Toeplitz matrices. Linear Algebra Appl., 243: 225–245 (1996).

[379]   I. Gohberg, S. Goldberg, N. Krupnik, Hilbert-Carleman and regularized determinants for linear operators. Integral Equations and Operator Theory, 27, 10–47 (1997).

[380]   Y. Eidelman, I. Gohberg, Fast inversion algorithms for diagonal plus semiseparable matrices. Integral Equations and Operator Theory, 27, 165–183 (1997).

[381]   J.A. Ball, I. Gohberg, M.A. Kaashoek, Nudelman interpolation and the band method. Integral Equations and Operator Theory, 27, 253–284 (1997).

[382]   I. Gohberg, H.J. Landau, Prediction for two processes and the Nehari problem. Journal of Fourier Analysis and Applications, vol. 3, 43–62 (1997).

[383]   C. Foias, A.E. Frazho, I. Gohberg, M.A. Kaashoek, A time-variant version of the commutant lifting theorem and nonstationary interpolation problems. Integral Equations and Operator Theory, 28, 158–190 (1997).

[384]   C. Foias, A.E. Frazho, I. Gohberg, M.A. Kaashoek, Parametrization of all solutions of the three chains completion problem. Integral Equations and Operator Theory, vol. 29, 455–490 (1997).

[385]   Y. Eidelman, I. Gohberg, Inversion formulas and linear complexity algorithm for diagonal plus semiseparable matrices. Comput. Math. Appl. 33: 4, 69–79 (1997).

[386]   I. Gohberg, V. Olshevsky, The fast generalized Parker-Traub algorithm for inversion of Vandermonde and related matrices. J. Complexity 13: (2), 208–234 (1997).

[387]   C. Foias, A.E. Frazho, I. Gohberg, M.A. Kaashoek, The maximum principle for the three chains completion problem. Integral Equations and Operator Theory, vol. 30, 67–82 (1998).

[388]   D. Alpay, I. Gohberg, Inverse problem for Sturm-Liouville operators with rationals reflection coefficient. Integral Equations and Operator Theory, vol. 30, 316–325 (1998).

[389]   I. Gohberg, M.A. Kaashoek, A.L. Sakhnovich, Sturm-Liouville systems with rational Weyl functions: Explicit formulas and applications. Integral Equations and Operator Theory, vol. 30, 338–377 (1998).

[390]   Y. Eidelman, I. Gohberg, A look-ahead block Schur algorithm for diagonal plus semiseparable matrices. Computers Math. Applic., vol. 35, 25–34 (1998).

[391]   I. Gohberg, M.A. Kaashoek, F. van Schagen, Operator blocks and quadruples of subspaces: Classification and the eigenvalue completion. Linear Algebra Appl., 269: 65–89 (1998).

[392]   I. Gohberg, Mark Grigorievich Krein: Recollections Integral Equations and Operator Theory, 30 (2), 123–134 (1998).

[393]   I. Gohberg, M.A. Kaashoek, A.L. Sakhnovich, Pseudo-canonical systems with rational Weyl functions: Explicit formulas and applications. J. Diff. Equations 146 (2), 375–398 (1998).

[394]   R.L. Ellis, I. Gohberg, D.C. Lay, Singular values of positive pencils and applications. Operator Theory: Advances and Applications, vol. 106, Birkhauser Verlag, Basel, 131–146 (1998).

[395] A. Dijksma, I. Gohberg, Heinz Langer and his work. Operator Theory: Advances and Applications, vol. 106, Birkhauser Verlag, Basel, 1–22 (1998).

[396] R.L. Ellis, I. Gohberg, Inversion formulas for infinite generalized Toeplitz matrices. Integral Equations and Operator Theory, 32, 29–64 (1998).

[397] A. Ben-Artzi, I. Gohberg, Singular numbers of contractions in spaces with an indefinite metric and Yamamoto"s theorem. Linear Algebra and its Applications, 290, 31–48 (1999).

[398] Y. Eidelman, I. Gohberg, A new class of structured matrices. Integral Equations and Operator Theory, 34, 293–363 (1999).

[399] Y. Eidelman, I. Gohberg, Linear complexity inversion algorithms for a class of structured matrices. Integral Equations and Operator Theory, 35, 28–52 (1999).

[400] I. Gohberg, M.A. Kaashoek, State space methods for analysis problems involving rational matrix functions. Progress in Systems and Control Theory, vol. 25, Birkhauser Verlag, 93–109 (1999).

[401] I. Gohberg, Vladimir Maz'ya: Friend and Mathematician. Recollections. Operator Theory: Advances and Applications, vol. 109, Birkhauser Verlag, Basel, 1–5 (1999).

Operator Theory:
Advances and Applications, Vol. 124
© 2001 Birkhäuser Verlag Basel/Switzerland

On the Spectral Theory of Degenerate Quadratic Operator Pencils

Vadim Adamyan, Reinhard Mennicken
and Vjacheslav Pivovarchik

Dedicated to Israel Gohberg on the occasion of his seventieth birthday

This paper deals with quadratic operator pencils

$$L(\lambda) = \lambda^2 A + \lambda B + C,$$

where A, B are non-invertible non-negative operators and C is a selfadjoint operator, which is bounded from below and has a compact resolvent. Such pencils naturally arise in stability problems of mechanics and resistive magnetohydrodynamics. Under certain assumptions on A, B and C the description of the spectrum of the pencil L is given.

1 Introduction

In this paper we study operator pencils

$$L(\lambda) = \lambda^2 A + \lambda B + C$$

whose coefficients A, B and C are 2×2 operator matrices acting in the orthogonal sum $\mathcal{H}_1 \oplus \mathcal{H}_2$ of Hilbert spaces $\mathcal{H}_1$, $\mathcal{H}_2$. Spectral problems for quadratic operator pencils occur in various problems of mathematical physics. The existing spectral theory of quadratic operator pencils deals mostly with pencils having a bounded or boundedly invertible coefficient A and a selfadjoint coefficient C. There are only some partial results related to *degenerate* pencils with noninvertible A and B. At any rate a substantial spectral theory of degenerate quadratic pencils is important for the solution of the stability problem for small oscillations of resistive plasmas in hydrodynamic approximation (see [Li], [T]).

Consider an equilibrium configuration of a non-ideal plasma surrounded by a perfectly conducting wall. Let us denote the corresponding density, pressure, magnetic field and current density by ρ_0, p_0, $\mathbf{B}_0$, $\mathbf{J}_0$ respectively. Perturbed quantities characterizing small oscillations of the plasma near the equilibrium state can be written in the form

$$
\begin{aligned}
\rho(\mathbf{x}, t) &= \rho_0(\mathbf{x}) + \varepsilon \rho_1(\mathbf{x}, t); \\
\mathbf{v}(\mathbf{x}, t) &= \varepsilon \mathbf{v}_1(\mathbf{x}, t); \\
p(\mathbf{x}, t) &= p_0(\mathbf{x}) + \varepsilon p_1(\mathbf{x}, t); \\
\mathbf{B}(\mathbf{x}, t) &= \mathbf{B}_0(\mathbf{x}) + \varepsilon \mathbf{B}_1(\mathbf{x}, t).
\end{aligned}
$$

The spectral problem related to such oscillations has the form

$$(1.1) \qquad\qquad L_r \mathbf{u} = \lambda \mathbf{u},$$

where $\mathbf{u} = (\rho_1, \mathbf{v}_1, p_1, \mathbf{B}_1)$ is an eight component column vector, L_r is the closure of the differential operator L_{r0} defined by the expression

$$L_{r0} \begin{pmatrix} \rho_1 \\ \mathbf{v}_1 \\ p_1 \\ \mathbf{B}_1 \end{pmatrix} = \begin{pmatrix} -(\mathbf{v}_1 \cdot \nabla)\rho_0 - \rho_0 \mathrm{div}\mathbf{v}_1 \\ \rho_0^{-1}(\mathrm{grad}\,p_1 - \mathrm{rot}\mathbf{B}_1 \times \mathbf{B}_0/\mu - \mathrm{rot}\mathbf{B}_0 \times \mathbf{B}_1/\mu) \\ -(\mathbf{v}_1 \cdot \nabla)p_0 - \gamma p_0 \mathrm{div}\mathbf{v}_1 + 2(\gamma-1)(\eta/\mu)\,\mathrm{rot}\mathbf{B}_1 \cdot \mathrm{rot}\mathbf{B}_0 \\ \mathrm{rot}(\mathbf{v}_1 \times \mathbf{B}_0) - \mathrm{rot}(\eta \mathrm{rot}\mathbf{B}_1) \end{pmatrix},$$

on smooth vector functions satisfying the boundary conditions

$$\mathbf{n} \cdot \mathbf{B}_{1|S_p} = 0, \; \mathbf{n} \cdot \mathbf{v}_{1|S_p} = 0, \; \mathbf{n} \times \mathrm{rot}\mathbf{B}_{1|S_p} = 0$$

on the conducting wall S_p. Here μ, γ and η are the magnetic permeability, the specific heat ratio and the resistivity, respectively. Consider, for example, the simple case of a resistive gravitating plasma layer $\{x, y, z : 0 \le x \le d, 0 \le y \le d_y, 0 \le z \le d_z\}$ bounded by the perfectly conducting rigid planes $x = 0, x = d$. Under the assumptions that the equilibrium functions ρ_0, p_0, $\mathbf{B}_0, \mathbf{J}_0$ as well as the resistivity η_0 depend only on x and that ρ_0, $\mathbf{B}_0, \mathbf{J}_0$ are connected by the ideal force-balance condition, the spectral problem (1.1) can be reduced to that for the degenerate quadratic pencil

$$M(\lambda) = \lambda^2 M_2 + \lambda M_1 + M_0,$$

where M_0, M_1, M_2 are selfadjoint operators in the Hilbert space $L^2(0, d) \oplus L^2(0, d)$. For a certain twice continuously differentiable real function F and Sturm–Liouville differential operators Δ_2 and Δ_0 defined on functions satisfying Dirichlet boundary conditions, M_0, M_1, M_2 are given by the expressions

$$M_0 = \frac{1}{\mu}\begin{pmatrix} -FF'' + k^2\rho_0'g & -iF'' \\ iF'' & \Delta_0 \end{pmatrix},$$

$$M_1 = \frac{1}{\mu\eta}\begin{pmatrix} F^2 & iF \\ -iF & 1 \end{pmatrix},$$

$$M_2 = \begin{pmatrix} \Delta_2 & 0 \\ 0 & 0 \end{pmatrix},$$

where k is the modulus of a wave vector, g is the gravitational constant.

In this example zero is an eigenvalue of infinite multiplicity for both operator coefficients at λ^2 and λ of the quadratic pencil M.

There are also important stability problems generating quadratic operator pencils for which the coefficients at λ^2 and λ have densely defined but unbounded inverses. Consider for instance the equation

$$(1.2) \qquad \beta\frac{\partial^4 u}{\partial x^4} + \frac{\partial^3}{\partial t \partial x^2}\left(\alpha\frac{\partial^2 u}{\partial x^2}\right) + \frac{\partial}{\partial x}\left(g\frac{\partial u}{\partial x}\right) + k\frac{\partial u}{\partial t} - \rho\frac{\partial^2 u}{\partial t^2} = 0$$

which describes small transverse vibrations of a viscoelastic thin beam of unit length with external and internal damping (the so-called Kelvin–Voigt material). Here $x \in [0, 1]$ is the coordinate along the beam, $t \geq 0$ is the time, $u(x, t)$ is the transverse displacement at position x and time t. The function $\alpha(x) \geq 0$ describes the internal damping, $k(x) \geq 0$ is the coefficient of the external damping (viscous friction), and $g(x)$ is responsible for the forces of contraction or tension. The function $\rho(x) \geq 0$ is the density. Let us assume that $\rho(x)$ is equal to zero only on a set of measure zero. For the sake of simplicity we assume also that k is continuous, g is continuously differentiable, and α is twice continuously differentiable. We assume that ρ belongs to $\mathbf{L}^2(0, 1)$. Let us choose the boundary conditions corresponding to clamped ends of the beam:

$$(1.3) \qquad u(0, t) = \frac{\partial u(x, t)}{\partial x}\Big|_{x=0} = u(1, t) = \frac{\partial u(x, t)}{\partial x}\Big|_{x=1} = 0.$$

Substituting $u(x, t) = y(\lambda, x)e^{\lambda t}$ into (1.2) and (1.3) we obtain the following boundary problem

$$(1.4) \qquad \begin{cases} y^{iv} + (gy')' + \lambda((\alpha y'')'' + ky) - \lambda^2 \rho y = 0, \\ y(\lambda, 0) = y'(\lambda, 0) = y(\lambda, 1) = y'(\lambda, 1) = 0. \end{cases}$$

Introduce the following operators acting in $\mathbf{L}^2(0, 1)$:

$$(1.5) \qquad \begin{cases} Ay = \rho y, \\ \mathcal{D}(A) = \{y \in \mathbf{L}^2(0, 1) : \rho y \in \mathbf{L}^2(0, 1)\}; \end{cases}$$

$$(1.6) \qquad \begin{cases} By = (\alpha y'')'' + ky, \\ \mathcal{D}(B) = \{y \in \mathbf{W}_4^2(0, 1) : y(0) = y'(0) = y(1) = y'(1) = 0\}; \end{cases}$$

$$(1.7) \qquad \begin{cases} Cy = y^{(iv)}, \\ \mathcal{D}(C) = \mathcal{D}(B); \end{cases}$$

$$(1.8) \qquad \begin{cases} Dy = (gy')', \\ \mathcal{D}(D) = \{y \in \mathbf{W}_2^2(0, 1) : y(0) = y(1) = 0\}. \end{cases}$$

Then the spectral problem (1.4) is reduced to that for the quadratic operator pencil

$$(1.9) \qquad L(\lambda) = \lambda^2 A + \lambda B + C + D.$$

This example yields a good motivation for studying abstract spectral problems for non-selfadjoint quadratic pencils of the form (1.9), where all coefficients are in general unbounded operators acting in some Hilbert space, C is positive and invertible, A, B and D are selfadjoint operators with domains containing the domain of C, so that by definition $\mathcal{D}(L(\lambda)) = \mathcal{D}(C)$ for all $\lambda \in \mathbb{C}$. In addition, taking into account that the functions α, g and ρ in the last example are not necessary strongly positive, the operators A, B and D in general are not invertible. The case with invertible A was considered in details in [Pi1], [Pi2], [Pi3], [LS], [GS], [S], [AP] and a particular case of a noninvertible A in [BP].

To develop a comprehensive operator approach to the stability problems for dissipative oscillatory systems, including those arising in the above mentioned examples, we single out and study a special class of quadratic operator pencils

$$(1.10) \qquad L(\lambda) = \lambda^2 A + \lambda B + C$$

in the orthogonal sum $\mathcal{H}_1 \oplus \mathcal{H}_2$ of Hilbert spaces $\mathcal{H}_1, \mathcal{H}_2$ with coefficients A, B, C naturally given as 2×2 block-operator matrices and subject to the following conditions:

1) A admits a representation

$$A = \begin{pmatrix} A_{11} & 0 \\ 0 & 0 \end{pmatrix},$$

where A_{11} is a selfadjoint operator in $\mathcal{H}_1$ such that $A_{11} \gg 0$ and A_{11}^{-1} is compact;

2) $B = B' + B''$, where

$$B' = \begin{pmatrix} B'_{11} & 0 \\ 0 & 0 \end{pmatrix},$$

and B'_{11} is a non-negative operator in $\mathcal{H}_1$ such that $\mathcal{D}(A_{11}) \subset \mathcal{D}(B'_{11})$ and there exist a bounded operator $R \geq 0$ and a compact operator K satisfying the relation

$$B'_{11} = (R + K)A_{11},$$

and

$$B'' = \begin{pmatrix} B''_{11} & B''_{12} \\ B''_{21} & B''_{22} \end{pmatrix} \geq 0$$

is a bounded operator in $\mathcal{H}_1 \oplus \mathcal{H}_2$ such that $B''_{22} > 0$;

3) $C = C' + C''$, where

$$C' = \begin{pmatrix} 0 & 0 \\ 0 & C'_{22} \end{pmatrix}$$

and C'_{22} is a selfadjoint operator in $\mathcal{H}_2$ such that $C'_{22} >> 0$ and C'^{-1}_{22} is compact, C'' is a densely defined symmetric operator in $\mathcal{H}_1 \oplus \mathcal{H}_2$, $\mathcal{D}(A_{11}) \oplus \mathcal{D}(C'_{22}) \subset \mathcal{D}(C'')$ and there exist a compact operator T in $\mathcal{H}_1 \oplus \mathcal{H}_2$ satisfying the relation

$$C'' = T \begin{pmatrix} A_{11} & 0 \\ 0 & C'_{22} \end{pmatrix}$$

on $\mathcal{D}(A_{11}) \oplus \mathcal{D}(C'_{22})$.

If we restrict our considerations to the simplest stability problems of resistive magnetohydrodynamics, we might take $B' = 0$ and assume that B'' has the form

$$B'' = \begin{pmatrix} Q_1^* G Q_1 & Q_1^* G Q_2 \\ Q_2^* G Q_1 & Q_2^* G Q_2 \end{pmatrix}$$

with some bounded operators Q_1 and G in $\mathcal{H}_1$, some bounded operator Q_2 from $\mathcal{H}_1$ to $\mathcal{H}_2$ and $G > 0$. Observe that even in this case spectral problems for quadratic pencils

$$L(\lambda) = \begin{pmatrix} \lambda^2 A_{11} & 0 \\ 0 & 0 \end{pmatrix} + \lambda \begin{pmatrix} Q_1^* G Q_1 & Q_1^* G Q_2 \\ Q_2^* G Q_1 & Q_2^* G Q_2 \end{pmatrix} + \begin{pmatrix} 0 & 0 \\ 0 & C'_{22} \end{pmatrix} + C''$$

in $\mathcal{H}_1 \oplus \mathcal{H}_2$ correspond to such problems for operator pencils in $\mathcal{H}_1$, which are at least cubic in the spectral parameter λ.

The aim of this paper is to compile some basic facts on the spectra of quadratic operator pencils of type (1.10) satisfying the conditions 1), 2) and 3). The domain of the pencil L, pointwise defined by $\mathcal{D}(L(\lambda)) = \mathcal{D}(A_{11}) \oplus \mathcal{D}(C'_{22})$, is λ-independent. Section 2 is devoted to the description of the essential spectrum and the set of normal eigenvalues of the pencil L. Section 3 contains some auxiliary results concerning the operators $L(\lambda)$ for $\lambda \in (0, \infty)$. In Section 4 it is proved that the sum of the geometric multiplicities of all eigenvalues of the pencil L in the open right half-plane coincides with the corresponding sum of their algebraic multiplicities and equals the dimension of the maximal negative subspace of the quadratic form $(L(0)x, x) = (Cx, x)$, $x \in \mathcal{D}(A_{11}) \oplus \mathcal{D}(C'_{22})$.

2 General Description of $\sigma(L)$

If $L(\lambda)$ is a bijection from $\mathcal{D}(A_{11}) \oplus \mathcal{D}(C'_{22})$ to $\mathcal{H}_1 \oplus \mathcal{H}_2$ and $L(\lambda)^{-1}$ is bounded, we say as usually that $L(\lambda)$ has a bounded inverse. We call

$$\rho(L) := \{\lambda \in \mathbb{C} : L(\lambda) \text{ has a bounded inverse}\}$$

the *resolvent set* of L and

$$\sigma(L) := \mathbb{C}\backslash\rho(L)$$

the *spectrum* of L. The point spectrum of L is defined by

$$\sigma_p(L) := \{\lambda \in \mathbb{C} : L(\lambda) \text{ is not injective}\}.$$

A point $\lambda_0 \in \sigma_p(L)$ is called an *eigenvalue* of L, and a vector $x_0 \neq 0$ such that

$$L(\lambda_0)x_0 = 0$$

is called an *eigenvector* of L with respect to the eigenvalue λ_0. The dimension of the kernel $\mathcal{N}(\lambda_0)$ of the operator $L(\lambda_0)$ is called the *geometric multiplicity* of the eigenvalue λ_0. The vectors $x_0, x_1, \ldots, x_{m-1}$ form a *chain of associated vectors* to the eigenvector $x_0 \neq 0$ if

$$L(\lambda_0)x_1 + (2\lambda_0 A + B)x_0 = 0, \quad \text{if } m > 1$$

and

$$L(\lambda_0)x_k + (2\lambda_0 A + B)x_{k-1} + Ax_{k-2} = 0, \quad k = 2, \ldots, m-1, \quad \text{if } m > 2.$$

The number m is called the *length* of this chain. The algebraic multiplicity of an eigenvalue is defined to be the greatest value of the sum of the lengths of chains corresponding to linearly independent eigenvectors. An *isolated* eigenvalue λ_0 of L is called *normal* if λ_0 has finite algebraic multiplicity and the range of the operator $L(\lambda_0)$ has at most finite codimension what is equivalent to the fact that $L(\lambda_0)$ is Fredholmian. By $\sigma_d(L)$ we denote the set of all normal eigenvalues, which is called the *discrete spectrum* of L. The set

$$\sigma_{ess}(L) := \sigma(L)\backslash\sigma_d(L)$$

is called the *essential spectrum* of L.

Set

$$
\tilde{L}(\lambda) := I + \frac{1}{\lambda} \begin{pmatrix} R & 0 \\ 0 & 0 \end{pmatrix} + \frac{1}{\lambda} \begin{pmatrix} K & 0 \\ 0 & 0 \end{pmatrix}
$$

(2.1)
$$
+ (\lambda B'' + C'') \begin{pmatrix} \frac{1}{\lambda^2} A_{11}^{-1} & 0 \\ 0 & C_{22}'^{-1} \end{pmatrix}.
$$

By our assumptions $\tilde{L}(\lambda)$ is a bounded operator for any $\lambda \in \mathbb{C}\backslash\{0\}$. The equality

(2.2)
$$
L(\lambda) = \tilde{L}(\lambda) \begin{pmatrix} \lambda^2 I & 0 \\ 0 & I \end{pmatrix} \begin{pmatrix} A_{11} & 0 \\ 0 & C_{22}' \end{pmatrix}
$$

shows that $\sigma(L)\backslash\{0\} = \sigma(\tilde{L})$. Moreover, $\lambda_0 \in \mathbb{C}\backslash\{0\}$ is a normal eigenvalue of L if and only if λ_0 is a normal eigenvalue of $\tilde{L}$ and $L(\lambda_0)$ is not Fredholmian if and only if $\tilde{L}(\lambda_0)$ is not Fredholmian.

Set

(2.3)
$$
\tilde{L}_0(\lambda) := I + \frac{1}{\lambda} \begin{pmatrix} R & 0 \\ 0 & 0 \end{pmatrix}.
$$

Evidently,

$$
\sigma(\tilde{L}_0) \subset [-\|R\|, 0].
$$

Theorem 2.1 *Let A, B, C satisfy the assumptions* 1)–3). *Then*

$$
\sigma_{ess}(L)\backslash\{0\} = \sigma_{ess}(\tilde{L}_0) \subset [-\|R\|, 0] \quad and
$$

$$
\sigma_d(L)\backslash\{0\} = \sigma(L)\backslash\sigma_{ess}(\tilde{L}_0) \cup \{0\}.
$$

In particular, the non-real points of the open right half-plane belong to the resolvent set of L.

Proof: According to (2.1) and by the assumptions 1)–3),

(2.4)
$$
\tilde{L}(\lambda) = I + \frac{1}{\lambda} \begin{pmatrix} R & 0 \\ 0 & 0 \end{pmatrix} + Q(\lambda) = \tilde{L}_0(\lambda) + Q(\lambda),
$$

where

$$
Q(\lambda) := \frac{1}{\lambda} \begin{pmatrix} K & 0 \\ 0 & 0 \end{pmatrix} + (\lambda B'' + C'') \begin{pmatrix} \frac{1}{\lambda^2} A_{11}^{-1} & 0 \\ 0 & C_{22}'^{-1} \end{pmatrix}
$$

is a holomorphic operator function on $\mathbb{C}\backslash\{0\}$ with compact coefficients. It follows from (2.4) that $\widetilde{L}(\lambda)$ is not Fredholmian if and only if $\widetilde{L}_0(\lambda)$ is not Fredholmian. Let

$$F(\lambda) := \widetilde{L}_0(\lambda)^{-1} Q(\lambda), \quad \lambda \notin \sigma(\widetilde{L}_0) \cup \{0\}.$$

It follows from (2.2) and the representation

$$(2.5) \qquad \widetilde{L}(\lambda) = \widetilde{L}_0(\lambda)(I + F(\lambda)), \quad \lambda \notin \sigma(\widetilde{L}_0) \cup \{0\},$$

that $\lambda_0 \in \mathbb{C}\backslash\sigma(\widetilde{L}_0) \cup \{0\}$ is an eigenvalue of $\widetilde{L}$ (and hence of L) of geometric multiplicity $n = n(\lambda_0)$ if and only if the equation

$$(2.6) \qquad x + F(\lambda_0)x = 0, \quad x \in \mathcal{H},$$

has n linearly independent solutions. Note that F is a holomorphic operator function in $\mathbb{C}\backslash\sigma(\widetilde{L}_0) \cup \{0\}$ with compact values and that $\mathbb{C}\backslash\sigma(\widetilde{L}_0) \cup \{0\}$ is a connected set. Therefore for all $\lambda_0 \in \mathbb{C}\backslash\sigma(\widetilde{L}_0) \cup \{0\}$ except for some isolated points the number $n(\lambda_0)$ of linearly independent solutions of (2.6) is a constant n_0 and at the mentioned isolated points $n_0 < n < \infty$ (see [GK], [GGK], Ch. XI, Corollary 8.4). Besides, the algebraic multiplicity of each eigenvalue $\lambda_0 \in \mathbb{C}\backslash\sigma(\widetilde{L}_0) \cup \{0\}$ of $I + F$ is finite (see e.g. [M]). In particular, if $n(\lambda_0) = 0$ for at least one $\lambda_0 \in \mathbb{C}\backslash\sigma(\widetilde{L}_0) \cup \{0\}$, then $n_0 = 0$ and the operator $I + F(\lambda)$ is invertible for almost all $\lambda \in \mathbb{C}\backslash\sigma(\widetilde{L}_0) \cup \{0\}$ except for a set of isolated points, which consists of the normal eigenvalues of $I + F$. Let us prove that for the given operator function F we indeed have $n_0 = 0$. To this end it suffices to show that there exist a point $\lambda \in \mathbb{C}\backslash\sigma(\widetilde{L}_0) \cup \{0\}$, which is not an eigenvalue of L. The last claim is evident since by the assumptions 1), 2) and 3), we have

$$\|L(\lambda)x\| \geq |\operatorname{Im}(L(\lambda)x, x)| \geq |\tau|((2\mu A + B)x, x) > 0$$

for any $\lambda = \mu + i\tau$, $\mu > 0$, $\tau \neq 0$ and all $x \in \mathcal{D}(A_{11}) \oplus \mathcal{D}(C_{22}')$ with $x \neq 0$. Hence the non-real points of the open right half-plane belong to the resolvent set of L, i.e., the eigenvalues of L in the open right half-plane only may lie on the positive half-axis.

Suppose now that $\lambda_0 \in \sigma(\widetilde{L}_0)\backslash\sigma_{ess}(\widetilde{L}_0) \cup \{0\}$ is a normal eigenvalue of $\widetilde{L}_0$. Evidently, $\operatorname{Im}\lambda_0 = 0$. We write P_0 for the orthogonal projector onto the null-space of $\widetilde{L}_0(\lambda_0)$. The operator P_0 has finite rank. Set

$$\widehat{L}_0(\lambda) = I + \frac{1}{\lambda}\begin{pmatrix} R & 0 \\ 0 & 0 \end{pmatrix} + P_0.$$

The operator $\widehat{L}_0(\lambda)$ is invertible at λ_0 and at each point λ of some neighborhood of λ_0. According to (2.1) and the assumptions 1)–3),

$$(2.7) \qquad \widetilde{L}(\lambda) = \widehat{L}_0(\lambda) + Q(\lambda) - P_0 = \widehat{L}_0(\lambda)(I + \widetilde{G}(\lambda)),$$

where $\widetilde{G}(\lambda) = \widehat{L}_0(\lambda)^{-1}(Q(\lambda) - P_0)$ is an operator function with compact values, which is holomorphic in some neighborhood of λ_0. Using the representation (2.7) we now proceed as above to show that λ_0 is not an accumulation point of eigenvalues of L. $\qquad\qquad\square$

3 The Spectrum of the Operators $L(\lambda)$ for λ on the Positive Real Axis

Proposition 3.1 *For $\lambda \in (0, \infty)$ the operator $L(\lambda)$, defined on $\mathcal{D}(L(\lambda)) = \mathcal{D}(A_{11}) \oplus \mathcal{D}(C'_{22})$, is selfadjoint, bounded from below and has a discrete spectrum.*

Proof: Let $\lambda \in (0, \infty)$. By the assumptions 1) and 3) the operator

$$\lambda^2 A + C' = \begin{pmatrix} \lambda^2 A_{11} & 0 \\ 0 & C'_{22} \end{pmatrix}$$

is selfadjoint and strictly positive, i.e., $\lambda^2 A + C' \gg 0$. Again by assumption 3) we have

$$C'' = T \begin{pmatrix} A_{11} & 0 \\ 0 & C'_{22} \end{pmatrix} = T_\lambda (\lambda^2 A + C'),$$

where

$$T_\lambda = T \begin{pmatrix} \frac{1}{\lambda^2} I & 0 \\ 0 & I \end{pmatrix},$$

which is compact since T has this property. This representation shows that C'' is $\lambda^2 A + C'$-compact which implies, see [K], that the operator

$$\lambda^2 A + C = (\lambda^2 A + C') + C''$$

is selfadjoint and bounded from below. Since $B \geq 0$, $L(\lambda)$ is bounded from below.

Next we prove the selfadjointness of the operator $L(\lambda)$. Choose $m_\lambda \geq 0$ such that $L(\lambda) + m_\lambda \cdot I \gg 0$. It is easy to show that

$$(3.1) \qquad (L(\lambda) + m_\lambda \cdot I)(\lambda^2 A + C')^{-1} = (I + K_\lambda)\widetilde{L}_0(\lambda)$$

where

$$K_\lambda = \left[\lambda \begin{pmatrix} K & 0 \\ 0 & 0 \end{pmatrix} + (\lambda B'' + C'' + m_\lambda \cdot I)(\lambda^2 A + C')^{-1} \right] \tilde{L}_0(\lambda)^{-1}$$

and $\tilde{L}_0(\lambda)$ is the operator defined in (2.3). By assumption 2) K is compact and by assumptions 1) and 3) the inverse of $\lambda^2 A + C'$ is compact. It follows that K_λ is a compact operator. Since $L(\lambda) + m_\lambda$ is injective, the same statement holds for $I + K_\lambda$. Hence $I + K_\lambda$ is also surjective, which implies by (3.1) that $L(\lambda) + m_\lambda \cdot I$ is surjective. By von Neumann's Theorem the selfadjointness of $L(\lambda) + m_\lambda \cdot I$ and thus of $L(\lambda)$ follows.

The last assertion remains to be proved. From (3.1) it follows that

$$(L(\lambda) + m_\lambda \cdot I)^{-1} = (\lambda^2 A + C')^{-1} \tilde{L}_0(\lambda)^{-1}(I + K_\lambda)^{-1}$$

is a compact operator since $\lambda^2 A + C'$ has a compact inverse. Hence the spectrum of $L(\lambda) + m_\lambda \cdot I$ is discrete which yields the discreteness of the spectrum of $L(\lambda)$. $\qquad \square$

Theorem 3.2 *Suppose that the operators A, B and C fulfill the assumptions* 1), 2) *and* 3). *Set*

$$\tilde{A} := A + C' = \begin{pmatrix} A_{11} & 0 \\ 0 & C'_{22} \end{pmatrix} \qquad (\gg 0)$$

and

$$\beta := \min \sigma(\tilde{A}) \quad (> 0).$$

Then for every q, $0 < q < 1$, there is a positive number α such that $L(\lambda) \geq q\beta I$, for all $\lambda \geq \alpha$. Further, $L(\lambda)^{-\frac{1}{2}}$ (and hence $L(\lambda)^{-1}$) converges strongly to zero as λ tends to infinite, i.e.,

$$(3.2) \qquad\qquad L(\lambda)^{-\frac{1}{2}} x \longrightarrow 0 \qquad as \qquad \lambda \to \infty$$

for all $x \in \mathcal{H}_1 \oplus \mathcal{H}_2$.

Proof: From the proof of Proposition 3.1 we know that $\tilde{A} + C''$ is selfadjoint and bounded from below since C'' is $\tilde{A}$-compact. This also yields the relation

$$(3.3) \qquad\qquad \sigma_{ess}(\tilde{A} + C'') = \sigma_{ess}(\tilde{A}) = \emptyset.$$

By the definition of β the number of eigenvalues of $\tilde{A} + C''$ less than β counted according to their multiplicities is finite which means that the spectral subspace

$\mathcal{H}_-(\beta)$ of the operator $\widetilde{A} + C''$ corresponding to the interval $(-\infty, \beta)$ is finite-dimensional. Let P_- denote the orthogonal projection of $\mathcal{H}_1 \oplus \mathcal{H}_2$ onto $\mathcal{H}_-(\beta)$ and v_0 be the smallest eigenvalue of $\widetilde{A} + C''$. Set

$$\widehat{L}(\lambda) := L(\lambda) + (\beta - v_0)P_-.$$

Since $\widetilde{A} + C'' \geq v_0 I$, $B \geq 0$ and $(\beta - v_0)P_- \geq \max\{\beta - v_0, 0\}I$, we conclude that

$$\widehat{L}(1) \geq \beta I.$$

Because the domains of the operators $L(\lambda)$ and thus of $\widehat{L}(\lambda)$ are λ-independent, we obtain the estimate

$$(3.4) \qquad \begin{cases} \widehat{L}(\lambda) = (\lambda^2 - 1)A + (\lambda - 1)B + \widehat{L}(1) \\ \qquad \geq (\lambda - 1)(A + B'') + \beta I \geq \beta I \end{cases}$$

for all $\lambda \geq 1$. The assumptions 1) and 2) imply that $A + B'' \geq 0$. Let $E(t)$, $-\infty < t < \infty$, denote its spectral function. From (3.4) we conclude that

$$(3.5) \qquad \begin{aligned} (\widehat{L}(\lambda)^{-1}x, x) &\leq ([(\lambda - 1)(A + B'') + \beta I]^{-1}x, x) \\ &= \int_0^\infty \frac{1}{(\lambda - 1)t + \beta} d(E(t)x, x) \end{aligned}$$

holds for arbitrary $x \in \mathcal{H}_1 \oplus \mathcal{H}_2$ and $\lambda \geq 1$. The positivity of $A + B''$ yields the relation $E(0-) = E(0) = 0$ which implies that

$$(3.6) \qquad \lim_{\lambda \to \infty} (\widehat{L}(\lambda)^{-1}x, x) = 0$$

or, equivalently,

$$(3.7) \qquad \lim_{\lambda \to \infty} \widehat{L}(\lambda)^{-\frac{1}{2}}x = 0$$

for all $x \in \mathcal{H}_1 \oplus \mathcal{H}_2$. By definition

$$L(\lambda) = \widehat{L}(\lambda) - (\beta - v_0)P_- = \widehat{L}(\lambda)^{\frac{1}{2}}(I - S(\lambda))\widehat{L}(\lambda)^{\frac{1}{2}}$$

where

$$S(\lambda) = (\beta - v_0)\widehat{L}(\lambda)^{-\frac{1}{2}}P_-\widehat{L}(\lambda)^{-\frac{1}{2}}$$

is a bounded selfadjoint operator of fine rank. If $v_0 \geq \beta$, then $L(\lambda) = \hat{L}(\lambda)$ and the proof is complete. Otherwise, let $e_1, \ldots, e_l$ be an orthonormal basis of $\mathcal{H}_-(\beta)$. Then P_- has the representation

$$P_- x = \sum_{s=1}^{l} (x, e_s) e_s, \qquad x \in \mathcal{H}_1 \oplus \mathcal{H}_2.$$

It follows that

$$|S(\lambda) x| = (\beta - v_0) \left| \sum_{s=1}^{l} (\hat{L}(\lambda)^{-\frac{1}{2}} x, e_s) \hat{L}(\lambda)^{-\frac{1}{2}} e_s \right|$$

$$\leq (\beta - v_0) \sum_{s=1}^{l} |\hat{L}(\lambda)^{-\frac{1}{2}} e_s|^2 |x|$$

and thus

$$|S(\lambda)| \leq (\beta - v_0) \sum_{s=1}^{l} |\hat{L}(\lambda)^{-\frac{1}{2}} e_s|^2.$$

Let $0 < q < 1$. Choose a positive number α such that

$$(\beta - v_0) \sum_{s=1}^{l} |\hat{L}(\lambda)^{-\frac{1}{2}} e_s|^2 \leq 1 - q \quad (< 1)$$

for all $\lambda \geq \alpha$. We conclude that for such λ

$$(L(\lambda) x, x) = ((I - S(\lambda)) \hat{L}(\lambda)^{-\frac{1}{2}} x, \hat{L}(\lambda)^{-\frac{1}{2}} x) \geq q(\hat{L}(\lambda) x, x)$$

for all $x \in \mathcal{D}(L(\lambda)) = \mathcal{D}(A_{11}) \times \mathcal{D}(C'_{22})$, i.e.,

(3.8) $$L(\lambda) \geq q\hat{L}(\lambda) \geq q\beta I$$

which proves the first assertion of Theorem 3.2. It follows that

$$(L(\lambda)^{-1} x, x) \leq \frac{1}{q} (\hat{L}(\lambda)^{-1} x, x) \leq \frac{1}{q\beta} (x, x)$$

and hence

(3.9) $$|L(\lambda)^{-\frac{1}{2}} x| \leq \frac{1}{\sqrt{q}} |\hat{L}(\lambda)^{-\frac{1}{2}} x| \leq \frac{1}{\sqrt{q\beta}} |x|$$

for $x \in \mathcal{H}_1 \oplus \mathcal{H}_2$, $\lambda \geq \alpha$. The relation (3.7) yields the assertion (3.2). Finally, by (3.9)

$$|L(\lambda)^{-1}x| \leq \frac{1}{\sqrt{q\beta}}|L(\lambda)^{-\frac{1}{2}}x|, \qquad x \in \mathcal{H}_1 \oplus \mathcal{H}_2, \ \lambda \geq \alpha$$

which completes the proof of Theorem 3.2. $\qquad\qquad\qquad\qquad\qquad\square$

4 The Spectrum of the Pencil L in the Open Right Half-plane

By Theorem 2.1 the essential spectrum of the pencil L is contained in the interval $[-\|R\|, 0]$ and its discrete spectrum in the open right half-plane is either empty or located on the positive real axis. By Theorem 3.2 there is a positive number α such that $L(\lambda) \gg 0$ for all $\lambda \geq \alpha$. Hence, the spectrum of L in the open right half-plane (if not empty) consists of a sequence (μ_j) of normal eigenvalues which converges to zero if it is an infinite sequence. We enumerate these eigenvalues in such a way that

(4.1) $$\alpha > \mu_1 > \mu_2 > \mu_3 > \cdots > 0.$$

Proposition 4.1 *Let* $x \in \mathcal{D}(A_{11}) \oplus \mathcal{D}(C'_{22})$, $x \neq 0$, *and define*

$$\varphi(\lambda) := (L(\lambda)x, x), \qquad \lambda \in [0, \infty).$$

Then φ *is a real-valued continuous function which is increasing in* $(0, \infty)$.

Proof: Consider

$$\varphi'(\lambda) = 2\lambda(Ax, x) + (Bx, x).$$

Since A and B are non-negative operators, $\varphi'(\lambda) \geq 0$ in $[0, \infty)$. Now let $\lambda > 0$ and suppose that $\varphi'(\lambda) = 0$. Then both (Ax, x) and (Bx, x) vanish. Set $x = (x_1, x_2)'$ with $x_1 \in \mathcal{D}(A_{11})$, $x_2 \in \mathcal{D}(C'_{22})$. Then $(A_{11}x_1, x_1) = 0$ whence $x_1 = 0$ since $A_{11} \gg 0$. From $(Bx, x) = 0$ it then follows that $(B''_{22}x_2, x_2) = 0$ whence $x_2 = 0$ since $B''_{22} > 0$. Thus we have the contradiction $x = 0$. $\qquad\qquad\square$

Remark 4.2 The eigenvalues μ_j of the pencil L are semi-simple, i.e., their algebraic multiplicities coincide with their geometric multiplicities.

Proof: Let $x_0, x_1 \in \mathcal{D}(A_{11}) \oplus \mathcal{D}(C'_{22})$, $x_0 \neq 0$, such that

(4.2) $$L(\mu_j)x_0 = 0, \qquad L(\mu_j)x_1 + L'(\mu_j)x_0 = 0.$$

Since $L(\mu_j)$ is a selfadjoint operator, (4.2) yields

$$(L'(\mu_j)x_0, x_0) = 0$$

which contradicts the first statement of Proposition 4.1 (with x_0 instead of x in the definition of φ).   □

Theorem 4.3 *Let the operators A, B and C fulfill the assumptions* 1), 2) *and* 3). *Let N_- denote the dimension of any maximal negative subspace of the quadratic form (Cx, x), $x \in \mathcal{D}(A_{11}) \oplus \mathcal{D}(C'_{22})$. Then the following statements hold:*

i) *The spectrum of the pencil L in the open right half-plane is empty if and only if $N_- = 0$, i.e., the quadratic form (Cx, x), $x \in \mathcal{D}(A_{11}) \oplus \mathcal{D}(C'_{22})$, is non-negative.*

ii) *If $0 < N_- \leq \infty$, then*

$$(4.3) \qquad\qquad \sum_j n_j = N_-,$$

where the natural numbers n_j denote the multiplicities of the eigenvalues μ_j.

Proof: For $\lambda \geq 0$, let $n_-(\lambda)$ denote the dimension of any maximal negative subspace of the quadratic form $(L(\lambda)x, x)$, $x \in \mathcal{D}(A_{11}) \oplus \mathcal{D}(C'_{22})$. In particular $n_-(0) = N_-$. Note that $n_-(\lambda)$ coincides with the dimension of the spectral subspace of the operator $L(\lambda)$ corresponding to its negative spectrum. By Proposition 3.1 $n_-(\lambda)$ is zero or a positive integer and by Proposition 4.1 n_- is a non-increasing function in $[0, \infty)$. According to Theorem 3.2 there is a positive number α such that $n_-(\lambda) = 0$ for $\lambda \geq \alpha$. Therefore, n_- is a piecewise constant function which has only a finite number of discontinuity points to the right of any λ_0; at these points the value of n_- drops by positive integers until $n_-(\lambda)$ becomes and remains zero.

First we show that the discontinuity points of n_- coincide with the eigenvalues of the pencil L and the relation

$$(4.4) \qquad\qquad n_-(\mu_j - 0) - n_-(\mu_j + 0) = n_j$$

holds.

Suppose that $\lambda_0 > 0$ is a discontinuity point of n_- and that at the same time the operator $L(\lambda_0)$ is invertible. Let us first suppose that $n_-(\lambda_0 - 0) > n_-(\lambda_0)$, i.e., $n_-(\lambda) > n_-(\lambda_0)$ for all $\lambda \in (0, \lambda_0)$. By $\mathcal{M}_\pm(\lambda_0)$ let us denote the spectral subspace of the operator $L(\lambda_0)$ corresponding to its positive and negative spectrum, respectively. From the invertibility of $L(\lambda_0)$ it follows that

$$(4.5) \qquad (L(\lambda_0)x, x) \geq \eta \|x\|^2$$

for all $x \in \mathcal{M}_+(\lambda_0) \cap (\mathcal{D}(A_{11}) \oplus \mathcal{D}(C'_{22}))$ where η denotes the smallest positive eigenvalue of $L(\lambda_0)$. By the supposition

$$n_-(\lambda) > n_-(\lambda_0) = \dim \mathcal{M}_-(\lambda_0) \qquad (\lambda \in (0, \lambda_0))$$

there is a vector $x_\lambda \in \mathcal{D}(A_{11}) \oplus \mathcal{D}(C'_{22})$ such that $x_\lambda \in \mathcal{M}_-(\lambda_0)^\perp = \mathcal{M}_+(\lambda_0)$ and at the same time

$$(4.6) \qquad (L(\lambda)x_\lambda, x_\lambda) < 0.$$

Choose some $\lambda_1 \in (0, \lambda_0)$. Since the operator $L(\lambda_1)$ is bounded from below, there is a positive number β_1 such that $L(\lambda_1) + \beta_1 I \gg 0$. Taking into account that L is a non-decreasing operator function, it follows that $L(\lambda) + \beta_1 I \gg 0$ for any $\lambda \geq \lambda_1$. Hence for such λ the operators $L(\lambda) + \beta_1 I$ are invertible and

$$(4.7) \qquad \|(L(\lambda) + \beta_1 I)^{-1}\| \leq (\nu_1 + \beta_1)^{-1},$$

where ν_1 denotes the smallest eigenvalue of $L(\lambda_1)$. Instead of L let us consider the bounded operator function G, defined by

$$(4.8) \qquad G(\lambda) := L(\lambda)(L(\lambda) + \beta_1 I)^{-1} = I - \beta_1(L(\lambda) + \beta_1 I)^{-1}$$

for $\lambda \in (\lambda_1, \infty)$. The operator function G is continuously differentiable in (λ_1, ∞), even holomorphic in an open complex neighbourhood of this interval. Note that the operator $G(\lambda_0)$ is invertible and that $\mathcal{M}_\pm(\lambda_0)$ are also the spectral subspaces of this operator corresponding to its positive and negative spectrum, respectively. Instead of (4.5) we have

$$(4.9) \qquad (G(\lambda_0)x, x) \geq \frac{\eta}{\eta + \beta_1} \|x\|^2$$

for all $x \in \mathcal{M}_+(\lambda_0) \cap (\mathcal{D}(A_{11}) \oplus \mathcal{D}(C'_{22}))$. The selfadjointness of $L(\lambda)$ yields the relation

$$
\begin{aligned}
(G(\lambda)y, y) &= ((L(\lambda) + \beta_1 I)^{-1} y, L(\lambda)y) \\
&= \frac{1}{\beta_1}(L(\lambda)y, y) - \frac{1}{\beta_1}((L(\lambda) + \beta_1 I)^{-1} L(\lambda)y, L(\lambda)y) \\
&\leq \frac{1}{\beta_1}(L(\lambda)y, y)
\end{aligned}
$$

for any y in the domain of $L(\lambda)$ whence

$$(4.10) \qquad\qquad (G(\lambda)x_\lambda, x_\lambda) < 0$$

for the vectors x_λ introduced above. Choose $(\lambda_0 - \lambda) > 0$ so small that

$$(4.11) \qquad\qquad \|G(\lambda) - G(\lambda_0)\| < \frac{\eta}{2(\eta + \beta_1)}.$$

Using (4.9) and (4.11) we conclude that

$$
\begin{aligned}
(G(\lambda)x_\lambda, x_\lambda) &= (G(\lambda_0)x_\lambda, x_\lambda) + ((G(\lambda) - G(\lambda_0))x_\lambda, x_\lambda) \\
&\geq \frac{\eta}{\eta + \beta_1} \|x_\lambda\|^2 - \|G(\lambda) - G(\lambda_0)\| \|x_\lambda\|^2 \\
&> \frac{\eta}{2(\eta + \beta_1)} \|x_\lambda\|^2 > 0
\end{aligned}
$$

which contradicts the inequality (4.10). Therefore $n_-(\lambda_0 - 0) = n_-(\lambda_0)$. By an analogous argumentation we obtain $n_-(\lambda_0 + 0) = n_-(\lambda_0)$. Note that for λ in a neighbourhood of λ_0 the operator $L(\lambda)$ is invertible so that the spectral subspace $\mathcal{M}_\pm(\lambda)$ are well-defined and $\mathcal{M}_-(\lambda)^\perp = \mathcal{M}_+(\lambda)$.

For an eigenvalue μ_j of the pencil L let us denote the spectral subspace of $L(\mu_j)$ corresponding to its positive and negative spectrum by $\mathcal{M}_\pm(\mu_j)$, respectively. Further denote the null space of $L(\mu_j)$ by $\mathcal{M}_0(\mu_j)$. By Proposition 4.1

$$(4.12) \qquad\qquad \mathcal{M}_-(\mu_j) \oplus \mathcal{M}_0(\mu_j) \subset \mathcal{M}_-(\lambda)$$

for $0 < \lambda < \mu_j$, and

$$\mathcal{M}_+(\mu_j) \oplus \mathcal{M}_0(\mu_j) \subset \mathcal{M}_+(\lambda)$$

and hence

$$(4.13) \qquad\qquad \mathcal{M}_-(\mu_j) \supset \mathcal{M}_-(\lambda)$$

for $\mu_j < \lambda$. It follows that

$$(4.14) \qquad\qquad n_-(\lambda) \geq n_-(\mu_j) + n_j$$

for $\lambda < \mu_j$ and

$$(4.15) \qquad\qquad n_-(\lambda) \leq n_-(\mu_j)$$

for $\lambda > \mu_j$. Suppose that

$$n_-(\lambda) > n_-(\mu_j) + n_j$$

for all positive numbers $\lambda < \mu_j$. Then for any such λ there exists a vector $x_\lambda \in \mathcal{D}(A_{11}) \oplus \mathcal{D}(C'_{22})$ such that

$$x_\lambda \in \mathcal{M}_-(\lambda) \cap (\mathcal{M}_-(\mu_j) \oplus \mathcal{M}_0(\mu_j))^\perp = \mathcal{M}_-(\lambda) \cap \mathcal{M}_+(\mu_j)$$

which implies that

$$(4.16) \qquad (L(\mu_j)x_\lambda, x_\lambda) \geq \eta \|x_\lambda\|^2, \qquad (L(\lambda)x_\lambda, x_\lambda) < 0$$

where here η denotes the smallest positive eigenvalue of $L(\mu_j)$. By the same argumentation as above we conclude that the inequalities (4.16) are incompatible for sufficiently small values of $(\mu_j - \lambda) > 0$.

Next we suppose that

$$n_-(\lambda) < n_-(\mu_j)$$

for all $\lambda > \mu_j$. Then for any $\lambda \in (\mu_j, \mu_{j-1})$ there is a vector $x_\lambda \in \mathcal{D}(A_{11}) \oplus \mathcal{D}(C'_{22})$ such that

$$x_\lambda \in \mathcal{M}_-(\mu_j) \cap \mathcal{M}_-(\lambda)^\perp = \mathcal{M}_-(\mu_j) \cap \mathcal{M}_+(\lambda)$$

which means that

$$(4.17) \qquad (L(\mu_j)x_\lambda, x_\lambda) \leq -\eta \|x_\lambda\|^2, \qquad (L(\lambda)x_\lambda, x_\lambda) > 0$$

where here $-\eta$ denotes the largest negative eigenvalue of $L(\mu_j)$. As above we infer that the inequalities (4.17) are incompatible if $\lambda - \mu_j > 0$ is sufficiently small.

Summarizing we conclude that equality holds in (4.12), (4.13), (4.14) and (4.15); in particular the relation (4.4) is proved. It follows that

$$n_-(\lambda) = \sum_{\mu_j > \lambda} n_j$$

for $\lambda > 0$ which implies that

$$N_- = n_-(0) \geq n_-(+0) = \sum_j n_j.$$

Finally suppose that $n_-(0) > n_-(0+)$. Consider an $n_-(0+) + 1$–dimensional negative subspace $\mathcal{N}_0$ of the quadratic form $(L(0)x, x) = (Cx, x), x \in \mathcal{D}(A_{11}) \oplus \mathcal{D}(C'_{22})$. Since $(L(\lambda), x, x)$ is a continuous function with respect to λ for a fixed element $x \in \mathcal{D}(A_{11}) \oplus \mathcal{D}(C'_{22})$ we infer that $\mathcal{N}_0 \subset \mathcal{M}_-(\lambda)$ for all sufficiently small $\lambda > 0$ which yields the contradiction $n_-(0+) \geq n_-(0+) + 1$. $\qquad \square$

Acknowledgements

This work was supported by the Deutsche Forschungsgemeinschaft, DFG, within a German-Ukrainian cooperation. V. Adamyan and V. Pivovarchik also wish to thank the University of Regensburg for its hospitality while this work was done.

References

[AP]   V.M. Adamyan and V.N. Pivovarchik, On the Spectra of Some Classes of Quadratic Operator Pencils, *Operator Theory: Adv. and Appl.* **106** (1998), 23–36.

[BP]   S.M. Barkar and V.N. Pivovarchik, On the Spectra of a Class of Operator Pencils in the Right Half-Plane (in Russian), Buletinul Academiei de Stiinte a Republicii Moldova, *Matematica* **1(7)** (1992), 17–28.

[GGK] I.C. Gohberg, S. Goldberg and M.A. Kaashoek, Classes of Linear Operators, volume I, *Operator Theory: Adv. and Appl.* **49**, Birkhäuser-Verlag Basel-Boston-New York, 1996.

[GK]   I.C. Gohberg and M.G. Krein, Introduction to the Theory of Linear Nonselfadjoint Operators, *Amer. Math. Soc.*, Providence, 1988.

[GS]   R.O. Griniv and A.A. Shkalikov, On Operator Pencils Arising in the Problem of Beam Oscillations with Internal Damping (in Russian), *Matem. Zametki* **56**, #2, (1994), 114–131; *English Transl. in Math. Notes* **56** (1994).

[K]   T. Kato, Perturbation Theory for Linear Operators, Springer-Verlag Berlin-Heidelberg-New York, 1966.

[KL]   M.G. Krein and H. Langer, On Some Mathematical Principles in the Linear Theory of Damped Oscillations of Continua I, II, *Integral Equations Operator Theory* **1** (1978), 364–399, 539–566.

[LS]   P. Lancaster and A.A. Shkalikov, Damped Vibrations of Beams and Related Spectral Problems, *Canadian Appl. Math. Quart.* **2**, #4 (1994), 45–90.

[Li]   A. Lifshitz, Magnetohydrodynamics and Spectral Theory, Kluwer Academic Publishers, Dordrecht/Boston/London, 1989.

[M]   A. Markus, On Holomorphic Operator Functions, *Doklady Akad. Nauk SSSR* **119**, #6 (1958), 1099–1102.

[Pi1]   V.N. Pivovarchik, A Problem Connected with Oscillations of Elastic Beams with Internal and Viscous Damping (in Russian), *Moscow Univ. Bulletin* **42**, (1987), 68–71.

[Pi2]   V.N. Pivovarchik, On the Spectrum of Certain Quadratic Pencils of Unbounded Operators (in Russian), *Function. Anal. i Ego Prilozhen.* **23**, #1 (1989), 80–81.

[Pi3]   V.N. Pivovarchik, On the Positive Spectra of a Class of Polynomial Operator Pencils, *Integral Equations Operator Theory* **19** (1994), 314–326.

[S]   A.A. Shkalikov, Operator Pencils Arising in Elasticity and Hydrodynamics: the Instability Index Formula, *Operator Theory: Adv. and Appl.* **87** (1996), 358–385.

[T]   H. Tasso, Linear and Nonlinear Stability in Resistive Magnetohydrodynamics, *Annalen Physics* **234** (1994), 211.

Vadim Adamyan
University of Odessa
Department of Theoretical Physics
vul. Dvorjanska 2
270026 Odessa
Ukraine
e-mail: vadamjan@m-vox.odessa.ua

Reinhard Mennicken
University of Regensburg
NWF I – Mathematik
D - 93040 Regensburg
Germany
e-mail: reinhard.mennicken@
      mathematik.uni-regensburg.de

Vjacheslav Pivovarchik
Odessa State Academy of Civil
    Engineering and Architecture
Department of Higher Mathematics
vul. Didrykhsona 4
2770028 Odessa
Ukraine
e-mail: v.pivovarchik@paco.net

Keywords. Quadratic operator pencils, spectrum, essential spectrum, normal eigenvalue.

Operator Theory:
Advances and Applications, Vol. 124
© 2001 Birkhäuser Verlag Basel/Switzerland

Two-sided Tangential Interpolation
for Hilbert-Schmidt Operator
Functions on Polydisks

Daniel Alpay, Vladimir Bolotnikov and Leiba Rodman

Dedicated with respect and affection to Professor Israel Gohberg on the occasion
of his 70-th birthday

Bitangential interpolation problems are formulated for the class of Hilbert-Schmidt operator-
valued functions, which are analytic on a polydisk and have square summable power series. A
procedure is described for reduction of the problems in d-disk to the analogous problems in
$(d-1)$-disk. Using this procedure, for the case of the bidisk, the minimal norm solutions are
explicitly described in terms of the interpolation data, and formulas for the general solution are
obtained.

1 Introduction

Tangential interpolation theory of matrix and operator functions is by now a well
developed topic, and many ways to approach tangential interpolation have florished
in the recent years; we mention here only the books [8], [11], [13], [14]. The subject
originated with the work of Schur [18]; see the above mentioned books and the
review paper [12] for many additional references and historical perspective.

In the context of tangential interpolation, it is of special interest to consider
Hilbert spaces of analytic operator functions as sets in which the interpolant
functions are sought. The Hilbert space structure allows one to study orthogonal
decompositions of the set of solutions, which in turn can be applied to tangential
interpolation with symmetries and to certain multipoint interpolation problems.
This line of investigation, with the standard Hardy space H_2 of the unit disk as the
Hilbert space of matrix functions, has been pursued in [2], [5], [6].

Interpolation problems of matrix and operator functions of several variables,
in particular, functions defined in the polydisk, seem to be much less studied,
although the interest in these problems is rising, and new techniques have become
available recently. We refer to [1], [9], and see the book [17] for the general theory
of analytic functions in polydisks. It turns out, not surprisingly, that tangential
interpolation of operator functions of several variables, in particular, interpolation
in Hilbert spaces of such functions, is much more involved than the one variable case. The investigation of tangential interpolation in Hilbert spaces of operator functions of several variables has been initiated in [4], where the one-sided

tangential interpolation problem for matrix-valued Hardy functions on the bidisk was studied, and continued in [7] with the study of one-sided interpolation for Hilbert-Schmidt operator-valued functions on the polydisk.

In the present paper we continue the line of [4] and [7]. Our main problem to study is *two-sided*, or *bitangential*, interpolation for Hilbert-Schmidt operator-valued functions defined on the polydisk $\mathbb{D}^d$:

$$\mathbb{D}^d = \{\mathbf{z}_d = (z_1, \ldots, z_d) \in \mathbb{C}^d : \ |z_k| < 1\}.$$

Here and throughout the paper we denote by $\mathbf{z}_k = (z_1, \ldots, z_k)$ a point in $\mathbb{C}^k$.

Let $\mathcal{H}$ be a separable Hilbert space with the inner product $\langle \cdot, \cdot \rangle_{\mathcal{H}}$ and the norm $\| \cdot \|_{\mathcal{H}}$, and let

(1.1)

$$H(\mathbf{z}_d) = \sum_{j_1,\ldots,j_d=0}^{\infty} H_{j_1,\ldots,j_d} z_1^{j_1} z_2^{j_2} \ldots z_d^{j_d},$$

$$H_{j_1,\ldots,j_d} \in \mathcal{H}, \quad \mathbf{z}_d = (z_1, \ldots, z_d) \in \mathbb{D}^d$$

be a square summable power series:

(1.2)
$$\sum_{j_1,\ldots,j_d=0}^{\infty} \|H_{j_1,\ldots,j_d}\|_{\mathcal{H}}^2 < \infty.$$

The series (1.1) represents a function $H(\mathbf{z}_d)$ which is defined and analytic in $\mathbb{D}^d$. The set of all $\mathcal{H}$-valued functions of the form (1.1) with square summable coefficients endowed with the inner product

(1.3)
$$\langle G, H \rangle = \sum_{j_1,\ldots,j_d=0}^{\infty} \langle G_{j_1,\ldots,j_d}, H_{j_1,\ldots,j_d} \rangle_{\mathcal{H}},$$

is a Hilbert space which we shall call the *Hardy space* $\mathcal{H}(\mathbb{D}^d)$. The boundary values of functions in $\mathcal{H}(\mathbb{D}^d)$ are well-defined almost everywhere on

$$\mathbb{T}^d = \{\mathbf{z}_d = (z_1, \ldots, z_d) \in \mathbb{C}^d : \ |z_k| = 1\},$$

and form square integrable $\mathcal{H}$-valued functions defined on $\mathbb{T}^d$.

Let $\mathcal{H}$ and $\mathcal{G}$ be two separable Hilbert spaces. We denote by $\mathbf{HS}^{\mathcal{G} \to \mathcal{H}}$ the class of Hilbert-Schmidt operators acting from $\mathcal{G}$ into $\mathcal{H}$, which will be considered as a (separable) Hilbert space under the standard inner product. Thus, $\mathbf{HS}^{\mathcal{G} \to \mathcal{H}}(\mathbb{D}^d)$ denotes the Hardy space of operator-valued functions which are analytic in $\mathbb{D}^d$ and whose values are Hilbert-Schmidt operators acting from $\mathcal{G}$ into $\mathcal{H}$. $\mathbf{HS}^{\mathcal{G}\to\mathcal{H}}(\mathbb{D}^d)$ is a Hilbert space with the inner product

$$(1.4) \quad \langle G, H\rangle_{\mathbf{HS}^{\mathcal{G}\to\mathcal{H}}(\mathbb{D}^d)}$$
$$:= \frac{1}{(2\pi)^d}\int_0^{2\pi}\cdots\int_0^{2\pi}\operatorname{Trace}(H(e^{it_1},\ldots,e^{it_d})^*$$
$$G(e^{it_1},\ldots,e^{it_d}))\,dt_1\ldots dt_d.$$

Throughout the paper all Hilbert spaces are assumed to be separable, and all operators are assumed to be linear and bounded. We denote by spec A the spectrum of an operator A. We consider the two-sided interpolation problem whose data set is an ordered collection

$$(1.5) \quad \Omega = \{C_+,\ C_j,\ D_j,\ A_j,\ B_j,\ B_-,$$
$$\Gamma_j \mid j = 1,\ldots,d\}.$$

The collection consists of Hilbert spaces $\mathcal{H}$, $\mathcal{G}$, $\mathcal{H}_j$ and $\mathcal{G}_j$ (the Hilbert spaces are suppressed in the data set (1.5)), $5d$ operators $A_j : \mathcal{H}_j \to \mathcal{H}_j$, $D_j : \mathcal{G}_j \to \mathcal{G}_j$, $\Gamma_j : \mathcal{G}_j \to \mathcal{H}_j$ $(j = 1,\ldots,d)$ and two additional operators $C_+ : \mathcal{G}_1 \to \mathcal{H}$ and $B_- : \mathcal{G} \to \mathcal{H}_d$ such that

$$(1.6) \quad \bigcup_{j=1}^d \operatorname{spec} A_j \subset \mathbb{D} \quad \text{and} \quad \bigcup_{j=1}^d \operatorname{spec} D_j \subset \mathbb{D}.$$

We assume that the spaces $\mathcal{H}_j$ and $\mathcal{G}_j$ are finite dimensional. In what follows, the symbol $\prod_{j=k}^{\curvearrowleft n} T_j$ denotes the left product of square matrices T_j:

$$\prod_{j=k}^{\curvearrowleft n} T_j \stackrel{\text{def}}{=} T_n T_{n-1}\ldots T_k.$$

Problem 1.1 Given a data set (1.5), and a positive number γ, find all functions $H \in \mathbf{HS}^{\mathcal{G}\to\mathcal{H}}(\mathbb{D}^d)$ satisfying the norm constraint

$$(1.7) \quad \|H\|_{\mathbf{HS}^{\mathcal{G}\to\mathcal{H}}(\mathbb{D}^d)} \leq \gamma,$$

the left-sided interpolation condition

$$(1.8) \quad \frac{1}{(2\pi i)^d}\int_{\mathbb{T}^d}\left(\prod_{j=1}^{\curvearrowleft d}(z_j I - A_j)^{-1}B_j\right)H(\mathbf{z}_d)\,dz_1\ldots dz_d = B_-,$$

the right-sided interpolation condition

$$(1.9) \qquad \frac{1}{(2\pi i)^d} \int_{\mathbb{T}^d} H(\mathbf{z}_d) \left(\overset{d}{\underset{j=1}{\overset{\frown}{\prod}}} C_j(z_j I - D_j)^{-1} \right) dz_d \ldots dz_1 = C_+,$$

and d two-sided interpolation conditions

$$\frac{1}{(2\pi i)^d} \int_{\mathbb{T}^d} \left(\overset{k}{\underset{j=1}{\overset{\frown}{\prod}}} (z_j I - A_j)^{-1} B_j \right) H(\mathbf{z}_d)$$

$$(1.10)$$

$$\times \left(\overset{d}{\underset{\ell=k}{\overset{\frown}{\prod}}} C_\ell(z_\ell I - D_\ell)^{-1} \right) dz_1 \ldots dz_{k-1} dz_d \ldots dz_k = \Gamma_k,$$

where $k = 1, \ldots, d$.

Note that the differential in (1.10) may be replaced by $dz_d \ldots dz_{k+1} dz_1 \ldots dz_k$ and it is equal to $dz_d dz_{d-1} \ldots dz_1$ for $k = 1$. Note also that without loss of generality matrices A_j and D_j can be assumed to be of Jordan form since the problem with the data set

$$(1.11) \qquad \begin{aligned} &\{C_+ S_1^{-1}, \; S_{j+1} C_j S_j^{-1}, \; S_j D_j S_j^{-1}, \; T_j A_j T_j^{-1}, \\ &\quad T_j B_j T_{j-1}^{-1}, \; T_d B_-, \; T_j \Gamma_j S_j^{-1} \}, \end{aligned}$$

where S_j, T_j are invertible operators, is equivalent to Problem 1.1.

It is easy to see that the set of functions in $\mathbf{HS}^{\mathcal{G} \to \mathcal{H}}(\mathbb{D}^d)$ satisfying the conditions (1.8)–(1.10) is convex. Therefore, there is a unique minimal norm solution $H_{\min}$ of Problem 1.1. Moreover, all solutions, if exist, are of the form $H = H_{\min} + G$, where G is a general solution of the corresponding homogeneous problem with a suitable norm constraint. Note that G is orthogonal to $H_{\min}$.

In this paper we obtain necessary and sufficient conditions for Problem 1.1 to be solvable, and in this case obtain also formulas for the general solution. To this end, we develop further the scheme used in [7], which is based on reduction to the case with fewer variables. In contrast with [7], where Problem 1.1 was studied also with respect to the operator-valued inner product (obtained by omitting the trace in the formula (1.4)), we consider here the scalar-valued inner product (1.4) only; see the remark after the statement of Theorem 4.6.

Observe that the set of interpolating points coincides with the spectra of the operators A_j and D_j (and the spectral condition (1.6) provides that all these points are in $\mathbb{D}^d$) while the directions at which the interpolant H has the preassigned values are determined by the operators B_+ and C_- from the left and from the right, respectively. We illustrate the latter remark by the Nevanlinna-Pick problem in the bidisk $\mathbb{D}^2$. For $d = 2$ conditions (1.8)–(1.10) are of the form

(1.12)
$$\frac{1}{(2\pi i)^2} \int_{|\zeta|=1} (\zeta I - A_2)^{-1} B_2$$
$$\left(\int_{|\xi|=1} (\xi I - A_1)^{-1} B_1 H(\xi, \zeta) d\xi \right) d\zeta = B_-$$

(1.13)
$$\frac{1}{(2\pi i)^2} \int_{|\xi|=1} \left(\int_{|\zeta|=1} H(\xi, \zeta) C_2 (\zeta I - D_2)^{-1} d\zeta \right) C_1$$
$$(\xi I - D_1)^{-1} d\xi = C_+$$

(1.14)
$$\frac{1}{(2\pi i)^2} \int_{|\xi|=1} (\xi - A_1)^{-1} B_1$$
$$\left(\int_{|\zeta|=1} (H(\xi, \zeta) C_2 (\zeta I - D_2)^{-1} d\zeta) \right) \cdot C_1 (\xi I - D_1)^{-1} d\xi = \Gamma_1$$

(1.15)
$$\frac{1}{(2\pi i)^2} \int_{|\zeta|=1} (\zeta I - A_2)^{-1} B_2$$
$$\left(\int_{|\xi|=1} (\xi I - A_1)^{-1} B_1 H(\xi, \zeta) d\xi \right) \cdot C_2 (\zeta I - D_2)^{-1} d\zeta = \Gamma_2.$$

Example 1.2 (The two-sided Nevanlinna-Pick problem). Let be given $n_L + n_R$ points $\mathbf{z}_k = (z_{1k}, z_{2k})$ and $\mathbf{w}_\ell = (\omega_{1\ell}, \omega_{2\ell})$ in $\mathbb{D}^2$, vectors a_k, $c_\ell \in \mathbb{C}^p$ and b_k, $d_\ell \in \mathbb{C}^q$, and scalars $\gamma_{k\ell}^{(1)}$ and $\gamma_{k\ell}^{(2)}$ $(k = 1, \ldots, n_L; \ \ell = 1, \ldots, n_R)$. Let in (1.12)–(1.15) $C_1 = I_{n_R}$, $B_2 = I_{n_L}$,

(1.16) $\quad (B_1, B_-) = \begin{pmatrix} a_1^* & b_1^* \\ \vdots & \vdots \\ a_{n_L}^* & b_{n_L}^* \end{pmatrix}$, $\qquad \begin{pmatrix} C_+ \\ C_2 \end{pmatrix} = \begin{pmatrix} c_1 & \cdots & c_{n_R} \\ d_1 & \cdots & d_{n_R} \end{pmatrix}$,

$$A_j = \begin{pmatrix} z_{j1} & & \\ & \ddots & \\ & & z_{jn_L} \end{pmatrix}, \qquad D_j = \begin{pmatrix} \omega_{j1} & & \\ & \ddots & \\ & & \omega_{jn_R} \end{pmatrix},$$

(1.17)
$$\Gamma_j = \left(\gamma_{k\ell}^{(j)} \right)$$

for $j = 1, 2$ (the blank spaces in (1.17) are assumed to be zeros). It is easily seen that conditions (1.12) and (1.13) reduce then to the left and right-sided Nevanlinna-Pick conditions

$$a_k^* H(z_{1k}, z_{2k}) = b_k^* \quad (k = 1, \ldots, n_L) \quad \text{and}$$
$$H(\omega_{1\ell}, \omega_{2\ell}) d_\ell = c_\ell \quad (\ell = 1, \ldots, n_R),$$

respectively. Using the residue theorem we conclude that condition (1.14) reduces to

$$\gamma_{k\ell}^{(1)} = \begin{cases} a_k^* \dfrac{H(z_{1k}, \omega_{2\ell}) - H(\omega_{1\ell}, \omega_{2\ell})}{z_{1k} - \omega_{1\ell}} d_\ell & \text{for } z_{1k} \neq \omega_{1\ell} \\ a_k^* H_{z_1}'(\omega_{1\ell}, \omega_{2\ell}) d_\ell & \text{for } z_{1k} = \omega_{1\ell} \end{cases}$$

and quite similarly, that (1.15) reduces to

$$\gamma_{k\ell}^{(2)} = \begin{cases} a_k^* \dfrac{H(z_{1k}, z_{2k}) - H(z_{1k}, \omega_{2\ell})}{z_{2k} - \omega_{2\ell}} d_\ell & \text{for } z_{2k} \neq \omega_{2\ell} \\ a_k^* H_{z_2}'(z_{1k}, z_{2\ell}) d_\ell & \text{for } z_{2k} = \omega_{2\ell} \end{cases}$$

Note that for the special choice (1.16) and (1.17) the Sylvester equation

$$(A_1 \Gamma_1 - \Gamma_1 D_1) + (A_2 \Gamma_2 - \Gamma_2 D_2) = B_- C_2 - B_1 C_+$$

(which is a necessary condition for existence of interpolants; see Lemma 2.1 in the next section) reduces to

$$(z_{1k} - \omega_{1\ell}) \gamma_{k\ell}^{(1)} + (z_{2k} - \omega_{2\ell}) \gamma_{k\ell}^{(2)}$$
$$= b_k^* d_\ell - a_k^* c_\ell \quad (k = 1, \ldots, n_L; \ \ell = 1, \ldots, n_R).$$

Returning to the general Problem 1.1, and assuming (in view of (1.11)) that A_j and D_j are matrices in Jordan forms, one can recast (1.8)–(1.10) as conditions involving values of H and of its partial derivatives of higher order at different prescribed points in the polydisk $\mathbb{D}^d$. In the case of one variable, all possible sets of these conditions that arise via (1.8)–(1.10) have been identified; see, e.g., Theorem 16.8.1 in [8] for the class of rational matrix functions. In particular, such sets are necessarily nested. For the many variables case, a complete description of all possible sets of conditions in terms of the values of interpolant and of its several partial derivatives at prescribed points, is an open, and seemingly very difficult, problem. We present two illustrative examples.

Example 1.3 Fix two distinct points $\lambda, \mu \in \mathbb{D}$. Let A_1 be the $(n+1) \times (n+1)$ lower triangular Jordan block with eigenvalue λ, let A_2 be the $(m+1) \times (m+1)$ lower triangular Jordan block with eigenvalue μ, and define the matrices $B_1 \in \mathbb{C}^{n+1}$, $B_2 \in \mathbb{C}^{(m+1)\times(n+1)}$ and $B_- \in \mathbb{C}^{m+1}$ as follows:

$$
B_1 = \begin{pmatrix} 1 \\ 0 \\ \vdots \\ 0 \end{pmatrix}, \quad
B_2 = \begin{pmatrix} 0 & \cdots & 0 & 1 \\ 0 & \cdots & 0 & 0 \\ \vdots & \ddots & \vdots & \vdots \\ 0 & \cdots & 0 & 0 \end{pmatrix}, \quad
B_- = \begin{pmatrix} b_0 \\ b_1 \\ \vdots \\ b_m \end{pmatrix}.
$$

Then, for $H \in \mathbf{HS}^{\mathbb{C}\to\mathbb{C}}(\mathbb{D}^2)$ the equation (1.8) amounts to the following conditions:

$$
\frac{\partial^{n+j}}{\partial z_1^n \partial z_2^j} H(\lambda, \mu) = n!\, j!\, b_j, \qquad j = 0, 1, \ldots, m.
$$

Example 1.4 Select four distinct points $\lambda_1, \lambda_2, \mu_1, \mu_2 \in \mathbb{D}$, and let

$$
A_1 = \begin{pmatrix} \lambda_1 & 0 \\ 0 & \lambda_2 \end{pmatrix}, \quad
A_2 = \begin{pmatrix} \mu_1 & 0 \\ 0 & \mu_2 \end{pmatrix}, \quad
B_1 = \begin{pmatrix} 1 \\ 1 \end{pmatrix}, \quad
B_2 = \begin{pmatrix} a & b \\ c & d \end{pmatrix},
$$

where a, b, c, d are complex numbers. The left-hand side of (1.8) (with $d = 2$) is equal to

$$
\begin{pmatrix} aH(\lambda_1, \mu_1) + bH(\lambda_2, \mu_1) \\ cH(\lambda_1, \mu_2) + dH(\lambda_2, \mu_2) \end{pmatrix}.
$$

Thus, in this example the interpolation condition (1.8) can be recast in terms of certain combinations of the values of H at the four points $(\lambda_j, \mu_k) \in \mathbb{T}^2$, $j, k = 1, 2$.

In the matrix case (all Hilbert spaces involved are finite dimensional) and for one variable, Problem 1.1 is well-known and was extensively studied. Without the norm constraint (1.7), but with an arbitrary domain in the complex plane replacing the open unit disk, it is known as the *two sided Lagrange-Sylvester tangential interpolation*. With the norm constraint (using the H_∞ norm), but still with respect to an arbitrary domain, it is called the *Nudelman interpolation*, and in the particular case when the domain is the unit disk, it is the *generalized Nevanlinna-Pick interpolation*. See the book [8] for a comprehensive exposition of these and related two sided interpolation problems in one variable, in the class of rational matrix functions.

Observe that Problem 1.1 is a two sided interpolation problem in a reproducing kernel Hilbert space of operator-valued functions. A general theory of tangential interpolation in the context of reproducing kernels and Hilbert modules was developed in [3]. As a particular case of this theory, for the Hardy space of one variable, formulas for solutions of one-sided interpolation problem have been obtained in [3].

Finally, we note a limitation of the interpolation conditions (1.8)–(1.10). As the referee pointed out, a natural interpolation condition for functions of several variables would be interpolation along an arbitrary subvariety. (The conditions (1.8)–(1.10) correspond to certain subvarieties of dimension zero.) This "sub-variety interpolation" in one variable has been treated in the context of zero-pole, or homogeneous (in the terminology of [8]), interpolation in [10]. As far as we know, it had not been attacked yet for the several variables case. It does not seem likely that the integral form representation of the interpolation conditions would be appropriate there, one reason being that subvarieties of positive dimension in a complex vector space are neither compact nor bounded.

We describe briefly the contents of the paper section by section. In Section 2 we derive necessary conditions (which later will be proved also sufficient) for solvability of Problem 1.1, without the norm constraint. Section 3 is auxillary: there we recall the solution of the one-variable problem obtained in [5], and present other preliminary material. Our main results are given in Section 4, in which the procedure of reducing the number of variables, and its full justification, is described. The "other side" variation of the main Theorem 4.6, and the particular cases of one-sided interpolation problems are given in Section 5. Finally, Section 6 contains a complete description of the solutions of Problem 1.1 in the special case of two variables.

A challenging problem, which is beyond the scope of this paper, is to extend, if possible, the analysis and results to the case when $\mathcal{H}_j$ and $\mathcal{G}_j$ are infinite dimensional Hilbert spaces.

2 Necessary Conditions

To start with, we derive necessary conditions (which later prove to be also sufficient) for existence of solutions H of the interpolation problem (1.8)–(1.10), i.e., Problem 1.1 with the norm constraint (1.7) omitted.

Making use of the expansion (1.1) leads to the explicit expression of integrals in the left hand sides of (1.8) and (1.9) in terms of Taylor coefficients of the function H as

$$\sum_{j_1,\ldots,j_d=0}^{\infty} A_d^{j_d} B_d \ldots A_1^{j_1} B_1 H_{j_1,\ldots,j_d} = B_-$$

and

$$\sum_{j_1,\ldots,j_d=0}^{\infty} H_{j_1,\ldots,j_d} C_d D_d^{j_d} \ldots C_1 D_1^{j_1} = C_+,$$

respectively. From these equalities it follows that the conditions

(2.1) $\operatorname{Ran} B_- \subseteq \operatorname{span}\{\operatorname{Ran} (A_d^{j_d} B_d \ldots A_1^{j_1} B_1); \ j_1,\ldots,j_d = 0,1,\ldots\}$

(we denote here and elsewhere in the paper the range of an operator X by Ran X) and

$$(2.2) \qquad \text{Ker } C_+ \supseteq \bigcap_{j_1,\dots,j_d=0}^{\infty} \text{Ker } (C_d D_d^{j_d} \dots C_1 D_1^{j_1})$$

are necessary for Problem 1.1 to be solvable. Equations (1.10) lead to additional necessary conditions:

$$(2.3) \qquad \text{Ker } \Gamma_k \supseteq \bigcap_{j_k,\dots,j_d=0}^{\infty} \text{Ker } (C_d D_d^{j_d} \dots C_k D_k^{j_k}),$$

$$\text{Ran } \Gamma_k \subseteq \text{span}\{\text{Ran } (A_k^{j_k} B_k \dots A_1^{j_1} B_1);$$

$$(2.4) \qquad j_1, \dots, j_k = 0, 1, \dots\},$$

for $k = 1, \dots, d$.

Still another necessary condition for Problem 1.1 to be solvable is established in the following lemma.

Lemma 2.1 *Let conditions* (1.8)–(1.10) *are satisfied for a function* $H \in \mathbf{HS}^{\mathcal{G} \to \mathcal{H}}$ ($\mathbb{D}^d$). *Then the following Sylvester identity holds:*

$$
\left(\overset{d}{\underset{j=2}{\overset{\frown}{\prod}}} B_j \right) (A_1 \Gamma_1 - \Gamma_1 D_1)
$$

$$(2.5) \qquad + \sum_{k=2}^{d-1} \left(\overset{d}{\underset{j=k+1}{\overset{\frown}{\prod}}} B_j \right) (A_k \Gamma_k - \Gamma_k D_k) \left(\overset{k-1}{\underset{\ell=1}{\overset{\frown}{\prod}}} C_\ell \right)$$

$$
+ (A_d \Gamma_d - \Gamma_d D_d) \overset{d-1}{\underset{j=1}{\overset{\frown}{\prod}}} C_j = B_- \overset{d}{\underset{j=1}{\overset{\frown}{\prod}}} C_j - \overset{d}{\underset{j=1}{\overset{\frown}{\prod}}} B_j C_+.
$$

Proof: It follows from (1.8)–(1.10), on account of equalities

$$(2.6) \qquad A_k(zI - A_k)^{-1} = z(zI - A_k)^{-1} - I$$

and

$$(2.7) \qquad (zI - D_k)^{-1} D_k = z(zI - D_k)^{-1} - I,$$

that

(2.8)
$$A_1\Gamma_1 - \Gamma_1 D_1 = -B_1 C_+ + \frac{1}{(2\pi i)^d} \int_{\mathbb{T}^d} (z_1 I - A_1)^{-1} B_1 H(\mathbf{z}_d)$$

$$\cdot \overset{d}{\underset{j=2}{\overset{\frown}{\prod}}} C_j (z_j I - D_j)^{-1} C_1 \, dz_d \ldots dz_1,$$

(2.9)
$$A_d \Gamma_d - \Gamma_d D_d = B_- C_d - \frac{1}{(2\pi i)^d} \int_{\mathbb{T}^d}$$

$$B_d \overset{d-1}{\underset{j=1}{\overset{\frown}{\prod}}} (z_j I - A_j)^{-1} B_j H(\mathbf{z}_d)$$

$$\cdot C_d (z_d I - D_d)^{-1} \, dz_1 \ldots dz_d$$

and

$$A_k \Gamma_k - \Gamma_k D_k = \frac{1}{(2\pi i)^d} \int_{\mathbb{T}^d} \left(\overset{k}{\underset{j=1}{\overset{\frown}{\prod}}} (z_j I - A_j)^{-1} B_j \right) H(\mathbf{z}_d)$$

$$\cdot \left(\overset{d}{\underset{j=k+1}{\overset{\frown}{\prod}}} C_j (z_j I - D_j)^{-1} \right) C_k \, dz_1 \ldots dz_k dz_d \ldots dz_{k+1}$$

$$- \frac{1}{(2\pi i)^d} \int_{\mathbb{T}^d} B_k \left(\overset{k-1}{\underset{j=1}{\overset{\frown}{\prod}}} (z_j I - A_j)^{-1} B_j \right) H(\mathbf{z}_d)$$

$$\cdot \left(\overset{d}{\underset{j=k}{\overset{\frown}{\prod}}} C_j (z_j I - D_j)^{-1} \right) dz_1 \ldots dz_k dz_d \ldots dz_{k+1}$$

for $k = 2, \ldots, d - 1$. The Sylvester identity (2.5) follows easily from the d last equalities. $\square$

We say that the interpolation data (1.5) is *admissible* if the conditions (2.1)–(2.5) hold true. It will be proved later that these conditions are also sufficient for solvability of (1.8)–(1.10).

3 Two-sided Interpolation Problem for Hardy Functions

In this section we recall some results on two-sided interpolation for Hardy functions of one variable. An ordered collection

$$(3.1) \qquad \widetilde{\Omega} = \{C_+, \ C_-, \ D, \ A, \ B_+, \ B_-, \ \Gamma\}$$

of operators $A : \mathcal{H}_L \to \mathcal{H}_L$, $D : \mathcal{H}_R \to \mathcal{H}_R$, $\Gamma : \mathcal{H}_R \to \mathcal{H}_L$, $C_+ : \mathcal{H}_R \to \mathcal{H}_1$, $C_- : \mathcal{H}_R \to \mathcal{H}_2$, $B_+ : \mathcal{H}_1 \to \mathcal{H}_L$ and $B_- : \mathcal{H}_2 \to \mathcal{H}_L$ such that

$$(3.2) \qquad \operatorname{spec} A \cup \operatorname{spec} D \subset \mathbb{D},$$

is called *admissible* if

$$(3.3) \qquad A\Gamma - \Gamma D = B_- C_- - B_+ C_+$$

and the following inclusions are satisfied:

$$(3.4) \qquad \begin{aligned} \operatorname{Ran} B_- &\subseteq \operatorname{span}\{\operatorname{Ran} (A^\ell B_+); \ell = 0, 1, \ldots\}, \\ \operatorname{Ker} C_+ &\supseteq \bigcap_{j=0}^{\infty} \operatorname{Ker} (C_- D^j), \end{aligned}$$

$$(3.5) \qquad \begin{aligned} \operatorname{Ran} \Gamma &\subseteq \operatorname{span}\{\operatorname{Ran} (A^\ell B_+); \ell = 0, 1, \ldots\}, \\ \operatorname{Ker} \Gamma &\supseteq \bigcap_{j=0}^{\infty} \operatorname{Ker} (C_- D^j). \end{aligned}$$

Again, we assume that $\mathcal{H}_L$ and $\mathcal{H}_R$ are finite dimensional.

Problem 3.1 Given admissible data set $\widetilde{\Omega}$, find all functions $H \in \mathbf{HS}^{\mathcal{H}_2 \to \mathcal{H}_1}(\mathbb{D})$ satisfying interpolation conditions

$$(3.6) \qquad \frac{1}{2\pi i} \int_{|z|=1} (zI_{\mathcal{H}_L} - A)^{-1} B_+ H(z) dz = B_-,$$

$$(3.7) \qquad \frac{1}{2\pi i} \int_{|z|=1} H(z) C_- (zI_{\mathcal{H}_R} - D)^{-1} dz = C_+,$$

$$(3.8) \qquad \frac{1}{2\pi i} \int_{|z|=1} (zI_{\mathcal{H}_L} - A)^{-1} B_+ H(z) C_- (zI_{\mathcal{H}_R} - D)^{-1} dz = \Gamma.$$

Note that the identity (3.3) as well as inclusions (3.4) and (3.5) follow immediately from (3.6)–(3.8) and are therefore necessary conditions for Problem 3.1 to be solvable. It turns out that these conditions are also sufficient:

Lemma 3.2 *If the equality (3.3) and the inclusions (3.4), (3.5) are satisfied, then Problem 3.1 has a solution.*

Proof: Denote

$$C = \operatorname{span}\{\operatorname{Ran}(A^{\ell} B_{+}); \ell = 0, 1, \ldots\},$$

$$\mathcal{O} = \bigcap_{j=0}^{\infty} \operatorname{Ker}(C_{-} D^{j}).$$

Clearly, C is A-invariant, and $\mathcal{O}$ is D-invariant. With respect to the orthogonal decompositions

$$\mathcal{H}_{L} = C \oplus C^{\perp}, \qquad \mathcal{H}_{R} = \mathcal{O} \oplus \mathcal{O}^{\perp},$$

and on account of inclusions (3.4) and (3.5), the seven matrices of (3.1) have the following block forms:

$$C_{\pm} = \begin{bmatrix} 0 & C_{\pm 0} \end{bmatrix}, \; B_{\pm} = \begin{bmatrix} B_{\pm 0} \\ 0 \end{bmatrix}, \; A = \begin{bmatrix} A_{0} & * \\ 0 & * \end{bmatrix},$$

$$D = \begin{bmatrix} * & * \\ 0 & D_{0} \end{bmatrix}, \; \Gamma = \begin{bmatrix} 0 & \Gamma_{0} \\ 0 & 0 \end{bmatrix},$$

for some matrices $C_{\pm 0}$, $B_{\pm 0}$, A_{0}, D_{0}, and Γ_{0}. Using these block forms, it is easy to see that the equations (3.6)–(3.8) are satisfied if and only if the equations (3.6)–(3.8) in which $C_{\pm 0}$, $B_{\pm 0}$, A_{0}, D_{0}, and Γ_{0} are substituted for $C_{\pm}$, $B_{\pm}$, A, D, and Γ respectively, are satisfied. Moreover, the Sylvester equation $A_{0}\Gamma_{0} - \Gamma_{0} D_{0} = B_{-0} C_{-0} - B_{+0} C_{+0}$ holds. In other words, we have reduced the proof to the case when $C = \mathcal{H}_{L}$ and $\mathcal{O} = \{0\}$. Further reduction, replacing B_{-} and B_{+} by their restrictions to the orthogonal complements of Ker B_{-} and of Ker B_{+}, respectively, allows us to consider only the case of finite dimensional $\mathcal{H}_{1}$ and $\mathcal{H}_{2}$. But then, existence of a rational function $H \in HS^{\mathcal{H}_{2} \to \mathcal{H}_{1}}(\mathbb{D})$ that solves Problem 3.1 is well-known (see, e.g., Theorem 16.10.1 in [8]).   □

The arguments used in the proof of Lemma 3.2 have been used before, see, e.g., the proof of Lemma 16.9.1 of [8].

The results that will be presented below in this section have been obtained in [5], for finite dimensional $\mathcal{H}_{1}$ and $\mathcal{H}_{2}$. The same proofs work also for the case when $\mathcal{H}_{j}$ are separable Hilbert spaces. We assume from now on in this section that an admissible data set (3.1) is given.

Due to the spectral condition (3.2), the Stein equations

(3.9) $\mathbb{P}_{L} - A\mathbb{P}_{L}A^{*} = B_{+}B_{+}^{*}$ and $\mathbb{P}_{R} - D^{*}\mathbb{P}_{R}D = C_{-}^{*}C_{-}$

have unique solutions $\mathbb{P}_L : \mathcal{H}_L \to \mathcal{H}_L$ and $\mathbb{P}_R : \mathcal{H}_R \to \mathcal{H}_R$ given by the converging series

$$(3.10) \qquad \mathbb{P}_L = \sum_{k=0}^{\infty} A^k B_+ B_+^* A^{*k} \quad \text{and} \quad \mathbb{P}_R = \sum_{k=0}^{\infty} D^{*k} C_-^* C_- D^k,$$

respectively, which are positive semidefinite.

Given a positive semidefinite matrix $\mathbb{P}$, the Moore-Penrose pseudoinverse matrix $\mathbb{P}^{[-1]}$ (see [16], [15, Section 12.8]) is uniquely defined by the conditions

$$\mathbb{P}^{[-1]} \mathbf{P}_{\text{Ran}\mathbb{P}} = \mathbf{P}_{\text{Ran}\mathbb{P}} \mathbb{P}^{[-1]} = \mathbb{P}^{[-1]},$$

$$(3.11)$$

$$\mathbb{P}^{[-1]} \mathbb{P} = \mathbb{P}\mathbb{P}^{[-1]} = \mathbf{P}_{\text{Ran}\mathbb{P}},$$

where $\mathbf{P}_{\text{Ran}\mathbb{P}}$ is the orthogonal projection on the range of $\mathbb{P}$.

Theorem 3.3 *Let $\mathbb{P}_L^{[-1]}$ and $\mathbb{P}_R^{[-1]}$ be the Moore-Penrose pseudoinverses of the matrices $\mathbb{P}_L$ and $\mathbb{P}_R$ defined in (3.10). Then:*

1. *The operator-valued functions*

$$\Theta_L(z) = I_{\mathcal{H}_1} + (z-1)B_+^*(I_{\mathcal{H}_L} - zA^*)^{-1}$$

$$(3.12)$$

$$\mathbb{P}_L^{[-1]}(I_{\mathcal{H}_L} - A)^{-1}B_+$$

$$\Theta_R(z) = I_{\mathcal{H}_2} + (z-1)C_-(I_{\mathcal{H}_R} - D)^{-1}$$

$$(3.13)$$

$$\mathbb{P}_R^{[-1]}(I_{\mathcal{H}_R} - zD^*)^{-1}C_-^*$$

satisfy the equalities

$$I_{\mathcal{H}_1} - \Theta_L(z)\Theta_L(\omega)^* = (1 - z\overline{\omega})B_+^*(I_{\mathcal{H}_L} - zA^*)^{-1}$$

$$(3.14)$$

$$\mathbb{P}_L^{[-1]}(I_{\mathcal{H}_L} - \overline{\omega}A)^{-1}B_+$$

and

$$I_{\mathcal{H}_2} - \Theta_R(\omega)^*\Theta_R(z) = (1 - z\overline{\omega})C_-(I_{\mathcal{H}_R} - \overline{\omega}D)^{-1}\mathbb{P}_R^{[-1]}$$

$$(3.15)$$

$$(I_{\mathcal{H}_R} - zD^*)^{-1}C_-^*$$

for every $z, \omega \in \mathbb{D}$.

2. The operator-valued functions

$$(3.16) \qquad H_L(z) = B_+^*(I_{\mathcal{H}_L} - zA^*)^{-1}\mathbb{P}_L^{[-1]}B_-,$$

$$(3.17) \qquad H_R(z) = C_+\mathbb{P}_R^{-1}(I_{\mathcal{H}_R} - zD^*)^{-1}C_-^*$$

belong to $\mathbf{HS}^{\mathcal{H}_2 \to \mathcal{H}_1}(\mathbb{D})$, satisfy interpolation conditions (3.6) and (3.7), respectively, and their norms are given by

$$\|H_L\|^2_{\mathbf{HS}^{\mathcal{H}_2 \to \mathcal{H}_1}(\mathbb{D})} = \mathrm{Trace}\,(B_-^*\mathbb{P}_L^{[-1]}B_-),$$

$$\|H_R\|^2_{\mathbf{HS}^{\mathcal{H}_2 \to \mathcal{H}_1}(\mathbb{D})} = \mathrm{Trace}\,(C_+\mathbb{P}_R^{[-1]}C_+^*).$$

It follows from (3.14) and (3.15) that the functions $\Theta_L(z)$ and $\Theta_R(z)$ are *inner* in $\mathbb{D}$. Note that in general, the functions H_L and H_R are not solutions of Problem 3.1. However, H_L satisfies the left-sided interpolation condition (3.6) and has the minimal norm among all $\mathbf{HS}^{\mathcal{H}_2 \to \mathcal{H}_1}(\mathbb{D})$ functions which satisfy the condition (3.6). Similarly, H_R is the minimal norm solution of the right-sided problem (3.7).

It is readily seen that the set of all the solutions of Problem 3.1 is a convex subset of the Hilbert space $\mathbf{HS}^{\mathcal{H}_2 \to \mathcal{H}_1}(\mathbb{D})$ and therefore, there exists the unique solution $H_{\min}$ with the minimal $\mathbf{HS}^{\mathcal{H}_2 \to \mathcal{H}_1}(\mathbb{D})$-norm. This minimal solution is of the form

$$(3.18) \qquad H_{\min}(z) = H_R(z) + \widehat{H}_L(z)\Theta_R(z)$$

where H_R is defined in (3.17),

$$\widehat{H}_L(z) = B_+^*(I - zA^*)^{-1}\mathbb{P}_L^{[-1]}\widehat{B}_-$$

and where

$$(3.19) \qquad \widehat{B}_- = B_- + \{\Gamma(I - D) - B_-C_-\}\mathbb{P}_R^{[-1]}(I - D^*)^{-1}C_-^*.$$

The sum in (3.18) is orthogonal with respect to the inner product in $\mathbf{HS}^{\mathcal{H}_2 \to \mathcal{H}_1}(\mathbb{D})$. By Theorem 3.3, the norm of the function $\widehat{H}_R$ is equal to

$$\|\widehat{H}_L\|^2_{\mathbf{HS}^{\mathcal{H}_2 \to \mathcal{H}_1}(\mathbb{D})} = \mathrm{Trace}\,\widehat{B}_-^*\mathbb{P}_L^{[-1]}\widehat{B}_-,$$

and since the representation (3.18) is orthogonal and the function Θ_R is inner,

$$\|H_{\min}\|^2_{\mathbf{HS}^{\mathcal{H}_2 \to \mathcal{H}_1}(\mathbb{D})} = \|H_R\|^2_{\mathbf{HS}^{\mathcal{H}_2 \to \mathcal{H}_1}(\mathbb{D})} + \|\widehat{H}_L\|^2_{\mathbf{HS}^{\mathcal{H}_2 \to \mathcal{H}_1}(\mathbb{D})}$$

$$(3.20) \qquad\qquad = \mathrm{Trace}\,(C_-\mathbb{P}_R^{[-1]}C_-^* + \widehat{B}_-^*\mathbb{P}_L^{[-1]}\widehat{B}_-).$$

Theorem 3.4 *All solutions to Problem 3.1 are parametrized by the formula*

$$(3.21) \qquad H(z) = H_{\min}(z) + \Theta_L(z)h(z)\Theta_R(z)$$

where $H_{\min}$, Θ_L *and* Θ_R *are the functions defined by (3.18), (3.12) and (3.13) respectively and h is a free parameter in* $\mathbf{HS}^{\mathcal{H}_2 \to \mathcal{H}_1}(\mathbb{D})$. *The representation (3.21) is orthogonal with respect to the inner product in* $\mathbf{HS}^{\mathcal{H}_2 \to \mathcal{H}_1}(\mathbb{D})$, *and (since* Θ_L *and* Θ_R *are inner),*

$$\|H\|^2_{\mathbf{HS}^{\mathcal{H}_2 \to \mathcal{H}_1}(\mathbb{D})} = \|H_{\min}\|^2_{\mathbf{HS}^{\mathcal{H}_2 \to \mathcal{H}_1}(\mathbb{D})} + \|h\|^2_{\mathbf{HS}^{\mathcal{H}_2 \to \mathcal{H}_1}(\mathbb{D})}.$$

Theorem 3.4, together with (3.21), also describes solutions of the interpolation problem 3.1 with the additional norm restriction:

Theorem 3.5 *There exists a solution H of Problem 3.1 satisfying*

$$(3.22) \qquad \|H\|_{\mathbf{HS}^{\mathcal{H}_2 \to \mathcal{H}_1}(\mathbb{D})} \leq \gamma$$

if and only if

$$\gamma \leq (\text{Trace } (C_- \mathbb{P}_R^{[-1]} C_-^* + \widehat{B}_-^* \mathbb{P}_L^{[-1]} \widehat{B}_-))^{\frac{1}{2}}.$$

If this condition holds, then all solutions of Problem 3.1 satisfying (3.22) are described by the formula (3.21) in which $h \in \mathbf{HS}^{\mathcal{H}_2 \to \mathcal{H}_1}(\mathbb{D})$ *is subject to*

$$\|h\|^2_{\mathbf{HS}^{\mathcal{H}_2 \to \mathcal{H}_1}(\mathbb{D})} \leq \gamma^2 - \text{Trace } (C_- \mathbb{P}_R^{[-1]} C_-^* + \widehat{B}_-^* \mathbb{P}_L^{[-1]} \widehat{B}_-).$$

There is also another orthogonal representation of the minimal solution, in which the roles of Θ_L and Θ_R are interchanged, namely,

$$(3.23) \qquad H_{\min}(z) = H_R(z) + \widehat{H}_L(z)\Theta_R(z),$$

where

$$\widehat{H}_R(z) = \widehat{C}_+ \mathbb{P}_R^{[-1]}(I - zD^*)^{-1} C_-^*,$$

and where

$$(3.24) \qquad \widehat{C}_+ = C_+ + B_+^*(I - A^*)^{-1} \mathbb{P}_L^{[-1]}\{(I - A)\Gamma - B_+ C_+\}.$$

Using the formula (3.23), an alternative version of Theorem 3.5 can be obtained in which

$$(3.25) \qquad \text{Trace } (B_-^* \mathbb{P}_L^{[-1]} B_- + \widehat{C}_+ \mathbb{P}_R^{[-1]} \widehat{C}_+^*)$$

is used in place of

$$(3.26) \qquad \text{Trace } (C_- \mathbb{P}_R^{[-1]} C_-^* + \widehat{B}_*^* \mathbb{P}_L^{[-1]} \widehat{B}_-).$$

Note the expressions (3.25) and (3.26) are equal.

4 Reducing the Number of Variables

We reduce here Problem 1.1 to two two-sided tangential interpolation problems for Hilbert-Schmidt operator-valued functions: the first problem of one variable, and the second problem of $d - 1$ variables. Given a separable Hilbert space $\mathcal{H}$, we define inductively the Hilbert spaces $\mathcal{H}(k), k = 0, 1, \ldots$ by setting $\mathcal{H}(0) = \mathcal{H}$, and $\mathcal{H}(k)$ the Hilbert space of square summable sequences $\{h_j\}_{j=0}^\infty, h_j \in \mathcal{H}(k - 1)$:

$$(4.1) \quad \mathcal{H}(k) = \left\{ \{h_j\}_{j=0}^\infty : h_j \in \mathcal{H}(k - 1) \quad \text{and} \quad \sum_{j=0}^\infty \|h_j\|_{\mathcal{H}(k-1)}^2 < \infty \right\}.$$

Let $\mathcal{H}$ be a (separable) Hilbert space, and for fixed $z \in \mathbb{D}$, let $E_1(z) : \mathcal{H}(1) \to \mathcal{H}$ be the operator defined by the rule

$$(4.2) \qquad E_1(z) : \mathbf{h} = \{h_j\}_{j=0}^\infty \to \sum_{j=0}^\infty z^j h_j \qquad h_j \in \mathcal{H}.$$

One verifies (see [7] for more details) that

$$(4.3) \qquad E_1(z) E_1(\omega)^* = \frac{1}{1 - z\bar{\omega}} I_{\mathcal{H}}$$

and in particular, $E_1(z) E_1(z)^* = \frac{1}{1-|z|^2} I_{\mathcal{H}}$, which implies that the operator $E_1(z)$ is bounded. It admits the matrix representation

$$(4.4) \qquad E_1(z) = (I_{\mathcal{H}} \quad z I_{\mathcal{H}} \quad z^2 I_{\mathcal{H}} \quad \cdots)$$

and is an operator-valued function which is well defined and analytic in $\mathbb{D}$. Starting with $E_1(z)$, we define inductively the operator-valued functions $E_k(z)$ which, for every fixed $z \in \mathbb{D}$, map $\mathcal{H}(k)$ into $\mathcal{H}(k - 1)$ by the rule

$$(4.5) \quad E_k(z) : \{f_j\}_{j=0}^\infty \to \{E_{k-1}(z) f_j\}_{j=0}^\infty \qquad f_j \in \mathcal{H}(k - 1), \ z \in \mathbb{D}.$$

For $k \geq 2$, the adjoint operator is defined by

$$E_k(z)^* : \{h_j\}_{j=0}^\infty \to \{E_{k-1}(z)^* h_j\}_{j=0}^\infty \qquad h_j \in \mathcal{H}(k - 2), \ z \in \mathbb{D}$$

and, making use of (4.3), we obtain by induction that

$$(4.6) \qquad E_k(z) E_k(\omega)^* = \frac{1}{1 - z\bar{\omega}} I_{\mathcal{H}(k-1)}.$$

Functions $E_k(z)$ admit recursive matrix representations

$$(4.7) \qquad E_k(z) = \mathrm{diag}(E_{k-1}(z), \ E_{k-1}(z), \ E_{k-1}(z), \ \ldots)$$

and are well defined and analytic in $\mathbb{D}$. The next lemma is taken from [7].

Lemma 4.1 *Let $\mathcal{H}$ and $\mathcal{G}$ be two separable Hilbert spaces and let E_k be the operator-valued function defined by (4.7). Then the operator $\mathbf{M}_{E_k}$ of multiplication by $E_k(z_1)$ is a unitary operator from $\mathbf{HS}^{\mathcal{G} \to \mathcal{H}(k)}(\mathbb{D}^{d-1})$ onto $\mathbf{HS}^{\mathcal{G} \to \mathcal{H}(k-1)}(\mathbb{D}^d)$: any function $H \in \mathbf{HS}^{\mathcal{G} \to \mathcal{H}(k-1)}(\mathbb{D}^d)$ admits a representation of the form*

$$H(z_1, \ldots, z_d) = E_k(z_1) G(z_2, \ldots, z_d),$$

where G is a (uniquely defined) function in $\mathbf{HS}^{\mathcal{G} \to \mathcal{H}(k)}(\mathbb{D}^{d-1})$ such that

$$\|H\|_{\mathbf{HS}^{\mathcal{G} \to \mathcal{H}(k-1)}(\mathbb{D}^d)} = \|G\|_{\mathbf{HS}^{\mathcal{G} \to \mathcal{H}(k)}(\mathbb{D}^{d-1})}.$$

Applying repeatedly Lemma 4.1, we obtain:

Lemma 4.2 *Let $\mathcal{H}$ and $\mathcal{G}$ be two separable Hilbert spaces and let*

$$(4.8) \qquad \mathbf{E}(\mathbf{z}_{d-1}) = E_1(z_1) E_2(z_2) \ldots E_{d-1}(z_{d-1}) : \mathcal{H}(d-1) \to \mathcal{H}$$

for every fixed $\mathbf{z}_{d-1} = (z_1, \ldots z_{d-1}) \in \mathbb{D}^{d-1}$. The operator $\mathbf{M}_{\mathbf{E}}$ of multiplication by $\mathbf{E}(\mathbf{z}_{d-1})$ is a unitary operator from $\mathbf{HS}^{\mathcal{G} \to \mathcal{H}(d-1)}(\mathbb{D})$ onto $\mathbf{HS}^{\mathcal{G} \to \mathcal{H}}(\mathbb{D}^d)$: any function $H \in \mathbf{HS}^{\mathcal{G} \to \mathcal{H}}(\mathbb{D}^d)$ admits a representation of the form

$$(4.9) \qquad H(\mathbf{z}_d) = \mathbf{E}(\mathbf{z}_{d-1}) F(z_d),$$

where F is a (uniquely defined) function in $\mathbf{HS}^{\mathcal{G} \to \mathcal{H}(d-1)}(\mathbb{D})$ such that

$$\|H\|_{\mathbf{HS}^{\mathcal{G} \to \mathcal{H}}(\mathbb{D}^d)} = \|F\|_{\mathbf{HS}^{\mathcal{G} \to \mathcal{H}(d-1)}(\mathbb{D})}.$$

Note the following property of $E_k(z_k)$:

$$\frac{1}{(2\pi i)^k} \int_{\mathbb{T}^k} (E_1(z_1) \ldots E_k(z_k))^* (E_1(z_1) \ldots E_k(z_k))$$

$$(4.10)$$

$$dz_1 \ldots dz_k = I_{\mathcal{H}(k)},$$

for $k = 1, 2, \ldots, d-1$. The formula (4.10) follows easily by induction on k using (4.7). In particular, for $k = d - 1$ the formula (4.10) (in view of (4.8)) gives:

$$(4.11) \qquad \frac{1}{(2\pi i)^{d-1}} \int_{\mathbb{T}^{(d-1)}} \mathbf{E}(\mathbf{z}_{d-1})^* \mathbf{E}(\mathbf{z}_{d-1}) dz_1 \ldots dz_{d-1} = I_{\mathcal{H}(d-1)}.$$

Lemma 4.2 will enable us to reduce Problem 1.1 to analogous problems with fewer variables.

Since the first multiplier $\mathbf{E}$ in the representation (4.9) is of the standard form, all the information about the function H is contained in the second factor $F(z_d)$. The idea is to reformulate Problem 1.1 in terms of the operator-valued function F of one variable. To this end let $\mathbf{B}_+ : \mathcal{H}(d-1) \to \mathcal{H}_d$ be the operator defined by

$$\mathbf{B}_+ = \frac{1}{(2\pi i)^{d-1}} B_d \int_{\mathbb{T}(\epsilon)^{d-1}} \left(\overset{d-1}{\underset{j=1}{\frown}\prod} (z_j I - A_j)^{-1} B_j \right)$$

$$(4.12)$$
$$\mathbf{E}(\mathbf{z}_{d-1}) dz_1 \ldots dz_{d-1},$$

where $\mathbb{T}(\epsilon) = \{z \in \mathbb{C} : |z| = \epsilon\}$, and $0 < \epsilon < 1$ is chosen so that $\operatorname{spec} A_j$ is inside $\mathbb{T}(\epsilon)$ for $j = 1, \ldots, d$. (For shorthand, in the sequel we shall omit ϵ in formulas such as (4.12)).

Lemma 4.3 *Let* $\mathbf{E}$ *and* $\mathbf{B}_+$ *be defined by* (4.8) *and* (4.12), *respectively. Then*

$$(4.13) \qquad \mathbf{B}_+ \mathbf{E}(\mathbf{z}_{d-1})^* = B_d \overset{d-1}{\underset{j=1}{\frown}\prod} (I - \bar{z}_j A_j)^{-1} B_j$$

for every choice of $\mathbf{z}_{d-1} \in \mathbb{D}^{d-1}$, *and*

$$(4.14) \qquad \mathbf{B}_+ \mathbf{B}_+^* = B_d \sum_{j_1, \ldots, j_{d-1}=0}^{\infty} \left(\overset{d-1}{\underset{k=1}{\frown}\prod} A_k^{j_k} B_k \right) \left(\overset{d-1}{\underset{k=1}{\frown}\prod} A_k^{j_k} B_k \right)^*.$$

Moreover, denoting by $\mathbb{Z}_+$ *the set of non-negative integers, we have*

$$\operatorname{span}\{\operatorname{Ran} (A_d^{j_d} B_d \ldots A_1^{j_1} B_1); \ j_\alpha \in \mathbb{Z}_+\}$$

$$(4.15)$$
$$= \operatorname{span}\{\operatorname{Ran} (A_d^\ell \mathbf{B}_+); \ \ell \in \mathbb{Z}_+\}.$$

Proof: The proof follows a general line of argument used in [7]. To show (4.13) we start with the equality

$$(4.16) \qquad \mathbf{E}(\mathbf{z}_{d-1})\mathbf{E}(\mathbf{w}_{d-1})^* = \prod_{k=1}^{d-1} \frac{1}{1 - z_k\bar{\omega}_k} \, I_{\mathcal{H}},$$

which follows immediately from (4.3), (4.6) and (4.8). Take arbitrary $h \in \mathcal{H}$ and apply operators from each side of (4.12) to $\mathbf{E}(\mathbf{z}_{d-1})^*h$. Making use of (4.16) and taking advantage of the residue calculus, we get

$$\mathbf{B}_+\mathbf{E}(\mathbf{z}_{d-1})^*h = \frac{1}{(2\pi i)^{d-1}} B_d \int_{\mathbb{T}^{d-1}} \left(\overset{d-1}{\underset{j=1}{\frown}} \prod (\xi_j I - A_j)^{-1} B_j \right)$$

$$\mathbf{E}(\xi_1, \ldots, \xi_{d-1}) \times \mathbf{E}(z_1, \ldots, z_{d-1})^*h \, d\xi_1 \ldots d\xi_{d-1}$$

$$= \frac{1}{(2\pi i)^{d-1}} B_d \int_{\mathbb{T}^{d-1}} \left(\overset{d-1}{\underset{j=1}{\frown}} \prod (\xi_j I - A_j)^{-1} \frac{B_j}{1 - \xi_j\bar{z}_j} \right)$$

$$h \, d\xi_1 \ldots d\xi_{d-1}$$

$$= B_d \left(\overset{d-1}{\underset{j=1}{\frown}} \prod \frac{1}{2\pi i} \int_{\mathbb{T}} (\xi_j I - A_j)^{-1} \frac{B_j}{1 - \xi_j\bar{z}_j} \, d\xi_j \right) h$$

$$= B_d \overset{d-1}{\underset{j=1}{\frown}} \prod (I - \bar{z}_j A_j)^{-1} B_j h,$$

which proves (4.13), since h is arbitrary. Next, applying Lemma 4.2 for the function $H(z_1, \ldots, z_d) = \mathbf{E}(z_1, \ldots, z_{d-1})\mathbf{B}_+^*$ and taking into account (4.13) we get

$$\mathbf{B}_+\mathbf{B}_+^* = [H, H]_{\mathbf{HS}^{\mathcal{H}_d \to \mathcal{H}}(\mathbb{D}^d)}$$

$$= \frac{1}{(2\pi i)^{d-1}} \int_{\mathbb{T}^{d-1}} \mathbf{B}_+\mathbf{E}(\xi_1, \ldots, \xi_{d-1})^*$$

$$(4.17) \qquad \mathbf{E}(\xi_1, \ldots, \xi_{d-1}) \, \mathbf{B}_+^* d\xi_1 \ldots d\xi_{d-1}$$

$$= \frac{1}{(2\pi)^{d-1}} \int_0^{2\pi} \cdots \int_0^{2\pi} B_d \left(\prod_{j=1}^{\overset{d-1}{\frown}} (I - e^{-it_j} A_j)^{-1} B_j \right)$$

$$\times \left(\prod_{j=1}^{\overset{d-1}{\frown}} B_j^* (I - e^{it_j} A_j^*)^{-1} \right) B_d^* \, dt_1 \ldots dt_{d-1}.$$

Expanding the integrand function in the latter integral as the multiple Fourier series it is readily seen that the right hand side of (4.17) coincides with the multiple series in the right hand side of (4.14).

To verify (4.15), we first observe that the right-hand side of (2.1) can be written in an alternative form:

$$\text{span}\{\text{Ran}\,(A_d^{j_d} B_d \ldots A_1^{j_1} B_1) : \ j_\alpha \in \mathbb{Z}_+\}$$

(4.18)
$$= \text{span}\{\text{Ran}\,((z_d I - A_d)^{-1} B_d \ldots (z_1 I - A_1)^{-1} B_1)$$

$$: z_1, \ldots, z_d \in \Omega\}.$$

Here Ω is a fixed subset of $\mathbb{C} \backslash (\cup_{j=1}^d \text{spec}\, A_j)$ that contains at least $m = \max_j \dim \mathcal{H}_j$ distinct complex numbers. The inclusion $\supseteq$ in (4.18) is clear because each $(z_k I - A_k)^{-1}$ is a linear combination of $I, A_k, \ldots, A_k^{m-1}$. For the converse inclusion, select m distinct points $\lambda_1, \ldots, \lambda_m$ in Ω. Using the Cayley-Hamilton theorem, it is easy to see that for any nonnegative integer j, A_k^j is a linear combination of $(\lambda_1 I - A_k)^{-1}, \ldots, (\lambda_m I - A_k)^{-1}$. Therefore, for every fixed d-tuple of nonnegative integer indices $j_1, \ldots, j_d$ the range of $A_d^{j_d} B_d A_{d-1}^{j_{d-1}} B_{d-1} \ldots A_1^{j_1} B_1$ is contained in the right hand side of (4.18). The same arguments lead to the equality

$$\text{span}\{\text{Ran}\,(A_d^j \mathbf{B}_+) : \ j \in \mathbb{Z}_+\}$$

(4.19)
$$= \text{span}\{\text{Ran}\,((z_d I - A_d)^{-1} \mathbf{B}_+) : \ z_d \in \Omega\}.$$

We note also another equality

$$\text{Ran}\, \mathbf{B}_+$$

(4.20)
$$= \text{span}\left\{ \text{Ran} \left(\prod_{j=1}^{\overset{d-1}{\frown}} (z_j I - A_j)^{-1} B_j \right) : z_1, \ldots, z_{d-1} \in \Omega \right\}.$$

Indeed, the inclusion $\supseteq$ in (4.20) follows from (4.12) upon approximating the integral in (4.13) by Riemann sums and replacing each term $(\xi_k I - A_k)^{-1}$ by a suitable linear combination of $(\lambda_1 I - A_k)^{-1}, \ldots, (\lambda_m I - A_k)^{-1}$ (as before, $\lambda_1, \ldots, \lambda_m$ are distinct complex numbers in Ω). The converse containment $\subseteq$ in (4.20) follows from (4.14), in view of which we only have to show that Ran $(A_{d-1}^{j_{d-1}} B_{d-1} \ldots A_1^{j_1} B_1)$ is contained in the right hand side of (4.20). But this is immediate upon writing each $A_k^{j_k}$ as a linear combination of $(\lambda_1 I - A_k)^{-1}, \ldots, (\lambda_m I - A_k)^{-1}$. Finally, combining (4.18), (4.19), and (4.20), we obtain (4.15). $\qquad\square$

Let $H \in \mathbf{HS}^{\mathcal{G} \to \mathcal{H}}(\mathbb{D}^d)$ of the form (4.9) be a solution of Problem 1.1. Substituting (4.9) into the left hand sides of the interpolation conditions (1.8) and (1.10) (for $k = d$), and taking (4.13) into account, we obtain

$$(4.21) \qquad \frac{1}{2\pi i} \int_{|\zeta|=1} (\zeta I - A_d)^{-1} \mathbf{B}_+ F(\zeta) d\zeta = B_-,$$

$$(4.22) \qquad \frac{1}{2\pi i} \int_{|\zeta|=1} (\zeta I_{\mathcal{H}_L} - A_d)^{-1} \mathbf{B}_+ F(\zeta) C_- (\zeta I - D_d)^{-1} d\zeta = \Gamma_d.$$

In order to apply the results from Section 2 and to involve interpolation conditions (1.9) and (1.10) we introduce the right sided condition

$$(4.23) \qquad \frac{1}{2\pi i} \int_{|\zeta|=1} F(\zeta) C_d (\zeta I - D_d)^{-1} d\zeta = X$$

for F, where X is a bounded operator acting from $\mathcal{H}_R$ into $\mathcal{H}(d-1)$. We associate with X the operator-valued function

$$(4.24) \quad \widehat{\mathbf{X}}(\mathbf{z}_{d-1}) = \mathbf{E}(\mathbf{z}_{d-1}) X = \frac{1}{2\pi i} \int_{|\zeta|=1} H(\mathbf{z}_{d-1}, \zeta) C_d (\zeta I - D_d)^{-1} d\zeta,$$

which belongs to $\mathbf{HS}^{\mathcal{G}_d \to \mathcal{H}}(\mathbb{D}^{d-1})$. It is easily seen from (4.11) that X is reconstructed from $\widehat{\mathbf{X}}(\mathbf{z}_{d-1})$ via

$$(4.25) \quad X = \frac{1}{2\pi} \int_0^{2\pi} \mathbf{E}(e^{it_1}, \ldots, e^{it_{d-1}})^* \widehat{\mathbf{X}}(e^{it_1}, \ldots, e^{it_{d-1}}) dt_1 \ldots dt_{d-1}.$$

Note also the equality

$$(4.26) \qquad \mathbf{B}_+ X = \Gamma_d D_d - A_d \Gamma_d + B_- C_d,$$

which follows readily from (2.9) by definitions of $\mathbf{B}_+$ and X. Let

$$(4.27) \qquad \mathbb{P}_L = \sum_{\ell=0}^{\infty} A_d^{\ell} \mathbf{B}_+ \mathbf{B}_+^* A_d^{*\ell} \quad \text{and} \quad \mathbb{P}_R = \sum_{k=0}^{\infty} D_d^{*k} C_d^* C_d D_d^k.$$

Then

$$(4.28) \qquad \mathrm{Ker}\, \mathbb{P}_R = \bigcap_{k \geq 0} \mathrm{Ker}\, C_d D_d^k = \bigcap_{\zeta \in \mathbb{T}} \mathrm{Ker}\, C_d (I - \zeta D_d)^{-1}$$

and

$$\mathrm{Ran}\, \mathbb{P}_R = \mathrm{span}\, \{\mathrm{Ran}\, ((I - \zeta D_d^*)^{-1} C_d^*) : \zeta \in \mathbb{T}\}$$
$$= \mathrm{span}\, \{\mathrm{Ran}\, (D_d^{*k} C_d^*) : k \in \mathbb{Z}_+\}.$$

It follows immediately from (4.23) and (4.24) that

$$(4.29) \qquad \widehat{\mathbf{X}}(\mathbf{z}_{d-1})\big|_{\mathrm{Ker}\, \mathbb{P}_R} \equiv X|_{\mathrm{Ker}\, \mathbb{P}_R} = 0.$$

It turns out that conditions (1.9) and (1.10) may be easily expressed in terms of the associated function $\widehat{\mathbf{X}}$. Indeed, substituting (4.9) into the left hand side of the interpolation condition (1.8), and using formulas (2.9) and (4.24), we obtain

$$\frac{1}{(2\pi i)^{d-1}} \int_{\mathbb{T}^{d-1}} \overset{d-1}{\underset{j=1}{\frown}} \prod (z_j I - A_j)^{-1} B_j \widehat{\mathbf{X}}(\mathbf{z}_{d-1})\, dz_1 \ldots dz_{d-1}$$

$$(4.30)$$
$$= \Gamma_d D_d - A_d \Gamma_d + B_- C_d.$$

Analogously, in view of (4.24) and (4.29), the interpolation condition (1.9) leads to

$$\frac{1}{(2\pi i)^{d-1}} \int_{\mathbb{T}^{d-1}} \widehat{\mathbf{X}}(\mathbf{z}_{d-1}) \mathbf{P}_{\mathrm{Ran}\, \mathbb{P}_R} \left(\overset{d-1}{\underset{j=1}{\frown}} \prod C_j (z_j I - D_j)^{-1} \right)$$

$$(4.31)$$
$$dz_{d-1} \ldots dz_1 = C_+,$$

and conditions (1.10) take the form

$$\frac{1}{(2\pi i)^{d-1}} \int_{\mathbb{T}^{d-1}} \left(\overset{k}{\underset{j=1}{\frown}} \prod (z_j I - A_j)^{-1} B_j \right) \widehat{\mathbf{X}}(\mathbf{z}_{d-1}) \mathbf{P}_{\mathrm{Ran}\, \mathbb{P}_R}$$

$$(4.32)$$
$$\cdot \left(\overset{d-1}{\underset{j=k}{\frown}} \prod C_j (z_j I - D_j)^{-1} \right) dz_1 \ldots dz_{k-1} dz_{d-1} \ldots dz_k = \Gamma_k$$

for $k = 1, \ldots, d - 1$. Recall that $\mathbb{P}_{\mathrm{Ran}\mathbb{P}_R}$ stands for the orthogonal projection on $\mathrm{Ran}\mathbb{P}_R$.

Making use of Lemma 4.2 we conclude that Problem 1.1 reduces to the following interpolation problem in one variable:

Problem 4.4 Given data set (1.5) find all functions $F \in \mathbf{HS}^{\mathcal{G} \to \mathcal{H}(d-1)}(\mathbb{D})$ satisfying interpolation conditions (4.21)–(4.23) and the norm constraint

$$(4.33) \qquad \|F\|_{\mathbf{HS}^{\mathcal{G} \to \mathcal{H}(d-1)}(\mathbb{D})} \le \gamma,$$

where γ is a preassigned positive number and $X : \mathcal{G}_d \to \mathcal{H}(d - 1)$ is a bounded operator defined via (4.25) from a function $\widehat{\mathbf{X}} \in \mathbf{HS}^{\mathcal{G}_d \to \mathcal{H}}(\mathbb{D}^{d-1})$ which in its turn, satisfies the interpolation conditions (4.29)–(4.32).

More precisely:

Remark 4.5 A function $H \in \mathbf{HS}^{\mathcal{G} \to \mathcal{H}}(\mathbb{D}^d)$ of the form (4.9) is a solution of Problem 1.1 if and only if the function F in the representation (4.9) is a solution of Problem 4.4.

We now state one of the main results of the paper: reduction of the original problem to analogous interpolation problems with fewer variables.

Theorem 4.6 *Let be given the data* (1.5), *and assume that the necessary conditions* (2.1)–(2.5) *are satisfied. Let* $\mathbb{P}_L^{[-1]}$ *and* $\mathbb{P}_R^{[-1]}$ *be the Moore-Penrose pseudo-inverses of the matrices* $\mathbb{P}_L$ *and* $\mathbb{P}_R$, *respectively, given by* (4.27), *and define*

$$(4.34) \qquad \begin{aligned} \Theta_L(z) &= I_{\mathcal{H}(d-1)} + (z - 1)\mathbf{B}_+^*(I - zA_d^*)^{-1} \\ &\quad \mathbb{P}_L^{[-1]}(I - A_d)^{-1}\mathbf{B}_+, \end{aligned}$$

where $\mathbf{B}_+$ *is given by* (4.12),

$$(4.35) \qquad \Theta_R(z) = I_{\mathcal{G}} + (z - 1)C_-(I - D_d)^{-1}\mathbb{P}_R^{[-1]}(I - zD_d^*)^{-1}C_d^*$$

and

$$(4.36) \qquad \begin{aligned} \widehat{\mathbf{B}}_d &= \mathbf{B}_- + \{\Gamma_d(I - D_d) - B_-C_d\}\mathbb{P}_R^{[-1]} \\ &\quad (I - D_d^*)^{-1}C_d^* : \mathcal{G} \to \mathcal{H}_d. \end{aligned}$$

A function $H \in \mathbf{HS}^{\mathcal{G} \to \mathcal{H}}(\mathbb{D}^d)$ is a solution to Problem 1.1 if and only if it is of the form

(4.37)
$$H(\mathbf{z}_d) = \widehat{\mathbf{X}}(\mathbf{z}_{d-1}) \mathbb{P}_R^{[-1]}(I - z_d D_d^*)^{-1} C_d^* + B_+^*$$
$$\overbrace{\prod_{j=1}^{d}}^{d} B_j^*(I - z_j A_j^*)^{-1} \mathbb{P}_L^{[-1]} \widehat{B}_d \Theta_R(z_d)$$
$$+ E(\mathbf{z}_{d-1}) \Theta_L(z_d) f(z_d) \Theta_R(z_d),$$

where f is an arbitrary function in $\mathbf{HS}^{\mathcal{G} \to \mathcal{H}(d-1)}(\mathbb{D})$, and where $\widehat{\mathbf{X}}$ is any function in $\mathbf{HS}^{\mathcal{G}_d \to \mathcal{H}}(\mathbb{D}^{d-1})$ which satisfies interpolation conditions (4.29)–(4.32) and the norm constraint

(4.38)
$$\| (\mathbb{P}_R^{[-1]})^{1/2} \widehat{\mathbf{X}}^* \|_{\mathbf{HS}^{\mathcal{G}_d \to \mathcal{H}}(\mathbb{D}^{d-1})}^2$$
$$\leq \gamma^2 - \mathrm{Trace}\, (\widehat{B}_d^* \mathbb{P}_L^{[-1]} \widehat{B}_d) - \| f \|_{\mathbf{HS}^{\mathcal{G} \to \mathcal{H}(d-1)}(\mathbb{D})}^2.$$

It will follow from Lemma 4.7 below that the parameters f and $\widehat{\mathbf{X}}$ in Theorem 4.6 are independent. In other words, there is one-to-one correspondence between the solutions $H(\mathbf{z}_d)$ of Problem 1.1 and the pairs $(f, \widehat{\mathbf{X}})$, where f and $\widehat{\mathbf{X}}$ satisfy the properties specified in the theorem.

The formula (4.37) also shows that our approach in the case of several variables does not yield nice formulas for the operator-valued inner product version of Problem 1.1, in contrast with [7]. The difficulty is that the operator of multiplication by an inner operator-valued function on the right is not isometric with respect to the operator-valued inner product

$$[G,\ H] = \frac{1}{(2\pi)^d} \int_0^{2\pi} \cdots \int_0^{2\pi} H(e^{it_1}, \ldots, e^{it_d})^*$$
$$G(e^{it_1}, \ldots, e^{it_d})\, dt_1 \ldots dt_d.$$

For that reason, we do not consider this inner product in the present paper.

Proof: By Remark 4.5, H is a solution of Problem 1.1 if and only if the function F in representation (4.9) is a solution of Problem 4.4. Conditions (4.21)–(4.23) form a two sided interpolation problem with the data set

(4.39)
$$\widetilde{\Omega}_F = \{X,\ C_d,\ D_d,\ A_d,\ \mathbf{B}_+,\ \mathbf{B}_-,\ \Gamma_d\}.$$

We verify that the data set (4.39) is admissible (as defined in Section 3). The inclusion

$$\mathrm{Ran}\, \mathbf{B}_- \subseteq \mathrm{span}\{\mathrm{Ran}\, (A^\ell \mathbf{B}_+);\ \ell = 0, 1, \ldots\}$$

follows from (2.1), (4.15) and the first inclusion in (3.4). The definition (4.23) of X implies

$$\mathrm{Ker}\, X \supseteq \bigcap_{j=0}^{\infty} \mathrm{Ker}\,(C_d D_d^j).$$

Next, (2.3) with $k = d$ gives

$$\mathrm{Ker}\, \Gamma_d \supseteq \bigcap_{j=0}^{\infty} \mathrm{Ker}\,(C_d D_d^j).$$

Using (4.15) again, we obtain from (2.4):

$$\mathrm{Ran}\, \Gamma_d \subseteq \mathrm{span}\{\mathrm{Ran}\,(A^\ell \mathbf{B}_+);\ \ell = 0, 1, \ldots\}.$$

Finally, the Sylvester equation

$$A_d \Gamma_d - \Gamma_d D_d = B_- C_d - \mathbf{B}_+ X$$

follows from (4.26). By the spectral condition (1.6), the series (4.27) converge and the positive semidefinite matrices $\mathbb{P}_L$ and $\mathbb{P}_R$ are the unique solutions of the Stein equations

$$(4.40) \qquad \mathbb{P}_L - A_d \mathbb{P}_L A_d^* = \mathbf{B}_+ \mathbf{B}_+^* \quad \text{and} \quad \mathbb{P}_R - D_d^* \mathbb{P}_R D_d = C_d^* C_d,$$

respectively. By Theorem 3.3, $\Theta_L(z)$ and $\Theta_R(z)$ are inner in $\mathbb{D}$. Furthermore, by Theorem 3.4, all functions $F \in \mathbf{HS}^{\mathcal{G} \to \mathcal{H}(d-1)}(\mathbb{D})$ that satisfy conditions (4.21)–(4.23) are parametrized by the formula

$$(4.41) \qquad F(z) = F_R(z) + \widehat{F}_L(z)\Theta_R(z) + \Theta_L(z) f(z)\Theta_R(z),$$

where f is an arbitrary parameter from $\mathbf{HS}^{\mathcal{G} \to \mathcal{H}(d-1)}(\mathbb{D})$,

$$(4.42) \qquad \begin{aligned} F_R(z) &= X \mathbb{P}_R^{[-1]}(I - z D_d^*)^{-1} C_d^*, \\ \widehat{F}_L(z) &= \mathbf{B}_+^*(I - z A_d^*)^{-1} \mathbb{P}_L^{[-1]} \widehat{B}_d \end{aligned}$$

and $\widehat{B}_d : \mathcal{G} \to \mathcal{H}_d$ is the operator defined in (4.36). By Theorem 3.4, the summands in (4.41) are mutually orthogonal with respect to the inner product of $\mathbf{HS}^{\mathcal{G} \to \mathcal{H}(d-1)}(\mathbb{D})$, and therefore the sum of the two first terms in the right hand side of (4.41) presents the minimal norm solution of Problem 4.4. Making use of

Theorem 3.3 we evaluate norms of all the terms in the left hand side of (4.41) and obtain

$$\|F\|^2_{\mathbf{HS}^{\mathcal{G}\to\mathcal{H}(d-1)}(\mathbb{D})} = \mathrm{Trace}\,(X\mathbb{P}_R^{[-1]}X^* + \widehat{B}_d^*\mathbb{P}_L^{[-1]}\widehat{B}_d)$$

(4.43)

$$+\|f\|^2_{\mathbf{HS}^{\mathcal{G}\to\mathcal{H}(d-1)}(\mathbb{D})}.$$

Thus, F of the form (4.41) satisfies the norm constraint (4.33) if and only if

$$\mathrm{Trace}\,(X\mathbb{P}_R^{[-1]}X^*) \le \gamma^2 - \mathrm{Trace}\,(\widehat{B}_d^*\mathbb{P}_L^{[-1]}\widehat{B}_d)$$

(4.44)

$$-\|f\|^2_{\mathbf{HS}^{\mathcal{G}\to\mathcal{H}(d-1)}(\mathbb{D})}.$$

Now compute:

$$
\begin{aligned}
H(\mathbf{z}_d) &= \text{(by (4.9))} \quad = \mathbf{E}(\mathbf{z}_{d-1})F(z_d) = \quad \text{(by (4.41))} \\
&= \mathbf{E}(\mathbf{z}_{d-1})F_R(z_d) + \mathbf{E}(\mathbf{z}_{d-1})\widehat{F}_L(z_d)\Theta_R(z_d) \\
&\quad +\mathbf{E}(\mathbf{z}_{d-1})\Theta_L(z_d)f(z_d)\Theta_R(z_d),
\end{aligned}
$$

which in view of (4.24) and (4.42), is equal to

$$\widehat{\mathbf{X}}(\mathbf{z}_{d-1})\mathbb{P}_R^{[-1]}(I - z_d D_d^*)^{-1}C_d^* + \mathbf{E}(\mathbf{z}_{d-1})\widehat{F}_L(z_d)\Theta_R(z_d)$$

$$+\mathbf{E}(\mathbf{z}_{d-1})\Theta_L(z_d)f(z_d)\Theta_R(z_d).$$

Using (4.42) and (4.13) we get (4.37). If $\widehat{\mathbf{X}}$ satisfies (4.30)–(4.32), then the function H of the form (4.37) satisfies conditions (1.8)–(1.10), by Remark 4.5. It remains to note that the norm constraint (4.38) is equivalent to (4.44) (use the left equality in (4.24) to verify this). □

The next question is: are the parameters $\widehat{\mathbf{X}}(\mathbf{z}_{d-1})$ and $f(z_d)$ in (4.37) independent and are they uniquely defined by a solution $H(\mathbf{z}_d)$ of the problem? The answer is affirmative:

Lemma 4.7 *Let $H \in \mathbf{HS}^{\mathcal{G}\to\mathcal{H}}(\mathbb{D}^d)$ be a solution of Problem 1.1 which admits two representations of the form* (4.37):

$$H(\mathbf{z}_d) = \widehat{\mathbf{X}}_i(\mathbf{z}_{d-1})\mathbb{P}_R^{[-1]}(I - z_d D_d^*)^{-1}C_d^*$$

(4.45)

$$+\overset{d}{\underset{j=1}{\frown}}\prod B_j^*(I - z_j A_j^*)^{-1}\mathbb{P}_L^{[-1]}\widehat{B}_d\Theta_R(z_d)$$

$$+\mathbf{E}(\mathbf{z}_{d-1})\Theta_L(z_d)f_i(z_d)\Theta_R(z_d) \qquad (i = 1, 2).$$

Then

(4.46) $f_1(z_d) \equiv f_2(z_d) \quad and \quad \widehat{\mathbf{X}}_1(\mathbf{z}_{d-1}) \equiv \widehat{\mathbf{X}}_2(\mathbf{z}_{d-1}).$

Proof: Consider the difference between the two representations in (4.45):

$$(\widehat{\mathbf{X}}_1(\mathbf{z}_{d-1}) - \widehat{\mathbf{X}}_2(\mathbf{z}_{d-1}))\mathbb{P}_R^{[-1]}(I - z_d D_d^*)^{-1}C_d^*$$

$$+ \mathbf{E}(\mathbf{z}_{d-1})\Theta_L(z_d)(f_1(z_d) - f_2(z_d))\Theta_R(z_d) = 0,$$

which is equivalent, on account of (4.24) and in view of Lemma 4.2, to

$$(X_1 - X_2)\mathbb{P}_R^{[-1]}(I - z_d D_d^*)^{-1}C_d^*$$

(4.47)

$$= \Theta_L(z_d)(f_2(z_d) - f_1(z_d))\Theta_R(z_d).$$

Since Θ_R is rational and inner in $\mathbb{D}$, it follows by the symmetry principle that

$$\Theta_R(z)^{-1} = \Theta_R(\bar{z}^{-1})^* = I_{\mathcal{G}} - (z - 1)C_d(zI - D_d)^{-1}\mathbb{P}_R^{[-1]}$$

(4.48)

$$(I - D_d^*)^{-1}C_d^*.$$

At this point it is convenient to recall some well-known properties of the positive semidefinite solution $\mathbb{P}_R$ of the Stein equation

(4.49) $$\mathbb{P}_R - D_d^*\mathbb{P}_R D_d = C_d^*C_d.$$

Namely:

(a) Ker $\mathbb{P}_R$ is D_d-invariant, and Ker $\mathbb{P}_R = \bigcap_{k=0}^{\infty}$ Ker $(C_d D_d^k)$;

(b) denoting by $\mathbf{P}_{\mathrm{Ran}\mathbb{P}_R}$ (resp., $\mathbf{P}_{\mathrm{Ker}\mathbb{P}_R}$) the orthogonal projection on Ran $\mathbb{P}_R$ (resp., on Ker $\mathbb{P}_R$), we have:

(4.50) $$\mathbf{P}_{\mathrm{Ran}\mathbb{P}_R}(I - D_d^*)^{-1}C_d^* = (I - D_d^*)^{-1}C_d^*$$

and

(4.51) $$\mathbf{P}_{\mathrm{Ker}\mathbb{P}_R}(zI - D_d)^{-1}\mathbb{P}_R^{[-1]} = 0;$$

(c) the following equality holds for every $z \notin \mathrm{spec}\, D_d \cup (\mathrm{spec}\, D_d^*)^{-1}$:

$$(I - zD_d^*)^{-1}C_d^*C_d(zI - D_d)^{-1} = \mathbb{P}_R(zI - D_d)^{-1}$$

(4.52)

$$+ (I - zD_d^*)^{-1}D_d^*\mathbb{P}_R.$$

We now postmultiply both sides of (4.47) by $\Theta_R(z)^{-1}$, writing z for z_d in (4.47). The left hand side, after the postmultiplication and ignoring for the time being the factor $(X_1 - X_2)$, and using (4.48), transforms as follows (the second equality below holds due to (4.52)):

$$\mathbb{P}_R^{[-1]}(I - zD_d^*)^{-1}C_d^*[I - (z-1)C_d(zI - D_d)^{-1}\mathbb{P}_R^{[-1]}$$

$$(I - D_d^*)^{-1}C_d^*] = \mathbb{P}_R^{[-1]}(I - zD_d^*)^{-1}C_d^*$$

$$-(z-1)\mathbb{P}_R^{[-1]}(I - zD_d^*)^{-1}C_d^*C_d(zI - D_d)^{-1}\mathbb{P}_R^{[-1]}$$

(4.53) $\qquad (I - D_d^*)^{-1}C_d^* = \mathbb{P}_R^{[-1]}(I - zD_d^*)^{-1}C_d^* - (z-1)\mathbb{P}_R^{[-1]}$

$$\cdot\{\mathbb{P}_R(zI - D_d)^{-1} + (I - zD_d^*)^{-1}D_d^*\mathbb{P}_R\}\mathbb{P}_R^{[-1]}(I - D_d^*)^{-1}C_d^*$$

$$= \mathbb{P}_R^{[-1]}(I - zD_d^*)^{-1}C_d^* - (z-1)\cdot\{\mathbb{P}_{\mathrm{Ran}\mathbb{P}_R}(zI - D_d)^{-1}\mathbb{P}_R^{[-1]}$$

$$+ \mathbb{P}_R^{[-1]}(I - zD_d^*)^{-1}D_d^*\mathbb{P}_{\mathrm{Ran}\mathbb{P}_R}\}(I - D_d^*)^{-1}C_d^*.$$

The second $\mathbb{P}_{\mathrm{Ran}\mathbb{P}_R}$ term can be omitted in (4.53) in view of (b). Thus, (4.53) is equal to

$$\mathbb{P}_R^{[-1]}(I - zD_d^*)^{-1}\{(I - D_d^*) - (z-1)D_d^*\}(I - D_d^*)^{-1}C_d^*$$

$$-(z-1)\mathbb{P}_{\mathrm{Ran}\mathbb{P}_R}(zI - D_d)^{-1}\mathbb{P}_R^{[-1]}(I - D_d^*)^{-1}C_d^*$$

$$= \{\mathbb{P}_R^{[-1]} - (z-1)\mathbb{P}_{\mathrm{Ran}\mathbb{P}_R}(zI - D_d)^{-1}\mathbb{P}_R^{[-1]}\}(I - D_d^*)^{-1}C_d^*$$

$$= \{(zI - D_d) - (z-1)\mathbb{P}_{\mathrm{Ran}\mathbb{P}_R}\}(zI - D_d)^{-1}\mathbb{P}_R^{[-1]}(I - D_d^*)^{-1}C_d^*$$

$$= \{(I - D_d) + (z-1)\mathbb{P}_{\mathrm{Ker}\mathbb{P}_R}\}(zI - D_d)^{-1}\mathbb{P}_R^{[-1]}(I - D_d^*)^{-1}C_d^*.$$

In view of (b), this expression is equal to

$$(I - D_d)(zI - D_d)^{-1}\mathbb{P}_R^{[-1]}(I - D_d^*)^{-1}C_d^*.$$

Using the above computation, postmultiplying both sides of (4.47) by $\Theta_R(z)^{-1}$ eventually yields:

$$(X_1 - X_2)(zI - D_d)^{-1}(I - D_d)\mathbb{P}_R^{[-1]}(I - D_d^*)^{-1}C_d^*$$

(4.54)

$$= \Theta_L(z)(f_2(z) - f_1(z)).$$

Since spec $D_d \subset \mathbb{D}$, we have the expansion

(4.55) $\qquad (e^{it}I - D_d)^{-1} = \displaystyle\sum_{k=1}^{\infty} e^{-ikt}D_d^{k-1}.$

We consider both sides of (4.54) as functions of the variable $z \in \mathbb{T}$ with values in the Banach algebra of bounded linear operators acting from $\mathcal{G}$ into $\mathcal{H}(d-1)$.

Applying (4.54) to a fixed vector $y_1 \in \mathcal{G}$, and taking the inner product with a fixed vector $y_2 \in \mathcal{H}(d-1)$, we see (in view of (4.55)) that the scalar function

$$\langle\{(X_1 - X_2)(zI - D_d)^{-1}(I - D_d)\mathbb{P}_R^{[-1]}(I - D_d^*)^{-1}C_d^*\}y_1, y_2\rangle_{\mathcal{H}(d-1)}$$

belongs to the orthogonal complement of the Hardy space H_2 in the Lebesgue L_2 space of scalar functions on the unit circle. However, the function

$$\langle \Theta_L(z)\,(f_2(z) - f_1(z))\,y_1,\ y_2\rangle$$

obviously belongs to H_2, and therefore we obtain

$$\langle\{(X_1 - X_2)(zI - D_d)^{-1}(I - D_d)\mathbb{P}_R^{[-1]}(I - D_d^*)^{-1}C_d^*\}y_1, y_2\rangle = 0$$

and

$$\langle \Theta_L(z)\,(f_2(z) - f_1(z))\,y_1,\ y_2\rangle = 0.$$

Since y_1, y_2 are arbitrary, in fact

$$(X_1 - X_2)(zI - D_d)^{-1}(I - D_d)\mathbb{P}_R^{[-1]}(I - D_d^*)^{-1}C_d^*$$

$$\equiv \Theta_L(z)(f_2(z) - f_1(z)) \equiv 0.$$

Because $\Theta_L(z)$ is inner, $f_1(z) \equiv f_2(z)$, which proves the first relation in (4.46). Setting

$$T = \mathbf{P}_{\mathrm{Ran}\mathbb{P}_R}(I - D_d)\mathbb{P}_R^{[-1]}\left(I - D_d^*\right)^{-1}$$

we obtain then, in view of (b):

$$(X_1 - X_2)\,(zI - D_d)^{-1}TC_d^* \equiv 0,$$

which is equivalent to

(4.56) $$(X_1 - X_2)\,D_d^k TC_d^* = 0 \qquad (k \geq 0).$$

Premultiplying the left hand side of (4.49) by T and postmultiplying by T^*, we obtain using (c) (with $z = 1$):

(4.57)
$$\mathbf{P}_{\mathrm{Ran}\mathbb{P}_R}(I - D_d)\mathbb{P}_R^{[-1]}\{\mathbb{P}_R(I - D_d)^{-1} + (I - D_d^*)^{-1}D_d^*\mathbb{P}_R\}$$
$$\mathbb{P}_R^{[-1]}(I - D_d^*)\mathbf{P}_{\mathrm{Ran}\mathbb{P}_R}$$
$$= \mathbf{P}_{\mathrm{Ran}\mathbb{P}_R}(I - D)\{\mathbf{P}_{\mathrm{Ran}\mathbb{P}_R}(I - D)^{-1}\mathbb{P}_R^{[-1]}$$
$$+ \mathbb{P}_R^{[-1]}(I - D_d^*)^{-1}D_d^*\mathbf{P}_{\mathrm{Ran}\mathbb{P}_R}\} \cdot (I - D_d^*)\mathbf{P}_{\mathrm{Ran}\mathbb{P}_R}.$$

Using the equalities

$$\mathbf{P}_{\mathrm{Ran}\mathbb{P}_R}(I - D_d)\mathbf{P}_{\mathrm{Ran}\mathbb{P}_R} = \mathbf{P}_{\mathrm{Ran}\mathbb{P}_R}(I - D_d),$$

$$\mathbf{P}_{\mathrm{Ran}\mathbb{P}_R}(I - D_d^*)\mathbf{P}_{\mathrm{Ran}\mathbb{P}_R} = (I - D_d^*)\mathbf{P}_{\mathrm{Ran}\mathbb{P}_R},$$

which in turn follow from (a), the expression (4.57) after some simple algebra takes the form

$$\mathbf{P}_{\mathrm{Ran}\mathbb{P}_R}(\mathbb{P}_R^{[-1]} - D_d\mathbb{P}_R^{[-1]}D_d^*)\mathbf{P}_{\mathrm{Ran}\mathbb{P}_R}$$

$$= \mathbb{P}_R^{[-1]} - \mathbf{P}_{\mathrm{Ran}\mathbb{P}_R}D_d\mathbb{P}_R^{[-1]}D_d^*\mathbf{P}_{\mathrm{Ran}\mathbb{P}_R}.$$

We obtain that $\mathbb{P}_R^{[-1]}$ satisfies the equation

$$(4.58) \qquad \mathbb{P}_R^{[-1]} - \mathbf{P}_{\mathrm{Ran}\mathbb{P}_R}D_d\mathbb{P}_R^{[-1]}D_d^*\mathbf{P}_{\mathrm{Ran}\mathbb{P}_R} = TC_d^*C_dT^*.$$

(A similar argument has been used in the proof of Lemma 6.3 of [6].) Therefore, by the property (a) applied to (4.58):

$$(4.59) \qquad \begin{aligned} \mathrm{Ran}\,\mathbb{P}_R^{[-1]} &= \mathrm{span}\,\{\mathrm{Ran}\,(\mathbf{P}_{\mathrm{Ran}\mathbb{P}_R}D_d)^k TC_d^* : \ k \in \mathbb{Z}_+\} \\ &\subseteq \mathrm{span}\,\{\mathrm{Ran}\,(D_d^k TC_d^*) : \ k \in \mathbb{Z}_+\}. \end{aligned}$$

By definition of the Moore-Penrose pseudoinverse, $\mathbb{P}_R^{[-1]}$ maps bijectively the subspace $\mathrm{Ran}\mathbb{P}_R \subseteq \mathcal{H}_R$ onto itself. Consequently, the inclusion in (4.59) is equivalent to

$$\mathrm{Ran}\,\mathbb{P}_R \subseteq \mathrm{span}\,\{\mathrm{Ran}\,(D_d^k TC_d^*) : \ k \in \mathbb{Z}_+\}$$

and implies, on account of (4.56),

$$(X_1 - X_2)|_{\mathrm{Ran}\mathbb{P}_R} = 0.$$

The latter equality together with (4.29) leads to $X_1 = X_2$ which being multiplied by $\mathbf{E}(\mathbf{z}_{d-1})$ on the left implies the second relation in (4.46). $\square$

Note that the interpolation problem (4.30)–(4.32) is of the same type as Problem 1.1 (without the norm constraint), having the data set

$$(4.60) \qquad \begin{aligned} \Omega_{\widehat{\mathbf{X}}} &= \{C_+, \ \widetilde{C}_j, \ D_j, \ A_j, \ B_j, \ \Gamma_d D_d - A_d\Gamma_d + B_-C_d, \ \Gamma_j \ | \\ & \quad j = 1, \ldots, d-1\}, \end{aligned}$$

where

$$\tilde{C}_{d-1} = \mathbf{P}_{\mathrm{Ran}\mathbb{P}_R} C_{d-1} \quad \text{and} \quad \tilde{C}_j = C_j \quad (j = 1, \ldots, d - 2).$$

Lemma 4.8 *If the data set (1.5) is admissible, then $\Omega_{\tilde{X}}$ is admissible as well.*

Proof: The proof amounts to verifying the Sylvester equation

$$\left(\overset{\overset{d-1}{\frown}}{\prod_{j=2}} B_j \right) (A_1 \Gamma_1 - \Gamma_1 D_1)$$

$$+ \sum_{k=2}^{d-2} \left(\overset{\overset{d-1}{\frown}}{\prod_{j=k+1}} B_j \right) (A_k \Gamma_k - \Gamma_k D_k) \left(\overset{\overset{k-1}{\frown}}{\prod_{\ell=1}} C_\ell \right)$$

(4.61)
$$+ (A_{d-1} \Gamma_{d-1} - \Gamma_{d-1} D_{d-1}) \mathbf{P}_{\mathrm{Ran}\mathbb{P}_R} \overset{\overset{d-1}{\frown}}{\prod_{j=1}} C_j$$

$$= (\Gamma_d D_d - A_d \Gamma_d + B_- C_d) \mathbf{P}_{\mathrm{Ran}\mathbb{P}_R} \overset{\overset{d-1}{\frown}}{\prod_{j=1}} C_j - \overset{\overset{d-1}{\frown}}{\prod_{j=1}} B_j C_+,$$

and the following inclusions:

(4.62)
$$\mathrm{Ran}\,(\Gamma_d D_d - A_d \Gamma_d + B_- C_d) \subseteq$$
$$\mathrm{span}\{\mathrm{Ran}\,(A_{d-1}^{j_{d-1}} B_{d-1} \ldots A_1^{j_1} B_1) : \ j_1, \ldots, j_{d-1} \in \mathbb{Z}_+\},$$

(4.63) $$\mathrm{Ker}\, C_+ \supseteq \bigcap_{j_1, \ldots, j_{d-1}=0}^{\infty} \mathrm{Ker}\,(\mathbf{P}_{\mathrm{Ran}\mathbb{P}_R} C_{d-1} D_{d-1}^{j_{d-1}} C_{d-2} \ldots C_1 D_1^{j_1}),$$

and

(4.64) $$\mathrm{Ker}\, \Gamma_k \supseteq \bigcap_{j_k, \ldots, j_d=0}^{\infty} \mathrm{Ker}\,(\mathbf{P}_{\mathrm{Ran}\mathbb{P}_R} C_{d-1} D_{d-1}^{j_{d-1}} C_{d-2} \ldots C_k D_k^{j_k}),$$

(4.65) $$\mathrm{Ran}\, \Gamma_k \subseteq \mathrm{span}\{\mathrm{Ran}\,(A_k^{j_k} B_k \ldots A_1^{j_1} B_1) : \ j_1, \ldots, j_k \in \mathbb{Z}_+\}$$

for $k = 1, \ldots, d - 1$. It follows from (4.26) and (4.29) that

$$(4.66) \qquad (\Gamma_d D_d - A_d \Gamma_d + B_- C_d) \mathbf{P}_{\text{Ker} \mathbb{P}_R} = 0.$$

Therefore,

$$(\Gamma_d D_d - A_d \Gamma_d + B_- C_d) \mathbf{P}_{\text{Ran} \mathbb{P}_R} = \Gamma_d D_d - A_d \Gamma_d + B_- C_d$$

and then (4.61) follows immediately from (2.5). Next, in view of (4.12),

$$\text{Ran } \mathbf{B}_+ X \subseteq \text{Ran } \mathbf{B}_+ \subseteq \text{span}\{\text{Ran } (A_{d-1}^{j_{d-1}} B_{d-1} \ldots A_1^{j_1} B_1)$$

$$: \quad j_1, \ldots, j_{d-1} \in \mathbb{Z}_+\},$$

which, by (4.26), is equivalent to (4.62). Now (4.28) yields

$$\text{Ker } (\mathbf{P}_{\text{Ran} \mathbb{P}_R} C_{d-1} D_{d-1}^{j_{d-1}} \ldots C_1 D_1^{j_1})$$

$$= \bigcap_{j_d=0}^{\infty} \text{Ker } (C_d D_d^{j_d} C_{d-1} D_{d-1}^{j_{d-1}} \ldots C_1 D_1^{j_1}),$$

and then (4.63) and (4.64) follow from (2.2) and (2.3), respectively. Finally, conditions (4.65) are contained in (2.4). $\qquad \square$

Following Lemma 4.8, one can further reduce the interpolation problem (4.30)–(4.32), with or without the norm constraints, to an analogous problem in $d - 2$ variables, and continuing in this way, the original Problem 1.1 will reduce eventually to the one variable problem, to which the results of [5] (also exposed in Section 3) are applicable. However, the formulas for the solution set which could be obtained in that way are too cumbersome to be presented here in full generality. In Section 5 we give such formulas for the bidisk, i.e., for $d = 2$. In the next section the particular case of one-sided problems is considered. These have been studied already in [7], under the additional hypotheses that $B_j = I$, $j = 2, \ldots, d$ for the left-sided problem and $C_j = I$, $j = 1, \ldots, d - 1$ for the right-sided problem. For this case also, explicit formulas for the solution set are presented.

We conclude this section with the criterion for solvability of Problem 1.1 (without the norm constraint) that was promised in Section 2.

Theorem 4.9 *The interpolation problem* (1.8)–(1.10) *admits a solution* $H \in$ $\mathbf{HS}^{\mathcal{G} \to \mathcal{H}}(\mathbb{D}^d)$ *if and only if the conditions* (2.1)–(2.5) *hold true.*

Proof: Use the induction on d. The case $d = 1$ is covered in Lemma 3.2. Assuming the theorem has been already proved for $d - 1$, it immediately follows for d by combining Theorem 4.6 and Lemma 4.8. $\qquad \square$

5 Variations and Particular Cases

The reduction established in Theorem 4.6 is not symmetric with respect to the right and to the left interpolation conditions. To formulate the dual result, in which the roles of left and right conditions are interchanged, we need some preparation.

Starting with the Hilbert space $\mathcal{G}$ we define inductively the Hilbert spaces $\mathcal{G}(k)$, $k = 0, 1, \ldots$ by setting $\mathcal{G}(0) = \mathcal{G}$, and $\mathcal{G}(k)$ the Hilbert space of square summable sequences $\{g_j\}_{j=0}^{\infty}$, $g_j \in \mathcal{G}(k-1)$, as in (4.1). Now we define inductively the operator-valued functions $\widetilde{E}_k(z)$ by

$$
\widetilde{E}_1(z) = \begin{pmatrix} I_{\mathcal{G}}, \\ z I_{\mathcal{G}} \\ \vdots \end{pmatrix}, \quad \widetilde{E}_k(z) = E_k(z) = \mathrm{diag}(\widetilde{E}_{k-1}(z), \ \widetilde{E}_{k-1}(z), \ \ldots)
$$

(cf. (4.4), (4.7)). The function $\widetilde{E}_k$ is analytic in $\mathbb{D}$ and for every fixed $z \in \mathbb{D}$ the operator $\widetilde{E}_k(z)$ acts from $\mathcal{G}(k-1)$ into $\mathcal{G}(k)$.
Let

$$(5.1) \quad \widetilde{\mathbf{E}}(z_2, \ldots, z_d) = \widetilde{E}_{d-1}(z_2)\widetilde{E}_{d-2}(z_3) \ldots \widetilde{E}_1(z_d) : \mathcal{G} \rightarrow \mathcal{G}(d-1).$$

Note the following "right sided" analogue of Lemma 4.2.

Lemma 5.1 *Let $\mathcal{H}$ and $\mathcal{G}$ be two separable Hilbert spaces and let $\widetilde{\mathbf{E}}$ be defined by (5.1). Then the operator $\mathbf{M}_{\widetilde{\mathbf{E}}}$ of the right multiplication by $\widetilde{\mathbf{E}}(z_2, \ldots, z_d)$ is a unitary operator from $\mathbf{HS}^{\mathcal{H} \rightarrow \mathcal{G}(d-1)}(\mathbb{D})$ onto $\mathbf{HS}^{\mathcal{G} \rightarrow \mathcal{H}}(\mathbb{D}^d)$: any function $H \in \mathbf{HS}^{\mathcal{G} \rightarrow \mathcal{H}}(\mathbb{D}^d)$ admits a representation of the form*

$$(5.2) \qquad\qquad H(\mathbf{z}_d) = \widetilde{F}(z_1)\widetilde{\mathbf{E}}(z_2, \ldots, z_d),$$

where $\widetilde{F}$ is a (uniquely defined) function in $\mathbf{HS}^{\mathcal{H} \rightarrow \mathcal{G}(d-1)}(\mathbb{D})$ such that

$$\|H\|_{\mathbf{HS}^{\mathcal{G} \rightarrow \mathcal{H}}(\mathbb{D}^d)} = \|F\|_{\mathbf{HS}^{\mathcal{H} \rightarrow \mathcal{G}(d-1)}(\mathbb{D})}.$$

Let $\mathbf{C}_- : \mathcal{G}_1 \rightarrow \mathcal{G}(d-1)$ be the operator defined by

$$
\mathbf{C}_- = \frac{1}{(2\pi i)^{d-1}} \int_{\mathbb{T}^{d-1}} \widetilde{\mathbf{E}}(z_2, \ldots, z_d) \overset{d}{\overbrace{\prod_{j=2}}} C_j (z_j I - D_j)^{-1} dz_d \ldots dz_2,
$$

let $\widetilde{\mathbb{P}}_L$ and $\widetilde{\mathbb{P}}_R$ be solutions of the Stein equations

$$(5.3) \qquad \widetilde{\mathbb{P}}_L - A_1\widetilde{\mathbb{P}}_L A_1^* = B_1 B_1^* \quad \text{and} \quad \widetilde{\mathbb{P}}_R - D_1^*\widetilde{\mathbb{P}}_R D_1 = C_-^* C_-,$$

which are given via the converging series

$$\widetilde{\mathbb{P}}_L = \sum_{\ell=0}^{\infty} A_1^\ell B_1 B_1^* A_1^{*\ell} \quad \text{and} \quad \widetilde{\mathbb{P}}_R = \sum_{k=0}^{\infty} D_1^{*k} C_-^* C_- D_1^k,$$

and denote by $\widetilde{\mathbb{P}}_L^{[-1]}$ and $\widetilde{\mathbb{P}}_R^{[-1]}$ the Moore–Penrose pseudoinverses of $\widetilde{\mathbb{P}}_L$ and $\widetilde{\mathbb{P}}_R$, respectively. By Theorem 3.3 the functions

$$(5.4) \qquad \widetilde{\Theta}_L(z) = I_{\mathcal{H}} + (z-1)C_-(I - zA_1^*)^{-1}\widetilde{\mathbb{P}}_L^{[-1]}(I - A_1)^{-1}B_1$$

and

$$(5.5) \quad \widetilde{\Theta}_R(z) = I_{\mathcal{H}(d-1)} + (z-1)C_-(I - D_1)^{-1}\widetilde{\mathbb{P}}_R^{[-1]}(I - zD_1^*)^{-1}C_-,$$

are inner in $\mathbb{D}$.

Theorem 5.2 *Let be given the data* (1.5) *such that the necessary conditions* (2.1)–(2.5) *are satisfied. Let $\widetilde{\Theta}_L$ and $\widetilde{\Theta}_R$ be defined by* (5.4) *and* (5.5), *respectively, and let*

$$(5.6) \qquad \widehat{C}_1 = C_+ + B_1^*(I - A_1^*)^{-1}\widetilde{\mathbb{P}}_L^{[-1]}\{(I - A_1)\Gamma_1 - B_1 C_+\} : \mathcal{G}_1 \to \mathcal{H}.$$

A function $H \in \mathbf{HS}^{\mathcal{G}\to\mathcal{H}}(\mathbb{D}^d)$ is a solution to Problem 1.1 if and only if it is of the form

$$H(\mathbf{z}_d) = B_1^*(I - z_1 A_1^*)^{-1}\widetilde{\mathbb{P}}_L^{[-1]}\widehat{\mathbf{Y}}(z_2, \ldots, z_d) + \widetilde{\Theta}_L(z_1)\widehat{C}_1\widetilde{\mathbb{P}}_R^{[-1]}$$

$$(5.7)$$

$$\prod_{j=1}^{d}(I - z_j D_j^*)^{-1}C_j + \widetilde{\Theta}_L(z_1)h(z_1)\widetilde{\Theta}_R(z_1)\widetilde{E}(z_2, \ldots, z_d),$$

where h is an arbitrary function in $\mathbf{HS}^{\mathcal{H}\to\mathcal{G}(d-1)}(\mathbb{D})$, and where $\widehat{\mathbf{Y}}$ is any function in $\mathbf{HS}^{\mathcal{G}\to\mathcal{H}_1}(\mathbb{D}^{d-1})$ which satisfies interpolation conditions

$$(5.8) \qquad\qquad \mathbf{P}_{\mathrm{Ker}\mathbb{P}_L}\widehat{\mathbf{Y}}(z_2, \ldots z_d) \equiv 0,$$

$$\frac{1}{(2\pi i)^{d-1}} \int_{\mathbb{T}^{d-1}} \left(\prod_{j=2}^{\overset{d}{\curvearrowleft}} (z_j I - A_j)^{-1} B_j \right) \mathbf{P}_{\text{Ran}\widetilde{\mathbb{P}}_L}$$

(5.9)
$$\widehat{\mathbf{Y}}(z_2, \ldots, z_d)\, dz_2 \ldots dz_d = B_-,$$

$$\frac{1}{(2\pi i)^{d-1}} \int_{\mathbb{T}^{d-1}} \widehat{\mathbf{Y}}(z_2, \ldots, z_d) \prod_{j=2}^{\overset{d}{\curvearrowleft}} C_j (z_j I - D_j)^{-1} \, dz_d \ldots dz_2$$

(5.10)
$$= A_1 \Gamma_1 - \Gamma_1 D_1 + B_1 C_+,$$

$$\frac{1}{(2\pi i)^{d-1}} \int_{\mathbb{T}^{d-1}} \left(\prod_{j=2}^{\overset{k}{\curvearrowleft}} (z_j I - A_j)^{-1} B_j \right) \mathbf{P}_{\text{Ran}\widetilde{\mathbb{P}}_L}$$

(5.11)
$$\widehat{\mathbf{Y}}(z_2, \ldots, z_d) \cdot \left(\prod_{j=k}^{\overset{d}{\curvearrowleft}} C_j (z_j I - D_j)^{-1} \right)$$

$$dz_2 \ldots dz_{k-1} dz_d \ldots dz_k = \Gamma_k$$

for $k = 2, \ldots, d$ and the norm constraint

(5.12)
$$\| (\mathbb{P}_L^{[-1]})^{1/2} \widehat{\mathbf{Y}} \|_{\mathbf{HS}^{\mathcal{G} \to \mathcal{H}_1}(\mathbb{D}^{d-1})}^2$$
$$\leq \gamma^2 - \text{Trace}\, (\widehat{C}_1 \widetilde{\mathbb{P}}_L^{[-1]} \widehat{C}_1) - \| h \|_{\mathbf{HS}^{\mathcal{H} \to \mathcal{G}(d-1)}(\mathbb{D})}^2.$$

Proof: Given a function $H(\mathbf{z}_d) \in \mathbf{HS}^{\mathcal{G} \to \mathcal{H}}(\mathbb{D}^d)$, define $\tilde{H}(\mathbf{z}_d) \in \mathbf{HS}^{\mathcal{H} \to \mathcal{G}}(\mathbb{D}^d)$ by the rule

$$\tilde{H}(z_1, \ldots, z_d) = (H(\bar{z}_1, \ldots \bar{z}_d))^*.$$

Observe that

$$\| \tilde{H}(\mathbf{z}_d) \|_{\mathbf{HS}^{\mathcal{H} \to \mathcal{G}}(\mathbb{D}^d)} = \| H(\mathbf{z}_d) \|_{\mathbf{HS}^{\mathcal{G} \to \mathcal{H}}(\mathbb{D}^d)}.$$

Taking adjoints in the interpolation conditions (1.8)–(1.10), we see that $H(\mathbf{z}_d)$ is a solution of Problem 1.1 (with or without the norm constraint) if and only if

$\tilde{H}(\mathbf{z}_d)$ is a solution (with or without the norm constraint, as the case may be) of Problem 1.1 associated with the data set

(5.13) $\tilde{\Omega} = \{\tilde{C}_+, \ \tilde{C}_j, \ \tilde{D}_j, \ \tilde{A}_j, \ \tilde{B}_j, \ \tilde{B}_-, \ \tilde{\Gamma}_j \ | j = 1, \ldots, d\},$

where

$$\begin{aligned}
\tilde{C}_+ &= B_-^*, \quad \tilde{C}_j = B_{d-j+1}^*, \quad \tilde{D}_j = A_{d-j+1}^*, \quad \tilde{A}_j = D_{d-j+1}^*, \\
\tilde{B}_j &= C_{d-j+1}^*, \quad \tilde{B}_- = C_+^*, \quad \tilde{\Gamma}_j = \Gamma_{d-j+1}^*.
\end{aligned}$$

Observe that the data set (1.5) is admissible if and only if (5.13) is. Now by applying Theorem 4.6 to the interpolation problem associated with $\tilde{\Omega}$, and expressing the description for the interpolants $\tilde{H}(\mathbf{z}_d)$ in terms of $H(\mathbf{z}_d)$, we obtain Theorem 5.2. $\square$

In conclusion, we remark that the formula for solutions of the left sided interpolation problem (i.e., (1.8) with or without the norm constraint (1.7)) may be derived as a particular case of Theorem 4.6. Namely, the left sided interpolation is equivalent to Problem 1.1 in which $C_- = 0, C_+ = 0$, and $\Gamma_k = 0$ for $k = 1, \ldots, d$. Then it follows from (4.24) and (4.35) that $\hat{X}(\mathbf{z}_{d-1}) \equiv 0$ and $\Theta_R(z) \equiv I_{\mathcal{G}}$. Substituting these in (4.37) we get:

$$H(\mathbf{z}_d) = \overbrace{\prod_{j=1}^{d}}^{d} B_j^*(I - z_j A_j^*)^{-1} \mathbb{P}_L^{[-1]} \hat{B}_d + \mathbf{E}(\mathbf{z}_{d-1}) \Theta_L(\mathbf{z}_d) f(\mathbf{z}_d).$$

This is the formula obtained in [7] for the particular case when dimensions of $\mathcal{H}_j$ are equal and B_j are identity matrices for $j = 2, \ldots, d$. Analogously, the formula for the right-sided problem may be obtained as a particular case of Theorem 5.2 by setting $B_1 = 0, B_- = 0$, and $\Gamma_k = 0$ for $k = 1, \ldots, d$ in Problem 1.1.

6 The Case of the Bidisk

In this section we apply the analysis of Section 4 to the case of two variables, i.e., we shall describe all functions $H \in \mathbf{HS}^{\mathcal{G} \to \mathcal{H}}$ which satisfy interpolation conditions (1.12)–(1.15) and the norm constraint (1.7). We shall refer to this problem as Problem 6.1. In view of Theorem 4.9 we assume that the data set of Problem 6.1

(6.1) $\Omega = \{C_+, \ C_j, \ D_j, \ A_j, \ B_j, \ B_-, \ \Gamma_j \ | j = 1, 2\}$

is admissible, i.e., that the spectral conditions (1.6) are in force and the following Sylvester identity holds:

$$B_2(A_1\Gamma_1 - \Gamma_1 D_1) + (A_2\Gamma_2 - \Gamma_2 D_2)C_1 = B_- C_2 C_1 - B_2 B_1 C_+.$$

Moreover, we may assume without loss of generality that the positive semidefinite matrix

$$\mathbb{P}_{R,2} = \sum_{k=0}^{\infty} D_2^{*k} C_2^* C_2 D_2^k$$

is an orthogonal projection. If this is not so, consider an invertible operator $G : \mathcal{G}_2 \to \mathcal{G}_2$ such that $G^* \mathbb{P}_{R,2} G$ is an orthogonal projection (recall that $\mathcal{G}_j$ are assumed to be finite dimensional), and note that the interpolation problem with the data set

$$\Omega_1 = \{C_+, \ \tilde{C}_j, \ \tilde{D}_j, \ A_j, \ B_1, \ B_-, \ \tilde{\Gamma}_j \ | j = 1, 2\},$$

where

$$\begin{aligned}
\tilde{C}_1 &= G^{-1} C_1, \ \tilde{C}_2 = C_2 G, \ \tilde{D}_1 = D_1, \\
\tilde{D}_2 &= G^{-1} D_2 G, \ \tilde{\Gamma}_1 = \Gamma_1, \ \tilde{\Gamma}_2 = \Gamma_2 G,
\end{aligned}$$

is equivalent to (i.e., has the same set of solutions as) Problem 6.1. Thus we have

(6.2) $$\mathbb{P}_{R,2} = \mathbb{P}_{R,2}^{[-1]} = \mathbf{P}_{\mathrm{Ran} \mathbb{P}_{R,2}}.$$

Note the equality

(6.3) $$(\Gamma_2 D_2 - A_2 \Gamma_2 + B_- C_2)(I - \mathbb{P}_{R,2}) = 0,$$

which follows from (4.66) for $d = 2$ and $\mathbb{P}_R = \mathbb{P}_{R,2}$. Introduce also the matrices

$$\mathbb{P}_{R,1} = \sum_{j=0}^{\infty} D_1^{*j} \mathbb{P}_{R,2} D_1^j, \ \mathbb{P}_{L,1} = \sum_{k=0}^{\infty} A_1^k B_1 B_1^* A_1^{*k},$$

$$\mathbb{P}_{L,2} = \sum_{k=0}^{\infty} A_2^k B_+ B_+^* A_2^{*k},$$

where, according to (4.12),

(6.4) $$\mathbf{B}_+ = \frac{1}{2\pi i} \int_{\mathbb{T}} (zI - A_1)^{-1} B_1 E_1(z) dz,$$

and where $E_1(z)$ is the function given by (4.2). Define the following four inner functions in $\mathbb{D}$:

$$\Theta_{R,1}(z) = I_{\mathcal{G}_2} + (z - 1) \mathbb{P}_{R,2} C_1 (I - D_1)^{-1} \mathbb{P}_{R,1}^{[-1]}$$

(6.5)

$$(I - z D_1^*)^{-1} C_1^* \mathbb{P}_{R,2},$$

$$\Theta_{R,2}(z) = I_{\mathcal{G}} + (z-1)C_2(I-D_2)^{-1}\mathbb{P}_{R,2}$$

(6.6)
$$(I - zD_2^*)^{-1}C_2^*,$$

(6.7) $\Theta_{L,1}(z) = I_{\mathcal{H}} + (z-1)B_1^*(I-zA_1^*)^{-1}\mathbb{P}_{L,1}^{[-1]}(I-A_1)^{-1}B_1,$

(6.8) $\Theta_{L,2}(z) = I_{\mathcal{H}(1)} + (z-1)\mathbf{B}_+^*(I-zA_1^*)^{-1}\mathbb{P}_{L,2}^{[-1]}(I-A_2)^{-1}\mathbf{B}_+,$

and operators

$$\widehat{B}_1 = \Gamma_2 D_2 - A_2\Gamma_2 + B_-C_2 + \{\Gamma_1(I-D_1) - \Gamma_2 D_2$$
$$-A_2\Gamma_2 + B_-C_2\}\mathbb{P}_{R,1}^{[-1]}(I-D_1^*)^{-1}C_1^*\mathbb{P}_{R,2}$$

and

$$\widehat{B}_2 = B_- + \{\Gamma_2(I-D_2) - B_-C_2\}\mathbb{P}_{R,2}(I-D_2^*)^{-1}C_2^*.$$

It follows from (6.3) and (6.5) that

(6.9) $$\widehat{B}_1\Theta_{R,1}(z)(I-\mathbb{P}_{R,2}) \equiv 0.$$

By Theorem 4.6, all solutions H to Problem 6.1 are parametrized by the formula

$$H(z_1,z_2) = \widehat{\mathbf{X}}(z_1)\mathbb{P}_{R,2}(I-z_2D_2^*)^{-1}C_2^* + B_1^*(I-z_1A_1^*)^{-1}$$

(6.10)
$$\mathbb{P}_{L,2}^{[-1]}\widehat{B}_2\Theta_{R,2}(z_2) + E_1(z_1)\Theta_{L,2}(z_2)f(z_2)\Theta_{R,2}(z_2),$$

where f is an arbitrary function in $\mathbf{HS}^{\mathcal{G}\to\mathcal{H}(1)}(\mathbb{D})$, and where $\widehat{\mathbf{X}}$ is any function in $\mathbf{HS}^{\mathcal{G}_2\to\mathcal{H}}(\mathbb{D})$ which satisfies the interpolation conditions

(6.11) $$\widehat{\mathbf{X}}(z_1)(I-\mathbb{P}_{R,2}) \equiv 0,$$

(6.12) $$\frac{1}{2\pi i}\int_{\mathbb{T}}(z_1 I - A_1)^{-1}B_1\widehat{\mathbf{X}}(z_1)\,dz_1 = \Gamma_2 D_2 - A_2\Gamma_2 + B_-C_2,$$

(6.13) $$\frac{1}{2\pi i}\int_{\mathbb{T}}\widehat{\mathbf{X}}(z_1)\mathbb{P}_{R,2}C_1(z_1 I - D_1)^{-1}\,dz_1 = C_+,$$

(6.14) $$\frac{1}{2\pi i}\int_{\mathbb{T}}(z_1 I - A_1)^{-1}B_1\widehat{\mathbf{X}}(z_1)\mathbb{P}_{R,2}C_1(z_1 I - D_1)^{-1}\,dz_1 = \Gamma_1,$$

and the norm constraint

$$\|\mathbb{P}_{R,2}\widehat{\mathbf{X}}^*\|^2_{\mathbf{HS}^{\mathcal{G}_2 \to \mathcal{H}}(\mathbb{D})} \le \gamma^2 - \mathrm{Trace}\,(\widehat{B}_1^*\mathbb{P}^{[-1]}_{L,2}\widehat{B}_2) - \|f\|^2_{\mathbf{HS}^{\mathcal{G} \to \mathcal{H}^{(1)}}(\mathbb{D})}.$$

In view of (6.11) and (6.2), the latter condition may be written as

$$(6.15) \quad \|\widehat{\mathbf{X}}\|^2_{\mathbf{HS}^{\mathcal{G}_2 \to \mathcal{H}}(\mathbb{D})} \le \gamma^2 - \mathrm{Trace}\,(\widehat{B}_2^*\mathbb{P}^{[-1]}_{L,2}\widehat{B}_2) - \|f\|^2_{\mathbf{HS}^{\mathcal{G} \to \mathcal{H}^{(1)}}(\mathbb{D})}.$$

Relations (6.12)–(6.14) present the interpolation problem with the admissible data set

$$\Omega_{\widehat{\mathbf{X}}} = \{C_+, \; \mathbb{P}_{R,2}C_1, \; D_1, \; A_1, \; B_1, \; \Gamma_2 D_2 - A_2\Gamma_2 + B_-C_2, \; \Gamma_1\}.$$

By Theorem 3.4, all solutions to this problem are parametrized by the formula

$$\widehat{\mathbf{X}}(z) = C_+\mathbb{P}^{[-1]}_{R,1}(I - zD_1^*)^{-1}C_1\mathbb{P}_{R,2} + B_1^*(I - zA_1^*)^{-1}$$

(6.16)

$$\mathbb{P}^{[-1]}_{L,1}\widehat{B}_1\Theta_{R,1}(z) + \Theta_{L,1}(z)h(z)\Theta_{R,1}(z),$$

where $\Theta_{L,1}$ and $\Theta_{R,1}$ are the functions defined by (6.7) and (6.5) respectively and h is a free parameter in $\mathbf{HS}^{\mathcal{G}_2 \to \mathcal{H}}(\mathbb{D})$. The representation (6.16) is orthogonal with respect to the inner product in $\mathbf{HS}^{\mathcal{G} \to \mathcal{H}}(\mathbb{D})$, and since $\Theta_{L,1}$ and $\Theta_{R,1}$ are inner,

$$\|\widehat{\mathbf{X}}\|^2_{\mathbf{HS}^{\mathcal{G}_2 \to \mathcal{H}}(\mathbb{D})} = \mathrm{Trace}\,(C_+\mathbb{P}^{[-1]}_{R,1}C_+^* + \widehat{B}_1^*\mathbb{P}^{[-1]}_{L,1}\widehat{B}_1)$$

(6.17)

$$+\|h\|^2_{\mathbf{HS}^{\mathcal{G} \to \mathcal{H}}(\mathbb{D})}.$$

It follows from (6.9) that the function $\widehat{\mathbf{X}}$ of the form (6.16) satisfies the condition (6.11) if and only if the corresponding parameter h is subject to

$$(6.18) \qquad\qquad h(z)(I - \mathbb{P}_{R,2}) \equiv 0.$$

Upon substituting (6.16) into (6.10) and making use of (6.15) and (6.17), we arrive at the following result:

Theorem 6.1 *All solutions H to Problem 6.1 are parametrized by the formula*

$$H(z_1, z_2) = C_+\mathbb{P}^{[-1]}_{R,1}(I - z_1D_1^*)^{-1}C_1^*\mathbb{P}_{R,2}(I - z_2D_2^*)^{-1}C_2^*$$
$$+ B_1^*(I - z_1A_1^*)^{-1}\mathbb{P}^{[-1]}_{L,1}\widehat{B}_1\Theta_{R,1}(z_1)(I - z_2D_2^*)^{-1}C_2^*$$
(6.19)
$$+ B_1^*(I - z_1A_1^*)^{-1}\mathbb{P}^{[-1]}_{L,2}\widehat{B}_2\Theta_{R,2}(z_2)$$
$$+ \Theta_{L,1}(z_1)h(z_1)\Theta_{R,1}(z_1)\mathbb{P}_{R,2}(I - z_2D_2^*)^{-1}C_2^*$$
$$+ E_1(z_1)\Theta_{L,2}(z_2)f(z_2)\Theta_{R,2}(z_2),$$

where $\Theta_{L,j}$ *and* $\Theta_{R,j}$ *are the inner functions defined by (6.5)–(6.8) and* f *and* h *are arbitrary functions in* $\mathbf{HS}^{\mathcal{G}\to\mathcal{H}^{(1)}}(\mathbb{D})$ *and* $\mathbf{HS}^{\mathcal{G}_2\to\mathcal{H}}(\mathbb{D})$, *respectively, such that the condition (6.18) holds together with the norm constraint*

$$\|f\|^2_{\mathbf{HS}^{\mathcal{G}\to\mathcal{H}^{(1)}}(\mathbb{D})} + \|h\|^2_{\mathbf{HS}^{\mathcal{G}\to\mathcal{H}}(\mathbb{D})}$$

$$\leq \gamma^2 - \text{Trace}\,(\widehat{B}_2^* \mathbb{P}_{L,2}^{[-1]} \widehat{B}_2 + \widehat{B}_1^* \mathbb{P}_{L,1}^{[-1]} \widehat{B}_1 + C_+ \mathbb{P}_{R,1}^{[-1]} C_+^*).$$

The representation (6.19) is orthogonal with respect to the inner product in $\mathbf{HS}^{\mathcal{G}\to\mathcal{H}}(\mathbb{D})$ *and therefore, the sum of the three first summands in the righthand side of (6.19) presents the solution of Problem 6.1 with the possibly minimal norm. Therefore, Problem 6.1 with the admissible data set (6.1) is solvable if and only if*

$$\gamma \leq (\text{Trace}\,(\widehat{B}_2^* \mathbb{P}_{L,2}^{[-1]} \widehat{B}_2 + \widehat{B}_1^* \mathbb{P}_{L,1}^{[-1]} \widehat{B}_1 + C_+ \mathbb{P}_{R,1}^{[-1]} C_+^*))^{\frac{1}{2}}.$$

It follows easily from Theorem 6.1 that all solutions of the homogeneous analogue of the Problem 6.1, i.e., when the right-hand sides of equations (1.12)–(1.15) are all equal zero, are given by

$$H(z_1, z_2) = \Theta_{L,1}(z_1)h(z_1)\Theta_{R,1}(z_1)\mathbb{P}_{R,2}(I - z_2 D_2^*)^{-1} C_2^*$$

(6.20)

$$+ E_1(z_1)\Theta_{L,2}(z_2)f(z_2)\Theta_{R,2}(z_2).$$

Note that when specialized to the case of one variable, this formula is a Beurling-Lax formula for shift invariant subspaces of special type given by the homogeneous interpolation conditions. It is well-known that the usual Beurling-Lax formula fails in the context of more than one variable. Of course, the set of functions satisfying the homogeneous interpolation conditions (1.12)–(1.15) is invariant under multiplication by the variables z_1 and z_2. Therefore, formula (6.20) can be understood as an analogue of Beurling-Lax theorem for a very special type of shift invariant subspaces of $\mathbf{H}_2(\mathbb{D}^2)$.

A "dual" description of all solutions of the Problem 6.1 may be obtained by using Theorem 5.2 in place of Theorem 4.6.

Acknowledgements

We thank the referee for several useful remarks, one of which has led the authors to consider Problem 1.1 in its present form.

The research of LR is partially supported by NSF Grant DMS 9800704, and by a Faculty Research Assignment Grant of the College of William and Mary.

References

[1] J. Agler, On the representation of certain holomorphic functions defined on a polydisk, *Operator Theory: Advances and Applications*, Birkhäuser Verlag, Basel **48** (1990), 47–66.

[2] D. Alpay and V. Bolotnikov, Two sided interpolation for matrix functions with entries in the Hardy space, *Linear Algebra and its Applications* **223/224** (1995), 31–56.

[3] D. Alpay and V. Bolotnikov, On tangential interpolation in reproducing kernel Hilbert space modules and applications, *Operator Theory: Advances and Applications* **95** (1997), 37–68.

[4] D. Alpay and V. Bolotnikov, On the tangential interpolation problem for matrix-valued H_2-functions of two variables, *Proceedings of the Amer. Math. Soc.* **127** (1999), 1789–1799.

[5] D. Alpay, V. Bolotnikov and Ph. Loubaton, On two-sided residue interpolation for matrix-valued h_2-functions with symmetries, *Journal of Mathematical Analysis and Applications* **200** (1996), 76–105.

[6] D. Alpay, V. Bolotnikov and L. Rodman, Tangential interpolation with symmetries and two-points interpolation for matrix valued H_2-functions, *Integral Equations and Operator Theory* **32** (1998), 1–28.

[7] D. Alpay, V. Bolotnikov and L. Rodman, One-sided tangential interpolation for operator-valued Hardy functions on polydisks, *Integral Equations and Operator Theory* (to appear).

[8] J.A. Ball, I. Gohberg and L. Rodman, *Interpolation of Rational Matrix Functions*, OT **45**, Birkhäuser Verlag, Basel 1990.

[9] J.A. Ball, T. Trent, Unitary colligations, reproducing kernel Hilbert spaces and Nevanlinna-Pick interpolation in several variables, *J. of Functional Analysis* **157** (1998), 1–61.

[10] J.A. Ball and V. Vinnikov, Zero-pole interpolation for meromorphic matrix functions on an algebraic curve and transfer functions of 2D systems, *Acta Appl. Math.* **45** (1996), 239–316.

[11] H. Dym, *J* Contractive Matrix Functions, Reproducing Kernel Spaces and Interpolation, *CBMS Lecture Notes. Amer. Math. Soc.*, Rhodes Island 1989.

[12] H. Dym, Book review: The commutant lifting approach to interpolation problems, by Ciprian Foiaş and Arthur E. Frazho, *Bull. of the Amer. Math. Soc.* **31** (1994), 125–140.

[13] C. Foiaş and A.E Frazho, *The Commutant Lifting Approach to Interpolation Problems*, OT **44**, Birkhäuser Verlag, Basel 1990.

[14] C. Foiaş, A.E. Frazho, I. Gohberg and M.A. Kaashoek, *Metric Constrained Interpolation, Commutant Lifting, and Systems*, OT **100**, Birkhäuser Verlag, Basel 1998.

[15] P. Lancaster and M. Tismenetsky, *The Theory of Matrices, 2nd edition*, Academic Press, Orlando 1985.

[16] R.A. Penrose, A generalized inverse for matrices, *Proceedings of Cambridge Phil. Soc.* **51** (1955), 406–413.

[17] W. Rudin, *Function Theory in Polydiscs*, W.A. Benjamin, Inc., New York–Amsterdam, 1969.

[18]   I. Schur, Über die Potenzreihen, die im Innern des Einheitkreises Beschrankt sind, *Journal für die reine und angewandte Mathematik* **147** (1917), English translation in: I. Schur methods in operator theory and signal processing. Operator theory: Advances and Applications Birkhäuser Verlag, Basel OT **18** (1986), 31–59; 61–88.

D. Alpay
Department of Mathematics
Ben-Gurion University of the Negev
Beer-Sheva 84105
Israel

V. Bolotnikov and L. Rodman
Department of Mathematics
College of William and Mary
Williamsburg, VA 23187-8795
USA

1991 Mathematics Subject Classification. Primary: 47A56, 41A05; Secondary: 32A35.

Nonstationary Analogs of the Herglotz Representation Theorem: Realizations Centered at an Arbitrary Point

D. Alpay, A. Dijksma and Y. Peretz

Dedicated to Israel Gohberg on the occasion of his seventieth birthday

In this paper we prove generalized Herglotz representation theorems for bounded upper triangular operators with nonnegative real part when the base "point" (in fact a diagonal operator) is different from 0.

1 Introduction

In this paper we continue our study of the representation (or realization) of bounded upper triangular operators with nonnegative real part in terms of (co-)isometric and unitary colligations. In [ADP] we considered the case of realizations centered at the origin. Here we tackle the case of realizations centered at an arbitrary point (precise definitions will be given in the sequel).

Upper triangular operators with nonnegative real part are the natural non-stationary analogs of $p \times p$ matrix valued Carathéodory functions. By definition, the latter are functions which are analytic in the open unit disk $\mathbb{D}$ and have a nonnegative real part there. If ϕ is such a function, then the kernel

$$K_\phi(z, w) = \frac{\phi(z) + \phi(w)^*}{1 - zw^*}$$

is nonnegative on $\mathbb{D} \times \mathbb{D}$ (in the sense of reproducing kernels) and hence is the reproducing kernel of a uniquely determined reproducing kernel Hilbert space $\mathcal{L}_+(\phi)$. For $v \in \mathbb{D}$ the formulas

$$(A_v f)(z) = \frac{f(z) - f(v)}{z - v}$$

$$(B_v \xi)(z) = \frac{\phi(z) - \phi(v)}{z - v}\xi$$

$$C_v f = f(v)$$

$$D_v \xi = \phi(v)\xi$$

define an operator matrix

$$\begin{pmatrix} A_v & B_v \\ C_v & D_v \end{pmatrix} : \begin{pmatrix} \mathcal{L}_+(\phi) \\ \mathbb{C}^p \end{pmatrix} \longrightarrow \begin{pmatrix} \mathcal{L}_+(\phi) \\ \mathbb{C}^p \end{pmatrix}$$

such that

(1.1) $\phi(z) = D_v + (z - v)C_v(I_{\mathcal{L}_+(\phi)} - (z - v)A_v)^{-1}B_v, \quad z \in \mathbb{D}.$

When $v = 0$, formula (1.1) is a realization in the usual sense and the operator A_0 is coisometric. Furthermore, $B_0 = A_0 C_0^*$ and Re $D_0 = \frac{1}{2}C_0 C_0^*$. For general $v \in \mathbb{D}$, the colligation $(\mathcal{L}_+(\phi), \mathbb{C}^p, A_v, B_v, C_v, D_v)$ is called coisometric in the generalized sense that

(1.2) $A_v A_v^* = (I + vA_v)(I + vA_v)^*.$

Note that also

(1.3) $B_v = A_v(I + vA_v)^{-*}C_v^*,$

(1.4)   Re $D_v = \dfrac{1}{2(1 - |v|^2)}C_v C_v^*,$

and that (1.1) can be rewritten as

$$\phi(z) = i\mathrm{Im}\, D_v + \frac{1}{2}C_v(I + (z - v)A_v)$$

(1.5) $(I - (z - v)A_v)^{-1}(I + vA_v)^{-*}C_v^*$

$$+\frac{1}{2}C_v((1 - |v|^2)(I + vA_v)^* - I)(I + vA_v)^{-*}C_v^*.$$

When $v = 0$, this formula reduces to

$$\phi(z) = i\mathrm{Im}\, D_0 + \frac{1}{2}C_0(I + zA_0)(I - zA_0)^{-1}C_0^*.$$

Note that the operator $(I + vA_v)$ is given by

$$((I + vA_v)f)(z) = \frac{zf(z) - vf(v)}{z - v}.$$

There are also analogous isometric and unitary realizations. In the classical case of analytic functions such realizations centered at a point different from the origin are of particular importance when one leaves the setting of Hilbert spaces and assumes that the kernel K_ϕ has a finite number of negative squares. Then much of the Hilbert space structure remains, but one may loose analyticity at the origin; compare for example with [ADRS1] and [ADRS2]. The paper [FFGK1] and the book [FFGK2] present a method that allows one to transfer problems in a nonstationary setting into a stationary operator setting. This method helps to guess the nonstationary results from the original stationary (function theoretic) setting. It is not clear if and how this method applies to the problems considered in this paper.

The paper consists of four sections besides the introduction. In the second section we review some facts on the discrete nonstationary setting. In Section 3 we consider the coisometric case, while the isometric case is studied in Section 4. The last section is devoted to the unitary case.

The authors are grateful to the referee for the positive yet critical remarks.

2 Some Preliminaries

The nonstationary setting has been treated in a number of papers (see for example [ADD], [DD, Section 1]), and this section will be kept to a minimum. Let $\mathcal{M}$ be a separable Hilbert space, the so called "coefficient space". The set of bounded operators from the space $\ell^2_{\mathcal{M}}$ of square summable two sided sequences with components in $\mathcal{M}$ into itself is denoted by $\mathcal{X}$. The space $\ell^2_{\mathcal{M}}$ is taken with the standard inner product. We denote by Z the bilateral backward shift operator

$$(Zf)_i = f_{i+1}, \quad i = \ldots, -1, 0, 1, \ldots$$

where $f = (\ldots, f_{-1}, \boxed{f_0}, f_1, \ldots) \in \ell^2_{\mathcal{M}}$. It is unitary on $\ell^2_{\mathcal{M}}$. An element $A \in \mathcal{X}$ can be represented as an operator matrix (A_{ij}) with $A_{ij} = \pi^* Z^i A Z^{*j} \pi$, where π denotes the injection map: $u \in \mathcal{M} \mapsto (\ldots, 0, \boxed{u}, 0, \ldots) \in \ell^2_{\mathcal{M}}$. The operator matrix AZ is obtained from A by moving the columns of A one place to the right; ZA is obtained from A by moving the rows of A one place upward. We set

$$A^{(j)} = Z^{*j} A Z^j, \quad j = \ldots, -1, 0, 1, \ldots$$

The spaces of upper triangular, lower triangular, and diagonal operators will be denoted by $\mathcal{U}$, $\mathcal{L}$ and $\mathcal{D}$:

$$\mathcal{U} = \{A \in \mathcal{X} | A_{ij} = 0, \, i > j\}, \; \mathcal{L} = \{A \in \mathcal{X} | A_{ij} = 0, \, j > i\}, \; \mathcal{D} = \mathcal{U} \cap \mathcal{L}.$$

We associate to an upper triangular operator F two point evaluations as follows: for any $F \in \mathcal{U}$, there exists a unique sequence of operators $F_{[j]} \in \mathcal{D}$, $j = 0, 1, \ldots$, namely $(F_{[j]})_{ii} = F_{i-j,i}$, such that

$$F = \sum_{n=0}^{\infty} Z^n F_{[n]}$$

in the sense that $F - \sum_{j=0}^{n-1} Z^j F_{[j]} \in Z^n \mathcal{U}$. One can view this series as a formal "left" power series in Z and we define the left point evaluation of this power series at $W \in \mathcal{D}$ by

$$F^{\wedge}(W) = \sum_{n=0}^{\infty} W^{[n]} F_{[n]},$$

where

$$W^{[0]} = I, \quad W^{[n]} = W W^{(1)} W^{(2)} \ldots W^{(n-1)} = (W Z^*)^n Z^n, \quad n \geq 1.$$

Similarly, there exists a unique sequence of diagonal operators $F_{\{j\}} \in \mathcal{D}$, namely $(F_{\{j\}})_{ii} = F_{i,i+j}$ (so $F_{\{j\}} = Z^j F_{[j]} Z^{*j} = F_{[j]}^{(-j)}$), such that

$$F = \sum_{n=0}^{\infty} F_{\{n\}} Z^n$$

in the sense that $F - \sum_{j=0}^{n-1} F_{\{j\}} Z^j \in \mathcal{U} Z^n$. This series is a formal "right" power series in Z. For $W \in \mathcal{D}$ we set

$$W^{\{n\}} = Z^n (Z^* W)^n, \quad n = 0, 1, \ldots$$

and we define the right point evaluation of F at W by

$$F^{\triangle}(W) = \sum_{n=0}^{\infty} F_{\{n\}} W^{\{n\}}.$$

For $W \in \mathcal{D}$, it holds that

$$\ell_W := \lim_{n \to \infty} \|W^{[n]}\|^{1/n} = \lim_{n \to \infty} \|W^{\{n\}}\|^{1/n} = r_{sp}(W Z^*) = r_{sp}(Z^* W),$$

where $r_{sp}(V)$ stands for the spectral radius of V. If $\ell_W < 1$ then the series defining the left and right point evaluations converge in the uniform operator norm of $\mathcal{D}$. We set

$$\Omega = \{W \in \mathcal{D} | \ell_W < 1\}.$$

The set Ω contains in particular the diagonal operators of norm strictly less than 1.

In the sequel we will be concerned with the maps $W \mapsto F^{\wedge}(W)$ and $W \mapsto F^{\triangle}(W)$ for $W \in \mathcal{D}$; the operators $F^{\wedge}(W)$ and $F^{\triangle}(W)$ are diagonal and can be characterized as follows: For $D \in \mathcal{D}$, (1) the operator $(Z - W)^{-1}(F - D)$ belongs to $\mathcal{U}$ for $W \in \Omega$ if and only if $D = F^{\wedge}(W)$ and (2) the operator $(F - D)(Z - W)^{-1}$ belongs to $\mathcal{U}$ for $W \in \Omega$ if and only if $D = F^{\triangle}(W)$. Continuous analogs of these generalized point evaluations are introduced in [BGK1] and [BGK2]. We have used them in a forthcoming publication (see [ABDP]) to get the continuous analogs of the results of the present paper.

An operator $F = (F_{ij}) \in \mathcal{X}$ is a Hilbert–Schmidt operator if all its entries F_{ij} are Hilbert–Schmidt operators on $\mathcal{M}$ and $\sum_{ij} \mathrm{Tr}\, F_{ij}^* F_{ij} < \infty$, where Tr stands for trace. The set of these operators will be denoted by $\mathcal{X}_2$, it is a Hilbert space with respect to the inner product

$$\langle F, G \rangle_{\mathcal{X}_2} = \sum_{ij} \mathrm{Tr}\, G_{ij}^* F_{ij} < \infty.$$

The subspaces of upper triangular, lower triangular, and diagonal operators in $\mathcal{X}_2$ will be denoted by $\mathcal{U}_2$, $\mathcal{L}_2$, and $\mathcal{D}_2$.

Let $\Phi \in \mathcal{U}$ be a Carathéodory operator, that is, an operator with nonnegative real part: $\operatorname{Re} \Phi = \frac{\Phi + \Phi^*}{2} \geq 0$. The left multiplication operator $\mathcal{M}_\Phi^\ell : \mathcal{U}_2 \longrightarrow \mathcal{U}_2$ defined by $\mathcal{M}_\Phi^\ell(F) = \Phi F$ is well defined and bounded with $\|\mathcal{M}_\Phi^\ell\| \leq \|\Phi\|$. Similarly, the left multiplication operator $\mathcal{M}_{\Phi^*}^\ell : \mathcal{L}_2 \longrightarrow \mathcal{L}_2$ defined by $\mathcal{M}_{\Phi^*}^\ell(G) = \Phi^* G$ is well defined and bounded with $\|\mathcal{M}_{\Phi^*}^\ell\| \leq \|\Phi\|$. We will denote by $\mathcal{L}_\ell(\Phi)$ and $\mathcal{L}_\ell(\Phi^*)$ the operator ranges

$$\mathcal{L}_\ell(\Phi) = \operatorname{ran}(\mathcal{M}_\Phi^\ell + \mathcal{M}_\Phi^{\ell*})^{1/2} \quad \text{and} \quad \mathcal{L}_\ell(\Phi^*) = \operatorname{ran}(\mathcal{M}_{\Phi^*}^\ell + \mathcal{M}_{\Phi^*}^{\ell*})^{1/2}$$

equipped with the lifted norms

$$\|(\mathcal{M}_\Phi^\ell + \mathcal{M}_\Phi^{\ell*})^{1/2} F\|_{\mathcal{L}_\ell(\Phi)} = \|(I - P)F\|_{\mathcal{U}_2},$$

and

$$\|(\mathcal{M}_{\Phi^*}^\ell + \mathcal{M}_{\Phi^*}^{\ell*})^{1/2} F\|_{\mathcal{L}_\ell(\Phi)} = \|(I - P_*)F\|_{\mathcal{L}_2},$$

where P (P_*) is the orthogonal projection from $\mathcal{U}_2$ onto $\ker(\mathcal{M}_\Phi^\ell + \mathcal{M}_\Phi^{\ell*})^{1/2}$ (from $\mathcal{L}_2$ onto $\ker(\mathcal{M}_{\Phi^*}^\ell + \mathcal{M}_{\Phi^*}^{\ell*})^{1/2}$, respectively). The spaces $\mathcal{L}_\ell(\Phi)$ and $\mathcal{L}_\ell(\Phi^*)$ are Hilbert spaces (see [ADP, Section 3]) and will be the state spaces for the coisometric and isometric realizations. For the unitary realization another space is needed, namely the analog of the classical "2 × 2 kernel" and this space will be recalled in Section 5. The theory developed in this paper also has a dual variant associated with the right multiplication operators $\mathcal{M}_\Phi^r F = F\Phi$ and $\mathcal{M}_{\Phi^*}^r F = F\Phi^*$. We shall not consider this here. We refer to [P] for more details on Schur and Carathéodory triangular operators and nonstationary systems.

3 The Coisometric Representation

In the sequel we fix $V \in \mathcal{D}$ with $\|V\| < 1$ to be the "point" at which the representations are to be centered. The results in the case $V = 0$ were proved in our previous paper [ADP].

Theorem 3.1 *Let Φ be a bounded Carathéodory operator. The formulas*

$$(3.1) \qquad \mathbf{T}_V(F) = (FZ - (FZ)^\Delta(V))(Z - V)^{-1},$$
$$(3.2) \qquad \widetilde{\mathbf{T}}_V(F) = (FZ - (FZ)^\Delta(V^{(1)}))(Z - V^{(1)})^{-1},$$
$$(3.3) \qquad \mathbf{A}_V(F) = (F - F^\Delta(V))(Z - V)^{-1},$$
$$(3.4) \qquad \mathbf{B}_V(E) = (\Phi E - (\Phi E)^\Delta(V))(Z - V)^{-1},$$
$$(3.5) \qquad \mathbf{C}_V(F) = F^\Delta(V),$$
$$(3.6) \qquad \mathbf{D}_V(E) = (\Phi E)^\Delta(V)$$

define bounded operators $\mathbf{T}_V, \tilde{\mathbf{T}}_V$ *and* $\mathbf{A}_V : \mathcal{L}_\ell(\Phi) \longrightarrow \mathcal{L}_\ell(\Phi)$, $\mathbf{B}_V : \mathcal{D}_2 \longrightarrow \mathcal{L}_\ell(\Phi)$, $\mathbf{C}_V : \mathcal{L}_\ell(\Phi) \longrightarrow \mathcal{D}_2$, *and* $\mathbf{D}_V : \mathcal{D}_2 \longrightarrow \mathcal{D}_2$. *The operators* $\mathbf{T}_V$ *and* $\tilde{\mathbf{T}}_V$ *are invertible. The colligation* $\mathcal{V}_V = (\mathcal{L}_\ell(\Phi), \mathcal{D}_2, \mathbf{T}_V, \tilde{\mathbf{T}}_V, \mathbf{A}_V, \mathbf{B}_V, \mathbf{C}_V, \mathbf{D}_V)$ *is coisometric in the sense that*

$$(3.7) \qquad\qquad \mathbf{A}_V \mathbf{A}_V^* = \mathbf{T}_V \mathbf{T}_V^*,$$

and the following equalities hold:

$$(3.8) \qquad\qquad \mathbf{B}_V = \mathbf{A}_V \tilde{\mathbf{T}}_V^{-*} \mathbf{C}_V^*,$$

$$(3.9) \qquad\qquad \operatorname{Re} \mathbf{D}_V = \frac{1}{2} \mathbf{C}_V (I_{\mathcal{L}_\ell(\Phi)} - \mathcal{M}_V^r \mathcal{M}_V^{r\,*}) \mathbf{C}_V^*.$$

The colligation $\mathcal{V}_V$ *is closely outerconnected, that is,*

$$\bigcap_{\lambda \in \mathbb{D}} \ker \mathbf{C}_V (\tilde{\mathbf{T}}_V - \lambda \mathbf{A}_V)^{-1} = \{0\}.$$

In the sequel for $W \in \mathcal{D}$ *with* $\|W\| < 1$ *we shall use the function* $\mathcal{F}_W : \mathcal{D}_2 \to \mathcal{D}_2$ *defined by*

$$\mathcal{F}_W(E) = (\Phi E)^\triangle(W).$$

For a formula for the adjoint of $\mathcal{F}_W$ in terms of Φ see (4.4) in the next section.

Theorem 3.2 *In terms of the operators in the colligation* $\mathcal{V}_V$ *of Theorem* 3.1, Φ *has the following representation. For all* $E \in \mathcal{D}_2$ *and all* $W \in \mathcal{D}$ *with* $\|W\| < 1$ *it holds that*

$$(3.10) \qquad \begin{aligned} (\Phi E)^\triangle(W) = (\mathbf{D}_V + \mathbf{C}_V(\mathcal{M}_W^r - \mathcal{M}_V^r) \\ (\mathbf{T}_V - \mathbf{A}_V \mathcal{M}_W^r)^{-1} \mathbf{B}_V)(E), \end{aligned}$$

or equivalently,

$$(3.11) \qquad \begin{aligned} (\Phi E)^\triangle(W) = \Big(& i \operatorname{Im} \mathbf{D}_V + \frac{1}{2} \mathbf{C}_V (I_{\mathcal{L}_\ell(\Phi)} + (\mathcal{M}_W^r - \mathcal{M}_V^r) \mathbf{A}_V) \\ & \times (I_{\mathcal{L}_\ell(\Phi)} - (\mathcal{M}_W^r - \mathcal{M}_V^r) \mathbf{A}_V)^{-1} \tilde{\mathbf{T}}_V^{-*} \mathbf{C}_V^* \\ & + \frac{1}{2} \mathbf{C}_V ((I_{\mathcal{L}_\ell(\Phi)} - \mathcal{M}_V^r \mathcal{M}_{V*}^r) \tilde{\mathbf{T}}_V^* - I_{\mathcal{L}_\ell(\Phi)}) \tilde{\mathbf{T}}_V^{-*} \mathbf{C}_V^* \Big) (E). \end{aligned}$$

Moreover, for all $W, U \in \mathcal{D}$ *of norm strictly less than* 1,

$$(3.12) \qquad \begin{aligned} \mathcal{F}_W + \mathcal{F}_U^* = \mathbf{C}_V(\tilde{\mathbf{T}}_V - \mathcal{M}_W^r \mathbf{A}_V)^{-1} \\ (I_{\mathcal{L}_\ell(\Phi)} - \mathcal{M}_W^r \mathcal{M}_{U*}^r)(\tilde{\mathbf{T}}_V^* - \mathbf{A}_V^* \mathcal{M}_{U*}^r)^{-1} \mathbf{C}_V^*. \end{aligned}$$

The colligation $\mathcal{V}_V$ will be called *the coisometric colligation for* Φ *centered at* V; compare formulas (3.7)–(3.9) with (1.2)–(1.4). Formulas (3.10) and (3.11) will be called the *coisometric representation of* Φ; compare these formulas with (1.1) and (1.5). In the proof below we also show that

$$(3.13) \qquad \mathbf{T}_V = I_{\mathcal{L}_\ell(\Phi)} + \mathbf{A}_V \mathcal{M}_V^r,$$

$$(3.14) \qquad \widetilde{\mathbf{T}}_V = I_{\mathcal{L}_\ell(\Phi)} + \mathcal{M}_V^r \mathbf{A}_V,$$

and hence

$$(3.15) \qquad \begin{aligned} (\mathcal{M}_W^r - \mathcal{M}_V^r)(\mathbf{T}_V - \mathbf{A}_V \mathcal{M}_W^r)^{-1} \\ = (\widetilde{\mathbf{T}}_V - \mathcal{M}_W^r \mathbf{A}_V)^{-1}(\mathcal{M}_W^r - \mathcal{M}_V^r). \end{aligned}$$

When $V = 0$, we get back the *backward shift coisometric colligation centered at the origin*

$$(3.16) \qquad \mathcal{V}_0 = \begin{pmatrix} \mathbf{A}_0 & \mathbf{B}_0 \\ \mathbf{C}_0 & \mathbf{D}_0 \end{pmatrix} : \begin{pmatrix} \mathcal{L}_\ell(\Phi) \\ \mathcal{D}_2 \end{pmatrix} \longrightarrow \begin{pmatrix} \mathcal{L}_\ell(\Phi) \\ \mathcal{D}_2 \end{pmatrix}$$

in which the operators are given by

$$(3.17) \qquad \begin{aligned} \mathbf{A}_0(F) &= (F - F_{[0]})Z^{-1}, \\ \mathbf{B}_0(E) &= (\Phi - \Phi_{[0]})EZ^{-1}, \\ \mathbf{C}_0(F) &= F_{[0]}, \\ \mathbf{D}_0(E) &= \Phi_{[0]}E. \end{aligned}$$

This colligation has been studied in [ADP, Lemma 4.2 and Theorem 4.4].

Proof of Theorems 3.1 and 3.2: The starting point is the coisometric realization (3.16), (3.17). The space $\mathcal{L}_\ell(\Phi)$ is $\mathcal{M}_V^r$ invariant and $\mathbf{A}_0$ invariant. Since $\mathcal{M}_V^r$ is a strict contraction and $\mathbf{A}_0$ is coisometric, the operator $(I_{\mathcal{L}_\ell(\Phi)} - \mathcal{M}_V^r \mathbf{A}_0)^{-1}$ is well defined, and so are the operators

$$(3.18) \qquad \mathbf{T}_V = (I_{\mathcal{L}_\ell(\Phi)} - \mathbf{A}_0 \mathcal{M}_V^r)^{-1},$$

$$(3.19) \qquad \widetilde{\mathbf{T}}_V = (I_{\mathcal{L}_\ell(\Phi)} - \mathcal{M}_V^r \mathbf{A}_0)^{-1},$$

$$(3.20) \qquad \mathbf{A}_V = (I_{\mathcal{L}_\ell(\Phi)} - \mathbf{A}_0 \mathcal{M}_V^r)^{-1}\mathbf{A}_0 = \mathbf{A}_0(I_{\mathcal{L}_\ell(\Phi)} - \mathcal{M}_V^r \mathbf{A}_0)^{-1},$$

$$(3.21) \qquad \mathbf{B}_V = (I_{\mathcal{L}_\ell(\Phi)} - \mathbf{A}_0 \mathcal{M}_V^r)^{-1}\mathbf{B}_0,$$

$$(3.22) \qquad \mathbf{C}_V = \mathbf{C}_0(I_{\mathcal{L}_\ell(\Phi)} - \mathcal{M}_V^r \mathbf{A}_0)^{-1},$$

$$(3.23) \qquad \mathbf{D}_V = \mathbf{D}_0 + \mathbf{C}_0 \mathcal{M}_V^r(I_{\mathcal{L}_\ell(\Phi)} - \mathbf{A}_0 \mathcal{M}_V^r)^{-1}\mathbf{B}_0.$$

We take (3.18)–(3.23) as definitions and show that (3.1)–(3.14) are valid. Definition (3.20) implies that the right hand side of (3.13) coincides with that of (3.18) and the right hand side of (3.14) coincides with that of (3.19). For example, the latter equality follows from

$$I_{\mathcal{L}_\ell(\Phi)} + \mathcal{M}_V^r \mathbf{A}_V = I_{\mathcal{L}_\ell(\Phi)} + \mathcal{M}_V^r (I_{\mathcal{L}_\ell(\Phi)} - \mathbf{A}_0 \mathcal{M}_V^r)^{-1}$$
$$\mathbf{A}_0 = (I_{\mathcal{L}_\ell(\Phi)} - \mathcal{M}_V^r \mathbf{A}_0)^{-1}.$$

Now (3.20) and (3.17) also imply (3.3). Indeed, set $(I_{\mathcal{L}_\ell(\Phi)} - \mathbf{A}_0 \mathcal{M}_V^r)^{-1}$ $\mathbf{A}_0(F) = G$. Then

$$\mathbf{A}_0(F) = G - \mathbf{A}_0(GV).$$

Multiplying both sides by Z from the right, we obtain $F - F_{[0]} = G(Z - V) + (GV)_{[0]}$, so

$$G = (F - (F_{[0]} - (GV)_{[0]}))(Z - V)^{-1}.$$

Since G is upper triangular, we have $F^\triangle(V) = F_{[0]} - (GV)_{[0]}$, and hence

$$\mathbf{A}_V(F) = G = (F - F^\triangle(V))(Z - V)^{-1}.$$

Formulas (3.1) and (3.2) now follow easily. We prove (3.2):

$$\begin{aligned}
\widetilde{\mathbf{T}}_V(F) &= (I_{\mathcal{L}_\ell(\Phi)} + \mathcal{M}_V^r \mathbf{A}_V)(F) \\
&= F + (F - F^\triangle(V))(Z - V)^{-1} V \\
&= F + (F - F^\triangle(V))(Z - V)^{-1}(V - Z + Z) \\
&= F^\triangle(V) + (F - F^\triangle(V))(Z - V)^{-1} Z \\
&= F^\triangle(V) + (F - F^\triangle(V))Z(Z - V^{(1)})^{-1} \\
&= (F^\triangle(V)Z - F^\triangle(V)V^{(1)} + FZ - F^\triangle(V)Z)(Z - V^{(1)})^{-1} \\
&= (FZ - F^\triangle(V)V^{(1)})(Z - V^{(1)})^{-1} \\
&= (FZ - (FZ)^\triangle(V^{(1)}))(Z - V^{(1)})^{-1},
\end{aligned}$$

where we have used that $F^\triangle(V)V^{(1)}$ is diagonal and hence equal to $(FZ)^\triangle(V^{(1)})$. The formula (3.4) for $\mathbf{B}_V$ follows from

$$\begin{aligned}
\mathbf{B}_V(E) &= \mathbf{T}_V \mathbf{B}_0(E) \\
&= ((\Phi E - \Phi_{[0]} E) - (\Phi E - \Phi_{[0]} E)^\triangle(V))(Z - V)^{-1} \\
&= ((\Phi E - (\Phi_{[0]} E)^\triangle(V))(Z - V)^{-1}.
\end{aligned}$$

We turn to (3.5):

$$\begin{aligned}
\mathbf{C}_V(F) &= \mathbf{C}_0 \widetilde{\mathbf{T}}_V(F) \\
&= \mathbf{C}_0(F + \mathbf{A}_V(F)V) \\
&= \mathbf{C}_0(F + (F - F^\triangle(V))(Z - V)^{-1} V)
\end{aligned}$$

$$
\begin{aligned}
&= \mathbf{C}_0(F + (F - F^\triangle(V))(Z - V)^{-1}(V - Z + Z)) \\
&= \mathbf{C}_0(F^\triangle(V) + (F - F^\triangle(V))(Z - V)^{-1}Z) \\
&= F^\triangle(V).
\end{aligned}
$$

The formula (3.6) is proved as follows:

$$
\begin{aligned}
\mathbf{D}_V(E) &= \Phi_{[0]}E + \mathbf{C}_0\mathcal{M}_V^r(\Phi E - (\Phi E)^\triangle(V))(Z - V)^{-1} \\
&= \Phi_{[0]}E + \mathbf{C}_0(\Phi E - (\Phi E)^\triangle(V))(Z - V)^{-1}(V - Z + Z) \\
&= \Phi_{[0]}E + \mathbf{C}_0(-\Phi E + (\Phi E)^\triangle(V)) \\
&\quad + \mathbf{C}_0(\Phi E - (\Phi E)^\triangle(V)(Z - V)^{-1}Z) \\
&= (\Phi E)^\triangle(V).
\end{aligned}
$$

Since $\mathbf{A}_0$ is coisometric and by (3.18), (3.7) holds:

$$
\mathbf{A}_V\mathbf{A}_V^* = (I_{\mathcal{L}_\ell(\Phi)} - \mathbf{A}_0\mathcal{M}_V^r)^{-1}\mathbf{A}_0\mathbf{A}_0^*(I_{\mathcal{L}_\ell(\Phi)} - \mathcal{M}_V^r\mathbf{A}_0^*)^{-1} = \mathbf{T}_V\mathbf{T}_V^*.
$$

Formulas (3.8) and (3.9) were proved in [ADP, Theorem 4.4] for the case $V = 0$. They imply that

$$
\begin{aligned}
\mathbf{A}_V\widetilde{\mathbf{T}}_V^{-*}\mathbf{C}_V^* &= (I_{\mathcal{L}_\ell(\Phi)} - \mathbf{A}_0\mathcal{M}_V^r)^{-1}\mathbf{A}_0\mathbf{C}_0^* \\
&= (I_{\mathcal{L}_\ell(\Phi)} - \mathbf{A}_0\mathcal{M}_V^r)^{-1}\mathbf{B}_0 = \mathbf{B}_V,
\end{aligned}
$$

which implies (3.8), and that

$$
\begin{aligned}
\mathbf{D}_V &= i\mathrm{Im}\,\mathbf{D}_0 + \frac{1}{2}\mathbf{C}_0\mathbf{C}_0^* + \mathbf{C}_0\mathcal{M}_V^r(I_{\mathcal{L}_\ell(\Phi)} - \mathbf{A}_0\mathcal{M}_V^r)^{-1}\mathbf{A}_0\mathbf{C}_0^* \\
&= i\mathrm{Im}\,\mathbf{D}_0 + \frac{1}{2}\mathbf{C}_0(I_{\mathcal{L}_\ell(\Phi)} + \mathcal{M}_V^r\mathbf{A}_0)(I_{\mathcal{L}_\ell(\Phi)} - \mathcal{M}_V^r\mathbf{A}_0)^{-1}\mathbf{C}_0^*,
\end{aligned}
$$

which leads to the formula (3.9):

$$
\begin{aligned}
\mathrm{Re}\,\mathbf{D}_V &= \frac{1}{4}\mathbf{C}_0(I_{\mathcal{L}_\ell(\Phi)} - \mathcal{M}_V^r\mathbf{A}_0)^{-1} \\
&\quad \{(I_{\mathcal{L}_\ell(\Phi)} + \mathcal{M}_V^r\mathbf{A}_0)(I_{\mathcal{L}_\ell(\Phi)} - \mathbf{A}_0^*\mathcal{M}_{V*}^r) \\
&\quad + (I_{\mathcal{L}_\ell(\Phi)} - \mathcal{M}_V^r\mathbf{A}_0) \\
&\quad (I_{\mathcal{L}_\ell(\Phi)} + \mathbf{A}_0^*\mathcal{M}_{V*}^r)\}(I_{\mathcal{L}_\ell(\Phi)} - \mathbf{A}_0^*\mathcal{M}_{V*}^r)^{-1}\mathbf{C}_0^* \\
&= \frac{1}{2}\mathbf{C}_0(I_{\mathcal{L}_\ell(\Phi)} - \mathcal{M}_V^r\mathbf{A}_0)^{-1} \\
&\quad \{I_{\mathcal{L}_\ell(\Phi)} - \mathcal{M}_V^r\mathcal{M}_{V*}^r\}(I_{\mathcal{L}_\ell(\Phi)} - \mathbf{A}_0^*\mathcal{M}_{V*}^r)^{-1}\mathbf{C}_0^* \\
&= \frac{1}{2}\mathbf{C}_V\{I_{\mathcal{L}_\ell(\Phi)} - \mathcal{M}_V^r\mathcal{M}_{V*}^r\}\mathbf{C}_V^*.
\end{aligned}
$$

We now turn to the realization formula (3.10). By (3.14) and (3.20), we have for $W \in \mathcal{D}$,

$$
\widetilde{\mathbf{T}}_V - \mathcal{M}_W^r\mathbf{A}_V = (I_{\mathcal{L}(\Phi)} - \mathcal{M}_W^r\mathbf{A}_0)(I_{\mathcal{L}(\Phi)} - \mathcal{M}_V^r\mathbf{A}_0)^{-1}.
$$

Hence, for $W \in \mathcal{D}$ with $\|W\| < 1$ and $F \in \mathcal{L}_\ell(\Phi)$,

$$\mathbf{C}_V(\tilde{\mathbf{T}}_V - \mathcal{M}_W^r \mathbf{A}_V)^{-1}(F)$$

(3.24)

$$= \mathbf{C}_0(I_{\mathcal{L}_\ell(\Phi)} - \mathcal{M}_W^r \mathbf{A}_0)^{-1}(F) = F^\Delta(W).$$

The last equality holds because if $G = (I_{\mathcal{L}_\ell(\Phi)} - \mathcal{M}_W^r \mathbf{A}_0)^{-1}(F)$, then

$$F - G_{[0]} = (G - G_{[0]})Z^{-1}(Z - W),$$

which, since $(G - G_{[0]})Z^{-1} \in \mathcal{U}$, implies $\mathbf{C}_0(G) = G_{[0]} = F^\Delta(W)$. Hence, by (3.4),

$$\begin{aligned}
\mathbf{C}_V(&\tilde{\mathbf{T}}_V - \mathcal{M}_W^r \mathbf{A}_V)^{-1}(\mathcal{M}_W^r - \mathcal{M}_V^r)\mathbf{B}_V(E) \\
&= ((\Phi E - (\Phi E)^\Delta(V))(Z - V)^{-1}(W - V))^\Delta(W) \\
&= ((\Phi E - (\Phi E)^\Delta(V))(Z - V)^{-1}(W - Z + Z - V))^\Delta(W) \\
&= (\Phi E - (\Phi E)^\Delta(V))^\Delta(W) \\
&\quad -((\Phi E - (\Phi E)^\Delta(V))(Z - V)^{-1}(Z - W))^\Delta(W) \\
&= (\Phi E - (\Phi E)^\Delta(V))^\Delta(W) \\
&= (\Phi E)^\Delta(W) - (\Phi E)^\Delta(V) \\
&= (\Phi E)^\Delta(W) - \mathbf{D}_V(E),
\end{aligned}$$

which, on account of (3.15), implies (3.10):

$$\begin{aligned}
(\Phi E)^\Delta(W) &= (\mathbf{D}_V + \mathbf{C}_V(\tilde{\mathbf{T}}_V - \mathcal{M}_W^r \mathbf{A}_V)(\mathcal{M}_W^r - \mathcal{M}_V^r)\mathbf{B}_V)(E) \\
&= (\mathbf{D}_V + \mathbf{C}_V(\mathcal{M}_W^r - \mathcal{M}_V^r)(\mathbf{T}_V - \mathbf{A}_V\mathcal{M}_W^r)^{-1}\mathbf{B}_V)(E).
\end{aligned}$$

Finally, from (3.10), (3.9) and (3.8) we have that

$$\begin{aligned}
(\Phi E)^\Delta(W) &= \big(\mathbf{D}_V + \mathbf{C}_V(\mathcal{M}_W^r - \mathcal{M}_V^r) \\
&\quad (I_{\mathcal{L}_\ell(\Phi)} - \mathbf{A}_V(\mathcal{M}_W^r - \mathcal{M}_V^r))^{-1}\mathbf{B}_V\big)(E) \\
&= \Big(i\operatorname{Im}\mathbf{D}_V + \frac{1}{2}\mathbf{C}_V\{I_{\mathcal{L}_\ell(\Phi)} - \mathcal{M}_V^r\mathcal{M}_{V*}^r\}\mathbf{C}_V^* \\
&\quad + \mathbf{C}_V(\mathcal{M}_W^r - \mathcal{M}_V^r)(I_{\mathcal{L}_\ell(\Phi)} - \mathbf{A}_V(\mathcal{M}_W^r - \mathcal{M}_V^r))^{-1} \\
&\quad \mathbf{A}_V\tilde{\mathbf{T}}_V^{-*}\mathbf{C}_V^*\Big)(E) = \Big(i\operatorname{Im}\mathbf{D}_V + \frac{1}{2}\mathbf{C}_V(2(\mathcal{M}_W^r - \mathcal{M}_V^r)\mathbf{A}_V \\
&\quad + (I_{\mathcal{L}_\ell(\Phi)} - \mathcal{M}_V^r\mathcal{M}_{V*}^r)\tilde{\mathbf{T}}_V^*(I_{\mathcal{L}_\ell(\Phi)} - (\mathcal{M}_W^r - \mathcal{M}_V^r)\mathbf{A}_V)) \\
&\quad \times (I_{\mathcal{L}(\Phi)} - \mathbf{A}_V(\mathcal{M}_W^r - \mathcal{M}_V^r))^{-1}\mathbf{A}_V\tilde{\mathbf{T}}_V^{-*}\mathbf{C}_V^*\Big)(E),
\end{aligned}$$

and since

$$2(\mathcal{M}_W^r - \mathcal{M}_V^r)\mathbf{A}_V + (I_{\mathcal{L}_\ell(\Phi)} - \mathcal{M}_V^r \mathcal{M}_{V*}^r)\widetilde{\mathbf{T}}_V^*$$
$$(I_{\mathcal{L}_\ell(\Phi)} - (\mathcal{M}_W^r - \mathcal{M}_V^r)\mathbf{A}_V)$$
$$= (I_{\mathcal{L}_\ell(\Phi)} + (\mathcal{M}_W^r - \mathcal{M}_V^r)\mathbf{A}_V)$$
$$+((I_{\mathcal{L}_\ell(\Phi)} - \mathcal{M}_V^r \mathcal{M}_{V*}^r)\widetilde{\mathbf{T}}_V^* - I_{\mathcal{L}_\ell(\Phi)}))$$
$$(I_{\mathcal{L}_\ell(\Phi)} - (\mathcal{M}_W^r - \mathcal{M}_V^r)\mathbf{A}_V),$$

we obtain (3.11). The closely outerconnectedness of the colligation $\mathcal{V}_V$ follows from (3.24) with $W = \lambda I_{\mathcal{D}}$: If $F \in \cap_{\lambda \in \mathbb{D}} \mathbf{C}_V (\widetilde{\mathbf{T}}_V - \lambda \mathbf{A}_V)^{-1}$, then

$$\sum_{n=0}^{\infty} F_{\{n\}} \lambda^n = F^{\triangle}(\lambda I_{\mathcal{D}}) = \mathbf{C}_V (\widetilde{\mathbf{T}}_V - \lambda \mathbf{A}_V)^{-1}(F) = 0, \qquad \lambda \in \mathbb{D},$$

and hence each diagonal $F_{\{n\}} = 0$, so $F = 0$. It remains to prove (3.12). By (3.10), the left hand side of (3.12) is independent of $V \in \mathcal{D}$. Hence, it is equal to

$$(\mathbf{D}_0 + \mathbf{C}_0(I_{\mathcal{L}_\ell(\Phi)} - \mathcal{M}_W^r \mathbf{A}_0)^{-1} \mathbf{B}_0) + (\mathbf{D}_0^* + \mathbf{B}_0^*(I_{\mathcal{L}_\ell(\Phi)} - \mathbf{A}_0^* \mathcal{M}_{U*}^r)^{-1} \mathbf{C}_0)^*.$$

Straightforward calculations show that this expression equals

$$\mathbf{C}_0(I_{\mathcal{L}_\ell(\Phi)} - \mathcal{M}_W^r \mathbf{A}_0)^{-1}(I_{\mathcal{L}_\ell(\Phi)} - \mathcal{M}_W^r \mathcal{M}_{U*}^r)(I_{\mathcal{L}_\ell(\Phi)} - \mathbf{A}_0^* \mathcal{M}_{U*}^r)^{-1} \mathbf{C}_0^*,$$

which, by (3.24), coincides with the right hand side of (3.12).

In the next theorem we prove the uniqueness of the representation (3.10) of the Carathéodory operator Φ.

Theorem 3.3 *Let $\mathcal{V} = (\mathcal{H}, \mathcal{D}_2, T, \widetilde{T}, A, B, C, D)$ be a colligation such that*

$$(3.25) \qquad (\Phi E)^{\triangle}(W) = (D + C(\mathcal{M}_W^r - \mathcal{M}_V^r)(T - A\mathcal{M}_W^r)^{-1} B)(E)$$

for any $E \in \mathcal{D}_2$ and any $W \in \mathcal{D}$ with $\|W\| < 1$, where $\mathcal{H}$ is a right $\mathcal{D}$-module, and the operators of the colligation $\mathcal{V}$ satisfy

$$
\begin{aligned}
T &= I_{\mathcal{H}} + A\mathcal{M}_V^r \quad \text{and} \quad T \text{ is invertible,} \\
\widetilde{T} &= I_{\mathcal{H}} + \mathcal{M}_V^r A \quad (\text{so } \widetilde{T} \text{ is invertible}), \\
AA^* &= TT^*, \\
B &= A\widetilde{T}^{-*} C^*, \\
\mathrm{Re}\, D &= \frac{1}{2} C(I_{\mathcal{H}} - \mathcal{M}_V^r \mathcal{M}_{V*}^r) C^*.
\end{aligned}
$$

Assume that the colligation $\mathcal{V}$ is closely outerconnected, that is,

$$\bigcap_{\lambda \in \mathbb{D}} \ker C(\widetilde{T} - \lambda A)^{-1} = \{0\}.$$

Finally, assume that for $\lambda, \mu \in \mathbb{D}$,

(3.26)
$$\frac{\mathcal{F}_{\lambda I_\mathcal{D}} + \mathcal{F}^*_{\mu I_\mathcal{D}}}{1 - \lambda\mu^*} = C(\widetilde{T} - \lambda A)^{-1}(\widetilde{T}^* - \mu^* A^*)^{-1} C^*.$$

Then the colligations $\mathcal{V}$ and $\mathcal{V}_V$ are unitarily equivalent, that is, there exists a unitary map $\sigma : \mathcal{L}_\ell(\Phi) \longrightarrow \mathcal{H}$ such that

$$
\begin{pmatrix} T & 0 & 0 & 0 \\ 0 & \widetilde{T} & 0 & 0 \\ 0 & 0 & A & B \\ 0 & 0 & C & D \end{pmatrix} = \begin{pmatrix} \sigma & 0 & 0 & 0 \\ 0 & \sigma & 0 & 0 \\ 0 & 0 & \sigma & 0 \\ 0 & 0 & 0 & I_{\mathcal{D}_2} \end{pmatrix}
$$

(3.27)
$$
\times \begin{pmatrix} \mathbf{T}_V & 0 & 0 & 0 \\ 0 & \widetilde{\mathbf{T}}_V & 0 & 0 \\ 0 & 0 & \mathbf{A}_V & \mathbf{B}_V \\ 0 & 0 & \mathbf{C}_V & \mathbf{D}_V \end{pmatrix} \begin{pmatrix} \sigma^* & 0 & 0 & 0 \\ 0 & \sigma^* & 0 & 0 \\ 0 & 0 & \sigma^* & 0 \\ 0 & 0 & 0 & I_{\mathcal{D}_2} \end{pmatrix}.
$$

Proof: Note that the condition $AA^* = TT^*$ implies $\|T^{-1}A\| = 1$ and hence the inverse in (3.25) exists. We define the linear relation $\mathbf{R}$ on $\mathcal{L}_\ell(\Phi) \times \mathcal{H}$ as the linear span of all pairs of the form

$$((\widetilde{\mathbf{T}}_V^* - \mu^* \mathbf{A}_V^*)^{-1} \mathbf{C}_V^*(E), (\widetilde{T}^* - \mu^* A^*)^{-1} C^*(E)),$$

where μ runs through $\mathbb{D}$ and E through $\mathcal{D}_2$. Formula (3.12) applied to $W = \lambda I_\mathcal{D}$ and $U = \mu I_\mathcal{D}$ gives

$$\frac{\mathcal{F}_{\lambda I_\mathcal{D}} + \mathcal{F}^*_{\mu I_\mathcal{D}}}{1 - \lambda\mu^*} = \mathbf{C}_V(\widetilde{\mathbf{T}}_V - \lambda \mathbf{A}_V)^{-1}(\widetilde{\mathbf{T}}_V^* - \mu^* \mathbf{A}_V^*)^{-1} \mathbf{C}_V^*$$

and so by (3.26), the relation $\mathbf{R}$ is isometric. Since both colligations are closely outer connected, the relation has a dense domain and a dense range. Therefore the closure of $\mathbf{R}$ is the graph of a unitary operator σ from $\mathcal{L}_\ell(\Phi)$ onto $\mathcal{H}$ satisfying

$$\sigma((\widetilde{\mathbf{T}}_V^* - \mu^* \mathbf{A}_V^*)^{-1} \mathbf{C}_V^*(E)) = (\widetilde{T}^* - \mu^* A^*)^{-1} C^*(E), \quad E \in \mathcal{D}_2, \lambda \in \mathbb{D}.$$

Formula (3.27) can be proved much as in the proof of [ADP, Theorem 4.6]. We omit the details. $\square$

Remark 3.4 (1) Instead of (3.26), we could have assumed

$$C(\widetilde{T} - \lambda A)^{-1}(\widetilde{T}^* - \mu^* A^*)^{-1} C^*$$

$$= \mathbf{C}_V(\widetilde{\mathbf{T}}_V - \lambda \mathbf{A}_V)^{-1}(\widetilde{\mathbf{T}}_V^* - \mu^* \mathbf{A}_V^*)^{-1} \mathbf{C}_V^*,$$

or the stronger noncommutative condition

$$C(\tilde{T} - \mathcal{M}^r_W A)^{-1}(\tilde{T}^* - A^* \mathcal{M}^r_{U*})^{-1} C^*$$
$$= C_V(\tilde{T}_V - \mathcal{M}^r_W A_V)^{-1}(\tilde{T}^*_V - A^*_V \mathcal{M}^r_{U*})^{-1} C^*_V,$$

and obtain the same result. Moment type conditions of this kind appear in [BT] in the setting of analytic functions in the polydisk.

(2) If V is a colligation as above satisfying (3.27) for some unitary operator $\sigma : \mathcal{L}_\ell(\Phi) \longrightarrow \mathcal{H}$, then

$$(\Phi E)^\triangle(W) = (D + C(\mathcal{M}^r_W - \mathcal{M}^r_V)(T - A\mathcal{M}^r_W)^{-1} B)(E)$$

holds if in addition we have that σ is a $\mathcal{D}$-module isomorphism, that is, $\sigma \mathcal{M}^r_U = \mathcal{M}^r_U \sigma$ for every $U \in \mathcal{D}$.

4 The Isometric Representation

In this short section we mention without proofs the results related to the isometric colligation associated to Φ. We first recall the definitions of point evaluations for lower triangular operators. For $F \in \mathcal{U}$ and $W \in \Omega$,

$$F^\wedge(W^*)^* = \sum_{n=0}^\infty F^*_{[n]} Z^{*n}(ZW)^n, \quad F^\triangle(W^*)^* = \sum_{n=0}^\infty (WZ)^n Z^{*n} F^*_{\{n\}}.$$

This gives rise to the analogs of the left and the right point evaluations for lower triangular operators: An element $G \in \mathcal{L}$ can formally be written as a right and a left power series in Z^*, namely $G = \sum_{n=0}^\infty G_{[n]} Z^{*n}$ and $G = \sum_{n=0}^\infty Z^{*n} G_{\{n\}}$ for some diagonal operators $G_{[n]}$ and $G_{\{n\}}$. We define the right point evaluation of G at $W \in \mathcal{X}$ by

$$G^\vee(W) = \sum_{n=0}^\infty G_{[n]} Z^{*n}(ZW)^n = \sum_{n=0}^\infty G_{[n]} W^{*[n]*},$$

and the left point evaluation by

$$G^\triangledown(W) = \sum_{n=0}^\infty (WZ)^n Z^{*n} G_{\{n\}} = \sum_{n=0}^\infty W^{*\{n\}*} G_{\{n\}}.$$

The four transforms are related by conjugation in the sense that for any $F \in \mathcal{U}$, the operator F^* belongs to $\mathcal{L}$ and hence for $W \in \Omega$

$$F^\wedge(W^*)^* = F^{*\vee}(W), \qquad F^\triangle(W^*)^* = F^{*\triangledown}(W).$$

When $\mathcal{M} = \mathbb{C}$, then $W^{*[n]*} = W^{[n]}$ and $W^{*\{n\}*} = W^{\{n\}}$.

Let $\widehat{\mathcal{V}}_V = (\mathcal{L}_\ell(\Phi^*), \mathcal{D}_2, \widehat{\mathbf{T}}_V, \widehat{\widehat{\mathbf{T}}}_V, \widehat{\mathbf{A}}_V, \widehat{\mathbf{B}}_V, \widehat{\mathbf{C}}_V, \widehat{\mathbf{D}}_V)$ be the colligation defined by

$$
\begin{aligned}
(4.1) \quad
\widehat{\mathbf{T}}_V(H) &= (HZ^* - (HZ^*)^\vee(V^*))(Z^* - V^*)^{-1}, \\
\widehat{\widehat{\mathbf{T}}}_V(H) &= (HZ^* - (HZ^*)^\vee(V^{*(1)}))(Z^* - V^{*(1)})^{-1}, \\
\widehat{\mathbf{A}}_V(H) &= (H - H^\vee(V^*))(Z^* - V^*)^{-1}, \\
\widehat{\mathbf{B}}_V(E) &= (\Phi^* E - (\Phi^* E)^\vee(V^*))(Z^* - V^*)^{-1}, \\
\widehat{\mathbf{C}}_V(H) &= H^\vee(V^*), \\
\widehat{\mathbf{D}}_V(E) &= (\Phi^* E)^\vee(V^*).
\end{aligned}
$$

Then one can show as in the previous section that (1) the operators in $\widehat{\mathcal{V}}_V$ are bounded, (2) the formulas

$$
\widehat{\mathbf{T}}_V = I_{\mathcal{L}_\ell(\Phi^*)} + \widehat{\mathbf{A}}_V \mathcal{M}^r_{V*}, \qquad \widehat{\widehat{\mathbf{T}}}_V = I_{\mathcal{L}_\ell(\Phi^*)} + \mathcal{M}^r_{V*}\widehat{\mathbf{A}}_V,
$$

are valid and (3) the following two representations hold:

$$
(4.2) \quad
\begin{aligned}
(\Phi^* E)^\vee(W^*) &= (\widehat{\mathbf{D}}_V + \widehat{\mathbf{C}}_V(\mathcal{M}^r_{W*} - \mathcal{M}^r_{V*}) \\
&\quad (\widehat{\mathbf{T}}_V - \widehat{\mathbf{A}}_V \mathcal{M}^r_{W*})^{-1}\widehat{\mathbf{B}}_V)(E),
\end{aligned}
$$

and

$$
(4.3) \quad
\begin{aligned}
(\Phi^* E)^\vee(W^*) &= \Big(i\mathrm{Im}\,\widehat{\mathbf{D}}_V + \tfrac{1}{2}\widehat{\mathbf{C}}_V(I_{\mathcal{L}_\ell(\Phi^*)} + (\mathcal{M}^r_{W*} - \mathcal{M}^r_{V*})\widehat{\mathbf{A}}_V) \\
&\quad \times (I_{\mathcal{L}_\ell(\Phi^*)} - (\mathcal{M}^r_{W*} - \mathcal{M}^r_{V*})\widehat{\mathbf{A}}_V)^{-1}\widehat{\widehat{\mathbf{T}}}_V^{-*}\widehat{\mathbf{C}}_V^* \\
&\quad + \tfrac{1}{2}\widehat{\mathbf{C}}_V((I_{\mathcal{L}_\ell(\Phi^*)} - \mathcal{M}^r_{V*}\mathcal{M}^r_V)\widehat{\widehat{\mathbf{T}}}_V^{-*} - I_{\mathcal{L}_\ell(\Phi^*)})\widehat{\widehat{\mathbf{T}}}_V^{-*}\widehat{\mathbf{C}}_V^*\Big)(E),
\end{aligned}
$$

where the main operator $\widehat{\mathbf{A}}_V$ is coisometric in the sense that $\widehat{\mathbf{A}}_V\widehat{\mathbf{A}}_V^* = \widehat{\mathbf{T}}_V\widehat{\mathbf{T}}_V^*$. It can be shown (see [ADP, Section 5]) that the adjoint of the mapping $\mathcal{F}_W : \mathcal{D}_2 \to \mathcal{D}_2$ defined by $\mathcal{F}_W(E) = (\Phi E)^\triangle(W)$ is given by

$$
(4.4) \qquad\qquad \mathcal{F}_W^*(E) = (\Phi^* E)^\vee(W^*).
$$

Applying this result to the identities (4.2) and (4.3) we obtain the representations

$$
\begin{aligned}
(\Phi E)^\triangle(W) &= (\widehat{\mathbf{D}}_V^* + \widehat{\mathbf{B}}_V^*(\widehat{\mathbf{T}}_V^* - \mathcal{M}^r_{W^{(1)}}\widehat{\mathbf{A}}_V^*)^{-1}(\mathcal{M}^r_{W^{(1)}} - \mathcal{M}^r_V)\widehat{\mathbf{C}}_V^*)(E) \\
&= (\widehat{\mathbf{D}}_V^* + \widehat{\mathbf{B}}_V^*(\mathcal{M}^r_{W^{(1)}} - \mathcal{M}^r_V)(\widehat{\mathbf{T}}_V^* - \widehat{\mathbf{A}}_V \mathcal{M}^r_{W^{(1)}})^{-1}\widehat{\mathbf{C}}_V^*)(E),
\end{aligned}
$$

$$
\begin{aligned}
(\Phi E)^\triangle(W) &= (i\mathrm{Im}\,\widehat{\mathbf{D}}_V^* + \tfrac{1}{2}\widehat{\mathbf{C}}_V\widehat{\widehat{\mathbf{T}}}_V^{-1}(I_{\mathcal{L}_\ell(\Phi^*)} + \widehat{\mathbf{A}}_V^*(\mathcal{M}^r_{W^{(1)}} - \mathcal{M}^r_V)). \\
&\quad \times (I_{\mathcal{L}_\ell(\Phi^*)} - \widehat{\mathbf{A}}_V^*(\mathcal{M}^r_{W^{(1)}} - \mathcal{M}^r_V))^{-1}\widehat{\mathbf{C}}_V^* \\
&\quad + \tfrac{1}{2}\widehat{\mathbf{C}}_V\widehat{\widehat{\mathbf{T}}}_V^{-1}((I_{\mathcal{L}_\ell(\Phi^*)} - \mathcal{M}^r_{V*}\mathcal{M}^r_V)\widehat{\widehat{\mathbf{T}}}_V - I_{\mathcal{L}_\ell(\Phi^*)})\widehat{\mathbf{C}}_V^*)(E).
\end{aligned}
$$

Now the main operator $\widehat{\mathbf{A}}_V^*$ is isometric in the sense that $\widehat{\mathbf{A}}_V \widehat{\mathbf{A}}_V^* = \widehat{\mathbf{T}}_V \widehat{\mathbf{T}}_V^*$. The colligation

$$\widehat{\mathcal{V}}_V^* = (\mathcal{L}_\ell(\Phi^*), \mathcal{D}_2, \widehat{\widehat{\mathbf{T}}}_V^*, \widehat{\mathbf{T}}_V, \widehat{\mathbf{A}}_V^*, \widehat{\mathbf{C}}_V^*, \widehat{\mathbf{B}}_V^*, \widehat{\mathbf{D}}_V^*)$$

is called the *isometric colligation for* Φ *centered at* V, and the above formulas are called the *isometric representations of* Φ. The colligation $\widehat{\mathcal{V}}_V^*$ is closely inner-connected, that is,

$$\cap_{\lambda \in \mathbb{D}} \ker \widehat{\mathbf{B}}_V^* (\widehat{\mathbf{T}}_V^* - \lambda \widehat{\mathbf{A}}_V^*)^{-1} = \{0\},$$

and a uniqueness theorem holds which is analogous to Theorem 3.3.

5 The Unitary Representation

In this section we derive a Herglotz representation for a Carathéodory operator Φ in which the main operator is unitary in a generalized sense. We use the analysis for upper triangular contractions (or Schur operators) introduced in [AP] which we now briefly review: The space $\mathcal{U}_2$ is a reproducing kernel Hilbert space with reproducing kernel

$$\rho_W^{-1} = (I - ZW^*)^{-1} = \sum_0^\infty (ZW^*)^n = \sum_0^\infty Z^n W^{[n]*}$$

in the sense that for all $W \in \Omega$, $E \in \mathcal{D}_2$, and $F \in \mathcal{U}_2$, the operator $\rho_W^{-1} E \in \mathcal{U}_2$ and

$$\langle F, \rho_W^{-1} E \rangle_{\mathcal{U}_2} = \text{Tr } E^* F^\wedge(W).$$

Similarly, the space $\mathcal{L}_2$ is a reproducing kernel Hilbert space with reproducing kernel

$$\sigma_W^{-1} = (I - Z^* W^*)^{-1} = \sum_0^\infty Z^{*n} W^{*\{n\}}$$

in the sense that for all $W \in \Omega$, $E \in \mathcal{D}_2$, and $G \in \mathcal{L}_2$, the operator $\sigma_W^{-1} E \in \mathcal{L}_2$ and

$$\langle G, \sigma_W^{-1} E \rangle_{\mathcal{L}_2} = \text{Tr } E^* G^\nabla(W).$$

Let $S \in \mathcal{U}$ be a Schur operator. Then the operator $\mathcal{M}_S^\ell$ of left multiplication by S: $\mathcal{M}_S^\ell(F) = SF$ is a contraction from $\mathcal{U}_2$ in itself. By $\mathcal{H}_\ell(S) \subset \mathcal{U}_2$ we denote the operator range

$$\mathcal{H}_\ell(S) = \text{ran } \Gamma_S^{\frac{1}{2}}, \quad \Gamma_S = I_{\mathcal{U}_2} - \mathcal{M}_S^\ell \mathcal{M}_S^{\ell*},$$

with the lifted norm

$$\|\Gamma_S^{\frac{1}{2}} F\|_{\mathcal{H}_\ell(S)} = \|(I - P_S)F\|_{\mathcal{U}_2},$$

where P_S is the orthogonal projection in $\mathcal{U}_2$ onto ker Γ_S. The space $\mathcal{H}_\ell(S^*) \subset \mathcal{L}_2$ is defined in a similar way. The operator $(S^* - S^{*\vee}(W^*))(Z^* - W^*)^{-1} E$ belongs to $\mathcal{H}_\ell(S^*)$ for all $W \in \Omega$ and $E \in \mathcal{D}_2$, and the formula

$$\Lambda(\Gamma_S \rho_W^{-1} E) = (S^* - S^{*\vee}(W^*))(Z^* - W^*)^{-1} E^{(1)}, \quad W \in \Omega, \ E \in \mathcal{D}_2,$$

defines a contraction $\Lambda : \mathcal{H}_\ell(S) \longrightarrow \mathcal{H}_\ell(S^*)$. Its adjoint Λ^* is given by

$$\Lambda^*(\Gamma_{S^*} \sigma_W^{-1} E) = (S - S^\Delta(W^*))(Z - W^*)^{-1} E^{(-1)}, \quad W \in \Omega, \ E \in \mathcal{D}_2;$$

see [AP, Lemmas 6.1 and 6.2]. It follows that the operator

$$\Theta_S = \begin{pmatrix} I_{\mathcal{H}_\ell(S)} & \Lambda^* \\ \Lambda & I_{\mathcal{H}_\ell(S^*)} \end{pmatrix} : \begin{pmatrix} \mathcal{H}_\ell(S) \\ \mathcal{H}_\ell(S^*) \end{pmatrix} \longrightarrow \begin{pmatrix} \mathcal{H}_\ell(S) \\ \mathcal{H}_\ell(S^*) \end{pmatrix}$$

is nonnegative on the space $\mathcal{H}_\ell(S) \oplus \mathcal{H}_\ell(S^*)$. We denote by $\mathcal{D}_\ell(S)$ the operator range ran $\Theta_S^{\frac{1}{2}}$ in $\mathcal{H}_\ell(S) \oplus \mathcal{H}_\ell(S^*)$ endowed with the lifted norm. By [ADP, Lemma 3.2], the operator

$$U = \begin{pmatrix} M^\ell_{\frac{1}{\sqrt{2}}(I+\Phi)} & 0 \\ 0 & M^\ell_{\frac{1}{\sqrt{2}}(I+\Phi^*)} \end{pmatrix} : \begin{pmatrix} \mathcal{H}_\ell(S) \\ \mathcal{H}_\ell(S^*) \end{pmatrix} \longrightarrow \begin{pmatrix} \mathcal{L}_\ell(\Phi) \\ \mathcal{L}_\ell(\Phi^*) \end{pmatrix}$$

is unitary. Then $\Theta_\Phi := U\Theta_S U^*$ is a nonnegative map from $\mathcal{L}_\ell(\Phi) \oplus \mathcal{L}_\ell(\Phi^*)$ to itself. By $\mathcal{D}_\ell(\Phi)$ we denote the operator range ran $\Theta_\Phi^{\frac{1}{2}}$ provided with the lifted norm. From $\Theta_\Phi^{\frac{1}{2}} = U\Theta_S^{\frac{1}{2}} U^*$ we see that as sets $\mathcal{D}_\ell(\Phi) = U\mathcal{D}_\ell(S)$. The restriction $U\,|_{\mathcal{D}_\ell(S)}$ is a unitary mapping from $\mathcal{D}_\ell(S)$ onto $\mathcal{D}_\ell(\Phi)$. In [ADP] we showed that the colligation

$$(5.1) \qquad W_0 = \begin{pmatrix} \alpha_0 & \beta_0 \\ \gamma_0 & \delta_0 \end{pmatrix} : \begin{pmatrix} \mathcal{D}_\ell(\Phi) \\ \mathcal{D}_2 \end{pmatrix} \longrightarrow \begin{pmatrix} \mathcal{D}_\ell(\Phi) \\ \mathcal{D}_2 \end{pmatrix}$$

defined by

$$(5.2) \qquad \alpha_0 \begin{pmatrix} F \\ H \end{pmatrix} = \begin{pmatrix} (F - F_{[0]})Z^*, \\ HZ^* - F_{[0]} \end{pmatrix},$$

$$(5.3) \qquad \beta_0(E) = \begin{pmatrix} (\Phi - \Phi_{[0]})EZ^{-1}, \\ -(\Phi^* + \Phi_{[0]})E \end{pmatrix},$$

$$(5.4) \qquad \gamma_0 \begin{pmatrix} F \\ H \end{pmatrix} = F_{[0]},$$

$$(5.5) \qquad \delta_0(E) = \Phi_{[0]}E,$$

is well defined and that α_0 is unitary. $\mathcal{W}_0$ is called the *unitary colligation for* Φ *centered at* 0. We now formulate three theorems. The first one concerns the existence of a unitary colligation $\mathcal{W}_V$ for Φ centered at a point $V \in \mathcal{D}$ with $\|V\| < 1$. The second one gives the unitary representation of Φ, that is, the representation in terms of the operators in the colligation $\mathcal{W}_V$. The last theorem deals with the uniqueness of the representation.

In the sequel for $W \in \mathcal{D}$ with $\|W\| < 1$, we define the right multiplication operator $\mathcal{M}^r_{(W, W^{(1)})}$ by

$$\mathcal{M}^r_{(W, W^{(1)})} \begin{pmatrix} F \\ H \end{pmatrix} = \begin{pmatrix} FW \\ HW^{(1)} \end{pmatrix}.$$

By [ADP, Lemma 6.5] it is a strict contraction from $\mathcal{D}_\ell(\Phi)$ to itself.

Theorem 5.1 *The formulas*

$$(5.6) \qquad \tau_V \begin{pmatrix} F \\ H \end{pmatrix} = \begin{pmatrix} (FZ - (FZ)^\triangle(V))(Z - V)^{-1} \\ (H - (FZ)^\triangle(V))\sigma_{V*}^{-1} \end{pmatrix},$$

$$(5.7) \qquad \tilde{\tau}_V \begin{pmatrix} F \\ H \end{pmatrix} = \begin{pmatrix} (FZ - (FZ)^\triangle(V^{(1)}))(Z - V^{(1)})^{-1} \\ (H - (FZ)^\triangle(V^{(1)}))\sigma_{V*(1)}^{-1} \end{pmatrix},$$

$$(5.8) \qquad \alpha_V \begin{pmatrix} F \\ H \end{pmatrix} = \begin{pmatrix} (F - F^\triangle(V))(Z - V)^{-1} \\ (HZ^* - F^\triangle(V))\sigma_{V*}^{-1} \end{pmatrix},$$

$$(5.9) \qquad \beta_V(E) = \begin{pmatrix} (\Phi E - (\Phi E)^\triangle(V))(Z - V)^{-1} \\ -(\Phi^* E + (\Phi E)^\triangle(V))\sigma_{V*}^{-1} \end{pmatrix},$$

$$(5.10) \qquad \gamma_V \begin{pmatrix} F \\ H \end{pmatrix} = F^\triangle(V),$$

$$(5.11) \qquad \delta_V(E) = (\Phi E)^\triangle(V)$$

define bounded operators τ_V, $\tilde{\tau}_V$ *and* $\alpha_V : \mathcal{D}_\ell(\Phi) \to \mathcal{D}_\ell(\Phi)$, $\beta_V : \mathcal{D}_2 \to \mathcal{D}_\ell(\Phi)$, $\gamma_V : \mathcal{D}_\ell(\Phi) \to \mathcal{D}_2$, *and* $\delta_V : \mathcal{D}_2 \to \mathcal{D}_2$. *The operators* τ_V *and* $\tilde{\tau}_V$ *are invertible.*

The colligation $\mathcal{W}_V = (\mathcal{D}_\ell(\Phi), \mathcal{D}_2, \tau_V, \tilde{\tau}_V, \alpha_V, \beta_V, \gamma_V, \delta_V)$ is unitary in the sense that

(5.12)
$$\alpha_V \alpha_V^* = \tau_V \tau_V^*, \qquad \tilde{\alpha}_V^* \tilde{\alpha}_V = \tilde{\tau}_V^* \tilde{\tau}_V,$$

and the following equalities hold

(5.13)
$$\beta_V = \alpha_V \tilde{\tau}_V^{-*} \gamma_V^*,$$
$$\operatorname{Re} \delta_V = \frac{1}{2} \gamma_V (I_{\mathcal{D}_\ell(\Phi)} - M_{(V,V^{(1)})}^r M_{(V^*,V^{*(1)})}^r) \gamma_V^*,$$

(5.14)
$$\gamma_V^* = \alpha_V^* \tau_V^{-1} \beta_V,$$
$$\operatorname{Re} \delta_V^* = \frac{1}{2} \beta_V^* (I_{\mathcal{D}_\ell(\Phi)} - M_{(V^*,V^{*(1)})}^r M_{(V,V^{(1)})}^r) \beta_V.$$

· Finally, $\mathcal{W}_V$ is closely innerconnected, that is, $\mathcal{D}_\ell(\Phi)$ is the closed linear span of all the elements of the form

(5.15)
$$(\tilde{\tau}_V^* - \lambda^* \alpha_V^*)^{-1} \mathbf{C}_V^*(E), \quad (\tau_V^* - \mu \alpha_V)^{-1} \mathbf{B}_V(F),$$
$$\lambda, \mu \in \mathbb{D}, \ E, F \in \mathcal{D}_2.$$

Remark 5.2 The adjoints of the operators in (5.6)–(5.11) are given by

$$\tau_V^* \begin{pmatrix} F \\ H \end{pmatrix} = \begin{pmatrix} (F - (HZ^*)^\vee(V^*))\rho_{V(-1)}^{\Delta-1} \\ (HZ^* - (HZ^*)^\vee(V^*))(Z^* - V^*)^{-1} \end{pmatrix},$$

$$\tilde{\tau}_V^* \begin{pmatrix} F \\ H \end{pmatrix} = \begin{pmatrix} (F - (HZ^*)^\vee(V^{*(1)}))\rho_V^{\Delta-1} \\ (HZ^* - (HZ^*)^\vee(V^{*(1)}))(Z^* - V^{*(1)})^{-1} \end{pmatrix},$$

$$\alpha_V^* \begin{pmatrix} F \\ H \end{pmatrix} = \begin{pmatrix} (FZ - H^\vee(V^{*(1)}))\rho_V^{\Delta-1} \\ (H - H^\vee(V^{*(1)}))(Z^* - V^{*(1)})^{-1} \end{pmatrix},$$

$$\beta_V^* \begin{pmatrix} F \\ H \end{pmatrix} = -H^\vee(V^{*(1)}),$$

$$\gamma_V^*(E) = \begin{pmatrix} (\Phi E + (\Phi^* E)^\vee(V^{*(1)}))\rho_V^{\Delta-1} \\ -(\Phi^* E - (\Phi^* E)^\vee(V^{*(1)}))(Z^* - V^{*(1)})^{-1} \end{pmatrix},$$

$$\delta_V^*(E) = (\Phi^* E)^\vee(V^{*(1)}).$$

Theorem 5.3 *The unitary representation of* Φ *in terms of the operators in the unitary collation* $\mathcal{W}_V$ *from Theorem 5.1 takes the following form. For* $E \in \mathcal{D}$ *and* $W \in \mathcal{D}$ *with* $\|W\| < 1$,

$$
(\Phi E)^{\triangle}(W) = (\delta_V + \gamma_V(\mathcal{M}^r_{(W,W^{(1)})} - \mathcal{M}^r_{(V,V^{(1)})})
$$

(5.16)
$$
(\tau_V - \alpha_V \mathcal{M}^r_{(W,W^{(1)})})^{-1}\beta_V)(E),
$$

or equivalently,

$$
(\Phi E)^{\triangle}(W)\left(i\mathrm{Im}\,\delta_V + \frac{1}{2}\gamma_V(I_{\mathcal{D}_\ell(\Phi)} + (\mathcal{M}^r_{(W,W^{(1)})} - \mathcal{M}^r_{(V,V^{(1)})})\alpha_V)\right.
$$

(5.17)
$$
\times (I_{\mathcal{D}_\ell(\Phi)} - (\mathcal{M}^r_{(W,W^{(1)})} - \mathcal{M}^r_{(V,V^{(1)})})\alpha_V)^{-1}\widetilde{\tau}_V^{-*}\gamma_V^*
$$

$$
\left. +\frac{1}{2}\gamma_V((I_{\mathcal{D}_\ell(\Phi)} - \mathcal{M}^r_{(V,V^{(1)})}\mathcal{M}^r_{(V^*,V^{*(1)})})\widetilde{\tau}_V^{-*} - I_{\mathcal{D}_\ell(\Phi)}\widetilde{\tau}_V^{-*}\gamma_V^*\right)(E).
$$

Moreover, for $W, U \in \mathcal{D}$ *of norm strictly less than 1,*

$$
\begin{pmatrix} \mathcal{F}_W + \mathcal{F}_U^* & \mathcal{F}_W - \mathcal{F}_{U^*} \\ \mathcal{F}_{W^*}^* - \mathcal{F}_U^* & \mathcal{F}_{W^*}^* + \mathcal{F}_{U^*} \end{pmatrix}
$$

$$
= \begin{pmatrix} \gamma_V(\widetilde{\tau}_V - \mathcal{M}^r_{(W,W^{(1)})}\alpha_V)^{-1} & 0 \\ 0 & \beta_V^*(\tau_V^* - \mathcal{M}^r_{(W,W^{(1)})}\alpha_V^*)^{-1} \end{pmatrix}
$$

$$
\times \begin{pmatrix} I_{\mathcal{D}_\ell(\Phi)} - \mathcal{M}^r_{(W,W^{(1)})}\mathcal{M}^r_{(U^*,U^{*(1)})} & \mathcal{M}^r_{(W,W^{(1)})} - \mathcal{M}^r_{(U^*,U^{*(1)})} \\ \mathcal{M}^r_{(W,W^{(1)})} - \mathcal{M}^r_{(U^*,U^{*(1)})} & I_{\mathcal{D}_\ell(\Phi)} - \mathcal{M}^r_{(W,W^{(1)})}\mathcal{M}^r_{(U^*,U^{*(1)})} \end{pmatrix}
$$

$$
\times \begin{pmatrix} (\widetilde{\tau}_V^* - \alpha_V^*\mathcal{M}^r_{(U^*,U^{*(1)})})^{-1}\gamma_V^* & 0 \\ 0 & (\tau_V - \alpha_V \mathcal{M}^r_{(U^*,U^{*(1)})})^{-1}\beta_V \end{pmatrix}.
$$

Remark 5.4 For $W = \lambda I_{\mathcal{D}}$ and $U = \mu I_{\mathcal{D}}$, the noncommutative relation above reduces to the factorization (we write $\mathcal{F}_\lambda$ for $\mathcal{F}_{\lambda I_{\mathcal{D}}}$)

$$
\begin{pmatrix} \frac{\mathcal{F}_\lambda + \mathcal{F}_\mu^*}{1 - \lambda\mu^*} & \frac{\mathcal{F}_\lambda - \mathcal{F}_\mu^*}{\lambda - \mu^*} \\ \frac{\mathcal{F}_{\lambda^*}^* - \mathcal{F}_\mu}{\lambda - \mu^*} & \frac{\mathcal{F}_{\lambda^*}^* + \mathcal{F}_\mu^*}{1 - \lambda\mu^*} \end{pmatrix} = \begin{pmatrix} \gamma_V(\widetilde{\tau}_V - \lambda\alpha_V)^{-1} \\ \beta_V^*(\tau_V^* - \lambda\alpha_V^*)^{-1} \end{pmatrix}
$$

(5.18)
$$
\cdot ((\widetilde{\tau}_V^* - \mu^*\alpha_V^*)^{-1}\gamma_V^* \quad (\tau_V - \mu^*\alpha_V)^{-1}\beta_V).
$$

Theorem 5.5 *Let* $\mathcal{W} = (\mathcal{H}, \mathcal{D}_2, \tau, \tilde{\tau}, \alpha, \beta, \gamma, \delta)$ *be a colligation such that*

(5.19) $(\Phi E)^{\triangle}(W) = (\delta + \gamma(\mathcal{M}_W^r - \mathcal{M}_V^r)(\tau - \alpha\mathcal{M}_W^r)^{-1}\beta)(E)$

for any $E \in \mathcal{D}_2$ *and any* $W \in \mathcal{D}$ *with* $\|W\| < 1$, *where* $\mathcal{H}$ *is a right* $\mathcal{D}$-*module, and the operators of the colligation* $\mathcal{W}$ *satisfy*

$$\tau = I_{\mathcal{H}} + \alpha\mathcal{M}_V^r \quad \text{and} \quad \tau \text{ is invertible,}$$
$$\tilde{\tau} = I_{\mathcal{H}} + \mathcal{M}_V^r\alpha \quad \text{(so that } \tilde{\tau} \text{ is invertible),}$$
$$\alpha\alpha^* = \tau\tau^*,$$
$$\beta = \alpha\tilde{\tau}^{-*}\gamma^*,$$
$$\text{Re } \delta = \frac{1}{2}\gamma(I_{\mathcal{H}} - \mathcal{M}_V^r\mathcal{M}_{V^*}^r)\gamma^*.$$

Assume that the colligation $\mathcal{W}$ *is closely connected in the sense that the span of all operators of the form*

$$(\tilde{\tau}^* - \lambda^*\alpha^*)^{-1}\gamma^*(E), \quad (\tau - \mu\alpha)^{-1}\beta(G), \qquad \lambda, \mu \in \mathbb{D}, \ E, G \in \mathcal{D}_2,$$

is dense in $\mathcal{H}$. *Finally, assume that for all* $\lambda, \mu \in \mathbb{D}$,

$$
\begin{pmatrix} \frac{\mathcal{F}_\lambda + \mathcal{F}_\mu^*}{1 - \lambda\mu^*} & \frac{\mathcal{F}_\lambda - \mathcal{F}_\mu^*}{\lambda - \mu^*} \\ \frac{\mathcal{F}_\lambda^* - \mathcal{F}_\mu}{\lambda - \mu^*} & \frac{\mathcal{F}_\lambda^* + \mathcal{F}_\mu}{1 - \lambda\mu^*} \end{pmatrix} = \begin{pmatrix} \gamma(\tilde{\tau} - \lambda\alpha)^{-1} \\ \beta^*(\tau^* - \lambda\alpha^*)^{-1} \end{pmatrix}
$$

(5.20)

$$\cdot \left((\tilde{\tau}^* - \mu^*\alpha^*)^{-1}\gamma^* \quad (\tau - \mu^*\alpha)^{-1}\beta \right).$$

Then the colligations $\mathcal{W}$ *and* $\mathcal{W}_V$ *are unitarily equivalent, that is, there exists a unitary map* $\sigma : \mathcal{D}_\ell(\Phi) \longrightarrow \mathcal{H}$ *such that*

$$
\begin{pmatrix} \tau & 0 & 0 & 0 \\ 0 & \tilde{\tau} & 0 & 0 \\ 0 & 0 & \alpha & \beta \\ 0 & 0 & \gamma & \delta \end{pmatrix} = \begin{pmatrix} \sigma & 0 & 0 & 0 \\ 0 & \sigma & 0 & 0 \\ 0 & 0 & \sigma & 0 \\ 0 & 0 & 0 & I \end{pmatrix}
$$

$$
\begin{pmatrix} \tau_V & 0 & 0 & 0 \\ 0 & \tilde{\tau}_V & 0 & 0 \\ 0 & 0 & \alpha_V & \beta_V \\ 0 & 0 & \gamma_V & \delta_V \end{pmatrix} \begin{pmatrix} \sigma^* & 0 & 0 & 0 \\ 0 & \sigma^* & 0 & 0 \\ 0 & 0 & \sigma^* & 0 \\ 0 & 0 & 0 & I \end{pmatrix}.
$$

The complete proofs of these theorems are lengthy, so we show only samples of the calculations.

Proof of Theorem 5.1: We start with the formulas (5.1)–(5.5) for the case $V = 0$. Since α_0 is unitary and $\mathcal{M}^r_{(V,V^{(1)})}$ is a contraction the operator $(I_{\mathcal{D}_\ell(\Phi)} - \alpha_0 \mathcal{M}^r_{(V,V^{(1)})})$ is invertible and so the following operators are well defined:

$$
\begin{aligned}
\tau_V &= (I_{\mathcal{D}_\ell(\Phi)} - \alpha_0 \mathcal{M}^r_{(V,V^{(1)})})^{-1}, \\
\tilde{\tau}_V &= (I_{\mathcal{D}_\ell(\Phi)} - \mathcal{M}^r_{(V,V^{(1)})} \alpha_0)^{-1}, \\
\alpha_V &= (I_{\mathcal{D}_\ell(\Phi)} - \alpha_0 \mathcal{M}^r_{(V,V^{(1)})})^{-1} \alpha_0 \\
&= \alpha_0 (I_{\mathcal{D}_\ell(\Phi)} - \mathcal{M}^r_{(V,V^{(1)})} \alpha_0)^{-1}, \\
\beta_V &= (I_{\mathcal{D}_\ell(\Phi)} - \alpha_0 \mathcal{M}^r_{(V,V^{(1)})})^{-1} \beta_0, \\
\gamma_V &= \gamma_0 (I_{\mathcal{D}_\ell(\Phi)} - \mathcal{M}^r_{(V,V^{(1)})} \alpha_0)^{-1}, \\
\delta_V &= \delta_0 + \gamma_0 \mathcal{M}^r_{(V,V^{(1)})} (I_{\mathcal{D}_\ell(\Phi)} - \alpha_0 \mathcal{M}^r_{(V,V^{(1)})})^{-1} \beta_0.
\end{aligned}
$$

(5.21)

They coincide with the ones mentioned in the theorem. We show this for the main operator α_V. For $\genfrac{}{}{0pt}{}{F}{H} \in \mathcal{D}_\ell(\Phi)$, set

$$
\begin{pmatrix} X \\ Y \end{pmatrix} = (I_{\mathcal{D}_\ell(\Phi)} - \mathcal{M}^r_{(V,V^{(1)})} \alpha_0)^{-1} \begin{pmatrix} F \\ H \end{pmatrix}.
$$

A straightforward calculation gives

$$
(F - F_{[0]})(Z - V)^{-1} = (X - X_{[0]}) Z^*,
$$

$$
(HZ^* - X_{[0]})(I_{\mathcal{D}_\ell(\Phi)} - Z^* V)^{-1} = Y^* Z - X_{[0]}.
$$

The first equality implies that $X_{[0]} = F^\triangle(V)$ and hence

$$
\begin{aligned}
\alpha_V \begin{pmatrix} F \\ H \end{pmatrix} = \alpha_0 \begin{pmatrix} X \\ Y \end{pmatrix} &= \begin{pmatrix} (X - X_{[0]}) Z^* \\ YZ^* - X_{[0]} \end{pmatrix} \\
&= \begin{pmatrix} (F - F^\triangle(V))(Z - V)^{-1} \\ (HZ^* - F^\triangle(V)) \sigma_{V*}^{-1} \end{pmatrix},
\end{aligned}
$$

that is, (5.8) holds. The formulas (5.6) and (5.7) for τ_V and $\tilde{\tau}_V$ follow from

$$
\tau_V = I_{\mathcal{D}_\ell(\Phi)} + \alpha_V \mathcal{M}^r_{(V,V^{(1)})}, \qquad \tilde{\tau}_V = I_{\mathcal{D}_\ell(\Phi)} + \mathcal{M}^r_{(V,V^{(1)})} \alpha_V,
$$

the one for β_V follows from $\beta_V = \tau_V \beta_0$, and so on. The definitions of τ_V, $\tilde{\tau}_V$ and α_V together with α_0 being unitary imply that the colligation $\mathcal{W}_V$ is unitary in the described sense. We omit the proofs of the other equalities. It then remains to prove the closely connectedness of $\mathcal{W}_V$. This follows by applying the following formulas to the case where $W = \lambda I_D$ with $\lambda \in \mathbb{D}$: For ${}^F_H \in \mathcal{D}_\ell(\Phi)$ and $W \in \mathcal{D}$, $\|W\| < 1$, we have

(5.22)

$$\gamma_V (\tilde{\tau}_V - \mathcal{M}^r_{(W,W^{(1)})} \alpha_V)^{-1} \begin{pmatrix} F \\ H \end{pmatrix}$$

$$= \gamma_0 (I_{\mathcal{D}_\ell(\Phi)} - \mathcal{M}^r_{(V,V^{(1)})} \alpha_0)^{-1} \begin{pmatrix} F \\ H \end{pmatrix}$$

$$= F^\triangle(W),$$

$$\beta_V^* (\tau_V^* - \mathcal{M}^r_{(W^*,W^{*(1)})} \alpha_V^*)^{-1} \begin{pmatrix} F \\ H \end{pmatrix}$$

$$= \beta_0^* (I_{\mathcal{D}_\ell(\Phi)} - \mathcal{M}^r_{(W^*,W^{*(1)})} \alpha_0^*)^{-1} \begin{pmatrix} F \\ H \end{pmatrix}$$

$$= -H^\vee(W^{*(1)}).$$

Indeed, if ${}^{\,'F\,'}_H \in \mathcal{D}_\ell(\Phi)$ is orthogonal to the elements in (5.15), then for all $\lambda \in \mathbb{D}$, $F^\triangle(\lambda I_D) = 0$ and $H^\vee(\lambda^* I_D) = 0$, that is, $F = 0$ and $H = 0$.

Proof of Theorem 5.3: By the formulas (5.9) for β_V and (5.22), we have

$$\gamma_V (\mathcal{M}^r_{(W,W^{(1)})} - \mathcal{M}^r_{(V,V^{(1)})})(\tau_V - \alpha_V \mathcal{M}^r_{(W,W^{(1)})})^{-1} \beta_V(E)$$

$$= \gamma_V (\tilde{\tau}_V - \mathcal{M}_{(W,W^{(1)})} \alpha_V)^{-1}(\mathcal{M}^r_{(W,W^{(1)})} - \mathcal{M}^r_{(V,V^{(1)})})\beta_V(E)$$

$$= ((\Phi E - (\Phi E)^\triangle(V))(Z - V)^{-1}(W - V))^\triangle(V)$$

$$= ((\Phi E - (\Phi E)^\triangle(V))(Z - V)^{-1}(W - Z + Z - V))^\triangle(V)$$

$$= ((\Phi E - (\Phi E)^\triangle(V)))^\triangle(W)$$

$$= (\Phi E - (\Phi E)^\triangle(V))^\triangle(W)$$

$$= (\Phi E)^\triangle(W) - (\Phi E)^\triangle(V)$$

which readily implies the representation formula (5.16). The last identity in Theorem 5.3 can be proved by using the formulas

$$\mathcal{F}_W(E) = (\Phi E)^\triangle(W)$$

$$= (\delta_V + \gamma_V(\mathcal{M}^r_{(W,W^{(1)})} - \mathcal{M}^r_{(V,V^{(1)})})(\tau_V - \alpha_V \mathcal{M}^r_{(W,W^{(1)})})^{-1}\beta_V)(E)$$

$$= (\delta_V + \gamma_V(\tilde{\tau}_V - \mathcal{M}^r_{(W,W^{(1)})} \alpha_V)^{-1}$$

$$(\mathcal{M}^r_{(W,W^{(1)})} - \mathcal{M}^r_{(V,V^{(1)})})\beta_V)(E),$$

$$\mathcal{F}_W^*(E) = (\Phi^* E)^\vee(W^{*(1)})$$
$$= (\delta_V^* + \beta_V^*(\mathcal{M}_{(W^*, W^{*(1)})}^r - \mathcal{M}_{(V^*, V^{*(1)})}^r)$$
$$(\widetilde{\tau}_V^* - \alpha_V^* \mathcal{M}_{(W^*, W^{*(1)})}^r)^{-1} \gamma_V^*)(E)$$
$$= (\delta_V^* + \beta_V^*(\tau_V^* - \mathcal{M}_{(W^*, W^{*(1)})}^r \alpha_V^*)^{-1}$$
$$(\mathcal{M}_{(W^*, W^{*(1)})}^r - \mathcal{M}_{(V^*, V^{*(1)})}^r)\gamma_V^*)(E)$$

and by reducing the general case to the case $V = 0$ as in the proof of formula (3.12).

Proof of Theorem 5.5: Let $\mathbf{R} \subset \mathcal{D}_\ell(\Phi) \times \mathcal{H}$ be the relation spanned by the elements of the form

$$\left(((\widetilde{\tau}_V^* - \mu^* \alpha_V^*)^{-1} \gamma_V^* \ (\tau_V - \alpha_V \mu^*)^{-1} \beta_V) \begin{pmatrix} E \\ G \end{pmatrix} , \right.$$
$$\left. ((\widetilde{\tau}^* - \mu^* \alpha^*)^{-1} \gamma^* \ (\tau - \mu^* \alpha)^{-1} \beta) \begin{pmatrix} E \\ G \end{pmatrix} \right),$$

where E, G run through $\mathcal{D}_2$ and μ through $\mathbb{D}$. By (5.18) and (5.20), we have

$$\begin{pmatrix} \gamma_V (I_{\mathcal{D}_V(\Phi)} - \lambda \alpha_V)^{-1} \\ \beta_V^* (I_{\mathcal{D}_V(\Phi)} - \lambda \alpha_V^*)^{-1} \end{pmatrix}$$

(5.23) $$\qquad ((I_{\mathcal{D}_V(\Phi)} - \overline{\mu} \alpha_V^*)^{-1} \gamma_V^* (I_{\mathcal{D}_V(\Phi)} - \overline{\mu} \alpha_V)^{-1} \beta_V)$$

$$= \begin{pmatrix} \gamma(I_{\mathcal{H}} - \lambda \alpha)^{-1} \\ \beta^*(I_{\mathcal{H}} - \lambda \alpha^*)^{-1} \end{pmatrix} ((I_{\mathcal{H}} - \overline{\mu} \alpha^*)^{-1} \gamma^* (I_{\mathcal{H}} - \overline{\mu} \alpha)^{-1} \beta),$$

which implies that $\mathbf{R}$ is isometric. On account of the closely connectedness of $\mathcal{W}_V$ and $\mathcal{W}$, $\mathbf{R}$ has a dense domain and a dense range. Hence the closure of $\mathbf{R}$ is the graph of a unitary operator σ from $\mathcal{D}_\ell(\Phi)$ to $\mathcal{H}$ such that for all $E, G \in \mathcal{D}_2$ and $\mu \in \mathbb{D}$,

$$\sigma\left(((\widetilde{\tau}_V^* - \mu^* \alpha_V^*)^{-1} \gamma_V^* \ (\tau_V - \alpha_V \mu^*)^{-1} \beta_V) \begin{pmatrix} E \\ G \end{pmatrix} \right)$$

$$= ((\widetilde{\tau}^* - \mu^* \alpha^*)^{-1} \gamma^* \ (\tau - \mu^* \alpha)^{-1} \beta) \begin{pmatrix} E \\ G \end{pmatrix}.$$

Taking $\mu = 0$ we get $\beta = \sigma \beta_V$ and $\gamma = \gamma_V \sigma^*$. From (5.23) we conclude that

$$\begin{cases} \gamma_V \alpha_V^m \alpha_V^{*n} \gamma_V^* = \gamma \alpha^m \alpha^{*n} \gamma^*, \\ \gamma_V \alpha_V^m \alpha_V^n \beta_V = \gamma \alpha^m \alpha^n \beta, \\ \beta_V^* \alpha_V^{*m} \alpha_V^n \beta_V = \beta \alpha^{*m} \alpha^n \beta. \end{cases}$$

Hence

$$\begin{pmatrix} \gamma_V(I-\lambda\alpha_V)^{-1} \\ \beta_V^*(I-\lambda\alpha_V^*)^{-1} \end{pmatrix} \alpha_V \begin{pmatrix} \gamma_V(I-\mu^*\alpha_V)^{-1} \\ \beta_V^*(I-\mu^*\alpha_V^*)^{-1} \end{pmatrix}^*$$

$$= \begin{pmatrix} \gamma(I-\lambda\alpha)^{-1} \\ \beta^*(I-\lambda\alpha^*)^{-1} \end{pmatrix} \alpha \begin{pmatrix} \gamma(I-\mu^*\alpha)^{-1} \\ \beta^*(I-\mu^*\alpha^*)^{-1} \end{pmatrix}^*$$

$$= \begin{pmatrix} \gamma_V(I-\lambda\alpha_V)^{-1} \\ \beta_V^*(I-\lambda\alpha_V^*)^{-1} \end{pmatrix} \sigma^*\alpha\sigma \begin{pmatrix} \gamma_V(I-\mu^*\alpha_V)^{-1} \\ \beta_V^*(I-\mu^*\alpha_V^*)^{-1} \end{pmatrix}^*,$$

which implies that $\alpha = \sigma\alpha_V\sigma^*$. Finally, $\delta_V(E) = (\Phi E)^\Delta(V) = \delta(E)$ for all $E \in \mathcal{D}_2$.

References

[ABDP]   D. Alpay, V. Bolotnikov, A. Dijksma and Y. Peretz, A coisometric realization for triangular integral operators, *Operator Theory: Adv. Appl.*, Birkhäuser Verlag, Basel (preprint), to appear.

[ADD]   D. Alpay, P. Dewilde and H. Dym, Lossless inverse scattering and reproducing kernels for upper triangular operators, *Operator Theory: Adv. Appl.*, Birkhäuser Verlag, Basel **47** (1990), 61–133.

[ADP]   D. Alpay, A. Dijksma and Y. Peretz, Nonstationary analogs of the Herglotz representation theorem: the discrete case, *J. Functional Analysis* **66** (1999), 85–129.

[ADRS1] D. Alpay, A. Dijksma, J. Rovnyak and H. de Snoo, Schur functions, operator colligations, and reproducing kernel Pontryagin spaces, *Operator Theory: Adv. Appl.*, Birkhäuser Verlag, Basel **96** (1997).

[ADRS2] D. Alpay, A. Dijksma, J. Rovnyak and H. de Snoo, Realization and factorization in reproducing kernel Pontryagin spaces, *Operator Theory: Adv. Appl.*, Birkhäuser Verlag, Basel (preprint), to appear.

[AP]   D. Alpay and Y. Peretz, Realizations for Schur upper triangular operators, in: Dijksma, A., Gohberg, I., Kaashoek, M. and Mennicken, R., editors, Contributions to operator theory in spaces with an indefinite metric, *Operator Theory: Adv. Appl.*, Birkhäuser Verlag, Basel **106** (1998), 37–90.

[BGK1]   J. Ball, I. Gohberg and M.A. Kaashoek, Nevanlinna-Pick interpolation for time-varying input-output maps: the continuous case, in: Gohberg, I., editor, Time-variant systems and interpolation, *Operator Theory: Adv. Appl.*, Birkhäuser Verlag, Basel **56** (1992), 52–89.

[BGK2]   J. Ball, I. Gohberg and M.A. Kaashoek, Bitangential interpolation for input-output maps of time-varying systems: the continuous time case, *Integral Equations Operator Theory* **20** (1994), 1–43.

[BT]   J. Ball and T. Trent, Unitary colligations, reproducing kernel Hilbert spaces and Nevanlinna-Pick interpolation in several variables, *J. Functional Analysis* **157** (1998), 1–61.

[DD]   P. Dewilde and H. Dym, Interpolation for upper triangular operators, in: Gohberg, I., editor, Time-variant systems and interpolation, *Operator Theory: Adv. Appl.*, Birkhäuser Verlag, Basel **56** (1992), 153–260.

[FFGK1]   C. Foias, A.E. Frazho, I. Gohberg and M.A. Kaashoek, Discrete time-invariant interpolation as classical interpolation with an operator argument, *Integral Equations Operator Theory* **26** (1996), 371–403.

[FFGK2]   C. Foias, A.E. Frazho, I. Gohberg and M.A. Kaashoek, Metric constrained interpolation, commutant lifting and systems, *Operator Theory: Adv. Appl.*, Birkhäuser Verlag, Basel **100** (1998).

[P]   Y. Peretz, *Nonstationary de Branges-Rovnyak spaces and realizations of triangular operators*, Ph.D. thesis, Ben-Gurion University of the Negev, Beer-Sheva, Israel, 1999.

D. Alpay   A. Dijksma
Department of Mathematics   Department of Mathematics
Ben-Gurion University of the Negev   University of Groningen, POB 800
Beer-Sheva 84105   9700 AV Groningen
Israel   The Netherlands

Y. Peretz
Department of Mathematics
Ben-Gurion University of the Negev
Beer-Sheva 84105
Israel

1991 Mathematics Subject Classification. 47A48, 47H06.

Operator Theory:
Advances and Applications, Vol. 124
© 2001 Birkhäuser Verlag Basel/Switzerland

Characteristic Functions of Maximal Sectorial Operators

Yury Arlinskii

Dedicated to Professor Israel Gohberg on the occasion of his 70-th birthday

The analytic properties of the characteristic functions of maximal sectorial operators are studied and functional models are constructed.

1 Introduction

The concepts of the characteristic functions was introduced by M.S. Livsic for some classes of bounded and unbounded operators in Hilbert spaces. His ideas were developed by different authors in order to construct and study the functional and triangular models of linear operators in Hilbert and Krein spaces. A brief survey of results in this directions is presented in the monograph [17].

In this paper we study the characteristic function and construct models for a maximal sectorial operator. In order to define an operator node and its the characteristic function in the form similar to the case of bounded nonselfadjoint operator (see [8], [9], [19], [20]), it is convenient to use the concept of the rigged Hilbert spaces [7]. Such approach was proposed earlier by E.R. Tsekanovskii (see [24]) for proper nonselfadjoint extensions of a symmetric operator with equal defect numbers and, in particular [22], [23], for proper maximal accretive and maximal sectorial extensions of nonnegative symmetric operator. Characteristic functions of maximal sectorial extensions of nonnegative Hermitian operators were studied in [12], [13] by using the concepts of boundary value space and the corresponding Weyl function [11]. One of the results of [13] is the characterizations of proper m-sectorial extensions of a nonnegative Hermitian operator in terms of the corresponding "boundary" parameter and in terms of the characteristic functions. Using the characteristic function and its the fractional linear transformation we construct the D.N. Clark's type [10] functional model and models with diagonal real parts for a maximal sectorial operator. It should be noted that characteristic functions of the Cayley's transform of a maximal sectorial operator were studied in [1].

We will use the following notations: $\mathcal{L}(H_1, H_2)$ is the Banach space of all continuous linear operators acting from the Hilbert space H_1 into the Hilbert space H_2 and $\mathcal{L}(H) = \mathcal{L}(H, H)$. The domain, the range and the null-space of a linear operator S we denote by $\mathcal{D}(S)$ and $\mathcal{R}(S)$ and *Ker* S, respectively, $\rho(S)$ is the resolvent set of S.

Let H be a complex separable Hilbert space. According to T. Kato [16], a linear operator T acting in H is called maximal sectorial with a semiangle $\alpha \in [0, \pi/2)$ and with the vertex at the origin if T is maximal accretive and if the numerical range

$$W(T) = \{(Tu, u), u \in \mathcal{D}(T), \|u\| = 1\}$$

is contained in the sector of the complex plane: $\Theta(\alpha) = \{\lambda \in \mathbb{C} : |\arg \lambda| \le \alpha\}$.

Below such an operator will be called m-α-sectorial or, shortly, m-sectorial. The domain of an m-α-sectorial operator is dense in H and its resolvent set contains $\mathbb{C}\backslash\Theta(\alpha)$. As is well known [16], if T is an m-sectorial operator then the sesquilinear form $(T\cdot, \cdot)$ has the closure which we will denote by $T[u, v]$ and its domain by $\mathcal{D}[T]$. The following representations hold [16]:

$$T = T_R^{1/2}(I + iG)T_R^{1/2}, \quad T[u, v] = ((I + iG)T_R^{1/2}u, T_R^{1/2}v), \quad u, v \in \mathcal{D}[T],$$

where T_R is the "real part" of T, i.e. the nonnegative selfadjoint operator associated with the closed form $T_R[u, v] = (T[u, v] + \overline{T[v, u]})/2$ and $G = G^* \in \mathcal{L}(\overline{\mathcal{R}(T)})$. According to the First and Second Representation Theorems [16] we have equalities $\mathcal{D}[T] = \mathcal{D}[T_R] = \mathcal{D}(T_R^{1/2})$.

2 Operator Nodes and Characteristic Functions

Let T be an m-sectorial operator and let $B = T_R^{1/2}$. Denote by H_+ the domain $\mathcal{D}[T]$ with the inner product

$$(u, v)_+ = (u, v) + (Bu, Bv).$$

Then H_+ is a Hilbert space and let $H_+ \subseteq H \subseteq H_-$ be the corresponding rigged Hilbert space [7]. Remind that the "negative" Hilbert space H_- is dual to H_+ with respect to the form (u, v). Consider the operator B as an element of $\mathcal{L}(H_+, H)$. Its adjoint belongs in $\mathcal{L}(H, H_-)$ and coincides with B on H_+. Therefore, this operator which we denote by $\overline{B}$ is the continuation of B on H. Clearly, $\overline{B}h \in H$ if and only if $h \in H_+$, $\overline{B}$ is a nonnegative selfadjoint operator in H_- with the domain H and the operators $\mathfrak{J} = (I + \overline{B}B)^{-1}$, $\mathfrak{J} = (I + \overline{B}B)^{-1/2}$ are the canonical isometries [7] between H_-, H_+ and H_-, H respectively, i.e.

$$(f, g)_- = (f, \mathfrak{J}g) = (\mathfrak{J}f, \mathfrak{J}g)_+ = (\mathfrak{J}f, \mathfrak{J}g), \quad f, g \in H_-.$$

Moreover, the following relation holds

$$\mathcal{R}(\overline{B}) = \left\{ f \in H_- : \sup_{a>0} \|(\overline{B}B + aI)^{-1/2}f\|^2 < \infty \right\}.$$

Actually, let $\|(\overline{B}B + aI)^{-1/2}\mathfrak{J}^{-1}h\|^2 < C$ for $h \in H$ and all $a > 0$. Since

$$(\overline{B}B + aI)^{-1/2}\mathfrak{J}^{-1} = (\overline{B}B + I)^{1/2}(\overline{B}B + aI)^{-1/2},$$

we have $\|(\overline{B}B + aI)^{-1/2}h\|^2 = \|(T_R + aI)^{-1/2}h\|^2 < C$ and hence $h \in \mathcal{R}(B)$. Therefore, $f = \mathfrak{J}^{-1}h = (\overline{B}B + I)^{1/2}h \in \mathcal{R}(\overline{B})$.

Let $\overline{T} = \overline{B}(I + iG)B$ then $\overline{T} \in \mathcal{L}(H_+, H_-)$ and $T[u, v] = (\overline{T}u, v)$, $u, v \in H_+$. The adjoint operator $\overline{T}^{\times} \in \mathcal{L}(H_+, H_-)$ has the form $\overline{T}^{\times} = \overline{B}(I - iG)B$. We say that $\lambda \in \rho(\overline{T})$ if the operator $\overline{T} - \lambda I$ is a bijection of H_+ onto H_-.

Proposition 2.1 *The following identity holds:* $\rho(\overline{T}) = \rho(T)$.

Proof: The inclusion $\rho(\overline{T}) \subseteq \rho(T)$ is evident. Let $Re\, z < 0$, then for $u \in H_+$ we have

$$\|(\overline{T} - zI)u\|_- = \sup_{v \in H_+} \frac{|((\overline{T} - zI)u, v)|}{\|v\|_+}$$

$$\geq \frac{|((\overline{T} - zI)u, u)|}{\|u\|_+} = \frac{|((I + iG)Bu, Bu) - z\|u\|^2|}{\|u\|_+}$$

$$\geq \frac{|\|Bu\|^2 - Re\, z\|u\|^2|}{\|u\|_+} \geq c\|u\|_+$$

with $c = max\{1, -Re\, z\}$. Since $\mathcal{R}((\overline{T} - zI)) \supseteq H$, it follows that $z \in \rho(\overline{T})$. If $\lambda \in \rho(T)$ then for every z, $Re\, z < 0$ the operator

$$(\overline{T} - zI)^{-1} + (\lambda - z)(T - \lambda I)^{-1}(\overline{T} - zI)^{-1}$$

is the inverse to $\overline{T} - \lambda I$. Hence, $\lambda \in \rho(\overline{T})$ and $\rho(T) \subseteq \rho(\overline{T})$. □

Note the Hilbert identity

(2.1)
$$(\overline{T} - \lambda I)^{-1} = (\overline{T} - zI)^{-1} + (\lambda - z)$$
$$(T - \lambda I)^{-1}(\overline{T} - zI)^{-1}, \lambda, z \in \rho(T)$$

and

(2.2)
$$(\overline{T} - \lambda I)^{-1}\mathcal{R}(\overline{B}) = \mathcal{R}(B) \cap H_+, \lambda \in \rho(T).$$

Definition 2.2 Let E be a Hilbert space, J be a fundamental symmetry of E, i.e. $J = J^* = J^{-1}$ and let $K \in \mathcal{L}(E, H_-)$, $\mathcal{R}(K) \subseteq \mathcal{R}(\overline{B})$. If the relation

$$(2.3) \qquad\qquad \overline{T} - \overline{T}^{\times} = 2i\, K J K^{\times}$$

holds then $\{T, E, K, J\}$ is called the operator node of an m-sectorial operator T.

Definition 2.3 The function

$$W(\lambda) = I - 2i\, J K^{\times} (\overline{T} - \lambda I)^{-1} K$$

will be called the characteristic function of the operator node $\{T, E, K, J\}$.

Remark 2.4 Definitions 2.2 and 2.3 coincide by the form with related one for bounded operators [8], [9], [19], [20]. For proper extensions of Hermitian operators by using biextensions and rigged Hilbert spaces the definitions of the operator node and the characteristic functions belong to E.R. Tsekanovskii [24], [22], [23] (see also [6]). The definitions of E.R. Tsekanovskii depend on the choice of a biextension of the operator T. Other definitions for such a case by using boundary triplets and Weyl functions were given by V. Derkach and M. Malamud in [11], [13], [12]. It should be noted that the characteristic functions in definitions of E.R. Tsekanovskii and V. Derkach – M. Malamud is defined up to J-unitary factors [6], [11].

If $H_B = \overline{\mathcal{R}(B)}$ then there exist a Hilbert space E, a fundamental symmetry J in E and an operator $\Gamma \in \mathcal{L}(E, H_B)$ such that $G = \Gamma J \Gamma^*$ [8]. Let $K = \overline{B}\Gamma$ then $\{T, E, K, J\}$ forms an operator node of T. In such a case the characteristic function takes the form

$$W(\lambda) = I - 2i\, J \Gamma^* B (\overline{T} - \lambda I)^{-1} \overline{B}\Gamma.$$

This is a general form of characteristic functions of nodes because if $\mathcal{R}(K) \subseteq \mathcal{R}(\overline{B})$ then [14] there exists $\Gamma \in \mathcal{L}(E, H_B)$ such that $K = \overline{B}\Gamma$.

Theorem 2.5 *Let $\{T, E, K, J\}$ be a node of an m-α-sectorial operator T. Then the characteristic function $W(\lambda)$ has the following properties*

1) *$W(\lambda)$ is a holomorphic operator-function on the domain $\rho(T)$;*

2) *$s - \lim_{\lambda \to \infty}\{W(\lambda),\ \lambda \in \mathbb{C} \backslash \Theta(\gamma)\} = I$ for every $\gamma \in (\alpha, \pi/2)$ and $\lim_{\mathrm{Re}\, z \to -\infty} \|W(z) - I\| = 0$;*

3) *there exists a strong limit $s - \lim_{\lambda \to 0}\{W(\lambda),\ \lambda \in \mathbb{C} \backslash \Theta(\gamma)\}$ for every $\gamma \in (\alpha, \pi/2)$;*

4) *the operator-function*

$$K(\lambda, \mu) = \frac{W^*(\mu) J W(\lambda) - J}{2i\, (\overline{\mu} - \lambda)}$$

is a positive definite kernel in $\rho(T)$;

5) *the operator-function*

$$Z(\lambda, \mu) = \frac{1}{2i}(W^*(\mu) - I)J + \lambda K(\lambda, \mu)$$

is an α-sectorial kernel in the following sense

$$\left| Im \sum_{i,j=1}^{n} (Z(\lambda_i, \lambda_j) f_i, f_j)_E \right| \leq \tan \alpha \ Re \sum_{i,j=1}^{n} (Z(\lambda_i, \lambda_j) f_i, f_j)_E$$

for every choice of points $\lambda_1, \ldots, \lambda_n \in \rho(T)$ and vectors $f_1, \ldots, f_n \in E$.

Proof: The first statement is evident. In view of sectoriality we have the estimates

$$\|(T - \lambda I)^{-1}\| \leq \frac{c_\gamma}{|\lambda|}, \ \lambda \in \mathbb{C} \backslash \Theta(\gamma),$$

$$\|(\overline{T} - zI)^{-1} f\|_+ \leq \frac{\|f\|_-}{-Re\, z}, \ -Re\, z > 1, \ f \in H_-.$$

It follows for all $f \in H$ the following relations in the space H

$$\lim_{\lambda \to \infty} \{(T - \lambda I)^{-1} f, \ \lambda \in \mathbb{C} \backslash \Theta(\gamma)\} = 0,$$

$$\lim_{\lambda \to \infty} \{\lambda (T - \lambda I)^{-1} f, \ \lambda \in \mathbb{C} \backslash \Theta(\gamma)\} = -f$$

Using (2.1) for every $h \in H_-$ we obtain the equality

$$\lim_{\lambda \to \infty} \{\|(\overline{T} - \lambda I)^{-1} h\|, \ \lambda \in \mathbb{C} \backslash \Theta(\gamma)\} = 0.$$

Since for $f \in H$

$$\begin{aligned}
\|B(T - \lambda I)^{-1} f\|^2 &= Re\,(T(T - \lambda I)^{-1} f, (T - \lambda I)^{-1} f) \\
&= Re\,(f, (T - \lambda I)^{-1} f) + Re\,\lambda \|(T - \lambda I)^{-1} f\|^2,
\end{aligned}$$

then

$$\lim_{\lambda \to \infty} \{\|B(T - \lambda I)^{-1} f\|, \ \lambda \in \mathbb{C} \backslash \Theta(\gamma)\} = 0.$$

Taking into account the relation established in [3] for $u \in H_+$

$$\lim_{\lambda \to \infty} \{\lambda B(T - \lambda I)^{-1} u, \ \lambda \in \mathbb{C} \backslash \Theta(\gamma)\} = -Bu$$

from

$$B(\overline{T} - \lambda I)^{-1} = B(\overline{T} - zI)^{-1} + (\lambda - z)B(T - \lambda I)^{-1}(\overline{T} - zI)^{-1}$$

we get

$$\lim_{\lambda \to \infty} \{\|B(\overline{T} - \lambda I)^{-1}h\|, \ \lambda \in \mathbb{C}\backslash\Theta(\gamma)\} = 0, \ h \in H_-.$$

Thus

$$\lim_{\lambda \to \infty} \{\|(\overline{T} - \lambda I)^{-1}h\|_+, \ \lambda \in \mathbb{C}\backslash\Theta(\gamma)\} = 0, \ h \in H_-.$$

It proves statement 2).
Further we will use the relation in H [2]

$$\lim_{\lambda \to 0} \{B(T - \lambda I)^{-1}Bf, \lambda \in \mathbb{C}\backslash\Theta(\gamma)\}$$
$$= (I + iG)^{-1}f, \ f \in H_+ \cap H_B.$$

Hence for $h \in H_+$ from the identity

$$B(\overline{T} - \lambda I)^{-1}\overline{B}h = B(\overline{T} - zI)^{-1}\overline{B}h$$

$$+(\lambda - z)B(T - \lambda I)^{-1}(\overline{T} - zI)^{-1}\overline{B}h$$

and from (2.2) we get that in H there exists

$$\lim_{\lambda \to 0} \{B(\overline{T} - \lambda I)^{-1}\overline{B}h, \ \lambda \in \mathbb{C}\backslash\Theta(\gamma)\}.$$

and we proved statement 3).
   Using (2.1) and (2.3) from the definition of the characteristic function we obtain
the relation

(2.4) $W^*(\mu)JW(\lambda) - J = 2i(\overline{\mu} - \lambda)K^\times(\overline{T}^\times - \overline{\mu}I)^{-1}(\overline{T} - \lambda I)^{-1}K.$

Therefore

$$K(\lambda, \mu) = \frac{W^*(\mu)JW(\lambda) - J}{2i(\overline{\mu} - \lambda)}$$

is a positive definite kernel in $\rho(T)$. Further we have

$$\frac{1}{2i}(W^*(\mu) - I)J + \lambda K(\lambda, \mu)$$

$$= K^\times(\overline{T}^\times - \overline{\mu}I)^{-1}K + \lambda K^\times(\overline{T}^\times - \overline{\mu}I)^{-1}(\overline{T} - \lambda I)^{-1}K$$

$$= K^\times(\overline{T}^\times - \overline{\mu}I)^{-1}\overline{T}(\overline{T} - \lambda I)^{-1}K.$$

Since the operator $\overline{T}$ is an α-sectorial we get that $Z(\lambda, \mu)$ is an α-sectorial kernel. □

Observe that if an m-sectorial operator T has the semiangle α then for every $x < y < 0$ operators

$$iJ(W(x) - I) = 2K^\times(\overline{T} - xI)^{-1}K$$

and

$$i(W^*(x)JW(y) - J) = 2(y - x)K^\times(\overline{T}^\times - xI)^{-1}(\overline{T} - yI)^{-1}K$$

are bounded sectorial with the same semiangle α in E. It follows that $iJ(W(0) - I)$ is an m-sectorial with the semiangle α.

Remark 2.6 Using different approaches the existence of strong limit values $W(0)$ and $W(-\infty)$ for characteristic functions of m-sectorial operators were proved in [1], [13], [12], [22]. In [13] and [12] is proved that the condition:

the operator $i(W^*(-\infty)JW(0) - J)$ is bounded sectorial

is necessary and sufficient for the operator T to be a proper m-sectorial extension of positive Hermitian operator.

Let T is an m-sectorial operator in H and $H^{(1)}$ is an invariant subspace under the resolvent $(T - \lambda I)^{-1}$, $Re\,\lambda < 0$. Then $T_1 = T|H^{(1)}$ and $T_2 = (T^*|H^{(2)})^*$ are m-sectorial operators in $H^{(1)}$ and in $H^{(2)} = H \ominus H^{(1)}$ respectively. Moreover, the following relations hold [3]:

$$(2.5) \qquad \mathcal{D}[T] = \mathcal{D}[T_1] \oplus \mathcal{D}[T_2], \quad \mathcal{R}(T_R^{1/2}) = \mathcal{R}(T_{1R}^{1/2}) \oplus \mathcal{R}(T_{2R}^{1/2}),$$

$$(2.6) \qquad \begin{aligned} T[u, v] &= T_1[P^{(1)}u, P^{(1)}v] + T_2[P^{(2)}u, P^{(2)}v] \\ &\quad + 2(XT_{2R}^{1/2}P^{(2)}u, T_{1R}^{1/2}P^{(1)}v), \end{aligned}$$

where $u, v \in \mathcal{D}[T]$, $P^{(1)}$, $P^{(2)}$ are orthogonal projections in H onto $H^{(1)}$ and $H^{(2)}$, respectively and

$$(2.7) \qquad X \in \mathcal{L}(\overline{\mathcal{R}(T_{2R}^{1/2})}, \overline{\mathcal{R}(T_{2R}^{1/2})}), \quad \|X\| < 1.$$

Let $B_j = T_{jR}^{1/2}$, $H_+^{(j)} = \mathcal{D}(B_j)$, $j = 1, 2$, $\tilde{B} = B_1 P^{(1)} + B_2 P^{(2)}$ and $H_+^{(j)} \subseteq H^{(j)} \subseteq H_-^{(j)}$ are corresponding rigged Hilbert spaces. The spaces $H_+^{(1)}$ and $H_+^{(2)}$ are orthogonal in H_+ with respect to the inner product

$$(u, v)_{\tilde{B}} = (\tilde{B}u, \tilde{B}v) + (u, v), \quad u, v \in H_+.$$

and the corresponding norm is equivalent to the $(+)$ - norm.

Denote by $P_+^{(j)}$ the orthogonal projections onto $H_+^{(j)}$ in H_+. Then its adjoints coincide with orthogonal projections $P_-^{(j)}$ onto $H_-^{(j)}$ in H_-. Let $\{T, E, K, J\}$ be a node of T and let $K_j = P_-^{(j)} K$, $j = 1, 2$. Then $\{T_1, K_1, J, E\}$ and $\{T_2, K_2, J, E\}$ are the nodes for T_1 and T_2 respectively. Actually, it is clear that

$$\overline{T}_j - \overline{T}_j^\times = 2i K_j J K_j^\times, \quad j = 1, 2.$$

Let us verify that $\mathcal{R}(K_j) \subseteq \mathcal{R}(\overline{B}_j)$. Since

$$\|Bu_j\|^2 = \operatorname{Re}(\overline{T}u_j, u_j) = \operatorname{Re}(\overline{T}_j u_j, u_j) = \|B_j u_j\|^2,$$

$$u_j \in H_+^{(j)}, \quad j = 1, 2$$

we have $Bu_j = U_j B_j u_j$, where U_j are isometries. Therefore

$$P_-^{(j)} \overline{B}h = \overline{B}_j U^* h, \quad h \in H, \ j = 1, 2.$$

Hence $K_j = P_-^{(j)} \overline{B}\Gamma = \overline{B}_j U_j^* \Gamma$.

Observe that

$$(2.8) \qquad \overline{T} = \overline{T}_1 P_+^{(1)} + \overline{T}_2 P_+^{(2)} + 2i K_1 J K_2^\times P_+^{(2)}.$$

Thus, the operator $\overline{T}$ has the form of coupling of $\overline{T}_1$ and $\overline{T}_2$ which is similar to the coupling of bounded operators [9]. Let $W_j(\lambda)$ be the characteristic functions of the nodes $\{T_j, K_j, J, E\}$. Then one can easily check the equality

$$W(\lambda) = W_1(\lambda) W_2(\lambda).$$

Remark 2.7 The factorization of the characteristic function is an essential part for construction of triangular models of linear operators. For characteristic functions of proper extensions of Hermitian operators the factorization were established in [23] and [11]. In view of (2.5)–(2.7) the coupling of the form (2.8) of m-sectorial operators in general is not an m-sectorial operator, therefore the function $W(\lambda) = W_1(\lambda) W_2(\lambda)$ has no the property 5) of Theorem 2.5.

Next by usual way we define a simple part of the node and prove that the characteristic function defines a simple part up to an unitary equivalence.

Let $\{T, E, K, J\}$ be a node of T. Consider the following subspace in H:

$$H^{(0)} = cls\{(\overline{T} - \lambda I)^{-1}\mathcal{R}(K), \ Re\,\lambda < 0\}.$$

Then $H^{(0)}$ is an invariant subspace under the resolvent of T and hence the subspace $H^{(0)\perp} = H \ominus H^{(0)}$ is an invariant under the resolvent of T^*. Moreover, since $\mathcal{R}(\overline{T} - \overline{T}^\times) \subseteq \mathcal{R}(K)$ we get for all $g \in H^{(0)\perp}$ and all $z, Re < 0$ the equality

$$\overline{T}(T^* - zI)^{-1}g = T^*(T^* - zI)^{-1}g.$$

It follows that $H^{(0)\perp}$ reduces T, $T|(\mathcal{D}(T) \cap H^{(0)\perp} = B^2$ is a selfadjoint operator, $T^{(0)} = T|(\mathcal{D}(T) \cap H^{(0)}$ is an m-sectorial in the subspace $H^{(0)}$, $\mathcal{R}(K) \subseteq H_-^{(0)}$ and $W(\lambda) = W_0(\lambda)$, where $W_0(\lambda)$ is the characteristic function of the node $\{T^{(0)}, K, J, E\}$. This node is called the simple part of the the node $\{T, E, K, J\}$. If $H^{(0)} = H$ then the node $\{T, E, K, J\}$ will be called a simple.

In view of the relations

$$(\overline{T} - \lambda I)^{-1}K - (\overline{B}B - \lambda I)^{-1}K = -i(\overline{T} - \lambda I)^{-1}$$

$$KJK^\times(\overline{B}B - \lambda I)^{-1}K = -i(\overline{B}B - \lambda I)^{-1}KJK^\times(\overline{T} - \lambda I)^{-1}K$$

we get that also

(2.9) $$H^{(0)} = cls\{(\overline{B}B - \lambda I)^{-1}\mathcal{R}(K), \ Re\,\lambda < 0\}.$$

Let T_1 and T_2 are two unitary equivalent m-sectorial operators in Hilbert spaces H_1 and H_2 respectively, i.e. $UT_1 = T_2U$, where U is an isometry from H_1 onto H_2 such that $U\mathcal{D}(T_1) = \mathcal{D}(T_2)$. It follows readily that U maps $\mathcal{D}[T_1]$ onto $\mathcal{D}[T_2]$ and $T_1[u, v] = T_2[Uu, Uv]$ for all $u, v \in \mathcal{D}[T_1]$. The last relation immediately yields that U is an isometry from H_{1+} onto H_{2+}, $\overline{U} = U^{\times -1}$ is the continuation of U and is an isometry from H_{1-} onto H_{2-} and $\overline{U}\,\overline{T}_1 = \overline{T}_2U$.

Two nodes $\{T_j, K_j, J, E\}$, $j = 1, 2$ will be called unitary equivalent if there exists an isometry from H_1 onto H_2 such that

$$UT_1 = T_2U, \ \overline{U}K_1 = K_2.$$

Theorem 2.8 *Let $\{T_j, K_j, J, E\}$, $j = 1, 2$ be two nodes of m-sectorial operators T_1 and T_2 and $W_1(\lambda) = W_2(\lambda)$, $Re\,\lambda < 0$. Then the simple parts of the nodes are unitary equivalent.*

Proof: The equality $W_1(\lambda) = W_2(\lambda)$, $Re\,\lambda < 0$ yields

$$W_1^*(\mu)JW_1(\lambda) = W_2^*(\mu)JW_2(\lambda).$$

By (2.4) we have

$$((\overline{T}_1 - \lambda I)^{-1} K_1 h, (\overline{T}_1 - \mu I)^{-1} K_1 g)_{H_1}$$

$$= ((\overline{T}_2 - \lambda I)^{-1} K_2 h, (\overline{T}_2 - \mu I)^{-1} K_2 g)_{H_2}.$$

Define the operator $U : H_1^{(0)} \to H_2^{(0)}$ by the relation

$$U \left(\sum_i (\overline{T}_1 - \lambda_i I)^{-1} K_1 h_i \right) = \sum_i (\overline{T}_2 - \lambda_i I)^{-1} K_2 h_i.$$

The operator U is an isometry from $H_1^{(0)}$ onto $H_2^{(0)}$ and the Hilbert identity implies

$$U(T_1 - zI)^{-1} f = (T_2 - zI)^{-1} U f, \quad f \in H_1^{(0)}.$$

Consequently, $U T_1^{(0)} = T_2^{(0)} U$ and $\overline{U} \, \overline{T}_1^{(0)} = \overline{T}_2^{(0)} U$. Therefore $(\overline{T}_2^{(0)} - \lambda I)^{-1}$ $K_2 = U(\overline{T}_1^{(0)} - \lambda I)^{-1} K_1 = (\overline{T}_2^{(0)} - \lambda I)^{-1} \overline{U} K_1$. Hence, $K_2 = \overline{U} K_1$. □

Observe that the statement of Theorem 2.8 is well known for nodes and their characteristic functions of bounded and unbounded operators (see [8], [9], [11], [17], [20], [24]).

Let T be an m-sectorial operator and let

$$\mathcal{D}(A) = \{f \in \mathcal{D}(T) \cap \mathcal{D}(T^*) : Tf = T^* f\}, \ Af = Tf, \ f \in \mathcal{D}(A).$$

Then A is a closed nonnegative Hermitian operator and T is a proper extension of A, i.e. $T \supseteq A$ and $T^* \supseteq A$. We shall show that the characteristic function of T given by Definitions 2.2 and 2.3 coincides with one of the characteristic functions of T defined in [11] in the case when A is densely defined and T is transversal to the Friedrichs extension A_F of the operator A. In this case the von Neumann-Krein extension A_N of A is also transversal to A_F [4] and $\mathcal{D}[T] = \mathcal{D}[A_N] \supset \mathcal{D}(A^*)$ [21]. Let $H_+ = \mathcal{D}[T] = \mathcal{D}[A_N]$ and let $H_+ \subset H \subset H_-$ be the corresponding rigged Hilbert space. Choose the basic boundary value space of A [2], [4], i.e. the triplet $\{\mathcal{H}, G, \Gamma\}$ where $\mathcal{H}$ is a Hilbert space, $G : \mathcal{D}(A^*) \to \mathcal{H}$, $\Gamma \in \mathcal{L}(H_+, \mathcal{H})$ are such linear operators that $Ker \, G = \mathcal{D}(A_N)$, $Ker \, \Gamma = \mathcal{D}[A_F]$, $\mathcal{R}(\Gamma) = \mathcal{H}$ and

$$A_N[u, v] = (A^* u, v) - (Gu, \Gamma v)_{\mathcal{H}}, \ u \in \mathcal{D}(A^*), \ v \in \mathcal{D}[A_N].$$

Since the m-sectorial extension T of A is transversal to A_F, there exists an operator $W \in \mathcal{L}(\mathcal{H})$ which is sectorial and such that (see [5])

$$(2.10) \qquad T[u, v] = A_N[u, v] + (W \Gamma u, \Gamma v)_{\mathcal{H}}, \ u, v \in \mathcal{D}[T].$$

It follows that

$$((\overline{T} - \overline{T}^{\times})u, v) = ((W - W^*)\Gamma u, \Gamma v)_{\mathcal{H}}.$$

Let E be a Hilbert space, J be a fundamental symmetry in E and $L \in \mathcal{L}(E, \mathcal{H})$ such that $W - W^* = 2i W_R^{1/2} LJL^* W_R^{1/2}$, where $W_R = (W + W^*)/2$. It is easy to see from (2.10) that $\mathcal{R}(\Gamma^{\times} W_R^{1/2}) \subseteq \mathcal{R}(\overline{B})$, where $B = T_R^{1/2}$ and $\Gamma^{\times} \in \mathcal{L}(\mathcal{H}, H_-)$ is the adjoint to Γ. Since for $K = \Gamma^{\times} W_R^{1/2} L$ we have $\overline{T} - \overline{T}^{\times} = 2i K J K^{\times}$, the collection $\{T, E, K, J\}$ is an operator node of T in the sense of Definition 2.2.

Let N_{λ} be the defect subspace of A, $\lambda \in \rho(A_F)$, $\gamma(\lambda) = (\Gamma | N_{\lambda})^{-1}$ be the operator field and $M(\lambda) = G\gamma(\lambda)$ be the corresponding Weyl function [11].

Proposition 2.9 *The following equality holds for* $\lambda \in \rho(T) \cap \rho(A_F)$:

$$I - 2i J K^{\times}(\overline{T} - \lambda I)^{-1} K = I - 2i J L^* W_R^{1/2}(W - M(\lambda))^{-1} W_R^{1/2} L.$$

Proof: According the resolvent formula [11] we have

$$(T - \lambda I)^{-1} = (A_F - \lambda I)^{-1} + \gamma(\lambda)(W - M(\lambda))^{-1}$$

$$\gamma^*(\overline{\lambda}), \quad \lambda \in \rho(T) \cap \rho(A_F).$$

The condition $Ker\ \Gamma = \mathcal{D}[A] = \mathcal{D}[A_F]$ and the definition of the operator K yield that

$$K^{\times}(T - \lambda I)^{-1} = L^* W_R^{1/2}(W - M(\lambda))^{-1} \gamma^*(\overline{\lambda}).$$

Since $\gamma(\lambda) \in \mathcal{L}(\mathcal{H}, H_+)$, the above relation can be extended by the continuation on H_- and in view of $\Gamma\gamma(\lambda)e = e$ for every $e \in \mathcal{H}$ and $\lambda \in \rho(T) \cap \rho(A_F)$ we get the equalities $\gamma^*(\overline{\lambda})\Gamma^{\times} = I | \mathcal{H}$ and

$$W(\lambda) = I - 2i J K^{\times}(\overline{T} - \lambda I)^{-1} K$$

$$= I - 2i J L^* W_R^{1/2}(W - M(\lambda))^{-1} W_R^{1/2} L.$$

$\square$

Thus, the characteristic function of the node $\{T, E, K, J\}$ has the same form as in [11], [12], [13].

3 The Model of a Simple m-sectorial Operator

Next using the Clark's and Krein-Langer's method [10], [18] we construct a model of an m-sectorial operator.

Theorem 3.1 *Let E be a Hilbert space, J be a fundamental symmetry of E and let* $W(\lambda)$ *be an operator-function holomorphic on the left half-plane and possessing properties:*

1) $s - \lim_{x \to -\infty} W(x) = I$,

2) *there exists a strong limit* $s - \lim_{x \to -0} W(x)$,

3) *the operator-function*

$$K(\lambda, \mu) = \frac{W^*(\mu) J W(\lambda) - J}{2i(\bar{\mu} - \lambda)}$$

is a positive definite kernel,

4) *the operator-function*

$$Z(\lambda, \mu) = \frac{1}{2i}(W^*(\mu) - I)J + \lambda K(\lambda, \mu)$$

is an α-sectorial kernel.

Then there exist a Hilbert space H, an m-α-sectorial operator T in H and a simple node $\{T, E, K, J\}$ such that its the characteristic function coincides with $W(\lambda)$.

Proof: Since the function $Z(\lambda, \mu)$ is an α-sectorial kernel, the following estimate holds for every choice of λ_1, λ_2 and $e_1, e_2 \in E$ (see [3]):

$$|(Z(\lambda_1, \lambda_2)e_1, e_2)_E|^2 \le \frac{1}{\cos^2\alpha} Re\,(Z(\lambda_1, \lambda_1)e_1, e_1)_E$$

(3.1)

$$Re\,(Z(\lambda_2, \lambda_2)e_2, e_2)_E.$$

The operators $Z(\lambda, \lambda)$ are m-sectorial and the operators $K(\lambda, \lambda)$ are nonnegative selfadjoint for all λ in the left half-plane. The relation

(3.2) $$\qquad Z(x, x) - x K(x, x) = \frac{1}{2i}(W^*(x) - I)J, \ x < 0$$

and condition 2) of the theorem yield that

(3.3) $$\qquad s - \lim_{x \to -\infty} Z(x, x) = s - \lim_{x \to -\infty} x K(x, x) = 0.$$

Consider the linear manifold $\mathcal{L}$ of finite sums $\sum_i K(\lambda_i, z) f_i$ and define the inner product

$$\left(\sum_{i=1}^{n} K(\lambda_i, z) f_i, \sum_{j=1}^{m} K(\mu_j, z) g_j \right)_H = \sum_{i=1}^{n} \sum_{j=1}^{m} (K(\lambda_i, \mu_j) f_i, g_j)_E.$$

Let H be the closure of $\mathcal{L}$ with respect to this inner product. Then $H = H_K$ is the Hilbert space with the reproducing kernel $K(\lambda, \mu)$. Let

$$D_0 = \left\{ \sum_i K(\lambda_i, z)e_i, \ \sum_i e_i = 0 \right\}, \ T_0 \left(\sum_i K(\lambda_i, z)e_i \right)$$

$$= \sum_i \lambda_i K(\lambda_i, z)e_i.$$

Then for $h = \sum_i K(\lambda_i, z)e_i \in D_0$ we have

$$(T_0 h, h)_H = \left(\sum_i \lambda_i K(\lambda_i, z)e_i, \ \sum_i K(\lambda_i, z)e_i \right)_H$$

$$= \sum_{ij} (Z(\lambda_i, \lambda_j)e_i, e_j)_E.$$

It follows that T_0 is an α-sectorial operator.

For every λ the vector $\sum_i (\lambda_i - \lambda)^{-1}(K(\lambda_i, z) - K(\lambda, z))f_i$ belongs to D_0 and

$$(T_0 - \lambda I) \sum_i (\lambda_i - \lambda)^{-1}(K(\lambda_i, z) - K(\lambda_z))f_i$$

$$= \sum_i K(\lambda_i, z)f_i.$$

Let $\varphi = \sum_i K(\lambda_i, z)f_i$, then the vector $\varphi - K(x, z) \sum_i f_i$ belongs to D_0. The equality

$$\left\| K(x, z) \sum_i f_i \right\|_H^2 = \left(K(x, x) \sum_i f_i, \ \sum_i f_i, \right)_E$$

implies that $\lim_{x \to -\infty} (\varphi - K(x, z) \sum_i f_i) = \varphi$. Hence D_0 is dense in H. Let T be the closure of T_0. Taking into account that $(T_0 - \lambda I)D_0$ is dense in H we get that T is an m-α-sectorial operator in H.

Let us show that for all λ and all $f \in E$ the vector $\varphi = K(\lambda, z)f$ belongs to $\mathcal{D}[T]$. Put $\varphi_x = \varphi - K(x, z)f, \ x < 0$. Then $\varphi_x \in D_0, \ \lim_{x \to -\infty} \varphi_x = \varphi$ and

$$T[\varphi_x - \varphi_y] = (Z(x, x)f, f)_E + (Z(y, y)f, f)_E$$
$$- (Z(x, y)f, f)_E - (Z(y, x)f, f)_E.$$

From (3.1), (3.3) it follows that $\lim_{x,y \to -\infty} T[\varphi_x - \varphi_y] = 0$. Consequently [16], $\varphi \in \mathcal{D}[T]$ and moreover

$$\begin{aligned} T[\varphi] &= \lim_{x \to -\infty} T[\varphi_x] = \lim_{x \to -\infty} [(Z(\lambda, \lambda)f, f)_E \\ &+ (Z(x, x)f, f)_E - (Z(x, \lambda)f, f)_E \\ &- (Z(\lambda, x)f, f)_E] = (Z(\lambda, \lambda)f, f)_E. \end{aligned}$$

Hence

(3.4) $\qquad T[K(\lambda, z)f, K(\mu, z)g] = (Z(\lambda, \mu)f, g)_E, \ \ f, g \in E.$

Let $H_+ = \mathcal{D}[T]$ and let $H_+ \subseteq H \subseteq H_-$ is the corresponding rigged Hilbert space. Since the operator T is associated with the form $T[u, v]$ and $K(\lambda_1, z)f - K(\lambda_2, z)f \in D_0 \subseteq \mathcal{D}(T)$ then for all $v \in \mathcal{D}[T]$ we get

$$\begin{aligned} T[K(\lambda_1, z)f - K(\lambda_2, z)f, v] &= (T(K(\lambda_1, z)f - K(\lambda_2, z)f), v)_H \\ &= (\lambda_1 K(\lambda_1, z)f - \lambda_2 K(\lambda_2, z)f, v). \end{aligned}$$

It follows that the expression

$$T[K(\lambda, z)f, v] - \lambda(K(\lambda, z)f, v)_H$$

does not depend on λ and is an antilinear continous functional on H_+ for every $f \in E$. Therefore, there exists an operator $K \in \mathcal{L}(E, H_-)$ such that

$$(Kf, v)_H = T[K(\lambda, z)f, v] - \lambda(K(\lambda, z)f, v)_H, \ \ f \in E, v \in H_+.$$

Thus, by the definition

$$(\overline{T} - \lambda I)^{-1} Kf = K(\lambda, z)f, \ \ \mathrm{Re}\,\lambda < 0.$$

Further we have for $f, g \in E$

$$\begin{aligned} (Kg, (\overline{T} - \lambda I)^{-1} Kf)_H &= (Kg, K(\lambda, z)f,)_H \\ &= \overline{T}[K(\mu, z)g, K(\lambda, z)f] - (K(\lambda, z)f, K(\mu, z)g)_H \\ &= (Z(\mu, \lambda)g, f)_E - \mu(K(\mu, \lambda)g, f)_E \\ &= ((2i)^{-1}(W^*(\lambda) - I)Jg, f)_E. \end{aligned}$$

Hence

$$I - 2i\,JK^{\times}(\overline{T} - \lambda I)^{-1} K = W(\lambda).$$

Let us show that $\mathcal{R}(K) \subseteq \mathcal{R}(\overline{B})$. In view of (2.2) it is sufficiently to prove that $(\overline{T} - \lambda I)^{-1} Kh = K(\lambda, z)h \in \mathcal{R}(T_R^{1/2})$ for every $h \in E$ and fixed λ. Since the operator T is an m-sectorial, a vector $g \in H$ belongs to $\mathcal{R}(T_R^{1/2})$ if and only if the function $((T - xI)^{-1}g, g)_H$ is bounded when $x \to -0$ (see [2]). Let $g = K(\lambda, z)h$. Then

$$((T - xI)^{-1}g, g)_H = (x - \lambda)^{-1} ((K(x, \lambda)h, h)_E - (K(\lambda, \lambda)h, h)_E).$$

By definition we have

$$K(x, \lambda) = \frac{1}{\lambda} Z(x, \lambda) - \frac{1}{2i\lambda}(W^*(\lambda) - I)J.$$

Relations (3.1), (3.2) and condition 2) of Theorem imply that the function $(K(x, \lambda) h, h)_E$ is bounded when $x \to -0$. Thus the vector $g = K(\lambda, z)h$ belongs to $\mathcal{R}(T_R^{1/2})$ and therefore, $Kh \in \mathcal{R}(\overline{B})$.

Our next goal is to prove the equality $2i\, KJK^\times = \overline{T} - \overline{T}^\times$. It is clear from (3.4) that

$$((\overline{T} - \overline{T}^\times)(\overline{T} - \lambda I)^{-1}Kf, (\overline{T} - \mu I)^{-1}Kg)_H$$
$$= (Z(\lambda, \mu)f, g)_E - (Z^*(\mu, \lambda)f, g)_E.$$

Taking into account that $2i\, JK^\times(\overline{T}-\lambda)^{-1}K = I-W(\lambda)$ and $K^*(\mu, \lambda) = K(\lambda, \mu)$ we get

$$2i(KJK^\times(\overline{T} - \lambda I)^{-1}Kf, (\overline{T} - \mu I)^{-1}Kg)_H$$
$$= (Z(\lambda, \mu)f, g)_E - (Z^*(\mu, \lambda)f, g)_E.$$

Hence

$$\left((\overline{T} - \overline{T}^\times)\left(\sum_i K(\lambda_i, z)f_i\right), \left(\sum_i K(\lambda_i, z)f_i\right)\right)_H$$
$$= \left(2i\, KJK^\times\left(\sum_i K(\lambda_i, z)f_i\right), \left(\sum_i K(\lambda_i, z)f_i\right)\right)_H.$$

Since finite sums $\sum_i K(\lambda_i, z)f_i$ are dense in H_+ (and in H by the definition), we get that $2i\, KJK^\times = \overline{T} - \overline{T}^\times$ and that $\{T, E, K, J\}$ forms a simple node which the characteristic function coincides with $W(\lambda)$. □

4 Fractional Linear Transformation of Characteristic Functions

Let $\{T, E, K, J\}$ be a node of an m-α-sectorial operator T. Consider the operator-function

$$V(\lambda) = K^\times(\overline{B}B - \lambda I)^{-1}K.$$

This function is defined and holomorphic on $\rho(B^2) \supseteq \mathbb{C}\backslash R_+$ and belongs to the class S of Nevanlinna functions [15], i.e. $V(x) \geq 0$ for $x < 0$. Moreover, since the function $V(\lambda)$ can be represented in the form

$$V(\lambda) = \Gamma^* B(\overline{B}B - \lambda I)^{-1}\overline{B}\Gamma = \Gamma^* B^2 (B^2 - \lambda I)^{-1}\Gamma,$$

it possesses properties (see the Proof of Theorem 2.5)

1) $s - \lim_{\lambda \to \infty}\{V(\lambda),\ \lambda \in \mathbb{C}\backslash\Theta(\gamma)\} = 0$ for every $\gamma \in (0, \pi/2)$;

2) there exists a strong limit $s - \lim_{\lambda \to 0}\{V(\lambda),\ \lambda \in \mathbb{C}\backslash\Theta(\gamma)\}$ for every $\gamma \in (0, \pi/2)$.

It should be noted that if A is a bounded sectorial operator in E then $(A + i\eta J)^{-1} \in \mathcal{L}(E)$ for all real $\eta \neq 0$. In fact, if $\lim_{n\to\infty}(A + i\eta J)e_n = 0$ and $\|e_n\| = 1$ for all n then $\lim_{n\to\infty}((A + i\eta J)e_n, e_n)_E = 0$ and in view of sectoriality we have $\lim_{n\to\infty} Re\,(Ae_n, e_n)_E = 0$. Since the operator A is bounded it follows $\lim_{n\to\infty} Ae_n = 0$ and therefore $\lim_{n\to\infty} Je_n = \lim_{n\to\infty} e_n = 0$. Contradiction with $\|e_n\| = 1$.

Since the operators $iJ(W(\lambda) - I)$ are bounded sectorial for $Re\,\lambda < 0$ and $W(\lambda) + I = -iJ(iJ(W(\lambda) - I) + 2iJ)$ then $(W(\lambda) + I)^{-1} \in \mathcal{L}(E)$. As in the case of nonselfadjoint bounded operator [8], the function $V(\lambda)$ is a fractional linear transformation of the characteristic function $W(\lambda)$:

$$V(\lambda) = iJ(W(\lambda) + I)^{-1}(W(\lambda) - I),\ \lambda \in \rho(T) \cap \rho(B^2)$$

and

$$W(\lambda) = -(V(\lambda) - iJ)^{-1}(V(\lambda) + iJ).$$

Remark 4.1 In [11] the Krein-Langer's method is used for a solution to the inverse problem of the theory of characteristic functions of almost solvable proper extensions of symmetric operator. The model Hilbert space in [11] has been defined as a Hilbert space H_V with a reproducing kernel

$$V(\lambda, \mu) := \frac{V(\lambda) - V^*(\mu)}{\lambda - \overline{\mu}}.$$

The connection between model spaces H_K in the Theorem 3.1 and H_V is given by the following relation

$$\frac{W^*(\mu)JW(\lambda) - J}{2i(\overline{\mu} - \lambda)} = (I - iV^*(\mu)J)^{-1}\frac{V(\lambda) - V^*(\mu)}{\lambda - \overline{\mu}}(I + iJV(\lambda))^{-1}.$$

The existence of limit values $V(0)$ and $V(-\infty)$ for fractional linear transformation of Derkach-Malamud's and Tsekanovskii's characteristic functions of a proper m-sectorial extension of a nonnegative Hermitian operator was proved in [12], [13], [22], therefore the next Theorem 4.2 is known. We give its proof for the sake of completeness.

Theorem 4.2 *Let E be a Hilbert space, J be a fundamental symmetry in E and let $V(\lambda) \in \mathcal{L}(E)$ belongs to the class S of Nevanlinna functions. If $V(\lambda)$ possesses properties*

1) *$s - \lim_{x \to -\infty} V(x) = 0$,*

2) *there exists a strong limit $s - \lim_{x \to -0} V(x)$ then the operator-function*

$$W(\lambda) = -(V(\lambda) - iJ)^{-1}(V(\lambda) + iJ), \ Re\,\lambda < 0$$

is the characteristic function of an operator node of some m-sectorial operator.

Proof: Since $V(\lambda) \in S$, the following integral representation holds [15]:

$$V(\lambda) = A + \int_0^\infty \frac{d\Omega(t)}{t - \lambda},$$

where $\Omega(t) \in \mathcal{L}(E)$ is a selfadjoint nondecreasing operator-function. In view of condition 1) of Theorem we have $A = s - \lim_{x \to -\infty} V(x) = 0$ and from condition 2) it follows that the integral

$$\int_0^\infty \frac{d\Omega(t)e}{t} := V(0)e$$

converges for every $e \in E$.

Let

$$\Sigma(t) = \int_0^t \frac{d\Omega(\tau)}{\tau}, \ t \geq 0.$$

Since $\Sigma(t)$ is a nondecreasing operator-function, according to M.A. Naimark's theorem there exist a Hilbert space H, an orthogonal resolution of the identity $E(t)$ in H and an operator $\Gamma \in \mathcal{L}(E, H)$ such that $\Sigma(t) = \Gamma^* E(t)\Gamma$.

Put

$$B = \int_0^\infty \sqrt{t}\,dE(t), \ G = \Gamma J\Gamma^*,$$

$$T[u, v] = ((I + iG)Bu, Bv), \ u, v \in \mathcal{D}(B).$$

Observe that

$$V(\lambda) = \Gamma^* B^2 (B^2 - \lambda I)^{-1} \Gamma$$

and that the sesquilinear form $T[u, v]$ is a closed sectorial with the semiangle $\alpha = \arctan \|G\| = \arctan \|V^{1/2}(0) J V^{1/2}(0)\|$ (see also [13]).

Let $H_+ = \mathcal{D}(B)$ is a Hilbert space with the graph norm, $H_+ \subseteq H \subseteq H_-$ is the corresponding rigged Hilbert space and let $\overline{B} \in \mathcal{L}(H, H_-)$ is the continuation of B. If T is the m-sectorial operator associated with the form $T[u, v]$ then $\overline{T} = \overline{B}(I + iG)B$. Put $K = \overline{B}\Gamma$ then $\{T, E, K, J\}$ forms an operator node of T. If $W(\lambda)$ is the characteristic function of this node then for its fractional linear transformaion we have

$$i J (W(\lambda) + I)^{-1}(W(\lambda) - I) = \Gamma^* B^2 (B^2 - \lambda I)^{-1} \Gamma$$
$$= V(\lambda), \quad \lambda \in \rho(T) \cap \rho(B^2).$$

Thus, $W(\lambda) = -(V(\lambda) - i J)^{-1}(V(\lambda) + i J), \ Re \, \lambda < 0.$ $\qquad\square$

Another model for an m-sectorial operator can be constructed by the following way (see also [11] for proper extensions of a Hermitian operator). Consider the Hilbert space $\mathcal{H} = \mathcal{L}_2(E, d\Omega)$ (see [7]) of all functions $f(t), t \in \mathbb{R}_+$ with values in E and such that

$$\|f(t)\|_{\mathcal{H}}^2 := \int_0^\infty (d\Omega(t) f(t), f(t))_E < \infty.$$

Put

$$Bf(t) = \sqrt{t} f(t), \ \mathcal{D}(B) = \left\{ f(t) \in \mathcal{H} : \int_0^\infty (t d\Omega(t) f(t), f(t))_E < \infty \right\}$$

and put $\mathcal{H}_+ = \mathcal{D}(B)$ with the corresponding graph norm. The operator B is nonnegative selfadjoint operator in $\mathcal{H}$ [7]. Then

$$\mathcal{H}_- = \left\{ f(t) : \int_0^\infty \frac{(d\Omega(t) f(t), f(t))_E}{t + 1} < \infty \right\}.$$

Define the operator $F \in \mathcal{L}(\mathcal{H}_+, E)$ by the equality

$$Ff = \int_0^\infty d\Omega(t) f(t)$$

and consider the sesquilinear form on $\mathcal{H}_+$:

$$S[f_1, f_2] = (Bf_1, Bf_2)_{\mathcal{H}} + i(J Ff_1, Ff_2)_E$$

Since for all $f = f(t) \in \mathcal{H}_+$ and all $e \in E$ we have

$$|(Ff, e)|_E^2 \leq \int_0^\infty (t d\Omega(t) f(t), f(t))_E \int_0^\infty \frac{d(\Omega(t)e, e)_E}{t},$$

the form $S[f_1, f_2]$ is a sectorial and a closed. Let S be the associated m-sectorial operator. It is clear that $\overline{S}f = \overline{B}Bf + iF^\times JFf$ where $F^\times e = e(t) := e$, $e \in E$ and thus

$$\mathcal{D}(S) = \left\{ \sqrt{t}f(t) \in \mathcal{L}_2(E, d\Omega) : tf(t) + iJ \int_0^\infty d\Omega(t)f(t) \in \mathcal{L}_2(E, d\Omega) \right\},$$

$$Sf(t) = tf(t) + iJ \int_0^\infty d\Omega(t)f(t), \quad f(t) \in \mathcal{D}(S).$$

If $K_0 = F^\times$, then $\{S, E, K_0, J\}$ is an operator node of S.

Put

$$\mathcal{H}^{(0)} = cls \left\{ \frac{e}{t - \lambda}, \ e \in E \right\}.$$

Then in view of (2.9) the subspace $\mathcal{H}^{(0)} \subseteq \mathcal{H}$ reduces S, $S_0 = S|\mathcal{H}^{(0)}$ is the m-sectorial operator and the operator node $\{S_0, E, K_0, J\}$ is the simple part of the node $\{S, E, K_0, J\}$. Thus, the following statement holds:

Theorem 4.3 *Let $\{T, E, K, J\}$ be a node of an m-sectorial operator T. Then there exist a Hilbert space E, a fundamental symmetry J in E and a nondecreasing operator-function $\Omega(t) \in \mathcal{L}(E)$, $t \in \mathbb{R}_+$ with the property*

$$\int_0^\infty \frac{(d\Omega(t)e, e)}{t} < \infty, \ e \in E$$

such that the simple part of the node $\{T, E, K, J\}$ is unitary equivalent to the node $\{S_0, E, K_0, J\}$ where the space $\mathcal{H}^{(0)}$ and the operators S_0, K_0 are defined above.

References

[1] Y.M. Arlinskii, Characteristic Functions of Operators of the Class $C(\alpha)$, *Izv. Vuzov. Matematika*. no. 2 (1991), 13–21. (Russian)

[2] Y.M. Arlinskii, Extremal Extensions of Sectorial Linear Relations, *Mat. Studii*. **7**, no. 1 (1997), 81–96.

[3] Y.M. Arlinskii, On Functions Connected with Sectorial Operators and their Extensions, *Int. Equat. Oper. Theory* **33**, no. 2 (1999), 125–152.

[4] Y.M. Arlinskii, Positive Bounary Value Spaces and Sectorial Extensions of Non-negative Symmetric Operator, *Ukrainian Math. Journ.* **40**, no. 1 (1988), 8–15. (Russian)

[5] Y.M. Arlinskii, Maximal Sectorial Extensions and Associated with them Closed Forms, *Ukrain. Math. Journ.* **48**, no. 6 (1996), 723–738.

[6] Y.M. Arlinskii and E.R. Tsekanovskii, Regular Biextensions of Unbounded Operators, Donetsk University, *Deposited in VINITI, N 2876-79* (1979), 73. (Russian)

[7] Yu.M. Berezanskii, Expansions in eigenfunction of selfadjoint operators, *Amer. Math. Soc. Providence* 1968.

[8] M.S. Brodskii, Triangular and Jordan representations of linear operators, *Nauka, Moscow* 1969. (Russian)

[9]   M.S. Brodskii and M.S. Livsic, Spectral Analysys of Nonself-Adjoint Operators and Intermediate Systems, *Uspekhi Mat. Nauk.* **13**, no. 1 (1958), 3–85. (Russian)

[10]   D.N. Clark, On Model for Non-Contractions, *Acta Sci. Math.* **36** (1974), 5–16.

[11]   V.A. Derkach and M.M. Malamud, Characteristic Functions of Almost Solvable Extensions of Hermitian Operators, *Ukrainian Math. Journ.* **44**, no. 4 (1992), 435–459. (Russian)

[12]   V.A. Derkach and M.M. Malamud, Nonselfadjoint Extensions of an Hermitian Operator and their Characteristic Functions, *J. Math. Sci.* **97**, no. 5 (1999), 4420–4455.

[13]   V.A. Derkach, M.M. Malamud and E.R. Tsekanovskii, Sectorial Extensions of Positive Operator and a Characteristic Function, *Ukrainian Mat. Journ.* **41**, no. 2 (1989), 151–158. (Russian)

[14]   P.A. Fillmore and J.P. Williams, On operator ranges, *Advances in Math.* **7** (1971), 254–281.

[15]   I.S. Kac and M.G. Krein, *R-functions – analityc functions mapping the upper half-plane into inself*, Suplement I to the Russian edition of F.V. Atkinson: Discrete and continuous boundary problems. Mir, Moscow 1968.

[16]   T. Kato, *Perturbation theory for linear operators*, Springer-Verlag, 1966.

[17]   A. Kuzhel, *Characteristic functions and models of nonself-adjoint operators*, Kluwer Academic Publishers. Dodrecht/Boston/London, 1996.

[18]   M.G. Krein and H. Langer, Uber die Q-Function eines Π-Hermiteschen Operators in Raum Π_κ, *Acta Sci. Math. Szeged* **34** (1973), 191–230.

[19]   M.S. Livsic, On a Spectral Decomposition of Linear Nonself-Adjoint Operators, *Mat. Sbornik.* **34** (1954), 144–199. (Russian)

[20]   M.S. Livsic and A.A. Yantsevich, *The theory of operator colligations in Hilbert spaces*, Kharkov University, Kharkov, 1971. (Russian)

[21]   M.M. Malamud, On some Classes of Extensions of Hermitian Operators with Gaps, *Ukrainian Math. Journ.* **44**, no. 2 (1992), 215–234. (Russian)

[22]   E.R. Tsekanovskii, Characteristic Function and Sectorial Boundary Problems, *Trudy Inst. Mat. (Novosibirsk) Issled. Geom. Mat. Anal.* **7** (1987), 180–194. (Russian)

[23]   E.R. Tsekanovskii, Triangular Models of Unbounded Accretive Operators and the Regular Factorization of their Characteristic Operator Functions, *Dokl. Akad. Nauk USSR* **297**, no. 3 (1987), 552–556. (Russian)

[24]   E.R. Tsekanovskii and Y.L. Shmulyan, The Theory of Biextensions of Operators in Hilbert Spaces, *Uspekhi Mat. Nauk.* **32**, no. 5 (1977), 69–124. (Russian)

Yury Arlinskii
EastUkrainian State University
Department of Mathematics
Kvartal Molodyozhny 20-A
348034 Lugansk
Ukraine

1991 Mathematics Subject Classification. Primary: 47A45; Secondary: 47B44.

Operator Theory:
Advances and Applications, Vol. 124
© 2001 Birkhäuser Verlag Basel/Switzerland

Similarity between Kreĭn Space Bicontractions and Hilbert Space Contractions

T.Ya. Azizov, A.I. Barsukov, A. Dijksma and P. Jonas

Dedicated to Israel Gohberg on the occasion of his 70-th birthday

Criteria for the similarity of a bicontraction T on a Kreĭn space $\mathcal{K}$ to a Hilbert space contraction are given in terms of power boundedness and when $\dim \mathcal{K}^- < \infty$, in terms of the location of the spectrum and a uniform bound on the growth of the resolvent of T.

1 Introduction

Two operators T and T_0 on a Hilbert space $\mathcal{H}$ are called *similar* if there is a bounded and boundedly invertible operator S on $\mathcal{H}$ such that $T = S^{-1}T_0S$. Similar operators have the same spectrum and the same subdivisons of the spectrum such as point, continuous and approximate spectrum; see, for example, [H2, Problem 75]. In this paper we investigate when T is similar to a contraction T_0. If it is and $T = S^{-1}T_0S$, then $T^n = S^{-1}T_0^nS$ and hence $\|T^n\| \le \|S^{-1}\|\|T_0\|^n\|S\| \le \|S^{-1}\|\|S\|$, which proves that T is *power bounded*: there is a constant M such that for all n

$$\|T^n\| \le M.$$

S.R. Foguel in [F] gave an example to prove that the converse is not true. By the von Neumann inequality for contractions: for all polynomials $p(z)$

$$\|p(T_0)\| \le \sup_{|z|=1} |p(z)|,$$

if $T = S^{-1}T_0S$, then T is *polynomially bounded*: there is a constant M such that for all polynomials $p(z)$

$$\|p(T)\| \le M \sup_{|z|=1} |p(z)|.$$

A. Lebow in [Le] showed that the operator in Foguel's example is not polynomially bounded and subsequently P.R. Halmos in [H1, Problem 6] discussed the question: Is every polynomially bounded operator similar to a contraction? In 1984 V.I. Paulsen proved that T is similar to a contraction if and only if T is *completely polynomially bounded*: there is a constant M such that for every square matrix valued function $P(z)$ whose entries are polynomials

$$\|P(T)\| \le M \sup_{|z|=1} \|P(z)\|_e,$$

where if $P(z)$ is $k \times k$ matrix valued

$$\|P(T)\| = \sup_{0 \neq h \in \mathcal{H}^k} \frac{\|P(T)h\|}{\|h\|}, \quad \|P(z)\|_e = \sup_{0 \neq x \in \mathbb{C}^k} \frac{\|P(z)x\|}{\|x\|},$$

and $\mathbb{C}^k$ is equipped with the Euclidean norm; see [Pa1], [Pa2], and for another proof also [Pi1]. This criterion can be applied to give another proof of the theorem in [SF, Chapter I, Section 11, and Chapter II, Section 8] that operators which admit a unitary ρ-dilation, $\rho > 0$, are similar to a contraction. Indeed, let T be a bounded operator on a Hilbert space $\mathcal{H}$ and for $\rho > 0$ let U be a unitary operator in a Hilbert space $\mathcal{G}$ which is a ρ-dilation of T. This means that $\mathcal{H}$ is a closed subspace of $\mathcal{G}$ and that

$$T^n = \rho \, P_{\mathcal{H}} U^n |_{\mathcal{H}}, \quad n = 1, 2, \ldots.$$

Then for any k and any $k \times k$ matrix valued function $P(z)$ whose entries are polynomials in z, we have

$$P(T) = \rho \, P_{\mathcal{H}^k} P(U)|_{\mathcal{H}^k} + (1 - \rho) \, P(0).$$

As U is unitary, U is completely polynomially bounded with bound 1:

$$\|P(U)\| \leq \sup_{|z|=1} \|P(z)\|_e,$$

and it follows that

$$\|P(T)\| \leq (\rho + |\rho - 1|) \sup_{|z| \leq 1} \|P(z)\|_e = (\rho + |\rho - 1|) \sup_{|z|=1} \|P(z)\|_e.$$

(The equality follows from the maximum modulus principle applied to the polynomial $[P(z)x, y], x, y \in \mathcal{H}^k$.) Hence T is completely polynomially bounded, and so by Paulsen's criterion, it is similar to a contraction. For example, all operators T whose numerical radius

$$w(T) := \sup_{0 \neq h \in \mathcal{H}} \frac{[Th, h]}{\|h\|^2}$$

satisfies the inequality $w(T) \leq 1$ are similar to contractions: These operators are precisely those which admit a unitary ρ-dilation with $\rho = 2$; see [SF, Chapter 1, Proposition 11.2].

Finally, G. Pisier in [Pi2] showed that polynomially boundedness does not imply completely polynomially boundedness, hence the answer to Halmos' question is negative.

The similarity problem was raised originally by B. Sz.-Nagy in [SN1] and [SN2]. In the first paper the following theorem is proved (see also [DK, Chapter 1, Section 6]).

Theorem 1.1 *Assume T is a bounded and boundedly invertible operator on a Hilbert space $\mathcal{H}$ and the powers of T and T^{-1} are uniformly bounded by c, then there is a bounded and boundedly invertible operator S on $\mathcal{H}$ with $\max\{\|S\|,$ $\|S^{-1}\|\} \leq c$ such that STS^{-1} is unitary on $\mathcal{H}$.*

In the second paper it is proved that if T is compact then T is similar to a contraction if and only if it is power bounded. In this paper instead of the class of compact operators we consider the class of bicontractions on a Kreĭn space. One of our main results, Theorem 2.10, is that if T is a bicontraction on a Kreĭn space, then T is similar to a Hilbert space contraction if and only if T is power bounded. The proof of the theorem is based on Lemma 2.1, which gives necessary and sufficient conditions for an operator T to be similar to a contraction in terms of the existence of a dilation of T with certain properties. To demonstrate the usefulness of the lemma we reprove the theorem that if T is the sum of a compact operator and an operator whose spectrum lies in the open unit disc, then T is similar to a contraction if and only if it is power bounded; see Theorem 2.2 and Corollary 2.3. If T is a contraction on a Pontryagin space (here a Kreĭn space with finite negative index, so that automatically T is a bicontraction) we formulate this power bounded-ness condition in terms of the location of the spectrum of T: $\sigma(T) \subset \overline{\mathbb{D}}$ and the uniform bound on the growth of the resolvent of T:

$$\sup\{(|\mu| - 1)\|(T - \mu)^{-1}\| : |\mu| > 1\} < \infty;$$

see Theorem 3.7.

Our results can be carried over to maximal dissipative operators on a Pontryagin space either by the same method or by using the Cayley transform. Theorem 3.8 gives criteria for such an operator to be similar to a maximal dissipative operator on a Hilbert space. E.W. Packel in [P] gave an example of a closed and densely defined operator A on a Hilbert space with property (d) in Theorem 3.8 which is not similar to a Hilbert space dissipative operator. The Kreĭn space version of Theorem 3.8 is more complicated and will be published elsewhere.

We assume that the reader is familiar with the geometry of indefinite inner product spaces and the corresponding operator theory; see the books [AI] and [IKL], and also [A] and [DR]. We briefly recall some of the notions, also in order to make clear the notations used in this paper. A Kreĭn space $(\mathcal{K}, [\cdot, \cdot])$ is a complex linear space with an inner product $[\cdot, \cdot]$ such that $\mathcal{K}$ admits a fundamental decomposition $\mathcal{K} = \mathcal{K}^+ [+] \mathcal{K}^-$ in which $(\mathcal{K}^{\pm}, \pm[\cdot, \cdot])$ are Hilbert spaces and $\mathcal{K}^+ \perp \mathcal{K}^-$. The operator $J = P^+ - P^-$, where $P^{\pm}$ is the orthogonal projection onto $\mathcal{K}^{\pm}$, is called the fundamental symmetry corresponding to this fundamental decomposition. The positive/negative index $\dim \mathcal{K}^{\pm}$ is independent of such decompositions. The Hilbert space J-inner products $(\cdot, \cdot) := [J\cdot, \cdot]$ which depend on J, give rise to equivalent norms, and by definition the Kreĭn space $\mathcal{K}$ is endowed with this norm topology. Whenever we use a norm on $\mathcal{K}$ we tacitly assume that this is one of these norms and denote the corresponding fundamental symmetry by J. The adjoint of an operator A on a Kreĭn or Hilbert space will be denoted by A^*; the adjoint of A with respect to a J-inner product will be denoted by $A^\times$.

By definition a contraction T on a Kreĭn space $\mathcal{K}$ is a bounded operator T on $\mathcal{K}$ with the property that $[Tx, Tx] \leq [x, x]$ for all $x \in \mathcal{K}$. A bicontraction T on $\mathcal{K}$ is a contraction whose adjoint T^* is a contraction also. In this paper a dissipative operator A on $\mathcal{K}$ is a closed densely defined operator A for which $\mathrm{Im}\,[Ax, x] \geq 0$, $x \in \mathrm{dom}\,A$. In the paper we use dilations of operators as considered in for example [GGK] and [N]; for a discussion of dilations in a Kreĭn space setting we refer to [DR].

In the sequel $\mathbb{R}$ and $\mathbb{C}$ are the sets of real and complex numbers, $\mathbb{D}$ is the open unit disk, $\mathbb{T}$ the unit circle, $\mathbb{E}$ the complement of $\mathbb{D} \cup \mathbb{T}$ in $\mathbb{C}$, $\mathbb{C}^\pm$ the open upper/lower half plane in $\mathbb{C}$. The closure of a set S is denoted by $\overline{S}$. Finally, $\varrho(T), \sigma(T), \sigma_c(T)$, and $\sigma_p(T)$ stand for the resolvent set, the spectrum, the continuous spectrum and the point spectrum of an operator T.

2 Power Bounded Operators

In this section we show that if T is a Kreĭn space bicontraction then T is similar to a Hilbert space contraction if and only if it is power bounded, and that in this case power boundedness can also be expressed in terms of certain T-invariant subspaces; see Theorem 2.10. We shall use the following lemma.

Lemma 2.1 *Let T be a bounded operator on a Hilbert space $\mathcal{H}$. The following statements are equivalent:*

(i) *T is similar to a contraction T_0 on $\mathcal{H}$: $T = S^{-1}T_0S$.*

(ii) *There is a dilation $\widehat{W}$ of T defined on a Hilbert space $\mathcal{K}$ which is similar to a unitary operator U on $\mathcal{K}$: $\widehat{W} = \widehat{S}^{-1}U\widehat{S}$.*

(iii) *There is a dilation W of T defined on a Hilbert space which is boundedly invertible and such that W and W^{-1} are power bounded.*

If (i)–(iii) are valid the bounded and boundedly invertible operators S and $\widehat{S}$ can be chosen so that

$$\|S\| \leq \|\widehat{S}\|, \quad \max\{\|\widehat{S}\|, \|\widehat{S}^{-1}\|\} \leq \sup_{n \in \mathbb{Z}} \|W^n\|.$$

Proof: (i) $\Rightarrow$ (ii). If $T = S^{-1}T_0S$ and T_0 is a contraction in $\mathcal{H}$, let U be a unitary dilation of T_0: so U is of the form

$$U = \begin{bmatrix} T_0 & 0 & * \\ * & * & * \\ 0 & 0 & * \end{bmatrix} : \begin{pmatrix} \mathcal{H} \\ \mathcal{H}_1 \\ \mathcal{H}_2 \end{pmatrix} \to \begin{pmatrix} \mathcal{H} \\ \mathcal{H}_1 \\ \mathcal{H}_2 \end{pmatrix},$$

where $\mathcal{H}_1$ and $\mathcal{H}_2$ are Hilbert spaces and every $*$ stands for some bounded operator which we do not further specify. Set $\widehat{S} = \mathrm{diag}\,(S, I_{\mathcal{H}_1}, I_{\mathcal{H}_2})$. Then $\widehat{W} = \widehat{S}^{-1}U\widehat{S}$

is a dilation of T satisfying (ii) with $\mathcal{K} = \mathcal{H} \oplus \mathcal{H}_1 \oplus \mathcal{H}_2$, and $\|S\| \leq \|\widehat{S}\|$ holds.
(ii) $\Rightarrow$ (i). Assume there is an operator $\widehat{W}$ of the form

$$\widehat{W} = \begin{bmatrix} T & 0 & * \\ * & * & * \\ 0 & 0 & * \end{bmatrix} : \begin{pmatrix} \mathcal{H} \\ \mathcal{H}_1 \\ \mathcal{H}_2 \end{pmatrix} \to \begin{pmatrix} \mathcal{H} \\ \mathcal{H}_1 \\ \mathcal{H}_2 \end{pmatrix},$$

where $\mathcal{H}_1, \mathcal{H}_2$ are Hilbert spaces, such that $\widehat{W} = \widehat{S}^{-1} U \widehat{S}$ for some unitary operator U on $\mathcal{K} = \mathcal{H} \oplus \mathcal{H}_1 \oplus \mathcal{H}_2$. Then

$$\begin{bmatrix} T & 0 \\ * & * \end{bmatrix} = \widehat{S}^{-1} U \widehat{S}|_{\mathcal{H} \oplus \mathcal{H}_1},$$

and if we set $S_1 = \widehat{S}|_{\mathcal{H} \oplus \mathcal{H}_1} : \mathcal{H} \oplus \mathcal{H}_1 \to \mathrm{ran}\, S_1$ then U maps $\mathrm{ran}\, S_1$ into itself and we have

$$\begin{bmatrix} T & 0 \\ * & * \end{bmatrix} = S_1^{-1} U_1 S_1,$$

where $U_1 := U|_{\mathrm{ran}\, S_1}$ is an isometry. Hence $T^* = S_1^* U_1^* (S_1^*)^{-1}|_{\mathcal{H}}$. Let $S_2 = (S_1^*)^{-1}|_{\mathcal{H}} : \mathcal{H} \to \mathrm{ran}\, S_2$, then $S_2 T^* = U_1^* S_2$ implies that $T_2 := U_1^*|_{\mathrm{ran}\, S_2}$ is a contraction from $\mathrm{ran}\, S_2$ into itself and we have $T^* = S_2^{-1} T_2 S_2$. Finally, let $S_2 = U_0 |S_2|$ be the polar decomposition of S_2 where U_0 is unitary and $|S_2|$ acts on $\mathcal{H}$. Then $T^* = |S_2|^{-1} U_0^* T_2 U_0 |S_2|$ and so $T = S^{-1} T_0 S$ where $S = |S_2|^{-1}$ and $T_0 = U_0^* T_2^* U_0$ is a contraction on $\mathcal{H}$.

By Theorem 1.1, (ii) and (iii) are equivalent and the last norm estimate at the end of the lemma holds. $\qquad\square$

As a first application of the lemma we give another proof of a theorem of G.C. Rota in [R].

Theorem 2.2 *Let T be a bounded operator on a Hilbert space $\mathcal{H}$. If $\sigma(T) \subset \mathbb{D}$, then T is similar to a contraction on $\mathcal{H}$.*

Proof: Denote by $\mathcal{H}_+$ and $\mathcal{H}_-$, the spectral subspaces for $I - T^*T$ in $\mathcal{H}$ corresponding to the intervals $(0, \infty)$ and $(-\infty, 0)$, and by P_+ and P_- the orthogonal projections in $\mathcal{H}$ onto these subspaces. Let V_2 on $\ell^2(\mathcal{H}_+)$ be the forward shift:

$$V_2(x_1, x_2, x_3, \ldots) = (0, x_1, x_2, \ldots),$$

and let $V_{21} : \mathcal{H} \to \ell^2(\mathcal{H}_+)$ be defined by

$$V_{21} h = (|I - T^*T|^{1/2} P_+ h, 0, 0, \ldots).$$

Then

$$V = \begin{bmatrix} T & 0 \\ V_{21} & V_2 \end{bmatrix} : \begin{pmatrix} \mathcal{H} \\ \ell^2(\mathcal{H}_+) \end{pmatrix} \to \begin{pmatrix} \mathcal{H} \\ \ell^2(\mathcal{H}_+) \end{pmatrix}$$

is a dilation of T on $\mathcal{G} := \mathcal{H} \oplus \ell^2(\mathcal{H}_+)$. For $f = \begin{pmatrix} h \\ x \end{pmatrix}$, $h \in \mathcal{H}$, $x \in \ell^2(\mathcal{H}_+)$,

$$
\begin{aligned}
\|Vf\|^2 &= \|Th\|^2 + [(I - T^*T)P_+h, h] + \|V_2x\|^2 \\
&= [T^*Th, h] + [(I - T^*T)h, h] - [(I - T^*T)P_-h, h] + \|x\|^2 \\
&\geq \|h\|^2 + \|x\|^2 = \|f\|^2,
\end{aligned}
$$

which shows that V is an expansive operator. We now prove that V is also power bounded. Recall the two formulas for the spectral radius r_T of T:

$$
r_T = \lim_{n \to \infty} \|T^n\|^{1/n} = \max\{|\lambda| \, \lambda \in \sigma(T)\}.
$$

The assumption $\sigma(T) \subset \mathbb{D}$ implies that $r_T < 1$ and hence $\sum_{n=0}^{\infty} \|T^n\| < \infty$. From

$$
V^n = \begin{bmatrix} T^n & 0 \\ V_{21n} & V_2^n \end{bmatrix}, \quad V_{21n} = \sum_{k=0}^{n-1} V_2^k V_{21} T^{n-k-1}, \quad n = 1, 2, \ldots,
$$

we obtain the uniform bound

$$
\begin{aligned}
\|V^n\| &\leq \|T^n\| + \|V_{21n}\| + \|V_2^n\| \\
&\leq \max_{n \in \mathbb{N}} \|T^n\| + \|V_{21}\| \sum_{k=0}^{\infty} \|T^k\| + 1.
\end{aligned}
$$

Thus for every $f \in \mathcal{G}$, the sequence $\|V^n f\|$ is nondecreasing and bounded, hence the limit $\lim_{n \to \infty} \|V^n f\|$ exists and by the polarisation formula so does the limit

$$
\langle f, g \rangle := \lim_{n \to \infty} [V^n f, V^n g], \quad f, g \in \mathcal{G}.
$$

The form $\langle f, g \rangle$ defines an inner product on $\mathcal{G}$ whose norm is equivalent to the original norm on $\mathcal{G}$. Hence $\mathcal{G}$ with this new inner product is again a Hilbert space, and in this space V is an isometry. Let $R := \sqrt{G}$, where G is the Gram operator such that $\langle f, g \rangle = (Gf, g)$ for all $f, g \in \mathcal{G}$. Then $V_0 := RVR^{-1}$ is an isometric operator on $\mathcal{G}$ and so

$$
U := \begin{bmatrix} V_0 & P_{\ker V_0^*} \\ 0 & V_0^* \end{bmatrix}
$$

is a unitary operator on $\mathcal{G} \oplus \mathcal{G}$. It is now easy to verify that with $\widehat{S} = \text{diag}\,(R, R^{-1})$ the operator $\widehat{W} = \widehat{S}^{-1} U \widehat{S}$ satisfies (ii) of Lemma 2.1. Thus the theorem follows from this lemma. $\qquad\square$

Corollary 2.3 *Let T_1 be a bounded operator on a Hilbert space $\mathcal{H}$ with $\sigma(T_1) \subset \mathbb{D}$, and let K be a compact operator on $\mathcal{H}$. Then $T = T_1 + K$ is similar to a contraction on $\mathcal{H}$ if and only if T is power bounded.*

We briefly sketch the proof of the "if" part, as the argument is the same as in [SN2] and [R], where the case T is compact, that is, $T_1 = 0$, is considered. The power boundedness of T implies that $\sigma(T) \subset \overline{\mathbb{D}}$. By the compactness of K the accumulation points of $\sigma(T)$ lie in $\sigma(T_1) \subset \mathbb{D}$, hence $\Lambda := \sigma(T) \cap \mathbb{T}$ is a finite set and $\overline{\sigma(T)\backslash\Lambda} = \sigma(T)\backslash\Lambda \subset \mathbb{D}$. If we decompose T according to its spectrum on $\mathbb{D}$ and its spectrum in $\mathbb{T}$, we find that T is the sum of an operator satisfying the conditions of Theorem 2.2 and an operator T' on a finite dimensional space. The operator T' has a matrix representation as a direct sum of Jordan blocks with eigenvalues of modulus 1. Since T' is power bounded, each such Jordan block is power bounded and hence a scalar. It follows that T' is diagonizable, and T is similar to a contraction.

Corollary 2.4 *Let T be a bounded operator on a Hilbert space $\mathcal{H}$ and let r_T be its spectral radius. For every $\varepsilon > 0$ there exists a bounded operator T_ε on $\mathcal{H}$ which is similar to T and has norm $\|T_\varepsilon\| \leq r_T + \varepsilon$.*

Proof: Consider the operator $\frac{1}{r_T+\varepsilon}T$. Its spectrum lies in $\mathbb{D}$, so on account of Theorem 2.2, it is similar to a contraction S_ε, say, on $\mathcal{H}$. Take $T_\varepsilon = (r_T + \varepsilon)S_\varepsilon$. $\qquad\square$

The norm $\|T\|$ and the spectral radius r_T of a bounded operator T only satisfy the inequality $r_T \leq \|T\|$, but are otherwise "independent". The corollary implies that there is an inner product on $\mathcal{H}$ such that the norm of T is close to its spectral radius.

The following theorem is also an application of Lemma 2.1.

Theorem 2.5 *Let T be a power bounded bicontraction in a Kreĭn space $\mathcal{K}$: $\|T^n\| \leq c < \infty$, $n = 1, 2, \ldots$. Then there is a bounded and boundedly invertible operator S with $\max\{\|S\|, \|S^{-1}\|\} \leq \sqrt{4c^2 + 3}$ such that STS^{-1} is a Hilbert space contraction.*

Proof: Let $\mathcal{H} = \ell^2(\mathcal{D})$, the Hilbert space of square summable sequences with entries in $\mathcal{D} := \overline{\operatorname{ran}}(J_\mathcal{K} - T^\times J_\mathcal{K}T)$, where $J_\mathcal{K}$ is a fundamental symmetry on $\mathcal{K}$. Consider the Kreĭn space $\mathcal{G} = \mathcal{K} \oplus \mathcal{H}$ with fundamental symmetry $J_\mathcal{G} = J_\mathcal{K} \oplus I$ and the operator

$$V = \begin{bmatrix} T & 0 \\ V_{21} & V_2 \end{bmatrix}$$

with $V_{21} : \mathcal{K} \to \mathcal{H}$ defined by

$$V_{21}x = ((J_\mathcal{K} - T^\times J_\mathcal{K}T)^{1/2}x, 0, 0, \ldots)$$

and $V_2 : \mathcal{H} \to \mathcal{H}$ defined by

$$V_2(x_1, x_2, x_3, \ldots) = (0, x_1, x_2, \ldots).$$

Then V is an isometric dilation of T, which is minimal in the sense that $\mathcal{G} = \overline{\text{span}}\,\{V^n \mathcal{K}\}_{n=0}^{\infty}$. As T is a bicontraction it follows from for example [DR, Theorem 3.1.3] that V is a bicontraction also. Since

$$
V^n = \begin{bmatrix} T^n & 0 \\ V_{21n} & V_2^n \end{bmatrix}, \quad V_{21n} := \sum_{k=0}^{n-1} V_2^k V_{21} T^{n-k-1}, \quad n = 1, 2, \ldots,
$$

is an isometric operator, for $z = x + y$, $x \in \mathcal{K}$, $y \in \mathcal{H}$, we have

$$
\begin{aligned}
(V^n z, V^n z) &= (T^n x, T^n x) + (V_{21n} x + V_2^n y, V_{21n} x + V_2^n y) \\
&= (T^n x, T^n x) - [T^n x, T^n x] + [T^n x, T^n x] \\
&\quad + (V_{21n} x + V_2^n y, V_{21n} x + V_2^n y) \\
&= ((I - J_{\mathcal{K}}) T^n x, T^n x) + [V^n z, V^n z] \\
&= ((I - J_{\mathcal{K}}) T^n x, T^n x) + [z, z] \\
&\leq (2c^2 + 1)(z, z).
\end{aligned}
$$

Hence the operators V^n are uniformly bounded and

$$
\|V^n\| \leq \sqrt{2c^2 + 1}.
$$

Therefore V^* is a power bounded bicontraction too and

$$
\|V^{*n}\| \leq \sqrt{2c^2 + 1}.
$$

Let $U^* : \mathcal{F} \to \mathcal{F}$ be a minimal isometric dilation of V^* constructed in the same way as V is constructed from T. Here $\mathcal{F}$ is a Kreĭn space with a fundamental decomposition $\mathcal{F} = \mathcal{F}^+[+]\mathcal{F}^-$, where $\mathcal{F}^- = \mathcal{G}^-$ and $\mathcal{G}^+ \subset \mathcal{F}^+$. Then from [DR, Theorem 3.1.5] it follows that U is a unitary dilation of T (with the minimality property $\mathcal{F} = \overline{\text{span}}\,\{U^n \mathcal{K}\}_{n=-\infty}^{\infty}$, but we do not need this here), and from the above we have that U^* is a power bounded operator with

$$
\|U^{*n}\| \leq \sqrt{4c^2 + 3}.
$$

Therefore U is a power bounded unitary operator with the same uniform estimate for the norm of its powers. Since $U^* = U^{-1}$, all negative and positive powers of U are uniformly bounded. Thus the theorem follows from Lemma 2.1. $\qquad\square$

Remark 2.6 Let T be a power bounded bicontractive isometry in a Kreĭn space. Then from the proofs of Theorem 2.5 and Lemma 2.1 it follows that there is a bounded and boundedly invertible operator S such that $S T S^{-1}$ is a Hilbert space isometry. Indeed, in the notation of the proof of Theorem 2.5, $V = T$ and U is an extension of T:

$$
U = \begin{bmatrix} T & * \\ 0 & * \end{bmatrix}
$$

By Theorem 1.1, U is similar to a Hilbert space unitary operator U_1: $U = S_1^{-1} U_1 S_1$. Hence $T = S_1^{-1} U_1 S_1|_{\mathcal{H}}$. Using the polar decomposition for $S_1|_{\mathcal{H}}$ we obtain the result as in the proof of Lemma 2.1.

The following theorem is a special case of a theorem due to R.S. Phillips, (see [Ph] and also [AI, Section 2, Corollary 5.20]).

Theorem 2.7 *For a unitary operator U on a Kreĭn space $\mathcal{K}$ the following statements are equivalent:*

(a) *U is power bounded.*

(b) *There exist a fundamental decomposition $\mathcal{K} = \mathcal{K}_+[+]\mathcal{K}_-$ in which the summands $\mathcal{K}_\pm$ are U-invariant.*

(c) *U is similar to a Hilbert space unitary operator.*

In Theorem 2.10 below we shall show a connection between "similarity" and the existence of uniformly definite T-invariant subspaces. It is a generalization of Theorem 2.7 to bicontractions. On the other hand Theorem 2.7 will used in the proof of Theorem 2.10. In the sequel, by $\mathfrak{M}^+(\mathcal{K})$ and $\mathfrak{M}^-(\mathcal{K})$ we denote the sets of maximal nonnegative and maximal nonpositive subspaces of $\mathcal{K}$, and by $\overset{\circ}{\mathfrak{M}}{}^+(\mathcal{K})$ and $\overset{\circ}{\mathfrak{M}}{}^-(\mathcal{K})$ the sets of maximal uniformly positive and maximal uniformly negative subspaces of $\mathcal{K}$. The proof of the "if" parts of the following lemma are due to S.A. Khoroshavin (private communication).

Lemma 2.8 *Let T be a contraction in the Kreĭn space $\mathcal{K}$ and let V in the Kreĭn space $\mathcal{G}$ be an isometric dilation of T. Then*
sadflakdsfasldfh

(i) *T has an invariant subspace in $\mathfrak{M}^-(\mathcal{K})$ if and only if V has an invariant subspace in $\mathfrak{M}^-(\mathcal{G})$.*

(ii) *If V has an invariant subspace in $\overset{\circ}{\mathfrak{M}}{}^-(\mathcal{G})$, then T has an invariant subspace in $\overset{\circ}{\mathfrak{M}}{}^-(\mathcal{K})$, but the converse does not hold in general.*

Proof: Assume $\mathcal{L}_{\mathcal{G}}^- \in \mathfrak{M}^-(\mathcal{G})$ is V-invariant. Then there is a contraction K from the Hilbert space $(\mathcal{G}^-, -[\cdot, \cdot]_{\mathcal{G}})$ to the Hilbert space $(\mathcal{G}^+, [\cdot, \cdot]_{\mathcal{G}})$ such that $\mathcal{L}_{\mathcal{G}}^-$ is the graph of K:
$$\mathcal{L}_{\mathcal{G}}^- = \{x^- + Kx^- \mid x^- \in \mathcal{G}^-\}.$$

Denote by $P_{\mathcal{K}}^{\mathcal{G}}$ the orthogonal projection in $\mathcal{G}$ onto $\mathcal{K}$. Then $P_{\mathcal{K}}^{\mathcal{G}} \mathcal{L}_{\mathcal{G}}^-$ is the graph of $P_{\mathcal{K}}^{\mathcal{G}} K|_{\mathcal{K}^-}$:
$$P_{\mathcal{K}}^{\mathcal{G}} \mathcal{L}_{\mathcal{G}}^- = \{x^- + P_{\mathcal{K}}^{\mathcal{G}} Kx^- \mid x^- \in \mathcal{K}^-\} =: \mathcal{L}_{\mathcal{K}}^-,$$

and $\mathcal{L}_{\mathcal{K}}^-$ is a T-invariant maximal nonpositive subspace of $\mathcal{K}$.

Conversely, let $\mathcal{L}_{\mathcal{K}}^-\in\mathfrak{M}^-(\mathcal{K})$ be a T-invariant subspace. Consider the V-invariant subspace $\mathcal{F}:=\mathcal{L}_{\mathcal{K}}^-\oplus(\mathcal{G}^+\ominus\mathcal{K}^+)$. Since the uniformly positive subspace $\mathcal{G}^+\ominus\mathcal{K}^+$ is V-invariant, the operator $V|_{\mathcal{F}}$ satisfies the conditions of [AADM, Theorem 2.1 and Remark 2.2] and therefore it has an invariant subspace $\mathcal{L}_{\mathcal{G}}^-\in\mathfrak{M}^-(\mathcal{F})\subset\mathfrak{M}^-(\mathcal{G})$ and this subspace is the desired V-invariant subspace.

The proof of the "if" statement in part (ii) is similar to the proof of the "if" statement in part (i): "maximal uniform negative" is equivalent to $\|K\|<1$. That the converse does not hold in general is shown by the following example. Let $\mathcal{H}$ be a Hilbert space and consider the bicontractive operator

$$T=\begin{bmatrix} I & 0 \\ 0 & -2I \end{bmatrix}$$

in the Kreĭn space $\mathcal{K}=\mathcal{H}\,[+]\,\mathcal{H}$ with fundamental symmetry

$$J=\begin{bmatrix} I & 0 \\ 0 & -I \end{bmatrix}.$$

Clearly $\{0\}\,[+]\,\mathcal{H}$ is a T-invariant maximal uniformly negative subspace. Let V in the Kreĭn space $\mathcal{G}$ be its minimal isometric dilation. Then V is bicontractive also. Assume that V has a V-invariant maximal uniformly negative subspace $\mathcal{G}^-$. Then $V\mathcal{G}^-=\mathcal{G}^-$ and hence $V^*\mathcal{G}^-=V^*V\mathcal{G}^-=\mathcal{G}^-$, which implies that the orthogonal complement $\mathcal{G}^+$ of $\mathcal{G}^-$ is also V-invariant. Hence V satisfies condition (2) of Theorem 2.10 below. In the proof of this theorem we only use the first statement of part (ii) of the lemma, so we may apply the theorem and conclude that V is power bounded. But then T is also power bounded, which it is not. Hence in $\mathcal{G}$ there is no V-invariant maximal uniformly negative subspace. $\qquad\square$

If V is an isometry on the Kreĭn space $\mathcal{K}$, then the subspaces $\ker V^{*k}$ are regular and the projections onto these subspaces are given by $Q_k=I-V^kV^{*k}$. With an isometry V on the Kreĭn space $\mathcal{K}$ we associate two subspaces:

$$\mathcal{R}:=\cap_{k=1}^\infty \mathrm{ran}\,V^k,\quad \mathcal{L}:=\mathcal{R}^\perp=\overline{\mathrm{span}}\,\{\ker V^{*k}\}_{k=0}^\infty=\overline{\cup_{k=0}^\infty\ker V^{*k}}.$$

Note that $V\mathcal{R}=\mathcal{R}$ and $V\mathcal{L}\subset\mathcal{L}$. If $\mathcal{R}$ is a regular subspace or, equivalently, $\mathcal{L}$ is a regular subspace, then $\mathcal{K}=\mathcal{R}\,[+]\,\mathcal{L}$ and $V|_{\mathcal{R}}$ is unitary. In the proof of the theorem below we shall use these notations and apply the following lemma (see [DR, Lemma 1.1.9]).

Lemma 2.9 *Let $\mathcal{N}_1,\mathcal{N}_2,\dots$ be regular subspaces of a Kreĭn space $\mathcal{K}$ such that $\mathcal{N}_1\supset\mathcal{N}_2\dots$. Assume that the projections $P_1,P_2,\dots$ of $\mathcal{K}$ onto the subspaces $\mathcal{N}_1,\mathcal{N}_2,\dots$ are uniformly bounded. Then*

$$\mathcal{N}=\cap_{n=1}^\infty\mathcal{N}_n,\quad \mathcal{M}=\overline{\mathrm{span}}\,\{\mathcal{N}_n^\perp\}_{n=1}^\infty$$

are regular subspaces of $\mathcal{K}$ with $\mathcal{N}=\mathcal{M}^\perp$. If P is the projection of $\mathcal{K}$ onto $\mathcal{N}$, then $P_n\to P$ in the strong operator topology as $n\to\infty$.

Theorem 2.10 *Let T be a bicontraction in a Kreĭn space $\mathcal{K}$. Then the following statements are equivalent.*

(1) *T is a power bounded operator.*

(2) *T has a maximal uniformly negative invariant subspace $\mathcal{L}_-$ and a maximal uniformly positive invariant subspace $\mathcal{L}_+$ and $T|_{\mathcal{L}_-}$ is a power bounded operator.*

(3) *T is similar to a Hilbert space contraction.*

Proof: (1)$\Rightarrow$ (2). Let V be a minimal isometric dilation of T. Then (see the Proof of Theorem 2.5) V is a power bounded operator. Therefore V^* is power bounded too. Hence the orthogonal projections $Q_k = I - V^k V^{*k}$ onto ker V^{*k} are uniformly bounded. By Lemma 2.9 with $P_n = I - Q_n$, $\mathcal{L}$ and $\mathcal{R}$ are both regular subspaces and if Q is the projection of $\mathcal{K}$ onto $\mathcal{L}$ then $Q_n \to Q$, as $n \to \infty$. Since the V^k are bicontractions, ker V^{*k}, $k = 1, 2, \ldots$, are uniformly positive subspaces (see [AI, Section 2.4, 4.14]) and then $\mathcal{L}$ is a uniformly positive subspace. Hence $\overset{\circ}{\mathfrak{M}}{}^-(\mathcal{R}) \subset \overset{\circ}{\mathfrak{M}}{}^-(\mathcal{G})$. Consider the operator $V|_{\mathcal{R}}$. It is a power bounded unitary operator in the Kreĭn space $\mathcal{R}$. By Theorem 2.7, there exists a fundamental decomposition

$$(2.1) \qquad\qquad \mathcal{R} = \mathcal{R}_+ [+] \mathcal{R}_-$$

of $\mathcal{R}$ such that $\mathcal{R}_\pm$ is V-invariant. The subspace $\mathcal{R}_- \in \overset{\circ}{\mathfrak{M}}{}^-(\mathcal{G})$ and by Lemma 2.8 the subspace $\mathcal{L}_- := P_{\mathcal{K}}^{\mathcal{G}} \mathcal{R}_- \in \overset{\circ}{\mathfrak{M}}{}^-(\mathcal{K})$ is T-invariant. Power boundedness of T implies the same property for T^* and therefore T^* also has an invariant subspace, say, $\mathcal{M}_- \in \overset{\circ}{\mathfrak{M}}{}^-(\mathcal{K})$. Then $\mathcal{L}_+ := \mathcal{M}_-^\perp \in \overset{\circ}{\mathfrak{M}}{}^+(\mathcal{K})$ is T-invariant. The power boundedness of T implies that of $T|_{\mathcal{L}_-}$.

(2)$\Rightarrow$(3). Let $\mathcal{L}_\pm \in \overset{\circ}{\mathfrak{M}}{}^\pm(\mathcal{K})$ be T-invariant. From a result of M.G. Kreĭn and Yu. L. Shmulyan [KSh] (see also [AI, Section 1, Corollary 8.18]) it follows that $\mathcal{K} = \mathcal{L}_+ \dotplus \mathcal{L}_-$. Being a restriction of T the operator $T|_{\mathcal{L}_-}$ is power bounded and expansive with respect to the Hilbert space norm $(-[x_-, x_-])^{\frac{1}{2}}$ and since also $T\mathcal{L}_- = \mathcal{L}_-$, there exists a Hilbert scalar product $(\cdot, \cdot)_1$ such that $T|\mathcal{L}_-$ is a $(\cdot, \cdot)_1$-unitary operator (see the proof of Theorem 2.2, or invoke Theorem 1.1, which is allowed since T^{-1} exists and is a contraction in the norm $(-[x_-, x_-])^{\frac{1}{2}}$). Note that $T|\mathcal{L}_+$ is a $[\cdot, \cdot]$-contraction. Thus T is a Hilbert space contraction with respect to the norm $\{x, x\}^{\frac{1}{2}} = ([x_+, x_+]_1 + (x_-, x_-)_1)^{\frac{1}{2}}$, $x = x_+ + x_-$, $x_\pm \in \mathcal{L}^\pm$, that is, T as an operator in $\mathcal{K}$ with the Hilbert space norm $\| \cdot \|$ is similar to a contraction.

(3)$\Rightarrow$(1) Every operator which is similar to a Hilbert space contraction is power bounded, in particular, a bicontraction T in a Kreĭn space too. $\qquad\square$

3 Pontryagin Space Contractions

In this section we prove some criteria for the similarity between a Pontryagin space contraction and a Hilbert space contraction. By the same method and sometimes by using the Cayley transform we obtain the corresponding results for dissipative operators.

We begin with a slight variation of a result due to A.S. Markus [M] with almost the same proof. The reflexive case is due to N. Dunford; see [DS, Corollary VII.7.5].

Lemma 3.1 *Let A be a linear operator in a Banach space $\mathcal{B}$ and let $\lambda_0 \in \sigma(A)$. Assume there is a sequence $\{\lambda_n\}$ in $\varrho(A)$ with $\lambda_n \to \lambda_0$ such that*

$$(3.1) \qquad \alpha := \sup_n |\lambda_n - \lambda_0| \, \|(A - \lambda_n I)^{-1}\| < \infty.$$

Then

$$(3.2) \qquad \ker(A - \lambda_0 I) \cap \overline{\operatorname{ran}}(A - \lambda_0 I) = \{0\},$$

and

$$\mathcal{L} := \ker(A - \lambda_0 I) + \overline{\operatorname{ran}}(A - \lambda_0 I)$$

is a closed subspace. If $\mathcal{B}$ is a reflexive space then $\mathcal{L} = \mathcal{B}$.

Proof: Let $z \in \ker(A - \lambda_0 I)$ and $y \in \operatorname{ran}(A - \lambda_0 I)$, $y = (A - \lambda_0 I)x$. Then

$$(\lambda_n - \lambda_0)(A - \lambda_n I)^{-1}(y - z) = (\lambda_n - \lambda_0)x + (\lambda_n - \lambda_0)^2 (A - \lambda_n)^{-1}x + z.$$

Hence, on account of (3.1),

$$\|(\lambda_n - \lambda_0)x + (\lambda_n - \lambda_0)^2 (A - \lambda_n)^{-1}x + z\| \le \alpha \|y - z\|.$$

Let $n \to \infty$, then again by (3.1), we obtain

$$(3.3) \qquad \|z\| \le \alpha \|y - z\|.$$

Since the expression on the left does not depend on y, this inequality is true for all $y \in \overline{\operatorname{ran}}(A - \lambda_0 I)$ and $z \in \ker(A - \lambda_0 I)$. If $y = z$ then $z = 0$ and $y = 0$, that is, (3.2) is valid.

To prove that $\mathcal{L}$ is a closed subspace, assume $y_n \in \overline{\operatorname{ran}}(A - \lambda_0 I)$, $z_n \in \ker(A - \lambda_0 I)$ and $y_n - z_n \to w$, say. Then $\{y_n - z_n\}$ is a fundamental sequence in $\mathcal{L}$. From (3.3) it follows that $\{z_n\}$ and $\{y_n\}$ are fundamental sequences too. Therefore $z_n \to z$ and $y_n \to y$, say, with $z \in \ker(A - \lambda_0 I)$ and $y \in \overline{\operatorname{ran}}(A - \lambda_0 I)$. Then $w = y - z$, that is, $w \in \mathcal{L}$, and $\mathcal{L}$ is a closed subspace.

Assume that $\mathcal{B}$ is a reflexive space. If $\mathcal{L} \ne \mathcal{B}$ then there exists $f \in \mathcal{B}^*$ which is orthogonal to $\mathcal{L}$ and therefore $f \in \ker(A^* - \lambda_0 I) \cap \overline{\operatorname{ran}}(A^* - \lambda_0 I)$. But A^* also satisfies (3.1) and therefore $f = 0$. Hence $\mathcal{L} = \mathcal{B}$. $\square$

Remark 3.2 If $\lambda_0 \in \sigma_p(A)$ satisfies the conditions of Lemma 3.1, then by (3.2), λ_0 is a semi-simple eigenvalue, that is, ker $(A - \lambda_0 I)$ coincides with the root space of A at λ_0.

Corollary 3.3 *Let T be a contraction in a Kreĭn space $\mathcal{K}$ and let $\mu_0 \in \sigma_p(T) \cap \mathbb{T}$. Assume there exists a sequence $\{\mu_n\}$ in $\varrho(T)$ with $\mu_n \to \mu_0$ such that*

$$\sup_n |\mu_n - \mu_0| \, \|(T - \mu_n I)^{-1}\| < \infty.$$

Then ker $(T - \mu_0 I)$ is a regular subspace, that is,

$$\mathcal{K} = \ker (T - \mu_0 I) [+] \ker (T - \mu_0 I)^{\perp},$$

and both subspaces on the right are T-invariant.

Proof: Since T is a contraction and $|\mu_0| = 1$, we have (see, for example, [AI, Section 2.4, Exercise 19]) that ker $(T - \mu_0 I) \subset$ ker $(T^* - \overline{\mu_0} I) = $ ran $(T - \mu_0 I)^{\perp}$. Hence $\overline{\text{ran}}\, (T - \mu_0 I) \subset$ ker $(T - \mu_0 I)^{\perp}$. From Lemma 3.1 it follows that

$$\mathcal{K} = \ker (T - \mu_0 I) + \overline{\text{ran}}\, (T - \mu_0 I).$$

This equality implies $\overline{\text{ran}}\, (T - \mu_0 I) = $ ker $(T - \mu_0 I)^{\perp}$, and therefore ker $(T - \mu_0 I)$ is regular and T-invariant and so is its orthogonal complement. □

Corollary 3.4 *Let A be a dissipative operator in a Krein space and let $\lambda_0 \in \sigma_p(A) \cap \mathbb{R}$. Assume there exists a sequence $\{\lambda_n\}$ in $\varrho(A)$ with $\lambda_n \to \lambda_0$ such that*

$$\sup_n |\lambda_n - \lambda_0| \, \|(A - \lambda_n I)^{-1}\| < \infty.$$

Then ker $(A - \lambda_0 I)$ is a regular subspace, that is,

$$\mathcal{K} = \ker (A - \lambda_0 I) [+] \ker (A - \lambda_0 I)^{\perp}.$$

Proof: By [AI, Section 2.2, Corollary 2.16] we have the inclusion ker $(A - \lambda_0) \subset$ ker $(A^* - \lambda_0)$ and now the proof is the same as the proof of Corollary 3.3. □

If T is a contraction in a Pontryagin space, Corollary 3.3 has a converse. In the proof of this converse we use that T has a unitary dilation U in a Pontryagin space with the same negative index (see, for example, [AI, Section 5.3] and the references given there, and also [DR]) and we apply to U the spectral theory of unitary operators in a Pontryagin space (see [L], and see [KL] for the case of selfadjoint Pontryagin space operators).

Theorem 3.5 *Let T be a contraction in a Pontryagin space and let $\mu_0 \in \mathbb{T}$. Then ker $(T - \mu_0 I)$ is a regular subspace if and only if there is a sequence $\{\mu_n\}$ in $\varrho(T)$ with $\mu_n \to \mu_0$ such that*

(3.4)
$$\sup_n |\mu_n - \mu_0| \, \|(T - \mu_n I)^{-1}\| < \infty$$

(and this sequence can be chosen in $\mathbb{E}$).

Proof: The "if" part follows from Corollary 3.3 if also $\mu_0 \in \sigma_p(T)$; if $\mu_0 \notin \sigma_p(T)$, then $\ker(T - \mu_0 I) = \{0\}$. For the converse let T be a contraction in the Pontryagin space Π and assume that $\ker(T - \mu_0 I)$ is a regular subspace of Π. Then T is a bicontraction in Π and there exist Hilbert spaces $\mathcal{H}_0$ and $\mathcal{H}_1$ and a unitary operator U on the orthogonal sum space $\widetilde{\Pi} = \mathcal{H}_0 [+] \Pi [+] \mathcal{H}_1$ with block decomposition of the form

$$(3.5) \qquad U = \begin{bmatrix} U_0 & 0 & 0 \\ * & T & 0 \\ * & * & U_1 \end{bmatrix} : \begin{pmatrix} \mathcal{H}_0 \\ \Pi \\ \mathcal{H}_1 \end{pmatrix} \rightarrow \begin{pmatrix} \mathcal{H}_0 \\ \Pi \\ \mathcal{H}_1 \end{pmatrix},$$

where $U_0 \in L(\mathcal{H}_0)$ is the adjoint of an isometric shift, $U_1 \in L(\mathcal{H}_1)$ is an isometric shift and the unspecified operators indicated by a $*$ are bounded.

First we assume that $\mu_0 \notin \sigma_p(T)$. Then, since U_0 and U_1 do not have eigenvalues on $\mathbb{T}$, $\mu_0 \notin \sigma_p(U)$. There exists a definitizing polynomial p for U which is nonnegative on $\mathbb{T}$ such that all zeros of p are eigenvalues of U (see [IK, §18]). Then $p(\mu_0) > 0$. If E is the spectral function of U, there exists an open arc Δ of $\mathbb{T}$ containing μ_0 such that either $E(\Delta)\widetilde{\Pi} = \{0\}$ or $E(\Delta)\widetilde{\Pi}$ is a Hilbert space (see [L]). In the first case we have $\mu_0 \in \varrho(U)$. In the second case $U_\Delta := U|_{E(\Delta)\widetilde{\Pi}}$ is a Hilbert space unitary operator and

$$\sup\{(|\mu| - 1)\|(U_\Delta - \mu I)^{-1}\| : \mu \in \mathbb{E}\} < \infty.$$

Hence if $\{\mu_n\} \subset \mathbb{E}$ is a sequence with

$$(3.6) \qquad \mu_n \rightarrow \mu_0 \quad \text{and} \quad \sup_n |\mu_n - \mu_0|(|\mu_n| - 1)^{-1} < \infty$$

we have

$$\sup_n |\mu_n - \mu_0|\|(U_\Delta - \mu_n I)^{-1}\| < \infty.$$

Therefore in both cases, for every sequence $\{\mu_n\} \subset \mathbb{E}\backslash(\sigma(U) \cup \sigma(T))$ satisfying (3.6) we have

$$(3.7) \qquad \sup_n |\mu_n - \mu_0|\|(U - \mu_n I)^{-1}\| < \infty.$$

Since Π and $\widetilde{\Pi}$ are Pontryagin spaces, $\mathbb{E} \cap (\sigma(U) \cup \sigma(T))$ is a finite set and sequences $\{\mu_n\}$ with the property (3.7) exist.

From (3.5) it follows that

$$PU^n|_\Pi = T^n, \qquad PU^{-n}|_\Pi = T^{*n}, \qquad n = 0, 1, 2, \ldots,$$

where P is the orthogonal projection in $\widetilde{\Pi}$ onto Π. Hence, for all μ in a neighbourhood of ∞, we have

$$(3.8) \qquad P(U - \mu I)^{-1}|_\Pi = (T - \mu I)^{-1}.$$

We see by analytic continuation that (3.8) holds for all $\mu \in \mathbb{E} \backslash (\sigma(U) \cup \sigma(T))$. Then every sequence $\{\mu_n\}$ as in (3.7) has the desired property.

Assume now that $\ker(T - \mu_0 I) \neq \{0\}$. By hypothesis it is a regular subspace and therefore Π admits the orthogonal decomposition

$$\Pi = \ker(T - \mu_0 I)\,[+]\,\ker(T - \mu_0 I)^{\perp},$$

and both subspaces on the right are T-invariant. By the preceding case, for the operator $T|_{\ker(T-\mu_0 I)^{\perp}}$ there exists a sequence $\{\mu_n\}$ of the desired kind. It is easy to see that this sequence also satisfies (3.4). $\qquad\square$

Using the Cayley transform we obtain the following corollary to Theorem 3.5.

Corollary 3.6 *Let A be a maximal dissipative operator in a Pontryagin space and let $\lambda_0 \in \mathbb{R} \cap \sigma_1(\mathbb{A})$. Then $\ker(A - \lambda_0 I)$ is regular if and only if there is a sequence $\{\lambda_n\}$ in $\varrho(A)$ with $\lambda_n \to \lambda_0$ such that*

$$\sup_n |\lambda_n - \lambda_0| \|(A - \lambda_n I)^{-1}\| < \infty.$$

Proof: Indeed, the maximality of A implies that $\mathbb{C}^- \subset \varrho(A)$ except for at most finitely many points. Choose a $z \in \varrho(A) \cap \mathbb{C}^-$ and set $T = (A - \bar{z})(A - z)^{-1}$. Then the conditions of the corollary imply that T satisfies the hypotheses of Theorem 3.5 with $\mu_0 = (\lambda_0 - \bar{z})/(\lambda_0 - z)$ and $\mu_n = (\lambda_n - \bar{z})/(\lambda_n - z)$. $\qquad\square$

Corollary 3.6 and Theorem 3.5 do not hold in a Kreĭn space setting. For example, let $\mathcal{H}$ be a Hilbert space and let B be a positive compact operator on $\mathcal{H}$ with $\sigma(B) = \{\beta_k\}_1^{\infty}$, $\beta_1 \geq \beta_2 \geq \cdots \to 0$. Define on $\mathcal{H} \times \mathcal{H}$ the inner product

$$(3.9) \qquad \left[\begin{pmatrix} x_1 \\ x_2 \end{pmatrix}, \begin{pmatrix} y_1 \\ y_2 \end{pmatrix} \right] = \left(\begin{bmatrix} 0 & I \\ I & 0 \end{bmatrix} \begin{pmatrix} x_1 \\ x_2 \end{pmatrix}, \begin{pmatrix} y_1 \\ y_2 \end{pmatrix} \right)_{\mathcal{H} \times \mathcal{H}}$$

and let $\mathcal{K}$ be the Kreĭn space $(\mathcal{H} \times \mathcal{H}, [\cdot, \cdot])$. We consider the operator

$$A = \begin{bmatrix} 0 & I \\ B & 0 \end{bmatrix},$$

which is a positive bounded operator on $\mathcal{K}$. Since $\lambda_0 = 0 \notin \sigma_p(A)$ we have $\mathcal{K} = \overline{\operatorname{ran}} A$, but there are no sequences $\{\lambda_n\} \subset \varrho(A)$ such that $\lambda_n \to \lambda_0 = 0$ and

$$\sup_n |\lambda_n| \|(A - \lambda_n I)^{-1}\| < \infty.$$

Indeed, it follows from the equality

$$(A - \lambda_n I)^{-1} = \begin{bmatrix} \lambda_n (B - \lambda_n^2 I)^{-1} & (B - \lambda_n^2 I)^{-1} \\ B(B - \lambda_n^2 I)^{-1} & \lambda_n (B - \lambda_n^2 I)^{-1} \end{bmatrix}$$

that

$$|\lambda_n| \|(A - \lambda_n I)^{-1}\| \geq |\lambda_n| \|(B - \lambda_n^2 I)^{-1}\|$$

$$= |\lambda_n| \sup_k \frac{1}{|\beta_k - \lambda_n^2|} \geq \frac{1}{|\lambda_n|} \to \infty, \quad n \to \infty.$$

Let $U := (A + iI)(A - iI)^{-1}$ be the Cayley transform of A. Then U is a Kreĭn space unitary operator. It follows that there are no sequences $\{\mu_n\} \subset \varrho(U)$ such that $\mu_n \to \mu_0 = -1$ and

$$\sup_n |\mu_n - \mu_0| \|(U - \mu_n I)^{-1}\| < \infty.$$

Now we can formulate and prove the main results of this section. Recall that if T is a contraction in a Pontryagin space, then $|\lambda| = |\mu| = 1$, $\lambda \neq \mu$, $x \in \ker(T - \lambda I)$, $y \in \ker(T - \mu I)$ implies $[x, y] = 0$ (see, for example, [AI], Section 2.4, Exercise 22]). Hence there are no more than a finite number of eigenvalues of T in $\mathbb{T}$ for which there exists a nonnegative eigenvector.

Theorem 3.7 *Let T be a Pontryagin space contraction and let $\mu_1, \mu_2, \ldots \mu_m$ be the points in $\mathbb{T}$ for which $\ker(T - \mu_j I)$ contains at least one nonpositive nonzero vector, $j = 1, 2, \ldots, m$. The following statements are equivalent:*

(a) $\sigma(T) \subset \overline{\mathbb{D}}$ *and for each $j = 1, 2, \ldots, m$, there is a sequence $\{\mu_{jn}\}$ in $\varrho(T) \cap \mathbb{E}$ with $\mu_{jn} \to \mu_j$ such that*

$$\sup_n |\mu_{jn} - \mu_j|(|\mu_{jn}| - 1)^{-1} < \infty,$$

$$\sup_n (|\mu_{jn}| - 1)\|(T - \mu_{jn}I)^{-1}\| < \infty.$$

(b) $\sigma(T) \subset \overline{\mathbb{D}}$ *and for each $j = 1, 2, \ldots, m$, there is a sequence $\{\mu_{jn}\}$ in $\varrho(T)$ with $\mu_{jn} \to \mu_j$ such that*

$$\sup_n |\mu_{jn} - \mu_j|\|(T - \mu_{jn}I)^{-1}\| < \infty.$$

(c) *T is similar to a Hilbert space contraction.*

(d) $\sigma(T) \subset \overline{\mathbb{D}}$ *and*

$$\sup\{(|\mu| - 1)^k \|(T - \mu I)^{-k}\| : k \in \mathbb{N}, \, \mu \in \mathbb{E}\} < \infty.$$

(e) $\sigma(T) \subset \overline{\mathbb{D}}$ *and*

$$\sup\{(|\mu| - 1)\|(T - \mu I)^{-1}\| : \mu \in \mathbb{E}\} < \infty.$$

Moreover, if T is isometric or unitary in a Pontryagin space, then (a), (b), (d), *and* (e) *hold if and only if T is similar to a isometric or unitary operator in a Hilbert space, respectively.*

Proof: The implication (a) $\Rightarrow$ (b) is evident.

We prove the implication (b) $\Rightarrow$ (c). Assume that T acts in the Pontryagin space Π with negative index κ. It is well known (see, for example, [IKL, Section 11] and [AI, Section 2.2]) that the contraction T has a κ-dimensional nonpositive invariant subspace $\mathcal{L}$, say, such that $\sigma(T|_{\mathcal{L}}) \subset \overline{\mathbb{E}}$. Since $\sigma(T) \subset \overline{\mathbb{D}}$ we have $\sigma(T|_{\mathcal{L}}) \subset \mathbb{T}$, hence $\sigma(T|_{\mathcal{L}}) \subset \{\mu_1, \mu_2, \ldots, \mu_m\}$. On account of Remark 3.2 applied to T, $\mathcal{L} \subset$ span $\{\ker (T - \mu_j I) : j = 1, 2, \ldots, m\} =: \mathcal{K}_1$. From Corollary 3.3 it follows that $\mathcal{K}_1$ is a regular T-invariant subspace with negative index κ. Since $\ker (T - \mu_j I)$, $j = 1, \ldots, m$, is also T^*-invariant, $\mathcal{K}_2 := \mathcal{K}_1^{\perp}$ is a T-invariant Hilbert subspace. By construction $T|_{\mathcal{K}_1}$ is similar to a Hilbert space unitary operator and $T|_{\mathcal{K}_2}$ is a Hilbert space contraction, which is isometric/unitary if T is isometric/unitary. Hence T is similar to a Hilbert space contraction, which again is isometric/unitary if T is isometric/unitary.

The implication (c) $\Rightarrow$ (d) follows from the fact that for a Hilbert space contraction W, say, we have $\sigma(W) \subset \overline{\mathbb{D}}$ and

$$\sup\{(|\mu| - 1)^k \|(W - \mu I)^{-k}\| : k \in \mathbb{N}, \mu \in \mathbb{E}\} = 1.$$

Indeed, if $T = S^{-1} W S$ for some bounded and boundedly invertible operator S, then $(T - \mu I)^{-k} = S^{-1}(W - \mu I)^{-k} S$ and hence the supremum in (d) is less than or equal to $\|S^{-1}\| \|S\|$.

The implications (d) $\Rightarrow$ (e) $\Rightarrow$ (a) are evident. $\qquad\square$

In a similar way one can prove the following result.

Theorem 3.8 *Let A be a maximal dissipative operator in a Pontryagin space and let $\lambda_1, \lambda_2, \ldots, \lambda_m$ be the points in $\mathbb{R}$ for which $\ker (A - \lambda_j I)$ contains at least one nonpositive nonzero vector, $j = 1, 2, \ldots, m$. The following statements are equivalent:*

(a) $\sigma(A) \subset \overline{\mathbb{C}^+}$ *and for each $j = 1, 2, \ldots, m$, there is a sequence $\{\lambda_{jn}\}$ in $\varrho(A) \cap \mathbb{C}^-$ with $\lambda_{jn} \to \lambda_j$ such that*

$$\sup_n |\lambda_{jn} - \lambda_j| |\operatorname{Im} \lambda_{jn}|^{-1} < \infty,$$

$$\sup_n |\operatorname{Im} \lambda_{jn}| \|(A - \lambda_{jn} I)^{-1}\| < \infty.$$

(b) $\sigma(A) \subset \overline{\mathbb{C}^+}$ *and for each $j = 1, 2, \ldots, m$, there is a sequence $\{\lambda_{jn}\}$ in $\varrho(A)$ with $\lambda_{jn} \to \lambda_j$ such that*

$$\sup_n |\lambda_{jn} - \lambda_j| \|(A - \lambda_{jn} I)^{-1}\| < \infty.$$

(c) *A is similar to a maximal dissipative operator on a Hilbert space.*

(d) $\sigma(A) \subset \overline{\mathbb{C}^+}$ *and*

$$\sup\{|\operatorname{Im}\lambda|^k \|(A-\lambda I)^{-k}\| : k \in \mathbb{N}, \lambda \in \mathbb{C}^-\} < \infty.$$

(e) $\sigma(A) \subset \overline{\mathbb{C}^+}$ *and*

$$\sup\{|\operatorname{Im}\lambda| \|(A-\lambda I)^{-1}\| : \lambda \in \mathbb{C}^-\} < \infty.$$

Moreover, if A is also symmetric or selfadjoint in a Pontryagin space, then (a), (b), (d), and (e) hold if and only if A is similar to a symmetric and maximal dissipative or selfadjoint operator in a Hilbert space, respectively.

Remark 3.9 What remains true with respect to conditions (c), (d), and (e) when A is a selfadjoint operator in a Kreĭn space? The answer is as follows: In this case (d) is equivalent to the statement

(c') *A is similar to a selfadjoint operator in a Hilbert space,*

and evidently (d) implies (e), but the converse is not true. Since (d) clearly holds when A is a selfadjoint operator in a Hilbert space, (c) implies (d). We now prove the converse. From $(A-\bar\lambda I)^{-k} = J((A-\lambda I)^{-k})^\times J$ we obtain

$$\|(A-\lambda I)^{-k}\| = \|((A-\lambda I)^{-k})^\times\|$$
$$= \|J((A-\lambda I)^{-k})^\times J\| = \|(A-\bar\lambda I)^{-k}\|$$

and hence (d) implies

$$\sup\{|\operatorname{Im}\lambda|^k \|(A-\lambda I)^{-k}\| : k \in \mathbb{N}, \lambda \in \mathbb{C}\backslash\mathbb{R}\} < \infty.$$

By [HP, Theorems 12.3.1 and 12.3.2], iA is the generator of a uniformly bounded C_0-group $V(t)$ on $\mathbb{R}$. By a generalization to groups of a theorem of B. Sz.-Nagy (see [DK, Chapter I, Theorem 6.2] and also [AI, Section 2.5, Theorem 5.18]), there is a bounded and boundedly invertible operator S such that $V(t) = S^{-1}U(t)S$, where $U(t)$ is a unitary group on a Hilbert space. It follows that there is a selfadjoint operator B on this Hilbert space, such that

$$iAx = S^{-1}\lim_{t\to 0}\frac{U(t)-I}{t}Sx = S^{-1}(iB)Sx,$$

that is, (c') holds.

We give an example to show that the implication (e) $\Rightarrow$ (d) is not true: Let M be a compact operator in the Hilbert space $\mathcal{H}$ for which

$$\sup\{|\operatorname{Im}\lambda|^k\|(M - \lambda I)^{-k}\| : k \in \mathbb{N},\ \lambda \in \mathbb{C}\backslash\mathbb{R}\} = \infty,$$

but

$$\sup\{|\operatorname{Im}\lambda|\|(M - \lambda I)^{-1}\| : \lambda \in \mathbb{C}\backslash\mathbb{R}\} < \infty.$$

Such an operator has been given by A.S. Markus in [M]. The selfadjoint operator

$$A = \begin{bmatrix} M & 0 \\ 0 & M^\times \end{bmatrix}$$

on the Kreĭn space $(\mathcal{H} \times \mathcal{H}, [\cdot,\cdot])$ with the inner product (3.9) clearly satisfies (e), but it does not satisfy (d).

Acknowledgements

The research of T.Ya. Azizov was supported by DFG grant 436 RUS 17/102/96, INTAS project 93-02449-ext and by NWO NB 61-432. The research of A.I. Barsukov was supported by the Russian Foundation for Basic Research RFBR 99–01–00391.

References

[A]   T. Ando, *Linear operators on Krein spaces*, Hokkaido University, Research Institute of Applied Electricity, Division of Applied Mathematics, Sapporo, 1979.

[AADM] R. Arocena, T.Ya. Azizov, A. Dijksma and S.A.M. Marcantognini, On commutant lifting with finite defect, *J. Operator Theory* **35** (1996), 117–132.

[AI]   T.Ya. Azizov and I.S. Iokhvidov, *Foundation of the theory of linear operators in spaces with an indefinite metric*, Nauka, Moscow, 1986 (Russian); English translation: *Linear operators in spaces with an indefinite metric*, Wiley, New York, 1989.

[DK]   Ju.L. Daleckii and M.G. Kreĭn, Stability of solutions of differential equations in Banach space, *Transl. Amer. Math. Soc.* vol. 43, Providence, R.I. 1974.

[DR]   M.A. Dritschel and J. Rovnyak, Extension theorems for contraction operators on Kreĭn spaces, *Operator Theory: Adv. Appl.*, vol. 47 Birkhäuser, Basel, 1990, 221–305.

[DS]   N. Dunford and J.T. Schwartz, *Linear operators, part I: General theory*, Interscience Publishers, Inc., New York, 1957.

[F]   S.R. Foguel, A counterexample to a problem of Sz. -Nagy, *Proc. Amer. Math. Soc.* **15** (1964), 788–790.

[GK]   I.C. Gohberg and M.G. Kreĭn, *Introduction to the theory of linear nonselfad-joint operators*, Nauka, Moscow, 1965 (Russian); English translation: *Transl. Amer. Math. Soc.*, vol. 18, *Amer. Math. Soc.*, Providence, R.I., 1969.

[GGK]   I. Gohberg, S. Goldberg and M.A. Kaashoek, Classes of linear operators, volume 2, *Operator Theory: Adv. Appl.*, vol. 63, Birkhäuser, Basel, 1990.

[H1]   P.R. Halmos, Ten problems in Hilbert space, *Bulletin Amer. Math. Soc.* **76** (1970), 887–933.

[H2]   P.R. Halmos, *A Hilbert space problem book*, Graduate texts in mathematics, vol. 19, second edition, Springer Verlag, Berlin, 1982.

[HP]   E. Hille and R.S. Phillips, Functional Analysis and semi-groups, *Amer. Math. Soc. Coll. Publ.*, vol. XXXI, revised edition, Providence, R.I. 1957.

[IK]   I.S. Iohvidov and M.G. Kreĭn, Spectral theory of operators in spaces with an indefinite metric, II, *Trudy Moskovsk. Matem. Obshestva* **8** (1959), 413–496 (Russian); English translation: *Amer. Math. Soc. Transl.* (2) **34** (1963), 283–373.

[IKL]   I.S. Iohvidov, M.G. Kreĭn and H. Langer, Introduction to the spectral theory of operators in spaces with an indefinite metric, *Mathematical Research*, vol. 9, Akademie-Verlag, Berlin 1982.

[KL]   M.G. Kreĭn and H. Langer, On the spectral function of a selfadjoint operator in a space with an indefinite metric, *Doklady Akad. Nauk SSSR* **1** (1963), 152, 39–42 (Russian); English translation: *Soviet Math. Doklady* **4** (1963), 1236–1239.

[KSh]   M.G. Kreĭn, Yu.L. Shmulyan, J-polar representation of plus-operators, *Mat. Issledovan.* **1** (1966), 172–210. (Russian); English translation: *Amer. Math. Soc. Transl.* (2) (1969), 115–143.

[L]   H. Langer, *Spektraltheorie linearer Operatoren in J-Räumen und einige Anwendungen auf die Schar $L(\lambda) = \lambda^2 + \lambda B + C$*, Habilitationschrift, Technische Universität Dresden, 1965.

[Le]   A. Lebow, A power-bounded operator that is not polynomially bounded, *Michigan Math. J.* **15** (1968), 397–399.

[M]   A.S. Markus, Sufficient conditions for the completeness of the system of root vectors of a linear operator in a Banach space, *Matem. Sb.* **70** (1966), no. 4, 526–561.

[N]   N.K. Nikolskii, *Treatise on the shift operator*, Springer Verlag, Berlin, 1986.

[P]   E.W. Packel, A semigroup analogue of Foguel's counterexample, *Proc. Amer. Math. Soc.* **21** (1969), 240–244.

[Pa1]   V.I. Paulsen, Every completely polynomially bounded operator is similar to a contraction, *J. Functional Analysis* **55** (1984), 1–17.

[Pa2]   V.I. Paulsen, *Completely bounded maps and dilations*, Pitman research notes in mathematics, vol. 146, Longman, 1986.

[Ph]   R.S. Phillips, The extension of dual subspaces invariant under an algebra, *Proc. Internat. Sympos. Linear Spaces, Jerusalem*, 1960, Pergamon Press, 1961, 366–398.

[Pi1]   G. Pisier, *Similarity problems and completely bounded maps*, Lecture notes in mathematics, vol. 1618, Springer-Verlag, Berlin 1995.

[Pi2]   G. Pisier, A polynomially bounded operator on a Hilbert space which is not similar to a contraction, *J. Amer. Math. Soc.* **10** (1997), 351–369.

[R]   G.C. Rota, On models for linear operators, *Comm. Pure Appl. Math.* **13** (1960), 469–472.

[SN1]   B. Sz.-Nagy, On uniformly bounded linear transformations in Hilbert space, *Acta Sci. Math. Szeged* **11** (1946/48), 152–157.

[SN2]   B. Sz.-Nagy, Completely continuous operators with uniformly bounded iterates, *Magyar Tud. Akad. Mat. Kutató Int. Közl* **4** (1959), 89–93.

[SF]   B. Sz.-Nagy and C. Foiaş, *Harmonic analysis of operators on Hilbert space*, North-Holland, Amsterdam, 1970.

T.Ya. Azizov
Department of Mathematics
Voronezh State University
Universitetskaja pl. 1
394693 Voronezh
Russia

A.I. Barsukov
Department of Mathematics
Voronezh State University
Universitetskaja pl. 1
394693 Voronezh
Russia

A. Dijksma
Department of Mathematics
University of Groningen
P.O Box 800
9700 AV Groningen
The Netherlands

P. Jonas
Department of Mathematics
University of Potsdam
P.O. Box 601553
D–14415 Potsdam
Germany

1991 Mathematics Subject Classification. Primary: 47B50; Secondary: 47A15.

Operator Theory:
Advances and Applications, Vol. 124
© 2001 Birkhäuser Verlag Basel/Switzerland

Classification of Cyclic Invariant Subspaces of Jordan Operators

Hari Bercovici and Thomas Smotzer

Dedicated to Israel Gohberg on the occasion of his 70th anniversary

Let T be an operator of class C_0 and $\mathcal{M}$ a cyclic invariant subspace for T. We show that the weakly quasiaffine orbit of $\mathcal{M}$ is determined by the quasisimilarity classes of the restriction $T|\mathcal{M}$ and of the compression $T_{\mathcal{M}^\perp} = (T^*|\mathcal{M}^\perp)^*$.

1 Introduction

Let T be a bounded linear operator on a complex Hilbert space $\mathcal{H}$, and let $\mathcal{M}$ and $\mathcal{N}$ be invariant subspaces for T. We say $\mathcal{M}$ is a quasiaffine transform of $\mathcal{N}$ if there exists a quasiaffinity X (i.e. a bounded one-to-one operator with dense range) commuting with T such that $(X\mathcal{M})^- = \mathcal{N}$. We indicate this relation by $\mathcal{M} \prec \mathcal{N}$ or $\mathcal{N} \succ \mathcal{M}$. The relation '$\prec$' is generally not an equivalence relation. There are two ways to generate a related equivalence relation, one stronger and the other weaker than '$\prec$'. Namely, we say that $\mathcal{M}$ and $\mathcal{N}$ are quasisimilar, and we write $\mathcal{M} \sim \mathcal{N}$, if $\mathcal{M} \prec \mathcal{N}$ and $\mathcal{M} \succ \mathcal{N}$. We say that $\mathcal{M}$ and $\mathcal{N}$ are weakly quasisimilar, and we write $\mathcal{M} \sim_w \mathcal{N}$ if there exist invariant subspaces $\mathcal{P}_0, \mathcal{P}_1, \ldots, \mathcal{P}_n$ for T such that $\mathcal{P}_0 = \mathcal{M}$, $\mathcal{P}_n = \mathcal{N}$, and for each $i = 1, 2, \ldots, n$ we have either $\mathcal{P}_{i-1} \prec \mathcal{P}_i$ or $\mathcal{P}_{i-1} \succ \mathcal{P}_i$. In earlier work [3] we showed that the quasisimilarity class of $\mathcal{M}$ only depends on the Jordan models of $T|\mathcal{M}$ and $T_{\mathcal{M}^\perp} = P_{\mathcal{M}^\perp}T|\mathcal{M}^\perp$ provided that T is a uniform Jordan operator. This is generally not true if T is an operator of class C_0. Examples are provided in [4] when T is nilpotent on a finite dimensional space. It was also noted there in the nilpotent case that the result is true if $T|\mathcal{M}$ is cyclic. In this paper we will prove an analogous result for arbitrary Jordan operators T. More precisely, if $\mathcal{M}$ and $\mathcal{N}$ are cyclic spaces for T such that $T|\mathcal{M} \sim T|\mathcal{N}$ and $T_{\mathcal{M}^\perp} \sim T_{\mathcal{N}^\perp}$, we will show that there exists a cyclic space $\mathcal{P}$ for T such that $\mathcal{M} \prec \mathcal{P}$ and $\mathcal{N} \prec \mathcal{P}$. In particular, $\mathcal{M}$ and $\mathcal{N}$ are weakly quasisimilar. We don't know whether $\mathcal{M}$ and $\mathcal{N}$ need to be quasisimilar. This seems to depend on finer analytic techniques.

The remainder of this paper is organized as follows. Section 2 recalls basic terminology and several useful results. Finally, Section 3 contains our classification of cyclic spaces.

2 Preliminaries

We recall here some basic facts from the theory of operators of class C_0. Most of the results in this section are well-known and can be found in [1] or [5]. Denote by H^∞ the Banach algebra of bounded analytic functions defined in the complex unit disk $\mathbb{D}$. We shall also denote by $\mathbb{C}$ the space of complex numbers and $\mathbb{T}$ the boundary of the complex unit disk. Let $\mathcal{H}$ be a Hilbert space, and let $L(\mathcal{H})$ denote the algebra of bounded linear operators on $\mathcal{H}$. An operator $T \in L(\mathcal{H})$ is of class C_0 if there exists a homomorphism $\Phi : H^\infty \to L(\mathcal{H})$ with the following properties:

(i) $\|\Phi(u)\| \le \|u\|$ for $u \in H^\infty$;

(ii) $\Phi(\chi) = T$ where $\chi(\lambda) = \lambda$, $\lambda \in \mathbb{D}$;

(iii) for every $h \in \mathcal{H}$ the map $u \mapsto \Phi(u)h$ is continuous if H^∞ is given its weak* topology and $\mathcal{H}$ its weak topology; and

(iv) Φ has nontrivial kernel.

The usual notation for $\Phi(u)$ is $\Phi(u) = u(T)$; this is the *Sz.-Nagy—Foias functional calculus* associated with T. The kernel of Φ has the form θH^∞ for some inner function θ which is uniquely determined up to a constant factor of absolute value one. This function is called the *minimal function* of T and is denoted m_T. If T is of class C_0 and $f \in \mathcal{H}$ then we will let $\mathcal{H}[f]$ denote the closed subspace generated by $\{T^n f : n \ge 0\}$, i.e., $\mathcal{H}[f] = \bigvee_{n=0}^\infty T^n f$. The restriction $T|\mathcal{H}[f]$ is also of class C_0. We will denote by m_f the minimal function $m_{T|\mathcal{H}_f}$. For two inner functions θ and ϕ we will write $\theta|\phi$ if θ divides ϕ. Recall that $\theta|\phi$ if $\phi = \theta\psi$ for some inner function ψ. We will write $\theta \equiv \phi$ if $\theta|\phi$ and $\phi|\theta$, and we will denote by $\theta \wedge \phi$ the greatest common inner divisor of θ and ϕ.

We now define Jordan blocks which are in some sense the building blocks of operators of class C_0. Let H^2 denote the set of functions of the form $f(\lambda) = \sum_{n=0}^\infty a_n \lambda^n$ for $\lambda \in \mathbb{D}$ where $\|f\|^2 = \sum_{n=0}^\infty |a_n|^2 < \infty$. If $u \in H^\infty$ and $f \in H^2$ then $uf \in H^2$ and $\|uf\| \le \|u\|_\infty \|f\|$. We can hence define the shift operator $S \in L(H^2)$ by $Sf = \chi f$, $f \in H^2$. If θ is an inner function then θH^2 is invariant for S, so $\mathcal{H}(\theta) = H^2 \ominus \theta H^2$ is invariant for S^*. We define the *Jordan block* $S(\theta) \in L(\mathcal{H}(\theta))$ by $S(\theta)^* = S^*|\mathcal{H}(\theta)$ or, equivalently, $S(\theta) = P_{\mathcal{H}(\theta)} S|\mathcal{H}(\theta)$, where $P_\mathcal{M}$ denotes the orthogonal projection onto the closed space $\mathcal{M}$. The operator $S(\theta)$ is of class C_0 and it has minimal function θ. The following statement contains some of the basic properties of Jordan blocks.

Proposition 2.1 *Let $\theta \in H^\infty$ be an inner function.*

(i) *The adjoint $S(\theta)^*$ is unitarily equivalent to $S(\tilde{\theta})$ where $\tilde{\theta}$ is defined by $\tilde{\theta}(\lambda) = \overline{\theta(\bar{\lambda})}$ for $\lambda \in \mathbb{D}$.*

(ii) *If ϕ is an inner divisor of θ then $\phi H^2 \ominus \theta H^2$ is invariant for $S(\theta)$. More precisely,*

$$\phi H^2 \ominus \theta H^2 = \mathrm{ran}[\phi(S(\theta))] = \ker[(\theta/\phi)(S(\theta))].$$

(iii) *If ϕ is an inner divisor of θ then $S(\theta)|\phi H^2 \ominus \theta H^2$ is unitarily equivalent to $S(\theta/\phi)$ while $S(\theta)^*|H^2 \ominus \phi H^2 = S(\phi)^*$.*

(iv) *For any inner function $u \in H^\infty$, the operator $S(\theta)|[\operatorname{ran} u(S(\theta))]^-$ is unitarily equivalent to $S(\theta/u \wedge \theta)$.*

(v) *A vector $h \in \mathcal{H}(\theta)$ is cyclic for $S(\theta)$ (i.e. $\bigvee_{n=0}^\infty S(\theta)^n h = \mathcal{H}(\theta)$) if and only if $\theta \wedge h \equiv 1$.*

A more general family of operators of class C_0 are the *separably acting Jordan operators*. These operators are of the form $\bigoplus_{j=0}^\infty S(\theta_j)$, where $\{\theta_j : j \geq 0\}$ is a sequence of inner functions satisfying the conditions $\theta_{j+1}|\theta_j$ for $j \geq 0$. Recall that the operators T and T' are *quasisimilar* if there exist quasiaffinities $X : \mathcal{H} \to \mathcal{H}'$ and $Y : \mathcal{H}' \to \mathcal{H}$ such that $XT = T'X$ and $YT' = TY$. (A *quasiaffinity* is a continuous linear operator with dense range and zero kernel.) We shall write $T \sim T'$ if T and T' are quasisimilar.

Theorem 2.2 *For every operator T of class C_0 acting on a separable Hilbert space there exists a unique Jordan operator T' such that $T \sim T'$.*

An analogous result is true in the nonseparable case (cf. [1]) but we will not need it in this paper. The Jordan operator T' given above is called the *Jordan model* of T.

A vector $x \in \mathcal{H}$ such that $m_x \equiv m_T$ is said to be *maximal* for T.

Theorem 2.3 *Let $T \in L(\mathcal{H})$ be an operator of class C_0. Then the set $\{x : m_f \equiv m_T\}$ is a dense G_δ in $\mathcal{H}$. More generally, if $\mathcal{X}$ is a Banach space, and $A : \mathcal{X} \to \mathcal{H}$ is a bounded linear operator such that $A\mathcal{X}$ is a cyclic set for T, the set $\{x \in \mathcal{X} : m_{Ax} \equiv m_T\}$ is a dense G_δ in $\mathcal{X}$.*

The next useful result is known as the splitting principle.

Theorem 2.4 *Let $T \in L(\mathcal{H})$ be an operator of class C_0, $h \in \mathcal{H}$ a maximal vector, $\mathcal{K} = \mathcal{H}[h]$. Then there exists an invariant subspace $\mathcal{M}$ for T such that $\mathcal{K} \vee \mathcal{M} = \mathcal{H}$ and $\mathcal{K} \cap \mathcal{M} = \{0\}$.*

Another useful result is the following.

Proposition 2.5 *Let $T \in L(\mathcal{H})$ be an operator of class C_0 with Jordan model $\bigoplus_{j=0}^\infty S(\theta_j)$, and $x \in \mathcal{H}$ a maximal vector for T. Set $\mathcal{L} = \bigvee\{T^n x : n \geq 0\}$ and $\mathcal{H}' = \mathcal{H} \ominus \mathcal{L}$. Then the Jordan model of $T_{\mathcal{H}'}$ is $\bigoplus_{j=1}^\infty S(\theta_j)$.*

An operator $T \in L(\mathcal{H})$ is said to have finite multiplicity if there exists a finite set $F \subset \mathcal{H}$ such that $\mathcal{H} = \bigvee\{T^n F : n \geq 0\}$. Such a set F is called a cyclic set for T. The smallest cardinality of a cyclic set is called the multiplicity of T and is denoted by μ_T. If $\mu_T = 1$ then T is said to be multiplicity-free.

Proposition 2.6 *Let $T \in L(\mathcal{H})$ be an operator of class C_0 with Jordan model $\bigoplus_{i=0}^{\infty} S(\theta_i)$. Then $\mu_T \leq n$ if and only if $\theta_n \equiv 1$.*

Let T be an operator of class C_0 with multiplicity n, and let $\bigoplus_{j=0}^{n-1} S(\theta_j)$ be the Jordan model of T. The determinant function of T is defined by $\det(T) \equiv \theta_0 \theta_1 \ldots \theta_{n-1}$. The following result from [1] gives an important property of the determinant function.

Theorem 2.7 *Let $T \in L(\mathcal{H})$ be an operator of class C_0 with finite multiplicity, $\mathcal{H}' \subset \mathcal{H}$ an invariant subspace for T, and $\mathcal{H}'' = \mathcal{H} \ominus \mathcal{H}'$. Then $\det(T) \equiv \det(T|\mathcal{H}')\det(T_{\mathcal{H}''})$.*

The following result is a consequence of Theorem 2.3. We will only use it for a finite family of functions.

Theorem 2.8 *Let $\{f_j : j \geq 0\}$ be a bounded sequence of functions in H^2, and let θ be an inner function. The set of those $\{a_j\}$ in ℓ^1 satisfying the relation*

$$\left(\sum_{j=0}^{\infty} a_j f_j \right) \wedge \theta \equiv \left(\bigwedge_{j=0}^{\infty} f_j \right) \wedge \theta$$

is a dense G_δ in ℓ^1.

The following useful proposition is from [1]. Recall that an operator of class C_0 with Jordan model $\bigoplus_{j=0}^{\infty} S(\theta_j)$ has property (P) if the greatest common inner divisor of the functions $\{\theta_j : j \geq 0\}$ is one.

Proposition 2.9 *Assume that $T \in L(\mathcal{H})$, $T' \in L(\mathcal{H}')$, $T'' \in L(\mathcal{H}'')$ are operators of class C_0, A and B are such that $AT' = TA$, $BT'' = TB$ and $A\mathcal{H}' \subset (B\mathcal{H}'')^-$. If in addition $T|(B\mathcal{H}'')^-$ has property (P) then*

(i) $(A^{-1}(B\mathcal{H}''))^- = \mathcal{H}'$; *and*

(ii) $(A\mathcal{H}' \cap B\mathcal{H}'')^- \supset A\mathcal{H}'$.

Corollary 2.10 *Let T, T', T'', A, and B satisfy the hypotheses of Proposition 2.9. Then $A^{-1}(B\mathcal{H}'')$ contains a dense set of maximal vectors.*

Proof: Let $\mathcal{K} \subset \mathcal{H}' \oplus \mathcal{H}''$ denote the subspace of vectors $\{f \oplus g : Af = Bg\}$. By Proposition 2.9(i) the minimal function for $T' \oplus T''|\mathcal{K}$ is $m_{T'}$. Let P_1 denote the projection of $\mathcal{H}' \oplus \mathcal{H}''$ onto $\mathcal{H}'$ and P_2 denote the projection of $\mathcal{H}' \oplus \mathcal{H}''$ onto $\mathcal{H}''$. It follows that $A P_1(T' \oplus T'')k = B P_2(T' \oplus T'')k$ for every vector in $k \in \mathcal{K}$. Thus $P_1(\mathcal{K}) = \{f \in \mathcal{H}' : Af \in B\mathcal{H}''\} = A^{-1}(B\mathcal{H}'')$ is dense in $\mathcal{H}'$ by Proposition 2.9(i). Since the dense set $A^{-1}(B\mathcal{H}'')$ is the range of the Hilbert space $\mathcal{K}$ under the continuous operator P_1, it follows from Theorem 2.3 that the set of $k \in K$ such that $m_{P_1 k} \equiv m_{T'}$ is a dense G_δ in $\mathcal{K}$. Thus $A^{-1}(B\mathcal{H}'')$ contains a dense set of maximal vectors. $\qquad\square$

We will also need the following lemma.

Lemma 2.11 *Let* $\xi_0, \xi_1, \ldots \in H^2 \backslash \{0\}$ *be a sequence of functions, then there exist an outer function* $v \in H^2$ *and positive constants* $c_0, c_1, c_2, \ldots$ *such that* $|\xi_j| \leq c_j |v|$ *for* $j = 0, 1, 2, \ldots$.

Proof: To define an outer function, we need only to define the absolute value of the outer function on the boundary of the unit disk. We set $|v| = \sum_{j=0}^{\infty} \varepsilon_j |\xi_j|$ where $\varepsilon_j > 0$ and $\sum_{j=0}^{\infty} \varepsilon_j \|\xi_j\| < \infty$. The lemma holds with $c_j \leq 1/\varepsilon_j$. $\qquad\square$

In the proof of the following result we use normalized arclength measure $dm = |d\xi|/2\pi$ on the unit circle.

Lemma 2.12 *If* $f \in H^2$ *satisfies* $\|f - 1\|_{H^2} < \varepsilon < 1/4$ *then there exists an outer function* ω *such that* $\|\omega f\|_\infty \leq 1$ *and* $\omega(0) f(0) \geq 1 - 4\varepsilon$.

Proof: The function ω is defined by the requirements that $\omega(0) f(0) \geq 0$ and for $\xi \in \mathbb{T}$,

$$|\omega(\xi)| = \begin{cases} 1 & \text{if } |f(\xi)| \leq 1 \\ \frac{1}{|f(\xi)|} & \text{if } |f(\xi)| > 1. \end{cases}$$

The condition that $\|\omega f\|_\infty \leq 1$ is obviously satisfied. To verify the second condition in the statement observe first that $|f(0) - 1| \leq \|f - 1\|_{H^2} < \varepsilon$ so that $|f(0)| \geq 1 - \varepsilon$. On the other hand,

$$|\omega(0)| = \exp\left(-\int_{\{\xi : |f(\xi)| > 1\}} \log |f(\xi)| \, dm(\xi) \right)$$

$$\geq \exp\left(-\int_{\{\xi : |f(\xi)| > 1\}} (|f(\xi)| - 1) \, dm(\xi) \right).$$

Since

$$\int_{\{\xi : |f(\xi)| > 1\}} (|f(\xi)| - 1) dm(\xi)$$

$$\leq \int_{\{\xi : |f(\xi)| > 1\}} (|f(\xi)| - 1)(|f(\xi)| + 1) dm(\xi)$$

$$\leq \int_{\{\xi : |f(\xi)| > 1\}} (|f(\xi)|^2 - 1) dm(\xi)$$

$$= \int_{\{\xi : |f(\xi)| > 1\}} (|f(\xi) - 1|^2 + 2\text{Re}(1 - f(\xi))) dm(\xi)$$

$$\leq \varepsilon^2 + 2 \left(\int_{\{\xi : |f(\xi)| > 1\}} |f - 1|^2 dm(\xi) \right)^{1/2}$$

$$\leq \varepsilon^2 + 2\varepsilon \leq 3\varepsilon,$$

(where we used the Schwarz inequality) we have

$$|\omega(0)| \geq e^{-3\varepsilon} \geq 1 - 3\varepsilon,$$

and

$$\omega(0) f(0) = |\omega(0) f(0)| \geq (1 - 3\varepsilon)(1 - \varepsilon) \geq 1 - 4\varepsilon,$$

as claimed.   □

3 Cyclic Spaces

In this section T will denote a fixed Jordan operator, say $T = \bigoplus_{j=0}^{\infty} S(\theta_j)$, acting on $\mathcal{H} = \bigoplus_{j=0}^{\infty} \mathcal{H}(\theta_j)$. In order to simplify our notation for cyclic spaces, let us start with a vector $\xi = \bigoplus_{j=0}^{\infty} \xi_j \in \bigoplus_{j=0}^{\infty} H^2$. Associated with ξ there will be a space $\mathcal{H}[\xi] = \bigvee_{n=0}^{\infty} T^n P_{\mathcal{H}} \xi$ where, of course, $P_{\mathcal{H}} \xi = \bigoplus_{j=0}^{\infty} P_{\mathcal{H}(\theta_j)} \xi_j$. Analogously, there are spaces $\mathcal{H}[\xi_j] \subset \mathcal{H}(\theta_j)$ generated by $P_{\mathcal{H}(\theta_j)} \xi_j$, that is $\mathcal{H}[\xi_j] = \bigvee_{k=0}^{\infty} S(\theta_j)^k P_{\mathcal{H}(\theta_j)} \xi_j$. These spaces have the form $\phi_j H^2 \ominus \theta_j H^2$, where $\phi_j = \xi_j \wedge \theta_j$.

Definition 3.1 We say the vector $\xi \in \mathcal{H}$ is *befitting* if for every $j = 0, 1, 2, \ldots$ we have $\phi_{j+1} | \phi_j$ and $\frac{\theta_{j+1}}{\phi_{j+1}} | \frac{\theta_j}{\phi_j}$.

It may be interesting to note that ξ is befitting if and only if the space $\bigoplus_{j=0}^{\infty} (\phi_j H^2 \ominus \theta_j H^2)$ is hyperinvariant for T (cf. [1] Proposition 2.1, Chapter 4). Our first task will be to show that for every cyclic space $\mathcal{M}$ there is an invertible operator $X \in \{T\}'$ such that $X\mathcal{M}$ is generated by a befitting vector.

Theorem 3.2 *For every vector $\xi \in \mathcal{H}$ there exists an invertible operator X in the commutant $\{T\}'$ such that $X\xi$ is befitting. The set of such operators X is a dense G_δ in the open set of invertible operators in $\{T\}'$.*

Proof: Denote by $\mathcal{K}$ the hyperinvariant subspace of T generated by ξ, that is $\mathcal{K} = \bigvee \{Y\xi : Y \in \{T\}'\}$. As mentioned above, this space can be written as $\bigoplus_{j=0}^{\infty} (\phi_j H^2 \ominus \theta_j H^2)$, with the functions ϕ_j satisfying the conditions in Definition 3.1. Denote by P_j the projection of $\mathcal{H}$ onto its jth component $H^2 \ominus \theta_j H^2$, and observe that for every $Y \in \{T\}'$ the vector $P_j Y\xi$ belongs to $\phi_j H^2 \ominus \theta_j H^2$, and in fact $\phi_j H^2 \ominus \theta_j H^2$ is the closure of $\{P_j Y\xi : Y \in \{T\}'\}$. By Theorem 2.3 applied to the map $A : Y \mapsto P_j Y\xi$, the set of those $Y \in \{T\}'$ for which $P_j Y\xi$ is cyclic in $\phi_j H^2 \ominus \theta_j H^2$ is a dense G_δ in $\{T\}'$. It follows that the set of those $X \in \{T\}'$ which are invertible and $P_j X \xi$ is cyclic in $\phi_j H^2 \ominus \theta_j H^2$ for *every* j is also a dense G_δ (in the set of invertibles). To conclude the proof it suffices to remark that $X\xi$ is befitting if X is in this dense G_δ. $\qquad\square$

Corollary 3.3 *Given a sequence $\{\xi_n : n \geq 0\}$ of vectors in $\mathcal{H}$, there exists an invertible operator $X \in \{T\}'$ such that $X\xi_n$ is befitting for every n.*

Interestingly, ξ is a befitting vector for T, the Jordan models of $T|\mathcal{H}[\xi]$ and $T_{\mathcal{H}[\xi]^\perp}$ only depend on the functions ϕ_j and θ_j for $j \geq 0$. Conversely the ϕ_j can be calculated from these Jordan models.

Theorem 3.4 *Let ξ be a befitting vector for T, where $\xi = \bigoplus_{j=0}^\infty \xi_j$, and let $\phi_j = \xi_j \wedge \theta_j$, $j = 0, 1, 2, \ldots$. Denote by $S(\alpha_0)$ and $\bigoplus_{j=0}^\infty S(\beta_j)$ the Jordan models of $T|\mathcal{H}[\xi]$ and $T_{\mathcal{H}[\xi]^\perp}$, respectively. Then*

(1) $\alpha_0 \equiv \theta_0/\phi_0$ *and* $\beta_j \equiv \theta_{j+1}\phi_j/\phi_{j+1}$ *for* $j \geq 0$; *and*

(2) $\phi_0 \equiv \theta_0/\alpha_0$, $\phi_j \equiv \dfrac{\theta_0\theta_1\ldots\theta_j}{\alpha_0\beta_0\beta_1\ldots\beta_{j-1}}$ *for* $j \geq 1$. *In particular*

$$\alpha_0\beta_0\beta_1 \ldots \beta_{j-1}|\theta_0\theta_1 \ldots \theta_j$$

for $j \geq 1$.

Proof: Observe first that (2) is a consequence of (1). The divisibility assertion in (2) could also be deduced from the analogue of Sigal's inequalities proved in [2]. We proceed now to prove (1). Since $T|\mathcal{H}[\xi]$ is cyclic, α_0 is simply the minimal function of $T|\mathcal{H}[\xi]$. Thus α_0 is the greatest common divisor of all inner functions ψ satisfying $\psi(T)\xi = 0$. This is equivalent to $\psi\xi \in \bigoplus_{j=0}^\infty \theta_j H^2$, that is $\psi\xi_j \in \theta_j H^2$ for all j. Now $\psi\xi_j \in \theta_j H^2$ if and only if $\psi\frac{\xi_j}{\phi_j} \in \frac{\theta_j}{\phi_j}H^2$ and, since $\frac{\xi_j}{\phi_j} \wedge \frac{\theta_j}{\phi_j} \equiv 1$, this happens only when $\frac{\theta_j}{\phi_j}|\psi$. Thus α_0 is the least common multiple of $\{\theta_j/\phi_j : j \geq 0\}$ which is θ_0/ϕ_0 by the befitting property of ξ.

We will set up now a process which can be continued inductively to calculate $\beta_0, \beta_1, \ldots$. Denote $\xi' = 0 \oplus (\bigoplus_{j=1}^\infty \xi_j)$, $\mathcal{H}' = \{0\} \oplus (\bigoplus_{j=1}^\infty \mathcal{H}(\theta_j))$, and let T' denote the restriction of T to its reducing subspace $\mathcal{H}'$. The orthogonal projection $P_{\mathcal{H}'}|\mathcal{H} : \mathcal{H} \to \mathcal{H}'$ maps $\mathcal{H}[\xi]$ into a dense linear submanifold of $\mathcal{H}[\xi'] = \bigvee_{n=0}^\infty T'^n\xi' = \bigvee_{n=0}^\infty T^n\xi'$. The Jordan model of $T'|\mathcal{H}[\xi']$ is seen to be $S(\theta_1/\phi_1)$ via the same calculation we did for α_0, and we will eventually show that the Jordan model of $T'_{\mathcal{H}'\ominus\mathcal{H}[\xi']}$ is $\bigoplus_{j=1}^\infty S(\beta_j)$. Granted this fact, we will only need to calculate β_0 to finish the proof. We start by showing that

(3.5) $\qquad \mathcal{H}[\xi] \cap (\mathcal{H}(\theta_0) \oplus \{0\} \oplus \cdots) = (\theta_1\phi_0/\phi_1 H^2 \ominus \theta_0 H^2) \oplus \{0\} \oplus \cdots$

and

(3.6) $\mathcal{H}[\xi] \vee (\mathcal{H}(\theta_0) \oplus \{0\} \oplus \cdots) = \{x \in \mathcal{H} : P_{\mathcal{H}'} x \in \mathcal{H}[\xi']\}.$

Equality (3.5) follows from the fact that the left hand side is the kernel of the operator $Z = P_{\mathcal{H}'}|\mathcal{H}[\xi] : \mathcal{H}[\xi] \to \mathcal{H}[\xi']$. As noted above, Z has dense range and $Z(T|\mathcal{H}[\xi]) = (T'|\mathcal{H}[\xi'])Z$. Therefore

$$\det(T|\ker Z) \equiv \det(T|\mathcal{H}[\xi])/\det(T'|\mathcal{H}[\xi'])$$

(3.7)

$$\equiv (\theta_0/\phi_0)/(\theta_1/\phi_1) = (\theta_0\phi_1)/(\theta_1\phi_0).$$

On the other hand, clearly $\ker Z$ must have the form $(\psi H^2 \ominus \theta_0 H^2) \oplus \{0\} \oplus \cdots$ for some divisor ψ of θ_0. The determinant calculation then indicates that $\frac{\theta_0}{\psi} \equiv \frac{\theta_0\phi_1}{\theta_1\phi_0}$, and this yields (3.5).

The inclusion '$\subset$' in (3.6) is trivial. To prove the opposite inclusion, consider a vector $x \in \mathcal{H}$ such that $P_{\mathcal{H}'} x \in \mathcal{H}_{\xi'}$. Clearly then $P_{\mathcal{H}'}\mathcal{H}[x] \subset \mathcal{H}[\xi']$ so that we can define an operator $Y : \mathcal{H}[x] \to \mathcal{H}[\xi']$ by $Y = P_{\mathcal{H}'}|\mathcal{H}[x]$; clearly $Y(T|\mathcal{H}[x]) = (T|\mathcal{H}[\xi'])Y$. Now $Y\mathcal{H}[x] \subset (Z\mathcal{H}[\xi])^-$, where $Z : \mathcal{H}[\xi] \to \mathcal{H}[\xi']$ was considered above. Proposition 2.9 implies that $\mathcal{H}[x] = (Y^{-1}(Z\mathcal{H}[\xi]))^-$ so that $x = \lim_{n\to\infty} x_n$, where $x_n = u_n + v$ with $u_n \in \mathcal{H}[\xi]$ and $v \in \ker P_{\mathcal{H}'} = \mathcal{H}(\theta_0) \oplus \{0\} \oplus \cdots$. Thus x belongs to the left hand side of (3.6), as desired.

We are now ready to calculate β_0 which is the minimal function of $T_{\mathcal{H}\ominus\mathcal{H}[\xi]}$. Note that $u(T_{\mathcal{H}\ominus\mathcal{H}[\xi]}) = 0$ if and only if $u(T)\mathcal{H} \subset \mathcal{H}[\xi]$. If $u(T)\mathcal{H} \subset \mathcal{H}[\xi]$ then, in particular,

$$P_{\mathcal{H}(\theta_0)}u \oplus 0 \oplus \cdots = u(T)(P_{\mathcal{H}(\theta_0)}1 \oplus 0 \oplus \cdots) \in \mathcal{H}[\xi].$$

Therefore (3.5) implies that $\theta_1\phi_0/\phi_1$ divides u. To conclude that $\beta_0 \equiv \theta_1\phi_0/\phi_1$ it suffices now to show that $(\theta_1\phi_0/\phi_1)(T)\mathcal{H} \subset \mathcal{H}[\xi]$, and to do this it suffices to show that

$$(\theta_1\phi_0/\phi_1)(T)(0 \oplus \cdots \oplus 0 \oplus P_{\mathcal{H}(\theta_j)}1 \oplus 0 \oplus \cdots) \in \mathcal{H}[\xi]$$

for $j = 0, 1, 2, \ldots$. For $j = 0$ this follows from (3.5) as above, while for $j \geq 1$, $\theta_j|\theta_1|(\theta_1\phi_0/\phi_1)$ so in fact

$$(\theta_1\phi_0/\phi_1)(T)(0 \oplus \cdots \oplus 0 \oplus P_{\mathcal{H}(\theta_j)}1 \oplus 0 \oplus \cdots) = 0.$$

We conclude by showing that the Jordan model of $T_{\mathcal{H}'\ominus\mathcal{H}[\xi']}$ is $\bigoplus_{j=1}^{\infty} S(\beta_j)$. To do this we consider the operator $W : \mathcal{H} \ominus \mathcal{H}[\xi] \to \mathcal{H}' \ominus \mathcal{H}[\xi']$ defined by $W = P_{\mathcal{H}'\ominus\mathcal{H}[\xi']}|\mathcal{H} \ominus \mathcal{H}[\xi]$. Since $P_{\mathcal{H}'\ominus\mathcal{H}[\xi']}\mathcal{H}[\xi] = \{0\}$, it follows easily that W is surjective and $WT_{\mathcal{H}\ominus\mathcal{H}[\xi]} = T_{\mathcal{H}'\ominus\mathcal{H}[\xi']}W$. It will suffice to show that $T_{\mathcal{H}\ominus\mathcal{H}[\xi]}|\ker W \sim S(\beta_0)$. Indeed, if that were true, the splitting lemma yields a subspace $\mathcal{M}$ invariant for $T_{\mathcal{H}\ominus\mathcal{H}[\xi]}$ such that $\mathcal{M} \cap \ker W = \{0\}$, $\mathcal{M} \vee \ker W = \mathcal{H} \ominus \mathcal{H}[\xi]$, and $T_{\mathcal{H}\ominus\mathcal{H}[\xi]}|\mathcal{M} \sim \bigoplus_{j=1}^{\infty} S(\beta_j)$. Now, $W|\mathcal{M}$ is a quasiaffinity showing that $T_{\mathcal{H}\ominus\mathcal{H}[\xi]}|\mathcal{M} \prec T_{\mathcal{H}'\ominus\mathcal{H}[\xi']}$, and this yields the desired Jordan model. Let us thus calculate

$$
\begin{aligned}
\ker W &= \{x \in \mathcal{H} \ominus \mathcal{H}[\xi] : P_{\mathcal{H}'}x \in \mathcal{H}_{\xi'}\} \\
&= P_{\mathcal{H}\ominus\mathcal{H}[\xi]}\{x \in \mathcal{H} : P_{\mathcal{H}'}x \in \mathcal{H}[\xi']\} \\
&= [P_{\mathcal{H}\ominus\mathcal{H}[\xi]}(\mathcal{H}(\theta_0) \oplus \{0\}\ldots)]^{-}
\end{aligned}
$$

by (3.6). Now $P_{\mathcal{H}\ominus\mathcal{H}[\xi]}T = T_{\mathcal{H}\ominus\mathcal{H}[\xi]}P_{\mathcal{H}\ominus\mathcal{H}[\xi]}$, and the above calculation indicates that $T_{\mathcal{H}\ominus\mathcal{H}[\xi]}|\ker W$ is a cyclic operator. To conclude the proof we must show that the determinant of this cyclic operator is β_0. To do this observe that

$$
\ker W = [\mathcal{H}[\xi] \vee (\mathcal{H}(\theta_0) \oplus \{0\} \oplus \cdots)] \ominus \mathcal{H}[\xi]
$$

so that

$$
\begin{aligned}
\det(T_{\mathcal{H}\ominus\mathcal{H}[\xi]}|\ker W) &\equiv \frac{\det(T|[\mathcal{H}[\xi] \vee (\mathcal{H}(\theta_0) \oplus \{0\} \oplus \cdots)])}{\det(T|\mathcal{H}[\xi])} \\
&\equiv \frac{\det(T|[\mathcal{H}[\xi] \vee (\mathcal{H}(\theta_0) \oplus \{0\} \oplus \cdots)])}{\theta_0/\phi_0}
\end{aligned}
$$

by Theorem 2.7. Similarly,

$$
\begin{aligned}
&\det(T|[\mathcal{H}[\xi] \vee (\mathcal{H}(\theta_0) \oplus \{0\} \oplus \cdots)]) \\
&\equiv \frac{\det(T|\mathcal{H}[\xi]) \cdot \det(T|[\mathcal{H}(\theta_0) \oplus \{0\} \oplus \cdots])}{\det(T|[\mathcal{H}[\xi] \cap (\mathcal{H}(\theta_0) \oplus \{0\} \oplus \cdots)])} \\
&\equiv \frac{\theta_0\theta_0/\phi_0}{\theta_0\phi_1/\theta_1\phi_0} = \theta_0\theta_1/\phi_1,
\end{aligned}
$$

where we used (3.7). The last two equations can now be combined to yield the desired value $\det(T_{\mathcal{H}\ominus\mathcal{H}[\xi]}|\ker W) \equiv \beta_0$. $\qquad\square$

An immediate consequence of Theorem 3.4 is as follows.

Corollary 3.8 *Let ξ and η be two befitting vectors for T, where $\xi = \bigoplus_{j=0}^{\infty}\xi_j$ and $\eta = \bigoplus_{j=0}^{\infty}\eta_j$. The following are equivalent:*

(1) $T|\mathcal{H}[\xi] \sim T|\mathcal{H}[\eta]$ *and* $T_{\mathcal{H}[\xi]^{\perp}} \sim T_{\mathcal{H}[\eta]^{\perp}}$; *and*

(2) $\xi_j \wedge \theta_j \equiv \eta_j \wedge \theta_j$ *for* $j = 0, 1, 2, \ldots$.

Condition (1) above is related to the classification of invariant subspaces as follows.

Proposition 3.9 *Let M and N be weakly quasisimilar invariant subspaces for T.*

(1) *We always have $T|M \sim T|N$.*

(2) *If $T|M$ has property (P), then we also have $T_{M^\perp} \sim T_{N^\perp}$.*

Proof: It suffices to prove this proposition in case $M \prec N$. Assume therefore that $X \in \{T\}'$ is a quasiaffinity satisfying $(XM)^- = N$. Then $Y = X|M :$ $M \to N$ is also a quasiaffinity and $Y(T|M) = (T|N)Y$ so that $T|M \prec T|N$. However, '$\prec$' is reflexive for operators of class C_0, and this proves (1). The operator $Z : M^\perp \to N^\perp$ defined by $Z = P_{N^\perp}X|M^\perp$ obviously has dense range and $ZT_{M^\perp} = T_{N^\perp}Z$. To conclude the proof of (2), it will suffice to show that Z is one-to-one if $T|M$ has property (P). To do that observe first that $T|N$ also has property (P). Consider a vector $\xi \in \ker Z$, so that $\xi \in M^\perp$ and $X\xi \in N$. Then the operator $W = X|H[\xi] : H[\xi] \to H$ satisfies $W(T|H[\xi]) = (T|N)W$, and $WH[\xi] \subset (YM)^-$. By Proposition 2.9, $W^{-1}(YM)$ is dense in $H[\xi]$. However, $W^{-1}(YM) = M \cap H[\xi]$, and we conclude that $H[\xi] \subset M$. Since $\xi \in M^\perp$, we must have $\xi = 0$. $\square$

The conclusion of (2) is not true if $T|M$ is not assumed to have property (P). An easy example is obtained by choosing $T = 0$ on an infinite-dimensional space H, and X some noninvertible quasiaffinity. Then there exists a subspace $M \subset H$ of codimension one such that $(XM)^- = H$. The space M can be constructed as $M = \{h\}^\perp$, where h is any vector not in the range of X^*.

The following result gives a method for constructing weakly quasisimilar subspaces.

Lemma 3.10 *Consider a vector $\xi = \bigoplus_{j=0}^\infty \xi_j \in H$.*

(1) *If $v \in H^\infty$ is such that $v \wedge \theta_0 \equiv 1$ then $H[\xi] = H[v(T)\xi]$.*

(2) *If $u_0, u_1, \ldots \in H^\infty$ are such that $\sup\{\|u_j\|_\infty : j \geq 0\} < \infty$ and $u_j \wedge \theta_j \equiv 1$ for $j = 1, 2, \ldots$, we have $H[\xi] \prec H[\eta]$, where $\eta = \bigoplus_{j=0}^\infty u_j(S(\theta_j))\xi_j$.*

Proof: Assertion (1) follows from the fact that

$$H[v(T)\xi] = (v(T)H[\xi])^- = (v(T|H[\xi])H[\xi])^-,$$

and $v(T|H[\xi])$ is a quasiaffinity because $v \wedge \theta_0 \equiv 1$. For (2), the desired quasi-affinity X satisfying $(XH[\xi])^- = H[\eta]$ is given by $X = \bigoplus_{j=0}^\infty u_j(S(\theta_j))$, and again $u_j(S(\theta_j))$ is a quasiaffinity because $u_j \wedge \theta_j \equiv 1$. $\square$

We are now ready for our classification result. Observe that $H[v(T)\xi] = H[v\xi]$ according to our convention regarding the notation of cyclic spaces.

Theorem 3.11 *Let $\mathcal{M}$ and $\mathcal{N}$ be cyclic subspaces for T. The following are equivalent:*

(1) *There exists a cyclic subspace $\mathcal{P}$ for T such that $\mathcal{M} \prec \mathcal{P}$ and $\mathcal{N} \prec \mathcal{P}$;*

(2) *$\mathcal{M}$ and $\mathcal{N}$ are weakly quasisimilar; and*

(3) *$T|\mathcal{M} \sim T|\mathcal{N}$ and $T_{\mathcal{M}^\perp} \sim T_{\mathcal{N}^\perp}$.*

Proof: The implication (1)$\Rightarrow$(2) is obvious, and (2)$\Rightarrow$(3) follows from Proposition 3.9. We will prove that (3)$\Rightarrow$(1). Assume therefore that $T|\mathcal{M} \sim T|\mathcal{N}$ and $T_{\mathcal{M}^\perp} \sim T_{\mathcal{N}^\perp}$. By Proposition 3.3, we may assume that both $\mathcal{M}$ and $\mathcal{N}$ are generated by befitting vectors, say $\mathcal{M} = \mathcal{H}[\xi]$ and $\mathcal{N} = \mathcal{H}[\eta]$ with $\xi = \bigoplus_{j=0}^{\infty} \xi_j$ and $\eta = \bigoplus_{j=0}^{\infty} \eta_j$. Corollary 3.8 implies that $\xi_j \wedge \theta_j \equiv \eta_j \wedge \theta_j$ for all j. By Lemma 2.11, there exists an outer function $a \in H^2$ such that the quotients $1/a$, ξ_j/a, and η_j/a are bounded for all j. Setting $b = 1/a$ we see that

$$\mathcal{H}[\xi] = \mathcal{H}[b\xi] \text{ and } \mathcal{H}[\eta] = \mathcal{H}[b\eta]$$

by Lemma 3.10.(1). Denote $\phi_j = \xi_j \wedge \theta_j \equiv (b\xi_j) \wedge \theta_j$, and observe that the vector $b\xi_j$ can be replaced by $b\xi_j + \lambda_j\theta_j$, $\lambda_j \in \mathbf{C}$, without changing the space $\mathcal{H}[b\xi] = \mathcal{H}[\xi]$. Choosing the constants λ_j so that

$$\left(\frac{b\xi_j}{\phi_j} + \lambda_j \frac{\theta_j}{b_j} \right) \wedge \theta_0 \equiv 1$$

(this can be done by Theorem 2.8 because $(\xi_j/\phi_j) \wedge (\theta_j/\phi_j) \equiv 1$), we may assume that $b\xi_j = \phi_j x_j$ with x_j bounded and $x_j \wedge \theta_0 \equiv 1$ for all j. Likewise, we may assume that $b\eta_j = \phi_j y_j$ with y_j bounded and $y_j \wedge \theta_0 \equiv 1$.

Fix now j, and note that $x_j(S(\theta_0))$ and $y_j(S(\theta_0))$ are both quasiaffinities so, in particular,

$$y_j(S(\theta_0))\mathcal{H}(\theta_0) \subset (x_j(S(\theta_0))\mathcal{H}(\theta_0))^-.$$

By Corollary 2.10 we can find a function $f_j^{(0)} \in \mathcal{H}(\theta_0)$ which is a cyclic vector for $S(\theta_0)$, and $g_j \in \mathcal{H}(\theta_0)$ such that

$$x_j(S(\theta_0))f_j^{(0)} = y_j(S(\theta_0))g_j,$$

and

$$\|f_j^{(0)} - P_{\mathcal{H}(\theta_0)}1\| < 4^{-j}.$$

If we set

$$f_j = f_j^{(0)} + P_{\theta_0 H^2}1$$

we will have the relations

$$x_j f_j - y_j g_j \in \theta_0 H^2, \; f_j \wedge \theta_0 \equiv g_j \wedge \theta_0 \equiv 1,$$

and

$$\|f_j - 1\| < 4^{-j}.$$

An application of Lemma 2.12 shows that there is an outer function ω_j such that the function $u_j = \omega_j f_j$ satisfies $u_j(0) = |u_j(0)| \geq 1 - 4^{-j+1}$, and $\|u_j\|_\infty \leq 1$. Note that we also have:

$$u_j x_j - \omega_j y_j g_j \in \theta_0 H^2, \quad u_j \wedge \theta_0 \equiv \omega_j g_j \wedge \theta_0 \equiv 1.$$

Observe that the product $u = \prod_{j=0}^{\infty} u_j$ converges, $\|u\|_\infty \leq 1$, and $u \wedge \theta_0 \equiv 1$. Upon setting $v_j = (u/u_j)\omega_j g_j$, we have

$$u x_j - v_j y_j \in \theta_0 H^2, \quad \text{and } u \wedge \theta_0 \equiv v_j \wedge \theta_0 \equiv 1$$

for all j. Choose now an outer function c and constants $\varepsilon_j \in (0,1)$ such that cv_j is in H^∞ for all j, and $\|\varepsilon_j cv_j\|_\infty \leq 1$. To prove (1), we set

$$\mathcal{P} = \mathcal{H}[\oplus_{j=0}^{\infty} \varepsilon_j cv_j \phi_j y_j].$$

Since $\varepsilon_j cu\phi_j x_j - \varepsilon_j cv_j \phi_j y_j \in \theta_0 H^2 \subset \theta_j H^2$, we have

$$\begin{aligned}
\mathcal{M} &= \mathcal{H}[b\xi] = \mathcal{H}[cub\xi] \\
&= \mathcal{H}[\oplus_{j=0}^{\infty} cu\phi_j x_j] \prec \mathcal{H}[\oplus_{j=0}^{\infty} \varepsilon_j cu\phi_j x_j] \\
&= \mathcal{H}[\oplus_{j=0}^{\infty} \varepsilon_j cv_j \phi_j y_j] = \mathcal{P}
\end{aligned}$$

where we also used Lemma 3.10. The same lemma yields

$$\mathcal{N} = \mathcal{H}[b\eta] = \mathcal{H}[\oplus_{j=0}^{\infty} \phi_j y_j] \prec \mathcal{H}[\oplus_{j=0}^{\infty} \varepsilon_j cv_j \phi_j y_j] = \mathcal{P}$$

because $\varepsilon_j cv_j$ are uniformly bounded. $\square$

The result can be improved if T has finite multiplicity.

Proposition 3.12 *If T has finite multiplicity, and $\mathcal{M}$, $\mathcal{N}$ are cyclic for T, then $\mathcal{M} \sim_w \mathcal{N}$ if and only if $\mathcal{M} \sim \mathcal{N}$.*

Proof: Assume $T = S(\theta_0) \oplus \cdots \oplus S(\theta_{n-1})$, where n is the multiplicity of T. Clearly it suffices to show that $[\xi] \prec [\eta]$ if $[\xi]$ and $[\eta]$ are befitting and $\xi_j \wedge \theta_j \equiv \eta_j \wedge \theta_j$ for $j = 0, 1, \ldots, n-1$. As in the preceeding proof, replace ξ_j by $\xi_j + \lambda_j \theta_j$, where this time we require that

$$\left(\frac{\xi_j}{\phi_j} + \lambda_j \frac{\theta_j}{\phi_j}\right) \wedge \theta_0 \equiv 1.$$

Thus $\xi_j = \phi_j u_j$ with $u_j \wedge \theta_0 \equiv 1$. Further, replacing ξ_j by $a\xi_j$ with a outer (and using Lemma 3.10.(1)) we may assume that u_j is bounded, $j = 0, 1, \ldots, n-1$.

Similarly, we may assume that $\eta_j = \phi_j v_j$ with $v_j \in H^\infty$ and $v_j \wedge \theta_0 \equiv 1$. If we set now $u = u_0 u_1 u_2 \ldots u_{n-1}$, we have

$$\mathcal{H}[\xi] \prec \mathcal{H}[\oplus_{j=0}^\infty u\phi_j] = \mathcal{H}[\oplus_{j=0}^\infty \phi_j] \prec \mathcal{H}[\eta],$$

where we used Lemma 3.10.(2) twice, (note that $u\phi_j/\xi_j = (u_0 u_1 \ldots u_{n-1})/u_j$ and $\eta_j/\phi_j = v_j$ are bounded), and Lemma 3.10.(1) for the equality in the middle. $\square$

Acknowledgements

The first named author was supported in part by a grant from the National Science Foundation. The second author expresses his gratitude to the Department of Mathematics of Indiana University for its kind hospitality while this paper was written.

References

[1]   H. Bercovici, Operator Theory and Arithmetic in H^∞, *Amer. Math. Soc.*, Providence, Rhode Island 1988.

[2]   H. Bercovici, W.S. Li and T. Smotzer, Classical linear algebra inequalities for the Jordan models of C_0 operators, *Linear Alg. Appl.* **251** (1997), 341–350.

[3]   H. Bercovici and T. Smotzer, Quasisimilarity of invariant subspaces for uniform Jordan operators of infinite multiplicity, *J. Funct. Anal.* **140** (1996), 87–99.

[4]   W.S. Li and V. Müller, Invariant subspaces of nilpotent operators and LR-sequences, *preprint*, 1997, 32.

[5]   B. Sz.-Nagy and C. Foias, Harmonic Analysis of Operators on Hilbert Space, North-Holland, Amsterdam 1970.

Hari Bercovici
Mathematics Department
Indiana University
Bloomington, Indiana 47405

Thomas Smotzer
Department of Mathematics and Statistics
Youngstown State University
Youngstown, OH 44555

1991 Mathematics Subject Classification. Primary: 47A45; Secondary: 47A15.

Operator Theory:
Advances and Applications, Vol. 124
© 2001 Birkhäuser Verlag Basel/Switzerland

Elliptic Spectral Problems of Higher Order with Eigenparameter Dependent Boundary Conditions

Paul Binding and Rostyslav Hryniv

Dedicated to Israel Gohberg on the occasion of his 70-th birthday

Let A be a uniformly elliptic differential operator of order $2m$ on a smooth bounded domain $\Omega \subset \mathbb{R}^n$. We consider the spectral problem $Au = \lambda u$ subject to boundary conditions $\mathcal{B}_0 u = \lambda \mathcal{B}_1 u$, $\mathcal{B}_1' u = 0$, where $\mathcal{B}_0$, $\mathcal{B}_1$, and $\mathcal{B}_1'$ are sets of boundary operators. The problem is recast in the form $\mathcal{A}\mathbf{u} = \lambda\mathbf{u}$ in an appropriate Hilbert space, and the essential spectrum of the corresponding operator $\mathcal{A}$ is found.

1 Introduction

Let Ω be a bounded connected domain in $\mathbb{R}^n$ with an infinitely smooth boundary $\partial\Omega$ and let A be a uniformly elliptic differential operator of order $2m$ with infinitely smooth coefficients on $\overline{\Omega}$. Let $r \leq m$ be a fixed natural number and $\mathcal{B}_0 = \{B_{01}, \ldots, B_{0r}\}$, $\mathcal{B}_1 = \{B_{11}, \ldots, B_{1r}\}$, and $\mathcal{B}_1' = \{B_{1,r+1}, \ldots, B_{1m}\}$ be sets of boundary operators of orders less than $2m$ with infinitely smooth coefficients such that $\mathcal{B}_0$ and $\mathcal{B}_1 \cup \mathcal{B}_1'$ are normal and $\mathcal{B}_1 \cup \mathcal{B}_1'$ covers A. Moreover, we assume that the problem $\{A, \mathcal{B}_1 \cup \mathcal{B}_1', \Omega\}$ is regular elliptic and that the mapping $H^{2m}(\Omega) \ni u \mapsto (Au - \lambda u, B_{11}u, \ldots, B_{1m}u)$ is injective for at least one $\lambda \in \mathbb{C}$ (see, e.g., [LM, Ch.II] for all the definitions).

We consider the spectral problem

$$(1.1) \qquad\qquad Au = \lambda u \quad \text{in } \Omega,$$

$$(1.2) \qquad\qquad \mathcal{B}_0 u = \lambda \mathcal{B}_1 u \quad \text{on } \partial\Omega,$$

$$(1.3) \qquad\qquad \mathcal{B}_1' u = 0 \quad \text{on } \partial\Omega,$$

where we denote $\mathcal{B}_0 u = (B_{01}u, \ldots, B_{0r}u)^T$ etc.

Put $\mathbb{L}_2(\partial\Omega) := (L_2(\partial\Omega))^r$, $\mathcal{H} := L_2(\Omega) \times \mathbb{L}_2(\partial\Omega)$, and $H_{\mathcal{B}}^{2m}(\Omega) := \{u \in H^{2m}(\Omega) \mid \mathcal{B}_1' u = 0\}$, and define the operator $\mathcal{A}_0 : \mathcal{H} \to \mathcal{H}$ on

$$\mathcal{D}(\mathcal{A}_0) = \left\{ \begin{pmatrix} u \\ \mathcal{B}_1 u \end{pmatrix} \Big| \, u \in H_{\mathcal{B}}^{2m}(\Omega) \right\}$$

by

$$\mathcal{A}_0 \begin{pmatrix} u \\ \mathcal{B}_1 u \end{pmatrix} = \begin{pmatrix} Au \\ \mathcal{B}_0 u \end{pmatrix}.$$

Then problem (1.1)–(1.3) can be recast in the operator form

$$\mathcal{A}_0 \begin{pmatrix} u \\ \mathcal{B}_1 u \end{pmatrix} = \lambda \begin{pmatrix} u \\ \mathcal{B}_1 u \end{pmatrix}.$$

It is therefore natural to define the spectrum of problem (1.1)–(1.3) as that of the closure $\mathcal{A}$ of the operator $\mathcal{A}_0$. Observe that, in general, the spectrum of $\mathcal{A}$ is not discrete unless some additional requirements are imposed (see, e.g., [ES, Hi, KY] and the references therein). The main objective of the present note is to identify the essential spectrum of the operator $\mathcal{A}$ and hence of the original problem (1.1)–(1.3). A problem with $m = 1$ was considered in detail in [BHLN].

2 Preliminaries

Throughout the paper we denote by

$$H^s(\Omega), \quad H^t(\partial\Omega), \qquad s, t \in \mathbb{R},$$

the Sobolev spaces of functions in the domain Ω and on its boundary $\partial\Omega$, respectively;

$$\| \cdot \|_s \quad \text{and} \quad \langle \cdot \rangle$$

are the norms in $H^s(\Omega)$ and $L_2(\partial\Omega)$, respectively, see [LM]. If T is a linear operator in a Hilbert space X, we shall write $\mathcal{D}(T)$, $\rho(T)$, and $\sigma(T)$ for domain, resolvent set, and spectrum of T, respectively.

Our assumptions about the problem $\{A, \mathcal{B}_1 \cup \mathcal{B}'_1, \Omega\}$ imply the following statements.

Lemma 2.1 ([Ag]). *Denote by A_0 the operator in $L_2(\Omega)$ defined on*

$$\mathcal{D}(A_0) = \{u \in H^{2m}(\Omega) \mid \mathcal{B}_1 u = 0,\ \mathcal{B}'_1 u = 0\}$$

by $A_0 u = Au$. Then A_0 has a discrete spectrum.

Put $m_k := \operatorname{ord} B_{1k}$, $M_k = \operatorname{ord} B_{0k}$, $k = 1, \ldots, r$, and for any real s set

$$\mathbb{H}^{s-\overline{m}}(\partial\Omega) := \prod_{k=1}^{r} H^{s-m_k}(\partial\Omega), \qquad \mathbb{H}^{s-\overline{M}}(\partial\Omega) := \prod_{k=1}^{r} H^{s-M_k}(\partial\Omega).$$

Lemma 2.2 (see [LM, Section 2.7.3]). *Fix $\lambda_0 \in \rho(A_0)$ and $s \geq 0$. Then for any set of elements $g_k \in H^{s-1/2-m_k}(\partial\Omega)$ the problem*

$$Au - \lambda_0 u = 0, \tag{2.1}$$

$$B_{1k}u = g_k, \qquad k = 1, \ldots, r, \tag{2.2}$$

$$B_{1k}u = 0, \qquad k = r+1, \ldots, m, \tag{2.3}$$

has a unique solution $u = u_0$. Moreover, $u_0 \in H^s(\Omega)$ and the operator $D(\lambda_0)$: $(g_1, \ldots, g_r)^T \mapsto u_0$ acts boundedly from $\mathbb{H}^{s-1/2-\overline{m}}(\partial\Omega)$ to $H^s(\Omega)$.

Corollary 2.3 $D(\lambda_0)$ *is compact as an operator from $\mathbb{L}_2(\partial\Omega)$ to $L_2(\Omega)$.*

Lemma 2.4 *For any $\lambda_0 \in \rho(A_0)$, the operator $\mathcal{B}_0 D(\lambda_0)$ defined on $\mathcal{B}_1 H_{\mathcal{B}}^{2m}(\Omega)$ is closable as an operator in $\mathbb{L}_2(\partial\Omega)$.*

Proof: Note that $\mathcal{B}_0 D(\lambda_0)$ is well defined on $\mathcal{B}_1 H_{\mathcal{B}}^{2m}(\Omega)$. Indeed, for $v \in H_{\mathcal{B}}^{2m}(\Omega)$ the solution of problem (2.1)–(2.3) with $g_k = B_{1k}v, k = 1, \ldots, r$, is

$$u = D(\lambda_0)\mathcal{B}_1 v = v - (A_0 - \lambda_0 I)^{-1}(A - \lambda_0 I)v \in H_{\mathcal{B}}^{2m}(\Omega),$$

whence $\mathcal{B}_0 D(\lambda_0)\mathcal{B}_1 v = \mathcal{B}_0 u \in \mathbb{L}_2(\partial\Omega)$.

Suppose that a sequence $(\hat{g}_n) \subset \mathcal{B}_1 H_{\mathcal{B}}^{2m}(\Omega)$ is such that $\hat{g}_n \to 0$ and $\mathcal{B}_0 D(\lambda_0)\hat{g}_n \to \hat{g}_0$ in $\mathbb{L}_2(\partial\Omega)$ as $n \to \infty$; we have to prove that $\hat{g}_0 = 0$. It follows from [LM, Theorems 2.7.3 and 2.7.4] that, for any real $s \geq 0$, the mapping $\mathcal{B}_0 D(\lambda_0)$ defines a bounded operator from $\mathbb{H}^{s-1/2-\overline{m}}(\partial\Omega)$ into $\mathbb{H}^{s-1/2-\overline{M}}(\partial\Omega)$. Take $s = 1/2$; since the embeddings $\mathbb{L}_2(\partial\Omega) \hookrightarrow \mathbb{H}^{-\overline{m}}(\partial\Omega)$ and $\mathbb{L}_2(\partial\Omega) \hookrightarrow \mathbb{H}^{-\overline{M}}(\partial\Omega)$ are continuous, we have $\hat{g}_n \to 0$ in $\mathbb{H}^{-\overline{m}}(\partial\Omega)$, whence $\mathcal{B}_0 D(\lambda_0)\hat{g}_n \to 0$ in $\mathbb{H}^{-\overline{M}}(\partial\Omega)$ and $\hat{g}_0 = 0$. The proof is complete. □

Denote by $\mathcal{F}(\lambda_0)$ the closure of $\mathcal{B}_0 D(\lambda_0)$ in $\mathbb{L}_2(\partial\Omega)$.

Lemma 2.5 *For any $\lambda_1, \lambda_2 \in \rho(A_0)$, $\mathcal{F}(\lambda_1) - \mathcal{F}(\lambda_2)$ is a compact operator in $\mathbb{L}_2(\partial\Omega)$.*

Proof: It suffices to show that $\mathcal{B}_0 D(\lambda_1) - \mathcal{B}_0 D(\lambda_2)$ is compact on $\mathcal{B}_1 H_{\mathcal{B}}^{2m}(\Omega)$. Let $v \in H_{\mathcal{B}}^{2m}(\Omega)$ and $u_k = D(\lambda_k)\mathcal{B}_1 v, k = 1, 2$. Then $u = u_1 - u_2$ solves the problem

$$Au - \lambda_1 u = (\lambda_1 - \lambda_2)u_2,$$

$$B_1 u = B_1' u = 0.$$

Since $\lambda_1 \in \rho(A_0)$ and $u_2 \in L_2(\Omega)$, this implies that

$$u = (A_0 - \lambda_1 I)^{-1}(\lambda_1 - \lambda_2)u_2 \in H_{\mathcal{B}}^{2m}(\Omega),$$

i.e., that

$$(2.4) \qquad D(\lambda_1) - D(\lambda_2) = (\lambda_1 - \lambda_2)(A_0 - \lambda_1 I)^{-1} D(\lambda_2).$$

In particular, we find that $D(\lambda_1) - D(\lambda_2)$ is bounded from $\mathbb{L}_2(\partial\Omega)$ to $H^{2m}(\Omega)$, as

$$\|u\|_{2m} \leq C\|u_2\|_0 = C\|D(\lambda_2)\mathcal{B}_1 v\|_0 \leq C'\langle \mathcal{B}_1 v \rangle.$$

It remains to note that $\mathcal{B}_0 : H^{2m}(\Omega) \to \mathbb{L}_2(\partial\Omega)$ is compact, and the proof is complete. $\qquad\qquad\qquad\square$

Corollary 2.6 *The domain* $\mathcal{D}(\mathcal{F}(\lambda))$ *and the essential spectrum* $\sigma_{ess}(\mathcal{F}(\lambda))$ *of* $\mathcal{F}(\lambda)$ *do not depend on* $\lambda \in \rho(A_0)$.

3 The Essential Spectrum of $\mathcal{A}$

With the results of Section 2 in hand, we are in a position to prove that the operator $\mathcal{A}_0$ is closable and describe its closure. In the sequel, we denote by $\mathcal{D}_\mathcal{F}$ and Σ the domain and the essential spectrum of (any of) $\mathcal{F}(\lambda)$, $\lambda \in \rho(A_0)$; this is justified by Corollary 2.6.

Lemma 3.1 *The operator* $\mathcal{A}_0$ *is closable.*

Proof: It suffices to show that $\mathcal{A}_0 - \lambda\mathcal{I}$ is closable for some $\lambda \in \mathbb{C}$, where $\mathcal{I}$ is the identity operator in $\mathcal{H}$.

Fix an arbitrary $\lambda_0 \in \rho(A_0)$ (see Lemma 2.1) and suppose that a sequence $\mathbf{u}_n := (u_n, \mathcal{B}_1 u_n)^T \in \mathcal{D}(\mathcal{A}_0)$ is such that $\mathbf{u}_n \to 0$ and $(\mathcal{A}_0 - \lambda_0\mathcal{I})\mathbf{u}_n \to (v, \hat{g})^T$ in $\mathcal{H}$ as $n \to \infty$. In particular, $\mathcal{B}_1 u_n \to 0$ and $\mathcal{B}_0 u_n \to \hat{g}$ in $\mathbb{L}_2(\partial\Omega)$.

Set $v_n := D(\lambda_0)\mathcal{B}_1 u_n \in H_\mathcal{B}^{2m}(\Omega)$; then $(A - \lambda_0 I)v_n = 0$ and $\mathcal{B}_1 v_n = \mathcal{B}_1 u_n$. Therefore $u_n - v_n \in \mathcal{D}(A_0)$, $A_0(u_n - v_n) \to v$ and $u_n - v_n \to 0$ in $L_2(\Omega)$ by Lemma 2.2. Closedness of A_0 then implies that $v = 0$.

Now we can conclude by [Ag, Theorem 1.1] that the sequence $(u_n - v_n)$ converges to zero in $H^{2m}(\Omega)$, whence $\mathcal{B}_0 v_n = \mathcal{B}_0 D(\lambda_0)\mathcal{B}_1 u_n \to \hat{g}$ in $\mathbb{L}_2(\partial\Omega)$. Thus Lemma 2.4 implies that $\hat{g} = 0$ and hence $\mathcal{A}_0$ is closable. $\qquad\square$

Theorem 3.2 *The closure* $\mathcal{A}$ *of* $\mathcal{A}_0$ *is given on*

$$(3.1) \qquad \mathcal{D}(\mathcal{A}) := \left\{ \begin{pmatrix} v + D(\lambda)\hat{g} \\ \hat{g} \end{pmatrix} \,\middle|\, v \in \mathcal{D}(A_0), \ \hat{g} \in \mathcal{D}_\mathcal{F} \right\}$$

by

$$(3.2) \qquad \mathcal{A}\begin{pmatrix} v + D(\lambda)\hat{g} \\ \hat{g} \end{pmatrix} = \begin{pmatrix} A_0 v + \lambda D(\lambda)\hat{g} \\ \mathcal{B}_0 v + \mathcal{F}(\lambda)\hat{g} \end{pmatrix}.$$

Here λ *is an arbitrary point of* $\rho(A_0)$.

Proof: Denote by $\tilde{\mathcal{A}}$ the operator acting on domain (3.1) according to (3.2). It is well defined in the sense that both $\mathcal{D}(\tilde{\mathcal{A}})$ and $\tilde{\mathcal{A}}\mathbf{u}$ for $\mathbf{u} \in \mathcal{D}(\tilde{\mathcal{A}})$ are independent of $\lambda \in \rho(A_0)$. The first statement follows from the fact that by (2.4) $D(\lambda_1)\hat{g} - D(\lambda_2)\hat{g} \in \mathcal{D}(A_0)$ for any $\hat{g} \in \mathbb{L}_2(\partial\Omega)$ and any $\lambda_1, \lambda_2 \in \rho(A_1)$. For the second statement, set $u = v + D(\lambda)\hat{g}$, where $v \in \mathcal{D}(A_0)$ and $\hat{g} \in \mathcal{D}_{\mathcal{F}}$. Then by Lemma 2.2 we have (in the distribution sense) $\hat{g} = \mathcal{B}_1 u$, $A_0 v + \lambda D(\lambda)\hat{g} = A_0 v + A D(\lambda)\hat{g} = Au$, and $\mathcal{B}_0 v + \mathcal{F}(\lambda)\hat{g} = \mathcal{B}_0 u$; hence (3.2) amounts to

$$\tilde{\mathcal{A}}\begin{pmatrix} u \\ \mathcal{B}_1 u \end{pmatrix} = \begin{pmatrix} Au \\ \mathcal{B}_0 u \end{pmatrix}$$

in the sense of distributions. Moreover, since both A_0 and $\mathcal{F}(\lambda)$ are closed, $\tilde{\mathcal{A}}$ is easily shown to be closed, too.

If we take $\hat{g} \in \mathcal{B}_1 H_{\mathcal{B}}^{2m}(\Omega)$, then all the equalities above hold in the classical sense, which implies that $A_0 \subset \tilde{\mathcal{A}}$ and $\mathcal{A} \subset \tilde{\mathcal{A}}$. Conversely, for any $\hat{g} \in \mathcal{D}_{\mathcal{F}}$ there exists a sequence $(\hat{g}_n) \subset \mathcal{B}_1 H_{\mathcal{B}}^{2m}(\Omega)$ such that $\hat{g}_n \to \hat{g}$ and $\mathcal{B}_0 D(\lambda)\hat{g}_n \to \mathcal{F}(\lambda)\hat{g}$ in $\mathbb{L}_2(\partial\Omega)$ as $n \to \infty$. Put $u_n := v + D(\lambda)\hat{g}_n$; then $\mathbf{u}_n := (u_n, \hat{g}_n)^T$ belongs to $\mathcal{D}(A_0)$, $\mathbf{u}_n \to (v + D(\lambda)\hat{g}, \hat{g})^T$ and

$$A_0 \mathbf{u}_n = \begin{pmatrix} Av + \lambda D(\lambda)\hat{g}_n \\ \mathcal{B}_0 v + \mathcal{B}_0 D(\lambda)\hat{g}_n \end{pmatrix} \to \begin{pmatrix} Av + \lambda D(\lambda)\hat{g} \\ \mathcal{B}_0 v + \mathcal{F}(\lambda)\hat{g} \end{pmatrix}$$

in $\mathcal{H}$ as $n \to \infty$, which implies that $\tilde{\mathcal{A}} \subset \mathcal{A}$ and hence $\mathcal{A} = \tilde{\mathcal{A}}$. $\square$

Next we study the essential spectrum of $\mathcal{A}$. We do this in two stages, the first being as follows.

Lemma 3.3 *Fix an arbitrary $\lambda_0 \in \rho(A_0)$ and denote by $\widehat{\mathcal{A}}$ the closure of the operator $\widehat{\mathcal{A}}_0$ defined on $\mathcal{D}(A_0)$ by*

$$\widehat{\mathcal{A}}_0 \begin{pmatrix} u \\ \mathcal{B}_1 u \end{pmatrix} = \begin{pmatrix} Au \\ \mathcal{F}(\lambda_0)\mathcal{B}_1 u \end{pmatrix}.$$

Then $\sigma_{ess}(\widehat{\mathcal{A}}) = \Sigma$.

Proof: Note that the operator $\widehat{\mathcal{A}}$ is given by formulae (3.1) and (3.2) without the term $\mathcal{B}_0 v$ in the latter; the proof is similar to that of Theorem 3.2.

Suppose first that $\Sigma \neq \mathbb{C}$. Let $\widehat{\mathcal{A}}' := A_0 \oplus \mathcal{F}(\lambda_0)$ and take any $\lambda \in \rho(\widehat{\mathcal{A}}') = \rho(A_0) \cap \rho(\mathcal{F}(\lambda_0))$. Then λ is not an eigenvalue of $\widehat{\mathcal{A}}$; otherwise for the corresponding (nonzero) eigenvector $(v + D(\lambda)\hat{g}, \hat{g})^T \in \mathcal{D}(\widehat{\mathcal{A}})$ we would have

$$\begin{aligned} Av - \lambda v &= 0, \\ (\mathcal{F}(\lambda_0) - \lambda)\hat{g} &= 0, \end{aligned}$$

whence $\lambda \in \sigma(\mathcal{F}(\lambda_0))$ or $\lambda \in \sigma(A_0)$, a contradiction.

For an arbitrary $f \in L_2(\Omega)$ and $\hat{g} \in \mathbb{L}_2(\partial\Omega)$ then

$$(\widehat{\mathcal{A}} - \lambda\mathcal{I})^{-1}\begin{pmatrix} f \\ \hat{g} \end{pmatrix} - (\widehat{\mathcal{A}'} - \lambda\mathcal{I})^{-1}\begin{pmatrix} f \\ \hat{g} \end{pmatrix} = \begin{pmatrix} w \\ 0 \end{pmatrix},$$

where w solves the problem

$$\begin{aligned} Aw - \lambda w &= 0, \\ \mathcal{B}_1 w &= (\mathcal{F}(\lambda_0) - \lambda)^{-1}\hat{g}, \\ \mathcal{B}_1' w &= 0. \end{aligned}$$

Therefore $w = D(\lambda)(\mathcal{F}(\lambda_0) - \lambda)^{-1}\hat{g}$ and

$$(\widehat{\mathcal{A}} - \lambda\mathcal{I})^{-1} - (\widehat{\mathcal{A}'} - \lambda\mathcal{I})^{-1} = \begin{pmatrix} 0 & D(\lambda)(\mathcal{F}(\lambda_0) - \lambda)^{-1} \\ 0 & 0 \end{pmatrix}$$

is a compact operator by Corollary 2.3. It follows from [Ka, Theorem IV.5.35] that $\sigma_{ess}(\widehat{\mathcal{A}}) = \sigma_{ess}(\widehat{\mathcal{A}'}) = \Sigma$.

If $\Sigma = \mathbb{C}$, we take any $\lambda \in \rho(A_0)$. By Corollary 2.6 then $\lambda \in \sigma_{ess}(\mathcal{F}(\lambda))$ and hence there exists a (Weyl) sequence $(\hat{g}_n) \subset \mathcal{D}_{\mathcal{F}}$ such that $\langle \hat{g}_n \rangle = 1$,

$$\mathcal{F}(\lambda)\hat{g}_n - \lambda\hat{g}_n \to 0,$$

but $(\hat{g}_n)$ does not contain any convergent subsequence. Then $(u_n, \hat{g}_n)^T$ with $u_n := D(\lambda)\hat{g}_n$ is a Weyl sequence for the operator $\widehat{\mathcal{A}} - \lambda\mathcal{I}$. As a result, $\lambda \in \sigma_{ess}(\widehat{\mathcal{A}})$ and $\sigma_{ess}(\widehat{\mathcal{A}}) = \mathbb{C}$.   □

The second stage is to characterize the essential spectrum of $\mathcal{A}$ by relating it to that of $\widehat{\mathcal{A}}$.

Theorem 3.4 $\sigma_{ess}(\mathcal{A}) = \Sigma$.

Proof: For any $(u, \mathcal{B}_1 u)^T \in \mathcal{D}(A_0)$ we have

$$\mathcal{A}\begin{pmatrix} u \\ \mathcal{B}_1 u \end{pmatrix} - \widehat{\mathcal{A}}\begin{pmatrix} u \\ \mathcal{B}_1 u \end{pmatrix} = \begin{pmatrix} 0 \\ B_0(u - D(\lambda_0)\mathcal{B}_1 u) \end{pmatrix}.$$

Put $w := u - D(\lambda_0)\mathcal{B}_1 u$; then w solves the problem

$$\begin{aligned} Aw - \lambda_0 w &= Au - \lambda_0 u \in L_2(\Omega), \\ \mathcal{B}_1 w = \mathcal{B}_1' w &= 0, \end{aligned}$$

and hence $w \in \mathcal{D}(A_0) \subset H^{2m}(\Omega)$, $B_0 w \in \mathbb{H}^{1/2}(\partial\Omega) := \prod_{k=1}^r H^{1/2}(\partial\Omega)$. Since the embedding $\mathbb{H}^{1/2}(\partial\Omega) \hookrightarrow L_2(\partial\Omega)$ is compact, $\mathcal{A} - \widehat{\mathcal{A}}$ is compact relative to $\mathcal{A}$, whence $\sigma_{ess}(\mathcal{A}) = \sigma_{ess}(\widehat{\mathcal{A}})$ by [Ka, Theorem IV.5.35]. The statement now follows from Lemma 3.3.   □

The set Σ, which by the previous theorem gives the essential spectrum of the operator $\mathcal{A}$, is defined only implicitly in terms of the original operators. To conclude this section, we examine a case of common occurrence (cf. Example 4.1 below), where the connection is more explicit.

Corollary 3.5 *Suppose that there exists a representation* $\mathcal{B}_0 = \mathfrak{A}\mathcal{B}_1 + \mathcal{C}$, *where* $\mathfrak{A}$: $\mathbb{L}_2(\partial\Omega) \to \mathbb{L}_2(\partial\Omega)$ *is some linear operator and* ord $\mathcal{B}_1 >$ ord $\mathcal{C}$ *componentwise. Then* $\sigma_{ess}(\mathcal{A}) = \sigma_{ess}(\mathfrak{A})$.

Proof: Under the assumptions of the lemma, for any $\hat{g} \in \mathcal{B}_1 H_{\mathcal{B}}^{2m}(\Omega)$ and any $\lambda \in \rho(A_0)$, the following relation holds:

$$\mathcal{B}_0 D(\lambda)\hat{g} = \mathfrak{A}\mathcal{B}_1 D(\lambda)\hat{g} + \mathcal{C}D(\lambda)\hat{g} = \mathfrak{A}\hat{g} + \mathcal{C}D(\lambda)\hat{g}.$$

Note that $\mathcal{C}D(\lambda)$ is a compact operator, and the claim follows. $\square$

Remark 3.6 A posteriori $\mathfrak{A}$, being a compact perturbation of $\mathcal{F}(\lambda)$, is closed.

4 Examples

Example 4.1 (cf. [BHLN]). Consider the spectral problem

(4.1) $$-\Delta u = \lambda u \quad \text{in } \Omega,$$
(4.2) $$B_0 u = \lambda B_1 u \quad \text{on } \partial\Omega$$

with $B_j = \alpha_j \frac{\partial}{\partial\nu} + \beta_j \gamma$, $j = 0, 1$, in a bounded smooth domain $\Omega \subset \mathbb{R}^n$. Here γ and $\frac{\partial}{\partial\nu}$ are the trace and the normal derivative, respectively, α_j and β_j, $j = 0, 1$, are smooth complex-valued functions on $\partial\Omega$, and α_1 does not vanish on $\partial\Omega$.

It is easily seen that the problem $\{-\Delta, B_1, \Omega\}$ is regular elliptic and hence Theorem 3.4 applies. Moreover, we have

$$B_0 = \frac{\alpha_0}{\alpha_1} B_1 - \frac{\alpha_0\beta_1 - \alpha_1\beta_0}{\alpha_1}\gamma,$$

whence the operator $\mathfrak{A}$ of Corollary 3.5 is the multiplication operator by the function α_0/α_1 in $L_2(\partial\Omega)$. Therefore the essential spectrum of problem (4.1)–(4.2) coincides with the set

$$\Sigma := \left\{ \frac{\alpha_0(x)}{\alpha_1(x)} \,\Big|\, x \in \partial\Omega \right\}.$$

Example 4.2 Consider the spectral problem

$$(4.3) \qquad\qquad \Delta^2 u = \lambda u \qquad \text{in } \Omega,$$

$$(4.4) \qquad\qquad \frac{\partial^2 u}{\partial \nu^2} = \lambda \gamma u \qquad \text{on } \partial\Omega,$$

$$(4.5) \qquad\qquad \gamma u = \lambda \frac{\partial u}{\partial \nu} \qquad \text{on } \partial\Omega$$

in a bounded smooth domain $\Omega \subset \mathbb{R}^n$.

The operator A_0 of Lemma 2.1 is the Dirichlet biharmonic operator in the domain Ω, hence A_0 is selfadjoint and uniformly positive and we can take $\lambda_0 = 0$ in Lemma 2.2. Now the operator $\mathcal{F}(0) := \overline{B_0 D(0)}$ acts in $\mathbb{L}_2(\partial\Omega)$ and has the matrix form

$$\mathcal{F}(0) = \begin{pmatrix} F_{11} & F_{12} \\ I & 0 \end{pmatrix},$$

where F_{11} and F_{12} are pseudodifferential operators of orders 2 and 1, respectively (see, e.g., [VG, Ko]).

It follows from the results of [ALMS] that, for any $\lambda \in \rho(F_{11})$,

$$\sigma_{ess}(\mathcal{F}(0)) = \sigma_{ess}(-(F_{11} - \lambda I)^{-1} F_{12}).$$

Since $(F_{11} - \lambda I)^{-1} F_{12}$ is a pseudodifferential operator of order -1 and hence compact in $L_2(\partial\Omega)$ [Ho], we have

$$\Sigma = \{0\},$$

whence the essential spectrum of problem (4.3)–(4.5) consists of the singleton $\{0\}$.

Acknowledgements

Research of the first author was supported by NSERC of Canada. The second author acknowledges appointment as a Post Doctoral Fellow of the Pacific Institute for the Mathematical Sciences at the University of Calgary.

References

[Ag]   S. Agmon, On the eigenfunctions and on the eigenvalues of general elliptic boundary value problems, *Comm. Pure Appl. Math.* **15** (1962), 119–147.

[ALMS]   F.V. Atkinson, H. Langer, R. Mennicken and A.A. Shkalikov, The essential spectrum of some matrix operators, *Math. Nachr.* **167** (1994), 5–20.

[BHLN] P. Binding, R. Hryniv, H. Langer and B. Najman, Elliptic eigenvalue problems with eigenparameter dependent boundary conditions, *submitted.*

[ES]   J. Ercolano and M. Schechter, Spectral theory for operators generated by elliptic boundary problems with eigenvalue parameter in boundary conditions, I, *Comm. Pure Appl. Math.* **18** (1965), 83–105.

[Hi]   T. Hintermann, Evolution equations with dynamic boundary conditions, *Proc. Roy. Soc. Edinburgh* **113A** (1989), 43–60.

[Ho]   L. Hörmander, *The Analysis of Linear Partial Differential Operators, III. Pseudo-Differential Operators*, Grundlehren der Mathematischen Wissenschaften, Band 274, Springer-Verlag, Berlin 1994.

[Ka]   T. Kato, *Perturbation Theory for Linear Operators*, 2nd ed., Grundlehren der Mathematischen Wissenschaften, Band 132, Springer-Verlag, Berlin-New York 1976.

[Ko]   A.N. Kozhevnikov, Spectral problems for pseudodifferential systems that are elliptic in the sense of Douglis-Nirenberg, and their applications (Russian), *Math. USSR Sb. (N.S.)* **92(134)** (1973), 60–88.

[KY]   A. Kozhevnikov and S. Yakubov, On operators generated by elliptic boundary problems with a spectral parameter in boundary conditions, *Integral Equations Operator Theory* **23** (1995), 205–231.

[LM]   J.-L. Lions and E. Magenes, *Non-Homogeneous Boundary Value Problems and Applications, I*, Grundlehren der Mathematischen Wissenschaften, Band 181, Springer-Verlag, Berlin-Heildelberg-New York 1972.

[VG]   B.R. Vainberg and V.V. Grushin, Uniformly nonelliptic problems. II (Russian), *Math. USSR Sb. (N.S.)* **73(115)** (1967), 126–159.

Paul Binding
Department of Mathematics
   and Statistics
University of Calgary
Calgary, AB
Canada T2N 1N4

Rostyslav Hryniv
Institute for Applied Problems
   of Mechanics and Mathematics
3b Naukova str.
290601 Lviv
Ukraine

1991 Mathematics Subject Classification. Primary: 35P05; Secondary: 35J40, 47A10, 47G30.

Operator Theory:
Advances and Applications, Vol. 124
© 2001 Birkhäuser Verlag Basel/Switzerland

About Scattering on the Ring

V.B. Bogevolnov, A.B. Mikhailova,
B.S. Pavlov and A.M. Yafyasov

This paper is dedicated to Professor I. Gohberg, on the occasion of his 70-th birthday. His trust in the applied value of the spectral theory of operators was our support and inspiration

The mathematical model of a simplest quasi-one-dimensional quantum network constructed of relatively narrow waveguides (the width of the waveguide is less than the de Broghlie wavelength of the electron in the material) is developed. This model allows to reduce the problem of calculating the current through the quantum network to the construction of scattered waves for some Schrödinger equation on the corresponding one-dimensional graph. We consider a graph consisting of a compact part and few semiinfinite rays attached to it via some boundary condition depending on a parameter β (analog of the inverse exponential "mass" e^{-bH} of a potential barrier H separating the rays from the compact part of the graph). This parameter regulates the connection between the rays and the compact part. Spectral properties of the Schrödinger operator on this graph are described with a special emphasis on the resonance case when the Fermi level in the rays coincides with one of eigenvalues of the nonperturbed Schrödinger operator on the ring. An explicit expression is obtained for the scattering matrix in the resonance case for weakening connection $\beta \to 0$ between the rays and the compact part.

1 Introduction

The spectral properties of the Schrödinger equation on graphs, see [3] and the most complete bibliography there, possess new interesting properties, which never appear for one-dimensional Schrödinger Operator on the real axis. For instance, the reflection coefficient on a homogeneous ring (length 2π) with one semi-infinite ray attached to it by "zero-current condition" (see Section 2) reveals a periodic behaviour in momentum at infinity:

$$S(k) = \frac{i - tgk\pi}{i + tgk\pi}.$$

In particular it does not approach 1 when $k \to \infty$.

The most important of the characteristic features which sharply distinguish the Schrödinger equation on graph from the Schrödinger equation on real axis is absence of a global solution of Cauchy problem: the solution exists only on the edge containing the initial point, but generally cannot be continued in a unique way across the neighboring node with 3 or more edges adjacent to it. In this respect the

Schrödinger equation on a graph takes an *intermediate position between ordinary and partial Schrödinger equations on a corresponding domain*, see also [4].

The modern interest to the investigation of spectral properties of the Schrödinger Operators on graph (see [1], [2], [3]) is partially motivated by the fact that despite the absence of "global" solutions of Cauchy problem we still may describe the whole set of solutions of the corresponding homogeneous differential equation on graph as a spline of solutions of Cauchy problem for ordinary differential equations on edges with proper boundary conditions at the nodes.

On the other hand the one-dimensional scalar Schrödinger equation on a graph is distinguished from a system of differential equations on a real axis because of *locality* of the corresponding potential: even if we assume that the solutions on different edges are different components of one vector function, we see, that the potential should be represented by some diagonal matrix: one scalar function $q_l(x)$ on each edge of the graph. In this representation all essential information which permits continuation of the solution from one edge to another is encoded in the boundary conditions at the nodes, see [1], [5]. Unfortunately this approach to the spectral theory of the Schrödinger Operator on graphs via Operator Extensions looks still too general to reveal the characteristic properties of differential operators on graphs approximating smooth manifolds.

In the present paper we consider the one-dimensional Schrödinger equation on a graph constructed of a compact part Γ_0 represented as a sum of oriented edges joined at the nodes with some self-adjoint boundary conditions connecting the boundary values of the wave function at the incident edges and with a finite number N of semi-infinite rays Γ_s : $0 < x_s < \infty$, $s = 1, \ldots, N$, attached to the compact subgraph at the *vertices* $x = a_1, \ldots, a_N$ which are *inner points* of some edges of the compact part Γ_0

$$-u_0'' + qu_0 = \lambda u_0,$$
$$-u_s'' = \lambda u_s, \ s = 1, 2, \ldots, N,$$

with proper boundary conditions at the vertices. We assume that only one ray is attached to each vertex. These boundary conditions correspond to selecting of Lagrangian planes of some symplectic boundary form (see for instance [2], [6], [8]). We assume that the potential $q(x)$ is a real bounded measurable function on the compact part Γ_0 and vanishes on the rays, $q(x_s) = 0$, $s = 1, 2, \ldots, N$. We choose the boundary conditions such that the component u_0 of the total wave function on the compact part Γ_0 is a continuous function and the boundary conditions connect the values of it and the jump of its derivative at the inner point – vertex – a_s of the oriented edge (arc) in Γ_0 to the boundary values $u_s(0)$, $u_s'(0)$ of the component of the wave function on the ray attached to a_s.

The object we get in this way

$$\Gamma_0 + \sum_{s=1}^{N} \Gamma_s$$

is a special sort of graph where the inner *nodes* of it with general self-adjoint boundary conditions and *vertices* a_s with special boundary conditions are in fact the elements of similar nature. Still we prefer to distinguish them as nodes and vertices, because of the special role of boundary conditions assigned to vertices.

The simplest but still nontrivial graph which possesses the features mentioned above is just a ring with few rays attached to it. Further we call our graph just "ring" but in fact the whole analysis is valid for any compact graph with few semi-infinite rays attached to it *as described above*.

We consider below a one-parameter family of *special* boundary conditions (see Section 2) which correspond to the weakening connection ($\beta \to 0$) between the rays and the compact part Γ_0. One can show that these boundary conditions simulate the interaction between real quantum wires when the rays are joined to the ring non directly but are connected to it via quantum tunnelling through the potential barrier with the total "mass" proportional to $\ln \frac{1}{|\beta|}$. Our analysis shows that the resonance properties of the corresponding scattering matrices are defined by the properties of eigenfunctions of the Schrödinger operator on the compact part Γ_0. In particular for the resonance case when $\lambda = k^2$ is a simple eigenvalue of the Schrödinger operator on the compact part Γ_0 the transmission coefficients $S_{st}(k)$ for pairs of rays attached at the points a_s, a_t are essentially defined by the products $\varphi_\lambda(a_s)\varphi_\lambda(a_t)$ of values of the corresponding eigenfunction at the vertices:

$$S_{st}(\lambda) = \frac{2}{\sum_r |\varphi_\lambda(a_r)|^2} \varphi_\lambda(a_s)\varphi_\lambda(a_t) + O(|\beta|^2).$$

Of course the limit value of the transmission coefficient for $\beta \to 0$ is not a continuous function of energy near the point λ, so, though the limit value of it for $\beta = 0$ is finite, practically the average value of it over Fermi distribution tends to zero when $\beta \to 0$ for any (positive) value of absolute temperature.

The last formula shows that the quantum current from one ray to another in *the resonance situation* when Fermi level in the rays is equal to some eigenvalue of the Schrödinger operator of the compact part can be controlled by the classical electric field applied to the ring. The physical meaning and technical implementation of this phenomenon will be discussed in following publications.

2 Schrödinger Operator on the Graph

In this section we collect several facts about graphs formulated in a convenient form. We use them in following sections.

Consider the Schrödinger operator defined by the differential expression

$$l_0 u_0 = -u_0'' + q(x)u_0$$

on the "ring" Γ_0 with real bounded measurable potential q and some general self-adjoint boundary conditions at the nodes of Γ_0. We assume that few semi-infinite rays $\Gamma_s, 0 \leq x_s < \infty$, are attached to the "ring" at the vertices $a_1, a_2, \ldots, a_s, \ldots, a_N \subset \Gamma_0$, $x_s|_{a_s} = 0$, the vertices being the inner points of some oriented edges (arcs) of Γ_0 where the wave functions is smooth one:

$$u_0(a_s - 0) = u_0(a_s + 0) = u_0(a_s), \quad u_0'(a_s - 0) = u_0'(a_s + 0).$$

We relate the Schrödinger operator L_0 with the Schrödinger operators L_s on the rays defined by the differential expressions

$$l_s u_s = -u_s'',$$

restricting all of them onto the subspace of all smooth functions vanishing near the vertices a_s, $s = 1, \ldots, N$ and then extending them with the boundary conditions connecting the jump of the derivative $[u_0']|_{a_s}$ of the *continuous* function u_0 on the corresponding oriented edge of Γ_0 with the boundary values $u_s(0)$, $u_s'(0)$ of the component of the wave function u_s on the ray Γ_s at the corresponding vertex:

$$u_0'(a_s + 0) - u_0'(a_s - 0) = [u_0']|_{a_s},$$

(2.1)
$$u_0(a_s - 0) = u_0(a_s + 0) = u_0(a_s),$$

(2.2)
$$\begin{pmatrix} [u_0']|_{a_s} \\ u_s(0) \end{pmatrix} = B_s \begin{pmatrix} u_0(a_s) \\ -u_s'(0) \end{pmatrix}$$

generally by some Hermitian matrix

$$B_s = B_s^* = \begin{pmatrix} \beta_{00}^s & \beta_{01}^s \\ \beta_{10}^s & \beta_{11}^s \end{pmatrix}.$$

Further we assume that

$$B_s = B_s^* = \begin{pmatrix} \beta_{00}^s & \beta_{01}^s \\ \beta_{10}^s & \beta_{11}^s \end{pmatrix} := \begin{pmatrix} 0 & \beta \\ \bar{\beta} & 0 \end{pmatrix},$$

β is the same for all vertices a_s. We call these boundary conditions *special* boundary conditions. Choosing $\beta = 1$ we receive "zero-current condition", but choosing $\beta \to 0$ we get the sequence of scattering problems with weakening connection between the compact part Γ_0 and the rays.

Theorem 2.1 *The operator $\mathcal{L}$ defined in $L_2(\Gamma_0) \oplus \sum_{s=1}^N L_2(\Gamma_s)$ by the differential expression $l_0 \oplus \sum_{s=1}^N l_s$ is essentially self-adjoint in the domain D_0 consisting of all smooth functions defined on the graph Γ which satisfy the boundary conditions (2.1), (2.2).*

Proof: One can easily check the symmetry of this operator just integrating by parts: for $u, v \in D_0$ due to the boundary conditions (2.1), (2.2) we have

$$\langle \mathcal{L}u, v \rangle - \langle u, \mathcal{L}v \rangle = 0.$$

On the other hand the adjoint operator $\mathcal{L}^*$ is defined in $L_2(\Gamma_0) \oplus \sum_{s=1}^{N} L_2(\Gamma_s)$ by the same differential expression $l_0 \oplus \sum_{s=1}^{N} l_s$ on the domain consisting of L_2 – functions with square-integrable first and second derivatives which satisfy the same boundary conditions at the vertex. Really for $u \in D_0$, $v \in D_0^*$ we have zero *boundary form*

$$\langle \mathcal{L}u, v \rangle - \langle u, \mathcal{L}^*v \rangle = 0.$$

Denoting by $[f]|_a$ the jump $f(a+0) - f(a-0)$ of the function f at the vertex a and by $\{f\}|_a$ the mean value $\frac{f(a-0)+f(a+0)}{2}$ of it we can represent the boundary form as follows:

$$\langle \mathcal{L}^*u, v \rangle - \langle u, \mathcal{L}^*v \rangle$$
$$= \sum_{s=1}^{N}([u_0']|_{a_s}\overline{\{v_0\}}|_{a_s} - \{u_0\}|_{a_s}\overline{[v_0']}|_{a_s})$$
$$+ \sum_{s=1}^{N}(\{u_0'\}|_{a_s}\overline{[v_0]}|_{a_s} - [u_0]|_{a_s}\overline{\{v_0'\}}|_{a_s})$$
$$+ \sum_{s=1}^{N}(u_s'(0)\overline{v_s(s)} - u_s(0)\overline{v_s'(s)}).$$

The special choice of functions $u_0, u_1, \ldots, u_N$ satisfying conditions $u_0(a_s) = u_0'(a_s) = u_s(0) = u_s'(0) = 0$ for each vertex except a_t and $u_0(a_t) = [u_0']|_{a_t} = u_s(0) = u_s'(0) = 0$, $\{u_0'\}|_{a_t} = 1$ permits to deduce from the vanishing boundary form that v_0 should be continuous at the vertex $[v_0]|_{a_s} = 0$. Then expressing $[u_0']|_{a_t}$, $u_t(0)$ in terms of $u_0(a_t)$, $u_t'(0)$ via the boundary conditions we deduce from the independence of the initial values $u_0(a_t)$, $u_t'(0)$ on the ray Γ_s that the boundary values of the element $v \in D_0^*$ at the vertex a_s satisfy the same boundary condition as the boundary values of $u \in D_0$. Then using the smoothness of v, $v \in W_2^2(\Gamma)$ we deduce[1] from integration by parts that $\mathcal{L}^*$ is symmetric and hence it coincides with the closure of $\mathcal{L}$:

$$\mathcal{L}^* \subseteq (\mathcal{L}^*)^* = \overline{\mathcal{L}} = (\overline{\mathcal{L}}^*)^* \subseteq \mathcal{L}^*.$$

This implies the essential self-adjointness

$$\overline{\mathcal{L}} = \mathcal{L}^*.$$

$\square$

---

[1] Sobolev class W_2^2 is embedded into the class C_1 of all continuous and continuously differentiable functions on each component Γ_s, $s = 0, 1, \ldots, N$ of the graph Γ hence the integration by parts with elements $v \in D_0^*$ is possible.

The resolvent kernel (Green function) of the operator $\mathcal{L}$ can be obtained as a solution of the corresponding inhomogeneous equation

$$\mathcal{L}g = \lambda g + \delta(x - \xi).$$

We shall represent it via the resolvent kernel (Green function) of the *nonperturbed* operator $\mathcal{L}^0 := \mathcal{L}_0^0 \oplus \sum_{s=1}^{N} \mathcal{L}_s^0$ which is defined by the same differential expression in $L_2(\Gamma_0) \oplus \sum_{1 \leq s \leq N} L_2(\Gamma_s)$ with the self-adjoint boundary conditions at the nodes of the ring Γ_0 and separating homogeneous boundary conditions $\beta = 0$ at the vertices:

$$[u_0']|_{a_s} = 0, \quad u_s(0) = 0, \quad s = 1, 2, \ldots, N.$$

This operator is a limit of the sequence of operators corresponding to weakening family of boundary conditions (2.1), (2.2) when $|\beta| \longrightarrow 0$. We assume that the eigenvalues, eigenfunctions and the resolvent kernel $g_0^0(x, \xi, \lambda)$ of the component $\mathcal{L}_0^0$ of the nonperturbed operator $\mathcal{L}^0$ on the ring Γ_0

$$-\frac{d^2 g_0^0(x, \xi, \lambda)}{dx^2} + q(x)g_0^0(x, \xi, \lambda) = \lambda g_0^0(x, \xi, \lambda) + \delta(x - \xi)$$

are known.

Theorem 2.2 *The spectrum of the operator $\mathcal{L}_0^0$ is discrete and the resolvent kernel of it is represented as a sum of an absolutely and uniformly convergent series*

$$g_0^0(x, \xi, \lambda) = \sum_{s=1}^{\infty} \frac{\varphi_s(x)\varphi_s(\xi)}{\lambda_s - \lambda},$$

where $\{\varphi_s\}$ are the normalized eigenfunctions of $\mathcal{L}_0^0$

$$\mathcal{L}_0^0 \varphi_s = \lambda_s \varphi_s, \quad |\varphi_s|_{L_2(\Gamma_0)} = 1.$$

The system $\{\varphi_s\}$ of all eigenfunctions is automatically orthogonal and complete if the spectrum of $\mathcal{L}_0^0$ is simple. In the case of multiple spectrum a normalized orthogonal system of eigenfunctions may be chosen as well.

When constructing the Green function of the perturbed operator $\mathcal{L}$ we use the fact that the Green function $g_0^0(x, \xi, \lambda)$ satisfies the homogeneous Schrödinger equation on Γ_0

$$l_0 g_0^0(x, \xi, \lambda) = \lambda g_0^0(x, \xi, \lambda), \ x \neq \xi,$$

and the boundary condition at the point ξ:

$$[g_0']|_\xi = -1.$$

The essential part of the Proof of the Theorem 2.2 - the convergence of the spectral series for the Green function g_0^0 may be obtained from embedding theorems (see also [1]). It is worth to note here that the regular asymptotics of eigenvalues at infinity is generally absent in this case because of mixing terms corresponding to nonconmeasurable edges as the following simple example shows.

Example 2.3 Consider a ring $q = 0$, $0 \le x \le 2\pi$ with nodes at the points a_1, a_2, $0 < a_1 - a_2 = \Delta < \pi$, and the boundary conditions $[u_0']|_{a_s} = \beta u_0|_{a_s}$. The resolvent kernel $g(x, \xi, \lambda)$ on the ring with these boundary conditions is represented as a linear combination of the Green functions $G(x, \xi, \lambda)$ on the "empty ring" with no boundary conditions:

$$G(x, s, \lambda) = -\frac{\cos(x - \pi - s)\sqrt{\lambda}}{2\sqrt{\lambda} \sin \pi \sqrt{\lambda}}, \quad |x - \pi - s| < \pi,$$

in the form $g(x, \xi, \lambda) = G(x, \xi, \lambda) + u_1 G(x, a_1, \lambda) + u_2 G(x, a_2, \lambda)$ where u_s, $s = 1, 2$ may be found from the linear system

$$-u_1 = \beta[G(a_1, \xi, \lambda) + u_1 G(a_1, a_1, \lambda) + u_2 G(a_1, a_2, \lambda)],$$

$$-u_2 = \beta[G(a_2, \xi, \lambda) + u_1 G(a_2, a_1, \lambda) + u_2 G(a_2, a_2, \lambda)],$$

with the determinant

$$\det \begin{pmatrix} \frac{\cos \pi \sqrt{\lambda}}{2\sqrt{\lambda} \sin \pi \sqrt{\lambda}} - \beta^{-1} & \frac{\cos(\pi - \Delta)\sqrt{\lambda}}{2\sqrt{\lambda} \sin \pi \sqrt{\lambda}} \\ \frac{\cos(\pi - \Delta)\sqrt{\lambda}}{2\sqrt{\lambda} \sin \pi \sqrt{\lambda}} & \frac{\cos \pi \sqrt{\lambda}}{2\sqrt{\lambda} \sin \pi \sqrt{\lambda}} - \beta^{-1} \end{pmatrix}$$

which vanishes if

$$\cos \pi \sqrt{\lambda} - 2\sqrt{\lambda}\beta^{-1} \sin \pi \sqrt{\lambda} = \pm \cos(\pi - \Delta)\sqrt{\lambda}.$$

If Δ and π are not conmeasurable then the set of zeroes of the determinant is "disordered" as a set of roots of a sum of two periodic functions with nonconmeasurable periods.

Further we use the fact that generally the component $g_0(x, \xi, \lambda)$ of the Green function of the "perturbed operator" $\mathcal{L}$ on the compact subgraph Γ_0 may be found as a linear combination of Green functions of the nonperturbed operator $\mathcal{L}_0^0$ attached to the pole ξ and the nodes. One can easily see that in the generic case when all eigenvalues of the operator $\mathcal{L}_0^0$ are simple we must distinguish the following situations:

1. For a given eigenvalue λ_0 the corresponding eigenfunction φ_0 of $\mathcal{L}_0^0$ vanishes at *all vertices* a_s. In this case the function φ_0 being continued as identical zero onto all rays satisfies the boundary conditions (2.1, 2.2) hence the continued function is an eigenfunction of the perturbed operator. It is obviously orthogonal to the subspace of absolutely continuous spectrum of the

perturbed operator in both cases when $\lambda_0 < 0$ or $\lambda_0 > 0$. In the second case λ_0 proves to be *imbedded* eigenvalue. The existence of imbedded eigenvalues (even for compactly-supported potentials) is a characteristic feature of Schrödinger operators on graphs.

2. For a given eigenvalue λ_0 of $\mathcal{L}_0^0$ there exists at least one vertex a_s such that the corresponding eigenfunction φ_0 does not vanish at a_s, $\varphi_0(a_s) \neq 0$. In this case the spectral point λ_0 will not be the eigenvalue of the perturbed operator at least for small values of β, i.e. in the case of weakly connected inner Γ_0 and outer channels Γ_s. We shall give the proof of this statement in the next section as a corollary of more general statement on "compensation of singularities". We shall show also that for negative λ_0 and weakly connected channels there exists a negative eigenvalue of the perturbed operator close to it:

$$\lambda_\beta = \lambda_0 + O(|\beta|^2),$$

and for positive λ_0 there exists a resonance of the perturbed operator close to λ_0.

We finish this section with a general statement concerning the representation of the resolvent of the perturbed operator. We can assume now that neither of eigenfunctions of the nonperturbed operator $\mathcal{L}_0^0$ vanishes *at all vertices* thus neither of eigenfunctions of $\mathcal{L}_0^0$ remains an eigenfunction of the perturbed operator.

Theorem 2.4 *The component $g_0 := g_0(x, \xi, \lambda)$, $x, \xi \in \Gamma_0$ of the Green function of the perturbed operator $\mathcal{L}$ in Γ_0 is represented in terms of the Green function g_0^0 of the nonperturbed operator the following way:*

$$g_0(x, \xi, \lambda) = \sum_{s=1}^{N} u_s g_0^0(x, a_s, \lambda) + g_0^0(x, \xi, \lambda),$$

where $\{u_s = u_s(\xi, \lambda)\}|_{s=1}^N$ are defined as solutions of the following linear algebraic system

$$\sum_r [g_0^0(a_s, a_r, \lambda) + \delta_{sr}(-ik|\beta|^2)^{-1}] u_r + g_0^0(a_s, \xi) = 0,$$

where $\lambda = k^2$, $\Im k > 0$. The spectrum of the perturbed operator $\mathcal{L}$ consists of all singularities of the Green function in the complex plane of the spectral parameter λ. In particular the absolutely continuous spectrum of $\mathcal{L}$ fills the positive half-axis $\lambda \geq 0$ with a constant multiplicity N. The eigenvalues $\lambda_r = k_r^2$, $\Im k_r > 0$ and *resonances* $\lambda_r = k_r^2$, $\Im k_r < 0$ *of the operator* $\mathcal{L}$ *are found as roots of the following dispersion equation in upper* $\Im k > 0$ *and lower* $\Im k < 0$ *half-planes respectively:*

$$(2.3) \qquad det(g_0^0(a_s, a_r, \lambda) + \delta_{sr}(-ik|\beta|^2)^{-1}) = 0.$$

Proof: Being solutions of the homogeneous equation $\mathcal{L}g = \lambda g$ the components of the complete Green function g of the perturbed operator on the rays Γ_s coincide generally with exponentials:

$$g_s(x_s, \xi, \lambda) = b_s e^{ikx_s}, \quad x_s > 0, \quad k = \sqrt{\lambda}, \quad \xi \in \Gamma_0.$$

Then due to the boundary conditions at each vertex $a = a_1, a_2, \ldots, a_N$ we have

$$(2.4) \qquad [g_0']|a_s = -ik|\beta|^2 g_0(a_s).$$

This implies the announced linear algebraic system for the coefficients u_s:

$$-u_s = \left(\sum_{r=1}^{N} u_r g_0^0(a_s, a_r) + g_0^0(a_s, \xi) \right) (-ik|\beta|^2).$$

Thus we get for the vector $\vec{u} = (u_1, \ldots, u_N)$ the representation

$$\vec{u} = - \left\{ G + i \frac{1}{k|\beta|^2} I \right\}^{-1} \vec{g}(\xi),$$

where $G = \{g_0^0(a_s, a_r)\}$ is an operator in the corresponding auxillary *channel-space*[2] and

$$(g_0^0(a_1, \xi), g_0^0(a_2, \xi), \ldots, g_0^0(a_N, \xi)) = \vec{g}(\xi) \in E.$$

Hence we have the following expression for the component of the Green function of the perturbed operator

$$(2.5) \qquad g_0(x, \xi, \lambda) = -\vec{g}(x) \left\{ G + \frac{i}{k|\beta|^2} \right\}^{-1} \vec{g}(\xi) + g_0^0(x, \xi, \lambda),$$

where $\vec{g}(x) = (g_0^0(x, a_1), \ldots, g_0^0(x, a_N))$ and $\vec{g}(\xi) = (g_0^0(a_1, \xi), g_0^0(a_2, \xi), \ldots, g_0^0(a_N, \xi))$.

We postpone the proof of the statement about zeroes of the determinant of the matrix $G + \frac{i}{k|\beta|^2}$ to the following Section 3 where we prove that all singularities of the resolvent kernel of the perturbed operator appear from these zeroes of the determinant, if neither of eigenfunctions φ_l of $\mathcal{L}$ vanishes at all vertices, $\sum_{s=1}^{N} |\varphi_l(a_s)|^2 > 0$. Modulo this important statement this is the end of the proof. $\qquad \square$

---

[2] The precise meaning of this space will be clarified later when we discuss scattering matrix.

In the following section we continue the discussion of the properties of the resolvent of the perturbed operator beginning from the formula (2.5).

Note that all roots of the equation (2.3) in upper halfplane $\Im k > 0$ which lie on the *physical sheet* of the spectral variable λ are situated on the imaginary axis $k = i\kappa$ ($0 < \kappa < \infty$) and correspond to the negative eigenvalues of $\mathcal{L}$. The roots of the dispersion equation (2.3) in the lower halfplane $\Im k_s < 0$ which lie on the *nonphysical sheet* are called resonances because of the special role they play in the description of asymptotic properties of solutions of the corresponding nonstationary equation (see [10])

$$\frac{1}{i}\frac{\partial \psi}{\partial t} = \mathcal{L}\psi,$$

$$\psi|_{t=0} = \psi_0.$$

The solution of this equation may be represented by the Riesz integral of the resolvent $R_\lambda f(x) = \int g(x, \xi, \lambda) f(\xi) d\xi$

$$e^{i\mathcal{L}t}\psi_0 = -\frac{1}{2\pi i}\int_{\Gamma_{\mathcal{L}}} e^{i\lambda t} R_\lambda d\lambda \psi_0$$

on some contour $\Gamma_{\mathcal{L}}$ on the physical sheet of the spectral variable around the spectrum $\sigma(\mathcal{L})$ of $\mathcal{L}$. The resonances become involved if we may deform this contour to the lower halfplane (see [10]). The spectral analysis of resonances is developed in [9] where the corresponding hyperbolic equation:

$$u_{tt} + \mathcal{L}u = 0$$

is considered. In our situation the similar analysis can be developed as well. For an asymptotic analysis of the Riesz integrals when $t \to \infty$ the description of the poles of the resolvent – the roots of the dispersion equation (2.3) both in the upper $\Im k > 0$ and the lower halfplane $\Im k < 0$ is required. We can perform the corresponding analysis for the family (sequence) of perturbed operators $\mathcal{L}_\beta$ which corresponds to weakening connection between the ring and the rays, $\beta \to 0$. The limit operator coincides with the nonperturbed operator

$$\mathcal{L}^0 = \mathcal{L}_0^0 \oplus \sum_{s=1}^{N} \mathcal{L}_s^0.$$

One can prove (see Theorem 3.1 in the next section) that

$$\mathcal{L}_\beta \longrightarrow \mathcal{L}^0$$

in a sense of the uniform convergence of resolvents

$$(\mathcal{L}_\beta - \lambda I)^{-1} \longrightarrow (\mathcal{L}^0 - \lambda I)^{-1}$$

on each compact of the complement of the spectrum $\sigma(\mathcal{L}^0)$ of the limit operator $\mathcal{L}^0$ in the complex plane.

3 Weakening Connection Limit in Resonance Case

We assume now that the nonperturbed operator has a simple spectrum and neither of its eigenfunctions φ_l vanishes at all vertices, $\sum_{s=1}^{N} |\varphi_l(a_s)|^2 := |\vec{\varphi_l}|^2 > 0$. In this section we investigate the asymptotic behaviour of the resolvent kernel $g_\beta(x, \xi, \lambda)$ and the scattering matrix $S_\beta(\lambda)$ for weakening connection $\beta \to 0$ in both nonresonance $\lambda \in \sigma(\mathcal{L}_0^0)$ and resonance case $\lambda \overline{\in} \sigma(\mathcal{L}_0^0)$, i.e. when λ coincides with one of eigenvalues of $\mathcal{L}_0^0$ or not. The next statement has a general nature and is valid for any resolvent-like matrices. It is based on an observation concerning the inverse matrix near the pole. For other important facts concerning Operator Matrices see for instance in [7].

Theorem 3.1 *Consider a sequence of operators $\mathcal{L}_\beta$ which corresponds to the vanishing connection between the ring and the rays: $\beta \to 0, \epsilon \to 0$. The resolvents of them*

$$(\mathcal{L}_\beta - \lambda I)^{-1}$$

converge uniformly to the resolvent of the nonperturbed operator $\mathcal{L}_0^0 \oplus \sum_{s=1}^{N} \mathcal{L}_s^0$ on each compact subset Ω of the complement of the spectrum of the nonperturbed operator. Besides, if λ_0 is an eigenvalue of the nonperturbed operator $\mathcal{L}^0$, then for sufficiently small β it can't be an eigenvalue of the perturbed operator $\mathcal{L}_\beta$ but there exist an eigenvalue of the perturbed operator (for $\lambda_0 < 0$) or resonance (for $\lambda_0 > 0$) in a $|\beta|^2$-neighborhood of it.

Proof: Consider the case when $x, \xi \in \Gamma_0$. We use the representation of the Green function of the perturbed operator derived in Theorem 2.2:

$$(3.1) \qquad g(x, \xi, \lambda) = -\vec{g}(x) \left\{ G + \frac{i}{k|\beta|^2} \right\}^{-1} \vec{g}(\xi) + g_0^0(x, \xi, \lambda).$$

The singularities of the resolvent of the nonperturbed operator $\mathcal{L}_0^0$ are present in both terms of the expression for the perturbed resolvent kernel. But in fact they compensate each other. To verify this statement we consider the last representation for sufficiently small values of β in a small neighborhood of the given eigenvalue λ_0 of the operator $\mathcal{L}_0^0$. At first sight the leading terms of the operator

$$G(\lambda) + i \frac{1}{k|\beta|^2}$$

are

$$\frac{\vec{\varphi_0}\vec{\varphi_0}}{\lambda_0 - \lambda} + i \frac{1}{k|\beta|^2}.$$

Here $\vec{\varphi}_0\vec{\varphi}_0$ is a matrix combined of values of the eigenfunction φ_0 at the points a_s, a_t. It is proportional to the projection operator P_0 in the N-dimensional auxiliary channel-space E:

$$\vec{\varphi}_0\vec{\varphi}_0 = \sum_s |\varphi_0(a_s)|^2 P_0 = |\vec{\varphi}_0|^2 P_0.$$

In fact it is slightly more convenient to write down the leading terms as orthogonal decomposition in two orthogonal subspaces $P_0 E + (I - P_0)E := P_0 E + P_0^\perp E$. Separating from the matrix $G_{st} = \sum_{l=0}^{\infty} \frac{\varphi_l(a_s)\varphi_l(a_t)}{\lambda_l-\lambda}$ the singular term at the point λ_0

$$G = \frac{|\vec{\varphi}_0|^2 P_0}{\lambda_0 - \lambda} + \sum_{\lambda_l \neq \lambda_0} \frac{|\vec{\varphi}_l|^2 P_l}{\lambda_l - \lambda}$$

$$:= \frac{|\vec{\varphi}_0|^2 P_0}{\lambda_0 - \lambda} + K_0,$$

and decomposing the expression in orthogonal sum we get the following formula for the denominator:

$$G(\lambda) + i\frac{1}{k|\beta|^2} = \left(\frac{|\vec{\varphi}_0|^2}{\lambda_0 - \lambda} + i\frac{1}{k|\beta|^2} \right)$$

(3.2)

$$P_0 + i\frac{1}{k|\beta|^2} P_0^\perp + K_0.$$

Note that the leading term of the last expression – the diagonal matrix

$$\begin{pmatrix} (\frac{|\vec{\varphi}_0|^2}{\lambda_0-\lambda} + i\frac{1}{k|\beta|^2}) P_0 & 0 \\ 0 & i\frac{1}{k|\beta|^2} P_0^\perp \end{pmatrix} := \Delta$$

is invertible

$$\Delta^{-1} = k|\beta|^2 \begin{pmatrix} \frac{\lambda_0-\lambda}{k|\beta|^2|\vec{\varphi}_0|^2+i(\lambda_0-\lambda)} P_0 & 0 \\ 0 & -i P_0^\perp \end{pmatrix}$$

and the inverse is holomorphic with respect to the variable $k = \sqrt{\lambda}$ in a small neighborhood of k_0, $\lambda_0 = k_0^2$ for all sufficiently small β. Then the inverse of $G + \frac{i}{k|\beta|^2}$ can be calculated as

$$k|\beta|^2 \left(\frac{\lambda_0 - \lambda}{k|\beta|^2|\vec{\varphi}_0|^2 + i(\lambda_0 - \lambda)} P_0 - i P_0^\perp \right) \times (I + K_0\Delta^{-1})^{-1}.$$

Consider the first term of the expression (3.1) for the Green function of the perturbed problem. The left and right factors of it

$$\vec{g}(x) = \sum_l \frac{\varphi_l(x)\bar{\varphi}_l}{\lambda_l - \lambda} = \frac{\varphi_0(x)\bar{\varphi}_0}{\lambda_0 - \lambda} + \sum_{\lambda_l \neq \lambda_0} \frac{\varphi_l(x)\bar{\varphi}_l}{\lambda_l - \lambda},$$

$$\vec{g}(\xi) = \sum_l \frac{\varphi_l(\xi)\bar{\varphi}_l}{\lambda_l - \lambda} = \frac{\varphi_0(\xi)\bar{\varphi}_0}{\lambda_0 - \lambda} + \sum_{\lambda_l \neq \lambda_0} \frac{\varphi_l(\xi)\bar{\varphi}_l}{\lambda_l - \lambda}$$

obviously have singularities at the eigenvalue λ_0 with the factors $\bar{\varphi}_0$ in front of them. Then a direct calculation of singularities of the whole expression shows that only first order term remains, since $P_0^\perp \bar{\varphi}_0 = 0$ and the coefficient in front of it is $-\varphi_0(x)\varphi_0(\xi)$. Combining this singularity $-\frac{\varphi_0(x)\varphi_0(\xi)}{\lambda_0 - \lambda}$ with the corresponding term in $g_0^0(x, \xi, \lambda)$ we see that both singular terms compensate each other. Thus we see that in the case when $\bar{\varphi}_0 \neq 0$ the inner component of the Green function of the perturbed operator is a holomorphic function at the eigenvalue λ_0 for sufficiently weak connection between the ring and the rays. On the other hand a new singularity from the denominator $G + i\frac{1}{k|\beta|^2}$ may appear. If $k_0^2 = \lambda_0 < 0$ then for small β the denominator has zero eigenvalue for some pure imaginary value of k close to k_0. This follows from the orthogonal decomposition (3.2)

$$\left[G(\lambda) + i\frac{1}{k|\beta|^2} \right] \vec{u} = P_0 \left(\frac{|\bar{\varphi}_0|^2}{\lambda_0 - \lambda} + i\frac{1}{k|\beta|^2} \right)$$

(3.3)
$$P_0\vec{u} + P_0 K_0 P_0\vec{u} + P_0 K_0 P_0^\perp \vec{u}$$

$$+ P_0^\perp K_0 P_0\vec{u} + P_0^\perp i\frac{1}{k|\beta|^2} P_0^\perp \vec{u} + P_0^\perp K_0 P_0^\perp \vec{u} = 0.$$

It follows from the operator version of Rouché's theorem [11] that the solution of the last equation (3.3) is close to the solution of the equation combined of leading terms and for small β

$$\lambda_\beta : = k_\beta^2 \approx \lambda_0 - i\sqrt{\lambda_0}|\bar{\varphi}_0|^2|\beta|^2,$$

$$\vec{u}_\beta \approx \bar{\varphi}_0.$$

Then the corresponding solutions of the Schrödinger equation may be obtained as

$$u(x) = \langle \vec{g}(x), \vec{u}_\beta \rangle.$$

If $\lambda_0 < 0$, then $\lambda_\beta := k_\beta^2 < 0$, $k_\beta = i\kappa$, $\kappa > 0$, hence the exponentials continuing the solution u from Γ_0 onto the rays Γ_s are square integrable and the total solution of the Schrödinger equation $\mathcal{L}u = \lambda u$ on the whole graph is a square-integrable function, i.e. is an eigenfunction of the operator $\mathcal{L}$. The finiteness of the total number of negative eigenvalues follows directly from the analyticity of the matrix $G + i\frac{1}{k|\beta|^2}$.

Vice versa if $\lambda_0 > 0$ then $\Im k_\beta < 0$ hence the corresponding solution u_β of the Schrödinger equation is exponentially growing at least on some rays. So it is not an eigenfunction but a resonance solution – "a resonance state". The total number of resonances is infinite which can be derived from the asymptotic behaviour of the matrix $G + i\frac{1}{k|\beta|^2}$ at infinity. The corresponding analysis will be done elsewhere. $\qquad\square$

If we take into account that $k = \sqrt{\lambda}$ we see that zero is a branching point (with respect to λ) of the operator - function $[G(\lambda) + i\frac{1}{k|\beta|^2}]^{-1}$ and the positive axis is a cut with different values of the resolvent kernel on different shores of it.

To accomplish the study of the non-resonance case we can now formulate the following statement concerning general spectral properties of the operator $\mathcal{L}$.

Theorem 3.2 *The spectrum of the operator $\mathcal{L}$ consists of an absolutely continuous branch $(0, \infty)$ multiplicity N and a finite number of negative eigenvalues. The eigenfunctions of an absolutely continuous spectrum are given by N families of scattered waves which serve as solutions of the homogeneous equation $\mathcal{L}\psi = k^2\psi$. For the components of the scattered wave ψ iniciated by the plane wave approaching from infinity on the ray Γ_t we have on the rays Γ_s the "asymptotics":*

$$\psi_s = \begin{cases} S_{st}e^{-ikx_s}, & s \neq t, \\ e^{ikx_t} + S_{tt}e^{-ikx_t}, & s = t, \end{cases}$$

and for the component on the ring we have

$$\mathcal{L}^0\psi_0 = k^2\psi_0$$

with the boundary conditions at the vertices:

$$[\psi_0']|_{a_s} = -\beta\psi_s'|_0,$$
$$\psi_s(0) = \bar\beta\psi_0(a_s),$$

$$\psi_0 = \sum_{s=1}^{N} u_s g_0^0(x, a_s) = \langle \vec{g}(x), \vec{u}\rangle.$$

These eigenfunctions are orthogonal and normalized in $L_2(\Gamma_0)$. The scattering matrix S_{st} in the ansatz above as well as the components of the scattered waves on Γ_0 are defined from the boundary conditions as:

$$S = \frac{G - \frac{i}{k|\beta|^2}}{G + \frac{i}{k|\beta|^2}}$$

$$\vec{u} = \frac{\beta}{G + \frac{i}{k|\beta|^2}}\vec{v}_t,$$

where $\vec{v}_t = \{\delta_{st}\}$.

Proof: The proof of this statement can be obtained with use of the standard Riesz techniques of contour integration of the resolvent basing on the asymptotic formulae for solutions of the homogeneous equation (see also [6]). Though it is not entirely equivalent to the techniques for one-dimensional Schrödinger operator we omit the essential part of the proof here and calculate only the expressions for transmission and reflection coefficients and the component of the scattered wave on the compact subgraph. Other properties of scattered waves and scattering matrix will be discussed elsewhere.

For the scattered wave initiated from the ray Γ_t we have the following anzatz for the component on the ring Γ_0

$$\psi_0 = \sum_{r=1}^{N} u_r g(x, a_r),$$

and the linear algebraic system for the coefficients u_r:

$$\begin{cases} -u_s = -\beta \psi_s'(0), \\ \psi_s(0) = \bar{\beta} \sum_{r=1}^{N} g(a_s, a_r) u_r, & s = 1, 2, \ldots, N, \end{cases}$$

where

$$\begin{cases} \psi_t = e^{ikx_t} + S_{tt} e^{-ikx_t} \\ \psi_s = S_{st} e^{-ikx_s}, & s \neq t. \end{cases}$$

Eliminating the variables of exteriors channels we obtain

$$\begin{cases} -u_s = ik\beta(-\delta_{st} + S_{st}) \\ \delta_{st} + S_{st} = \bar{\beta} \sum_{r=1}^{N} g(a_s, a_r) u_r \end{cases}$$

or

(3.4)
$$\begin{cases} \vec{u} = i\beta k(I - S)\vec{v}_t \\ \bar{\beta} G\vec{u} = (I + S)\vec{v}_t. \end{cases}$$

It gives immediately

$$(ik|\beta|^2 G - I)\vec{u} = 2ik\beta\vec{v}_t.$$

Then from the system (3.4) we get the expression for the scattering matrix. $\quad\Box$

In the remaining part of our paper we analyze a special but practically important situation when the energy λ of the scattered wave coincides with some eigenvalue

λ_0 of the nonperturbed operator $\mathcal{L}^0$. Following [3] we call this situation a *resonance case*. In this case we use the block-representation of the operator $G - \frac{i}{k|\beta|^2}$ with respect to the orthogonal decomposition of the auxillary channel-space E used in the Proof of the Theorem 3.1

$$G(\lambda) - i\frac{1}{k|\beta|^2} = \left(\frac{|\vec{\varphi}_0|^2}{\lambda_0 - \lambda} - i\frac{1}{k|\beta|^2} \right)$$

(3.5)

$$P_0 - i\frac{1}{k|\beta|^2} P_0^\perp + K_0 := \Delta_- + K_0.$$

Further we use the notations

$$\left(\frac{|\vec{\varphi}_0|^2}{\lambda_0 - \lambda} + i\frac{1}{k|\beta|^2} \right) := (+).$$

It is obvious that $(+) \approx \beta^{-2}$, $\beta \to 0$ provided $\lambda \neq \lambda_0$.

Theorem 3.3 *The scattering matrix of the ring with few rays attached to it via the weakening boundary condition $\beta \to 0$ has the following asymptotics near the simple resonance eigenvalue λ_0* [3]:

$$-I - 2ik|\beta|^2 K_0 + (I + ik|\beta|^2 K_0)\frac{2k|\beta|^2|\vec{\varphi}_0|^2}{k|\beta|^2|\vec{\varphi}_0|^2 + i(\lambda_0 - \lambda)} P_0 + O(|\beta|^4).$$

Proof: The leading terms of the denominator in the expression for the scattering matrix derived in the last theorem are represented near the resonance eigenvalue by the diagonal matrix in the orthogonal decomposition of the auxillary space $E = P_0 E + P_0^\perp E$.

$$\begin{pmatrix} (\frac{|\vec{\varphi}_0|^2}{\lambda_0 - \lambda} + i\frac{1}{k|\beta|^2})P_0 & 0 \\ 0 & i\frac{1}{k|\beta|^2}P_0^\perp \end{pmatrix} = \begin{pmatrix} (+)P_0 & 0 \\ 0 & i\frac{1}{k|\beta|^2}P_0^\perp \end{pmatrix}; := \Delta,$$

$$\Delta^{-1} = \begin{pmatrix} \frac{1}{(+)}P_0 & 0 \\ 0 & -ik|\beta|^2 P_0^\perp \end{pmatrix}.$$

Hence we can write the expression for the scattering matrix as

$$(\Delta_- + K_0)\Delta^{-1}(I + K_0\Delta^{-1})^{-1}.$$

The product of the first and the second factors gives

$$\Delta_-\Delta^{-1} + K_0\Delta^{-1} = -I + \frac{2k|\beta|^2|\vec{\varphi}_0|^2}{k|\beta|^2|\vec{\varphi}_0|^2 + i(\lambda_0 - \lambda)} P_0 + K_0\Delta^{-1}.$$

---

[3] Mr. M. Harmer noticed, that similar statement remains true for multiple eigenvalues as well.

The last factor is represented in form of convergent series for small values of β, $\lambda_0 - \lambda$:

$$I - K_0 \Delta^{-1} + O(\beta^4)$$

which gives for the scattering matrix the approximate expression

$$S(k) = -I + \frac{2k|\beta|^2|\vec{\varphi}_0|^2}{k|\beta|^2|\vec{\varphi}_0|^2 + i(\lambda_0 - \lambda)} P_0$$
$$+ 2k|\beta|^2 K_0 \left(\frac{\lambda_0 - \lambda}{k|\beta|^2|\vec{\varphi}_0|^2 + i(\lambda_0 - \lambda)} P_0 - i P_0^{\perp} \right)$$
$$+ O(\beta^4) = I - 2ik|\beta|^2 K_0 + (I + ik|\beta|^2 K_0)$$
$$\frac{2k|\beta|^2|\vec{\varphi}_0|^2}{k|\beta|^2|\vec{\varphi}_0|^2 + i(\lambda_0 - \lambda)} P_0 + O(|\beta|^4).$$

$\square$

In particular for λ close to λ_0 and $\beta \to 0$ we have the following approximate expression for the transmission coefficient for weakly connected rays:

$$S_{s,t}(\lambda_0) = \frac{2k|\beta|^2}{k|\beta|^2|\vec{\varphi}|^2 + i(\lambda_0 - \lambda)} \varphi(a_s)\varphi(a_t) + O(\beta^2), \quad s \neq t.$$

where the second term is uniformly small when $\beta \to 0$, but the first one exhibits a nonuniform behaviour in dependence on ratio $(\lambda_0 - \lambda)/\beta^2$, see below.

Remark: The last formula being applied formally to the case $\lambda = \lambda_0$ shows, that the transmission coefficient is approximately equal to

$$S_{s,t}(\lambda_0) = \frac{2}{|\vec{\varphi}|^2} \varphi(a_s)\varphi(a_t) + O(\beta^2).$$

This looks surprising for $\beta = 0$ since it gives a nonzero transmission coefficient for zero connection. Actually it means that the transmission coefficients are not continuous with respect to the energy λ uniformly in β. The physically significant values of the transmission coefficient may be obtained via averaging with respect to Fermi distribution

$$\rho(\lambda, T) = \frac{1}{1 + e^{\frac{\lambda - \lambda_f}{\kappa T}}}$$

$$: \overline{|S_{ij}(T)|^2} = \int |S_{ij}(\sqrt{\lambda})|^2 \left| \frac{d\rho(\lambda, T)}{d\lambda} \right| d\lambda.$$

Here λ_f is the Fermi-level in the wires.

One may consider two different cases: $\kappa T \langle \langle \lambda_0 | \vec{\varphi} |^2 | \beta |^2$ and $\kappa T \rangle \rangle \lambda_0 | \vec{\varphi} |^2 | \beta |^2$ which correspond to averaging on intervals $| \lambda - \lambda_0 | < \kappa T$, $\kappa T \langle \langle \lambda_0 | \vec{\varphi} |^2 | \beta |^2$ and $| \lambda - \lambda_0 | < \kappa T$, $\kappa T \rangle \rangle \lambda_0 | \vec{\varphi} |^2 | \beta |^2$. In the first case we still have:

$$\overline{|S_{ij}(T)|^2} \approx \frac{2 |\varphi(a_s) \varphi(a_t)|^2}{|\vec{\varphi}|^4}$$

but in the second case, when $\lambda_0 | \vec{\varphi} |^2 | \beta |^2$ is small comparing with κT we have:

$$\overline{|S_{ij}(T)|^2} \approx 4 \frac{|\varphi(a_s) \varphi(a_t)|^2}{|\vec{\varphi}|^4} \frac{1}{1 + \frac{\kappa^2 T^2}{\lambda_0 |\beta|^4 |\vec{\varphi}|^4}}.$$

Hence for small β and positive absolute temperature the averaged transmission coefficient is small, according to natural physical expectations.

Acknowledgements

We are grateful to the Commission of the European Communities for financial support in the frame of EC-Russia Exploratory Collaborative Activity under EU ESPRIT Project 28890 NTCONGS and partial support from the Russian Academy of Sciences (Grant RFFI 97 - 01 - 01149). One of us (B.P) proudly recognizes the support from the Royal Society of New Zealand (Marsden Fund Grant 3368152).
    We are greatful to professor V. Oleinik for important discussions.

References

[1]   N.I. Gerasimenko and B.S. Pavlov, Scattering problems on compact graphs, *Theor. Math. Phys.* **74** (1988), 230.

[2]   S.P. Novikov, Schrödinger operators on graphs and symplectic geometry, *in: The Arnol'dfest (Proceedings of the Fields Institute Conference in Honour of the 60th Birthday of Vladimir I. Arnol'd), eds. E. Bierstone, B. Khesin, A. Khovanskii and J. Marsden* (Fields Institute Communications Series), to appear.

[3]   V. Kostrykin and R. Schrader, Kirchhoff's rule for quantum wires, *J. Phys. A: Math. Gen.* **32** (1999), 595–630.

[4]   Y. Melnikov and B. Pavlov, Quantum Scattering on Graphs, (Manuscript).

[5]   B. Carlson, Inverse Eigenvalue Problems on Directed Graphs, *Transactions of the AMS* (vol. 351, no. 10) (1999), 4069–4088.

[6]   B.S. Pavlov, A model of zero-radius potential with internal structure, *Theoret. and Math. Phys.* **59** (3) (1984), 544.

[7]   R. Mennicken and A. Shkalikov, Spectral Decomposition of Symmetric Operator Matrices, *Mathematische Nachrichten* **179** (1996), 259–273.

[8]   M. Faddeev and B. Pavlov, Scattering by resonator with the small opening, *J. Sov. Math.* **27** (1984), 2527.

[9]   P. Lax and R. Phillips, *Scattering theory for automorphic functions*, Princeton University press and University of Tokyo press, Princeton, New Jersey 1976.

[10]   M. Reed and B. Simon, *Methods of modern mathematical physics*, Academic press, New York, London 1972.

[11]   I.S. Gohberg and E.I. Sigal, Operator extension of the theorem about logarithmis residue and Rouché's theorem, *Mat. sbornik* **84** (1971), 607.

V.B. Bogevolnov
Department of Solid State Electronics
Institute for Physics
St. Petersburg State University
Ulianovskaya 1
Petrodvorets
St. Petersburg, 198904
Russia

A.B. Mikhailova
Department of Solid State Electronics
Institute for Physics
St. Petersburg State University
Ulianovskaya 1
Petrodvorets
St. Petersburg, 198904
Russia

B.S. Pavlov
Department of Mathematics
University of Auckland
Private Bag 92019
Auckland
New Zealand

A.M. Yafyasov
Department of Solid State Electronics
Institute for Physics
St. Petersburg State University
Ulianovskaya 1
Petrodvorets
St. Petersburg, 198904
Russia

Operator Theory:
Advances and Applications, Vol. 124
© 2001 Birkhäuser Verlag Basel/Switzerland

Dilation of Generalized Toeplitz Kernels on Lexicographic $\Gamma \times \mathbb{R}$

Ramón Bruzual and Marisela Domínguez

To Professor Israel Gohberg, on the occasion of his 70-th birthday

Let Γ be a locally compact abelian ordered group. We prove a dilation theorem for positive definite generalized Toeplitz kernels on the group $\Gamma \times \mathbb{R}$ with the lexicographic order. The proof is constructive and it is based on a technique which combines special properties of the lexicographic order, geometrical facts of $\Gamma \times \mathbb{R}$ and local semigroups of isometries.

1 Introduction

Let Ω be an abelian group and let $L(\mathcal{H})$ be the space of the bounded linear operators on the Hilbert space $\mathcal{H}$. A $L(\mathcal{H})$-valued kernel is a function $k : \Omega \times \Omega \to L(\mathcal{H})$.

The kernel k is said to be positive definite if for any positive integer n and any $x_1, \ldots, x_n \in \Omega$ and $h_1, \ldots, h_n \in \mathcal{H}$ we have

$$\sum_{i,j=1}^{n} \langle k(x_i, y_j)h_i, h_j \rangle \geq 0,$$

and k is called a Toeplitz kernel if there exists a function $C : \Omega \to L(\mathcal{H})$ such that $k(x, y) = C(x - y)$ for all $x, y \in \Omega$.

As it is well known positive definite Toeplitz kernels appears in many problems of mathematical analysis.

Several problems in analysis led Cotlar and Sadosky to introduce the so-called generalized Toeplitz kernels, as kernels defined in $\mathbb{Z} \times \mathbb{Z}$ or in $\mathbb{R} \times \mathbb{R}$ (see [6]), where $\mathbb{Z}$ is the set of the integers and $\mathbb{R}$ is the set of the real numbers. These kernels satisfy a condition more general than being Toeplitz in their domain. It is supposed that there are four functions $(C_{\alpha\beta})_{\alpha,\beta=1,2}$ such that

$$k(x, y) = C_{\alpha\beta}(x - y)$$

whenever $(x, y) \in \mathbb{Z}_\alpha \times \mathbb{Z}_\beta$, where $\mathbb{Z}_1 = \{n \in \mathbb{Z} : n \geq 0\}$, $\mathbb{Z}_2 = \{n \in \mathbb{Z} : n < 0\}$. An analogous definition can be given for $\mathbb{R} \times \mathbb{R}$.

Let $(\Omega, +)$ be an abelian group with neutral element 0. Ω is an ordered group if there exists a set $\Omega_1 \subset \Omega$ such that: $\Omega_1 + \Omega_1 = \Omega_1$, $\Omega_1 \bigcap (-\Omega_1) = \{0\}$, $\Omega_1 \bigcup (-\Omega_1) = \Omega$. In this case if $x, y \in \Omega$ we write $x \leq y$ if $y - x \in \Omega_1$.

An analogue of the canonical partition $\mathbb{Z} = \mathbb{Z}_1 \cup \mathbb{Z}_2$ makes sense in general ordered groups $(\Omega, +)$. Thus the notion of generalized Toeplitz kernels can be defined in this kind of groups (see [8]).

Let Ω be an abelian ordered locally compact group, $\Omega_1 = \{\omega \in \Omega : \omega \geq 0\}$, $\Omega_2 = \{\omega \in \Omega : \omega < 0\}$ and let $(\mathcal{H}_1, \mathcal{H}_2)$ be a pair of Hilbert spaces.

Definition 1.1 A generalized Toeplitz kernel on $(\Omega, \mathcal{H}_1, \mathcal{H}_2)$ is a kernel C such that there exist four weakly continuous functions $C_{\alpha\beta} : \Omega_\alpha - \Omega_\beta \to L(H_\alpha, H_\beta)$ for $\alpha, \beta = 1, 2$ such that

$$C(x, y) = C_{\alpha\beta}(x - y) \quad \text{if} \quad (x, y) \in \Omega_\alpha \times \Omega_\beta$$

and $C_{12}(\omega) = C_{21}(-\omega)^*$ for all $\omega \in \Omega_1$.

As it is natural a generalized Toeplitz kernel C on $(\Omega, \mathcal{H}_1, \mathcal{H}_2)$ is said to be positive definite if

$$\sum_{\alpha,\beta=1}^{2} \sum_{(x,y)\in\Omega_\alpha\times\Omega_\beta} \langle C_{\alpha\beta}(x - y)h_\alpha(x), h_\beta(y)\rangle_{\mathcal{H}_\beta} \geq 0$$

for all pair of functions $h_\alpha : \Omega_\alpha \to \mathcal{H}_\alpha$ with finite support.

It is known that for the case $\Omega = \mathbb{Z}$ or $\Omega = \mathbb{R}$ the following dilation result holds: For every pair $(\mathcal{H}_1, \mathcal{H}_2)$ of Hilbert spaces and for every positive definite generalized Toeplitz kernel C on $(\Omega, \mathcal{H}_1, \mathcal{H}_2)$, there exist a Hilbert space $\mathcal{G}$, two bounded operators $\tau_\alpha : \mathcal{H}_\alpha \to \mathcal{G}$, $\alpha = 1, 2$, and a strongly continuous unitary representation $(U_\omega)_{\omega\in\Omega}$ of Ω on $L(\mathcal{G})$ such that:

$$C_{\alpha\beta}(\omega) = \tau_\beta^* U_\omega \tau_\alpha \text{ for all } \omega \in \Omega_\alpha - \Omega_\beta \text{ for } \alpha, \beta = 1, 2.$$

(see [1], [7] and [5]).

Let Γ be a locally compact abelian ordered group, we will consider an ordered group more general than $\mathbb{R}$, set $\Omega = \Gamma \times \mathbb{R}$ with the product topology and the lexicographic order, that is, we define in Ω the following order:

$$(\gamma_1, r_1) < (\gamma_2, r_2) \text{ if } r_1 < r_2 \text{ or if } r_1 = r_2 \text{ and } \gamma_1 < \gamma_2.$$

It is clear that Ω is a locally compact abelian ordered group.

In Section 2 we show that a multiplicative family of isometries can be associated in a natural way to a positive definite generalized Toeplitz kernel on $\Gamma \times \mathbb{R}$ and it is proved that if this family can be extended to a group of unitary operators then the kernel can be dilated.

In Section 3 we recall some results on local semigroups of isometries.

In Section 4 we develop a special technique that combines special properties of the lexicographic order, geometrical facts of $\Gamma \times \mathbb{R}$, the theory of local semigroups of isometries and the result of Section 2 in order to prove that a similar dilation result holds for positive definite generalized Toeplitz kernels on $\Omega = \Gamma \times \mathbb{R}$.

2 Multiplicative Families of Isometries Associated to Generalized Toeplitz Kernels on $\Gamma \times \mathbb{R}$

When we are working with a classical Toeplitz kernel the set of translations gives rise to a unitary group. A pre-Hilbert space and a family of isometric operators corresponds in a natural way to a positive definite generalized Toeplitz kernel. The properties of this family of isometries were the main motivation of our definition of multiplicative family of isometries (see also [2]). The result that we give in this section allows us to obtain the Proof of Theorem 4.1.

Let $\mathcal{E}$ be a Hilbert space and let Ω be an abelian locally compact ordered group, also let Ω_1 and Ω_2 be as before.

Definition 2.1 A multiplicative family of isometries on $\mathcal{E}$ is a family $(S_\omega, \mathcal{E}_\omega)_{\omega \in \Omega_1}$ such that:

i) For each $\omega \in \Omega_1$ we have that $\mathcal{E}_\omega$ is a closed subspace of $\mathcal{E}$, $S_\omega : \mathcal{E}_\omega \to \mathcal{E}$ is a linear isometry, $\mathcal{E}_y \subset \mathcal{E}_x$ if $x, y \in \Omega_1$ and $x < y$, $\mathcal{E}_0 = \mathcal{E}$, $S_0 = I_\mathcal{E}$.

ii) If $x, y \in \Omega_1$ then $S_y \mathcal{E}_{x+y} \subset \mathcal{E}_x$ and $S_{x+y} h = S_x S_y h$ for all $h \in \mathcal{E}_{x+y}$.

We shall say that $(S_\omega, \mathcal{E}_\omega)_{\omega \in \Omega_1}$ is strongly continuous if for all $x \in \Omega_1, h \in \mathcal{E}_x$ we have that the function $\omega \mapsto S_\omega h$ from $[0, x] = \{\omega \in \Omega : 0 \leq \omega \leq x\}$ to $\mathcal{E}$ is continuous.

We will associate a multiplicative family of isometries to a positive definite generalized Toeplitz kernel.

Let C be a positive definite generalized Toeplitz kernel on $(\Omega, \mathcal{H}_1, \mathcal{H}_2)$. Let $\mathcal{L}$ be the linear space of the pairs (f_1, f_2) such that $f_1 : \Omega_1 \to \mathcal{H}_1$ and $f_2 : \Omega_2 \to \mathcal{H}_2$ are functions with finite support.

For $(f_1, f_2), (g_1, g_2) \in \mathcal{L}$ we define

$$\langle (f_1, f_2), (g_1, g_2) \rangle_\mathcal{L} = \sum_{\alpha, \beta=1}^{2} \sum_{(x,y) \in \Omega_\alpha \times \Omega_\beta} \langle C_{\alpha\beta}(x - y) f_\alpha(x), g_\beta(y) \rangle_{\mathcal{H}_\beta}.$$

Then $\langle \, , \, \rangle_\mathcal{L}$ is a nonnegative sesquilinear form on $\mathcal{L}$. Let $\mathcal{E}$ be the Hilbert space obtained by completing $\mathcal{L}$ after the natural quotient. For $\omega \in \Omega_1$ set

$$\mathcal{L}_\omega = \{(f_1, f_2) \in \mathcal{L} \mid supp(f_2) \subset \{x \in \Omega_2 : x < -\omega\}\}.$$

We define the operator S_ω in the following way: let $(f_1, f_2) \in \mathcal{L}_\omega$, then $S_\omega(f_1, f_2) = (g_1, g_2)$ if

$$g_1(x) = \begin{cases} 0 & \text{if } 0 \leq x < \omega \\ f_1(x - \omega) & \text{if } x \geq \omega \end{cases}$$
$$g_2(x) = f_2(x - \omega) \text{ for } x \in \Omega_2.$$

It is easy to prove that $S_\omega : \mathcal{L}_\omega \to \mathcal{L}$ is an isometric operator. Let $\mathcal{E}_\omega$ be the closure of the set of the equivalence class of the elements of $\mathcal{L}_\omega$ in $\mathcal{E}$. S_ω gives rise to a linear isometry from $\mathcal{E}_\omega$ to $\mathcal{E}$, that we will denote by S_ω too and it is easy to check that $(S_\omega, \mathcal{E}_\omega)_{\omega \in \Omega_1}$ is a strongly continuous multiplicative family of isometries.

Suppose that Γ is a locally compact abelian ordered group and $\Omega = \Gamma \times \mathbb{R}$ with the lexicographic order.

Let $\{x_n\}$ be a sequence of real numbers such that $x_n < 0$ for all n and $\lim x_n = 0$. Let $h_2 \in \mathcal{H}_2$. Let us denote by δ_ω the characteristic function of the set $\{\omega\}$. Then we have that

$$\|(0, h_2\delta_{(0,x_n)}) - (0, h_2\delta_{(0,x_m)})\|_{\mathcal{L}}^2$$
$$= 2\langle C_{22}(0)h_2, h_2\rangle_{\mathcal{H}_2} - \langle C_{22}(x_n - x_m)h_2, h_2\rangle_{\mathcal{H}_2}$$
$$- \langle C_{22}(x_m - x_n)h_2, h_2\rangle_{\mathcal{H}_2}$$

Therefore $\{(0, h_2\delta_{(0,x_n)})\}$ is a Cauchy sequence in $\mathcal{L}$. The limit in $\mathcal{E}$ of this sequence will be denoted by $[(0, h_2\delta_{(0,0)})]$.

We have

Theorem 2.2 *Let Γ be a locally compact abelian ordered group and let $\Omega = \Gamma \times \mathbb{R}$ with the product topology and the lexicographic order.*

Let $(\mathcal{H}_1, \mathcal{H}_2)$ be a pair of Hilbert spaces and let C be a positive definite generalized Toeplitz kernel on $(\Omega, \mathcal{H}_1, \mathcal{H}_2)$.

If the multiplicative family of isometries associated to C can be extended to a strongly continuous group $(U_\omega)_{\omega \in \Omega}$ of unitary operators in a larger Hilbert space $\mathcal{G}$ then there exist two bounded operators $\tau_\alpha : \mathcal{H}_\alpha \to \mathcal{G}$, $\alpha = 1, 2$ such that:

$$C_{\alpha\beta}(\omega) = \tau_\beta^* U_\omega \tau_\alpha \text{ for all } \omega \in \Omega_\alpha - \Omega_\beta \text{ for } \alpha, \beta = 1, 2.$$

Proof: Let $[(f_1, f_2)]$ denote the equivalence class on $\mathcal{E}$ of the element $(f_1, f_2) \in \mathcal{L}$.

If $\omega \in \Omega_1$ and $h_1 \in \mathcal{H}_1$, then $[(h_1\delta_0, 0)] \in \mathcal{E}_\omega$ and $S_\omega[(h_1\delta_0, 0)] = [(h_1\delta_\omega, 0)]$.

If $\omega \in \Omega_2$ and $h_2 \in \mathcal{H}_2$, then $[(0, h_2\delta_\omega)] \in \mathcal{E}_{-\omega}$ and $S_{-\omega}[(0, h_2\delta_\omega)] = [(0, h_2\delta_0)]$.

Therefore

$$U_\omega[(h_1\delta_0, 0)] = [(h_1\delta_\omega, 0)] \ \forall \omega \in \Omega_1 \ \forall h_1 \in \mathcal{H}_1$$

$$U_\omega[(0, h_2\delta_0)] = [(0, h_2\delta_\omega)] \ \forall \omega \in \Omega_2 \ \forall h_2 \in \mathcal{H}_2.$$

Let $\tau_1 : \mathcal{H}_1 \to \mathcal{G}$ and $\tau_2 : \mathcal{H}_2 \to \mathcal{G}$ defined by

$$\tau_1(h_1) = [(h_1\delta_0, 0)] \quad \text{and} \quad \tau_2(h_2) = [(0, h_2\delta_0)].$$

We have that

$$\| \tau_1(h_1) \|_{\mathcal{G}}^2 = \langle C_{11}(0)h_1, h_1\rangle_{\mathcal{H}_1} \leq \| C_{11}(0) \| \| h_1 \|_{\mathcal{H}_1}^2$$

so we have that τ_1 is a bounded linear operator. In the same way it follows that τ_2 is a bounded linear operator.

Let $x, y \in \Omega_1, h_1, h'_1 \in \mathcal{H}_1$, then

$$
\begin{aligned}
\langle C_{11}(x-y)h_1, h'_1 \rangle_{\mathcal{H}_1} &= \langle [(h_1\delta_x, 0)], [(h'_1\delta_y, 0)] \rangle_{\mathcal{E}} \\
&= \langle U_x[(h_1\delta_0, 0)], U_y[(h'_1\delta_0, 0)] \rangle_{\mathcal{G}} \\
&= \langle U_x \tau_1(h_1), U_y \tau_1(h'_1) \rangle_{\mathcal{G}} \\
&= \langle U_{x-y} \tau_1(h_1), \tau_1(h'_1) \rangle_{\mathcal{G}} \\
&= \langle \tau_1^* U_{x-y} \tau_1(h_1), h'_1 \rangle_{\mathcal{H}_1}.
\end{aligned}
$$

Therefore $C_{11}(\omega) = \tau_1^* U_\omega \tau_1$ for all $\omega \in \Omega_1 - \Omega_1$.
In the same way we obtain:

$$
C_{\alpha\beta}(\omega) = \tau_\beta^* U_\omega \tau_\alpha \quad \text{for all } \omega \in \Omega_\alpha - \Omega_\beta \text{ for } \alpha, \beta = 1, 2.
$$

$\square$

3 Local Semigroups of Isometries

We recall some results and defintions given in [5], which will be used in the next section.

Let $0 < a \leq +\infty$ and let $I = [0, a)$.

A local semigroup of isometries on the Hilbert space $(\mathcal{H}, \langle \ , \ \rangle)$ is a family $(S_r, \mathcal{H}_r)_{r \in [0,a)}$ such that:

(i) $\mathcal{H}_r$ is a closed subspace of $\mathcal{H}$, $S_r : \mathcal{H}_r \to \mathcal{H}$ is an isometric operator, $\mathcal{H}_t \subset \mathcal{H}_r$ for $0 \leq r < t < a$ and $\mathcal{H}_0 = \mathcal{H}$, $S_0 = I_{\mathcal{H}}$.

(ii) If $r, t \in [0, a)$ are such that $r + t < a$ then $S_t \mathcal{H}_{r+t} \subset \mathcal{H}_r$ and $S_{r+t}h = S_r S_t h$ for all $h \in \mathcal{H}_{r+t}$.

(iii) $\bigcup_{r \in (x,a)} \mathcal{H}_r$ is dense in $\mathcal{H}_x$ for all $x \in [0, a)$.

(iv) If $r \in [0, a)$ and $f \in \mathcal{H}_r$ then the function $t \mapsto S_t h$ is continuous on $[0,r]$.

An infinitesimal generator can be associated to a local semigroup of isometries. It is defined

$$
D(A) = \left\{ h \in \bigcup_{r \in (0,a)} \mathcal{H}_r : \lim_{t \to 0^+} \frac{S_t h - h}{t} \text{ exist} \right\}
$$

and

$$
Ah = \lim_{t \to 0^+} \frac{S_t h - h}{t} \quad \text{for } h \in D(A).
$$

The following facts are proved:

(a) $D(A)$ is a linear dense manifold on $\mathcal{H}$.

(b) A is a skew-symmetric operator.

(c) Each skew-adjoint extension of A to a larger Hilbert space $\mathcal{G}$, which contains $\mathcal{H}$ as a closed subspace, is the infinitesimal generator of a strongly continuous group of unitary operators $(U_r)_{r\in(-\infty,+\infty)} \subset L(\mathcal{G})$ such that $U_r|_{\mathcal{H}_r} = S_r$ for all $r \in [0, a)$.

4 The Dilation Theorem

Theorem 4.1 *Let Γ be a locally compact abelian ordered group and consider $\Omega = \Gamma \times \mathbb{R}$ with the product topology and the lexicographic order.*

Let $(\mathcal{H}_1, \mathcal{H}_2)$ be a pair of Hilbert spaces and let C be a positive definite generalized Toeplitz kernel on $(\Omega, \mathcal{H}_1, \mathcal{H}_2)$.

Then there exist a Hilbert space $\mathcal{G}$, two bounded operators $\tau_\alpha : \mathcal{H}_\alpha \to \mathcal{G}$, $\alpha = 1, 2$ and a strongly continuous unitary representation $(U_\omega)_{\omega\in\Omega}$ of Ω on $L(\mathcal{G})$ such that:

$$C_{\alpha\beta}(\omega) = \tau_\beta^* U_\omega \tau_\alpha \text{ for all } \omega \in \Omega_\alpha - \Omega_\beta \text{ for } \alpha, \beta = 1, 2.$$

In order to prove this theorem we will need some auxiliary results .

Let $\mathcal{L}, \mathcal{E}, \langle\ ,\ \rangle_\mathcal{E}$, and $(S_\omega, \mathcal{E}_\omega)_{\omega\in\Omega_1}$ be as before. Again $[(f_1, f_2)]$ will denote the equivalence class on $\mathcal{E}$ of the element $(f_1, f_2) \in \mathcal{L}$.

Let $\omega_0 = (\gamma_0, r_0) \in \Omega_1$. Then $\mathcal{E}_{\omega_0}$ is equal to

$$Closure\{[(f_1, f_2)] : (f_1, f_2) \in \mathcal{L} \text{ and } supp(f_2) \subset \{(\gamma,r) \in \Omega_2$$
$$: (\gamma, r) < -\omega_0\}\} = Closure\{[(f_1, f_2)] : (f_1, f_2) \in \mathcal{L} \text{ and }$$
$$supp(f_2) \subset \{(\gamma,r) \in \Omega_2 : r < -r_0 \text{ or } r = -r_0,\ \gamma < -\gamma_0\}\}$$

The idea is to prove that $\mathcal{E}_{\omega_0}$ really is equal to

$Closure \{[(f_1, f_2)] : (f_1, f_2) \in \mathcal{L} \text{ and } supp(f_2) \subset \{(\gamma,r) \in \Omega_2 : r \leq -r_0\}\}$

which does not depend on γ_0.

This result will allow us to obtain a decomposition

$$S(\gamma, r) = S(\gamma, 0)S(0, r)$$

where the domain of $S(\gamma, 0)$ is the whole space $\mathcal{E}$ and the domain of $S(0, r)$ depends only on r.

Extending $S(\gamma, 0)$ and $S(0, r)$ in a suitable way we will obtain a unitary extension of the family.

Lemma 4.2 *Let $h_1 \in \mathcal{H}_1, h_2 \in \mathcal{H}_2, \omega_1, \omega_1' \in \Omega_1, \omega_2, \omega_2' \in \Omega_2$. Then $\|[(h_1\delta_{\omega_1} - h_1\delta_{\omega_1'}, h_2\delta_{\omega_2} - h_2\delta_{\omega_2'})]\|_\mathcal{E}^2 = \sum_{\alpha,\beta=1}^2 \langle(C_{\alpha\beta}(\omega_\alpha - \omega_\beta) - C_{\alpha\beta}(\omega_\alpha - \omega_\beta') - C_{\alpha\beta}(\omega_\alpha' - \omega_\beta) + C_{\alpha\beta}(\omega_\alpha' - \omega_\beta'))h_\alpha, h_\beta\rangle_{\mathcal{H}_\beta}$.*

In particular

$$\|[(h_1\delta_{(\gamma,r)}, 0)] - [(h_1\delta_{(\gamma,r_0)}, 0)]\|_{\mathcal{E}}^2$$
$$= 2\langle C_{11}(0,0)h_1, h_1\rangle_{\mathcal{H}_1} - \langle C_{11}(0, r - r_0)h_1, h_1\rangle_{\mathcal{H}_1}$$
$$-\langle C_{11}(0, r_0 - r)h_1, h_1\rangle_{\mathcal{H}_1}.$$

Proof:

$$\|[(h_1\delta_{\omega_1} - h_1\delta_{\omega_1'}, h_2\delta_{\omega_2} - h_2\delta_{\omega_2'})]\|_{\mathcal{E}}^2$$
$$= \langle [(h_1\delta_{\omega_1} - h_1\delta_{\omega_1'}, h_2\delta_{\omega_2} - h_2\delta_{\omega_2'})],$$
$$[(h_1\delta_{\omega_1} - h_1\delta_{\omega_1'}, h_2\delta_{\omega_2} - h_2\delta_{\omega_2'})]\rangle_{\mathcal{E}}$$

$$= \sum_{\alpha,\beta=1}^{2} \sum_{(x,y)\in\Omega_\alpha \times \Omega_\beta} \langle C_{\alpha\beta}(x-y)$$
$$(h_\alpha\delta_{\omega_\alpha}(x) - h_\alpha\delta_{\omega_\alpha'}(x)), h_\beta\delta_{\omega_\beta}(y) - h_\beta\delta_{\omega_\beta'}(y)\rangle_{\mathcal{H}_\beta}$$

$$= \sum_{(x,y)\in\Omega_1\times\Omega_1} \langle C_{11}(x-y)(h_1\delta_{\omega_1}(x)$$
$$-h_1\delta_{\omega_1'}(x)), h_1\delta_{\omega_1}(y) - h_1\delta_{\omega_1'}(y)\rangle_{H_1}$$
$$+ \sum_{(x,y)\in\Omega_1\times\Omega_2} \langle C_{12}(x-y)(h_1\delta_{\omega_1}(x)$$
$$-h_1\delta_{\omega_1'}(x)), h_2\delta_{\omega_2}(y) - h_2\delta_{\omega_2'}(y)\rangle_{\mathcal{H}_2}$$
$$+ \sum_{(x,y)\in\Omega_2\times\Omega_1} \langle C_{21}(x-y)(h_2\delta_{\omega_2}(x)$$
$$-h_2\delta_{\omega_2'}(x)), h_1\delta_{\omega_1}(y) - h_1\delta_{\omega_1'}(y)\rangle_{\mathcal{H}_1}$$
$$+ \sum_{(x,y)\in\Omega_2\times\Omega_2} \langle C_{22}(x-y)(h_2\delta_{\omega_2}(x)$$
$$-h_2\delta_{\omega_2'}(x)), h_2\delta_{\omega_2}(y) - h_2\delta_{\omega_2'}(y)\rangle_{\mathcal{H}_2}$$
$$= 2\langle C_{11}(0)h_1, h_1\rangle_{\mathcal{H}_1} - \langle C_{11}(\omega_1 - \omega_1')h_1, h_1\rangle_{\mathcal{H}_1}$$
$$-\langle C_{11}(\omega_1' - \omega_1)h_1, h_1\rangle_{\mathcal{H}_1}$$
$$+\langle C_{12}(\omega_1 - \omega_2)h_1, h_2\rangle_{\mathcal{H}_2} + \langle C_{12}(\omega_1' - \omega_2')h_1, h_2\rangle_{\mathcal{H}_2}$$
$$-\langle C_{12}(\omega_1 - \omega_2')h_1, h_2\rangle_{\mathcal{H}_2}$$
$$-\langle C_{12}(\omega_1' - \omega_2)h_1, h_2\rangle_{\mathcal{H}_2} + \langle C_{21}(\omega_2 - \omega_1)h_2, h_1\rangle_{\mathcal{H}_1}$$
$$+\langle C_{21}(\omega_2' - \omega_1')h_2, h_1\rangle_{\mathcal{H}_1}$$
$$-\langle C_{21}(\omega_2 - \omega_1')h_2, h_1\rangle_{\mathcal{H}_1} - \langle C_{21}(\omega_2' - \omega_1)h_2, h_1\rangle_{\mathcal{H}_1}$$
$$+2\langle C_{22}(0)h_2, h_2\rangle_{\mathcal{H}_2}$$
$$-\langle C_{22}(\omega_2 - \omega_2')h_2, h_2\rangle_{\mathcal{H}_2} - \langle C_{22}(\omega_2' - \omega_2)h_2, h_2\rangle_{\mathcal{H}_2}.$$

$\square$

Lemma 4.3 *Let* $h_1 \in \mathcal{H}_1$, $h_2 \in \mathcal{H}_2$.

(a) *If* $r_0 \in \mathbb{R}$ *and* $r_0 > 0$ *then* $\lim_{r \to r_0^+} [(h_1 \delta_{(\gamma,r)}, 0)] = [(h_1 \delta_{(\gamma,r_0)}, 0)] \ \forall \gamma \in \Gamma$.

(b) *If* $\gamma \in \Gamma$ *and* $\gamma \geq 0$ *then* $\lim_{r \to 0^+} [(h_1 \delta_{(\gamma,r)}, 0)] = [(h_1 \delta_{(\gamma,0)}, 0)]$.

(c) *If* $r_0 \in \mathbb{R}$ *and* $r_0 > 0$ *then* $\lim_{r \to -r_0^-} [(0, h_2 \delta_{(\gamma,r)})] = [(0, h_2 \delta_{(\gamma,-r_0)})]$ $\forall \gamma \in \Gamma$.

(d) *If* $\gamma \in \Gamma$ *and* $\gamma \leq 0$ *then* $\lim_{r \to 0^-} [(0, h_2 \delta_{(\gamma,r)})] = [(0, h_2 \delta_{(\gamma,0)})]$.

Proof: (a) From Lemma 4.2 we have that

$$\| [(h_1 \delta_{(\gamma,r)}, 0)] - [(h_1 \delta_{(\gamma,r_0)}, 0)] \|_{\mathcal{E}}^2$$
$$= 2 \langle C_{11}(0,0) h_1, h_1 \rangle_{\mathcal{H}_1} - \langle C_{11}(0, r - r_0) h_1, h_1 \rangle_{\mathcal{H}_1}$$
$$- \langle C_{11}(0, r_0 - r) h_1, h_1 \rangle_{\mathcal{H}_1}$$

which tends to 0 as $r \to r_0^+$ by means of the weak continuity of the triplet C.

(b) From Lemma 4.2 we have that

$$\| [(h_1 \delta_{(\gamma,r)}, 0)] - [(h_1 \delta_{(\gamma,0)}, 0)] \|_{\mathcal{E}}^2$$
$$= 2 \langle C_{11}(0,0) h_1, h_1 \rangle_{\mathcal{H}_1} - \langle C_{11}(0, r) h_1, h_1 \rangle_{\mathcal{H}_1}$$
$$- \langle C_{11}(0, -r) h_1, h_1 \rangle_{\mathcal{H}_1}$$

which tends to 0 as $r \to 0^+$ by means of the weak continuity of the triplet C.

(c) and (d) are analogous to (a) and (b). $\qquad\qquad\qquad\qquad$ □

Proposition 4.4 *Let* $(\gamma_0, r_0) \in \Omega_1 = (\Gamma \times \mathbb{R})_1$.

(a) *If* $r_0 > 0$ *then* $\mathcal{E}_{(\gamma_0, r_0)} = Closure \ \{ \ [(f_1, f_2)] : (f_1, f_2) \in \mathcal{L}$ *and* $supp$ $f_2 \subset \{ (\gamma, r) \in \Gamma \times \mathbb{R} : r \leq -r_0 \} \}$.

(b) *If* $r_0 = 0$ *then* $\mathcal{E}_{(\gamma_0, 0)} = \mathcal{E}$.

Proof:

(a) It is enough to prove that if $(\gamma, r) \in \Gamma \times \mathbb{R}$, $h_2 \in \mathcal{H}_2$ and $r \leq -r_0$ then $[(0, h_2 \delta_{(\gamma,r)})] \in \mathcal{E}_{(\gamma_0, r_0)}$.

If $r < -r_0$ then $(\gamma, r) < -(\gamma_0, r_0)$, therefore $[(0, h_2 \delta_{(\gamma,r)})] \in \mathcal{E}_{(\gamma_0, r_0)}$.

If $r = -r_0$ then

$$[(0, h_2 \delta_{(\gamma,-r_0)})] = \lim_{r \to -r_0^-} [(0, h_2 \delta_{(\gamma,r)})] \in \mathcal{E}_{(\gamma_0, r_0)}.$$

(b) It is sufficient to prove that if $(\gamma, r) \leq (0, 0)$, $h_2 \in \mathcal{H}_2$ then $[(0, h_2\delta_{(\gamma,r)})] \in \mathcal{E}_{(\gamma_0,0)}$.

If $r < 0$ then $(\gamma, r) < -(\gamma_0, 0)$, and therefore $[(0, h_2\delta_{(\gamma,r)})] \in \mathcal{E}_{(\gamma_0,0)}$.

If $r = 0$ then

$$[(0, h_2\delta_{(\gamma,0)})] = \lim_{r \to -0^-} [(0, h_2\delta_{(\gamma,r)})] \in \mathcal{E}_{(\gamma_0,0)}.$$

$\square$

Remark 4.5 Last proposition shows that $\mathcal{E}(\gamma, r)$ does not depend of γ if $(\gamma, r) \in (\Gamma \times \mathbb{R})_1$. Let $\mathcal{N}_r$ denote the space $\mathcal{E}(\gamma, r)$. It is important to observe that $\mathcal{N}_0 = \mathcal{E}$.

Proposition 4.6 *Let* $(\gamma_0, r_0) \in \Omega_1 = (\Gamma \times \mathbb{R})_1$.

(a) *If* $r_0 > 0$ *then*

$$Range(S_{(\gamma_0,r_0)}) = S_{(\gamma_0,r_0)}\mathcal{E}_{(\gamma_0,r_0)} = S_{(\gamma_0,r_0)}\mathcal{N}_{r_0}$$
$$= Closure\{[(f_1, f_2)] : (f_1, f_2) \in \mathcal{L} \text{ and}$$
$$supp f_1 \subset \{(\gamma, r) \in \Gamma \times \mathbb{R} : r \geq r_0\}\}.$$

(b) *If* $r_0 = 0$ *then* $Range\ (S_{(\gamma_0,0)}) = \mathcal{E}$.

Proof: (a) It is clear that

$$Range(S_{(\gamma_0,r_0)}) \subset Closure\{[(f_1, f_2)] : (f_1, f_2) \in \mathcal{L} \text{ and}$$
$$supp f_1 \subset \{(\gamma, r) \in \Gamma \times \mathbb{R} : r \geq r_0\}\}$$

In order to prove the equality it is sufficient to show that if $(\gamma, r) \in \Gamma \times \mathbb{R}$, $h_1 \in \mathcal{H}_1$ and $r \geq r_0$ then $[(h_1\delta_{(\gamma,r)}, 0)] \in Range(S_{(\gamma_0,r_0)})$.

If $r > r_0$ then $r - r_0 > 0$, so $(\gamma - \gamma_0, r - r_0) \in (\Gamma \times \mathbb{R})_1$, therefore

$$[(h_1\delta_{(\gamma,r)}, 0)] = S_{(\gamma_0,r_0)}[(h_1\delta_{\gamma-\gamma_0,r-r_0}), 0] \in Range(S_{(\gamma_0,r_0)}).$$

If $r = r_0$ then

$$[(h_1\delta_{(\gamma,r_0)}, 0)] = \lim_{r \to r_0^+} S_{(\gamma_0,r_0)}[(h_1\delta_{(\gamma-\gamma_0,r-r_0)}, 0)] \in \mathcal{E}_{(\gamma_0,r_0)}.$$

(b) It is enough to prove that if $(\gamma, r) \in (\Gamma \times \mathbb{R})_1$ and $h_1 \in \mathcal{H}_1$ then $[(h_1\delta_{(\gamma,r)}, 0)] \in S_{(\gamma_0,0)}\mathcal{E}_{(\gamma_0,0)}$.

If $r > 0$ then $(\gamma - \gamma_0, r) > (\gamma_0, 0)$, therefore

$$[(h_1\delta_{(\gamma,r)}, 0)] = S_{(\gamma_0,0)}[(h_1\delta_{(\gamma-\gamma_0,r)}, 0)] \in S_{(\gamma_0,0)}\mathcal{E}_{(\gamma_0,0)}.$$

If $r = 0$ then

$$[(h_1\delta_{(\gamma,r)}, 0)] = \lim_{r\to 0^+} S_{(\gamma_0,0)}[(h_1\delta_{(\gamma-\gamma_0,r)}, 0)] \in S_{(\gamma_0,0)}\mathcal{E}_{(\gamma_0,0)}.$$

$\square$

Lemma 4.7 *The following holds:*

(a) $(S_{(\gamma,0)})_{\gamma\in\Gamma_1}$ *is a semigroup of unitary operators in the Hilbert space* $\mathcal{E}$.

(b) *Set* $T_\gamma = S_{(\gamma,0)}$ *for* $\gamma \in \Gamma_1$ *and* $T_\gamma = S^*_{(-\gamma,0)}$ *for* $\gamma \in \Gamma_2$ *then* $(T_\gamma)_{\gamma\in\Gamma}$ *is a strongly continuous group of unitary operators.*

(c) $(S_{(0,r)}, \mathcal{N}_r)_{r\in[0,+\infty)}$ *is a local semigroup of isometries in the Hilbert space* $\mathcal{E}$.

(d) $T_\gamma(\mathcal{N}_r) = \mathcal{N}_r$ *and* $T_\gamma S_{(0,r)} = S_{(0,r)} T_\gamma|_{\mathcal{N}_r}$ *for every* $\gamma \in \Gamma$ *and for every* $r \in [0, +\infty)$.

(e) $S(\gamma, r) = T_\gamma S_{(0,r)} = S_{(0,r)} T_\gamma|_{\mathcal{N}_r}$ *for every* $(\gamma, r) \in \Omega_1$

Proof: From Propositions 4.4, 4.6 and Lemma 4.3 we obtain (a) and (b).

(c) The strong continuity of $(S_{(0,r)}, \mathcal{N}_r)$ follows from Lemma 4.2 so it is enough to show that if $r_0 \in [0, +\infty)$ then $\bigcup_{r>r_0} \mathcal{E}_{(0,r)}$ is dense in $\mathcal{E}_{(0,r_0)}$.

It is sufficient to prove:

$[(0, h_2\delta_{(\gamma,-r_0)})]$ is a limit point of $\bigcup_{r>r_0} \mathcal{E}_{(0,r)}$ for every $h_2 \in \mathcal{H}_2$, $\gamma \in \Gamma$ if $r_0 > 0$ and $[(0, h_2\delta_{(\gamma,0)})]$ is a limit point of $\bigcup_{r>0} \mathcal{E}_{(0,r)}$ for every $h_2 \in \mathcal{H}_2$, $\gamma \in \Gamma$.

The first assertion follows since

$$[(0, h_2\delta_{(\gamma,-r_0)})] = \lim_{r\to -r_0^-} [(0, h_2\delta_{(\gamma,r)})]$$

for every $\gamma \in \Gamma$ and the second is analogous.

(d) It is clear that we can suppose $r > 0$. First we shall see that

$$T_\gamma \mathcal{N}_r = \mathcal{N}_r \quad \text{for every } \gamma \geq 0.$$

If $\gamma \geq 0$ then

$$T_\gamma \mathcal{N}_r = S_{(\gamma,0)}\mathcal{N}_r = S_{(\gamma,0)}\mathcal{E}_{(\gamma,r)} \subset \mathcal{N}_r.$$

The equality follows from the following facts:

If $h_1 \in \mathcal{H}_1, \gamma' \in \Gamma y \ r' > 0$ then

$$[(h_1 \delta_{(\gamma',r')}, 0)] = S_{(\gamma,0)}[(h_1 \delta_{(\gamma+\gamma',r')}, 0)].$$

If $h_1 \in \mathcal{H}_1$ and $\gamma' \in \Gamma_1$ then

$$[(h_1 \delta_{(\gamma',0)}, 0)] = \lim_{r' \to 0^+} S_{(\gamma,0)}[(h_1 \delta_{(\gamma+\gamma',r')}, 0)].$$

If $h_2 \in \mathcal{H}_2, \gamma' \in \Gamma \ y \ r' < 0$ then

$$[(0, h_2 \delta_{(\gamma',r')})] = S_{(\gamma,0)}[(0, h_2 \delta_{(\gamma+\gamma',r')})].$$

If $h_2 \in \mathcal{H}_2$ and $\gamma' \in \Gamma_2$ then

$$[(0, h_2 \delta_{(\gamma',0)})] = \lim_{r' \to 0^-} S_{(\gamma,0)}[(0, h_2 \delta_{(\gamma+\gamma',r')})].$$

Since $T_\gamma \mathcal{N}_r = \mathcal{N}_r$ for every $\gamma \geq 0$ and $T_{-\gamma} = T_\gamma^{-1}$ the equality holds for every γ. Finally, it is clear that

$$T_\gamma S_{(0,r)} = S_{(0,r)} T_\gamma|_{\mathcal{E}_{(0,r)}} \quad \text{for every } \gamma \in \Gamma_1.$$

From $T_{-\gamma} = T_\gamma^{-1}$ the result follows.
(e) Let $(\gamma, r) \in \Omega_1$. If $\gamma \geq 0$ then

$$S(\gamma, r) = S(\gamma, 0)S(0, r) = T_\gamma S(0, r).$$

Suppose $\gamma < 0$. Then $r > 0$. Since $\bigcup_{s>r} \mathcal{N}_s$ is dense in $\mathcal{N}_r$ and $\bigcup_{s>0} \mathcal{N}_s$ is dense in $\mathcal{E}$ it is enough to show that

$$\langle S^*_{(-\gamma,0)} S_{(0,r)} f, g \rangle_{\mathcal{E}} = \langle S_{(\gamma,r)} f, g \rangle_{\mathcal{E}} \text{ for all}$$

$$f \in \bigcup_{s>r} \mathcal{N}_s \text{ and } g \in \bigcup_{s>0} \mathcal{N}_s.$$

Let $f \in \bigcup_{s>r} \mathcal{N}_s$ and $g \in \bigcup_{s>0} \mathcal{N}_s$. Let $\epsilon_0 > 0$ be such that $f \in \mathcal{N}_{r+3\epsilon_0}$ and $g \in \mathcal{N}_{2\epsilon_0}$. For $0 < \epsilon < \epsilon_0$ we have that

$$
\begin{aligned}
\langle S^*_{(-\gamma,0)} S_{(0,r+\epsilon)} f, g \rangle_{\mathcal{E}} &= \langle S_{(0,r+\epsilon)} f, S_{(-\gamma,0)} g \rangle_{\mathcal{E}} \\
&= \langle S_{(0,\epsilon)} S_{(0,r+\epsilon)} f, S_{(0,\epsilon)} S_{(-\gamma,0)} g \rangle_{\mathcal{E}} \\
&= \langle S_{(0,r+2\epsilon)} f, S_{(-\gamma,\epsilon)} g \rangle_{\mathcal{E}} \\
&= \langle S_{(\gamma,\epsilon)} S_{(0,r+2\epsilon)} f, S_{(\gamma,\epsilon)} S_{(-\gamma,\epsilon)} g \rangle_{\mathcal{E}} \\
&= \langle S_{(\gamma,r+3\epsilon)} f, S_{(0,2\epsilon)} g \rangle_{\mathcal{E}}.
\end{aligned}
$$

Letting $\epsilon \to 0^+$ we obtain the result. $\qquad \square$

Now we prove Theorem 4.1.

Proof: Let T_γ be as in Lemma 4.7. Then we have that

$$S_\omega = S_{(\gamma,r)} = T_\gamma S_{(0,r)} = S_{(0,r)} T_\gamma$$

for $\omega = (\gamma, r) \in \Omega_1$, where $(T_\gamma)_{\gamma \in \Gamma}$ is a strongly continuous group of unitary operators and $(S_{(0,r)}, \mathcal{N}_r)_{r \in [0,+\infty)}$ is a local semigroup of isometries.

Therefore to obtain the group $(U_\omega)_{\omega \in \Omega}$ it is sufficient to obtain a Hilbert space $\mathcal{G}$ containing $\mathcal{E}$ as a closed subspace, and two strongly continuous unitary groups $(\tilde{S}_r)_{r \in (-\infty,+\infty)} \subset L(\mathcal{G})$, $(\tilde{T}_\gamma)_{\gamma \in \Gamma} \subset L(\mathcal{G})$ such that:

$$\tilde{S}_r|_{\mathcal{N}_r} = S(0, r) \quad \text{for all } r \in [0, +\infty)$$

$$\tilde{T}_\gamma|_{\mathcal{E}} = T_\gamma \quad \text{for all } \gamma \in \Gamma$$

$$\tilde{S}_r \tilde{T}_\gamma = \tilde{T}_\gamma \tilde{S}_r \quad \text{for all } \gamma \in \Gamma, \quad r \in (-\infty, +\infty).$$

According to Section 3 $(S_{(0,r)}, \mathcal{N}_r)_{r \in [0,+\infty)}$ has an infinitesimal generator A, which is a skew-symmetric operator, and from Lemma 4.7 (d) it follows that $T_\gamma(domain(A)) \subset domain(A)$ and $T_\gamma A = A T_\gamma$.

Let τ_A be the Cayley transform of A. Then τ_A is an isometric operator with domain $D \subset \mathcal{E}$, range $R \subset \mathcal{E}$, $ker(\tau_A - I) = 0$ and $T_\gamma(D) \subset D$, $\tau_A T_\gamma = T_\gamma \tau_A$. Also each skew-selfadjoint extension to a larger Hilbert space of A is the infinitesimal generator of a group of unitary operators which extends $(S_{(0,r)})$. So it is clear that any unitary extension $\tilde{\tau}$ of τ_A to a larger Hilbert space such that $ker(\tilde{\tau} - I) = \{0\}$ is the cogenerator of a unitary extension of $(S_{(0,r)})$.

Since the unitary group $(\tilde{T}_\gamma)_{\gamma \in \Gamma} \subset L(\mathcal{G})$ commutes with the unitary group $(\tilde{S}_r)_{r \in (-\infty,+\infty)} \subset L(\mathcal{G})$ if and only if each $\tilde{T}_\gamma$ commutes with the cogenerator $\tilde{\tau}$ of $(\tilde{S}_r)_{r \in (-\infty,+\infty)}$ in order to finish our proof we need to show that there exists a Hilbert space $\mathcal{G}$ containing $\mathcal{E}$ as a closed subspace, a unitary operator $\tilde{\tau} \in L(\mathcal{G})$ and a strongly continuous group of unitary operators $(\tilde{T}_\gamma)_{\gamma \in \Gamma} \subset L(\mathcal{G})$ such that

$$\tilde{\tau}|_D = \tau_A \quad ker(\tilde{\tau} - I) = \{0\}$$

$$\tilde{T}_\gamma|_{\mathcal{E}} = T_\gamma|_{\mathcal{E}} = S_{(\gamma,0)} \quad \forall \gamma \in \Gamma$$

$$\tilde{\tau}\tilde{T}_\gamma = \tilde{T}_\gamma \tilde{\tau} \quad \forall \gamma \in \Gamma.$$

This construction can be done in several ways. A special simple construction is given in [2], page 329.   □

Acknowledgements

Both authors were partially supported by the CDCH of the Universidad Central de Venezuela.

References

[1]   R. Arocena, Generalized Toeplitz kernels and dilations of interwining operators, *Integral Equations and Operator Theory* **6** (1983), 759–778.

[2]   R. Arocena, On the Extension Problem for a class of translation invariant positive forms, *J. Operator Theory* **21** (1989), 323–347.

[3]   R. Arocena and M. Cotlar, Continuous generalized Toeplitz kernels in R, *Portugaliae Mathematica* **39** (1980), 419–434.

[4]   R. Arocena and M. Cotlar, Dilation of generalized Toeplitz kernels and some vectorial moment and weighted problems, *Lecture Notes in Math. Springer* **908** (1982), 169–188.

[5]   R. Bruzual, Local semigroups of contractions and some applications to Fourier representation theorems, *Int. Eq. and Op. Theory* **10** (1987), 780–801.

[6]   M. Cotlar and C. Sadosky, On the Helson-Szegö theorem and a related class of modified Toeplitz kernels, *Proc. Symp. Pure Math. AMS* **35-I** (1979), 383–407.

[7]   M. Cotlar and C. Sadosky, Prolongements des formes de Hankel généralisées en formes de Toeplitz, *C. R. Acad. Sci. Paris* **305** (1987), 167–170.

[8]   M. Domínguez, Interpolation and prediction problems for connected compact abelian groups, *Preprint* (1998), 1–15.

Ramón Bruzual
Universidad Central de Venezuela
Facultad de Ciencias
Escuela de Matemática
Apartado Postal 47686
Caracas 1041-A
Venezuela
e-mail: rbruzual@euler.ciens.ucv.ve
        rbruzual@reacciun.ve

Marisela Domínguez
Universidad Central de Venezuela
Facultad de Ciencias
Escuela de Matemática
Apartado Postal 47159
Caracas 1041-A
Venezuela
e-mail: mdomin@euler.ciens.ucv.ve

1991 Mathematics Subject Classification. Primary: 47A20; Secondary: 47D03.

Operator Theory:
Advances and Applications, Vol. 124
© 2001 Birkhäuser Verlag Basel/Switzerland

On Riccati Equations and Reproducing Kernel Spaces

Harry Dym

To Israel Gohberg: teacher, colleague and valued friend, with admiration and affection, on his seventieth

A class of finite dimensional reproducing kernel Krein spaces of vector valued rational functions $\mathcal{M}_X$ with an indefinite inner product that is defined in terms of a singular Hermitian matrix X is analyzed. It is shown that if X is positive semidefinite, then $\mathcal{M}_X$ is a reproducing kernel Hilbert space of the kind that originates in the work of L. de Branges if and only if X is a solution of an associated Riccati equation and a certain invariance condition (which is automatically fulfilled in some cases) is met. Analogous conclusions are obtained for the case when X is Hermitian and $\mathcal{M}_X$ is a Krein space.

1 Introduction

The purpose of these notes is to present some connections between Riccati equations and the special class of reproducing kernel spaces that originate in the work of de Branges. The material evolved in the course of the author's attempt to understand some of the recent literature on H_∞ control theory from more familiar grounds.

The most recent starting point (over the years there have been several) was the following theorem which appears in a slightly more general form (and slightly different notation) as Theorem 5.2 in the book of Kimura [Ki]. There are also a number of points of contact with the papers of [Gr] and [GGLD].

Theorem 1.1 *Let $(A, B) \in \mathbb{C}^{n \times n} \times \mathbb{C}^{n \times m}$ be a controllable pair such that the matrix A has no imaginary eigenvalues and let J be an $m \times m$ signature matrix. Then there exists a rational $m \times m$ matrix valued function (mvf) $\Theta(\lambda)$ that is J-inner with respect to the open right half plane Π_+ such that:*

1. *$(\lambda I_n - A)^{-1} B J \Theta(\lambda)$ is analytic in the closed right half plane $\overline{\Pi}_+$, and*

2. *the McMillan degree of $\Theta(\lambda)$ is equal to the number of eigenvalues of A in Π_+ counting multiplicities,*

if and only if the Riccati equation

$$(1.1) \qquad X A + A^* X - X 2\pi B J B^* X = 0$$

has a solution $X \geq 0$ such that the spectrum $\sigma(\widehat{A})$ of the matrix

$$(1.2) \qquad \widehat{A} = A - 2\pi B J B^* X$$

lies in the open left half plane Π_-. *Moreover, in this case the matrix*

(1.3) $$\Theta(\lambda) = I_m - 2\pi B^*(\lambda I_n + A^*)^{-1} X B J$$

achieves the stated goals.

The perhaps unfamiliar 2π is included in order to match the normalizations that are used in a number of the references that will be cited. It can be eliminated by renormalization.

The basic objective of these notes is to expose and exploit the reproducing kernel spaces that underlie theorems of this type and their analogues for the disc. Thus, for example, in the setting of Theorem 1.1, if

$$F(\lambda) = B^*(\lambda I_n + A^*)^{-1} \quad \text{and} \quad X \geq 0 ,$$

then the following conclusions (and more) will emerge from the subsequent analysis:

1. The space

   (1.4) $$\mathcal{M}_X = \{F(\lambda)Xu : u \in \mathbb{C}^n\}$$

   endowed with the inner product

   (1.5) $$\langle FXu, FXv \rangle = v^*Xu$$

   is a reproducing kernel Hilbert space (RKHS) with reproducing kernel (RK)

   (1.6) $$K_\omega(\lambda) = F(\lambda)XF(\omega)^* .$$

2. The RK can be expressed in the form

   (1.7) $$K_\omega(\lambda) = \frac{J - \Theta(\lambda)J\Theta(\omega)^*}{2\pi(\lambda + \bar{\omega})}$$

   if and only if X is a solution of the Riccati equation (1.1).

3. If $\sigma(A) \subset \Pi_+$ and X is a solution of the Riccati equation (1.1), then

   (1.8) $$\langle FXu, FXv \rangle_{\mathcal{M}_X} = v^*X\left\{\int_{-\infty}^{\infty} F(iv)^*JF(iv)dv\right\}Xu.$$

4. If $\sigma(A) \subset \Pi_+$, X is a solution of (1.1) and $\sigma(\widehat{A}) \subset \Pi_-$, then X is invertible and X^{-1} is the one and only solution of the Lyapunov equation

   (1.9) $$AX^{-1} + X^{-1}A^* = 2\pi BJB^*.$$

There are analogues of Theorem 1.1 for the disc case; see e.g., Theorem 3.2 of [KK1] and, for other variants and related issues, [KK2] and [IG]. The basic issue under study in this setting amounts to the following: Given a controllable pair (A, B), where the matrix A has no eigenvalues of modulus one, find a mvf $\Theta(\lambda)$ which is J-inner with respect to the open unit disc $\mathbb{D}$ such that

1. $(\lambda I_n - A)^{-1} B J \Theta(\lambda)$ is analytic in the closed unit disc $\overline{\mathbb{D}}$, and

2. the McMillan degree of $\Theta(\lambda)$ is equal to the number of eigenvalues of A in $\mathbb{D}$, counting multiplicities.

A number of analogues for this setting of the results described in formulas (1.4) to (1.9) will emerge from the analysis in Section 5. In particular, if $X \geq 0$, then:

1. The space $\mathcal{M}_X$ defined in (1.4), endowed with the inner product (1.5), but with

$$(1.10) \qquad F(\lambda) = B^*(I_n - \lambda A^*)^{-1},$$

is a RKHS with RK given by formula (1.6).

2. The RK can be expressed in the form

$$(1.11) \qquad K_\omega(\lambda) = \frac{J - \Theta(\lambda) J \Theta(\omega)^*}{1 - \lambda \overline{\omega}}$$

if and only if A meets the invariance condition

$$(1.12) \qquad A^*X = XX^\dagger A^*X$$

and X is a solution of the Riccati equation

$$(1.13) \qquad X - XAX^\dagger A^*X = XBJB^*X,$$

where $X^\dagger$ denotes the Moore-Penrose inverse of X.

3. If X satisfies the conditions (1.12) and (1.13), then the J-inner mvf $\Theta(\lambda) = \Theta_X(\lambda)$ that appears in formula (1.11) is unique up to a J-unitary constant multiplier on the right (see formula (5.4)).

4. If X satisfies the conditions (1.12) and (1.13) and if $\sigma(A) \subset \mathbb{D}$ and $(\lambda I_n - A)^{-1} B J \Theta_X(\lambda)$ is analytic in $\overline{\mathbb{D}}$, then X is invertible and X^{-1} is the one and only solution of the Stein equation

$$(1.14) \qquad X^{-1} - AX^{-1}A^* = BJB^*.$$

5. If X satisfies the conditions (1.12) and (1.13) and if $\sigma(A) \subset \mathbb{D}$, then

(1.15)
$$\langle FXu, FXv \rangle_{\mathcal{M}_X} = v^*X$$
$$\left\{ \frac{1}{2\pi} \int_0^{2\pi} F(e^{i\theta})^* J F(e^{i\theta}) d\Theta \right\} Xu.$$

6. If A and $I_n - BJB^*X$ are invertible, the Riccati equation (1.13) can be reexpressed as

(1.16)
$$A^*X(I_n - BJB^*X)^{-1}A = X.$$

Moreover, if X is a Hermitian solution of the Riccati equation (1.16), then the invariance condition (1.12) is automatically met. Furthermore, the mvf $(I_n - \lambda A)^{-1} BJ\Theta_X(\lambda)$ belongs to $H_\infty^{n \times m}(\mathbb{D})$ if and only if

(1.17)
$$\sigma(A^{-1}(I_n - BJB^*X)) \subset \mathbb{D}.$$

A number of the results cited above are available in the literature in one form or another, but they are obtained by other methods. In particular (1.16) and (1.17) are well known; see e.g., [KK1]. It should perhaps be emphasized that we do not assume that the matrix A is invertible in the first five statements.

As the material was developed, it became clear that analogous conclusions held for Hermitian solutions X of the Riccati equation and reproducing kernel Krein spaces (RKKS's). Accordingly, the analysis is presented in this more general setting. Even so, since we are dealing with finite dimensional spaces, the analysis is mostly elementary. However, it should be emphasized that this is only a beginning. We have considered the spaces underlying the Riccati equations associated with pole elimination. It remains to consider the analogous issues for zero elimination, which is harder. But that lies in the future.

The paper is organized as follows: A number of preliminary definitions and needed facts are summarized in Section 2. The third section explores the implications of invariance under the generalized backwards shift operator R_α that is defined by the rule

(1.18)
$$(R_\alpha f)(\lambda) = \frac{f(\lambda) - f(\alpha)}{\lambda - \alpha}.$$

Then in Section 4, the general theme that was outlined just above is developed in the finite dimensional RKKS setting with respect to the right half plane. Analogous developments for the disc are presented in Section 5.

Finally a word on notation: We shall use the symbols $\mathbb{R}$ and $\mathbb{C}$ for the real and complex numbers, as usual, $\mathbb{C}_+$ [resp. $\mathbb{C}_-$] for the open upper [resp. lower] half plane, Π_+ [resp. Π_-] for the open right [resp. left] half plane, $\mathbb{D} = \{\lambda \in \mathbb{C} : |\lambda| < 1\}$ and $\mathbb{E} = \{\lambda \in \mathbb{C} : |\lambda| > 1\}$.

The acronym mvf [resp. vvf] is short for matrix valued function [resp. vector valued function]. The symbol $H_2^m(\Omega_+)$ will be used to designate the set of $m \times 1$ vvf's that belong to the Hardy space of index 2 (i.e., square integrable) with respect to the region Ω_+, where Ω_+ is either $\mathbb{D}$, $\mathbb{C}_+$ or Π_+ and Ω_0 denotes the boundary of Ω_+: $\mathbb{T}$, $\mathbb{R}$ and $i\mathbb{R}$ respectively.

If A is a matrix, then $\mathcal{R}_A$ will denote its range, $\mathcal{N}_A$ its null space, A^* its conjugate transpose and, if it is square, then $\sigma(A)$ will denote its spectrum. For complex numbers ω, both $\overline{\omega}$ and ω^* will be used for the complex conjugate of ω. The symbol J stands for an $m \times m$ signature matrix, i.e., a constant matrix which is both unitary and selfadjoint with respect to the standard inner product in the space of $m \times 1$ complex vectors $\mathbb{C}^m$. The symbol $F^{\#}(\lambda)$ is defined in Table 2.1, below; $\mathbb{C}^{m \times n}$ denotes the $m \times n$ matrices with complex entries.

2 Preliminaries

A Krein space $\mathcal{K}$ of $m \times 1$ vector valued functions (vvf's) that is defined on a subset Ω of the complex plane $\mathbb{C}$ is said to be a reproducing kernel Krein space (RKKS) if there exists an $m \times m$ matrix valued function $K_\omega(\lambda)$ on $\Omega \times \Omega$ such that for every choice of $v \in \mathbb{C}^m$, $\omega \in \Omega$ and $f \in \mathcal{K}$ the following two conditions are met:

(2.1)   1. $K_\omega v \in \mathcal{K}$ (as a function of λ).

(2.2)   2. $\langle f, K_\omega v \rangle_{\mathcal{K}} = v^* f(\omega)$.

It is well known (and readily checked) that there is at most one mvf $K_\omega(\lambda)$ on $\Omega \times \Omega$ which meets these two conditions; it is referred to as the reproducing kernel (RK) of the RKKS $\mathcal{K}$.

In the present analysis we shall focus primarily on finite dimensional RKKS's of the form

$$\mathcal{M}_X = \{F(\lambda)Xu : \; u \in \mathbb{C}^n\},$$

that are constructed from the columns of the $m \times n$ mvf

(2.3) $F(\lambda) = V(M - \lambda N)^{-1}$,

where $V \in \mathbb{C}^{m \times n}$, $M, N, X \in \mathbb{C}^{n \times n}$, X is Hermitian and it is assumed throughout that:

(A1) The $n \times n$ matrix polynomial

$$G(\lambda) = M - \lambda N$$

is invertible in $\mathbb{C}$ except for at most n choices of λ.

(A2) The columns of $F(\lambda)$ are linearly independent in the sense that if $F(\lambda)u = 0$ for some $u \in \mathbb{C}^n$ in an open nonempty subset of $\mathbb{C}$ in which $G(\lambda)$ is invertible, then $u = 0$.

If $N = \pm I_n$ [resp. $M = \pm I_n$], then condition (A2) is equivalent to assuming that the pair (V, M) [resp. (V, N)] is observable. For a more extensive discussion of this circle of ideas for general $G(\lambda)$, see e.g., [AD3].

Theorem 2.1 *Let assumptions (A1) and (A2) be in force and let $X \in \mathbb{C}^{n \times n}$ be a nonzero Hermitian matrix. Then the linear space*

$$(2.4) \qquad \mathcal{M}_X = \{F(\lambda)Xu : \; u \in \mathbb{C}^n\}$$

is an RKKS with respect to the indefinite inner product

$$(2.5) \qquad \langle FXu, \; FXv \rangle_{\mathcal{M}_X} = v^* Xu.$$

The RK $K_\omega(\lambda)$ of this RKKS is given by the formula

$$(2.6) \qquad K_\omega(\lambda) = F(\lambda)XF(\omega)^*.$$

Proof: The first order of business is to check that the indicated inner product is well defined. But if

$$FXu_1 = FXu_2 \quad \text{and} \quad FXv_1 = FXv_2$$

for some choice of u_1, u_2, v_1, v_2 in $\mathbb{C}^n$, then assumption (A2) guarantees that

$$Xu_1 = Xu_2 \quad \text{and} \quad Xv_1 = Xv_2.$$

Thus,

$$\langle FXu_1, FXv_1 \rangle_{\mathcal{M}_X} = \langle FXu_2, FXv_2 \rangle_{\mathcal{M}_X}$$

as needed.

Next, upon setting

$$\mathcal{R}_X^+ = \text{span}\{\text{eigenvectors of } X \text{ with positive eigenvalues}\}$$

$$\mathcal{R}_X^- = \text{span}\{\text{eigenvectors of } X \text{ with negative eigenvalues}\}$$

and

$$\mathcal{M}_X^\pm = \{F(\lambda)Xu : u \in \mathcal{R}_X^\pm\},$$

it is easily verified that:

1. $\mathcal{M}_X = \mathcal{M}_X^+ + \mathcal{M}_X^-$.
2. $\mathcal{M}_X^+ \cap \mathcal{M}_X^- = \{0\}$.

3. $\mathcal{M}_X^+$ is a Hilbert space with respect to the inner product defined by the restriction of (2.5) to u, v in $\mathcal{R}_X^+$.

4. $\mathcal{M}_X^-$ is a Hilbert space with respect to the inner product defined by the negative of the restriction of (2.5) to u, v in $\mathcal{R}_X^-$.

5. $\mathcal{M}_X^+$ is orthogonal to $\mathcal{M}_X^-$ with respect to the indefinite inner product (2.5).

Thus, $\mathcal{M}_X$ is a Krein space. Finally, it is readily checked that if $K_\omega(\lambda)$ is defined by (2.6), then the two conditions (2.1) and (2.2) are met. Therefore, $\mathcal{M}_X$ is a RKKS and its RK is given by formula (2.6). $\qquad\square$

Remark 2.2 The proof of the preceding theorem remains valid if $\mathcal{R}_X^+$ is replaced by $\mathbb{C}^n \ominus \mathcal{R}_X^-$ in the definition of $\mathcal{M}_X^+$. In particular,

$$\langle FXu, FXu \rangle_{\mathcal{M}_X} = 0 \Longleftrightarrow Xu = 0 \Longleftrightarrow FXu = 0$$

for every $u \in \mathbb{C}^n \ominus \mathcal{R}_X^-$.

In the sequel we shall be particularly interested in the case when the RK $K_\omega(\lambda)$ of the RKKS $\mathcal{M}_X$ can be expressed in the form

$$(2.7) \qquad\qquad K_\omega(\lambda) = \frac{J - \Theta(\lambda) J \Theta(\omega)^*}{\rho_\omega(\lambda)},$$

where:

1. J is an $m \times m$ signature matrix (i.e., $J = J^*$ and $J J^* = I_m$).

2. $I_m/\rho_\omega(\lambda)$ is the RK for the RKHS $H_2^m(\Omega_+)$.

3. $\Theta(\lambda)$ is a rational $m \times m$ mvf.

If such a representation for $K_\omega(\lambda)$ exists, then the mvf $\Theta(\lambda)$ which appears in (2.7) is unique up to a right J-unitary constant factor, and we shall say that:

$\mathcal{M}_X$ is a de Branges space $\mathcal{H}(\Theta)$ if $X \geq 0$.

$\mathcal{M}_X$ is a dBK space $\mathcal{K}(\Theta)$ if X is Hermitian.

Both $K_\omega(\lambda)$ and $\Theta(\lambda)$ depend upon X. We do not always indicate this dependence in the notation in order to keep the typography simple. However, it should be kept in mind that:

1. The number of negative squares of the kernel $K_\omega(\lambda)$ is equal to the number of negative eigenvalues of X, counting multiplicities.

2. The mvf $\Theta(\lambda)$ in (2.7) will be J-inner if and only if $X \geq 0$.

3. The McMillan degree of $\Theta(\lambda)$ is equal to the rank of X.

Necessary and sufficient conditions for the RK of $\mathcal{M}_X$ to admit a representation of the form (2.7) are provided by the following theorem, which is an elaboration of a fundamental result that is due to de Branges [dB]. It is formulated in terms of the polynomials $a(\lambda)$ and $b(\lambda)$ that are given in Table 2.1 in order to obtain a statement which is applicable to each of the three classical choices of Ω_+. In terms of these polynomials,

$$\rho_\omega(\lambda) = a(\lambda)a(\omega)^* - b(\lambda)b(\omega)^*.$$

This formula, which serves to unify the analysis for all three of the classical choices of Ω_+, appears earlier in the work of Lev-Ari and Kailath [LAK] on fast algorithms and also in the work of a number of Russian mathematicians; see [Nu]. It was exploited extensively in a much higher degree of generality in the papers [AD1]–[AD3].

Table 2.1

Ω_+	$\mathbb{D}$	$\mathbb{C}_+$	Π_+
$a(\lambda)$	1	$\sqrt{\pi}(1-i\lambda)$	$\sqrt{\pi}(1+\lambda)$
$b(\lambda)$	λ	$\sqrt{\pi}(1+i\lambda)$	$\sqrt{\pi}(1-\lambda)$
$\rho_\omega(\lambda)$	$1-\lambda\omega^*$	$-2\pi i(\lambda-\omega^*)$	$2\pi(\lambda+\omega^*)$
Ω_0	$\mathbb{T}$	$\mathbb{R}$	$i\mathbb{R}$
$\langle f,g \rangle$	$\frac{1}{2\pi}\int_0^{2\pi} g(e^{i\theta})^* f(e^{i\theta})d\theta$	$\int_{-\infty}^{\infty} g(x)^* f(x)dx$	$\int_{-\infty}^{\infty} g(iy)^* f(iy)dy$
λ°	$1/\lambda^*$ if $\lambda \neq 0$	λ^*	$-\lambda^*$
$f^{\#}(\lambda)$	$f(\lambda^\circ)^*$	$f(\lambda^\circ)^*$	$f(\lambda^\circ)^*$
$ab'-ba'$	1	$2\pi i$	-2π

A set Δ is said to be symmetric with respect to Ω_0 (or $\rho_\omega(\lambda)$) if for every $\lambda \in \Delta$ (except 0 for $\Omega_0 = \mathbb{T}$) the point $\lambda^\circ \in \Delta$; note that $\rho_\omega(\omega^\circ) = 0$.

Theorem 2.3 *Let $\mathcal{K}$ be a RKKS of $m \times 1$ vector valued functions which are analytic in an open subset Δ of $\mathbb{C}$ that is symmetric with respect to Ω_0 and assume that $\Delta \cap \Omega_0 \neq \emptyset$. Then the reproducing kernel $K_\omega(\lambda)$ can be expressed in the form*

$$(2.8) \qquad K_\omega(\lambda) = \frac{J - \Theta(\lambda)J\Theta(\omega)^*}{\rho_\omega(\lambda)},$$

for some choice of $m \times m$ matrix valued function $\Theta(\lambda)$ which is analytic in Δ and some signature matrix J, if and only if the following two conditions hold:

1. *$\mathcal{K}$ is R_α invariant for every $\alpha \in \Delta$.*

2. *The structural identity*

$$\langle R_\alpha(bf), R_\beta(bg) \rangle_{\mathcal{H}} - \langle R_\alpha(af), R_\beta(ag) \rangle_{\mathcal{H}}$$

$$(2.9)$$

$$= |ab' - ba'|^2 g(\beta)^* J f(\alpha)$$

holds for every choice of α, β in Δ and f, g in $\mathcal{K}$.

Moreover, in this case, the function $\Theta(\lambda)$ which appears in (2.8) is unique up to a J unitary constant factor on the right; it can be taken equal to

$$(2.10) \qquad \Theta(\lambda) = I_m - \rho_\mu(\lambda) K_\mu(\lambda) J$$

for any point $\mu \in \Delta \cap \Omega_0$.

This formulation is adapted from [AD1]; see especially Theorems 4.1, 4.3 and 4.4. The restriction to the three choices of $a(\lambda)$ and $b(\lambda)$ specified earlier, permits some simplification in the presentation, because the terms $r(a, b; \alpha) f$ and $r(b, a; \alpha) f$ which intervene there are constant multiples of $R_\alpha(af)$ and $R_\alpha(bf)$, respectively.

The restriction $\Delta \cap \Omega_0 = \emptyset$ can be relaxed at the expense of a more sophisticated formulation. However, since we shall be dealing with finite dimensional spaces and rational functions, there is no need for this extra complication. The interested reader can refer to [AD1] for more information.

For the three cases of interest, the structural identity (2.9) can be reexpressed as:

$$(2.11) \qquad \langle (I + \alpha R_\alpha) f, (I + \beta R_\beta) g \rangle_K - \langle R_\alpha f, R_\beta g \rangle_K = g(\beta)^* J f(\alpha)$$

if $\Omega_+ = \mathbb{D}$,

$$\langle R_\alpha f, g \rangle_K - \langle f, R_\beta g \rangle_K - (\alpha - \beta^*)$$
$$(2.12) \qquad \langle R_\alpha f, R_\beta g \rangle_K = 2\pi i \; g(\beta)^* J f(\alpha)$$

if $\Omega_+ = \mathbb{C}_+$, and

$$\langle R_\alpha f, g \rangle_K + \langle f, R_\beta g \rangle_K + (\alpha + \beta^*)$$
$$(2.13) \qquad \langle R_\alpha f, R_\beta g \rangle_K = -2\pi \; g(\beta)^* J f(\alpha)$$

if $\Omega_+ = \mathbb{T}_+$.

Formula (2.12) appears in de Branges [dB]; formula (2.11) is equivalent to a formula which appears in Ball [Ba], who adapted de Branges' work to the disc, including an important technical improvement due to Rovnyak [Rov]. All three of these references deal with the Hilbert space case only.

The role of the two conditions in Theorem 2.2: R_α invariance and the structural identity, becomes particularly transparent when K is finite dimensional; see e.g., [D1] and [D2], where this theme is elaborated upon at great length, and [AD1]–[AD3]. The analysis developed there for the classical settings can be incorporated into the present formalism. It corresponds to the special case when X is invertible.

3 R_α Invariance

In this section we characterize those Hermitian matrices X for which the RKKS's $\mathcal{M}_X$ are R_α invariant for any one point α (and hence in fact every point α) at which $G(\alpha)$ is invertible.

Let $X^\dagger$ denote the Moore-Penrose inverse of X. Then, since X is Hermitian and hence admits a representation of the form

$$(3.1) \qquad\qquad X = U \begin{bmatrix} X_1 & 0 \\ 0 & 0 \end{bmatrix} U^*$$

with U unitary and X_1 both Hermitian and invertible, it follows that

$$(3.2) \qquad\qquad X^\dagger = U \begin{bmatrix} X_1^{-1} & 0 \\ 0 & 0 \end{bmatrix} U^*.$$

Thus, $X^\dagger$ commutes with X and

$$(3.3) \qquad\qquad X^\dagger X = X X^\dagger$$

is an orthogonal projection; it projects $\mathbb{C}^n$ onto $\mathcal{R}_X$.

Lemma 3.1 *Assume that $G(\alpha)$ is invertible and that either $M = \pm I_n$ or $N = \pm I_n$. Then the space $\mathcal{M}_X$ is R_α invariant if and only if*

$$(3.4) \qquad\qquad MX = XX^\dagger MX \quad \text{and} \quad NX = XX^\dagger NX.$$

Proof: By direct calculation,

$$(R_\alpha F)(\lambda)X = F(\lambda)NG(\alpha)^{-1}X.$$

Therefore, $\mathcal{M}_X$ will be R_α invariant if and only if there exists a matrix $Q_\alpha \in \mathbb{C}^{n \times n}$ such that

$$NG(\alpha)^{-1}X = X Q_\alpha.$$

Since $NG(\alpha)^{-1} = G(\alpha)^{-1}N$ in the present setting, this is the same as to require

$$(3.5) \qquad\qquad NX = G(\alpha)X Q_\alpha.$$

Let U be a unitary matrix such that

$$U^*XU = \begin{bmatrix} X_1 & 0 \\ 0 & 0 \end{bmatrix},$$

where X_1 is invertible, and let

$$U^*NU = \begin{bmatrix} N_{11} & N_{12} \\ N_{21} & N_{22} \end{bmatrix}, \quad U^*MU = \begin{bmatrix} M_{11} & M_{12} \\ M_{21} & M_{22} \end{bmatrix}$$

and $U^*Q_\alpha U = \begin{bmatrix} Q_{11} & Q_{12} \\ Q_{21} & Q_{22} \end{bmatrix}$

be block decompositions that are conformable with the block decomposition of X. Then (3.5) holds if and only if

$$\begin{bmatrix} N_{11}X_1 & 0 \\ N_{21}X_1 & 0 \end{bmatrix} = U^*G(\alpha)U \begin{bmatrix} X_1Q_{11} & X_1Q_{12} \\ 0 & 0 \end{bmatrix},$$

that is, if and only if $Q_{12} = 0$,

$$N_{11}X_1 = (M_{11} - \alpha N_{11})X_1Q_{11}$$

and

$$N_{21}X_1 = (M_{21} - \alpha N_{21})X_1Q_{11}.$$

The last two equations can be reexpressed as

(3.6) $$N_{11}X_1(I + \alpha Q_{11}) = M_{11}X_1Q_{11}$$

and

(3.7) $$N_{21}X_1(I + \alpha Q_{11}) = M_{21}X_1Q_{11}.$$

To complete the proof of necessity, there are two cases to consider:

1. N_{11} is invertible and $N_{21} = 0$.
   In this situation, Q_{11} must be invertible because its kernel is equal to zero:

$$Q_{11}u = 0 \implies N_{11}X_1(I + \alpha Q_{11})u = 0 \quad \text{(by (3.6))}$$
$$\implies N_{11}X_1u = 0$$
$$\implies u = 0 \quad \text{(since } N_{11} \text{ and } X_1 \text{ are invertible).}$$

   Therefore $M_{21} = 0$, by (3.7).

2. M_{11} is invertible and $M_{21} = 0$.
   In this setting, $I + \alpha Q_{11}$ must be invertible because its kernel is equal to zero:

$$(I + \alpha Q_{11})u = 0 \implies M_{11}X_1Q_{11}u = 0 \quad \text{(by (3.6))}$$
$$\implies Q_{11}u = 0 \quad \text{(since } M_{11} \text{ and } X_1 \text{ are invertible)}$$
$$\implies u = 0 \quad \text{(by the first line).}$$

   Therefore $N_{21} = 0$, by (3.7).

Thus, in both cases we have both $M_{21} = 0$ and $N_{21} = 0$. But that is equivalent to the stated condition (3.4). This completes the proof of necessity.

The proof of the sufficiency of condition (3.4) for R_α invariance is a straightforward calculation that is left for the reader. $\square$

4 The Right Half Plane Case

Throughout this section we shall assume that

$$(4.1) \qquad\qquad F(\lambda) = B^*(\lambda I_n + A^*)^{-1},$$

where $A \in \mathbb{C}^{n \times n}$, $B \in \mathbb{C}^{n \times m}$ and the pair (A, B) is controllable. This choice of $F(\lambda)$ is dictated by the identity

$$-F^\#(\lambda) J \Theta(\lambda) = (\lambda I_n - A)^{-1} B J \Theta(\lambda).$$

Theorem 4.1 *Let $F(\lambda)$ be given by (4.1) with (A, B) controllable. Then the RKKS M_X is a dBK space $\mathcal{K}(\Theta)$ if and only if the Hermitian matrix X is a solution of the Riccati equation*

$$(4.2) \qquad\qquad A^*X + XA = 2\pi XBJB^*X.$$

Moreover, in this case the mvf $\Theta(\lambda) = \Theta_X(\lambda)$ is uniquely determined by the formula

$$(4.3) \qquad\qquad \Theta_X(\lambda) = I_m - 2\pi B^*(\lambda I_n + A^*)^{-1} XBJ$$

up to a J-unitary constant multiplier on the right and the following two identities hold:

$$(4.4) \qquad\qquad X(\lambda I_n - A)^{-1} BJ \Theta_X(\lambda) = (\lambda I_n + A^*)^{-1} XBJ$$

$$(4.5) \qquad\qquad (\lambda I_n - A)^{-1} BJ \Theta_X(\lambda) = (\lambda I_n - \widehat{A})^{-1} BJ,$$

where

$$(4.6) \qquad\qquad \widehat{A} = A - 2\pi BJB^*X.$$

The mvf $\Theta_X(\lambda)$ is J-inner if and only if $X \geq 0$; it can be expressed in terms of $\widehat{A}$ as

$$(4.7) \qquad\qquad \Theta_X(\lambda) = I_m - 2\pi B^* X(\lambda I_n - \widehat{A})^{-1} BJ.$$

Proof: Suppose first that $\mathcal{M}_X$ is a dBK space. Then, in view of Theorem 2.3 and Lemma 3.1,

$$(4.8) \qquad A^*X = XX^\dagger A^*X$$

and the structural identity (2.13) holds.

Let A_1 be any matrix in $\mathbb{C}^{n \times n}$ which meets the equality

$$(4.9) \qquad A^*X = -XA_1.$$

(The existence of at least one such matrix is guaranteed by (4.8).) Then

$$(\alpha I_n + A^*)X = X(\alpha I_n - A_1)$$

and thus,

$$(\alpha I_n + A^*)^{-1}X = X(\alpha I_n - A_1)^{-1}$$

for every point $\alpha \in \mathbb{C}$ for which the two inverses exist. Let

$$f(\lambda) = F(\lambda)Xu \quad \text{and} \quad g(\lambda) = F(\lambda)Xv$$

for any choice of u, v in $\mathbb{C}^n$, and suppose that $\alpha, \beta \notin \sigma(-A^*) \cup \sigma(A_1)$. Then it is readily checked that:

$$(R_\alpha f)(\lambda) = -F(\lambda)(\alpha I_n + A^*)^{-1}Xu = -F(\lambda)X(\alpha I_n - A_1)^{-1}u.$$
$$(R_\beta g)(\lambda) = -F(\lambda)(\beta I_n + A^*)^{-1}Xv = -F(\lambda)X(\beta I_n - A_1)^{-1}v.$$
$$f(\alpha) = B^*(\alpha I_n + A^*)^{-1}Xu = B^*X(\alpha I_n - A_1)^{-1}u.$$
$$g(\beta) = B^*(\beta I_n + A^*)^{-1}Xv = B^*X(\beta I_n - A_1)^{-1}v.$$

Next, upon substituting these formulas into the structural identity (2.13) and invoking the inner product rule (2.5), we see that

$$\begin{aligned} v^*\{X(\alpha I_n - A_1)^{-1} &+ (\bar{\beta} I_n - A_1^*)^{-1}X \\ &- (\alpha + \bar{\beta})(\bar{\beta} I_n - A_1^*)^{-1}X(\alpha I_n - A_1)^{-1}\}u \\ &= 2\pi v^*\{(\bar{\beta} I_n - A_1^*)^{-1}XBJB^*X(\alpha I_n - A_1)^{-1}\}u. \end{aligned}$$

However, this last equality holds for every choice of u, $v \in \mathbb{C}^n$ if and only if

$$(\bar{\beta} I_n - A_1^*)X + X(\alpha I_n - A_1) - (\alpha + \bar{\beta})X = 2\pi XBJB^*X,$$

that is, if and only if

$$-A_1^*X - XA_1 = 2\pi XBJB^*X.$$

But, in view of formula (4.9), this last identity implies that X is a solution of the Riccati equation (4.2) and thus serves to complete the proof of the assertion that if $\mathcal{M}_X$ is a dBK space, then X is a solution of equation (4.2).

Conversely, if X is a solution of equation (4.2), then it is readily checked that (4.9) and the structural identity (2.13) hold and therefore, by Lemma 3.1 and Theorem 2.3, that $\mathcal{M}_X$ is a dBK space. Formula (4.3) for $\Theta(\lambda) = \Theta_X(\lambda)$ is obtained by letting $\mu \longrightarrow \infty$ along the imaginary axis in the general formula (2.10). The fact that $\Theta_X(\lambda)$ is J-inner if and only if $X \geq 0$ follows from (2.6) and (2.7). One direction is easy. The other exploits the fact that if $(A, \cdot B)$ is a controllable pair, then there exist a set of points $\omega_1, \ldots, \omega_n$ in the domain of analyticity of $F(\lambda)$ and a set of vectors $u_1, \ldots, u_n$ in $\mathbb{C}^m$ such that the $n \times n$ matrix

$$Y = [F(\omega_1)^* u_1 \quad \cdots \quad F(\omega_n)^* u_n]$$

is invertible. Now, if $\Theta_X(\lambda)$ is J inner, then the $n \times n$ matrix

$$Y^* X Y = [u_i^* K_{\omega_j}(\omega_i) u_j]$$

is positive semidefinite. Therefore $X \geq 0$ also, since Y is invertible.

Finally, when X is a solution of the Riccati equation (4.2), we may choose

$$\widehat{A} = A - 2\pi B J B^* X$$

and verify that

$$(\lambda I_n + A^*)^{-1} X = X(\lambda I_n - \widehat{A})^{-1}$$

for all points $\lambda \notin \sigma(-A^*) \cup \sigma(\widehat{A})$. The verification of formulas (4.7), (4.5) and (4.4) (in that order) is then straightforward. $\qquad\square$

Theorem 4.2 *Let X be a Hermitian solution of the Riccati equation* (4.2) *in the setting of Theorem 4.1 and suppose further that $\sigma(A) \subset \mathbb{H}_+$. Then:*

1. *The mvf*

$$X(\lambda I_n - A)^{-1} B J \Theta_X(\lambda)$$

   belongs to $H_\infty^{n \times m}(\mathbb{H}_+)$.

2. *The mvf*

$$(\lambda I_n - A)^{-1} B J \Theta_X(\lambda)$$

   belongs to $H_\infty^{n \times m}(\mathbb{H}_+)$ if and only if

   (4.10) $$\sigma(A - 2\pi B J B^* X) \subset \mathbb{H}_-.$$

3. *If condition* (4.10) *is in force, then X is invertible and X^{-1} is the one and only solution of the Lyapunov equation*

   (4.11) $$X^{-1} A^* + A X^{-1} = 2\pi B J B^*.$$

Proof: The first two statements are immediate from the preceding theorem. The third rests upon the observation that the null space $\mathcal{N}_X$ of X is invariant under the action of A:

$$Xu = 0 \Longrightarrow XAu = 0.$$

Thus, if $\mathcal{N}_X \neq \{0\}$, then there exists a vector $u \in \mathcal{N}_X$ which is an eigenvector of A. But then, if α is the corresponding eigenvalue,

$$\alpha u = Au = (A - 2\pi BJB^*X)u,$$

which is impossible, since

$$\sigma(A) \cap \sigma(A - 2\pi BJB^*X) = \emptyset$$

by assumption. This proves that X is invertible. The rest is plain sailing.   □

Our next objective is to show that if, in the setting of Theorem 4.1, X is a Hermitian solution of the Riccati equation (4.2) and $\sigma(A) \subset \Pi_+$, then

$$(4.12) \qquad \langle FXu, FXv \rangle_{\mathcal{M}_X} = v^*X \left\{ \int_{-\infty}^{\infty} F(iv)^* JF(iv)dv \right\} Xu$$

for every choice of u, v in $\mathbb{C}^n$. To this end it is useful to recall that when $\sigma(A) \subset \Pi_+$, the $n \times n$ matrix

$$(4.13) \qquad\qquad P = \int_{-\infty}^{\infty} F(iv)^* JF(iv)dv$$

can be reexpressed by the Plancherel theorem as

$$(4.14) \qquad\qquad P = 2\pi \int_{0}^{\infty} e^{-At} BJB^* e^{-A^*t} dt$$

and can also be characterized as the unique solution of the matrix equation

$$(4.15) \qquad\qquad AP + PA^* = 2\pi BJB^*.$$

Theorem 4.3 *Let X be a Hermitian solution of the Riccati equation (4.2) and let $\sigma(A) \subset \Pi_+$. Then*

$$(4.16) \qquad\qquad XPX = X$$

and formula (4.12) holds.

Proof: Multiply (4.15) on the left and on the right by X to obtain

$$XPA^*X + XAPX = X2\pi BJB^*X.$$

Comparing this with the Riccati equation (4.2), we see that

$$XPA^*X + XAPX = A^*X + XA,$$

or, equivalently, that

$$(XP - I_n)A^*X + XA(PX - I_n) = 0.$$

Thus, in view of (4.8), the matrix

$$W = XPX - X$$

is a solution of the equation

(4.17) $$WX^\dagger A^*X + XAX^\dagger W = 0.$$

We shall use equation (4.17) to show that

(4.18) $$XX^\dagger WX^\dagger X = 0.$$

Since

(4.19) $$(I_n - XX^\dagger)W = 0 = W(I_n - X^\dagger X),$$

this will prove that $W = 0$, as needed. Perhaps the easiest way to obtain (4.18) is via the conformable block decompositions

$$X = U \begin{bmatrix} X_1 & 0 \\ 0 & 0 \end{bmatrix} U^*, \quad W = U \begin{bmatrix} w_{11} & w_{12} \\ w_{21} & w_{22} \end{bmatrix} U^*$$

$$\text{and } A = U \begin{bmatrix} a_{11} & a_{12} \\ a_{21} & a_{22} \end{bmatrix} U^*,$$

where U is unitary and $X_1 = X_1^*$ is invertible. Then (4.17) implies that

(4.20) $$w_{11}(X_1a_{11}X_1^{-1})^* + (X_1a_{11}X_1)^{-1}w_{11} = 0.$$

Moreover, since A is a solution of the Riccati equation (4.2), $a_{12} = 0$. Therefore,

$$\sigma(X_1a_{11}X_1^{-1}) = \sigma(a_{11}) \subset \sigma(A) \subset \Pi_+$$

and hence $w_{11} = 0$ is the one and only solution of equation (4.20).   □

In the setting of Theorem 4.3, we can rewrite the inner product for $\mathcal{M}_X$ in terms of the J-inner product in $L_2^m(i\mathbb{R})$. In particular,

$$v^* F(\omega)Xu = \langle FXu, K_\omega v\rangle_{\mathcal{M}_X}$$

$$= \left\langle JFXu, \frac{J - \Theta J\Theta(\omega)^*}{\rho_\omega} v \right\rangle_{L_2^m}$$

for $\Theta(\lambda) = \Theta_X(\lambda)$ and every choice of $u \in \mathbb{C}^n$, $v \in \mathbb{C}^m$ and $\omega \in \Pi_+$. However, since I_m / ρ_ω is the reproducing kernel for $H_2^m(\Pi_+)$ and $\Theta \in H_\infty^{m \times m}(\Pi_+)$, we can split the last inner product into two pieces to obtain

$$v^* F(\omega) X u = \left\langle J F X u, \frac{J v}{\rho_\omega} \right\rangle_{L_2^m} - \left\langle J F X u, \frac{\Theta J \Theta(\omega)^* v}{\rho_\omega} \right\rangle_{L_2^m}$$

$$= v^* F(\omega) X u - \left\langle \Theta^* J F X u, \frac{J \Theta(\omega)^* v}{\rho_\omega} \right\rangle_{L_2^m} .$$

Therefore, the inner product on the right is equal to zero for every choice of $v \in \mathbb{C}^m$ and $\omega \in \Pi_+$. Thus, $\Theta^\# J F X u$ is orthogonal to $H_2^m(\Pi_+)$ for every $u \in \mathbb{C}^n$, that is to say, the mvf

$$-X F^\#(\lambda) J \Theta(\lambda) = X(\lambda I_n - A)^{-1} B J \Theta(\lambda)$$

must belong to $H_2^{n \times m}(\Pi_+)$. This serves to clarify the first conclusion in Theorem 4.2. It is not just a happy coincidence.

5 The Disc Case

In this section we choose

(5.1) $$F(\lambda) = B^*(I_n - \lambda A^*)^{-1},$$

where $A \in \mathbb{C}^{n \times n}$, $B \in \mathbb{C}^{n \times m}$ and the pair (A, B) is controllable. The choice of $F(\lambda)$ is dictated by the identity

$$F^\#(\lambda) J \Theta(\lambda) = \lambda(\lambda I_n - A)^{-1} B J \Theta(\lambda).$$

Theorem 5.1 *Let $F(\lambda)$ be given by (5.1) with (A, B) controllable. Then the RKKS $\mathcal{M}_X$ is a dBK space $\mathcal{K}(\Theta)$ if and only if*

(5.2) $$A^* X = X X^\dagger A^* X$$

and X is a solution of the Riccati equation

(5.3) $$X - X A X^\dagger A^* X = X B J B^* X.$$

If these two conditions are met, then $\Theta(\lambda) = \Theta_X(\lambda)$ is uniquely specified by the formula

(5.4) $$\Theta_X(\lambda) = I_m - (1 - \overline{\mu}\lambda) B^*(I_n - \lambda A^*)^{-1} X(I_n - \overline{\mu}A)^{-1} B J$$

up to a right J-unitary constant factor, where μ is any point on the unit circle $\mathbb{T}$ such that $\mu \notin \sigma(A)$. In this instance,

$$X(\lambda I_n - A)^{-1} B J \Theta_X(\lambda) = (I_n - \lambda A^*)^{-1}$$

(5.5)

$$(I_n - \mu A^*) X(\mu I_n - A)^{-1} B J.$$

Moreover, if there exists a matrix $\widehat{A} \in \mathbb{C}^{n \times n}$ such that

$$(5.6) \qquad A^*X = X\widehat{A} \quad \text{and} \quad I_n - BJB^*X = A\widehat{A},$$

then $\overline{\mu} \notin \sigma(\widehat{A})$ and

$$(5.7) \quad (\lambda I_n - A)^{-1} BJ\Theta_X(\lambda) = (I_n - \lambda\widehat{A})^{-1}(I_n - \mu\widehat{A})(\mu I_n - A)^{-1} BJ.$$

The mvf $\Theta_X(\lambda)$ is J-inner if and only if $X \geq 0$.

Proof: Suppose first that $\mathcal{M}_X$ is a dBK space. Then $\mathcal{M}_X$ is R_α invariant for every point α for which the matrix $I_n - \alpha A^*$ is invertible and the structural identity (2.11) holds. By Lemma 3.1, the presumed R_α invariance implies (5.2). To investigate the structural identity, let $A_1 \in \mathbb{C}^{n \times n}$ be any matrix which meets the condition

$$(5.8) \qquad\qquad\qquad A^*X = XA_1.$$

The condition (5.2) guarantees the existence of at least one such matrix A_1, and hence that

$$(I_n - \alpha A^*)^{-1} X = X(I_n - \alpha A_1)^{-1}$$

for every point $\alpha \in \mathbb{C}$ for which the two indicated inverses exist. Let

$$f(\lambda) = B^*(I_n - \lambda A^*)^{-1} Xu \quad \text{and} \quad g(\lambda) = B^*(I_n - \lambda A^*)^{-1} Xv$$

for any choice of u, v in $\mathbb{C}^n$ and take any two points α, β such that $(I_n - \alpha A^*)$, $(I_n - \beta A^*)$, $(I_n - \alpha A_1)$ and $(I_n - \beta A_1)$ are all invertible. Then the following formulas are easily verified:

$$
\begin{aligned}
(R_\alpha f)(\lambda) &= F(\lambda)A^*(I_n - \alpha A^*)^{-1} Xu = F(\lambda)XA_1(I_n - \alpha A_1)^{-1}u. \\
(R_\beta g)(\lambda) &= F(\lambda)A^*(I_n - \beta A^*)^{-1} Xv = F(\lambda)XA_1(I_n - \beta A_1)^{-1}v. \\
f(\alpha) &= B^*(I_n - \alpha A^*)^{-1} Xu = B^*X(I_n - \alpha A_1)^{-1}u. \\
g(\beta) &= B^*(I_n - \beta A^*)^{-1} Xv = B^*X(I_n - \beta A_1)^{-1}v.
\end{aligned}
$$

Substituting these evaluations into the structural identity (2.11) and calculating the requisite inner products via formula (2.5) it is readily checked that

$$X - A_1^*XA_1 = XBJB^*X.$$

But, by (5.8) and (5.2),

$$
\begin{aligned}
A_1^*XA_1 &= A_1^*A^*X \\
&= A_1^*XX^\dagger A^*X \\
&= XAX^\dagger A^*X.
\end{aligned}
$$

which, when substituted into the preceding identity, leads immediately to (5.3). This completes the proof that the conditions (5.2) and (5.3) are necessary in order for $\mathcal{M}_X$ to be a dBK space. The sufficiency of these conditions follows easily by running the arguments backwards and turning once again to Theorem 2.3.

The formula (5.4) for $\Theta(\lambda) = \Theta_X(\lambda)$ is an immediate consequence of the general formulas (2.6) and (2.10) applied to the present setting.

Next,

$$(\lambda I_n - A)^{-1} B J \Theta(\lambda)$$
$$= (\lambda I_n - A)^{-1} \{ I_n - (1 - \bar{\mu}\lambda) B J B^* (I_n - \lambda A^*)^{-1}$$
$$X (I_n - \bar{\mu}A)^{-1} \} B J$$
$$= (\lambda I_n - A)^{-1} \{ I_n - \bar{\mu}A - (1 - \bar{\mu}\lambda) B J B^*$$
$$(I_n - \lambda A^*)^{-1} X \} (I_n - \bar{\mu}A)^{-1} B J$$
$$= \boxed{1} + \boxed{2},$$

where

$$\boxed{1} \quad = (\lambda I_n - A)^{-1} \bar{\mu} (\lambda I_n - A)(I_n - \bar{\mu}A)^{-1} B J$$
$$= (\mu I_n - A)^{-1} B J$$

and

$$\boxed{2} \quad = (1 - \bar{\mu}\lambda)(\lambda I_n - A)^{-1} \{ I_n - B J B^*$$
$$(I_n - \lambda A^*)^{-1} X \} (I_n - \bar{\mu}A)^{-1} B J.$$

Now let A_1 be any $n \times n$ matrix which meets (5.8). Then, upon multiplying the term inside the curly brackets in $\boxed{2}$ by X, we obtain

$$X\{\cdots\} = (X - X B J B^* X - \lambda X A_1)(I_n - \lambda A_1)^{-1}$$
$$= (X A X^\dagger A^* X - \lambda X A_1)(I_n - \lambda A_1)^{-1}$$
$$= (X A X^\dagger X A_1 - \lambda X A_1)(I_n - \lambda A_1)^{-1}$$
$$= (X A A_1 - \lambda X A_1)(I_n - \lambda A_1)^{-1} \quad \text{(by (5.2))}$$
$$= X (A - \lambda I_n) A_1 (I_n - \lambda A_1)^{-1}.$$

Thus, as

$$X(\lambda I_n - A)^{-1} = X(\lambda I_n - A)^{-1} X^\dagger X,$$

it follows that

$$X(\lambda I_n - A)^{-1} X^\dagger X(\lambda I_n - A) = X$$

and hence that

$$X \cdot \boxed{2} = (1 - \overline{\mu}\lambda)X(\lambda I_n - A)^{-1}X^\dagger X(A - \lambda I_n)$$
$$A_1(I_n - \lambda A_1)^{-1}(I_n - \overline{\mu}A)^{-1}BJ$$
$$= -(1 - \overline{\mu}\lambda)XA_1(I_n - \lambda A_1)^{-1}(I_n - \overline{\mu}A)^{-1}BJ.$$

Putting it all together, we obtain

$$X(\lambda I_n - A)^{-1}BJ\Theta(\lambda)$$
$$= X(\mu I_n - A)BJ - (1 - \overline{\mu}\lambda)XA_1(I_n - \lambda A_1)^{-1}(I_n - \overline{\mu}A)^{-1}BJ$$
$$= X(I_n - \mu A_1)(I_n - \lambda A_1)^{-1}(\mu I_n - A)^{-1}BJ,$$

which is equivalent to formula (5.5).

The evaluation

$$X(I - BJB^*X) = XAA_1$$

played a key role in the verification of formula (5.5). To obtain formula (5.7) we shall need the same identity, but without the factor X on the left. This is achieved by invoking assumption (5.6).

We first show that if (5.6) is in force and $\mu \in \mathbb{T}$ but $\mu \notin \sigma(A)$, then $\overline{\mu} \notin \sigma(\widehat{A})$: If $\widehat{A}v = \overline{\mu}v$ for some $\overline{\mu} \in \mathbb{T}$ and $v \in \mathbb{C}^n$, then, by the first condition in (5.6),

$$(A^* - \overline{\mu}I_n)Xv = (A^*X - X\widehat{A})v = 0.$$

Therefore, since $\overline{\mu} \notin \sigma(A^*)$, $Xv = 0$. Thus, by the second condition in (5.6),

$$v = (I_n - BJB^*X)v = A\widehat{A}v = \overline{\mu}Av,$$

which implies that $v = 0$, since $\mu \notin \sigma(A)$.

Now, with the aid of (5.6), we can write

$$\boxed{2} = (1 - \overline{\mu}\lambda)(\lambda I_n - A)^{-1}$$
$$\{I_n - BJB^*X(I_n - \lambda\widehat{A})^{-1}\}(I_n - \overline{\mu}A)^{-1}BJ$$

and reexpress the term inside the curly brackets on the right as

$$\{\cdots\} = (I_n - BJB^*X - \lambda\widehat{A})(I_n - \lambda\widehat{A})^{-1}$$
$$= (A - \lambda I_n)\widehat{A}(I_n - \lambda\widehat{A})^{-1}.$$

Therefore,

$$\boxed{2} = -(1 - \overline{\mu}\lambda)\widehat{A}(I_n - \lambda\widehat{A})^{-1}(I_n - \overline{\mu}A)^{-1}BJ,$$

which, when added to the previous formula $\boxed{1}$ (which is still valid), leads readily to formula (5.7).

Finally, the claim that X is J-inner if and only if $X \geq 0$ is immediate from (2.6) and (2.7), just as before. $\square$

The next order of business is to explore the condition (5.6). If X is an invertible solution of the Riccati equation (5.3), then the invariance condition (5.2) is automatically in force, and every matrix $\widehat{A}$ which meets the first condition in (5.6) satisfies the second condition in (5.6) also. Therefore, it remains only to consider the case when X is not invertible.

Theorem 5.2 *Let X be a nonzero noninvertible Hermitian solution of the Riccati equation (5.3) and let $A \in \mathbb{C}^{n \times n}$ meet the invariance condition (5.2). Then there exists a matrix $\widehat{A} \in \mathbb{C}^{n \times n}$ which meets the two conditions in (5.6) if and only if*

$$(I_n - X^\dagger X)A(I_n - X^\dagger X) \text{ is an invertible map of}$$

(5.9)

$$(I_n - X^\dagger X)\mathbb{C}^n \text{ onto itself.}$$

If the full matrix A is invertible, then the condition (5.9) is automatically met and the matrix

(5.10) $$\widehat{A} = A^{-1}(I_n - BJB^*X)$$

satisfies both of the conditions in (5.6).

Proof: Suppose first that there exists a matrix $\widehat{A}$ which meets the conditions in (5.6). Then, the first of these conditions implies that

$$X^\dagger X \widehat{A}(I_n - X^\dagger X) = X^\dagger A^* X(I_n - X^\dagger X) = 0$$

and hence, upon combining this information with the second condition, we obtain the formula

$$\begin{aligned} I_n - X^\dagger X &= (I_n - X^\dagger X)(I_n - BJB^*X)(I_n - X^\dagger X) \\ &= (I_n - X^\dagger X)A\widehat{A}(I_n - X^\dagger X) \\ &= (I_n - X^\dagger X)A(I_n - X^\dagger X)\widehat{A}(I_n - X^\dagger X). \end{aligned}$$

Thus,

$$(I_n - X^\dagger X)A\big|_{(I_n - X^\dagger X)\mathbb{C}^n}$$

must be invertible, i.e., (5.9) holds.

Suppose next that (5.9) holds and let $\widehat{A} \in \mathbb{C}^{n \times n}$ be any matrix which meets the two conditions

(5.11) $$X^\dagger X \widehat{A}(I_n - X^\dagger X) = 0$$

and

(5.12) $$X \widehat{A} X^\dagger X = XX^\dagger A^* X.$$

Then, with the help of (5.2), it is readily checked that $\widehat{A}$ meets the first condition in (5.6). Furthermore, the Riccati equation (5.3) guarantees that

$$X(I_n - BJB^*X) = XAX^\dagger XX^\dagger A^*X$$
$$= XA\widehat{A}.$$

It remains to show that

$$(I_n - X^\dagger X)(I_n - BJB^*X) = (I_n - X^\dagger X)A\widehat{A},$$

or, equivalently, that

(5.13) $$I_n - X^\dagger X = (I_n - X^\dagger X)A\widehat{A}(I_n - XX^\dagger)$$

and

(5.14) $$-(I_n - XX^\dagger)BJB^*XX^\dagger = (I_n - X^\dagger X)A\widehat{A}XX^\dagger.$$

In view of (5.11), equation (5.13) is equivalent to the equation

$$I_n - X^\dagger X = \{(I_n - X^\dagger X)A(I_n - X^\dagger X)\}\{(I_n - X^\dagger X)\widehat{A}(I_n - X^\dagger X)\},$$

which is solvable by assumption. On the other hand, with the help of (5.12), equation (5.14) can be reexpressed as

$$-(I_n - XX^\dagger)BJB^*XX^\dagger$$
$$= (I_n - X^\dagger X)A(I_n - X^\dagger X)\widehat{A}X^\dagger X + (I_n - X^\dagger X)AX^\dagger X\widehat{A}X^\dagger X$$
$$= (I_n - X^\dagger X)A(I_n - X^\dagger X)\widehat{A}X^\dagger X + (I_n - X^\dagger X)AX^\dagger XX^\dagger A^*X$$
$$= (I_n - X^\dagger X)A(I_n - X^\dagger X)\widehat{A}X^\dagger X + (I_n - X^\dagger X)AX^\dagger A^*X.$$

Thus, equation (5.14) is equivalent to the equation

$$(I_n - X^\dagger X)A(I_n - X^\dagger X)\widehat{A}X^\dagger X$$
$$= -(I_n - X^\dagger X)BJB^*XX^\dagger - (I_n - X^\dagger X)AX^\dagger A^*X$$

for $\widehat{A}$, which is solvable, thanks to (5.9).

Finally, to complete the proof, we consider the case when A is invertible. Then, in view of the invariance assumption (5.2), A automatically satisfies condition (5.9). Moreover, since A satisfies (5.2) if and only if A^{-1} satisfies (5.2), it is readily seen that if $\widehat{A}$ is defined by (5.10), then

$$X\widehat{A} = XA^{-1}X^\dagger X(I_n - BJB^*X)$$
$$= XA^{-1}X^\dagger(X - XBJB^*X)$$

$$= XA^{-1}X^{\dagger}XAX^{\dagger}A^*X \quad \text{(by (5.3))}$$

$$= XA^{-1}AX^{\dagger}A^*X \quad \text{(by (5.2) applied to } A^{-1})$$

$$= XX^{\dagger}A^*X$$

$$= A^*X \quad \text{(by (5.2))}.$$

This serves to establish the first condition in (5.6). The second is selfevident. $\square$

Remark 5.3 Let X be Hermitian, nonzero and noninvertible. Then the space $\mathcal{M}_X$ only depends on $XX^{\dagger}A^*XX^{\dagger}$. Thus, we can adjust $(I_n - X^{\dagger}X)A(I_n - X^{\dagger}X)$ to be an invertible map of $(I_n - X^{\dagger}X)\mathbb{C}^n$ onto itself without affecting the space $\mathcal{M}_X$.

Theorem 5.4 *Let (A, B) be controllable, let X be a nonzero Hermitian solution of the Riccati equation (5.3) such that (5.2) is in force and suppose further that $\sigma(A) \subset \mathbb{D}$. Then:*

1. *The mvf*

$$X(\lambda I_n - A)^{-1}BJ\Theta_X(\lambda)$$

belongs to $H_{\infty}^{n \times m}(\mathbb{D})$.

2. *If $\widehat{A}$ is an $n \times n$ matrix that meets (5.6), then the mvf*

$$(\lambda I_n - A)^{-1}BJ\Theta_X(\lambda)$$

belongs to $H_{\infty}^{n \times m}(\mathbb{D})$ if and only if $\sigma(\widehat{A}) \subset \mathbb{D}$.

3. *If $\widehat{A}$ mets (5.6) and $\sigma(\widehat{A}) \subset \mathbb{D}$, then X is invertible and X^{-1} is the one and only solution of the Stein equation*

(5.15) $$X^{-1} - AX^{-1}A^* = BJB^*.$$

4. *If A is invertible, then*

$$(\lambda I_n - A)^{-1}BJ\Theta_X(\lambda)$$

belongs to $H_{\infty}^{n \times m}(\mathbb{D})$ if and only if

(5.16) $$\sigma(A^{-1}(I_n - BJB^*X)) \subset \mathbb{D}.$$

Moreover, this conclusion is valid even if $\sigma(A)$ is not a subset of $\mathbb{D}$ providing that $\sigma(A) \cap \mathbb{T} = \emptyset$.

Proof: Items (1) and (2) are immediate from Theorem 5.1.

To obtain (3), observe that the first condition in (5.6) implies that $\mathcal{N}_X$ is invariant under $\widehat{A}$: $Xu = 0 \Longrightarrow X\widehat{A}u = 0$. Therefore, if $\mathcal{N}_X \neq \{0\}$, then there must exist an eigenvector u of $\widehat{A}$ in $\mathcal{N}_X$. Thus, if α denotes the corresponding eigenvalue, then $\alpha \in \mathbb{D}$ and the second condition in (5.6) implies that

$$u = (I_n - BJB^*X)u = A\widehat{A}u = \alpha Au.$$

But this means that either $\alpha = 0$ and u is the zero vector, or $1/\alpha \in \sigma(A)$. Since none of these possibilities are viable, it follows that $\mathcal{N}_X = \{0\}$, i.e., X is invertible. Equation (5.15) is then an easy consequence of (5.3). It is well known that this equation has only one solution when $\sigma(A) \subset \mathbb{D}$.

Finally, item (4) is immediate from formula (5.7) in Theorem 5.1 and Theorem 5.2, which allows us to choose $\widehat{A}$ as in formula (5.10). $\square$

Theorem 5.5 *Let X be a Hermitian solution of the Riccati equation (5.3), let $A \in \mathbb{C}^{n \times n}$ meet the invariance condition (5.2) and suppose further that A and $I_n - BJB^*X$ are invertible. Then the Riccati equation can be reexpressed as*

$$(5.17) \qquad\qquad A^*X(I_n - BJB^*X)^{-1}A = X.$$

Moreover, if X is a Hermitian solution of the Riccati equation (5.17), then the invariance condition (5.2) is automatically met.

Proof: The Riccati equation (5.3) can be reexpressed as

$$(5.18) \qquad\qquad X(I_n - BJB^*X) = A_1^*XA_1,$$

where $A_1 \in \mathbb{C}^{n \times n}$ is any matrix for which

$$A^*X = XA_1.$$

Formula (5.17) is now easily obtained from (5.18) by choosing $A_1 = \widehat{A}$, where $\widehat{A}$ is given by (5.10).

Finally, if X is a Hermitian solution of (5.17), then

$$A^*X = XA^{-1}(I_n - BJB^*X),$$

which implies (5.2). $\square$

Theorem 5.6 *Let $F(\lambda)$ be given by (5.1) with $\sigma(A) \subset \mathbb{D}$ and (A, B) controllable, let*

$$(5.19) \qquad\qquad P = \frac{1}{2\pi} \int_0^{2\pi} F(e^{i\theta})^* JF(e^{i\theta})d\theta$$

and suppose that X is a nonzero Hermitian solution of the Riccati equation (5.3) such that (5.2) is in force. Then

(5.20) $$XPX = X$$

and

(5.21) $$\langle FXu, FXv \rangle_{\mathcal{M}_X} = v^* XPXu$$

for every choice of u and v in $\mathbb{C}^n$.

Proof: Since $\sigma(A) \subset \mathbb{D}$, we can write

$$F(e^{i\theta}) = \sum_{k=0}^{\infty} B^* A^{*k} e^{ik\theta}$$

and reexpress P as

(5.22) $$P = \sum_{k=0}^{\infty} A^k BJB^* A^{*k}.$$

It is now easy to verify (the well-known fact) that the matrix P can also be characterized as the unique solution of the Stein equation

(5.23) $$P - APA^* = BJB^*.$$

Multiplying both sides of equation (5.23) by X we see that

$$XPX - XAPA^*X = XBJB^*X$$

and hence, upon comparison with the Riccati equation (5.3), that

$$X - XAX^{\dagger}A^*X = XPX - XAPA^*X.$$

In view of (5.2) and the fact that $X^{\dagger}$ is the Moore-Penrose inverse of X, this last identity can be reexpressed as

$$X - XAX^{\dagger}XX^{\dagger}A^*X = XPX - XAX^{\dagger}XPXX^{\dagger}A^*X.$$

Therefore, the matrix

$$W = X - XPX$$

is a solution of the equation

$$W - (XAX^{\dagger})W(XAX^{\dagger})^* = 0.$$

Thus, as $\sigma(XAX^{\dagger}) \subset \mathbb{D}$, it follows that $W = 0$, i.e., (5.20) holds.

Finally, formula (5.21) is an immediate consequence of the definition (2.5) of the indefinite inner product in $\mathcal{M}_X$ and formula (5.20). $\qquad\square$

In the setting of Theorem 5.4, we can rewrite the indefinite inner product for $\mathcal{M}_X$ in terms of the J-inner product in $L_2^m(\mathbb{T})$ with respect to normalized Lebesgue measure when $\sigma(A) \subset \mathbb{D}$. In particular,

$$
v^* F(\omega) X u = \langle F X u, K_\omega v \rangle_{\mathcal{M}_X}
$$

$$
= \left\langle J F X u, \frac{J - \Theta J \Theta(\omega)^*}{\rho_\omega} v \right\rangle_{L_2^m}
$$

for $\Theta(\lambda) = \Theta_X(\lambda)$ and every $u \in \mathbb{C}^n$, $v \in \mathbb{C}^m$ and $\omega \in \mathbb{D}$. However, since I_m/ρ_ω is the reproducing kernel for $H_2^m(\mathbb{D})$ and $\Theta \in H_\infty^{m \times m}(\mathbb{D})$, the last inner product can be split into two pieces just as in the development at the tail end of the last section to obtain the conclusion that $\Theta^\# J F X u$ is orthogonal to $H_2^m(\mathbb{D})$ for every $u \in \mathbb{C}^n$. Therefore,

$$
X F^\#(\lambda) J \Theta(\lambda) = \lambda X (\lambda I_n - A)^{-1} B J \Theta(\lambda)
$$

belongs to $\lambda \cdot H_2^{n \times m}(\mathbb{D})$. This is what lies behind formula (5.5).

Acknowledgement

I wish to thank Shahar Nevo for reading the manuscript carefully and catching a number of misprints and a couple of unclear passages.

References

[AD1]   D. Alpay and H. Dym, On a new class of structured reproducing kernel spaces, *J. Funct. Anal.* **111** (1993), 1–28.

[AD2]   D. Alpay and H. Dym, On a new class of reproducing kernel spaces and a new generalization of the Iohvidov laws, *Linear Algebra Appl.* **178** (1993), 109–183.

[AD3]   D. Alpay and H. Dym, On a new class of realization formulas and their application, *Linear Algebra Appl.* **241–43** (1996), 3–84.

[Ba]   J.A. Ball, Models for non contractions, *J. Math. Anal. Appl.* **52** (1975), 240–254.

[dB]   L. de Branges, Some Hilbert spaces of analytic functions I, *Trans. Amer. Math. Soc.* **106** (1963), 445–468.

[D1]   H. Dym, Shifts, realizations and interpolation, redux, in: *Operator Theory and its Applications*, (A. Feintuch and I. Gohberg, eds.), *Oper. Theory Adv. Appl.* **OT73** Birkhäuser-Verlag, Basel, 1994, 182–243.

[D2]   H. Dym, A basic interpolation problem, in: *Holomorphic Spaces*, (S. Axler, J.E. McCarthy and D. Sarason, eds.), Cambridge University Press, Cambridge, 1998, 381–425.

[Gr]   M. Green, H_∞ controller synthesis by J-lossless coprime factorization, *SIAM J. Control Opt.* **30** (1992), 522–547.

[GGLD] M. Green, K. Glover, D.J.N. Limebeer and J.C. Doyle, A J-spectral factorization approach to H_∞ control, *SIAM J. Control Opt.* **28** (1990), 1350–1371.

[IG] P.A. Iglesias and K. Glover, State-space approach to discrete-time H_∞ control, *Int. J. Control* **54** (1991), 1031–1073.

[Ki] H. Kimura, *Chain Scattering Approach to H^∞-Control*, Birkhäuser, Boston, 1997.

[KK1] W. Kongprawechnon and H. Kimura, J-lossless conjugation and factorization for discrete-time systems, *Int. J. Control* **65** (1996), 867–884.

[KK2] W. Kongprawechnon and H. Kimura, J-lossless factorization and H^∞ control for discrete-time systems, *Int. J. Control* **70** (1998), 423–446.

[LAK] H. Lev-Ari and T. Kailath, Triangular factorization of structured Hermitian matrices, in: *I. Schur Methods in Operator Theory and Signal Processing*, (I. Gohberg, ed.), *Oper. Theory Adv. Appl.* **OT 18**, Birkhäuser, 1986, 301–324.

[Nu] A.A. Nudelman, Some generalizations of classical interpolation problems, in: *Operator Extensions, Interpolation of Functions and Related Topics*, (A. Gheondea, D. Timotin and F.-H. Vascilescu, eds.), *Oper. Theory Adv. Appl.* **OT 61**, Birkhäuser, 1993, 171–188.

[Rov] J. Rovnyak, *Characterization of spaces $\mathcal{K}(M)$*, unpublished manuscript, 1968.

Harry Dym
Department of Mathematics
The Weizmann Institute of Science
Rehovot 76100
Israel

1991 Mathematics Subject Classification. Primary: 30E05, 47A57; Secondary: 47A68 93B36, 93D05.

Operator Theory:
Advances and Applications, Vol. 124
© 2001 Birkhäuser Verlag Basel/Switzerland

A Status Report on the Asymptotic Behavior of Toeplitz Determinants with Fisher–Hartwig Singularities

Torsten Ehrhardt

Dedicated to I. Gohberg on the occasion of his 70-th birthday

The Fisher–Hartwig conjecture describes the asymptotic behavior of Toeplitz determinants for a certain class of singular generating functions. It has been proved in many cases and reformulated in others. Recently, the author proved the conjecture in all the cases in which it can be expected to be true. In the present paper we want to give an account of the latest developments in connection with this conjecture and describe the main ideas of the proof.

1 Introduction

In 1968, Fisher and Hartwig [FH] raised a conjecture which describes the asymptotic behavior of determinants of Toeplitz matrices for a certain class of singular generating functions. This conjecture is motivated by several applications in statistical physics and represents a generalization of the classical Szegö Limit Theorem. Since that time, the conjecture has been proved in many important cases, but it has also turned out that the conjecture fails in other cases.

In this paper we want to give a summary of the recent results on this conjecture and in particular a brief review of the author's dissertation [E], where this conjecture has been proved under certain smoothness assumptions in all the cases in which it can be expected to be true. We first start with recalling this conjecture and explaining what is known about it (see also [BS5, BT, ES, BS6]). In the subsequent sections we then present the main results of [E] and some ideas of their proof.

For a function $c \in L^1(\mathbb{T})$ defined on the unit circle $\mathbb{T}$ with Fourier coefficients

$$(1.1) \qquad c_n = \frac{1}{2\pi} \int_0^{2\pi} c(e^{i\theta}) e^{-in\theta} \, d\theta,$$

we consider the $n \times n$ Toeplitz matrices generated by c,

$$(1.2) \qquad T_n(c) = (c_{j-k}), \qquad 0 \le j, k \le n-1,$$

and their determinants $D_n(c) = \det T_n(c)$. Assume that c is of the form

$$(1.3) \qquad c(e^{i\theta}) = b(e^{i\theta}) \prod_{r=1}^{R} \omega_{t_r, \alpha_r, \beta_r}(e^{i\theta}),$$

where b is a smooth nonvanishing complex-valued function on the unit circle with winding number zero. The functions $\omega_{t_r,\alpha_r,\beta_r}$ are defined by

(1.4) $\omega_{t_r,\alpha_r,\beta_r}(t_r e^{i\theta}) = (2 - 2\cos\theta)^{\alpha_r} e^{i\beta_r(\theta - \pi)}$, $0 < \theta < 2\pi$,

and α_r, $\beta_r \in \mathbb{C}$ and $t_r \in \mathbb{T}$ are certain parameters. The function $\omega_{t_r,\alpha_r,\beta_r}$ is smooth and nonvanishing on $\mathbb{T}\backslash\{t_r\}$ and may have a singularity at t_r, which is a combination of two different kinds. The term $(2 - 2\cos\theta)^{\alpha_r}$ has a zero if $\operatorname{Re}\alpha_r > 0$, a pole if $\operatorname{Re}\alpha_r < 0$, and an oscillating discontinuity if $\operatorname{Re}\alpha_r = 0$ but $\operatorname{Im}\alpha_r \neq 0$. The term $\exp(i\beta_r(\theta - \pi))$ is a function with a jump discontinuity. The one-sided limits are $\exp(-i\beta_r\pi)$ and $\exp(i\beta_r\pi)$ as $\theta \to +0$ and $\theta \to 2\pi - 0$, respectively.

Then the conjecture of Fisher and Hartwig asserts that

(1.5) $D_n(c) \sim G[b]^n n^{\Omega} E$ as $n \to \infty$,

where $G[b]$, Ω and E are the following constants,

(1.6) $G[b] = \exp[\log b]_0$,

(1.7) $\Omega = \sum_{r=1}^{R}(\alpha_r^2 - \beta_r^2)$,

(1.8)
$$E = E[b]\prod_{r=1}^{R} b_+(t_r)^{-\alpha_r+\beta_r} b_-(t_r)^{-\alpha_r-\beta_r}$$
$$\times \prod_{1 \leq s \neq r \leq R}(1 - t_s/t_r)^{-(\alpha_r+\beta_r)(\alpha_s-\beta_s)}$$
$$\times \prod_{r=1}^{R} \frac{G(1 + \alpha_r + \beta_r)G(1 + \alpha_r - \beta_r)}{G(1 + 2\alpha_r)}.$$

Here $G(*)$ stands for the Barnes G-function [Ba, WW], which is an entire analytic function defined by

(1.9)
$$G(1 + z) = (2\pi)^{z/2} e^{-(z+1)z/2 - \gamma_E z^2/2}$$
$$\prod_{k=1}^{\infty}\left(\left(1 + \frac{z}{k}\right)^k e^{-z+z^2/(2k)}\right)$$

with γ_E being Euler's constant. The constant $E[b]$ is given by

(1.10) $E[b] = \exp\left(\sum_{k=1}^{\infty} k[\log b]_k [\log b]_{-k}\right)$

with $[\log b]_k$ refering to the k-th Fourier coefficient of the logarithm of b. The functions

$$(1.11) \qquad b_{\pm}(t) \; = \; \exp\left(\sum_{k=1}^{\infty} t^{\pm k}[\log b]_{\pm k}\right), \qquad t \in \mathbb{T},$$

as well as their inverses can be extended analytically to $\{z \in \mathbb{C} : |z| < 1\}$ and $\{z \in \mathbb{C} : |z| > 1\} \cup \{\infty\}$, respectively. Moreover, $b_{+}(0) = b_{-}(\infty) = 1$, and the representation $b(t) = b_{-}(t) G[b] b_{+}(t)$, $t \in \mathbb{T}$, represents the normalized canonical Wiener–Hopf factorization of b.

In support of their conjecture, some special cases were considered by Fisher and Hartwig themselves [FH] and later also by Lenard [L1, L2]. This work was considerably extended by Harold Widom [W1], who proved the conjecture under the assumptions $\mathrm{Re}\,\alpha_r > -1/2$ and $\beta_r = 0$ for all $1 \leq r \leq R$. He also treated the case of one singularity $(R = 1)$ with $|\mathrm{Re}\,\alpha_1| < 1/2$ and $|\mathrm{Re}\,\beta_1| < 1/2$, but did not evaluate the constant E. Using the same approach, Estelle L. Basor [B1] was able to consider the case $\mathrm{Re}\,\alpha_r > -1/2$ and $\mathrm{Re}\,\beta_r = 0$ for all $1 \leq r \leq R$. Moreover, she evaluated the constant E.

Later on, Estelle L. Basor [B2] treated the case where $\alpha_r = 0$ and $|\mathrm{Re}\,\beta_r| < 1/2$ for all $1 \leq r \leq R$, i.e., where the generating function has several jump discontinuities with conditions on their size. There and in the paper of Basor and Helton [BH], a "localization" or "separation" technique has been developed. Using this method they reduced the case of several jumps to the case of one jump by proving that for certain functions a and b the quotient $D_n(ab)/(D_n(a)D_n(b))$ converges to some nonzero constant. The localization idea represents a further advancement of the operator theoretic approach initiated by Widom.

The case of several jump discontinuities was independently treated by Albrecht Böttcher [Bö], who developed similar techniques. In 1985, Böttcher and Silbermann [BS3, BS5] were finally able to confirm the conjecture for parameters $|\mathrm{Re}\,\alpha_r| < 1/2$ and $|\mathrm{Re}\,\beta_r| < 1/2$ for all $1 \leq r \leq R$. In their approach, they also used Banach algebra methods.

Another partial case of the conjecture where either $\alpha_r + \beta_r = 0$ for all $1 \leq r \leq R$ or $\alpha_r - \beta_r = 0$ for all $1 \leq r \leq R$, was also settled by Böttcher and Silbermann [S, BS2]. In this case the generating function is said to be of analytic type. This notion can be explained as follows. Let η_{t_r,γ_r} and ξ_{t_r,δ_r} be the functions

$$(1.12) \qquad \eta_{t_r,\gamma_r}(t) \; = \; (1 - t/t_r)^{\gamma_r}, \qquad |t| \leq 1,\; t \neq t_r,$$

$$(1.13) \qquad \xi_{t_r,\delta_r}(t) \; = \; (1 - t_r/t)^{\delta_r}, \qquad |t| \geq 1,\; t \neq t_r,$$

where the branches of the analytic (in t) functions are chosen such that $\eta_{t_r,\gamma_r}(0) = \xi_{t_r,\delta_r}(\infty) = 1$. Then

$$(1.14) \qquad \omega_{t_r,\alpha_r,\beta_r}(t) \; = \; \eta_{t_r,\gamma_r}(t)\xi_{t_r,\delta_r}(t), \qquad t \in \mathbb{T}\backslash\{t_r\},$$

if $\gamma_r = \alpha_r + \beta_r$ and $\delta_r = \alpha_r - \beta_r$.

Further results are known if the generating function has only one singularity ($R = 1$). Böttcher and Silbermann [BS4] confirmed the conjecture for parameters that satisfy $\mathrm{Re}\,\alpha_1 \geq 0$, $\mathrm{Re}\,(\alpha_1 + \beta_1) > -1$ and $\mathrm{Re}\,(\alpha_1 - \beta_1) > -1$. Libby [Li] proved the conjecture for the case $\alpha_1 = 0$ and $|\mathrm{Re}\,\beta_1| < 5/2$. Finally, Silbermann and the author [ES] settled the problem completely, i.e., for all parameters α_1 and β_1 for which the conjecture makes sense and with b sufficiently smooth.

Despite the above cases, where the conjecture has been confirmed, there are examples where the asymptotic behavior of the Toeplitz determinants differs from the predictions of the conjecture.

The simplest exceptional cases to mention are those where the asymptotic formula (1.5) formally holds but with a constant $E = 0$. In fact, for continuous and sufficiently smooth nonvanishing functions c with a nonzero winding number one can obtain certain estimates for the asymptotics of $D_n(c)$ (see, e.g., [BS5]). Note that in the particular case $c(t) = \omega_{t_r,0,\varkappa}(t) = (-t/t_r)^\varkappa$ with $\varkappa \in \mathbb{Z}\setminus\{0\}$ one even has $D_n(c) = 0$. The special case of continuous and piecewise smooth but not entirely smooth functions was studied in more detail by Widom [W2].

More interesting examples are functions with zeros of integral order, i.e., where $\alpha_r + \beta_r$ and $\alpha_r - \beta_r$ are nonnegative integers for each $1 \leq r \leq R$. Böttcher and Silbermann [BS1] showed that then

$$(1.15) \qquad D_n(c) = G^n n^\Omega \left(\sum E_s T_s^n + o(1) \right) \qquad \text{as } n \to \infty,$$

where the summation is taken over a finite set, G, Ω, E_s and T_s are certain constants and $|T_s| = 1$. If $b(t) \equiv 1$, then c is even a rational function, and the determinant $D_n(c)$ can be evaluated explicitly [D].

It was long time believed that the above classes are the only examples where the Fisher–Hartwig conjecture breaks down. However, Basor and Tracy [BT] discovered a different class of counterexamples. The simplest one has the generating function

$$(1.16) \qquad c(e^{i\theta}) = \begin{cases} 1 & -\pi < \theta < 0 \\ -1 & 0 < \theta < \pi. \end{cases}$$

Then $D_n(c) = 0$ for n odd, and $D_n(c) \sim (i)^n n^{-1/2} 2^{1/2} G(1/2)^2 G(3/2)^2$ for n even and $n \to \infty$ (see also [BM1, BM2]). They pointed out that functions for which the Fisher–Hartwig conjecture is expected to fail possess at least two different representations of the form (1.3). Hence the original conjecture does not have a clear interpretation. They proposed a generalized conjecture and guessed that

$$(1.17) \qquad D_n(c) \sim \sum_\tau G(\tau)^n n^{\Omega(\tau)} E(\tau) \qquad \text{as } n \to \infty,$$

where $G(\tau)$, $\Omega(\tau)$ and $E(\tau)$ are constants defined by (1.6)–(1.8) relating to the different representations τ of c. The summation is taken over all representations (indexed by τ) for which the growth of the corresponding term is maximal.

In the following section we will make the Basor–Tracy conjecture more concrete. We are led to three cases, in which this conjecture can be formulated in specific ways.

2 On the Validity of the Fisher–Hartwig Conjecture

A function c defined on the unit circle will be called a *function of Fisher–Hartwig type* if it can be written in the form (1.3) with certain parameters $\alpha_r, \beta_r \in \mathbb{C}$ and $t_r \in \mathbb{T}$ and a smooth (i.e., infinitely differentiable) nonvanishing function b with winding number zero. We also assume that $2\alpha_r \notin \mathbb{Z}_-$ for all $1 \le r \le R$, where $\mathbb{Z}_- := \{-1, -2, -3, \ldots\}$.

Originally, the stronger assumption $\mathrm{Re}\,\alpha_r > -1/2$ for all $1 \le r \le R$ had been imposed, which ensures that $c \in L^1(\mathbb{T})$ and hence that the Fourier coefficients of c and the determinants $D_n(c)$ are well defined. In Section 4 we will explain how, under the assumption $2\alpha_r \notin \mathbb{Z}_-$, the generating function c can be replaced by a distribution in a reasonable way. In the case of one singularity this has already been done in [ES]. Note that $2\alpha_r \notin \mathbb{Z}_-$ also guarantees that the constant E given in (1.8) is well defined because $G(1 + z) = 0$ for the Barnes G-function if and only if $z \in \mathbb{Z}_-$.

A function of Fisher–Hartwig type possesses in general different representations of the form (1.3) with different parameters. Related to these representations there correspond different constants $G[b]$, Ω and E.

In order to analyze the relations between different representations, we have first to exclude trivial modifications by a factor $\omega_{t_r,0,\varkappa}(t) = (-t/t_r)^\varkappa$ with $\varkappa \in \mathbb{Z}$. A representation without such factors will be called a *proper representation*. Note that for a representation including such a factor with $\varkappa \ne 0$, the corresponding constant E equals zero.

Analyzing the singularities of a function of Fisher–Hartwig type, the location of which is determined by the parameters t_r, it is easy to see that the parameters α_r are uniquely determined, whereas the parameters β_r (relating to the jumps) are determined only up to an additive integer. More precisely, if (1.3) is a proper representation of c, then any other proper representation of c is given by

$$(2.1) \qquad c(t) = b^*(t) \prod_{r=1}^{R} \omega_{t_r,\alpha_r,\beta_r+n_r}(t),$$

where $n_r \in \mathbb{Z}$ for each $1 \le r \le R$, $\sum_{r=1}^{R} n_r = 0$ and $b^*(t) = b(t) \prod_{r=1}^{R}(-t_r)^{n_r}$. It follows that all proper representations of a function of Fisher–Hartwig type can be enumerated by R-tuples of integers $(n_1, \ldots, n_R) \in \mathbb{Z}^R$ the sum of which is zero. In the case of one singularity ($R = 1$), the proper representation is unique. But if the function has at least two singularities ($R \ge 2$), then there exists a countably infinite number of proper representations.

We will denote the different representation by a symbol τ, the corresponding parameters by α_r and $\beta_r^{(\tau)}$, and the constants by $G(\tau)$, $\Omega(\tau)$ and $E(\tau)$. Moreover, we define

$$(2.2) \qquad \gamma_r^{(\tau)} = \alpha_r + \beta_r^{(\tau)}, \qquad \delta_r^{(\tau)} = \alpha_r - \beta_r^{(\tau)}.$$

Note that the modulus of $G(\tau)$ is always the same because so is the modulus of the different functions b. Hence it makes sense to put $|G| := |G(\tau)|$.

Elaborating on the proposed asymptotic formula (1.17) of Basor and Tracy, i.e., asking when the growth of the terms in the sum is maximal, we introduce

$$(2.3) \qquad\qquad T = \{\tau \text{ proper} \mid E(\tau) \neq 0\},$$

$$(2.4) \qquad\qquad \Omega_{\max} = \max\{\operatorname{Re}\Omega(\tau) \mid \tau \in T\},$$

$$(2.5) \qquad\qquad T_{\max} = \{\tau \in T \mid \operatorname{Re}\Omega(\tau) = \Omega_{\max}\}.$$

Using these definitions, we can make the generalized conjecture more precise:

$$(2.6) \qquad D_n(c) \sim \sum_{\tau \in T_{\max}} G(\tau)^n n^{\Omega(\tau)} E(\tau) \qquad \text{as } n \to \infty.$$

Before proceeding, we consider the above definitions in more detail. Analyzing the product of the Barnes G-functions in the definition (1.8), it follows that

$$(2.7) \qquad \begin{aligned} T = \{&\tau \text{ proper} \mid \gamma_r^{(\tau)} \notin \mathbb{Z}_- \text{ and } \delta_r^{(\tau)} \notin \mathbb{Z}_- \\ &\text{for all } 1 \leq r \leq R\}. \end{aligned}$$

It can happen that the set T is empty. A function c for which this is the case will be called a *function of degenerate Fisher–Hartwig type*. If T is non-empty, then T may be finite or countably infinite, but in both cases one can show that a maximum $\Omega_{\max}$ exists and is attained on a finite non-empty set $T_{\max}$. We now distinguish the case where $T_{\max}$ is a singleton from the case where $T_{\max}$ contains at least two elements. In the first case, we call the function c a *function of unique Fisher–Hartwig type*, otherwise a *function of ambiguous Fisher–Hartwig type*.

Based on this classification we specify the generalized conjecture as follows:

Conjecture for the Degenerate Case.
If c is a function of degenerate Fisher–Hartwig type, then for each real ω we have

$$D_n(c) = O(|G|^n n^\omega) \qquad \text{as } n \to \infty.$$

Conjecture for the Unique Case.
If c is a function of unique Fisher–Hartwig type, then

$$D_n(c) = G^n n^\Omega E + o(|G|^n n^{\Omega_{\max}}) \qquad \text{as } n \to \infty,$$

where G, Ω and E correspond to the unique representation $\tau \in T_{\max}$.

Conjecture for the Ambiguous Case.
If c is a function of ambiguous Fisher–Hartwig type, then

$$D_n(c) = \sum_{\tau \in \mathcal{T}_{\max}} G(\tau)^n n^{\Omega(\tau)} E(\tau) + o(|G|^n n^{\Omega_{\max}}) \qquad as \ n \to \infty.$$

In [E], the first two conjectures have been proved, and we will sketch this proof later on. The third conjecture has not yet been proved. Notice that the conjecture for the unique case coincides with the original Fisher–Hartwig conjecture.

In Section 4 we will characterize functions of degenerate and unique Fisher–Hartwig type in terms of their representations. More precisely, we will single out a suitable representation for such a function and describe the conditions on the parameters of this representation.

In what follows, we will introduce some concepts which are needed for a generalization of the Fisher–Hartwig conjecture into another direction and for the proof of the main results as well.

3 Distributions on the Unit Circle

We are going to consider distributions on the unit circle. In particular, we define a product of distributions provided that these distributions satisfy certain conditions. We also introduce certain subsets of distributions.

Let $C^\infty(\mathbb{T})$ stand for the (topological) algebra of all infinitely differentiable complex-valued functions defined on the unit circle $\mathbb{T}$. A distribution on $\mathbb{T}$ is a linear continuous functional on $C^\infty(\mathbb{T})$. The set $\mathcal{D}'$ of all distributions on $\mathbb{T}$ is a linear topological space.

The Fourier coefficients of a distribution $\widehat{a} \in \mathcal{D}'$ are given by

$$(3.1) \qquad [\widehat{a}]_n = \widehat{a}(\chi_{-n}),$$

where $\chi_{-n} \in C^\infty(\mathbb{T})$ is the function $\chi_{-n}(t) = t^{-n}, t \in \mathbb{T}$.

There is a natural identification of functions in $L^1(\mathbb{T})$ with distributions in $\mathcal{D}'$. Given $f \in L^1(\mathbb{T})$, we associate the distribution $\widehat{f} \in \mathcal{D}'$ defined by

$$(3.2) \qquad \widehat{f}(g) = \frac{1}{2\pi} \int_0^{2\pi} f(e^{i\theta}) g(e^{i\theta}) \, d\theta, \qquad g \in C^\infty(\mathbb{T}).$$

The Fourier coefficients of the function f and the distribution $\widehat{f}$ are the same. We will call $\widehat{f}$ the *naturally associated distribution to* f.

Functions in $C^\infty(\mathbb{T})$ and distributions in $\mathcal{D}'$ can be characterized (and even be defined) in terms of their Fourier coefficients. For $f \in C^\infty(\mathbb{T})$ with Fourier coefficients f_n we have

$$(3.3) \qquad \sup_{n \in \mathbb{Z}} |f_n|(1 + |n|)^\mu < \infty$$

for all (arbitrarily large) $\mu \in \mathbb{R}$. For $\widehat{f} \in \mathcal{D}'$ with Fourier coefficients f_n, relation (3.3) holds for some (sufficiently small) $\mu \in \mathbb{R}$. Conversely, sequences f_n satisfying these conditions are the Fourier coefficients of some uniquely determined function in $C^\infty(\mathbb{T})$ or distribution in $\mathcal{D}'$, respectively.

A product of two distributions is in general not defined. However, it is if one of the distributions is a function in $C^\infty(\mathbb{T})$. Namely, the product of $a \in C^\infty(\mathbb{T})$ and $\widehat{b} \in \mathcal{D}'$ is defined by

$$(3.4) \qquad (a\widehat{b})(f) = \widehat{b}(af), \qquad f \in C^\infty(\mathbb{T}).$$

We need a generalization of this definition for two distributions which "behave" like smooth functions on $\mathbb{T}$ except at certain disjoint subset of $\mathbb{T}$.

Let K be a compact subset of $\mathbb{T}$. We denote by $C^\infty(\mathbb{T}\backslash K)$ the set of all infinitely differentiable functions defined on $\mathbb{T}\backslash K$ and by $C_K^\infty(\mathbb{T})$ the set of all functions in $C^\infty(\mathbb{T})$ which vanish identically on an open neighborhood of K. The product of a function $f \in C^\infty(\mathbb{T}\backslash K)$ with a function $g \in C_K^\infty(\mathbb{T})$ is a well defined function $fg \in C^\infty(\mathbb{T})$ by stipulating $(fg)(t) = 0$ for $t \in K$.

Let $\mathcal{D}'(K)$ stand for the set of all distributions $\widehat{a} \in \mathcal{D}'$ for which there exists a function $a \in C^\infty(\mathbb{T}\backslash K)$ such that

$$(3.5) \qquad \widehat{a}(f) = \frac{1}{2\pi} \int_0^{2\pi} a(e^{i\theta}) f(e^{i\theta})\, d\theta$$

for all $f \in C_K^\infty(\mathbb{T})$. The function $a \in C^\infty(\mathbb{T}\backslash K)$ is uniquely determined by $\widehat{a}$ and will be called the *smooth part of the distribution* $\widehat{a}$.

We are now prepared to define, under certain conditions, the product of two distributions.

Proposition 3.1 *Let M and N be compact subsets of $\mathbb{T}$ with $M \cap N = \emptyset$. Suppose in addition that $\widehat{a} \in \mathcal{D}'(M)$ has the smooth part $a \in C^\infty(\mathbb{T}\backslash M)$ and that $\widehat{b} \in \mathcal{D}'(N)$ has the smooth part $b \in C^\infty(\mathbb{T}\backslash N)$. Let $f_a \in C_M^\infty(\mathbb{T})$ and $f_b \in C_N^\infty(\mathbb{T})$ be functions for which $f_a + f_b = 1$. Then the product $\widehat{a}\widehat{b}$ defined by*

$$(3.6) \qquad (\widehat{a}\widehat{b})(g) = \widehat{a}(bf_bg) + \widehat{b}(af_ag), \qquad g \in C^\infty(\mathbb{T}),$$

is a distribution in $\mathcal{D}'(M \cup N)$ with smooth part $ab \in C^\infty(\mathbb{T}\backslash(M \cup N))$. The definition of $\widehat{a}\widehat{b}$ does not depend on the particular choice of f_a and f_b.

The definition (3.6) of the product $\widehat{a}$ and $\widehat{b}$ can be restated also as follows:

$$(3.7) \qquad \widehat{a}\widehat{b} = (bf_b)\widehat{a} + (af_a)\widehat{b}.$$

Here we use definition (3.4) and the fact that both $af_a \in C^\infty(\mathbb{T})$ and $bf_b \in C^\infty(\mathbb{T})$.

One can show that this multiplication of distributions is commutative and associative whenever it is defined.

For our purposes we need to single out some further classes of distributions. Let $\mathcal{D}'_+$ be the set of all distributions $\widehat{a} \in \mathcal{D}'$ for which the Fourier coefficients $[\widehat{a}]_n$ vanish for all $n < 0$. This linear topological subspace of $\mathcal{D}'$ becomes even an algebra when introducing a multiplication as follows. For distributions $\widehat{a} \in \mathcal{D}'_+$ and $\widehat{b} \in \mathcal{D}'_+$ with Fourier coefficients a_n and b_n, let $\widehat{c} = \widehat{a} \cdot \widehat{b} \in \mathcal{D}'_+$ be the (unique) distribution which has the Fourier coefficients

$$(3.8) \qquad\qquad c_n = \sum_{k=0}^{n} a_{n-k} b_k$$

for $n \geq 0$ and $c_n = 0$ for $n < 0$. Also this multiplication is commutative and associative. The unit element in $\mathcal{D}'_+$ is the distribution naturally associated to the function $e(t) \equiv 1$. The group of all invertible distributions in $\mathcal{D}'_+$ is denoted by $\mathcal{G}\mathcal{D}'_+$, and we write $\widehat{a}^{-1}$ for the inverse of a distribution $\widehat{a}$.

The following proposition shows that the product $\widehat{a} \cdot \widehat{b}$ coincides with $\widehat{ab}$ whenever both exist. Let $\mathcal{D}'_+(K) = \mathcal{D}'_+ \cap \mathcal{D}'(K)$.

Proposition 3.2 *Let K and M be compact subsets of $\mathbb{T}$. Suppose that $\widehat{a} \in \mathcal{D}'_+(K)$ has the smooth part $a \in C^\infty(\mathbb{T}\backslash K)$ and that $\widehat{b} \in \mathcal{D}'_+(M)$ has the smooth part $b \in C^\infty(\mathbb{T}\backslash M)$. Then $\widehat{a} \cdot \widehat{b} \in \mathcal{D}'_+(K \cup M)$, and $\widehat{a} \cdot \widehat{b}$ has the smooth part $ab \in C^\infty(\mathbb{T}\backslash(K \cup M))$. Moreover, if $K \cap M = \emptyset$, then $\widehat{a} \cdot \widehat{b} = \widehat{ab}$.*

The group of all invertible elements in $\mathcal{D}'_+(K)$ will be denoted by $\mathcal{G}\mathcal{D}'_+(K)$. One can show that $\widehat{a} \in \mathcal{G}\mathcal{D}'_+(K)$ if and only if $\widehat{a} \in \mathcal{G}\mathcal{D}'_+ \cap \mathcal{D}'(K)$ and the smooth part $a \in C^\infty(\mathbb{T}\backslash K)$ of $\widehat{a}$ is nonzero on all of $\mathbb{T}\backslash K$.

In addition to $\mathcal{D}'_+$, one can also consider the set $\mathcal{D}'_-$ of all distributions $\widehat{a} \in \mathcal{D}'$ for which the Fourier coefficients $[\widehat{a}]_n$ vanish for all $n > 0$. This is again a topological algebra with a similarly defined multiplication. One can introduce the sets $\mathcal{D}'_-(K)$, $\mathcal{G}\mathcal{D}'_-$ and $\mathcal{G}\mathcal{D}'_-(K)$, and the above results remain true with obvious modifications.

Finally, there exists a conjugation relation for both functions and distributions. Let f be a function and define

$$(3.9) \qquad\qquad \widetilde{f}(t) = f(1/t), \qquad t \in \mathbb{T}.$$

For a distribution $\widehat{a}$, let $\widetilde{\widehat{a}}$ denote the distribution defined by

$$(3.10) \qquad\qquad \widetilde{\widehat{a}}(f) = a(\widetilde{f}), \qquad f \in C^\infty(\mathbb{T}).$$

In terms of Fourier coefficients, this means that $[\widetilde{f}]_n = [f]_{-n}$ and $[\widetilde{\widehat{a}}]_n = [\widehat{a}]_{-n}$.

In connection with this conjugation relation, there exist relations between several of the above sets. We mention only the following two and leave others to the reader. We have $\widehat{a} \in \mathcal{D}'_+$ if and only if $\widetilde{\widehat{a}} \in \mathcal{D}'_-$. Moreover, if K is a compact subset of $\mathbb{T}$, then $\widehat{a} \in \mathcal{D}'(K)$ has the smooth part $a \in C^\infty(\mathbb{T}\backslash K)$ if and only if $\widetilde{\widehat{a}} \in \mathcal{D}'(\widetilde{K})$ has the smooth part $\widetilde{a} \in C^\infty(\mathbb{T}\backslash\widetilde{K})$, where $\widetilde{K} = \{t \mid 1/t \in K\}$.

4 Distributions of Fisher–Hartwig Type

In the original formulation of the Fisher–Hartwig conjecture, the Toeplitz determinants $D_n(c)$ are defined by a generating *function* c of the form (1.3). We want to replace this function c in a reasonable way by a *distribution* $\widehat{c}$ (see [E, Sect. 5]). Reasonable means that at least the Fourier coefficients of c and $\widehat{c}$ are the same if $\operatorname{Re}\alpha_r > -1/2$ for all $1 \leq r \leq R$. We will see shortly that such a distribution $\widehat{c}$ can be defined under the assumption $2\alpha_r \notin \mathbb{Z}_-$ for all $1 \leq r \leq R$.

First of all, we are going to define distributions with a "pure" Fisher–Hartwig singularity. Suppose $t_r \in \mathbb{T}$, $\alpha_r, \beta_r \in \mathbb{C}$ such that $2\alpha_r \notin \mathbb{Z}_-$. Let $\widehat{\omega}_{t_r,\alpha_r,\beta_r}$ be the (uniquely determined) distribution with Fourier coefficients

$$(4.1) \quad [\widehat{\omega}_{t_r,\alpha_r,\beta_r}]_n \; = \; (-1/t_r)^n \; \frac{\Gamma(1 + 2\alpha_r)}{\Gamma(1 + \alpha_r + \beta_r - n)\,\Gamma(1 + \alpha_r - \beta_r + n)}.$$

Moreover, for $\gamma_r \in \mathbb{C}$ and $\delta_r \in \mathbb{C}$, we introduce the distributions $\widehat{\eta}_{t_r,\gamma_r}$ and $\widehat{\xi}_{t_r,\delta_r}$ in terms of their Fourier coefficients

$$(4.2) \qquad [\widehat{\eta}_{t_r,\gamma_r}]_n \; = \; \begin{cases} (-1/t_r)^n \dbinom{\gamma_r}{n} & n \geq 0 \\[2mm] 0 & n < 0, \end{cases}$$

$$(4.3) \qquad [\widehat{\xi}_{t_r,\delta_r}]_n \; = \; \begin{cases} 0 & n > 0 \\[2mm] (-t_r)^{-n} \dbinom{\delta_r}{-n} & n \leq 0. \end{cases}$$

These definitions are correct because the sequences (4.1), (4.2) and (4.3) behave at most polynomially as $n \to \pm\infty$. The Fourier coefficients of the functions $\omega_{t_r,\alpha_r,\beta_r}(t)$, $\eta_{t_r,\gamma_r}(t)$ and $\xi_{t_r,\delta_r}(t)$ coincide with the values given in (4.1), (4.2) and (4.3) (see, e.g., [BS5, Chap. 6]). Therefore, $\widehat{\omega}_{t_r,\alpha_r,\beta_r}$, $\widehat{\eta}_{t_r,\gamma_r}$ and $\widehat{\xi}_{t_r,\delta_r}$ are the naturally associated distributions to the functions $\omega_{t_r,\alpha_r,\beta_r}(t)$, $\eta_{t_r,\gamma_r}(t)$ and $\xi_{t_r,\delta_r}(t)$, respectively, if $\operatorname{Re}\alpha_r > -1/2$, $\operatorname{Re}\gamma_r > -1$ and $\operatorname{Re}\delta_r > -1$.

One can show that the distribution $\widehat{\omega}_{t_r,\alpha_r,\beta_r}$ is contained $\mathcal{D}'(\{t_r\})$ and has the smooth part $\omega_{t_r,\alpha_r,\beta_r}$. The distribution $\widehat{\eta}_{t_r,\gamma_r}$ is contained $\mathcal{GD}'_+(\{t_r\})$, has the smooth part η_{t_r,γ_r} and the inverse $\widehat{\eta}_{t_r,-\gamma_r}$. The distribution $\widehat{\xi}_{t_r,\delta_r}$ is contained in $\mathcal{GD}'_-(\{t_r\})$, has the smooth part ξ_{t_r,δ_r} and the inverse $\widehat{\xi}_{t_r,-\delta_r}$.

Because of the properties of $\widehat{\omega}_{t_r,\alpha_r,\beta_r}$, the following product of distributions (in the sense of the previous section),

$$(4.4) \qquad\qquad \widehat{c} \; = \; b \prod_{r=1}^{R} \widehat{\omega}_{t_r,\alpha_r,\beta_r},$$

is well defined. Here $b \in C^\infty(\mathbb{T})$ is a nonvanishing function with winding number zero, and $2\alpha_r \notin \mathbb{Z}_-$ for all $1 \leq r \leq R$. Such a distribution $\widehat{c}$ will be called a *distribution of Fisher–Hartwig type*.

The distribution $\widehat{c}$ is contained in $\mathcal{D}'(K)$ with $K = \{t_1, \ldots, t_R\}$ and has the smooth part c with c given by (1.3). If $\operatorname{Re}\alpha_r > -1/2$ for all $1 \leq r \leq R$, then $\widehat{c}$ is the naturally associated distribution to c. This latter fact justifies to study the asymptotics of $D_n(\widehat{c})$ with the distribution $\widehat{c}$ instead of the function c.

A representation of the form (4.4) is in general not unique. In fact, the same relations hold between different representations of the form (4.4) as those established in the case of functions (see Section 2 and (2.1) in particular). The basis of this observation is that for $n_r \in \mathbb{Z}$,

$$(4.5) \qquad \widehat{\omega}_{t_r, \alpha_r, \beta_r + n_r} = \chi_{t_r, n_r}\, \widehat{\omega}_{t_r, \alpha_r, \beta_r},$$

where $\chi_{t_r, n_r} \in C^\infty(\mathbb{T})$ is the function defined by $\chi_{t_r, n_r}(t) = (-t/t_r)^{n_r}$, $t \in \mathbb{T}$.

The distributions $\widehat{\eta}_{t_r, \gamma_r}$ and $\widehat{\xi}_{t_r, \delta_r}$ are special cases of the distributions $\widehat{\omega}_{t_r, \alpha_r, \beta_r}$. More precisely,

  (a) if $\gamma_r = \alpha_r + \beta_r$, $0 = \alpha_r - \beta_r$ and $2\alpha_r \notin \mathbb{Z}_-$, then $\widehat{\omega}_{t_r, \alpha_r, \beta_r} = \widehat{\eta}_{t_r, \gamma_r}$,

  (b) if $0 = \alpha_r + \beta_r$, $\delta_r = \alpha_r - \beta_r$ and $2\alpha_r \notin \mathbb{Z}_-$, then $\widehat{\omega}_{t_r, \alpha_r, \beta_r} = \widehat{\xi}_{t_r, \delta_r}$.

Note that in these cases the exclusion $2\alpha_r \notin \mathbb{Z}_-$ corresponds to the exclusions $\gamma_r \notin \mathbb{Z}_-$ and $\delta_r \notin \mathbb{Z}_-$. For $\gamma_r \in \mathbb{Z}_-$ and $\delta_r \in \mathbb{Z}_-$, however, the distributions $\widehat{\eta}_{t_r, \gamma_r}$ and $\widehat{\xi}_{t_r, \delta_r}$ are still defined.

Nevertheless, one has to be careful when working with these distributions. For instance, whereas for functions the identity $\xi_{t_r, -1} = \chi_{t_r, 1}\, \eta_{t_r, -1}$ holds, the corresponding relation for distributions breaks down: $\widehat{\xi}_{t_r, -1} \neq \chi_{t_r, 1}\, \widehat{\eta}_{t_r, -1}$.

One can apply the same classification to distributions of Fisher–Hartwig type as to functions of Fisher–Hartwig type. We say that $\widehat{c}$ is a *distribution of degenerate (unique, ambiguous) Fisher–Hartwig type* if the smooth part c of $\widehat{c}$ is a function of degenerate (unique, ambiguous) Fisher–Hartwig type.

We cite the following characterizations of distributions of degenerate and unique Fisher–Hartwig type from [E, Sect. 5] without giving the proofs. These results are the promised descriptions in terms of conditions on the parameters of a suitable representation.

Proposition 4.1 *A distribution $\widehat{c}$ is of degenerate Fisher–Hartwig type if and only if it can be written in one of the following two forms:*

$$(4.6) \qquad \widehat{c} = t^\varkappa b \prod_{r=1}^{R} \widehat{\eta}_{t_r, \gamma_r},$$

where $\varkappa \in \mathbb{Z}$, $\varkappa > 0$, and $\gamma_r \in \mathbb{C}\backslash\mathbb{Z}_-$ for all $1 \leq r \leq R$, or

$$(4.7) \qquad \widehat{c} = t^\varkappa b \prod_{r=1}^{R} \widehat{\xi}_{t_r, \delta_r},$$

where $\varkappa \in \mathbb{Z}$, $\varkappa < 0$, and $\delta_r \in \mathbb{C}\backslash\mathbb{Z}_-$ for all $1 \leq r \leq R$. Here $b \in C^\infty(\mathbb{T})$ is a nonvanishing function with winding number zero.

Proposition 4.2 *A distribution $\widehat{c}$ is of unique Fisher–Hartwig type if and only if it can be written in the form*

$$(4.8) \qquad \widehat{c} = b \prod_{r \in M_0} \widehat{\omega}_{t_r,\alpha_r,\beta_r} \prod_{r \in M_+} \widehat{\eta}_{t_r,\gamma_r} \prod_{r \in M_-} \widehat{\xi}_{t_r,\delta_r},$$

where M_0, M_+ and M_- is a partition of the index set $\{1, \ldots, R\}$, $b \in C^\infty(\mathbb{T})$ is a nonvanishing function with winding number zero, and the parameters satisfy the following conditions:

(a) $\alpha_r, \beta_r \in \mathbb{C}$, $2\alpha_r \notin \mathbb{Z}_-$, $\alpha_r + \beta_r \notin \mathbb{Z}_- \cup \{0\}$, $\alpha_r - \beta_r \notin \mathbb{Z}_- \cup \{0\}$ *for* $r \in M_0$;

(b) $\gamma_r \in \mathbb{C}$, $\gamma_r \notin \mathbb{Z}_- \cup \{0\}$ *and* $\beta_r := \gamma_r/2$ *for* $r \in M_+$;

(c) $\delta_r \in \mathbb{C}$, $\delta_r \notin \mathbb{Z}_- \cup \{0\}$ *and* $\beta_r := -\delta_r/2$ *for* $r \in M_-$;

(d) $\varrho_{\max} - \varrho_{\min} < 1$, *where*

$$\varrho_{\max} = \max \{\operatorname{Re} \beta_r \mid r \in M_+ \cup M_0\},$$

$$\varrho_{\min} = \min \{\operatorname{Re} \beta_r \mid r \in M_- \cup M_0\}.$$

The representation (4.8) is the (unique) representation of $\widehat{c}$ contained in $\mathcal{T}_{\max}$.

Later on (as a slight generalization) we will also allow distributions $\widehat{\eta}_{t_r,\gamma_r}$ and $\widehat{\xi}_{t_r,\delta_r}$ with $\gamma_r \in \mathbb{Z}_-$ and $\delta_r \in \mathbb{Z}_-$ to appear in the products (4.6), (4.7) and (4.8). It is possible to formulate similar but more complicated conditions for the characterization of distributions of ambiguous Fisher–Hartwig type and their representations contained in $\mathcal{T}_{\max}$. Because we will not elaborate on the ambiguous case, the reader is referred to [E, Sect. 3] for details.

5 Asymptotics of Inverse Toeplitz Matrices and the Concept of $\mathcal{R}$-Convergence

The main results, which will be presented in the following section, will provide asymptotic information about two different objects. Firstly, they contain a description of the asymptotic behavior of Toeplitz determinants for distributions of unique Fisher–Hartwig type. Secondly, they contain a characterization of the asymptotic behavior of the inverses $T_n^{-1}(c)$ of Toeplitz matrices, where the generating distribution is also assumed to be of unique Fisher–Hartwig type. This characterization is of a very particular form and requires to introduce several sequences of operators and to define a certain concept of convergence [E, Sect. 10].

Let ℓ_μ^2 be the Hilbert space of all one-sided infinite sequences $\{x_n\}_{n=0}^\infty$ for which

$$(5.1) \qquad \|\{x_n\}_{n=0}^\infty\|_{\ell_\mu^2} = \left(\sum_{n=0}^\infty |x_n|^2 (1+n)^{2\mu} \right)^{1/2} < \infty,$$

where $\mu \in \mathbb{R}$. If $\mu_1 \geq \mu_2$, then $\ell_{\mu_1}^2 \subseteq \ell_{\mu_2}^2$ is a continuous and dense embedding.

Given a distribution $c \in \mathcal{D}'$ with Fourier coefficients c_n, the *Toeplitz operator* $T(c)$: $\ell^2_{\mu_1} \to \ell^2_{\mu_2}$ and the *Hankel operator* $H(c)$: $\ell^2_{\mu_1} \to \ell^2_{\mu_2}$ are defined by the infinite matrices

$$(5.2) \qquad T(c) = (c_{j-k}), \qquad 0 \le j, k < \infty,$$

$$(5.3) \qquad H(c) = (c_{j+k+1}), \qquad 0 \le j, k < \infty.$$

One can show that these operators are well defined and bounded for μ_1 sufficiently large and μ_2 sufficiently small.

Moreover, we also need the finite projection matrix P_n and the finite flip matrix W_n,

$$(5.4) \qquad P_n : \{x_k\}_{k=-\infty}^{\infty} \mapsto \{y_k\}_{k=-\infty}^{\infty}, \quad y_k = \begin{cases} x_k & \text{if } 0 \le k \le n-1 \\ 0 & \text{if } k \ge n, \end{cases}$$

$$(5.5) \qquad W_n : \{x_k\}_{k=-\infty}^{\infty} \mapsto \{y_k\}_{k=-\infty}^{\infty}, \quad y_k = \begin{cases} x_{n-k-1} & \text{if } 0 \le k \le n-1 \\ 0 & \text{if } k \ge n. \end{cases}$$

These linear bounded operators are defined on ℓ^2_μ. We have $P_n^2 = W_n^2 = P_n$ and $W_n = W_n P_n = P_n W_n$. We remark that

$$(5.6) \qquad T_n(c) = P_n T(c) P_n, \qquad T_n(\tilde{c}) = W_n T_n(c) W_n.$$

Next we introduce a certain type of convergence. Let H_1 and H_2 be Hilbert spaces. We consider sequences $\{C_n\}_{n=1}^{\infty}$, the elements of which are well defined linear bounded operators (or matrices) C_n : $H_1 \to H_2$ for all sufficiently large n. Let the notation $\mathcal{O}(\varrho)$ with $\varrho \in \mathbb{R}$ stand for the set of all such sequences for which

$$(5.7) \qquad \|C_n\|_{\mathcal{L}(H_1, H_2)} = O(n^\varrho) \qquad \text{as } n \to \infty.$$

The dependence of $\mathcal{O}(\varrho)$ on H_1 and H_2 will not be displayed in the notation.

Now let $H_1, H_2, \tilde{H}_1$ and $\tilde{H}_2$ be Hilbert spaces and $\varrho_0, \varrho_1, \varrho_2 \in \mathbb{R}$. We denote by $\mathcal{O}(\varrho_0, \varrho_1, \varrho_2)$ the set of all sequences $\{C_n\}_{n=1}^{\infty}$ of 2×2 block operators

$$(5.8) \qquad C_n = \begin{pmatrix} C_n^{(11)} & C_n^{(12)} \\ C_n^{(21)} & C_n^{(22)} \end{pmatrix} : H_1 \oplus \tilde{H}_1 \to H_2 \oplus \tilde{H}_2$$

for which $\{C_n^{(11)}\}, \{C_n^{(22)}\} \in \mathcal{O}(\varrho_0), \{C_n^{(12)}\} \in \mathcal{O}(\varrho_1), \{C_n^{(21)}\} \in \mathcal{O}(\varrho_2)$. We also use the notation $\mathcal{O}(\varrho_0, \varrho_1, \varrho_2)$ to denote any sequence of this type. In this sense,

$$(5.9) \qquad C_n = C + \mathcal{O}(\varrho_0, \varrho_1, \varrho_2)$$

means that $\{C_n - C\}_{n=1}^{\infty} \in \mathcal{O}(\varrho_0, \varrho_1, \varrho_2)$.

The set $\mathcal{O}(\varrho_0, \varrho_1, \varrho_2)$ is used along with the condition $\varrho_1 + \varrho_2 \leq \varrho_0 < 0$. The reason is that under this condition, one can perform a simple calculus with such sequences. For instance, suppose that

$$(5.10) \qquad A_n = \text{diag}(A, \tilde{A}) + \mathcal{O}(\varrho_0, \varrho_1, \varrho_2),$$

$$(5.11) \qquad B_n = \text{diag}(B, \tilde{B}) + \mathcal{O}(\varrho_0, \varrho_1, \varrho_2).$$

If these sequences are considered on *suitable* Hilbert spaces, then

$$(5.12) \qquad A_n B_n = \text{diag}(AB, \tilde{A}\tilde{B}) + \mathcal{O}(\varrho_0, \varrho_1, \varrho_2).$$

Moreover, if in addition both operators A and $\tilde{A}$ are invertible, then

$$(5.13) \qquad A_n^{-1} = \text{diag}(A^{-1}, \tilde{A}^{-1}) + \mathcal{O}(\varrho_0, \varrho_1, \varrho_2).$$

We are not going to show explicitly how this kind of calculus is employed to prove the main results. We only remark that it is an important tool in the proof of the separation theorem, which will be stated in Section 7.

Given $a \in \mathcal{D}'$, assume that $T_n(a)$ is invertible for sufficiently large n, and introduce the following sequences of operators of 2×2 block form:

$$(5.14) \qquad R_n(a) = \begin{pmatrix} T_n^{-1}(a) & T_n^{-1}(a)W_n \\ T_n^{-1}(\tilde{a})W_n & T_n^{-1}(\tilde{a}) \end{pmatrix},$$

$$(5.15) \qquad RH_n(a) = \begin{pmatrix} T_n^{-1}(a)P_nH(a) & T_n^{-1}(a)W_nH(\tilde{a}) \\ T_n^{-1}(\tilde{a})W_nH(a) & T_n^{-1}(\tilde{a})P_nH(\tilde{a}) \end{pmatrix},$$

$$(5.16) \qquad HR_n(a) = \begin{pmatrix} H(\tilde{a})P_nT_n^{-1}(a) & H(\tilde{a})W_nT_n^{-1}(\tilde{a}) \\ H(a)W_nT_n^{-1}(a) & H(a)P_nT_n^{-1}(\tilde{a}) \end{pmatrix},$$

$$(5.17) \qquad \begin{aligned} & HRH_n(a) = \\ & \begin{pmatrix} H(\tilde{a})P_nT_n^{-1}(a)P_nH(a) & H(\tilde{a})P_nT_n^{-1}(a)W_nH(\tilde{a}) \\ H(a)P_nT_n^{-1}(\tilde{a})W_nH(a) & H(a)P_nT_n^{-1}(\tilde{a})P_nH(\tilde{a}) \end{pmatrix} \\ & - \begin{pmatrix} T(\tilde{a}) & H(\tilde{a}\chi_{-n}) \\ H(a\chi_{-n}) & T(a) \end{pmatrix}. \end{aligned}$$

These sequences of operators are considered on $\ell_{\mu_1}^2 \oplus \ell_{\mu_1}^2 \to \ell_{\mu_2}^2 \oplus \ell_{\mu_2}^2$ with μ_1 sufficient large and μ_2 sufficiently small, which ensures the boundedness of the operators.

Now we are prepared to explain the main concept, called $\mathcal{R}$-convergence. It has been introduced in [E]. Let $a \in \mathcal{D}'$, $a_+ \in \mathcal{G}\mathcal{D}'_+$ and $a_- \in \mathcal{G}\mathcal{D}'_-$ be distributions, and let $\varrho_0, \varrho_1, \varrho_2 \in \mathbb{R}$. We say that the distribution a *effects $\mathcal{R}$-convergence with respect to* $[a_+, a_-]$ *and* $(\varrho_0, \varrho_1, \varrho_2)$ if there exist $\mu_1 \geq 0$ and $\mu_2 \leq 0$ such that

$$(5.18) \qquad R_n(a) = \text{diag}\,(T(a_+^{-1})T(a_-^{-1}),\ T(\widetilde{a}_-^{-1})T(\widetilde{a}_+^{-1})) + \mathcal{O}(\varrho_0, \varrho_1, \varrho_2),$$

$$(5.19) \quad RH_n(a) = \text{diag}\,(T(a_+^{-1})H(a_+),\ T(\widetilde{a}_-^{-1})H(\widetilde{a}_-)) + \mathcal{O}(\varrho_0, \varrho_1, \varrho_2),$$

$$(5.20) \quad HR_n(a) = \text{diag}\,(H(\widetilde{a}_-)T(a_-^{-1}),\ H(a_+)T(\widetilde{a}_+^{-1})) + \mathcal{O}(\varrho_0, \varrho_1, \varrho_2),$$

$$HRH_n(a) = -\text{diag}\,(T(\widetilde{a}_-)T(\widetilde{a}_+),$$

$$(5.21)$$

$$T(a_+)T(a_-)) + \mathcal{O}(\varrho_0, \varrho_1, \varrho_2),$$

where these sequences are considered on $\ell^2_{\mu_1} \oplus \ell^2_{\mu_1} \to \ell^2_{\mu_2} \oplus \ell^2_{\mu_2}$.

Obviously, if this statement holds for particular $\mu_1 = \mu_1^*$ and $\mu_2 = \mu_2^*$, then it also holds for all $\mu_1 \geq \mu_1^*$ and $\mu_2 \leq \mu_2^*$. Moreover, if the statement is true for particular $\varrho_0, \varrho_1, \varrho_2$, then it is also true when the values of these parameters are larger. Note that the block diagonal operators on the right hand side of (5.18)–(5.20) are well defined bounded operators on $\ell^2_{\mu_1} \oplus \ell^2_{\mu_1} \to \ell^2_{\mu_2} \oplus \ell^2_{\mu_2}$ if μ_1 is sufficiently large and μ_2 is sufficiently small.

We remark that while the sequences $HR_n(a)$ and $RH_n(a)$ themselves can be obtained from $R_n(a)$ by multiplying from the left or right with certain Hankel operators, the corresponding assertions (5.19) and (5.20) cannot be obtained from (5.18) in this way because the underlying Hilbert spaces do in general not "fit" to each other.

It is natural to ask for the "meaning" of the above sequences of operators and block diagonal operators. We cannot give here an exhaustive answer, but it might help to consider $R_n(a)$ in the special case of a smooth nonvanishing function a with winding number zero. Then it is well known that $T_n^{-1}(a) \to T^{-1}(a)$ strongly on ℓ^2. If one considers this sequence in $\ell^2_{\mu_1} \to \ell^2_{\mu_2}$ with suitable $\mu_1 > 0$ and $\mu_2 < 0$, then $T_n^{-1}(a)$ converges even in the norm and with a rate of convergence depending on the smoothness of a. Now, if one factors $a = a_+a_-$, then the inverse of $T(a)$ on ℓ^2 is precisely $T^{-1}(a) = T(a_+^{-1})T(a_-^{-1})$. If a_+ and a_- are distributions, then $T(a_+^{-1})T(a_-^{-1})$ cannot be understood as an inverse (because of the underlying Hilbert spaces), but it has apparently the same "structure" as the inverse of a Toeplitz operator. On the other hand, $T_n^{-1}(a)W_n \to 0$ weakly on ℓ^2, and if one considers weighted spaces, then the convergence is again in the norm. This explains to some extent the appearance of the block diagonal operator on the right hand side of (5.18). Notice that we may have $\varrho_1 \geq 0$ or $\varrho_2 \geq 0$ for distributions effecting $\mathcal{R}$-convergence (although still expecting $\varrho_1 + \varrho_2 \leq \varrho_0 < 0$).

For completeness sake we remark that the concept of $\mathcal{R}$-convergence has been introduced in [E] in a slightly different way. However, it has been proved there that both definitions are equivalent. Some more perhaps interesting implications of $\mathcal{R}$-convergences have also been stated there.

Finally, we agree on the following convention. We say that

(i) *a effects $\mathcal{R}$-convergence w.r.t. $[a_+, a_-]$ and $(-\infty, -\infty, -\infty)$ if and only if for each $\varrho \in \mathbb{R}$, a effects $\mathcal{R}$-convergence w.r.t. $[a_+, a_-]$ and $(\varrho, \varrho, \varrho)$;*

(ii) *a effects $\mathcal{R}$-convergence w.r.t. $[a_+, a_-]$ and $(-\infty, -\infty, \mu)$ if and only if for each $\varrho \in \mathbb{R}$, a effects $\mathcal{R}$-convergence w.r.t. $[a_+, a_-]$ and (ϱ, ϱ, μ);*

(iii) *a effects $\mathcal{R}$-convergence w.r.t. $[a_+, a_-]$ and $(-\infty, \mu, -\infty)$ if and only if for each $\varrho \in \mathbb{R}$, a effects $\mathcal{R}$-convergence w.r.t. $[a_+, a_-]$ and (ϱ, μ, ϱ).*

6 The Main Results

In this section we state the main results of the author's dissertation [E, Sect. 13]. We will describe the asymptotic behavior of Toeplitz determinants for distributions of unique Fisher–Hartwig type and we show that these distributions effect $\mathcal{R}$-convergence. Moreover, we give an estimate on the asymptotic behavior of Toeplitz determinants for distributions of degenerate Fisher–Hartwig type. Hence the first two conjectures of Section 2 are proved. In addition, the results about $\mathcal{R}$-convergence give us some information about certain aspects of the asymptotic behavior of the inverses of Toeplitz matrices.

Theorem 6.1 *Let $t_1, \ldots, t_R$ be distinct points on the unit circle, let M_0, M_+, M_+^*, M_- and M_-^* be a partition of the index set $\{1, \ldots, R\}$, and let $\widehat{c}$ be the distribution*

$$(6.1) \qquad \widehat{c} = b \prod_{r \in M_0} \widehat{\omega}_{t_r, \alpha_r, \beta_r} \prod_{r \in M_+ \cup M_+^*} \widehat{\eta}_{t_r, \gamma_r} \prod_{r \in M_- \cup M_-^*} \widehat{\xi}_{t_r, \delta_r}.$$

Assume that the following conditions are satisfied:

(a) *The function $b \in C^\infty(\mathbb{T})$ is nonvanishing and has winding number zero. The normalized canonical Wiener–Hopf factorization of b is $b(t) = b_+(t) Gb_-(t)$ with $b_\pm$ given by (1.11) and $G = G[b] \in \mathbb{C}\backslash\{0\}$ given by (1.6).*

(b) *$\alpha_r, \beta_r \in \mathbb{C}$, $2\alpha_r \notin \mathbb{Z}_-$, $\gamma_r := \alpha_r + \beta_r \notin \mathbb{Z}_- \cup \{0\}$, $\delta_r := \alpha_r - \beta_r \notin \mathbb{Z}_- \cup \{0\}$ for each $r \in M_0$.*

(c) *$\gamma_r \in \mathbb{C}$, $\gamma_r \notin \mathbb{Z}_- \cup \{0\}$, $\beta_r := \gamma_r/2$ for each $r \in M_+$.*

(d) *$\gamma_r \in \mathbb{Z}_-$ for each $r \in M_+^*$.*

(e) *$\delta_r \in \mathbb{C}$, $\delta_r \notin \mathbb{Z}_- \cup \{0\}$, $\beta_r := -\delta_r/2$ for each $r \in M_-$.*

(f) *$\delta_r \in \mathbb{Z}_-$ for each $r \in M_-^*$.*

(g) $\varrho_0 < 0$ *(or, equivalently, $\varrho_1 + \varrho_2 < 0$), where*

$$\varrho_1 = \max\{-1 - 2\operatorname{Re}\beta_r \mid r \in M_0 \cup M_-\},$$

$$\varrho_2 = \max\{-1 + 2\operatorname{Re}\beta_r \mid r \in M_0 \cup M_+\},$$

$$\varrho_0^* = \begin{cases} -1 & \text{if } M_0 \neq \emptyset \\ -\infty & \text{if } M_0 = \emptyset, \end{cases}$$

$$\varrho_0 = \max\{\varrho_0^*, \varrho_1 + \varrho_2\}.$$

Finally, we define the following distributions and constants:

(6.2) $$\widehat{c}_+ = Gb_+ \prod_{r \in M_0 \cup M_+ \cup M_+^*} \widehat{\eta}_{t_r, \gamma_r},$$

(6.3) $$\widehat{c}_- = b_- \prod_{r \in M_0 \cup M_- \cup M_-^*} \widehat{\xi}_{t_r, \delta_r},$$

(6.4) $$\Omega = \sum_{r \in M_0} (\alpha_r^2 - \beta_r^2),$$

$$E = E[b] \prod_{\substack{r \in M_0 \cup M_+ \cup M_+^* \\ s \in M_0 \cup M_- \cup M_-^* \\ r \neq s}} (1 - t_s/t_r)^{-\gamma_r \delta_s}$$

(6.5) $$\times \prod_{r \in M_0 \cup M_- \cup M_-^*} b_+(t_r)^{-\delta_r} \prod_{r \in M_0 \cup M_+ \cup M_+^*} b_-(t_r)^{-\gamma_r}$$

$$\times \prod_{r \in M_0} \frac{G(1 + \alpha_r + \beta_r)G(1 + \alpha_r - \beta_r)}{G(1 + 2\alpha_r)},$$

where $E[b]$ is the constant (1.10). *Then the following statements are true:*

(1) *The matrix $T_n(\widehat{c})$ is invertible for all sufficiently large n.*

(2) *The distribution $\widehat{c}$ effects $\mathcal{R}$-convergence with respect to $[\widehat{c}_+, \widehat{c}_-]$ and $(\varrho_0, \varrho_1, \varrho_2)$.*

(3) *If $\varrho_0 \neq -\infty$, then*

$$D_n(\widehat{c}) = G^n n^{\Omega} E(1 + O(n^{\varrho_0})) \qquad \text{as } n \to \infty.$$

(4) *If $\varrho_0 = -\infty$, then for each $\varrho \in \mathbb{R}$,*

$$D_n(\widehat{c}) = G^n E(1 + O(n^{\varrho})) \qquad \text{as } n \to \infty.$$

The preceding result confirms the Fisher–Hartwig conjecture for distributions of unique Fisher–Hartwig type. The assumptions on the parameters stated here and in Proposition 4.2 are the same. Indeed, $\varrho_1 = -1 - 2\varrho_{min}$ and $\varrho_2 = -1 + 2\varrho_{max}$, and hence the condition $\varrho_1 + \varrho_2 < 0$ is equivalent to $\varrho_{max} - \varrho_{min} < 1$. In addition, the previous theorem includes distributions $\widehat{\eta}_{t_r,\gamma_r}$ and $\widehat{\xi}_{t_r,\delta_r}$ with $\gamma_r, \delta_r \in \mathbb{Z}_-$.

The result for distributions of degenerate Fisher–Hartwig type is as follows. Again, in comparison with Proposition 4.1 we include distributions $\widehat{\eta}_{t_r,\gamma_r}$ and $\widehat{\xi}_{t_r,\delta_r}$ with $\gamma_r, \delta_r \in \mathbb{Z}_-$.

Theorem 6.2 *Let $\widehat{c}$ be a distribution that is of one of the following two forms:*

$$(6.6) \qquad \widehat{c} = t^{\varkappa} b \prod_{r=1}^{R} \widehat{\eta}_{t_r,\gamma_r},$$

where $\varkappa \in \mathbb{Z}$, $\varkappa > 0$ and $\gamma_r \in \mathbb{C}$ for each $1 \leq r \leq R$, or

$$(6.7) \qquad \widehat{c} = t^{\varkappa} b \prod_{r=1}^{R} \widehat{\xi}_{t_r,\delta_r},$$

where $\varkappa \in \mathbb{Z}$, $\varkappa < 0$ and $\delta_r \in \mathbb{C}$ for each $1 \leq r \leq R$. Assume that $b \in C^{\infty}(\mathbb{T})$ is a nonvanishing function with winding number zero, and let $G = G[b]$ be defined by (1.6). Then for each $\varrho \in \mathbb{R}$ we have

$$(6.8) \qquad D_n(\widehat{c}) = O(|G|^n n^{\varrho}) \qquad as\ n \to \infty.$$

7 Main Ideas of the Proof

We can give here only the main ideas of the proof. For details we must refer the reader to [E].

One of the main ideas is a so-called separation theorem, which will be stated below. The separation idea, i.e., to prove that under certain conditions the quotient $D_n(ab)/(D_n(a)D_n(b))$ converges to some nonzero constant, has already been used in previous papers (see, e.g., [B2, BH, Bö, BS5]). The separation theorem stated below applies, however, to a much broader class of functions (or distributions) than all previous separation results.

In order to describe the constant that appears in the separation theorem, consider first nonvanishing functions $a, b \in C^{\infty}(\mathbb{T})$ with winding number zero. Then

$$(7.1) \qquad \lim_{n \to \infty} \frac{D_n(ab)}{D_n(a)D_n(b)} = E(a,b)E(b,a),$$

where $E(a, b)$ is a constant, which is defined in terms of an operator determinant,

$$(7.2) \qquad E(a,b) = \det T^{-1}(a)T(ab)T^{-1}(b).$$

Note that $T^{-1}(a)T(ab)T^{-1}(b) = I + T^{-1}(a)H(a)H(\tilde{b})T^{-1}(b)$ with $H(a)H(\tilde{b})$ being trace class. Moreover, $E(a,b) = E(a_+, b_-)$ where a_+ and b_- are the Wiener–Hopf factors (see (1.11)). One can evaluate this constant:

$$(7.3) \qquad E(a,b) = \exp\left(\sum_{k=1}^{\infty} k[\log a]_k [\log b]_{-k}\right).$$

For our purposes, we need a generalization of this constant where the functions are replaced by distributions. Suppose that $\widehat{a}_+ \in \mathcal{GD}'_+(K)$ and $\widehat{b}_- \in \mathcal{GD}'_-(L)$, where K and L are disjoint compact subsets of the unit circle. Then one can show that the following limit exists:

$$(7.4) \qquad E(\widehat{a}_+, \widehat{b}_-) := \lim_{r \to 1-0} E(h_r\widehat{a}_+, h_r\widehat{b}_-).$$

Here $(h_r\widehat{c})(e^{i\theta}) = \sum_{n=-\infty}^{\infty} r^{|n|} e^{in\theta} [\widehat{c}]_n$ is the harmonic extension of a distribution. One can also give an alternative definition of (7.4) which is not based on a limit but relies on an operator determinant that involves some more complicated operators. However, for the purpose of the evaluation of the constant $E(\widehat{a}_+, \widehat{b}_-)$ for given concrete distributions, the definition (7.4) is quite useful because the limit can be computed (in principle) by means of (7.3).

The separation theorem deals (as already the main results) with the asymptotics of Toeplitz determinants on the one hand and the concept of $\mathcal{R}$-convergence on the other hand. This means that, in contrast to previous versions of separation results, the separation idea is also employed for the study of the asymptotic behavior of the inverses of Toeplitz matrices. The latter amounts to the following question: Suppose that the distributions $\widehat{a}$ and $\widehat{b}$ effect $\mathcal{R}$-convergence. Does it follow that the distribution $\widehat{a}\,\widehat{b}$ effects $\mathcal{R}$-convergence? It turns out that, under certain conditions, the answer is "yes".

We state the separation theorem immediately for the case of a product of several distributions. Of course, the crucial point is the case of a product of two distribution, from which the more general result can be derived by induction.

Theorem 7.1 *Given a finite index set I, let K_i, $i \in I$, be mutually disjoint compact subsets of $\mathbb{T}$. Moreover, let $\widehat{a}^{(i)} \in \mathcal{D}'(K_i)$, $\widehat{a}^{(i)}_+ \in \mathcal{GD}'_+(K_i)$ and $\widehat{a}^{(i)}_- \in \mathcal{GD}'_-(K_i)$ be distributions and denote their smooth parts by $a^{(i)}$, $a^{(i)}_+$ and $a^{(i)}_-$, respectively. Assume that $a^{(i)} = a^{(i)}_+ a^{(i)}_-$ in the sense of functions in $C^\infty(\mathbb{T} \backslash K_i)$. In addition, suppose that $\widehat{a}^{(i)}$ effects $\mathcal{R}$-convergence with respect to $[\widehat{a}^{(i)}_+, \widehat{a}^{(i)}_-]$ and $(\varrho_0^{(i)}, \varrho_1^{(i)}, \varrho_2^{(i)})$, where $\varrho_0^{(i)}, \varrho_1^{(i)}, \varrho_2^{(i)} \in \mathbb{R} \cup \{-\infty\}$ for each $i \in I$. Let*

$$(7.5) \qquad \widehat{a} = \prod_{i\in I} \widehat{a}^{(i)}, \qquad \widehat{a}_+ = \prod_{i\in I} \widehat{a}^{(i)}_+, \qquad \widehat{a}_- = \prod_{i\in I} \widehat{a}^{(i)}_-.$$

Finally, assume $\varrho_0 < 0$, where

(7.6) $\varrho_1 = \max\{\varrho_1^{(i)} \mid i \in I\}$, $\varrho_2 = \max\{\varrho_2^{(i)} \mid i \in I\}$,

(7.7) $\varrho_0^* = \max\{\varrho_0^{(i)} \mid i \in I\}$, $\varrho_0 = \max\{\varrho_0^*, \varrho_1 + \varrho_2\}$.

Then the following assertions hold:

(i) *The matrix $T_n(\widehat{a})$ is invertible for all sufficiently large n.*

(ii) *The distribution $\widehat{a}$ effects $\mathcal{R}$-convergence with respect to $[\widehat{a}_+, \widehat{a}_-]$ and $(\varrho_0, \varrho_1, \varrho_2)$.*

(iii) *As $n \to \infty$, we have*

$$\frac{D_n(\widehat{a})}{\prod_{i \in I} D_n(\widehat{a}^{(i)})} = E + O(n^{\varrho_0}),$$

where E is the (nonzero) constant given by

$$E = \prod_{\substack{r,s \in I \\ r \neq s}} E(\widehat{a}_+^{(r)}, \widehat{a}_-^{(s)}).$$

Most of the assumptions imposed in this separation theorem are quite natural. Not so easy to understand is, however, the meaning of the condition $\varrho_0 < 0$, which is equivalent to the condition $\varrho_1 + \varrho_2 < 0$ provided that $\varrho_0^* < 0$. The condition $\varrho_1 + \varrho_2 < 0$ is by no means satisfied for all collections of distribution $\widehat{a}^{(i)}$, $i \in I$, of Fisher–Hartwig type. However, it turns out that it is fulfilled if $\widehat{a} = \prod_{i \in I} \widehat{a}^{(i)}$ is a distribution of unique Fisher–Hartwig type and if this product is just the representation contained in $\mathcal{T}_{max}$. This is essentially the reason why we have been able to prove the main results in the stated generality. On the other hand, if $\widehat{a}$ is not a distribution of unique Fisher–Hartwig type or if the above representation is not the one contained in $\mathcal{T}_{max}$, then the assertions of the separation theorem can simply not be expected to be true. In other word, the condition $\varrho_1 + \varrho_2 < 0$ explains in some sense why the original Fisher–Hartwig conjecture sometimes holds and sometimes breaks down.

In regard to a proof of Theorem 6.1, the separation theorem reduces the asymptotic problems (i.e., the asymptotics of the Toeplitz determinants and the problem of $\mathcal{R}$-convergence) for a general distributions of unique Fisher–Hartwig type to the corresponding problems for the distributions which appear in the product of the given representation. Hence we have to determine the asymptotics of the Toeplitz determinants and to prove $\mathcal{R}$-convergence for these particular distributions.

These problems are fairly easy to settle for smooth nonvanishing functions b with winding number zero and for distributions of analytic type, $\widehat{\eta}_{t_r,\gamma_r}$ and $\widehat{\xi}_{t_r,\delta_r}$ with $\gamma_r, \delta_r \in \mathbb{C}$. We confine ourselves to state the results concerning $\mathcal{R}$-convergence (see [E, Sect. 12]).

Theorem 7.2 *Let* $\gamma_r, \delta_r \in \mathbb{C} \setminus (\mathbb{Z}_- \cup \{0\})$ *and* $t_r \in \mathbb{T}$. *Then*

(a) $\widehat{\eta}_{t_r,\gamma_r}$ *effects* $\mathcal{R}$-*convergence w.r.t.* $[\widehat{\eta}_{t_r,\gamma_r}, 1]$ *and* $(-\infty, -\infty, -1 + \operatorname{Re}\gamma_r)$,

(b) $\widehat{\xi}_{t_r,\delta_r}$ *effects* $\mathcal{R}$-*convergence w.r.t.* $[1, \widehat{\xi}_{t_r,\delta_r}]$ *and* $(-\infty, -1 + \operatorname{Re}\delta_r, -\infty)$.

Let $\gamma_r, \delta_r \in \mathbb{Z}_- \cup \{0\}$ *and* $t_r \in \mathbb{T}$. *Then*

(c) $\widehat{\eta}_{t_r,\gamma_r}$ *effects* $\mathcal{R}$-*convergence w.r.t.* $[\widehat{\eta}_{t_r,\gamma_r}, 1]$ *and* $(-\infty, -\infty, -\infty)$,

(d) $\widehat{\xi}_{t_r,\delta_r}$ *effects* $\mathcal{R}$-*convergence w.r.t.* $[1, \widehat{\xi}_{t_r,\delta_r}]$ *and* $(-\infty, -\infty, -\infty)$.

Let $b \in C^\infty(\mathbb{T})$ *be a nonvanishing function with winding number zero, which has the normalized canonical Wiener–Hopf factorization* $b(t) = b_-(t)Gb_+(t)$. *Then*

(e) b *effects* $\mathcal{R}$-*convergence w.r.t.* $[Gb_+, b_-]$ *and* $(-\infty, -\infty, -\infty)$.

The nontrivial problem arises with the distributions $\widehat{\omega}_{t_r,\alpha_r,\beta_r}$. Here the result is as follows [E, Sect. 11].

Theorem 7.3 *Let* $\alpha_r, \beta_r \in \mathbb{C}$ *and* $t_r \in \mathbb{T}$, *and assume that* $\gamma_r := \alpha_r + \beta_r \notin \mathbb{Z}_-$, $\delta_r := \alpha_r - \beta_r \notin \mathbb{Z}_-$ *and* $2\alpha_r \notin \mathbb{Z}_-$. *Then*

$$(7.8) \qquad D_n(\widehat{\omega}_{t_r,\alpha_r,\beta_r}) = \frac{G(1+\gamma_r)G(1+\delta_r)}{G(1+\gamma_r+\delta_r)}(1 + O(1/n))n^{\gamma_r\delta_r}$$

as $n \to \infty$. *Moreover, the distribution* $\widehat{\omega}_{t_r,\alpha_r,\beta_r}$ *effects* $\mathcal{R}$-*convergence with respect to* $[\widehat{\eta}_{t_r,\gamma_r}, \widehat{\xi}_{t_r,\delta_r}]$ *and* $(-1, -1 - 2\operatorname{Re}\beta_r, -1 + 2\operatorname{Re}\beta_r)$.

Now the proof of Theorem 6.1 follows by piecing together the results of Theorem 7.2 and Theorem 7.3 by means of Theorem 7.1.

Let us make some remarks on the (nontrivial) proof of Theorem 7.3. Fortunately, we can use a couple of operator identities, which are also quite interesting in themselves. Some of them have already been known (see [ES]).

We first observe that it suffices to consider the case $t_r = 1$. For convenience, let us write $\widehat{\omega}_{1,\alpha,\beta} = \xi_\delta\eta_\gamma$. Then a first important operator identity is

$$(7.9) \qquad T(\xi_\delta\eta_\gamma) = \Gamma_{\gamma,\delta}^{-1}M_\delta^{-1}T(\eta_\gamma)M_{\gamma+\delta}T(\xi_\delta)M_\gamma^{-1},$$

where $\Gamma_{\gamma,\delta} = \Gamma(1+\gamma)\Gamma(1+\delta)/G(1+\gamma+\delta)$ and $M_\alpha = \operatorname{diag}(\mu_0^{(\alpha)}, \mu_1^{(\alpha)}, \ldots)$ is an infinite diagonal matrix with $\mu_k^{(\alpha)} = \Gamma(1+\alpha+k)/(\Gamma(1+\alpha)k!)$. It follows that

$$(7.10) \qquad T_n(\xi_\delta\eta_\gamma) = \Gamma_{\gamma,\delta}^{-1}M_{\delta,n}^{-1}T_n(\eta_\gamma)M_{\gamma+\delta,n}T_n(\xi_\delta)M_{\gamma,n}^{-1},$$

where $M_{\alpha,n} = \mathrm{diag}\,(\mu_0^{(\alpha)}, \ldots, \mu_{n-1}^{(\alpha)})$ is an $n \times n$ diagonal matrix. Because the Toeplitz matrices $T_n(\eta_\gamma)$ and $T_n(\xi_\delta)$ are triangular matrices, we can easily evaluate the determinant and the inverse. Using the recurrence relation $G(1+z) = \Gamma(z)G(z)$ for the Barnes G-function [Ba, WW] we obtain the formula

$$(7.11) \quad D_n(\xi_\delta \eta_\gamma) = \frac{G(1+\gamma)G(1+\delta)}{G(1+\gamma+\delta)} \frac{G(1+n)G(1+\gamma+\delta+n)}{G(1+\gamma+n)G(1+\delta+n)},$$

which has already been known [BS3, BS5]. Note that the second term in this product behaves asymptotically as $n^{\gamma\delta}$. The inverse of $T_n(\xi_\delta \eta_\gamma)$ equals

$$(7.12) \qquad T_n^{-1}(\xi_\delta \eta_\gamma) = \Gamma_{\gamma,\delta} M_{\gamma,n} T_n(\xi_{-\delta}) M_{\gamma+\delta,n}^{-1} T_n(\eta_{-\gamma}) M_{\delta,n},$$

Using this formula, the asymptotic behavior of $T_n^{-1}(\xi_\delta \eta_\gamma)$ can be analyzed. After passing to the limit, where the operators are considered on appropriate spaces ℓ_μ^2, one obtains

$$(7.13) \qquad \Gamma_{\gamma,\delta} M_\gamma T(\xi_{-\delta}) M_{\gamma+\delta}^{-1} T(\eta_{-\gamma}) M_\delta = T(\eta_{-\gamma}) T(\xi_{-\delta}),$$

which is another operator identity. A more careful analysis gives

$$(7.14) \qquad T_n^{-1}(\xi_\delta \eta_\gamma) = T(\eta_{-\gamma}) T(\xi_{-\delta}) + \mathcal{O}(-1),$$

which is one part of the assertions in regard to the $\mathcal{R}$-convergence of $\xi_\delta \eta_\gamma$.

However, there appear also sequences involving Hankel operators. In this connection the following two identities,

$$(7.15) \qquad H(\xi_\gamma \eta_\delta) = \delta \Gamma_{\gamma,\delta}^{-1} M_{-\delta} H(\tau_\gamma) M_{\gamma+\delta} T(\xi_\delta) M_\gamma^{-1},$$
$$(7.16) \qquad H(\xi_\delta \eta_\gamma) = \gamma \Gamma_{\gamma,\delta}^{-1} M_\delta^{-1} T(\eta_\gamma) M_{\gamma+\delta} H(\tau_\delta) M_{-\gamma},$$

play a role. Here another Hankel operator,

$$(7.17) \qquad H(\tau_\alpha) = \left(-\frac{(j+k)!\,\Gamma(1+\alpha)}{\Gamma(2+\alpha+j+k)} \right), \qquad 0 \le j, k < \infty,$$

occurs. Indeed, from these formulas, it follows that

$$(7.18) \qquad H(\xi_\gamma \eta_\delta) T_n^{-1}(\xi_\delta \eta_\gamma) = \delta M_{-\delta} H(\tau_\gamma) T_n(\eta_{-\gamma}) M_{\delta,n},$$
$$(7.19) \qquad T_n^{-1}(\xi_\delta \eta_\gamma) H(\xi_\delta \eta_\gamma) = \gamma M_{\gamma,n} T_n(\xi_{-\delta}) H(\tau_\delta) M_{-\gamma},$$
$$(7.20) \quad H(\xi_\gamma \eta_\delta) T_n^{-1}(\xi_\delta \eta_\gamma) H(\xi_\delta \eta_\gamma) = \gamma\delta \Gamma_{\gamma,\delta}^{-1} M_{-\delta} H(\tau_\gamma) M_{\gamma+\delta,n} H(\tau_\delta) M_{-\gamma}.$$

Now we can pass to the limit and apply the formulas

$$\text{(7.21)} \qquad \delta M_{-\delta} H(\tau_\gamma) T(\eta_{-\gamma}) M_\delta \;=\; H(\eta_\delta) T(\xi_{-\delta}),$$

$$\text{(7.22)} \qquad \gamma M_\gamma T(\xi_{-\delta}) H(\tau_\delta) M_{-\gamma} \;=\; T(\eta_{-\gamma}) H(\eta_\gamma),$$

$$\text{(7.23)} \quad \gamma\delta \Gamma_{\gamma,\delta}^{-1} M_{-\delta} H(\tau_\gamma) M_{\gamma+\delta} H(\tau_\delta) M_{-\gamma} \;=\; T(\xi_\gamma \eta_\delta) - T(\eta_\delta) T(\xi_\gamma),$$

in order to prove the desired asymptotic behavior. In fact, one can show that

$$\text{(7.24)} \qquad H(\xi_\gamma \eta_\delta) T_n^{-1}(\xi_\delta \eta_\gamma) \;=\; H(\eta_\delta) T(\xi_{-\delta}) + \mathcal{O}(-1),$$

$$\text{(7.25)} \qquad T_n^{-1}(\xi_\delta \eta_\gamma) H(\xi_\delta \eta_\gamma) \;=\; T(\eta_{-\gamma}) H(\eta_\gamma) + \mathcal{O}(-1),$$

$$\text{(7.26)} \quad H(\xi_\gamma \eta_\delta) T_n^{-1}(\xi_\delta \eta_\gamma) H(\xi_\delta \eta_\gamma) \;=\; T(\xi_\gamma \eta_\delta) - T(\eta_\delta) T(\xi_\gamma) + \mathcal{O}(-1).$$

The sequences containing the matrices W_n, which appear in the definition of $\mathcal{R}$-convergence, can be analyzed also by using the above formulas. We will omit these computations.

Finally, let us sketch the proof of Theorem 6.2. Assume that $\widehat{c}$ is a distributions of degenerate Fisher–Hartwig type, which is without loss of generality of the form (6.6). Write $\widehat{c} = t^\varkappa \widehat{a}$, where $\varkappa > 0$. Then $\widehat{a}$ is a distribution of unique Fisher–Hartwig type. From Theorem 6.1, it follows that $\widehat{a}$ effects $\mathcal{R}$-convergence with respect to $[\widehat{a}_+, \widehat{a}_-]$ and $(-\infty, -\infty, \mu)$ with certain distributions $\widehat{a}_+$ and $\widehat{a}_-$ and a certain $\mu \in \mathbb{R} \cup \{-\infty\}$. Now one uses the identity

$$\text{(7.27)} \qquad (-1)^{\varkappa n} \frac{D_{n-\varkappa}(t^\varkappa \widehat{a})}{D_n(\widehat{a})} \;=\; \det P_\varkappa T_n^{-1}(\widehat{a}) W_n P_\varkappa,$$

which is a consequence of Cramer's rule. The asymptotic behavior of $R_n(\widehat{a})$ (more precisely, of its $(1, 2)$-entry) implies that $\| T_n^{-1}(\widehat{a}) W_n \| = O(n^\varrho)$ for all arbitrarily small $\varrho \in \mathbb{R}$. Because $\varkappa > 0$ is fixed, so behaves the determinant on the right hand side of the above formula. This combined with the asymptotics of $D_n(\widehat{a})$, which also follows from Theorem 6.1, gives the estimate for the asymptotics of $D_n(\widehat{c})$.

References

[Ba]   E.W. Barnes, The theory of the G-function, *Quart. J. Pure and Appl. Math.* **31** (1900), 264–313.

[B1]   E.L. Basor, Asymptotic formulas for Toeplitz determinants, *Trans. Amer. Math. Soc.* **239** (1978), 33–65.

[B2]   E.L. Basor, A localization theorem for Toeplitz determinants, *Indiana Univ. Math. J.* **28** (1979), 975–983.

[BH]   E.L. Basor and J.W. Helton, A new proof of the Szegö limit theorem and new results for Toeplitz operators with discontinuous symbols, *J. Operator Theory* **3** (1980), 23–29.

[BM1]   E.L. Basor and K.E. Morrison, The Fisher–Hartwig conjecture and Toeplitz eigenvalues, *Linear Algebra Appl.* **202** (1994), 129–142.

[BM2]   E.L. Basor and K.E. Morrison, The extended Fisher–Hartwig conjecture for symbols with multiple jump discontinuities, *Oper. Theory: Adv. Appl.* **71** (1994), 16–28.

[BT]   E.L. Basor and C.A. Tracy, The Fisher–Hartwig conjecture and generalizations, *Phys. A* **177** (1991), 167–173.

[Bö]   A. Böttcher, Toeplitz determinants with piecewise continuous generating function, *Z. Anal. Anw.* **1** (1982), 23–39.

[BS1]   A. Böttcher and B. Silbermann, The asymptotic behavior of Toeplitz determinants for generating functions with zeros of integral order, *Math. Nachr.* **102** (1981), 79–105.

[BS2]   A. Böttcher and B. Silbermann, Wiener–Hopf determinants with symbols having zeros of analytic type, *Seminar Analysis* **1982/83** (Karl-Weierstraß-Institut Berlin, 1983), 224–243.

[BS3]   A. Böttcher and B. Silbermann, Toeplitz matrices and determinants with Fisher–Hartwig symbols, *J. Funct. Anal.* **63** (1985), 178–214.

[BS4]   A. Böttcher and B. Silbermann, Toeplitz operators and determinants generated by symbols with one Fisher–Hartwig singularity, *Math. Nachr.* **127** (1986), 95–124.

[BS5]   A. Böttcher and B. Silbermann, *Analysis of Toeplitz operators*, Springer Verlag, Berlin 1990.

[BS6]   A. Böttcher and B. Silbermann, *Introduction to large truncated Toeplitz matrices*, Springer Verlag, Berlin 1998.

[D]   K.M. Day, Toeplitz matrices generated by the Laurent series expansion of an arbitrary rational function, *Trans. Amer. Math. Soc.* **206** (1975), 224–245.

[E]   T. Ehrhardt, *Toeplitz determinants with several Fisher–Hartwig singularities*, Dissertation, Technische Universität Chemnitz 1997.

[ES]   T. Ehrhardt and B. Silbermann, Toeplitz determinants with one Fisher–Hartwig singularity, *J. Funct. Anal.* **148** (1997), 229–256.

[FH]   M.E. Fisher and R.E. Hartwig, Toeplitz determinants: some applications, theorems and conjectures, *Adv. Chem. Phys.* **15** (1968), 333–353.

[L1]   A. Lenard, The momentum distribution in the ground state of the one-dimensional systems of impenetrable bosons, *J. Math. Phys.* **5** (1964), 930–943.

[L2]   A. Lenard, Some remarks on large Toeplitz determinants, *Pacific J. Math.* **42** (1972), 137–145.

[Li]   R.A. Libby, *Asymptotics of determinants and eigenvalue distribution for Toeplitz matrices associated with certain discontinuous symbols*, Ph.D. Thesis, Univ. of California Santa Cruz 1990.

[S]   B. Silbermann, The strong Szegö limit theorem for a class of singular generating functions. I, *Demonstr. Math.* **14:3** (1981), 647–667.

[WW]   E.T. Whittaker and G.N. Watson, *A course of modern analysis*, 4th ed., Cambridge Univ. Press, London, New York 1952.

[W1]   H. Widom, Toeplitz determinants with singular generating functions, *Amer. J. Math.* **95** (1973), 333–383.

[W2]   H. Widom, Eigenvalue distribution of nonselfadjoint Toeplitz matrices and the asymptotics of Toeplitz determinants in the case of nonvanishing index, *Oper. Theory: Adv. Appl.* **48** (1990), 387–421.

Torsten Ehrhardt
Fakultät für Mathematik
Technische Universität Chemnitz
09107 Chemnitz
Germany

1991 Mathematics Subject Classification. Primary: 47B35; Secondary: 15A15.

Operator Theory:
Advances and Applications, Vol. 124
© 2001 Birkhäuser Verlag Basel/Switzerland

On the Spectrum of Unbounded Off-diagonal
2×2 Operator Matrices in Banach Spaces

Volker Hardt and Reinhard Mennicken

We dedicate this paper to Israel Gohberg - an outstanding mathematician and man

In the product of two Banach spaces we study the spectrum of nonselfadjoint 2×2 off-diagonal operator matrices

$$\begin{pmatrix} 0 & T_1 \\ T_2 & 0 \end{pmatrix}$$

with unbounded entries T_1, T_2. We show that the spectral properties of such operator matrices are closely related to those of the product operators $T_2 T_1$ and $T_1 T_2$. The results on the spectra generalize well-known results on so-called supersymmetries, i.e., selfadjoint 2×2 off-diagonal operator matrices

$$\begin{pmatrix} 0 & T \\ T^* & 0 \end{pmatrix}$$

in the product of Hilbert spaces.

0 Introduction

In this note we consider operators defined by 2×2 block operator matrices of the form

(0.1)
$$\mathbf{T} = \begin{pmatrix} 0 & T_1 \\ T_2 & 0 \end{pmatrix}$$

where the entries T_1 and T_2 are closed linear operators in Banach spaces. The operator T_1 maps the Banach space B_2 to the Banach space B_1; conversely T_2 maps B_1 to B_2. Under appropriate assumptions on the operators T_1 and T_2 we will show that the spectrum and the non-discrete spectrum of $\mathbf{T}$ away from zero is equal to the square root of the spectrum and non-discrete spectrum, respectively, of the product operator $T_2 T_1$ away from zero:

(0.2)
$$\sigma(\mathbf{T}) \setminus \{0\} = \{\lambda \in \mathbb{C} : \lambda^2 \in \sigma(T_2 T_1)\} \setminus \{0\},$$

$$\sigma_{\mathrm{nd}}(\mathbf{T}) \setminus \{0\} = \{\lambda \in \mathbb{C} : \lambda^2 \in \sigma_{\mathrm{nd}}(T_2 T_1)\} \setminus \{0\}.$$

Connected with this formula is the relationship between the spectra of the operator products $T_2 T_1$ and $T_1 T_2$. We will prove that

(0.3)
$$\sigma(T_2 T_1)\backslash\{0\} = \sigma(T_1 T_2)\backslash\{0\},$$

$$\sigma_{nd}(T_2 T_1)\backslash\{0\} = \sigma_{nd}(T_1 T_2)\backslash\{0\}.$$

These questions are related to the relationships

(0.4)
$$(T_2 T_1)^* = T_1^* T_2^*, \quad (T_1 T_2)^* = T_2^* T_1^*$$

which in general do not hold.

The case of Hilbert spaces B_1, B_2 and a symmetric operator $\mathbf{T}$, i.e., $T_1 \subset T_2^*$ or equivalently $T_2 \subset T_1^*$, has recently been considered in [M]; see also [T] for the case of a selfadjoint operator $\mathbf{T}$, i.e., $T_1 = T_2^*$. In this case the non-discrete spectrum coincides with the essential spectrum.

The reason for considering operators of the type (0.1) is that many problems in mathematical physics can be described by block operator matrices

(0.5)
$$\mathbb{L} = \begin{pmatrix} A & B \\ C & D \end{pmatrix}$$

where the entries are unbounded operators. Physicists are in particular interested in the location of the essential spectrum. A successful approach to such problems has recently been developed in [ALMS] under certain assumptions on the entries A, B, C and D. The main assumptions are that A has compact resolvent and $\mathcal{D}(A) \subset \mathcal{D}(C)$, $\mathcal{D}(A^*) \subset \mathcal{D}(B^*)$, where $\mathcal{D}$ denotes the domain of an operator. In [ALMS] the key point consists in the consideration of the Frobenius Schur complement

$$D - C(A - \lambda_0)^{-1} B.$$

In numerous physical problems the assumptions of the [ALMS] paper are not fulfilled; for an example see [FMM1]. However, it might be possible to decompose the operator $\mathbb{L}$ in such a way that

(0.6)
$$\mathbb{L} = \mathbf{T} + \mathbf{M}$$

where $\mathbf{T}$ is an operator of the form (0.1) as e.g. in [FLMM] or a product $T_2 T_1$ of closed operators as e.g. in [FMM2], and $\mathbf{M}$ is a bounded operator. In the applications [FLMM], [FMM2] $\mathbf{T}$, i.e., T_1 and T_2 are differential operators in L_2-spaces and $\mathbf{M}$ is a bounded multiplication operator. In both applications the non-discrete spectrum away from zero of the product $T_2 T_1$ is empty, so that by the formulae (0.2), (0.3) also the non-discrete spectra of $\mathbf{T}$ and $T_1 T_2$ have the same property. In the Hilbert space case and a selfadjoint operator $\mathbf{T}$ the essential spectrum of $\mathbb{L}$ equals the essential spectrum of the bounded operator M_{11} defined by the decomposition

$$
\mathbf{M} = \begin{pmatrix} M_{11} & M_{12} \\ M_{21} & M_{22} \end{pmatrix} : \begin{matrix} N(\mathbf{T}) \\ \oplus \\ N(\mathbf{T})^\perp \end{matrix} \longrightarrow \begin{matrix} N(\mathbf{T}) \\ \oplus \\ N(\mathbf{T})^\perp \end{matrix} ,
$$

see [LM], Lemma 3.1, and thus may become explicitly calculable as in the paper [FMM2].

In Section 1 we consider the resolvent $(\mathbf{T} - \lambda)^{-1}$ for $\lambda \in \varrho(\mathbf{T}) \setminus \{0\}$ and derive necessary conditions on the product operators $T_2 T_1$, $T_1 T_2$, their adjoints and the operators $T_2^* T_1^*$, $T_1^* T_2^*$ for $\varrho(\mathbf{T}) \neq \emptyset$. In particular it is shown that λ^2 belongs to the resolvent sets of the operators $T_1 T_2$, $T_2 T_1$, $T_2^* T_1^*$ and $T_1^* T_2^*$. Further we obtain a representation of $(\mathbf{T} - \lambda)^{-1}$ in terms of the resolvents of the product operators $T_1 T_2$ and $T_2 T_1$.

In Section 2 we consider $\lambda^2 \in \varrho(T_2 T_1) \setminus \{0\}$ and show that the necessary conditions from Section 1 are also sufficient for the existence of the resolvent $(\mathbf{T} - \lambda)^{-1}$. The keypoint in the proof is the idea to consider a densely defined algebraic inverse of the linear operator $\mathbf{T} - \lambda$ and show that it has a bounded extension to the whole Banach space $B_1 \times B_2$ which is the topological inverse of $\mathbf{T} - \lambda$. The authors would like to point out that this idea of starting with an algebraic inverse and extending it to the topological inverse has successfully been used in [HMN] to characterize the essential spectrum of a selfadjoint operator generated by a 3 × 3-system of singular ordinary differential operators of mixed order.

In Section 3 we describe the connections between the resolvent sets, the spectra and the non-discrete spectra of the operators $\mathbf{T}$, $T_1 T_2$ and $T_2 T_1$.

In Section 4 we discuss the special case of symmetric operators $\mathbf{T}$ of type (0.1) presenting new proofs of the main results in [M] which better illustrate the underlying idea.

Section 5 contains an application to a 2 × 2 system of singular ordinary differential operators of mixed order.

1 From the Resolvent Set of T to the Resolvent Sets of the Products $T_2 T_1$ and $T_1 T_2$

Let B_1 and B_2 be Banach spaces. In the product Banach space $B := B_1 \times B_2$ we consider an operator which is defined by $\mathcal{D}(\mathbf{T}) = \mathcal{D}(T_2) \times \mathcal{D}(T_1)$ and

$$
(1.1) \qquad\qquad \mathbf{T} = \begin{pmatrix} 0 & T_1 \\ T_2 & 0 \end{pmatrix}.
$$

In the following it is always assumed that the operators T_1 and T_2 satisfy the following conditions:

a) T_1 is a densely defined closed linear operator from $\mathcal{D}(T_1) \subset B_2$ into B_1,

b) T_2 is a densely defined closed linear operator from $\mathcal{D}(T_2) \subset B_1$ into B_2.

It is clear that **T** is densely defined and closed.

Lemma 1.1 *Let the assumptions* a) *and* b) *be fulfilled and suppose that* $\lambda \in \varrho(\mathbf{T}) \setminus \{0\}$. *Then the operators* $T_2 T_1$ *and* $T_1 T_2$ *are closed and* λ^2 *belongs to their resolvent sets. Further the operators*

(1.2) $$(T_2 T_1 - \lambda^2)^{-1} T_2, \quad T_1 (T_2 T_1 - \lambda^2)^{-1} T_2$$

are bounded on their domain $\mathcal{D}(T_2)$ *and the operators*

(1.3) $$(T_1 T_2 - \lambda^2)^{-1} T_1, \quad T_2 (T_1 T_2 - \lambda^2)^{-1} T_1$$

are bounded on their domain $\mathcal{D}(T_1)$. *The resolvent of* **T** *at the point* λ *has the following representations:*

$$(\mathbf{T} - \lambda)^{-1} = \begin{pmatrix} -\frac{1}{\lambda} + \frac{1}{\lambda}\overline{T_1(T_2 T_1 - \lambda^2)^{-1} T_2} & T_1(T_2 T_1 - \lambda^2)^{-1} \\ (T_2 T_1 - \lambda^2)^{-1} T_2 & \lambda(T_2 T_1 - \lambda^2)^{-1} \end{pmatrix}$$

$$= \begin{pmatrix} \lambda(T_1 T_2 - \lambda^2)^{-1} & \overline{(T_1 T_2 - \lambda^2)^{-1} T_1} \\ T_2(T_1 T_2 - \lambda^2)^{-1} & -\frac{1}{\lambda} + \frac{1}{\lambda}\overline{T_2(T_1 T_2 - \lambda^2)^{-1} T_1} \end{pmatrix},$$

where the overlining by a bar denotes that we take the bounded extensions (closures) of the corresponding operators to the whole space B_1 *or* B_2, *respectively.*

Proof: i) Since $\lambda \in \varrho(\mathbf{T})$, the system of equations

(1.4)
$$\begin{aligned} T_1 f_2 - \lambda f_1 &= g_1, \\ T_2 f_1 - \lambda f_2 &= g_2, \end{aligned}$$

has a unique solution $(f_1, f_2) \in \mathcal{D}(T_2) \times \mathcal{D}(T_1)$ for any given vector $(g_1, g_2) \in B_1 \times B_2$. In particular, for $g_1 = 0$ we obtain

(1.5)
$$\begin{aligned} T_1 f_2 - \lambda f_1 &= 0, \\ T_2 f_1 - \lambda f_2 &= g_2. \end{aligned}$$

Since $\lambda \neq 0$ and $f_1 \in \mathcal{D}(T_2)$, the first equation in (1.5) shows that $f_2 \in \mathcal{D}(T_2T_1)$. Inserting $f_1 = \lambda^{-1}T_1f_2$ into the second equation in (1.5) yields that the equation

$$(1.6) \qquad \frac{1}{\lambda}(T_2T_1 - \lambda^2)f_2 = g_2$$

is uniquely solvable for any $g_2 \in B_2$ by some $f_2 \in \mathcal{D}(T_2T_1)$ whence

$$(1.7) \qquad f_2 = \lambda(T_2T_1 - \lambda^2)^{-1}g_2.$$

The algebraic inverse in (1.7) can be represented by the compressed resolvent of $\mathbf{T}$ at λ, i.e.,

$$(1.8) \qquad (T_2T_1 - \lambda^2)^{-1} = \frac{1}{\lambda}P_2(\mathbf{T} - \lambda)^{-1}J_2,$$

where J_l is the canonical embedding of B_l into B and P_l denotes the projection of B onto B_l ($l = 1, 2$). It follows that the algebraic inverse $(T_2T_1 - \lambda^2)^{-1}$ is a bounded operator whence T_2T_1 is closed and $\lambda^2 \in \varrho(T_2T_1)$.

ii) Analogously we conclude that T_1T_2 is a closed operator, $\lambda^2 \in \varrho(T_1T_2)$ and

$$(1.9) \qquad (T_1T_2 - \lambda^2)^{-1} = \frac{1}{\lambda}P_1(\mathbf{T} - \lambda)^{-1}J_1.$$

iii) We define the operator $\mathbf{R}^0(\lambda)$ in $B_1 \times B_2$ by $\mathcal{D}(\mathbf{R}^0(\lambda)) = \mathcal{D}(T_2) \times B_2$ and

$$(1.10) \qquad \begin{aligned} \mathbf{R}^0(\lambda) &:= \begin{pmatrix} R_{11}^0(\lambda) & R_{12}(\lambda) \\ R_{21}^0(\lambda) & R_{22}(\lambda) \end{pmatrix} \\ &:= \begin{pmatrix} -\frac{1}{\lambda} + \frac{1}{\lambda}T_1(T_2T_1 - \lambda^2)^{-1}T_2 & T_1(T_2T_1 - \lambda^2)^{-1} \\ (T_2T_1 - \lambda^2)^{-1}T_2 & \lambda(T_2T_1 - \lambda^2)^{-1} \end{pmatrix}. \end{aligned}$$

From i) we know that $\lambda^2 \in \varrho(T_2T_1)\backslash\{0\}$ and, therefore, the operator $\mathbf{R}^0(\lambda)$ is well-defined and $R_{12}(\lambda)$ and $R_{22}(\lambda)$ are bounded operators. Let $(x, y) \in \mathcal{D}(T_2) \times B_2$. We set

$$\begin{aligned} \begin{pmatrix} f \\ g \end{pmatrix} &:= \mathbf{R}^0(\lambda)\begin{pmatrix} x \\ y \end{pmatrix} \\ &= \begin{pmatrix} -\frac{1}{\lambda}x + \frac{1}{\lambda}T_1(T_2T_1 - \lambda^2)^{-1}T_2x + T_1(T_2T_1 - \lambda^2)^{-1}y \\ (T_2T_1 - \lambda^2)^{-1}T_2x + \lambda(T_2T_1 - \lambda^2)^{-1}y \end{pmatrix}. \end{aligned}$$

248 *Volker Hardt and Reinhard Mennicken*

Since $\mathcal{R}((T_2T_1 - \lambda^2)^{-1}) = \mathcal{D}(T_2T_1) \subset \mathcal{D}(T_1)$, where $\mathcal{R}$ denotes the range of an operator, and $x \in \mathcal{D}(T_2)$, it is obvious that $(f, g) \in \mathcal{D}(T_2) \times \mathcal{D}(T_1) = \mathcal{D}(\mathbf{T})$. Therefore we can apply $\mathbf{T} - \lambda$ to $\mathbf{R}^0(\lambda)(x, y)$ and obtain

$$(\mathbf{T} - \lambda)\mathbf{R}^0(\lambda)(x, y) = (x, y)$$

for arbitrary $(x, y) \in \mathcal{D}(T_2) \times B_2$. Since $\lambda \in \varrho(\mathbf{T})\backslash\{0\}$ it follows that

$$\mathbf{R}^0(\lambda) = (\mathbf{T} - \lambda)^{-1}|_{\mathcal{D}(T_2) \times B_2}.$$

Because T_2 is densely defined by assumption b) we conclude that the operator $\mathbf{R}^0(\lambda)$ has a unique bounded extension to the whole space $B_1 \times B_2$ which is given by $(\mathbf{T} - \lambda)^{-1}$. We denote this extension by $\mathbf{R}(\lambda)$, i.e.,

$$(1.11) \quad \mathbf{R}(\lambda) := \begin{pmatrix} R_{11}(\lambda) & R_{12}(\lambda) \\ R_{21}(\lambda) & R_{22}(\lambda) \end{pmatrix} := \overline{\mathbf{R}^0(\lambda)} = (\mathbf{T} - \lambda)^{-1}.$$

In particular, the operators

$$R_{11}^0(\lambda) = -\lambda^{-1} + \lambda^{-1}T_1(T_2T_1 - \lambda^2)^{-1}T_2,$$
$$R_{21}^0(\lambda) = (T_2T_1 - \lambda^2)^{-1}T_2$$

are bounded on their domain $\mathcal{D}(T_2)$ and have bounded extensions to the whole space B_1. Hence $T_1(T_2T_1 - \lambda^2)^{-1}T_2$ is also bounded on its domain $\mathcal{D}(T_2)$ and has a bounded extension to the whole space B_1.

iv) If we define the operator $\mathbf{S}^0(\lambda)$ in $B_1 \times B_2$ by $\mathcal{D}(\mathbf{S}^0(\lambda)) = B_1 \times \mathcal{D}(T_1)$ and

$$\mathbf{S}^0(\lambda) := \begin{pmatrix} S_{11}(\lambda) & S_{12}^0(\lambda) \\ S_{21}(\lambda) & S_{22}^0(\lambda) \end{pmatrix}$$

(1.12)

$$:= \begin{pmatrix} \lambda(T_1T_2 - \lambda^2)^{-1} & (T_1T_2 - \lambda^2)^{-1}T_1 \\ T_2(T_1T_2 - \lambda^2)^{-1} & -\frac{1}{\lambda} + \frac{1}{\lambda}T_2(T_1T_2 - \lambda^2)^{-1}T_1 \end{pmatrix}$$

it follows analogously that $\mathbf{S}^0(\lambda)$ is bounded on its domain and has a unique bounded extension to the whole space $B_1 \times B_2$ which is given by $(\mathbf{T} - \lambda)^{-1}$. We denote this extension by $\mathbf{S}(\lambda)$, i.e.,

$$\mathbf{S}(\lambda) := \overline{\mathbf{S}^0(\lambda)} = (\mathbf{T} - \lambda)^{-1}.$$

In particular the operators

$$S_{22}^0(\lambda) = -\lambda^{-1} + \lambda^{-1}T_2(T_1T_2 - \lambda^2)^{-1}T_1, \quad S_{12}^0(\lambda) = (T_1T_2 - \lambda^2)^{-1}T_1$$

and hence also the operator

$$T_2(T_1 T_2 - \lambda^2)^{-1} T_1$$

are bounded on their domain $\mathcal{D}(T_1)$ and thus have bounded extensions to the whole space B_2 which completes the Proof of Lemma 1.1. $\qquad \square$

Remark 1.2 Let the assumptions a) and b) be fulfilled. Then

(1.13) $\qquad \sigma_p(\mathbf{T}) \cup \sigma_p(-\mathbf{T}) = \{\lambda \in \mathbb{C} : \lambda^2 \in \sigma_p(T_2 T_1) \cup \sigma_p(T_1 T_2)\}.$

Proof: The relation (1.13) follows directly from $\mathcal{D}(\mathbf{T}^2) = \mathcal{D}(T_1 T_2) \times \mathcal{D}(T_2 T_1)$,

(1.14) $$\mathbf{T}^2 = \begin{pmatrix} T_1 T_2 & 0 \\ 0 & T_2 T_1 \end{pmatrix}$$

and

(1.15) $\qquad (\mathbf{T} + \lambda)(\mathbf{T} - \lambda) = \mathbf{T}^2 - \lambda^2 = (\mathbf{T} - \lambda)(\mathbf{T} + \lambda).$

$\qquad \square$

Corollary 1.3 *Let the assumptions* a) *and* b) *be fulfilled. Then*

$$\varrho(\mathbf{T}) \subset \{\lambda \in \mathbb{C} : \lambda^2 \in \varrho(T_1 T_2) \cap \varrho(T_2 T_1)\}.$$

Further, the point 0 *belongs to* $\varrho(\mathbf{T})$ *if and only if* $0 \in \varrho(T_1 T_2) \cap \varrho(T_2 T_1)$.

Proof: From Lemma 1.1 we know that

$$\varrho(\mathbf{T}) \backslash \{0\} \subset \{\lambda \in \mathbb{C} : \lambda^2 \in \varrho(T_1 T_2) \cap \varrho(T_2 T_1)\}.$$

Further the properties concerning the point 0 immediately follow from the formula (1.14) and the fact that $\mathbf{T}$ is invertible if and only if $\mathbf{T}^2$ is invertible. $\qquad \square$

Remark 1.4 It may happen that $0 \in \varrho(T_2 T_1)$, but $0 \in \sigma_p(T_1 T_2)$. For example, let B_2 be a proper closed complementable subspace of a Banach space B_1, T_1 the inclusion from B_2 into B_1 and T_2 a projection of B_1 onto B_2.

Since the operator $\mathbf{T}$ is densely defined, the adjoint operator $\mathbf{T}^*$ is well-defined in the dual space $B' = B_1' \times B_2'$ and given by the representation

$$\mathbf{T}^* = \begin{pmatrix} 0 & T_2^* \\ T_1^* & 0 \end{pmatrix}.$$

Lemma 1.5 *Let the assumptions* a) *and* b) *be fulfilled and suppose that* $\lambda \in \varrho(\mathbf{T}) \backslash \{0\}$. *Then the operators* $T_1^* T_2^*$ *and* $T_2^* T_1^*$ *are closed and* λ^2 *belongs to their resolvent set. The operators*

$$(T_1^* T_2^* - \lambda^2)^{-1} T_1^*, \quad T_2^* (T_1^* T_2^* - \lambda^2)^{-1} T_1^*$$

are bounded on their domain $\mathcal{D}(T_1^*)$ *and have bounded extensions to the whole space* B_1' *if* T_1^* *is densely defined. Under this additional assumption*

$$(\mathbf{T}^* - \lambda)^{-1}$$

(1.16)

$$= \begin{pmatrix} -\frac{1}{\lambda} + \frac{1}{\lambda} \overline{T_2^* (T_1^* T_2^* - \lambda^2)^{-1} T_1^*} & T_2^* (T_1^* T_2^* - \lambda^2)^{-1} \\ \overline{(T_1^* T_2^* - \lambda^2)^{-1} T_1^*} & \lambda (T_1^* T_2^* - \lambda^2)^{-1} \end{pmatrix}.$$

Analogously, the operators

$$(T_2^* T_1^* - \lambda^2)^{-1} T_2^*, \quad T_1^* (T_2^* T_1^* - \lambda^2)^{-1} T_2^*$$

are bounded on their domain $\mathcal{D}(T_2^*)$ *and have bounded extensions to the whole space* B_2' *if* T_2^* *is densely defined. Under this additional assumption*

$$(\mathbf{T}^* - \lambda)^{-1}$$

(1.17)

$$= \begin{pmatrix} \lambda (T_2^* T_1^* - \lambda^2)^{-1} & \overline{(T_2^* T_1^* - \lambda^2)^{-1} T_2^*} \\ T_1^* (T_2^* T_1^* - \lambda^2)^{-1} & -\frac{1}{\lambda} + \frac{1}{\lambda} \overline{T_1^* (T_2^* T_1^* - \lambda^2)^{-1} T_2^*}. \end{pmatrix}$$

Proof: The proof of this lemma is completely analogous to that of Lemma 1.1 taking into account the fact that $\varrho(\mathbf{T}^*) = \varrho(\mathbf{T})$. $\qquad \square$

Remark 1.6 If B_1 and B_2 are reflexive Banach spaces, then the operators T_1^* and T_2^* are densely defined under the assumptions a) and b), see e.g. [K], Chap. III, Theorem 5.29.

Corollary 1.7 *Let the assumptions* a) *and* b) *be fulfilled and suppose that* $\varrho(\mathbf{T}) \neq \emptyset$. *If* $T_2 T_1$ *is densely defined, then* $(T_2 T_1)^* = T_1^* T_2^*$. *An analogous statement holds for the product* $T_1 T_2$.

Proof: Let $\lambda \in \varrho(\mathbf{T}) \backslash \{0\}$ and suppose that $T_2 T_1$ is densely defined. By Lemma 1.1 λ^2 belongs to $\varrho(T_2 T_1)$ and thus also to $\varrho((T_2 T_1)^*)$. From Lemma 1.5 we know that $\lambda^2 \in \varrho(T_1^* T_2^*)$. Thus $\lambda^2 \in \varrho((T_2 T_1)^*) \cap \varrho(T_1^* T_2^*)$ which immediately implies that $(T_2 T_1)^* = T_1^* T_2^*$. $\qquad \square$

2 From the Resolvent Sets of the Products $T_2 T_1$ and $T_1 T_2$ to the Resolvent Set of T

In the previous section we have proved the inclusion

$$\varrho(\mathbf{T}) \backslash \{0\} \subset \{\lambda \in \mathbb{C} : \lambda^2 \in \varrho(T_2 T_1)\}$$

under the assumptions a) and b). This section deals with the inverse inclusion. To this end we make the following additional assumptions:

  c) $T_2 T_1$ is densely defined in B_2,

  d) the operator $T_1 (T_2 T_1 - \mu)^{-1} T_2$ is bounded on its domain $\mathcal{D}(T_2)$ for some $\mu \in \varrho(T_2 T_1)$,

  e) $(T_2 T_1)^* = T_1^* T_2^*$,

  f) T_1^* is densely defined in the dual space B_1'.

Notice that, under the assumptions a), b) and c), the conditions d) and e) hold if $\varrho(\mathbf{T}) \neq \emptyset$ as shown in Section 1.

Lemma 2.1 *If the assumptions* a), ..., f) *are fulfilled, then*

$$\{\lambda \in \mathbb{C} : \lambda^2 \in \varrho(T_2 T_1)\} \backslash \{0\} \subset \varrho(\mathbf{T}).$$

If, additionally, $T_1 T_2$ and T_2^* are densely defined, then also

$$\{\lambda \in \mathbb{C} : \lambda^2 \in \varrho(T_1 T_2)\} \backslash \{0\} \subset \varrho(\mathbf{T}).$$

Proof: Let $\lambda \in \mathbb{C} \backslash \{0\}$ such that $\lambda^2 \in \varrho(T_2 T_1)$. We consider the algebraic inverse of $(\mathbf{T} - \lambda)^{-1}$, i.e., we define the operator $\mathbf{R}^0(\lambda)$ as in (1.10). By assumption e)

$$\mathcal{R}(((T_2 T_1 - \lambda^2)^{-1})^*) = \mathcal{D}((T_2 T_1)^*) = \mathcal{D}(T_1^* T_2^*) \subset \mathcal{D}(T_2^*).$$

Hence, it follows from [ALMS], Proposition 3.1, that the operator $R_{21}^0(\lambda) = (T_2 T_1 - \lambda^2)^{-1} T_2$ is bounded on $\mathcal{D}(T_2)$. Since T_2 is densely defined, this operator has a unique bounded extension to the whole space B_1 which we denote by

$$R_{21}(\lambda) := \overline{R_{21}^0(\lambda)} = \overline{(T_2 T_1 - \lambda^2)^{-1} T_2}$$

as in Section 1. Further, choose $\mu \in \varrho(T_2 T_1)$ according to the assumption d). From the identity

$$T_1 (T_2 T_1 - \lambda^2)^{-1} T_2 = T_1 (T_2 T_1 - \mu)^{-1} T_2 + (\lambda^2 - \mu)$$

(2.1)

$$T_1 (T_2 T_1 - \mu)^{-1} R_{21}(\lambda)$$

we conclude that $T_1(T_2T_1 - \lambda^2)^{-1}T_2$ and hence $R_{11}^0(\lambda)$ are bounded operators on $\mathcal{D}(T_2)$. Again as in Section 1 we denote its bounded extension to the whole space B_1 by $R_{11}(\lambda)$, i.e.,

$$R_{11}(\lambda) := \overline{R_{11}^0(\lambda)} = -\lambda^{-1} + \lambda^{-1}\overline{T_1(T_2T_1 - \lambda^2)^{-1}T_2}.$$

It follows that $\mathbf{R}^0(\lambda)$ has a bounded extension to the whole space B which we denote by $\mathbf{R}(\lambda)$. By straightforward calculations as in Section 1 we obtain the relationship

$$(2.2) \qquad (\mathbf{T} - \lambda)\mathbf{R}(\lambda)|_{\mathcal{D}(T_2) \times B_2} = (\mathbf{T} - \lambda)\mathbf{R}^0(\lambda) = \mathrm{id}_B|_{\mathcal{D}(T_2) \times B_2}.$$

Since T_2 is densely defined in B_1, $\mathbf{T}$ is closed and $\mathbf{R}(\lambda)$ is bounded, it is easily seen that

$$(\mathbf{T} - \lambda)\mathbf{R}(\lambda) = \mathrm{id}_B,$$

i.e., $\mathbf{T} - \lambda$ is surjective and open. It remains to be shown that $\mathbf{T} - \lambda$ is injective. By the identity

$$N(\mathbf{T} - \lambda)^{\perp} = R(\mathbf{T}^* - \lambda)$$

where the orthogonal complement is taken with respect to the dual pair (B, B') (see [K], Chap. IV, Theorem 5.13), the injectivity of $\mathbf{T} - \lambda$ is equivalent to the surjectivity of $\mathbf{T}^* - \lambda$. We know that $\mathbf{T}^*$ is a closed operator in the Banach space $B' = B_1' \times B_2'$ with $\mathcal{D}(\mathbf{T}^*) = \mathcal{D}(T_1^*) \times \mathcal{D}(T_2^*)$ and has the following matrix representation

$$\mathbf{T}^* = \begin{pmatrix} 0 & T_2^* \\ T_1^* & 0 \end{pmatrix}.$$

Since $\lambda^2 \in \varrho(T_2T_1)$ and by assumption e), i.e., $(T_2T_1)^* = T_1^*T_2^*$, λ^2 belongs to the set $\varrho((T_2T_1)^*) = \varrho(T_1^*T_2^*)$. It follows that the algebraic inverse of $\mathbf{T}^* - \lambda$ is well-defined and given by

$$\tilde{\mathbf{R}}^0(\lambda) := \begin{pmatrix} \tilde{R}_{11}^0(\lambda) & \tilde{R}_{12}(\lambda) \\ \tilde{R}_{21}^0(\lambda) & \tilde{R}_{22}(\lambda) \end{pmatrix}$$

$$(2.3) \qquad := \begin{pmatrix} -\frac{1}{\lambda} + \frac{1}{\lambda}T_2^*((T_2T_1)^* - \lambda^2)^{-1}T_1^* & T_2^*((T_2T_1)^* - \lambda^2)^{-1} \\ ((T_2T_1)^* - \lambda^2)^{-1}T_1^* & \lambda((T_2T_1)^* - \lambda^2)^{-1} \end{pmatrix}$$

$$= \begin{pmatrix} -\frac{1}{\lambda} + \frac{1}{\lambda}T_2^*(T_1^*T_2^* - \lambda^2)^{-1}T_1^* & T_2^*(T_1^*T_2^* - \lambda^2)^{-1} \\ (T_1^*T_2^* - \lambda^2)^{-1}T_1^* & \lambda(T_1^*T_2^* - \lambda^2)^{-1} \end{pmatrix}.$$

In the next step we show that

$$(2.4) \qquad \mathbf{R}(\lambda)^*|_{\mathcal{D}(T_1^*) \times B_2'} = \tilde{\mathbf{R}}^0(\lambda).$$

With $\mathbf{R}(\lambda)$ also $\mathbf{R}(\lambda)^*$ is a bounded operator and has the following matrix representation

$$(2.5) \qquad \mathbf{R}(\lambda)^* = \begin{pmatrix} R_{11}(\lambda)^* & R_{21}(\lambda)^* \\ R_{12}(\lambda)^* & R_{22}(\lambda)^* \end{pmatrix}.$$

From the equations (1.10), (1.11), (2.3) and (2.5) it follows by straightforward calculations using well-known rules for adjoint operators that

$$R_{22}(\lambda)^* = (\lambda(T_2T_1 - \lambda^2)^{-1})^* = \lambda((T_2T_1)^* - \lambda^2)^{-1} = \tilde{R}_{22}(\lambda)$$

and, since $(T_2T_1 - \lambda^2)^{-1}$ is bounded,

$$R_{21}(\lambda)^* = R_{21}^0(\lambda)^* = ((T_2T_1 - \lambda^2)^{-1}T_2)^* = T_2^*((T_2T_1)^* - \lambda^2)^{-1}$$
$$= \tilde{R}_{12}(\lambda).$$

Further, for $x \in B_2$ and $x' \in \mathcal{D}(T_1^*) \subset B_1'$ we obtain that

$$\langle x, R_{12}(\lambda)^*x' \rangle = \langle R_{12}(\lambda)x, x' \rangle = \langle T_1(T_2T_1 - \lambda^2)^{-1}x, x' \rangle$$
$$= \langle (T_2T_1 - \lambda^2)^{-1}x, T_1^*x' \rangle = \langle x, ((T_2T_1)^* - \lambda^2)^{-1}T_1^*x' \rangle$$
$$= \langle x, \tilde{R}_{21}^0(\lambda)x' \rangle$$

and, hence, $R_{12}(\lambda)^*|_{\mathcal{D}(T_1^*)} = \tilde{R}_{21}^0(\lambda)$. By the same argumentation we obtain for $x \in \mathcal{D}(T_2)$ and $x' \in \mathcal{D}(T_1^*)$ that

$$\langle x, R_{11}(\lambda)^*x' \rangle = \langle R_{11}(\lambda)x, x' \rangle$$
$$= \langle (-\lambda^{-1} + \lambda^{-1}T_1(T_2T_1 - \lambda^2)^{-1}T_2)x, x' \rangle$$
$$= \langle x, (-\lambda^{-1} + \lambda^{-1}T_2^*((T_2T_1)^* - \lambda^2)^{-1}T_1^*)x' \rangle$$
$$= \langle x, \tilde{R}_{11}^0(\lambda)x' \rangle$$

since $R(((T_2T_1)^* - \lambda)^{-1}T_1^*) \subset \mathcal{D}(T_2^*)$. Thus, $R_{11}(\lambda)^*|_{\mathcal{D}(T_1^*)} = \tilde{R}_{11}^0(\lambda)$ and this finishes the proof of the relationship (2.4). Since $\mathbf{R}(\lambda)^*$ is bounded and T_1^* is densely defined in B_1' by assumption f), we conclude analogously as for $\mathbf{T} - \lambda$ that $\mathbf{T}^* - \lambda$ is surjective which completes the proof of the first assertion of Lemma 2.1.

For the proof of the second statement we show that the assumptions c)–f) are fulfilled if we interchange the roles of T_1 and T_2. The operators T_1T_2 and T_2^* are densely defined by assumption. Since $\varrho(\mathbf{T}) \neq \emptyset$ by the first statement it follows from Lemma 1.1 that $\varrho(T_1T_2) \neq \emptyset$, $(T_1T_2)^* = T_2^*T_1^*$ and the operator $T_2(T_1T_2 - \lambda^2)^{-1}T_1$ is bounded on its domain $\mathcal{D}(T_1)$. $\square$

Remark 2.2 In the Proof of Lemma 2.1 we have seen that the assumption e) yields the boundedness of $R_{21}^0(\lambda) = (T_2T_1 - \lambda^2)^{-1}T_2$ for all $\lambda^2 \in \varrho(T_2T_1)$ on its domain $\mathcal{D}(T_2)$ and that the assumption d) is fulfilled for all $\lambda \in \varrho(T_2T_1)$.

Corollary 2.3 *Under the assumptions* a)$, \ldots,$f)

$$\varrho(\mathbf{T}) \backslash \{0\} = \{\lambda \in \mathbb{C} : \lambda^2 \in \varrho(T_2T_1)\} \backslash \{0\}.$$

If, additionally, $T_1 T_2$ and T_2^* are densely defined, then also

$$\varrho(\mathbf{T}) \backslash \{0\} = \{\lambda \in \mathbb{C} : \lambda^2 \in \varrho(T_1 T_2)\} \backslash \{0\}$$

and thus

$$\varrho(T_2 T_1) \backslash \{0\} = \varrho(T_1 T_2) \backslash \{0\}.$$

Further, for all λ from this set, the relations

(2.6)
$$-\frac{1}{\lambda} + \frac{1}{\lambda} \overline{T_1 (T_2 T_1 - \lambda)^{-1} T_2} = (T_1 T_2 - \lambda)^{-1}$$

and

(2.7)
$$-\frac{1}{\lambda} + \frac{1}{\lambda} \overline{T_2 (T_1 T_2 - \lambda)^{-1} T_1} = (T_2 T_1 - \lambda)^{-1}$$

hold.

Proof: The relations (2.6) and (2.7) immediately follow from the representations of $(\mathbf{T} - \lambda)^{-1}$ in Lemma 1.1. $\qquad\square$

Our next aim is to give a sufficient condition for the condition d).

Let S be an operator in a Banach space X. We say that the resolvent of the operator S has a ray of minimal growth if there exist some $\Theta \in [0, 2\pi)$ and some $t_0 \geq 0$ such that

(2.8)
$$\gamma_\Theta := \{\lambda \in \mathbb{C} : \lambda = t e^{i\Theta} \ t \geq t_0\} \subset \varrho(S)$$

and there is a constant M such that

(2.9)
$$\|(S - \lambda)^{-1}\| \leq \frac{M}{1 + |\lambda|}$$

holds for all $\lambda \in \gamma_\Theta$.

The operator S is said to be of positive type if the estimate (2.9) holds for all $\lambda \in \mathbb{R}^-$. The operator $e^{i(\pi - \Theta)} (S - \mu_0)$ is of positive type if γ_Θ is a ray of minimal growth of the resolvent of S. For operators of positive type fractional powers are defined (see e.g. [KZPS], Chap. 4). Therefore fractional powers can also be defined for operators whose resolvents have rays of minimal growth.

Lemma 2.4 *Let the assumptions* a), b) *and* c) *be fulfilled. Suppose that the resolvent of the product operator $T_2 T_1$ has a ray γ_Θ of minimal growth. Let $0 < \tau < 1$ and assume that the inclusions*

(2.10)
$$\mathcal{D}(T_2^*) \supset \mathcal{D}(((T_2 T_1 - \mu_0)^*)^\tau) = \mathcal{D}(((T_2 T_1 - \mu_0)^\tau)^*)$$

and

(2.11) $$\mathcal{D}(T_1) \supset \mathcal{D}((T_2T_1 - \mu_0)^{1-\tau})$$

hold, where $\mu_0 = t_0 e^{i\Theta}$ *is chosen from the ray* γ_Θ. *Then the condition* d) *is satisfied.*

Proof: Notice first that $\mu_0 \in \varrho(T_2T_1)$. From well-known rules concerning fractional powers it follows that

(2.12) $$T_1(T_2T_1 - \mu_0)^{-1}T_2 = [T_1(T_2T_1 - \mu_0)^{-(1-\tau)}][(T_2T_1 - \mu_0)^{-\tau}T_2].$$

The inclusion (2.11) yields that

$$T_1(T_2T_1 - \mu_0)^{-(1-\tau)}$$

is a bounded operator on B_2. By [ALMS], Proposition 3.1, we conclude from (2.10) that

$$(T_2T_1 - \mu_0)^{-\tau}T_2$$

is a bounded operator on $\mathcal{D}(T_2)$. Hence the operator $T_1(T_2T_1 - \mu_0)^{-1}T_2$ is bounded on $\mathcal{D}(T_2)$ and its bounded extension to the whole space B_1 is given by

(2.13) $$\overline{T_1(T_2T_1 - \mu_0)^{-1}T_2} = T_1(T_2T_1 - \mu_0)^{-(1-\tau)}\overline{(T_2T_1 - \mu_0)^{-\tau}T_2},$$

i.e., the assumption d) is fulfilled. □

3 Main Results

A point $\lambda \in \sigma(\mathbf{T})$ is called an eigenvalue of $\mathbf{T}$ if there exists a vector $f \neq 0$ in $\mathcal{D}(\mathbf{T})$ such that $(\mathbf{T} - \lambda)f = 0$. Such a vector is called an eigenvector of $\mathbf{T}$ corresponding to λ. The set of all eigenvalues of $\mathbf{T}$ is called the point spectrum of $\mathbf{T}$ and is denoted by $\sigma_p(\mathbf{T})$. If $\lambda \in \sigma_p(\mathbf{T})$, then the subspace

$$\mathcal{L}_\lambda(\mathbf{T}) := \{x \in B : \exists n \in \mathbb{N}_0 \ (\mathbf{T} - \lambda)^n x = 0\}$$

is called the principal subspace of $\mathbf{T}$ corresponding to λ and dim $\mathcal{L}_\lambda(\mathbf{T})$ is called the algebraic multiplicity of λ. We denote the set of all eigenvalues of $\mathbf{T}$ which are isolated in $\sigma(\mathbf{T})$ and have finite algebraic multiplicity by $\sigma_d(\mathbf{T})$ and call this set the discrete spectrum of $\mathbf{T}$. Further we set

$$\sigma_{nd}(\mathbf{T}) = \sigma(\mathbf{T}) \backslash \sigma_d(\mathbf{T}).$$

Finally, an $L(B)$-valued function G is called finitely meromorphic at λ_0 if G has a pole at λ_0 and the coefficients of the principal part of its Laurent expansion at λ_0

are operators of finite rank, i.e., in some punctured neighbourhood of λ_0 we have an expansion

$$G(\lambda) = \sum_{\nu=-q}^{\infty} (\lambda - \lambda_0)^\nu G_\nu,$$

which converges in the operator norm on $L(B)$, such that $G_{-1}, \ldots, G_{-q}$ are finite rank operators. It is well-known that an eigenvalue λ_0 of $\mathbf{T}$ belongs to the set $\sigma_d(\mathbf{T})$ if and only if $(\mathbf{T} - \lambda)^{-1}$ is finitely meromorphic at λ_0 (see [GGK], Chap. XV.2).

Theorem 3.1 *Let the assumptions a)–f) be fulfilled. Then the following relations hold:*

   i) $\sigma(\mathbf{T})\setminus\{0\} = \{\lambda \in \mathbb{C} : \lambda^2 \in \sigma(T_2 T_1)\}\setminus\{0\}$.

   ii) $\sigma_{nd}(\mathbf{T})\setminus\{0\} = \{\lambda \in \mathbb{C} : \lambda^2 \in \sigma_{nd}(T_2 T_1)\}\setminus\{0\}$.

If additionally $T_1 T_2$ and T_2^* are densely defined, then we also have the following statements:

   iii) $\sigma(\mathbf{T})\setminus\{0\} = \{\lambda \in \mathbb{C} : \lambda^2 \in \sigma(T_1 T_2)\}\setminus\{0\}$.

   iv) $\sigma_{nd}(\mathbf{T})\setminus\{0\} = \{\lambda \in \mathbb{C} : \lambda^2 \in \sigma_{nd}(T_1 T_2)\}\setminus\{0\}$.

Proof: The relations i) and iii) are direct consequences of Corollary 2.3. Let $\mu \in \sigma_d(\mathbf{T})\setminus\{0\}$. Then $(\mathbf{T}-\lambda)^{-1}$ is finitely memomorphic in μ. From the representation of $\mathbf{R}(\lambda) = (\mathbf{T} - \lambda)^{-1}$ (see (1.10) and (1.11)) we know that

$$(T_2 T_1 - \lambda^2)^{-1} = \lambda^{-1} P_2 (\mathbf{T} - \lambda)^{-1} J_2.$$

Therefore $(T_2 T_1 - \lambda)^{-1}$ is also finitely meromorphic at μ^2, i.e., $\mu^2 \in \sigma_d(T_2 T_1)$. Conversely, let $\mu^2 \in \sigma_d(T_2 T_1)\setminus\{0\}$. Then $(T_2 T_1 - \lambda^2)^{-1}$ and hence $R_{22}(\lambda) = \lambda(T_2 T_1 - \lambda^2)^{-1}$ is finitely meromorphic in μ. For $\lambda_0 \in \varrho(\mathbf{T})\setminus\{0\}$

$$
\begin{aligned}
R_{12}(\lambda) &= T_1(T_2 T_1 - \lambda^2)^{-1} = T_1(T_2 T_1 - \lambda_0^2)^{-1} \\
&\quad + (\lambda^2 - \lambda_0^2) T_1 (T_2 T_1 - \lambda_0^2)^{-1}(T_2 T_1 - \lambda^2)^{-1} \\
&= R_{12}(\lambda_0) + (\lambda^2 - \lambda_0^2) R_{12}(\lambda_0)(T_2 T_1 - \lambda^2)^{-1}.
\end{aligned}
$$

Since $(T_2 T_1 - \lambda^2)^{-1}$ is finitely meromorphic at μ, also the operator function $R_{12}(\lambda)$ is finitely meromorphic there. Further, for $\lambda_0 \in \varrho(\mathbf{T})\setminus\{0\}$ we calculate

$$
\begin{aligned}
R_{21}^0(\lambda) &= (T_2 T_1 - \lambda^2)^{-1} T_2 = (T_2 T_1 - \lambda_0^2)^{-1} T_2 \\
&\quad + (\lambda^2 - \lambda_0^2)(T_2 T_1 - \lambda^2)^{-1}(T_2 T_1 - \lambda_0^2)^{-1} T_2 \\
&= R_{21}^0(\lambda_0) + (\lambda^2 - \lambda_0^2)(T_2 T_1 - \lambda^2)^{-1} R_{21}^0(\lambda_0).
\end{aligned}
$$

Since $R_{21}^0(\lambda_0)$ is bounded on $\mathcal{D}(T_2)$, it follows that

$$(3.1) \qquad R_{21}(\lambda) = R_{21}(\lambda_0) + (\lambda^2 - \lambda_0^2)(T_2 T_1 - \lambda^2)^{-1} R_{21}(\lambda_0)$$

and we conclude that $R_{21}(\lambda)$ is finitely meromophic at μ. Analogously we obtain

$$
\begin{aligned}
R_{11}^0(\lambda) &= -\lambda^{-1} + \lambda^{-1} T_1 (T_2 T_1 - \lambda_0^2)^{-1} T_2 \\
&\quad + \lambda^{-1}(\lambda^2 - \lambda_0^2) T_1 (T_2 T_1 - \lambda^2)^{-1} (T_2 T_1 - \lambda_0^2)^{-1} T_2 \\
&= -\lambda^{-1} + \lambda^{-1} T_1 (T_2 T_1 - \lambda_0^2)^{-1} T_2 \\
&\quad + \lambda^{-1}(\lambda^2 - \lambda_0^2) T_1 (T_2 T_1 - \lambda_0^2)^{-1} (T_2 T_1 - \lambda_0^2)^{-1} T_2 \\
&\quad + \lambda^{-1}(\lambda^2 - \lambda_0^2)^2 T_1 (T_2 T_1 - \lambda_0^2)^{-1} (T_2 T_1 - \lambda^2)^{-1} (T_2 T_1 - \lambda_0^2)^{-1} T_2.
\end{aligned}
$$

Since $T_1 (T_2 T_1 - \lambda_0^2)^{-1} T_2$ and $(T_2 T_1 - \lambda_0^2)^{-1} T_2$ are bounded operators on their domain $\mathcal{D}(T_2)$, and $T_1 (T_2 T_1 - \lambda_0^2)^{-1}$ is a bounded operator on B_2 we obtain the representation

$$
\begin{aligned}
(3.2) \qquad R_{11}(\lambda) &= \lambda^{-1}[\lambda_0 R_{11}(\lambda_0) + (\lambda^2 - \lambda_0^2) R_{12}(\lambda_0) R_{21}(\lambda_0) \\
&\quad + (\lambda^2 - \lambda_0^2)^2 R_{12}(\lambda_0)(T_2 T_1 - \lambda^2)^{-1} R_{21}(\lambda_0)]
\end{aligned}
$$

which shows that $R_{11}(\lambda)$ is also finitely meromorphic at μ. By Lemma 1.1 we conclude that $(\mathbf{T} - \lambda)^{-1}$ is finitely meromorphic at μ, i.e., $\mu \in \sigma_d(\mathbf{T})$. Thus the statement ii) is proved. The statement iv) can be proved in the same way. $\qquad \square$

4 The Hilbert Space Case

First we show that von Neumann's Theorem (see [K], Chap. V, Theorem 3.24) is an immediate consequence of Lemma 1.1.

Theorem 4.1 *Let* B_1, B_2 *be Hilbert spaces and* T *a densely defined closed linear operator from* B_2 *into* B_1. *Then* T^*T *and* TT^* *are selfadjoint operators.*

Proof: The operator $\mathbf{T}$ for $T_1 = T$ and $T_2 = T^*$ defined by formula (1.1) is selfadjoint and, hence, $\sigma(\mathbf{T}) \subset \mathbb{R}$. Lemma 1.1 yields that T^*T and TT^* are closed symmetric operators with defect indices $(0, 0)$ and thus selfadjoint. $\qquad \square$

The following results have been proved by M. Möller [M], c.f. also [T].

Theorem 4.2 *Let* B_1, B_2 *be Hilbert spaces and* T_1, T_2 *densely defined linear operators from* B_2 *into* B_1 *or from* B_1 *into* B_2, *respectively. Suppose that* $T_1 \subset T_2^*$ *and that* $T_2 T_1$ *is selfadjoint. Then the following relations hold:*

i) $\overline{T_2 T_1} = T_2 T_1$;

ii) *the operator* $\mathbf{T}$, *defined in* (0.1), *is essentially selfadjoint, i.e.,* $\overline{\mathbf{T}}$ *is selfadjoint;*

iii) $\sigma(\overline{\mathbf{T}}) \backslash \{0\} = \{\lambda \in \mathbb{C} : \lambda^2 \in \sigma(T_2 T_1)\} \backslash \{0\}$;

iv) $\sigma_{\text{ess}}(\overline{\mathbf{T}}) \backslash \{0\} = \{\lambda \in \mathbb{C} : \lambda^2 \in \sigma_{\text{ess}}(T_2 T_1)\} \backslash \{0\}$.

If, additionally, $T_1 T_2$ is densely defined, then also the following statements hold:

   v) $T_1 T_2$ is essentially selfadjoint and $\overline{T_1 T_2} = \overline{T_1}\,\overline{T_2}$;

   vi) $\sigma(T_2 T_1)\backslash\{0\} = \sigma(\overline{T_1}\,\overline{T_2})\backslash\{0\} = \sigma(\overline{T_1 T_2})\backslash\{0\}$;

   vii) $\sigma_{\text{ess}}(T_2 T_1)\backslash\{0\} = \sigma_{\text{ess}}(\overline{T_1}\,\overline{T_2})\backslash\{0\} = \sigma_{\text{ess}}(\overline{T_1 T_2})\backslash\{0\}$.

Proof: From the assumption $T_1 \subset T_2^*$ it follows that the operators T_1, T_2 are closable and $\overline{T_1} \subset T_2^*$, $\overline{T_2} \subset T_1^*$. Hence **T** is closable and $\overline{\mathbf{T}}$ is symmetric. Thus its square,

$$(4.1) \qquad\qquad \overline{\mathbf{T}}^2 = \begin{pmatrix} \overline{T_1}\,\overline{T_2} & 0 \\ 0 & \overline{T_2}\,\overline{T_1} \end{pmatrix},$$

is also symmetric which yields that both $\overline{T_1}\,\overline{T_2}$ and $\overline{T_2}\,\overline{T_1}$ are symmetric operators. We conclude that

$$T_2 T_1 \subset \overline{T_2}\,\overline{T_1} \subset (\overline{T_2}\,\overline{T_1})^* \subset (T_2 T_1)^*.$$

Since $T_2 T_1$ is selfadjoint, we may replace the above inclusions everywhere by equality which proves the assertion i).

   Obviously, the operators $\overline{T_1}, \overline{T_2}$ fulfill the assumptions a), b), c) and f). Further

$$(4.2) \qquad\qquad (T_2 T_1)^* = T_2 T_1 \subset T_1^* T_2^* \subset (T_2 T_1)^*$$

which, by statement i), yields the relation

$$(\overline{T_2}\,\overline{T_1})^* = \overline{T_1}^*\,\overline{T_2}^*$$

and hence the property e) for $\overline{T_1}, \overline{T_2}$. Finally we show that the operator

$$(4.3) \qquad\qquad \overline{T_1}(\overline{T_2}\,\overline{T_1} - \mu)^{-1}\overline{T_2} = \overline{T_1}(T_2 T_1 - \mu)^{-1}\overline{T_2}$$

is bounded on $\mathcal{D}(\overline{T_2})$ for $\mu \in \varrho(T_2 T_1)$ which means that the property d) holds with respect to $\overline{T_1}, \overline{T_2}$. Since $T_2 T_1$ is selfadjoint and a restriction of the operator $T_1^* T_1$, it follows that $T_2 T_1 \geq 0$ and thus its square root $(T_2 T_1)^{\frac{1}{2}}$ is well-defined. We consider the sesquilinear form

$$h[u, v] = (\overline{T_1}u, \overline{T_1}v) \qquad (u, v \in \mathcal{D}(\overline{T_1})).$$

It is obvious that h is a densely defined closed symmetric nonnegative form in the Hilbert space B_2 (see e.g. [K], Chap. VI). We conclude that for $u \in \mathcal{D}(T_2 T_1)$, $v \in \mathcal{D}(\overline{T_1})$ the relation

$$h[u, v] = (\overline{T_1}^*\overline{T_1}u, v) = (T_2 T_1 u, v)$$

holds which shows (see [K], Chap. VI, Theorem 2.1) that $T_2 T_1$ is the unique selfadjoint operator associated to the form h. By [K], Chap. VI, Theorem 2.23, the relation

$$\mathcal{D}(\overline{T_1}) = \mathcal{D}((T_2 T_1)^{\frac{1}{2}})$$

follows which implies that

$$\mathcal{D}(T_2^*) \supset \mathcal{D}((T_2 T_1)^{\frac{1}{2}}) = \mathcal{D}(((T_2 T_1)^*)^{\frac{1}{2}}).$$

Following the Proof of Lemma 2.4 with $\mu_0 = 0$, $\tau = \frac{1}{2}$ we obtain the boundedness of the operator (4.3) on $\mathcal{D}(\overline{T_2})$. (Notice that we need the ray of minimal growth of the resolvent of $T_2 T_1$ in Lemma 2.4 only for the definition of the fractional powers of $T_2 T_1 - \mu_0$ and its adjoint operator.)

We have shown that the operators $\overline{T_1}$, $\overline{T_2}$ fulfill the assumptions a)–f). From Lemma 2.1 we conclude by taking into account the selfadjointness of $T_2 T_1$ and the relation i) that the symmetric operator $\overline{T}$ has defect indices $(0, 0)$ and thus is selfadjoint which proves the assertion ii). The statements iii), iv) and the first equalities in vi), vii) immediately follow from Theorem 3.1. The relation $\overline{T_1 T_2} = \overline{T_1}\, \overline{T_2}$ remains to be proved: Since $\mathbf{T}$ is symmetric, $\mathbf{T}^2$ is symmetric and thus also

$$\mathbf{T}^2 = \begin{pmatrix} \overline{T_1 T_2} & 0 \\ 0 & \overline{T_2 T_1} \end{pmatrix} = \begin{pmatrix} \overline{T_1}\,\overline{T_2} & 0 \\ 0 & \overline{T_2}\,\overline{T_1} \end{pmatrix}.$$

Hence $\overline{T_1 T_2}$ is symmetric. Let $\lambda \in \varrho(\overline{\mathbf{T}}) \backslash \{0\}$; Lemma 1.1 yields that $\lambda^2 \in \varrho(T_2 T_1)$. It is easy to prove (see [M], Proposition 2.7) that the operator (4.3) for $\mu = \lambda^2$ maps $\mathcal{D}(\overline{T_1}\,\overline{T_2})$ into itself and the relation

$$(T_1 T_2 - \lambda^2) R_{11}^0(\lambda)|_{\mathcal{D}(T_1 T_2)} = I|_{\mathcal{D}(T_1 T_2)}$$

holds where $R_{11}^0(\lambda)$ denotes the operator defined in (1.10) with $\overline{T_1}$, $\overline{T_2}$ instead of T_1, T_2; observe that

$$R_{11}^0(\lambda)|_{\mathcal{D}(T_2)} = -\frac{1}{\lambda} + \frac{1}{\lambda} T_1 (T_2 T_1 - \lambda^2)^{-1} T_2.$$

As in (1.11) denote the bounded extension of $R_{11}^0(\lambda)$ by $R_{11}(\lambda)$. Since $\mathcal{D}(T_1 T_2)$ is dense in B_1, we obtain the relation

$$(\overline{T_1 T_2} - \lambda^2) R_{11}(\lambda) = I$$

which shows that $\overline{T_1 T_2} - \lambda^2$ is surjective. It follows that $\overline{T_1 T_2}$ has defect indices $(0, 0)$ and thus is selfadjoint which implies that $\overline{T_1 T_2} = \overline{T_1}\,\overline{T_2}$. $\qquad \square$

5 An Example

In this section we give an example demonstrating how our results can be used to calculate the essential spectrum for some nonselfadjoint system of singular differential operators of mixed order.

First let us state some notations. By $L^2((0, 1), \frac{1}{r})$ we denote the set of (equivalence classes) of measurable functions in $(0, 1)$ for which $\frac{1}{r}|f|^2 \in L^1(0, 1)$. $L^2((0, 1), \frac{1}{r})$ is a Hilbert space with respect to the scalar product

$$(f, g)_{\frac{1}{r}} := \int_0^1 f(r)\overline{g(r)}\frac{1}{r}dr \qquad (f, g \in L^2((0, 1), \frac{1}{r}))$$

and norm $|f|_{\frac{1}{r}} := \sqrt{(f, f)_{\frac{1}{r}}}$ $(f \in L^2((0, 1), \frac{1}{r}))$. For $m \in \mathbb{N}$ we denote the usual Sobolov space of order m by $H^m(0, 1)$, i.e.,

$$H^m(0, 1) := \{f \in L^2(0, 1) : f^{(j)} \in L^2(0, 1) \ j = 0, \dots, m\},$$

and define

$$H^m((0, 1), \frac{1}{r}) := \{f \in L^2((0, 1), \frac{1}{r}) : f^{(j)} \in L^2((0, 1), \frac{1}{r}) \ j = 0, \dots, m\}$$

where the derivatives are to be understood in the sense of distributions. The space $H^m((0, 1), \frac{1}{r})$ equipped with the scalar product

$$(f, g)_{H^m((0,1),\frac{1}{r})} := \sum_{j=0}^m (f^{(j)}, g^{(j)})_{\frac{1}{r}} \qquad (f, g \in H^m((0, 1), \frac{1}{r}))$$

is a Hilbert space. Set

$$H^1_{0,loc}((0, 1), \frac{1}{r}) := \{h \in L^2((0, 1), \frac{1}{r}) : \forall 0 < \varepsilon < 1 \quad h|_{(\varepsilon,1)} \in H^1(\varepsilon, 1)\}.$$

By ∂_r we denote the ordinary derivative with respect to the variable r.

Proposition 5.1 Let $f \in H^1((0, 1), \frac{1}{r})$. Then $\frac{1}{r}f$ and $r\partial_r \frac{1}{r}f$ are in $L^2((0, 1), \frac{1}{r})$ and the inequality $|\frac{1}{r}f|_{\frac{1}{r}} \leq |\partial_r f|_{\frac{1}{r}}$ holds.

For the proof we refer the reader to [FMM1], Proposition 2.1 and Corollary 2.2.

Let b_1 and b_2 be continuous not necessarily real-valued functions on $[0, 1]$. We assume that

(5.1) $b_1(0) \neq 0$

and that for some $0 < \sigma \leq 1$

$$(5.2) \qquad \max_{t \in [0, \sigma]} \frac{|\overline{b_1}(t) - b_2(t)|}{|b_1(t)|} < \frac{1}{2}.$$

We define an operator T_1 from $(L^2((0, 1), \frac{1}{r}))^2$ to $L^2((0, 1), \frac{1}{r})$ by

$$\mathcal{D}(T_1) := \{f \in (L^2((0, 1), \tfrac{1}{r}))^2 \ : \ f_1 \in H^1_{0, loc}((0, 1), \tfrac{1}{r}),$$
$$f_1(1) = 0, \partial_r f_1 - \tfrac{1}{r} b_1 f_2 \in L^2((0, 1), \tfrac{1}{r})\}$$

and

$$T_1 f := (\partial_r, \ -\tfrac{1}{r} b_1)(f_1, f_2)^t = \partial_r f_1 - \tfrac{1}{r} b_1 f_2,$$

where here and in the following, for an element $f \in (L^2((0, 1), \frac{1}{r}))^l$ we denote its j-th component by f_j. Further, we denote the canonical scalar product and norm in $(L^2((0, 1), \frac{1}{r}))^l$ by $(.\, , .)_{\frac{1}{r}}$ and $|.|_{\frac{1}{r}}$, respectively.

Proposition 5.2 *The operator T_1 is densely defined and closed.*

Proof: (Cf. [FMM2], Proposition 2.1). Since $(C_0^\infty(0, 1))^2 \subset \mathcal{D}(T_1)$, T_1 is densely defined in $(L^2((0, 1), \frac{1}{r}))^2$. Let $(f_n)_{n \in \mathbb{N}}$ be a sequence in $\mathcal{D}(T_1)$ which converges in $(L^2((0, 1), \frac{1}{r}))^2$ to some h and such that $(T_1 f_n)_{n \in \mathbb{N}}$ converges to some g in $L^2((0, 1), \frac{1}{r})$. Let $0 < \varepsilon < 1$. Then $f_{n,1}|_{(\varepsilon, 1)} \in H^1(\varepsilon, 1)$ and, therefore, $\partial_r f_{n,1}|_{(\varepsilon, 1)} \in L^2(\varepsilon, 1)$. Since $T_1 f_n$ is convergent it follows that $\partial_r f_{n,1}|_{(\varepsilon, 1)}$ converges in $L^2(\varepsilon, 1)$ and, therefore, $h_1 \in H^1_{0, loc}((0, 1), \frac{1}{r})$ and $f_{n,1}|_{(\varepsilon, 1)}$ converges in $H^1(\varepsilon, 1)$ to h_1. Since $H^1(\varepsilon, 1)$ is imbedded in $C([\varepsilon, 1])$ and $f_{n,1}(1) = 0$ for all $n \in \mathbb{N}$, it follows that $h_1(1) = 0$. Further, we conclude that

$$\partial_r h_1 - \tfrac{1}{r} b_1 h_2 = g \in L^2((0, 1), \tfrac{1}{r}).$$

Therefore $h \in \mathcal{D}(T_1)$ and $T_1 h = g$ which proves the closeness of T_1.   □

Since T_1 is densely defined and closed, the adjoint operator exists and is densely defined.

Proposition 5.3 *The adjoint operator T_1^* is given by $\mathcal{D}(T_1^*) = H^1((0, 1), \frac{1}{r})$ and*

$$T_1^* g = \begin{pmatrix} -r \partial_r \frac{1}{r} g \\ -\tfrac{1}{r} \overline{b_1} g \end{pmatrix} \qquad (g \in \mathcal{D}(T_1^*)).$$

Proof: (Cf. [FMM2], Proposition 2.3). It follows from Proposition 5.1 that

$$H^1((0, 1), \tfrac{1}{r}) = \{f \in L^2((0, 1), \tfrac{1}{r}) \ : \ \tfrac{1}{r} f \in L^2((0, 1), \tfrac{1}{r}),$$
$$(5.3)$$
$$r \partial_r \tfrac{1}{r} f \in L^2((0, 1), \tfrac{1}{r})\}.$$

By $\langle ., . \rangle_l$ we denote the canonical bilinearform on $(\mathcal{D}'(0, 1))^l \times (C_0^\infty(0, 1))^l$, where $\mathcal{D}'(0, 1)$ denotes the space of distributions on the interval $(0, 1)$. Let $g \in \mathcal{D}(T_1^*)$ and $f \in (C_0^\infty(0, 1))^2$. Then

$$\langle \tfrac{1}{r} T_1^* g, \bar{f} \rangle_2 = (T_1^* g, f)_{\frac{1}{r}} = (g, T_1 f)_{\frac{1}{r}} = (g, \partial_r f_1)_{\frac{1}{r}} - (g, \tfrac{1}{r} b_1 f_2)_{\frac{1}{r}}$$

$$= \int_0^1 \tfrac{1}{r} g \overline{\partial_r f_1} \, dr - \int_0^1 \tfrac{1}{r} g \tfrac{1}{r} \overline{b_1 f_2} \, dr$$

$$= \langle \tfrac{1}{r} g, \partial_r \overline{f_1} \rangle_1 + \langle -\tfrac{1}{r^2} \overline{b_1} g, \overline{f_2} \rangle_1$$

$$= -\langle \partial_r \tfrac{1}{r} g, \overline{f_1} \rangle_1 + \langle -\tfrac{1}{r^2} \overline{b_1} g, \overline{f_2} \rangle_1.$$

Hence, we have proved that

$$\langle \tfrac{1}{r} T_1^* g, \bar{f} \rangle_2 = \left\langle \tfrac{1}{r} \begin{pmatrix} -r \partial_r \tfrac{1}{r} g \\ -\tfrac{1}{r} b_1 g \end{pmatrix}, \bar{f} \right\rangle_2$$

in the sense of distributions. Since $f \in (C_0^\infty(0, 1))^2$ was arbitrary, it follows that

$$\begin{pmatrix} -r \partial_r \tfrac{1}{r} g \\ -\tfrac{1}{r} b_1 g \end{pmatrix} = T_1^* g$$

and thus belongs to $(L^2((0, 1), \tfrac{1}{r}))^2$. Since $b_1(0) \neq 0$ by assumption (5.1), we obtain from the second component that $\tfrac{1}{r} g \in L^2((0, 1), \tfrac{1}{r})$ and from the first one that $r \partial_r \tfrac{1}{r} g \in L^2((0, 1), \tfrac{1}{r})$. By the characterization (5.3) it follows that $g \in H^1((0, 1), \tfrac{1}{r})$.

Conversely, let $g \in L^2((0, 1), \tfrac{1}{r})$ be such that $\tfrac{1}{r} g$ and $r \partial_r \tfrac{1}{r} g$ are in $L^2((0, 1), \tfrac{1}{r})$. Then we infer for $f \in \mathcal{D}(T_1)$ that

$$(g, T_1 f)_{\frac{1}{r}} = \int_0^1 \tfrac{1}{r} g \, \overline{T_1 f} \, dr = \lim_{\varepsilon \searrow 0} \int_\varepsilon^1 \tfrac{1}{r} (g \partial_r \overline{f_1} - \tfrac{1}{r} g \overline{b_1 f_2}) \, dr$$

$$= \lim_{\varepsilon \searrow 0} \int_\varepsilon^1 \tfrac{1}{r} g \partial_r \overline{f_1} \, dr - \int_0^1 \tfrac{1}{r} (\tfrac{1}{r} g \overline{b_1 f_2}) \, dr$$

since $\tfrac{1}{r} g \in L^2((0, 1), \tfrac{1}{r})$. Integration by parts leads to

$$(g, T_1 f)_{\frac{1}{r}} = \lim_{\varepsilon \searrow 0} \left[\tfrac{1}{r} g \overline{f_1} \big|_\varepsilon^1 - \int_\varepsilon^1 \tfrac{1}{r} (r \partial_r \tfrac{1}{r} g) \overline{f_1} \, dr \right] - (\tfrac{1}{r} b_1 g, f_2)_{\frac{1}{r}}.$$

Since $g \in H^1((0, 1), \tfrac{1}{r})$, $f_1(1) = 0$, $r \partial_r \tfrac{1}{r} g \in L^2((0, 1), \tfrac{1}{r})$ we obtain that

$$(g, T_1 f)_{\frac{1}{r}} = \lim_{\varepsilon \searrow 0} (-\tfrac{1}{r} g \overline{f_1})(\varepsilon) + (T_1^* g, f)_{\frac{1}{r}}.$$

Since $\frac{1}{r}g \in L^2((0, 1), \frac{1}{r})$, it follows that $|\frac{1}{r}g\overline{f_1}| \in L^1((0, 1), \frac{1}{r})$ and

$$\int_0^\varepsilon |\frac{1}{r}g\overline{f_1}|\, dr \leq \varepsilon \int_0^\varepsilon \frac{1}{r}|\frac{1}{r}g\overline{f_1}|\, dr = \varepsilon o(1) \qquad (\varepsilon \to 0).$$

Hence we obtain

$$0 = \liminf_{\varepsilon \searrow 0} \frac{1}{\varepsilon}g(\varepsilon)\overline{f_1(\varepsilon)} = \lim_{\varepsilon \searrow 0}\frac{1}{\varepsilon}g(\varepsilon)\overline{f_1(\varepsilon)},$$

since we already know that the limit exists. $\qquad\qquad\qquad\qquad\qquad\square$

Now we define an operator T_2 from $L^2((0, 1), \frac{1}{r})$ into $(L^2((0, 1), \frac{1}{r}))^2$ by

$$\mathcal{D}(T_2) := H^1((0, 1), \frac{1}{r}) \quad \text{and} \quad T_2 g := \begin{pmatrix} -r\partial_r\frac{1}{r}g \\ -\frac{1}{r}b_2 g \end{pmatrix} \quad (g \in \mathcal{D}(T_2)).$$

Our aim is to calculate the essential spectrum of the operator

$$T_2 T_1 = \begin{pmatrix} -r\partial_r\frac{1}{r}\partial_r & r\partial_r\frac{b_1}{r^2} \\ -\frac{1}{r}b_2\partial_r & \frac{1}{r^2}b_2 b_1 \end{pmatrix}$$

which in general is not symmetric.

Proposition 5.4 *The operator T_2 is densely defined and closed.*

Proof: Since $C_0^\infty(0, 1)$ is included in $\mathcal{D}(T_2)$, T_2 is densely defined. From the inequality (5.2) we conclude that $b_2(0) \neq 0$. According to Proposition 5.3 T_2 is the adjoint of an operator $\hat{T}_2$ which is defined in an analogous way as T_1 with $\overline{b_2}$ instead of b_1. Hence, T_2 is closed. $\qquad\qquad\qquad\qquad\qquad\square$

Proposition 5.5 *It is the case that $\varrho(T_1 T_2) \neq \emptyset$ and that the operators T_1, T_2 fulfill the assumptions c)–f).*

Proof: We consider the operator $\mathbf{T}$ in $(L^2((0, 1), \frac{1}{r}))^3$ defined as in (1.1) with domain $\mathcal{D}(\mathbf{T}) = \mathcal{D}(T_2) \times \mathcal{D}(T_1)$. Further we introduce the operator $\tilde{T}_2$ by $\mathcal{D}(\tilde{T}_2) := \mathcal{D}(T_2) = \mathcal{D}(T_1^*) = H^1((0, 1), \frac{1}{r})$ and

$$\tilde{T}_2 := T_2 - T_1^* = \begin{pmatrix} 0 \\ -\frac{1}{r}(b_2 - \overline{b_1}) \end{pmatrix}.$$

Let $g \in H^1((0, 1), \frac{1}{r})$. By (5.2) we obtain

$$|\frac{1}{r}(b_2 - \overline{b_1})g|_{\frac{1}{r}} \leq |\chi_{[0,\sigma]}\frac{1}{r}(b_2 - \overline{b_1})g|_{\frac{1}{r}} + |(1 - \chi_{[0,\sigma]})\frac{1}{r}(b_2 - \overline{b_1})g|_{\frac{1}{r}}$$

$$\leq c_1 |T_1^* g|_{\frac{1}{r}} + c_2 |g|_{\frac{1}{r}},$$

where

$$c_1 := \max_{t\in[0,\sigma]} \frac{|\overline{b_1(t)} - b_2(t)|}{|b_1(t)|} < \frac{1}{2}, \quad c_2 := \frac{1}{\sigma} \max_{t\in[0,\sigma]} |\overline{b_1(t)} - b_2(t)|.$$

It follows that

$$\left| \begin{pmatrix} 0 & 0 \\ \tilde{T}_2 & 0 \end{pmatrix} \begin{pmatrix} g \\ f \end{pmatrix} \right|_{\frac{1}{r}} \leq c_1 |T_1^* g + \lambda f|_{\frac{1}{r}} + c_1 |\lambda| |f|_{\frac{1}{r}} + c_2 |g|_{\frac{1}{r}}$$

$$\leq c_1 \left| \begin{pmatrix} \lambda & T_1 \\ T_1^* & \lambda \end{pmatrix} \begin{pmatrix} g \\ f \end{pmatrix} \right|_{\frac{1}{r}} + (c_1 |\lambda| + c_2) \left| \begin{pmatrix} g \\ f \end{pmatrix} \right|_{\frac{1}{r}}$$

for all $(g, f)' \in \mathcal{D}(T_1^*) \times \mathcal{D}(T_1)$ and $\lambda \in \mathbb{C}$ which means that the operator $\begin{pmatrix} 0 & 0 \\ \tilde{T}_2 & 0 \end{pmatrix}$ is $\begin{pmatrix} \lambda & T_1 \\ T_1^* & \lambda \end{pmatrix}$-bounded with the constants $a = c_1 |\lambda| + c_2$ and $b = c_1$.

Choose $\lambda = i\beta$, $\beta > 0$, $2c_1 + \frac{1}{\beta} c_2 < 1$. Then $\begin{pmatrix} \lambda & T_1 \\ T_1^* & \lambda \end{pmatrix}$ has a bounded inverse,

$$\left| \begin{pmatrix} \lambda & T_1 \\ T_1^* & \lambda \end{pmatrix}^{-1} \right| \leq \frac{1}{\beta},$$

and hence

$$a \left| \begin{pmatrix} \lambda & T_1 \\ T_1^* & \lambda \end{pmatrix}^{-1} \right| + b \leq 2c_1 + \frac{1}{\beta} c_2 < 1.$$

This implies by [K], Chap. IV, Theorem 1.16 that for such λ

$$\mathbf{T} - \lambda = \begin{pmatrix} \lambda & T_1 \\ T_1^* & \lambda \end{pmatrix} + \begin{pmatrix} 0 & 0 \\ \tilde{T}_2 & 0 \end{pmatrix}$$

has a bounded inverse whence $\varrho(\mathbf{T}) \neq \emptyset$. Lemma 1.1 yields that $\varrho(T_1 T_2) \neq \emptyset$ and that (apart from the assumptions a) and b)) the conditions c)–f) are fulfilled.   □

Proposition 5.6 *The operator $T_1 T_2$ has a compact resolvent and hence $\sigma_{\text{ess}}(T_1 T_2) = \emptyset$.*

Proof: (Cf. [FMM2], Proposition 2.4). Obviously

$$\mathcal{D}(T_1 T_2) \subset \mathcal{D}(T_2) \subset H^1((0, 1), \tfrac{1}{r}),$$

where these three spaces are Banach spaces, $\mathcal{D}(T_1 T_2)$ and $\mathcal{D}(T_2)$ being equipped with the respective graph norms. Each of these three spaces is also continuously embedded into $L^2((0, 1), \frac{1}{r})$. Hence the embeddings

$$\mathcal{D}(T_1 T_2) \hookrightarrow \mathcal{D}(T_2) \hookrightarrow H^1((0, 1), \frac{1}{r})$$

are closed, and it follows from the Closed Graph Theorem that these embeddings are continuous. From [FMM1], Proposition 2.6 i) we know that the embedding $H^1((0, 1), \frac{1}{r}) \hookrightarrow L^2((0, 1), \frac{1}{r})$ is compact. Altogether, we conclude that $\mathcal{D}(T_1 T_2) \hookrightarrow L^2((0, 1), \frac{1}{r})$ is compact. Since $\varrho(T_1 T_2) \neq \emptyset$ by Proposition 5.5 and the resolvent $(T_1 T_2 - \lambda)^{-1}$ is bounded from $L^2((0, 1), \frac{1}{r})$ to $\mathcal{D}(T_1 T_2)$ by the Closed Graph Theorem, the proof is complete. □

Since $C_0^\infty(0, 1) \subset \mathcal{D}(T_1 T_2)$ and $(C_0^\infty(0, 1))^2 \subset \mathcal{D}(T_2^*)$ we conclude from Theorem 3.1 and Proposition 5.6:

Lemma 5.7 *Let the operators* T_1, T_2 *and* **T** *be defined as above. Then*

$$\sigma_{\text{ess}}(\mathbf{T}) = \{0\} \quad and \quad \sigma_{\text{ess}}(T_2 T_1) = \{0\}.$$

Proof: It only remains to be proved that $0 \in \sigma_{\text{ess}}(\mathbf{T})$ and $0 \in \sigma_{\text{ess}}(T_2 T_1)$. We observe that the assumption (5.1) (i.e., $b_1(0) \neq 0$) implies $b_1(t) \neq 0$ for $t \in [0, \tilde{\varepsilon}]$ with a suitable $\tilde{\varepsilon} \in (0, 1]$. From the definition of T_1 it follows that

$$\tilde{N} := \left\{ f \in (C_0^\infty(0, 1))^2 : \text{supp}(f) \subset (0, \tilde{\varepsilon}), \quad f_2 = \frac{1}{b_1} r \partial_r f_1 \right\}$$

is an infinite dimensional subspace of $N(T_1)$ and, hence, also of $N(T_2 T_1)$ which shows $0 \in \sigma_{\text{ess}}(T_2 T_1)$. $0 \in \sigma_{\text{ess}}(\mathbf{T})$ follows from $\{0\} \times \tilde{N} \subset N(\mathbf{T})$. □

References

[ALMS] F.V. Atkinson, H. Langer, R. Mennicken and A.A. Shkalikov, The essential spectrum of some matrix operators, *Math. Nachr.* **167** (1994), 5–20.

[FLMM] M. Faierman, A. Lifschitz, R. Mennicken and M. Möller, On the essential spectrum of a differentially rotating star, *ZAMM* **79** (1999), 739–755.

[FMM1] M. Faierman, R. Mennicken and M. Möller, The essential spectrum of a system of singular ordinary differential operators of mixed order. Part I: The general problem and an almost regular case, *Math. Nachr.* **208** (1999), 101–115.

[FMM2] M. Faierman, R. Mennicken and M. Möller, The essential spectrum of a system of singular ordinary differential operators of mixed order. Part II: The generalization of Kako's problem, *Math. Nachr.* **209** (1999), to appear.

[GGK] I.C. Gohberg, S. Goldberg and M.A. Kaashoek, *Classes of linear operators* vol. I, Birkhäuser-Verlag, Basel-Boston-Berlin 1990.

[HMN]   V. Hardt, R. Mennicken and S. Naboko, Systems of singular differential operators of mixed order and applications to 1-dimensional MHD problems, *Math. Nachr.* **205** (1999), 19–68.

[K]   T. Kato, *Perturbation theory for linear operators*, Springer-Verlag, New York 1966.

[KZPS]   M.A. Krasnoselskii, P.P. Zabreiko, E.I. Pustylnik and P.E. Sobolevskii, *Integral operators in spaces of summable functions*, Noordhoff Publishing, Leiden 1976.

[LM]   H. Langer and M. Möller, The essential spectrum of a non-elliptic boundary value problem, *Math. Nachr.* **178** (1996), 233–248.

[M]   M. Möller, On the essential spectrum of a class of operators in Hilbert spaces, *Math. Nachr.* **194** (1998), 185–196.

[T]   B. Thaller, *The Dirac equation*, Springer-Verlag, Berlin-Heidelberg-New York 1992.

[W]   J. Weidmann, *Linear operators in Hilbert spaces*, Springer-Verlag, New York-Heidelberg-Berlin 1980.

Volker Hardt and Rienhard Mennicken
University of Regensburg
Department of Mathematics
D-93040 Regensburg
Germany
e-mail: volker.hardt@mathematik.
uni-regensburg.de
e-mail: reinhard.mennicken@mathematik.
uni-regensburg.de

1991 Mathematics Subject Classification. Primary: 47A10, 47A25, 39B42.

Operator Theory:
Advances and Applications, Vol. 124
© 2001 Birkhäuser Verlag Basel/Switzerland

Multithreshold Spectral Phase Transitions
for a Class of Jacobi Matrices

J. Janas and S. Naboko

Dedicated to Professor Israel Gohberg on the occasion of his seventieth birthday

Consider an arbitrary periodic Jacobi matrix J_{per} with periodic weights c_n. A class of unbounded Jacobi matrices $A + cB$ is investigated, where B is a diagonal matrix and A is a Jacobi matrix with zero main diagonal and modulated weights λ_n (e.g. $\lambda_n = c_n n^{\alpha}$, $\alpha > 0$). Depending on whether the coupling constant $(-c)$ belongs to the absolutely continuous spectrum of J_{per} or not, the spectrum of $A + cB$ is either pure absolutely continuous or discrete. This gives us a class of examples with multithreshold spectral phase transition phenomena.

1 Introduction

Let W be a unilateral weighted shift operator defined in $l^2 = l^2(\mathbb{N})$ by $W e_n = 2\lambda_n e_{n+1}$, where e_n is the canonical basis in l^2, $\lambda_n \in (0, +\infty)$. Consider a sequence $\{q_n\} \subset \mathbb{R}$ and denote by Q the diagonal operator in l^2 defined by $Q e_n = q_n e_n$. The paper is concerned with spectral properties of the operator J given formally as $Re W + Q$. In the basis e_n its matrix form can written as

$$
\begin{pmatrix}
q_1 & \lambda_1 & 0 & 0 & \cdot & \cdot & \cdot \\
\lambda_1 & q_2 & \lambda_2 & 0 & \cdot & \cdot & \cdot \\
0 & \lambda_2 & q_3 & \lambda_3 & \cdot & \cdot & \cdot \\
\cdot & 0 & \lambda_3 & q_4 & \lambda_4 & \cdot & \cdot
\end{pmatrix}.
$$

Matrices of the above form are called Jacobi or tridiagonal [A], [S]. They induce essentially selfadjoint operators in l^2 (on the dense set of finite vectors) provided $\sum_k \lambda_k^{-1} = +\infty$, [Ber]. The main question considered in the papers [CK], [Ki], [JN], and [J N_1] was: when does the spectrum of J have a nontrivial absolutely continuous component. The present paper concerns the problem of determining conditions which guarantee <u>pure absolute</u> continuity of $\sigma(J)$ or its <u>discrete</u> character. Some sufficient conditions implying pure absolute continuity are formulated for general weights $\{\lambda_k\}$ and the diagonal $\{q_k\}$ and essentially use the notion of $\mathcal{D}^{k,r}$ classes of 2×2 matrices introduced by Stolz in his studies of discrete Schrödinger operators [St], [St1]. Besides the Stolz "smooth" method (§3) we develop here a sort of discrete WKB approach for finding the asymptotics of generalized eigenvectors [CC]. It allows to study the spectral structure of J. In turn the analysis of the discrete case requires a new approach also of WKB type. Applications to special classes of weights produce interesting relations with the spectrum of periodic Jacobi matrices. Conditions found for general weights and the diagonal are applied to a class of J's studied earlier in the framework of the representations of Lie groups by Masson-Repka [MR] and Edward [E] in the case of a discrete spectrum.

We explain their results from our point of view. In particular, for these weights a drastic spectral picture takes place: two threshold spectral phase transition under changing of coupling constant. Surely this phenomenon can be easily illustrated in the classical context. Consider for real δ the operator $L_\delta = (-\frac{d^2}{dx^2} + x) + \delta x$ acting in $L^2(0, \infty)$ with boundary condition $y(0) = 0$. Then for $\delta \leq -1$ the absolutely continuous spectrum of L_δ is equal to $\mathbb{R}$, and for $\delta > -1$ is discrete [RS], [RS1]. Therefore one threshold case can be constructed easily as the linear combination of an operator with discrete spectrum (the "diagonal" operator) and an operator with pure absolutely continuous spectrum ("weights"). In §6, and §7 a similar behaviour with multithreshold spectral phase transitions is studied for J with weights $\lambda_n = c_n n^\alpha$, where $\alpha > 0$ and $\{c_n\}$ is a periodic sequence $c_1, \ldots, c_N$ of nonzero numbers with period N, and the diagonal $q_n = \delta n^\alpha$. We should mention the numerous papers of mathematical physics [Ben], [CFKS], [DSS], [DS], [DS1], [Ko], [S1], [SZ], [St], [St1], which deal with transition phenomena, mostly in the stochastic case.

One aspect of the present paper and our previous ones is common however. Namely all papers are based on asymptotic analysis of products of transfer matrices and subordinacy theory due to Gilbert-Pearson [GP] and Khan-Pearson [KP]. For future use we now define the transfer matrix of J. Consider the infinite system of equations

(1.1) $\lambda_{n-1} u_{n-1} + \lambda_n u_{n+1} + q_n u_n = \lambda u_n, \quad \lambda \in \mathbb{R}, n = 2, 3, \ldots$

Denote

$$\vec{u}_n = \begin{pmatrix} u_{n-1} \\ u_n \end{pmatrix}.$$

Then (1.1) can be written as

$$\vec{u}_{n+1} = B_n \vec{u}_n, \quad n = 2, 3, \ldots,$$

where

$$B_n = \begin{pmatrix} 0 & 1 \\ -\frac{\lambda_{n-1}}{\lambda_n}, & \frac{\lambda - q_n}{\lambda_n} \end{pmatrix}$$

is called the transfer matrix of J. Our study of pure absolute continuity of $\sigma(J)$ is mainly based on the Khan-Pearson theory of subordinacy. This requires a detailed asymptotic analysis of the products $B_n \cdot B_{n-1} \ldots B_2$, as $n \to \infty$. Recall that a solution $\{u_n\}$, $n \in \mathbb{N}$ of (1.1) is said to be subordinated if for any linearly independent solution $\{v_n\}$ of (1.1) we have $\lim_{n \to \infty} (\sum_{k=1}^n |u_k|^2)[\sum_{k=1}^n |v_k|^2]^{-1} = 0$. See [KP].

2 Preliminaries

In what follows we shall use the following notations,

$$(2.1) \qquad \lambda_{n-1}/\lambda_n = 1 + \varepsilon_n, \quad (\lambda - q_n)/\lambda_n = w_n.$$

Given a sequence $\{D_s\}$ of 2×2 matrices the products $\prod_{k=n_1}^n D_k$, $n_1 \leq n$, are always understood below as: $D_n \cdot D_{n-1} \ldots D_{n_1}$ (so called chronological product). For a sequence $A = \{A(n)\}$ of complex 2×2 matrices, let

$$(\Delta A)(n) = A(n+1) - A(n), \quad \Delta^s A = \Delta(\Delta^{s-1} A), \quad s = 2, 3, \ldots,$$

and $\Delta^0 A = A$. G. Stolz introduced in [St] the class $\mathcal{D}^{k,r}$ (for $k \in \mathbb{N}$ and $r \in \{0, \ldots, k-1\}$) by

$$A = \{A(n)\} \in \mathcal{D}^{k,r} \text{ if and only if } \|(\Delta^j A)(\cdot)\| \in l^{\frac{k}{j+r}}, j = 1, \ldots, k-r.$$

We have the following result [St, Th4].

Proposition 2.1 (G. Stolz). *Let* $Z(n) = diag(z_+(n), z_-(n))$, *where* $\overline{z_+(n)} = z_-(n)$, $|z_\pm(n)| \to 1$, *as* $n \to \infty$, *Im* $z_+(n) \geq \delta > 0$ *for large* n, *and* $Z \in \mathcal{D}^{k,0}$ *for some* $k \in \{2, 3, \ldots\}$. *Suppose that* $A \in \mathcal{D}^{k,r}$ *for some* $r \in \{0, \ldots, k-2\}$, $A(n)$ *are invertible for all* n, $\|A(n)\|$ *and* $\|A(n)^{-1}\|$ *are bounded in* n, *and*

$$(2.2) \quad a) \; \overline{A(n)} = \begin{pmatrix} 0 & 1 \\ 1 & 0 \end{pmatrix} A(n) \begin{pmatrix} 0 & 1 \\ 1 & 0 \end{pmatrix} \text{ or } b) \; \overline{A(n)} = A(n) \begin{pmatrix} 0 & 1 \\ 1 & 0 \end{pmatrix}.$$

Then there exist $V \in \mathcal{D}^{k,0}$ *and* $C \in \mathcal{D}^{k,r+1}$ *such that* $C(n) \to I$ *as* $n \to \infty$,

$$Z(n)A(n)^{-1}A(n-1) = C(n)V(n)C(n)^{-1}$$

for large n,

$$\overline{C(n)} = \begin{pmatrix} 0 & 1 \\ 1 & 0 \end{pmatrix} C(n) \begin{pmatrix} 0 & 1 \\ 1 & 0 \end{pmatrix},$$

$V(n) = diag(v_+(n), v_-(n))$, $\overline{v_+(n)} = v_-(n)$, $|v_\pm(n)| \to 1$, *and Im* $v_+(n) \geq \delta' > 0$ *for large* n. *Finally,* $|v_+(n)|^2 = \frac{d_{n-1}}{d_n}|z_+(n)|^2$, *where* $d_n = det A(n)$.

Let us recall the definition of $\mathcal{D}^k$ $(k \in \mathbb{N})$, the class of slowly changing sequences, also due to Stolz [St]. A sequence $g = g(n)$ belongs to $\mathcal{D}^k$ if and only if g is bounded and $\Delta^s g \in l^{k/s}$, $s = 1, \ldots, k$. Here Δ is the forward difference operator (as above). The following lemma will be used in the next section.

Lemma 2.2 *Let* $x(\cdot), y(\cdot) \in \mathcal{D}^k$. *If* $x(n) \in [a, b]$, $y(n) \in [c, d]$ *and* $f \in C^k([a, b] \times [c, d])$, *then* $f(x(n), y(n)) \in \mathcal{D}^k$.

Proof: (Sketch). The proof follows by writing $(\Delta^s f)(x(\cdot), y(\cdot))$ as a suitable sum of integrals, e.g.,

$$(\Delta^2 f)(x(n), y(n)) = (\Delta f)(x(n+1), y(n+1))$$
$$-\Delta f(x(n), y(n)) = I_1 + I_2 + I_3 + I_4,$$

where with $a_1(n) = x(n), b_1(n) = x(n+1), c_1(n) = y(n)$, and $d_1(n) = y(n+1)$,

$$I_1 = \int_{a_1(n+1)}^{b_1(n+1)} \int_{c_1(n+1)}^{d_1(n+1)} f_{xy}(s, r)dsdr,$$

I_3 is defined similarly,

$$I_2 = -\int_{a_1(n)}^{a_1(n)+(\Delta a_1)(n)} f_x(s, y(n+1))ds$$
$$-\int_{b_1(n)}^{b_1(n)+(\Delta b_1)(n)} f_x(s, y(n+1))ds,$$

and I_4 is defined similarly.   □

3 Absolute Continuity of the Spectrum, Dominating Weights

Proposition 2.1 allows to find sufficient conditions on the weights which imply absolute continuity of the spectral measure of J.

Theorem 3.1 *If* $\lim_k \varepsilon_k = 0$, $\overline{\lim}_k |w_k| < 2$ *and* $\{\varepsilon_k\} \in \mathcal{D}^n$, $\{w_k\} \in \mathcal{D}^m$ *(see* (2.1)), *then the spectral measure of* J *is absolutely continuous.*

Proof: Let $u = \{u_k\}$ be a non-zero solution of (1.1). The Khan-Pearson theory [KP] and the generalized Behncke-Stolz Lemma [JN] reduce the absolute continuity problem to the estimate

$$(3.1) \qquad \qquad \|\vec{u}_{k+1}\|^2 \le C/\lambda_k, \quad C = C(\lambda) > 0,$$

which should hold for k sufficiently large.

Since $\vec{u}_{k+1} = B_k \ldots B_2 \vec{u}_2$ we have to analyze the product $B_k \ldots B_2$. Observe that $\sigma(B_k) = (z_+(k), z_-(k))$, where $z_+(k) = \frac{1}{2}(w_k + i\sqrt{\Delta_k})$, $\Delta_k := 4(1 + \varepsilon_k) - w_k^2$ and $z_-(k) = \overline{z_+(k)}$, for k sufficiently large. Hence

$$B_k = A(k)Z(k)A(k)^{-1},$$

where $Z(k) = diag(z_+(k), z_-(k))$, and

$$A(k) = \begin{pmatrix} 1 & 1 \\ z_+(k) & z_-(k) \end{pmatrix}$$

satisfies (2.2) (the second condition b)). Suppose that $n \geq m$. The above formulae and Lemma 2.2 imply that

$$A(\cdot) \in \mathcal{D}^{n,0}, \, Z(\cdot) \in \mathcal{D}^{n,0}.$$

Indeed, $z_+(k) = f(w_k, \varepsilon_k)$, where $f(x, y) := \frac{1}{2}(x + i\sqrt{4(1 + y) - x^2})$ for $|x| < 2 - \varepsilon$, $|y| < \varepsilon$ and some $\varepsilon > 0$. Choose ε so small that $4(1 + y) - x^2 \geq \rho > 0$ for x and y as above. Then $f(\cdot, \cdot)$ is C^∞ in the above rectangle and we can apply Lemma 2.2. Next, since $Im \, z_+(k) \geq \frac{1}{2}\sqrt{\rho}$ for large k, all assumptions of Proposition 2.1 are satisfied with $r = 0$. Hence after applying Proposition 2.1 $(n - 1)$ times, we obtain sequences of matrices $\{A_{n-1}(k)\} \in \mathcal{D}^{n,n-1}$ and $\{W_{n-1}(k)\} \in \mathcal{D}^{n,0}$ such that $W_{n-1}(k) = diag(w_+(k), w_-(k))$, $\overline{w_+(k)} = w_-(k)$, $|w_+(k)| \to 1$, $Im \, w_+(k) \geq \rho' > 0$ for large k, and n_0 is so large that the spectrum of B_k is elliptic [Har] for $k \geq n_0$.

$$\prod_{k=n_0}^{N} B_k = A(N)A_1(N) \ldots A_{n-1}(N)W_{n-1}(N)$$

$$\cdot A_{n-1}(N)^{-1}A_{n-1}(N-1)W_{n-1}(N-1)A_{n-1}(N-1)^{-1}$$

(3.2)
$$A_{n-1}(N-2) \ldots W_{n-1}(n_0 + 1)A_{n-1}(n_0 + 1)^{-1}$$

$$A_{n-2}(n_0)^{-1} \ldots A_1(n_0)^{-1}Z(n_0)A(n_0)^{-1}.$$

Here $A_s(N)$ plays the role of $C(n)$ from Proposition 2.1, $s = 1, \ldots, n - 1$. Since $\{A_{n-1}(\cdot)\} \in \mathcal{D}^{n,n-1}$,

$$A_{n-1}(s)^{-1}A_{n-1}(s - 1) = I + R_{n-1}(s),$$

where $\|R_{n-1}(\cdot)\| \in l^1$ (recall that $\|A_{n-1}(\cdot)^{-1}\|$ is uniformly bounded). We also have (by Proposition 2.1)

$$\prod_{k=n_0}^{N} |w_+(k)|^2 = det A_{n-2}(N - 1)[det A_{n-2}(n_0)]^{-1} det A_{n-3}(N - 1)$$

(3.3)
$$[det A_{n-3}(n_0)]^{-1} \ldots det A(N - 1)[det A(n_0)]^{-1} \prod_{k=n_0}^{N} |z_+(k)|^2$$

for n_0 sufficiently large. Formulae (3.2) and (3.3) imply that for n_0 sufficiently large

$$\|\vec{u}_{N+1}\|^2 \leq \|B_N \ldots B_{n_0}\|^2 \|B_{n_0-1} \ldots B_2\vec{u}_2\|^2$$

$$\leq C\|A(N)A_1(N)\ldots A_{n-1}(N)\|^2 \prod_{k=n_0+2}^{N} (1 + \|R_{n-1}(k)\|)$$

$$\times \prod_{k=n_0}^{N} |w_+(k)|^2 \leq C_1(n)C_{n_0} \prod_{k=n_0}^{N} |z_+(k)|^2 = C_1(n)C_{n_0}\frac{\lambda_{n_0-1}}{\lambda_N}.$$

Here C_{n_0}, $C_1(n)$ are some positive constants which depend only on n_0, n and $\|\vec{u}_2\|$. This completes the proof of the desired estimate (3.1). $\qquad\square$

Remark 3.2 Notice that from the results of [JN, Theorem 3.1] it is clear that J has pure absolutely continuous spectrum provided that $\lambda_n = n^\alpha(1 + \Delta_n)$, $\alpha \in (\frac{1}{2}, 1)$, $\{\Delta_n\} \in \mathcal{D}^2$ and $q_n \equiv 0$.

Corollary 3.3 *Suppose that λ_k and q_k are given by the following formulae*

$$\lambda_k = [p_m(k)]^{1/s}, \qquad q_k = \delta[\tilde{p}_{m_1}(k)]^{1/s},$$

where $p_m(\cdot)$, $\tilde{p}_{m_1}(\cdot)$ are polynomials of degrees m, m_1, respectively, and with leading coefficients equal to 1 $s, t \in \mathbb{R}_+$, $|\delta| < 2$, and $\frac{m}{s} = \frac{m_1}{t} \leq 1$. Then J defined by the above weights has pure absolutely continuous spectrum.

Proof: Due to the definition of ε_k and w_k we have

$$1 + \varepsilon_k = [p_m(k-1)/p_m(k)]^{1/s} = \left[1 + \frac{c}{k} + O\left(\frac{1}{k^2}\right)\right]^{1/s}$$

$$= 1 + \frac{c}{ks} + O\left(\frac{1}{k^2}\right).$$

It follows that $\varepsilon_k \in \mathcal{D}^1$. Similarly

$$w_k = \frac{q_k - \lambda}{\lambda_k} = \delta\left[1 + \frac{c_1}{k} + O\left(\frac{1}{k^2}\right)\right]^{1/s}$$

$$\cdot\left[1 + \frac{c_2}{k} + O\left(\frac{1}{k^2}\right)\right]^{-1/t} - \lambda k^{-m/s}\left[1 + O\left(\frac{1}{k}\right)\right]$$

$$= \delta\left[1 + \left(\frac{c_1}{s} - \frac{c_2}{t}\right)\frac{1}{k} + O\left(\frac{1}{k^2}\right)\right] - \lambda k^{-m/s} + O(k^{-1-\frac{m}{s}}),$$

and so $w_k \in \mathcal{D}^1$. $\qquad\square$

4 Discrete Spectrum, Dominating Diagonal

It turns out that situation changes drastically in the opposite case, i.e., when $\lim_n |q_n|/\lambda_n = \delta > 2$. We have

Theorem 4.1 *If* $\lim_n q_n = \infty$ *and* $\lim_n \inf q_n^2(\lambda_n^2 + \lambda_{n-1}^2)^{-1} = \frac{\delta^2}{2} > 2$, *then* J *has discrete spectrum.*

Proof: Set $B = \operatorname{Re} W$ and $A = diag(q_n)$.
Case 1 First assume that $q_n^2 \geq \frac{\delta^2}{2}(\lambda_n^2 + \lambda_{n-1}^2)$ for $n = 1, 2, \ldots$. Then for $u \in D(A)$ we have

$$
\begin{aligned}
\|Bu\|^2 &= \lambda_1^2|u_2|^2 + \sum_{n=2}^{\infty}|\lambda_{n-1}u_{n-1} + \lambda_n u_{n+1}|^2 \\
&\leq \lambda_1^2|u_2|^2 + 2\sum_{n=2}^{\infty}(\lambda_{n-1}^2|u_{n-1}|^2 + \lambda_n^2|u_{n+1}|^2) \\
&\leq 2\sum_{n=1}^{\infty}(\lambda_n^2 + \lambda_{n-1}^2)|u_n|^2 \leq 2\frac{2}{\delta^2}\sum_{n=1}^{\infty}q_n^2|u_n|^2 = \frac{4}{\delta^2}\|Au\|^2.
\end{aligned}
$$

Therefore ($\delta^2 > 4$) B is subordinated to A in the sense of Kato (see [K]) and so the resolvent $R(\cdot, J) = R(\cdot, A + B)$ is compact.

Case 2 If $q_n^2(\lambda_n^2 + \lambda_{n-1}^2)^{-1} > \delta/2$ for $n \geq n_o$, then we replace $\lambda_1, \ldots, \lambda_{n_o}$ in the definition of B by zeros. Denote this new B by $\hat{B}$. By the above Case 1 applied to $\hat{B}$ and A we know that $\hat{B} + A$ has discrete spectrum. Since $B - \hat{B}$ is a finite rank operator, $B + A$ must also have discrete spectrum. $\qquad\square$

5 Application to Edward's Weights. Two Threshold Spectral Phase Transition

Following Edward [E] we consider the special class of Jacobi matrices J with

$$(5.1) \quad \lambda_n = (\alpha n^2 + \beta n + \gamma)^{1/2}, \quad q_n = \delta n, \quad \text{where} \quad \alpha > 0, \ \beta^2 \geq 4\alpha\gamma.$$

In the case $|\delta| > 2\sqrt{\alpha}$ he proved that $\sigma(J)$ is discrete. Below we analyze the opposite case $|\delta| \leq 2\sqrt{\alpha}$. It turns out that $\sigma(J)$ is purely absolutely continuous for $|\delta| < 2\sqrt{\alpha}$. We simply can apply the general results from §3 (see Corollary 3.3). The border case $|\delta| = 2\sqrt{\alpha}$ requires extra assumptions and $\sigma(J)$ remains purely absolutely continuous (at least for a special class of weights). This is one of the reasons we decided to treat Edward's weights in a separate section. As we mentioned in the Introduction, weights given by (5.1) are also interesting because they appear naturally in the representation theory of $su(1, 1)$ Lie algebra. Moreover, the border case is really complicated for investigation (for example, the Stolz approach from §3 fails here). Therefore this situation requires a new technique. Except this paragraph we will avoid the border case in the paper.

Theorem 5.1 *Let λ_n and q_n be given by formula (5.1) with $\alpha > 0$ and $\beta^2 \geq 4\alpha\gamma$. We have the following description of $\sigma(J)$.*

a) *If $|\delta| < 2\sqrt{\alpha}$, then $\sigma(J)$ is pure absolutely continuous.*

b) *If $|\delta| > 2\sqrt{\alpha}$, then $\sigma(J)$ is discrete. (See also [E]).*

c) *If $\delta = 2\sqrt{\alpha}$ and $\gamma = \frac{\beta^2}{4\alpha} - \frac{\alpha}{16}$, then $\sigma(J)$ is pure absolutely continuous.*

Proof: The implication a) is an immediate consequence of Corollary 3.3 and b) follows by using Theorem 4.1.

The case c) can be shown by applying recent results due to Dombrowski and Pedersen [DP]. In fact, we can write $\lambda_n = \frac{1}{2}\sqrt{\alpha}\tilde{\lambda}_n$, $q_n = \frac{1}{2}\sqrt{\alpha}\tilde{q}_n$, where $\tilde{\lambda}_n = (4n^2 - cn + \gamma_1)^{1/2}$, $\tilde{q}_n = 4n$, $c = -\frac{4\beta}{\alpha}$, $\gamma_1 = \frac{c^2-4}{16}$. The last equality is due to the assumption on γ. Define $a_k = (k - \frac{c-2}{4})^{1/2}$. Then

$$\tilde{\lambda}_n = a_{2n}a_{2n-1}$$

and

$$a_{2n-1}^2 + a_{2n}^2 = 4n - 1 - \frac{c-2}{2}.$$

Direct computation shows that

$$(a_{2n}^2 + a_{2n+1}^2 - a_{2n-1}a_{2n} - a_{2n+1}a_{2n+2})$$

is bounded. Let $d_n := a_n - a_{n-1}$ and

$$g(x) = \ln\{[4x^2 - cx + \gamma_1]^{1/2} - [4(x-1)^2 - c(x-1) + \gamma_1]^{1/2}\}.$$

One can check that $\frac{d^2g}{dt^2}(x) > 0$ for x sufficiently large. It follows that

$$d_{n+1}^2 \leq d_n d_{n+2}, \quad \text{for } n \geq n_0.$$

Therefore we can apply Theorem 3.13 of [DP] and conclude that J has pure absolutely continuous spectrum. $\square$

Actually the Dombrowski-Pedersen results allow us to consider much more general weights than those given in item c) (at least for proving that $\sigma_{ac}(J) = \mathbb{R}$) but this requires tedious calculations.

6 Multithreshold Models and Periodic Jacobi Matrices. The Case of Absolutely Continuous Spectrum

In this paragraph, we concentrate on a class of weights given by

$$\lambda_n = c_n n^{\alpha},$$

where $0 < \alpha \leq 1$ and c_n is a periodic sequence of non-zero numbers of period N. The Jacobi matrices J we are going to study have the above weights and the diagonal $q_n = \delta n^{\alpha}$. Despite the simplicity of the formulae, spectral analysis of this class is nontrivial as will be shown below. It turns out that the spectral properties of J are strongly related to the position of $(-\delta)$ with respect to the interior of the pure absolute continuous spectrum of J_{per}, where J_{per} is Jacobi matrix with periodic weights $c_1, \ldots, c_N$ and vanishing diagonal. Since in the nondegeneric case $\sigma_{ac}(J_{per})$ consists of exactly N different intervals [He], the spectral picture of J will illustrate multithreshold spectral phase transition.

The method of proof is based on local, with respect to spectral parameter λ, reduction of the transfer matrix B_n for J to a matrix $\tilde{B}_n$ resembling the transfer matrix of the discrete Schrödinger operator. Recall that the transfer matrix of the discrete Schrödinger operator has the form

$$\begin{pmatrix} 0 & 1 \\ -1 & \lambda - q_n \end{pmatrix},$$

where $\{q_n\} \subset \mathbb{R}$ is the discrete potential. This is not the simplest approach to the spectral analysis of J but it works in the discrete and pure absolutely continuous cases. Before proceeding further observe that by the definition of $B_k = B_k(\lambda)$,

(6.1)
$$\lim_l B_k(\lambda) = \begin{pmatrix} 0 & 1 \\ -\frac{c_{j-1}}{c_j} & -\frac{\delta}{c_j} \end{pmatrix} =: F_j$$

provided $k = lN + j$, $1 \leq j \leq N$, and j is fixed. Hence we know the asymptotics of

$$\lim_{l \to \infty} \prod_{s=1}^{N} B_{lN+s}.$$

Define the following polynomial $P_N(\delta)$ (related to the asymptotic behaviour of B_n) by

(6.2)
$$P_N(\delta) := Tr \left[\prod_{n=1}^{N} \begin{pmatrix} 0 & 1 \\ -\frac{c_{n-1}}{c_n} & -\frac{\delta}{c_n} \end{pmatrix} \right].$$

Since $Tr AB = Tr BA$ the definition of $P_N(\delta)$ does not depend on the ordering of the above product. On the other hand for J_{per} (defined by $\lambda_n := c_n$, $q_n \equiv 0$)

$$d_{J_{per}}(\lambda) := Tr \left[\prod_{n=1}^{N} \begin{pmatrix} 0 & 1 \\ -\frac{c_{n-1}}{c_n} & \frac{\lambda}{c_n} \end{pmatrix} \right]$$

is the characteristic polynomial of J_{per} of order N in λ and is related to its spectral parameter. Namely $x \in Int\, \sigma_{ac}(J_{per})$ is equivalent to $|d_{J_{per}}(x)| < 2$, see [N]. Observe that

(6.3) $$d_{J_{per}}(\lambda) \equiv P_N(\delta)|_{\delta=-\lambda}.$$

Theorem 6.1 *If $(-\delta) \in Int\, \sigma_{ac}(J_{per})$, then J has pure absolutely continuous spectrum and $\sigma_{ac}(J) = \mathbb{R}$.*

Proof: The proof given below is relatively long but some parts of it will be used in the proof of Theorem 7.2. Additionally it provides the semiclassical asymptotics of the generalized eigenvectors in the elliptic case. Put $\mu_n := n^\alpha$ and define the sequence $\tilde{B}_n$ of matrices by

(6.4) $$\tilde{B}_n := \begin{pmatrix} 0 & 1 \\ -c_{n-1}c_n^{-1}\mu_n^{-1}(\mu_{n+1}\mu_{n-1})^{1/2} & \sqrt{\frac{\mu_{n+1}}{\mu_n}}(\frac{\lambda}{\mu_n} - \delta)c_n^{-1} \end{pmatrix}.$$

Note that the $\tilde{B}_n$'s resemble the transfer matrices of discrete Schrödinger operators. We have

(6.5) $$B_n = V_{n+1}^{-1} \tilde{B}_n V_n,$$

where $V_n = diag(\mu_{n-1}^{1/2}, \mu_n^{1/2})$. The above equality reduces the analysis of $\prod_{s=2}^{n} B_s$ to that of $\prod_{s=2}^{n} \tilde{B}_s$. As will be shown below all products $\prod_{s=2}^{n} \tilde{B}_s$ are uniformly bounded in norm provided $(-\delta) \in Int\, \sigma_{ac}(J_{per})$. An essential role in the analysis of these products is played by the condition: $\{(\mu_{n+1}\mu_{n-1})^{1/2}\mu_n^{-1} - 1\} \in l^1$.

We claim that $\tilde{B}_n$ can be written as the sum of a periodic matrix and a matrix of small norm. Indeed,

(6.6) $$\tilde{B}_s = F_s + H_s,$$

where

$$A_s = \begin{pmatrix} 0 & 1 \\ -c_{s-1}c_s^{-1} & -\delta c_s^{-1} \end{pmatrix} \quad \text{and } H_s := \tilde{B}_s - A_s.$$

Since

$$\sqrt{\frac{\mu_{s+1}}{\mu_s}} - 1 = \frac{1}{2\alpha}\frac{1}{s} + O\left(\frac{1}{s^2}\right)$$

and

$$\mu_s^{-1}\sqrt{\frac{\mu_{s+1}}{\mu_s}} = \frac{1}{s^\alpha} + O\left(\frac{1}{s^{1+\alpha}}\right) + O\left(\frac{1}{s^2}\right),$$

we have

$$H_s = \begin{pmatrix} 0 & 0 \\ 0 & \lambda c_s^{-1} \end{pmatrix}\frac{1}{s^\alpha} + \begin{pmatrix} 0 & 0 \\ 0 & -\delta(2\alpha c_s)^{-1} \end{pmatrix}\frac{1}{s} + R_s,$$

where $\{\|R_s\|\} \in l^1$. Let

$$C_l := \prod_{j=1}^{N} \tilde{B}_{(l-1)N+j}, \quad l = 2, 3, \ldots.$$

Using (6.6) and the form of H_s by opening brackets in the above product one can easily check that

$$C_l = \prod_{j=1}^{N} A_{(l-1)N+j} + \tilde{H}_l, \quad \text{where } \tilde{H}_l \in \mathcal{D}^{1,0}.$$

Observe that for $n = (l-1)N + j$

$$det\,\tilde{B}_n = \frac{c_{n-1}}{c_n}(1 + r_{jl}), \quad \text{where } \{r_{jl}\} \in l^1 \text{ in } \quad l, \text{ for } j = 1, \ldots, N.$$

Thus (due to the N-periodicity of c_n)

$$(6.7) \quad detC_l = det \prod_{j=1}^{N} \tilde{B}_{(l-1)N+j} = \prod_{j=1}^{N}(1 + r_{jl}) =: 1 + \tilde{R}_l, \{\tilde{R}_l\} \in l^1.$$

Using again the N-periodicity of c_n one can easily check that $C_\infty := \prod_{j=1}^{N} A_{(l-1)N+j}$ does not depend on l. Now $(-\delta) \in Int\sigma_{ac}(J_{per})$ implies that $|P_N(-\delta)| < 2$ (see (6.3)). But $P_N(-\delta) = TrC_\infty$ and so the eigenvalues $\mu_+(\infty)$ and $\mu_-(\infty)$ of C_∞ must be different because $detC_\infty = 1$.

Since $TrC_l = TrC_\infty + Tr\tilde{H}_l$ and $Tr\tilde{H}_l =: \Delta_l \in \mathcal{D}^1$, the eigenvalues

$$(6.8) \quad \mu_\pm(l) = \frac{1}{2}(TrC_\infty + \Delta_l) \pm \left[\frac{(TrC_\infty + \Delta_l)^2}{4} - detC_l\right]^{1/2}$$

of C_l are also different for l sufficiently large.

By a proper choice of $\mu_+(l)$ and $\mu_-(l)$ we have $\mu_\pm(l) \in \mathcal{D}^1$ (Lemma 2.2). Moreover, the same can be checked for the eigenvectors $\vec{e}_1(l), \vec{e}_2(l)$ of C_l (again by a proper choice of them, see Lemma 6.2 below) corresponding to $\mu_+(l)$ and $\mu_-(l)$ i.e., $\vec{e}_s(l) \in \mathcal{D}^1$ (coordinatewise). The last inclusions can be derived from the following general elementary fact on 2×2 matrices.

Lemma 6.2 *Let* $K_l = \begin{pmatrix} a_l & b_l \\ c_l & d_l \end{pmatrix}$ *be a sequence of* 2×2 *real matrices. Suppose that* $\lim_l K_l = K = \begin{pmatrix} a & b \\ c & d \end{pmatrix}$, $K_l \in \mathcal{D}^{1,0}$, $\sigma(K) = \{\mu_1, \mu_2\}$, μ_1, μ_2 *are non-zero numbers and* $\mu_1 \neq \mu_2$. *Then for a proper choice of the eigenvalues* $\mu_1(l), \mu_2(l)$ *of* K_l *and the corresponding normed eigenvectors* $\vec{e}_1(l), \vec{e}_2(l)$ *of* K_l *we have*

  i) $\mu_s(l) \in \mathcal{D}^1$, $s = 1, 2$,

  ii) $\vec{e}_s(l) \in \mathcal{D}^1$ *coordinatewise, and*

  iii) $\lim_{l \to \infty} \vec{e}_s(l) = \vec{e}_s$, $s = 1, 2$, *where* $K\vec{e}_s = \mu_s \vec{e}_s$.

Proof: (Sketch). Using the formulae for the eigenvalues of K_l and Lemma 2.2 we obtain i). Concerning ii) we have three cases.

1) $b \neq 0$. Then $b_l \neq 0$ for $l \gg 1$ and $\vec{e}_1(l) = \vec{e}_l = (x_l, y_l)$, here $x_l := [1 + (\mu_1(l) - a_l)^2 b_l^{-2}]^{-\frac{1}{2}}$, $y_l := (\mu_1(l) - a_l)b_l^{-1}x_l$. Applying again Lemma 2.2 we see that $\{x_l\}$ and $\{y_l\}$ belong to $\mathcal{D}^1$. Similar reasoning for $\vec{e}_2(l)$ shows the desired ii).

2) $b = 0$, $c \neq 0$ can be considered in the same way. It is enough to replace K_l by K_l^t.

3) $b = c = 0$. Then $a \neq d$ since the spectrum of K is simple. Let $\mu_1 \neq a$ ($\mu_1 \neq d$ can be treated in the same way), and put for $l \gg 1$

$$y_l := (1 + b_l^2(a_l - \mu_1(l))^{-2})^{-1/2}, \quad x_l := -b_l(a_l - \mu_1(l))^{-1}y_l.$$

It is clear that $x_l, y_l \in \mathcal{D}^1$ because $(a_l - \mu_1(l)) \to const \neq 0$, as $l \to \infty$.

iii) Follows from the above construction of $\vec{e}_s(l)$.   □

Diagonalization of $C_\infty + \tilde{H}_l$ leads to the equality:

(6.9) $$C_\infty + \tilde{H}_l = T_l \, diag \, (\mu_+(l), \mu_-(l))T_l^{-1},$$

where $T_l = (\vec{e}_1(l), \vec{e}_2(l))$. Here the eigenvectors are considered as columns. It is clear that $\{T_l\} \in \mathcal{D}^{1,0}$ and converges to an invertible matrix (the spectrum of C_∞ is simple). It follows that $\|\Delta(T_l(\cdot))\| \in l^1$. Thus $T_{l+1}^{-1}T_l = (T_l + \Delta(T_l))^{-1}T_l = I + \Gamma_l$ for l sufficiently large and $\{\|\Gamma_l\|\} \in l^1$. Therefore (using (6.9)) we have

$$\prod_{l=2}^{m}(C_\infty + \tilde{H}_l) = T_{m+1}\prod_{l=2}^{m}[(I + \Gamma_l)\,diag\,(\mu_+(l), \mu_-(l))]T_2$$

(6.10)

$$= T_{m+1}\prod_{l=2}^{m}[diag\,(\mu_+(l), \mu_-(l)) + \tilde{\Gamma}_l]T_2,$$

and again $\{\|\tilde{\Gamma}_l\|\} \in l^1$.

Since $\overline{\mu_+(l)} = \mu_-(l)$ for $l \gg 1$ and $\|T_m\|^{-1}$ is uniformly bounded, from (6.6) and (6.8) we know that the products $\prod_{l=1}^{m}\mu_+(l)$ are uniformly bounded and so, combining this with (6.10), we see that all solutions $\{u_k\}$ of (1.1) satisfy the estimate $|\vec{u}_m|^2 \le C\mu_m^{-1}$ for some $C > 0$. The factor u_m^{-1} comes from $V_m\mu_m^{-1} \to I$, as $m \to \infty$. The generalized Behncke-Stolz lemma completes the proof. $\square$

Remark 6.3 The above proof shows that one could deduce Theorem 6.1 by applying Proposition 2.1 (the Stolz "smooth" approach) to the products

$$\prod_{k=1}^{N}\tilde{B}_{lN+k} = C_{l+1}.$$

This is true because $C_{l+1} = C_\infty + \tilde{H}_{l+1}$ for some $\tilde{H}_{l+1} \in \mathcal{D}^{1,0}$, and the condition (2.2) b) is satisfied since the matrices C_{l+1} have real entries and the spectrum of C_{l+1} is almost elliptic (i.e., $\sigma(C_{l+1})$ is close to the unit circle) for l sufficiently large. Recall that we assumed $(-\delta) \in Int\,\sigma_{ac}(J_{per})$. Additionally by applying the semiclassical asymptotics for generalized eigenvectors (see §7, (7.4)) and the almost scalar form of V_n, we could obtain the absence of a subordinated solution straightforwardly without using the generalized Behncke-Stolz Lemma [Beh], [St], [JN].

Remark 6.4 Observe that in the proof of Theorem 6.2 we did not obtain exactly the transfer matrix of the discrete Schrödinger operator (see (6.3)). However, we could easily obtain the transfer matrix of it by a suitable change of V_n (but still keeping it diagonal). Moreover this can always be done provided $\{(\lambda_{n+1}\lambda_{n-1})^{1/2}\lambda_n^{-1} - 1\} \in l^1$, and the matrices V_n have the following simple asymptotic structure $V_n \sim \sqrt{\lambda_n}\begin{pmatrix} a & 0 \\ 0 & 1 \end{pmatrix}$, as $n \to \infty$, for some $a \in \mathbb{R}\backslash\{0\}$.

Remark 6.5 (Modulated Weights with Zero Diagonal). If $\lambda_n = c_n n^\alpha$, $\alpha > 0$, $q_n \equiv 0$ (i.e., $\delta = 0$) and $0 \in Int\sigma_{ac}(J_{per})$, then $\sigma(J)$ is pure absolutely continuous. This follows from Theorem 6.2.

Note that in the case N is odd 0 always belongs to $Int\, \sigma_{ac}(J_{per})$ and therefore $\sigma(J)$ is pure absolutely continuous. In fact, $d_{J_{per}}(0)$ equals the trace of an odd product of antidiagonal matrices, so it must be zero. On the other hand if N is even then one has

$$d_{J_{per}}(0) = Tr\left[\prod_{k=2}^{N}\left(\begin{matrix} -\frac{c_{k-1}}{c_k} & 0 \\ 0 & -\frac{c_k}{c_{k+1}} \end{matrix}\right)\right]$$

$$= (-1)^{\frac{N}{2}}\left(\frac{c_1}{c_2}\cdots\frac{c_{N-1}}{c_N} + \frac{c_2}{c_3}\cdots\frac{c_N}{c_1}\right).$$

Therefore $\sigma(J)$ is discrete provided $c_1 \cdot c_3 \ldots c_{N-1} \neq \pm c_2 c_4 \ldots c_N$ and this is the generic case. In other words, the spectrum of J with $q_n \equiv 0$ drastically depends on the evenness of N.

7 Multithreshold Models and Periodic Jacobi Matrices. The Case of Discrete Spectrum

The case $(-\delta)$ is not in $\sigma_{ac}(J_{per})$ is equivalent to the nonellipticity of C_∞, i.e., its eigenvalues are real and $\mu_+(\infty) > 1$, $\mu_-(\infty) < 1$. Therefore it is not suprising (at least from a qualitative point of view) that $\sigma(J)$ is discrete [CFKS], [DS1], [Ko], [SZ]. In the stochastic case this type of results was proved by Kotani [Ko]. The proof of the discreteness is based on a discrete variant of the Levinson theorem [CL]. This variant will allow to obtain the semiclassical asymptotics of solutions of (1.1). Actually we will only need the following special case of the Levinson result, which can be proved similarly as its continuous version [CL].

Proposition 7.1 (N. Levinson). *Let A, A_n, V_n and R_n be sequences of 2×2 matrices with real entries such that*

1) *A is a constant, invertible matrix with two different eigenvalues $\lambda_1 \neq \lambda_2$ and $A\vec{e}_i = \lambda_i \vec{e}_i$, $i = 1, 2$,*
2) *$\{\|V_{n+1} - V_n\|\} \in l^1$ and $\|V_n\| \to 0$, as $n \to \infty$, and*
3) *$\{\|R_n\|\} \in l^1$.*

Consider the infinite system of reccurence equations

$$(7.1) \qquad\qquad \vec{x}_{n+1} = (A + V_n + R_n)\vec{x}_n.$$

Then there exist two non-zero solutions $\vec{x}^{(1)}(n)$, $\vec{x}^{(2)}(n)$ of (7.1) such that

$$(7.2) \qquad \vec{x}^{(i)}(n) = \left[\prod_{k=p}^{n}\lambda_i(k)\right](\vec{e}_i + o(1)), \text{ as } n \to \infty, \quad i = 1, 2,$$

for some p sufficiently large. Here $\lambda_i(n)$ *are eigenvalues of* $A + V_n$ *chosen such that* $\lim_{n \to \infty} \lambda_i(n) = \lambda_i$, $i = 1, 2$.

In what follows we will apply Proposition 7.1 to the matrices $C_l = C_\infty + \tilde{H}_l$. More precisely, we apply it in the case $A = C_\infty$, $V_k = \tilde{H}_k$, $R_k \equiv 0$ because $\tilde{H}_k \in \mathcal{D}^{1,0}$ (see §6). Since in our case $det\, C_l \neq 0$ for all l we can replace p in formula (7.2) by 1. Recall that the eigenvalues of $(C_\infty + \tilde{H}_l)$ are $\mu_\pm(l)$. They are given by formula (6.8) in the nonelliptic case too. But now $\mu_+(l) > 1$ and $\mu_-(l) < 1$ for l sufficiently large.

Let $C_\infty \vec{e}_1 = \mu_-(\infty)\vec{e}_1$, $C_\infty \vec{e}_2 = \mu_+(\infty)\vec{e}_2$. Using (7.2) we can write for $n = (l-1)N + j$ the asymptotic formula for the solutions $\vec{u}_{n+1}^{(i)}$ of (1.1) as:

$$(7.3) \qquad \vec{u}_{n+1}^{(i)} = V_{n+1}^{-1} F_{j+1} \ldots F_2 \left[\prod_{k=1}^{l} \mu_\pm(k) \right] (\vec{e}_i + o(1)),$$

where F_s is given by (6.1). Evoking the definitions of V_n and F_j we can rewrite (7.3) as follows

$$(7.4) \qquad \begin{aligned} \vec{u}_{n+1}^{(i)} = \mu_n^{-\frac{1}{2}} \prod_{k=1}^{l} \mu_\pm(k) \begin{pmatrix} 0 & 1 \\ -\frac{c_j}{c_{j+1}} & -\frac{\delta}{c_{j+1}} \end{pmatrix} \\ \cdot \begin{pmatrix} 0 & 1 \\ -\frac{c_1}{c_2} & -\frac{\delta}{c_2} \end{pmatrix} (\vec{e}_i + o(1)), \end{aligned}$$

where $i = 1, 2$ and we choose $\mu_+(k)$ for $i = 1$, and $\mu_-(k)$ for $i = 2$.

Remark 7.2 Note that (7.4) also holds true in the elliptic case of §6 when $(-\delta) \in Int\, \sigma_{ac}(J_{per})$ (in this case $\mu_+(l)$, $\mu_-(l)$ are close to the unit circle). We can again apply Proposition 7.1. Moreover, in both cases (elliptic and nonelliptic) $\mu_+(l)$ $\mu_-(l) = det\, C_l \in \mathbb{R}$, even for complex λ and therefore $arg\,\mu_+(l) + arg\,\mu_-(l) = 0$. By the way the semiclassical asymptotics of the generalized eigenvectors in elliptic case is also of the form given by (7.4).

The asymptotics of the solutions given in (7.4) is useful in finding estimates of the Green function $G(k, n; \lambda) = ((J - \lambda)^{-1} e_k, e_n), k, n \in \mathbb{N}, \lambda \notin \sigma(J)$.

Let $u^D(\cdot, \lambda)$ respectively $u^N(\cdot, \lambda)$ be two Weyl solutions of (1.1) with boundary conditions given by $u^D(0) = 0$, $u^D(1) = 1$, and $u^N(0) = 1$, $u^N(1) = 0$, respectively. Then defining for $\lambda \in \mathbb{C} \backslash \mathbb{R}$ the unique Jost solution $\varphi_+^N(n, \lambda) = u^D(n, \lambda) + m(\lambda)u^N(n, \lambda)$ in l^2, one can write

$$(7.5) \qquad G(k, n; \lambda) = C(\lambda) u_D(k_<, \lambda) \varphi_+^N(k_>, \lambda),$$

where $k_< = min(k, n)$, $k_> = max(k, n)$, see [Ki], [LS]. Note that due to (7.4) we have

(7.6)
$$|u_D(n, \lambda)| \leq C_1 \mu_n^{-\frac{1}{2}} \prod_{k=1}^{n} \mu_+(k)$$

(7.7)
$$|\varphi_+^N(n, \lambda)| \leq C_2 \mu_n^{-\frac{1}{2}} \prod_{n=1}^{n} \mu_-(k),$$

for some positive constants C_1, C_2 which depend only on λ. Recall that in our situation only one solution of (1.1) belongs to l^2.

In turn the products $\prod_{s=1}^{n} \mu_+(s)\mu_-(s)$ are uniformly bounded (see (6.7) and (6.8)). Hence using (7.5), (7.6) and (7.7) we get

(7.8)
$$|G(k, n; \lambda)| \leq C(\mu_n\mu_k)^{-\frac{1}{2}} \prod_{s=n+1}^{k} \mu_-(s),$$

where $C = C_1(\lambda) > 0$ and $n \leq k$ but n is sufficiently large (here for $n = k$ the product $\prod_{s=n+1}^{k} \mu_-(s)$ is understood to be equal to 1). Since $\mu_-(s) \leq r < 1$ for $s \gg 1$, (7.8) implies that

(7.9)
$$|G(k, n; \lambda)| \leq C_1(\mu_n\mu_k)^{-\frac{1}{2}} r^{|k-n|}$$

for $n \leq k$ but n sufficiently large, say $n \geq n_o$. It is clear that (7.9) also holds for $n < n_o$ and $k \leq n$ (see (7.8)). The case $n \leq k$ follows due to the symmetry in (7.5) and so (7.9) is satisfied for all k, n.

In other words we have proved the following

Lemma 7.3 *Let J be a Jacobi matrix with the weights $\lambda_n = c_n n^\alpha$, $\alpha > 0$, where c_n is an N-periodic sequence, and the diagonal $q_n = \delta n^\alpha$. If $(-\delta) \notin \sigma_{ac}(J_{per})$ and $\lambda \in \mathbb{C}\backslash\sigma(J)$ (say λ is not real), then the Green function $G(k, n; \lambda)$ of J satisfies the estimate (7.9) with $\mu_n = n^\alpha$.*

Lemma 7.3 will be used below to prove

Theorem 7.4 *Let J be the Jacobi matrix with the same weights and the same diagonal as in Lemma 7.3. Assume that $(-\delta) \notin \sigma_{ac}(J_{per})$. Then J has discrete spectrum $\{\rho_n\}$, $(|\rho_n| \leq |\rho_{n+1}|)$ of multiplicity one and $\rho_n^{-1} = O(|\lambda_n|^{-1})$, as $n \to \infty$.*

Proof: Fix $\lambda \in \mathbb{C}\backslash\mathbb{R}$. Applying the estimate (7.9) (due to Lemma 7.3) we get $|G(k, n; \lambda)| \leq M(\mu_n\mu_k)^{-\frac{1}{2}} r^{|n-k|}$ for some $M > 0$ and $0 < r < 1$. Hence $G(k, n, \lambda)$ can be written as:

$$G(k, n; \lambda) = \mu_k^{-\frac{1}{2}} [F(k, n, \lambda) r^{|n-k|}] \mu_n^{-\frac{1}{2}},$$

where $|F(k, n, \lambda)| \leq M$, for all $k, n \in \mathbb{N}$. Therefore the operator G given by the matrix $\{G(k, n; \lambda)\}_{k,n \in \mathbb{N}}$ can be written as the product $G = ABA$, where A is the diagonal operator with the diagonal $\{\mu_n^{-1/2}\}_{n \in \mathbb{N}}$ and B is the operator in l^2 defined by the matrix $\{F(k, n, \lambda) r^{|k-n|}\}$.

Note that

(7.10) $$\sum_k |F(k, n, \lambda)| |r|^{|n-k|} \leq 2M(1-r)^{-1} \text{ for } n \in \mathbb{N}$$

and

(7.11) $$\sum_n |F(k, n, \lambda)| |r|^{|n-k|} \leq 2M(1-r)^{-1} \text{ for } k \in \mathbb{N}.$$

Hence B is bounded in l^2 (with $\|B\| \leq 2M(1-r)^{-1}$) [Hal]. Since A is compact ($\mu_n \to +\infty$, as $n \to \infty$), G must be compact also, and so the spectrum of J is discrete. To prove the estimate of the eigenvalues ρ_k of J observe that for the s-numbers $s_m(J)$ of J we have the inequalities [GK]

$$s_m(J)^{-1} \leq C s_m((J - \lambda)^{-1}) = C s_m(ABA)$$
$$\leq C\|B\| s_{[\frac{m}{2}]-1}(A)^2 \leq C_1 \|B\| \mu_m^{-1}.$$

$\square$

We conclude with

Remark 7.5 The semi-classical discrete methods can be applied to a much more general class of weights. We restricted our considerations in this paper to the special case $\lambda_n = n^\alpha$ in order to avoid technical difficulties. However, it was enough to expose general ideas. We plan to study the above mentioned classes of Jacobi matrices in a forthcoming paper.

Acknowledgements

The research of the first author was supported by grant PB 2 PO3A 002 13 of the Komitet Badan Naukowych, Warsaw.

The second author gratefully appreciates support of the Swedish Royal Academy of Sciences.

The authors thank Pani Bozena Skoczylas for her help in the preparation of the manuscript.

References

[A]   N.I. Akhiezer, *The classical moment problem*, Hafner, New York, 1965.
[Beh] H. Behncke, Absolute Continuity of Hamiltonians with von Neumann-Wigner Potentials II, *Manuscripta Math.* **71** (1991), 163–181.

[Ben]   F. Bentosela, R. Carmona, P. Duclos, B. Simon, B. Souillard and R. Weder, Schrödinger Operators with an Electric Field and Random or Deterministic Potentials, *Comm. Math. Phys.* **88** (1983), 387–397.

[Ber]   Yu.M. Berezanskii, *Expansions in Eigenfunctions of Selfadjoint Operators*, Naukova Dumka, Kiev 1965. (Russian)

[CK]   M. Christ and A. Kiselev, Absolutely continuous spectrum for one-dimensional Schrödinger operator with slowly decaying potential: some optimal results, *J. Amer. Math. Soc.* **11**:4 (1998), 771–797.

[CL]   E.A. Coddington and N. Levinson, *Theory of Ordinary Differential Equations*, Mc Graw-Hill 1955.

[CC]   O. Costin and R. Costin, Rigorous WKB for finite-order linear recurrence relations with smooth coefficients, *SIAM J. Math. Anal.* **27**:1 (1996), 110–134.

[CFKS]   H.L. Cycon, R.G. Froese, W. Kirsch and B. Simon, *Schrödinger Operators with Application to Quantum Mechanics and Global Geometry*, Springer-Verlag 1987.

[DSS]   F. Deylon, B. Simon and B. Souillard, From power law localized to extended states in a disordered system, *Ann. Inst. Henri Poincare* **42** (1985), 283.

[DS]   F. Deylon and B. Souillard, The rotation number for finite difference operators and its properties, *Comm. Math. Phys.* **89** (1983), 415.

[DS1]   F. Deylon and B. Souillard, Remark on the continuity of the density of states of ergodic finite difference operators, *Comm. Math. Phys.* **94** (1984), 289.

[D]   J. Dombrowski, Absolutely continuous measures for systems of orthogonal polynomials with unbounded recurrence coefficients, *Constr. Approx.* **8** (1992), 161–167.

[DP]   J. Dombrowski and S. Pedersen, Spectral measures and Jacobi matrices related to Laguerre-type systems of orthogonal polynomials, *Constr. Approx.* **13** (1997), 421–433.

[E]   J. Edward, Spectra of Jacobi matrices differential equations on the circle, and the $su(1, 1)$ Lie algebra, *SIAM J. Math. Anal.* **34**:3 (1993), 824–831.

[GS]   F. Gesztesy and B. Simon, M-functions and inverse spectral analysis for finite and semi-infinite Jacobi matrices, *J. Anal. Math.* **73** (1997), 267–297.

[GP]   D.J. Gilbert and D.B. Pearson, On subordinacy and analysis of the spectrum one-dimensional Schrödinger operators, *J. Math. Anal. Appl.* **128** (1987), 30–56.

[GK]   I. Gohberg and M.G. Krein, Introduction to the Theory of Linear Nonselfadjoint Operators in Hilbert space, *Amer. Math. Soc., Providence, R. I.* 1969.

[Hal]   P. Halmos, *A Hilbert space problem book*, Van Nostrand, Princeton 1967.

[Har]   P. Hartman, *Ordinary differential equations*, John Wiley 1964.

[He]   H. Heinig, Inversion of periodic Jacobi matrices, *Math. Islled., Chisinau* **1**:27 (1973), 180–200.

[JN]   J. Janas and S. Naboko, Jacobi matrices with power like weights-grouping in blocks approach, *J. of Func. Anal.* **166** (1999), 218–243.

[JN_1] J. Janas and S. Naboko, Asymptotics of generalized eigenvectors for unbounded Jacobi matrices with power-like weights, Pauli matrices commutation relations and Cesaro averaging, *Operator Theory: Advances and Applications, volume dedicated to M.G. Krein*, (to appear), 1999.

[K]   T. Kato, *Perturbation theory for linear operators*, 2nd ed., Springer, Berlin, Heidelberg, N.Y. 1980.

[KP]   S. Khan and D.B. Pearson, Subordinacy and Spectral Theory for Infinite Matrices, *Helv. Phys. Acta* **65** (1992), 505–527.

[Ki]   A. Kiselev, Absolutely continuous spectrum of one-dimensional Schrödinger operators and Jacobi matrices with slowly decaying potentials, *Comm. Math. Phys.* **179** (1996), 377–400.

[Ko]   S. Kotani, Ljapunov indices determine absolutely continuous spectra of stationary random one-dimensional Schrödinger operators, *Stochastic Analysis,* ed. by K. Ito, North Holland, Amsterdam **225** (1984).

[LS]   Y. Last and B. Simon, Eigenfunctions, transfer matrices and absolutely continuous spectrum of one-dimensional Schrödinger operators, *Invent. Math.* **135** (1999), 329–367.

[MR]   D. Masson and J. Repka, Spectral theory of Jacobi matrices in $l^2(\mathbb{Z})$ and the $su(1, 1)$ Lie algebra, *SIAM J. Math. Anal.* **22** (1991), 1131–1146.

[M]   N. Mott, *Electrons in disordered structures*, Russian Edition: Mir, Moscow 1969.

[N]   P.B. Najman, On the theory of periodic and limit-periodic Jacobi matrices, *Doklady AN USSR,* Russian **143** (1962), 277.

[O]   F.W. Olver, *Asymptotic and special functions*, New York, London, A.P. 1974.

[RS]   M. Reed and B. Simon, *Methods of Modern Math. Phys. III, Scattering Theory*, AP 1979.

[RS1]   M. Reed and B. Simon, *Methods of Modern Math. Phys. IV, Analysis of Operators*, AP 1978.

[S]   B. Simon, The classical moment problem as a selfadjoint finite difference operator, *Adv. Math.* **137** (1998), 82–203.

[S1]   B. Simon, Some Jacobi matrices with decaying potential and dense point spectrum, *Comm. Math. Phys.* **87** (1982), 253–258.

[SZ]   B. Simon and Y. Zhu, The Lyapunov exponents for Schrödinger operators with slowly oscillating potentials, *J. Funct. Anal.* **140** (1996), 541–556.

[St]   G. Stolz, Spectral Theory for Slowly Oscillating Potentials I. Jacobi Matrices, *Manuscripta Math.* **84** (1994), 245–260.

[St1]   G. Stolz, Spectral Theory for Slowly Oscillating Potentials II. Schrödinger Operators, *Math. Nachr.* **183** (1997), 275–294.

J. Janas
Institute of Mathematics
Polish Academy of Sciences
Cracow Branch
Sw. Tomasza 30
31-027 Cracow
Poland
e-mail: najanas@cyf-kr.edu.pl

S. Naboko
Department of Mathematical Physics
Institute of Physics
St. Petersburg University
Ulianovskaia 1
St. Petergoff
S. Petersburg
Russia, 198904
e-mail: naboko@snoopy.phys.spbu.ru

1991 Mathematics Subject Classification. Primary: 47B25; Secondary: 47B39.

Operator Theory:
Advances and Applications, Vol. 124
© 2001 Birkhäuser Verlag Basel/Switzerland

Positive Extensions and Diagonally Connected Patterns

M.A. Kaashoek and H.J. Woerdeman

Dedicated with respect and affection to Israel Gohberg on the occasion of his 70-th birthday

The band method with two coupled semi-band structures is used to give complete solutions to positive extension problems with an underlying pattern that is diagonally connected.

1 Introduction

The band method, which originated in the work of Dym and Gohberg [3], [4], is an abstract mathematical scheme, or more precisely any one of several closely related schemes, that offers a unified approach to various extension problems, in particular, positive and contractive extensions, and extensions with positive real part. Its numerous applications include solutions to the Carathéodory-Toeplitz and Nehari extension problems for matrix and operator valued functions, also the non-stationary versions of these problems, the stationary and non-stationary four block problems, extensions of Fredholm integral operators, bitangential Nevanlinna-Pick interpolation, extension problems for almost periodic matrix functions, etc. The method applies equally well to discrete and continuous problems. Much of the material pertaining to this method is found in books [18], [6]. For further details about the early history of the method see [7] and the references therein.

The band method, as it appears in the works of Gohberg, Kaashoek, and Woerdeman [7], [8], [9], [10], has three important features which for a positive completion problem can be summarized as follows: (1) by solving linear equations a particular positive completion, called the *band extension*, with a right and a left spectral factorization is constructed; (2) a linear fractional parametrization of all positive completions is obtained by using a right and a left spectral factorization of the band extension; (3) a maximum principle identifies the band extension. Later, it was shown in [11], [12] that some of the abstract results remain valid when certain algebraic restrictions were omitted, allowing for so-called "non-band" situations. In [5], a version of the abstract band method was developed that allowed, in some cases, to go beyond a typical Wiener algebra setting.

The incorporation of interpolation problems (such as Nevanlinna-Pick) in the general scheme of the abstract band method was done by Ball, Gohberg, and Kaashoek [1], [2], who combined ideas of the abstract band method and ideas of the

Grassmanian approach to interpolation. In Kaashoek and Zeinstra [14], by revising the scheme as presented in [2], the algebraic structure was further loosened to "semi-band structures", and in this context a Carathéodory-Toeplitz interpolation involving operator functions with operator arguments was solved. The most recent addition to the band method is Rodman, Spitkovsky and Woerdeman [16] where Toeplitz operators were introduced in an abstract algebra context and the existence of a positive band extension was directly related to the positivity of such an abstract Toeplitz operator.

In this paper we apply the scheme of [14] to give a complete solution to certain positive extension problems of non-band type involving partially given operator matrices whose underlying pattern is diagonally connected, a notion introduced in [11]. In the latter paper existence and maximum entropy results were obtained, while in the current paper we provide in addition a description of all solutions. The paper is organized as follows. In Section 2 we show how diagonally connected patterns can be decomposed so that they fit the underlying structure of the coupled semi-band structures developed in [14]. In Sections 3 and 4 we apply these results in the settings of finite and semi-infinite operator matrices, respectively.

2 Partitions with a Triangular Property and Linear Orders

Let Λ be a set. Recall the following elementary notions (cf., [17], §1.8). A *relation* R on Λ is a subset of $\Lambda \times \Lambda$, i.e., $R \subseteq \Lambda \times \Lambda$. For a relation R we define

$$\overleftarrow{R} = \{(i, j) : (j, i) \in R\}.$$

When $R = \overleftarrow{R}$ we say that R is *symmetric*. If R and S are relations on Λ, we define their *composition* $R \circ S$ to be the relation

$$R \circ S = \{(k, \ell) : \exists \, p \in \Lambda \text{ such that } (k, p) \in R \text{ and } (p, \ell) \in S\}.$$

A relation R is called *transitive* when $R \circ R \subseteq R$, and R is called *reflexive* if $\{(i, i) : i \in \Lambda\} \subseteq R$. When R is reflexive, symmetric and transitive, R is called an *equivalence relation*.

In what follows we shall use composition tables such as

$$\begin{array}{c|c} \circ & S \\ \hline R & P \end{array}$$

which means that $R \circ S \subseteq P$. A collection $\{A_i : i \in \alpha\}$ of subsets of a set Q is said to form a *partition* of Q if $\cup_{i \in \alpha} A_i = Q$ and $A_i \cap A_j = \emptyset$ for $i \neq j$.

In the following theorems we shall highlight how certain composition properties between relations on Λ interact with linear orderings $\leq$ on Λ. In subsequent sections these results will allow us to apply the results of [14] to a number of positive extension problems.

In analogy to [14] we shall say that a partition $\{S, D, \overleftarrow{S}\}$ of $\Lambda \times \Lambda$ has *the triangular property* if the following table holds

(2.1)

$\circ$	S	D	$\overleftarrow{S}$
S	S	S	$\Lambda \times \Lambda$
D	S	D	$\overleftarrow{S}$
$\overleftarrow{S}$	$\Lambda \times \Lambda$	$\overleftarrow{S}$	$\overleftarrow{S}$

Notice that the (1.3)-entry in table (2.1) means that there is no restriction on the composition $S \circ \overleftarrow{S}$. To understand the table (2.1) better, let us consider the special case when $\Lambda = \{1, \ldots, n\}$ and

$$S = \{(i, j) \mid j < i\}, \quad D = \{(i, i) \in \Lambda \times \Lambda \mid i \in \Lambda\}.$$

If we view $\Lambda \times \Lambda$ as an $n \times n$ array, then S corresponds to the lower triangular part of $\Lambda \times \Lambda$, the set D to its diagonal, and $\overleftarrow{S}$ to its upper triangular part. In this case the $(1, 1)$-entry in table (2.1), corresponds to the statement that the product of two lower triangular $n \times n$ matrices (i.e., $n \times n$ matrices with support in S) are again lower triangular. The other entries have analogous interpretations.

Theorem 2.1 *Let Λ be a set that allows a linear ordering. A partition $\{S, D, \overleftarrow{S}\}$ of $\Lambda \times \Lambda$ has the triangular property if and only if there exists a linear order $\leq$ on Λ so that*

   (i) $S \subseteq \{(i, j) \in \Lambda \times \Lambda : i \leq j\}$;

   (ii) $(i, j) \in S$, $k \leq i$ and $\ell \geq j$ imply $(k, \ell) \in S$;

   (iii) *D has a partition of the form $\{J_i \times J_i : i \in I\}$ where for each i the set $J_i \subseteq \Lambda$ is an interval in Λ with respect to $\leq$.*

Recall that a subset J of an ordered set $(\Lambda, \leq)$ is an *interval* if $i, j \in J$ and $i \leq k \leq j$ imply $k \in J$. For the definition of a linear order we refer to [17], §1.9.

Note that when one accepts the Axiom of Choice, or the weaker axiom that every set can be linearly ordered, one does not need to assume that Λ allows a linear order. We shall be applying the statements to subsets of $\mathbb{R}$, so that we will not need to depend on the Axiom of Choice.

Lemma 2.2 *Let $\{S, D, \overleftarrow{S}\}$ be a partition of $\Lambda \times \Lambda$ that has the triangular property. Then D is an equivalence relation.*

Proof: Let $i \in \Lambda$. Then $(i, i) \in S \cup D \cup \overleftarrow{S}$. If $(i, i) \in S$ or $(i, i) \in \overleftarrow{S}$, then $(i, i) \in S \cap \overleftarrow{S}$, contradicting that $S \cap \overleftarrow{S} = \emptyset$. Thus $(i, i) \in D$, for all $i \in \Lambda$, yielding that D is reflexive.

Since $D = (\Lambda \times \Lambda)\setminus(S \cup \overleftarrow{S})$ and $S \cup \overleftarrow{S}$ is symmetric, we get that D is symmetric. Lastly, from the multiplication table (2.1) we get that $D \circ D \subseteq D$, so that D is transitive. $\qquad\square$

Lemma 2.3 *Let* $\{S, D, \overleftarrow{S}\}$ *be a partition of* $\Lambda \times \Lambda$ *that has the triangular property. For* $i \in \Lambda$ *define*

$$(2.2) \qquad\qquad \Sigma_i := \{j \in \Lambda : (i, j) \in S\}.$$

Then for all i *and* j *in* Λ *we have that* $\Sigma_i \subseteq \Sigma_j$ *or* $\Sigma_j \subseteq \Sigma_i$.
  Moreover, $\Sigma_i = \Sigma_j$ *if and only if* $(i, j) \in D$ *and* $\Sigma_i \subsetneq \Sigma_j$ *if and only if* $(j, i) \in S$.

Note that the first observation, the Σ_i-s being linearly ordered by inclusion, was made earlier in Lemma 2.2 of [19].

Proof: We split the proof into four parts.

Part (a). In this part we show that $\Sigma_j \setminus \Sigma_i \neq \emptyset$ implies that $(i, j) \in \overleftarrow{S}$. Indeed, let $k \in \Sigma_j \setminus \Sigma_i$. Thus $(j, k) \in S$ and $(i, k) \notin S$. From $(i, k) \notin S$ we conclude that $(i, k) \in D \cup \overleftarrow{S}$, and hence $(k, i) \in D \cup S$, because $D = \overleftarrow{D}$ by Lemma 2.2. Thus

$$(j, i) \in S \circ (D \cup S) \subset S$$

by the multiplication table (2.1).

Part (b). In this part we show that $\Sigma_i \subseteq \Sigma_j$ or $\Sigma_j \subseteq \Sigma_i$. Assume not. Then both $\Sigma_j \setminus \Sigma_i$ and $\Sigma_i \setminus \Sigma_j$ are non-empty. By the result of the previous part this yields (i, j) and (j, i) are both in S. Thus $(i, j) \in S \cap \overleftarrow{S}$, which is impossible.

Part (c). In this part we show that $\Sigma_i = \Sigma_j$ if and only if $(i, j) \in D$. Take $(i, j) \in D$. Let $p \in \Sigma_i$. Thus $(j, i) \in D$, because $D = \overleftarrow{D}$ by Lemma 2.2, and $(i, p) \in S$. But then $(j, p) \in D \circ S \subset S$, and $p \in \Sigma_j$. Thus $(i, j) \in D$ yields $\Sigma_i \subset \Sigma_j$. Since $(i, j) \in D$ implies $(j, i) \in D$, we also have $\Sigma_j \subset \Sigma_i$, and therefore $\Sigma_i = \Sigma_j$.

Conversely, assume $\Sigma_i = \Sigma_j$. If $(i, j) \in S$, then $j \in \Sigma_i = \Sigma_j$, and thus $(j, j) \in S$. So $(j, j) \in S \cap \overleftarrow{S}$ which is impossible. Thus $(i, j) \notin S$. If $(i, j) \in \overleftarrow{S}$, then $(j, i) \in S$, and again we get a contradiction. Thus $(i, j) \in D$.

Part (d). In this part we show that $\Sigma_i \subsetneq \Sigma_j$ if and only if $(j, i) \in S$. Assume $\Sigma_i \subsetneq \Sigma_j$. In particular, $\Sigma_j \setminus \Sigma_i$ is non-empty, and hence by the result of Part (a) we have $(j, i) \in S$.

Conversely, assume $(j, i) \in S$, and take $p \in \Sigma_i$. Then $(i, p) \in S$, and hence $(j, p) \in S \circ S \subset S$, by the multiplication table (2.1). So $p \in \Sigma_j$. Thus $(j, i) \in S$ yields $\Sigma_i \subseteq \Sigma_j$. Furthermore, by the result of the previous part, $\Sigma_i \neq \Sigma_j$ since $(j, i) \notin D$. $\qquad\square$

Note that we in fact showed that

$$S = \{(i, j) : \Sigma_j \subsetneq \Sigma_i\},$$
$$D = \{(i, j) : \Sigma_i = \Sigma_j\},$$
$$\overleftarrow{S} = \{(i, j) : \Sigma_i \subsetneq \Sigma_j\}.$$

Proof of Theorem 2.1: First assume that the partition $\{S, D, \overleftarrow{S}\}$ of $\Lambda \times \Lambda$ has the triangular property. For $i \in \Lambda$ let $[i]$ denote the equivalence class of i with respect to the equivalence relation D (use Lemma 2.2), i.e.,

$$[i] = \{j \in \Lambda : (i, j) \in D\}.$$

Furthermore, let $\leq_\Lambda$ be any linear order on Λ. We define now the new linear order $\leq$ on Λ by

$$i \leq j \Leftrightarrow \begin{cases} i \leq_\Lambda j, & \text{when } [i] = [j], \\ \Sigma_j \subseteq \Sigma_i, & \text{when } [i] \neq [j], \end{cases}$$

where $\Sigma_k = \{p \in \Lambda : (k, p) \in S\}$, $k \in \Lambda$. Since $[i] = [j]$ is equivalent to $(i, j) \in D$, it follows from Lemma 2.3 that $i \leq j$ always implies that $\Sigma_j \subseteq \Sigma_i$. By Lemma 2.3 and the fact that $\leq_\Lambda$ is a linear order, every two elements i and j of Λ are comparable with respect to $\leq$. When $i = j$ we get that $i \leq j$, so $\leq$ is reflexive. Next, suppose that $i \leq j$ and $j \leq i$. If $(i, j) \in S$, we have that $[i] \neq [j]$, so $i \leq j$ implies $\Sigma_j \subseteq \Sigma_i$ and $j \leq i$ implies $\Sigma_i \subseteq \Sigma_j$. Thus $\Sigma_i = \Sigma_j$, giving $(i, j) \in D$. This contradicts $[i] \neq [j]$. Consequently, when $i \leq j$ and $j \leq i$ we must have that $[i] = [j]$. But then it follows from the anti-symmetry of $\leq_\Lambda$ that $i = j$. Thus $\leq$ is anti-symmetric.

Lastly, we need to check that $\leq$ is transitive, so suppose that $i \leq j$ and $j \leq k$. In case $[i] = [j] = [k]$, we get $i \leq k$ from the transitivity of $\leq_\Lambda$. When $[i] = [j] \neq [k]$, we have that $\Sigma_k \subsetneq \Sigma_j = \Sigma_i$, thus $i \leq k$. When $[i] \neq [j] = [k]$, we have that $\Sigma_k = \Sigma_j \subsetneq \Sigma_i$, thus $i \leq k$. Finally, if $[i] \neq [j] \neq [k]$ we get that $\Sigma_k \subsetneq \Sigma_j \subsetneq \Sigma_i$, thus $i \leq k$.

When $(i, j) \in S$ we have that $\Sigma_j \subsetneq \Sigma_i$, and thus $i \leq j$. This gives (i). To check (ii), let $(i, j) \in S$, $k \leq i$ and $\ell \geq j$. Then $\Sigma_j \subsetneq \Sigma_i$, $\Sigma_i \subseteq \Sigma_k$ and $\Sigma_\ell \subseteq \Sigma_j$. So $\Sigma_\ell \subsetneq \Sigma_k$, yielding that $(k, \ell) \in S$ by Lemma 2.3. Finally, to prove (iii) notice that $(i, j) \in D$ implies that

$$(i, j) \in [i] \times [i] \subset D \times D.$$

Hence in order to prove (iii) it suffices to show that each equivalence class of D is an interval. For this purpose let $J = [i]$, take $p \in J$, $q \in J$, and assume that $p \leq r \leq q$ for some $r \in \Lambda$. Then $\Sigma_p = \Sigma_q$ because $(p, q) \in D$, and $\Sigma_q \subseteq \Sigma_r \subseteq \Sigma_p$. So $\Sigma_p = \Sigma_r$, and we have $(p, r) \in D$. Thus $r \in [p] = J$, and therefore J is an interval.

Conversely, suppose that a linear order $\leq$ on Λ exists so that (i), (ii) and (iii) are satisfied. We need to check that the partition $\{S, D, \overleftarrow{S}\}$ satisfies the composition table (2.1).

The first paragraph of the proof of Lemma 2.2 shows that D is a reflexive relation. But then we see from (iii) that $\{J_i : i \in I\}$ is a partition of Λ. Since $(a, b) \in D$ if and only if $a \in J_i$ and $b \in J_i$ for some $i \in I$, we conclude that D is an equivalence relation on Λ. In particular, $D \circ D \subseteq D$.

In order to check that $S \circ S \subseteq S$, let $(i, j) \in S$ and $(j, \ell) \in S$. By (i) we thus get $j \leq \ell$. Applying (ii) (with $k = i$) now gives that $(i, \ell) \in S$.

For the inclusion $D \circ S \subseteq S$ let $(i, j) \in D$ and $(j, k) \in S$. If $i \leq j$, then (ii) implies that $(i, k) \in S$. Suppose $j \leq i$ and $j \neq i$. Note that we cannot have that $k \leq i$. Indeed, $k \leq i$ implies $j \leq k \leq i$ by (i). Since $[i] = [j]$ because $(i, j) \in D$, property (iii) would imply $k \in [j]$, contradicting $(j, k) \in S$. Thus $i \leq k$ and $i \neq k$. In particular, $(k, i) \notin S$. Also, $(k, i) \notin D$. Indeed, if $(k, i) \in D$, then $(k, j) \in D \circ D \subseteq D$ which contradicts the fact $(k, j) \in \overleftarrow{S}$. We conclude that $(k, i) \in \overleftarrow{S}$, and hence $(i, k) \in S$ as desired.

For the inclusion $S \circ D \subseteq S$, one argues analogously. The inclusions involving $\overleftarrow{S}$ follow from the inclusions with S.

Next, we prove the following theorem.

Theorem 2.4 *Let $\{S, D, \overleftarrow{S}\}$ be a partition of $\Lambda \times \Lambda$ that has the triangular property, and let Q_1 be a subset of S such that*

$$(2.3) \qquad\qquad (S \cup D) \circ Q_1 \subseteq Q_1.$$

Then there exists a linear order $\leq$ on Λ such that (i), (ii) *and* (iii) *in Theorem 2.1 and the following condition hold:*

(iv) $(i, j) \in Q_1$ *and* $k \leq i$ *imply* $(k, j) \in Q_1$.

Proof: For $i \in \Lambda$ let Σ_i be defined by (2.2). Put $T_i = \{j \in \Lambda : (i, j) \in Q_1\}$. First we show that $(i, j) \in D$ implies $T_i = T_j$. Indeed, let $(i, j) \in D$ and $k \in T_i$. Then $(i, k) \in Q_1$ by definition. Thus, by (2.3) and the symmetry of D, we get that $(j, k) \in D \circ Q_1 \subseteq Q_1$, and hence $k \in T_j$. Likewise, one proves $T_j \subseteq T_i$, and thus $T_i = T_j$.

Let $\leq$ be the ordering on Λ as defined in the first paragraph of the proof of Theorem 2.1. In particular, we have that $i \leq j$ implies $\Sigma_j \subseteq \Sigma_i$. We only need to show that (iv) holds. Let therefore $(i, j) \in Q_1$ and $k \leq i$. Thus $j \in T_i \subseteq \Sigma_i$. Suppose $(i, k) \in S$. Then $i \leq k$ by (i) in Theorem 2.1 and hence $i = k$. But for $i = k$ item (iv) is satisfied trivially. Assume $(i, k) \in D \cup \overleftarrow{S}$. If $(i, k) \in D$, then $j \in T_i = T_k$ by the result of the first paragraph of the proof, yielding $(k, j) \in Q_1$, proving (iv) in this case. If $(i, k) \in \overleftarrow{S}$, then $(k, i) \in S$. Thus by (2.3) we get that $(k, j) \in S \circ Q_1 \subseteq Q_1$. $\qquad\square$

The converse of Theorem 2.4 holds as well. In fact, the following stronger results holds.

Theorem 2.5 *Let* $\{S, D, \overleftarrow{S}\}$ *be a partition of* $\Lambda \times \Lambda$, *and* $Q_1 \subseteq S$ *so that* $D \circ Q_1 \subseteq Q_1$. *If there exists a linear order* $\leq$ *on* Λ *for which* (i), (ii) *and* (iii) *in Theorem 2.1 and* (iv) *in Theorem 2.4 hold. Then* $\{S, D, \overleftarrow{S}\}$ *has the triangular property and* (2.3) *holds. Moreover, letting*

$$\tilde{Q}_+^\circ = S, \quad \tilde{Q}_d = D, \quad \tilde{Q}_-^\circ = \overleftarrow{S},$$
$$\tilde{Q}_+ = S \cup D, \quad \tilde{Q}_- = \overleftarrow{S} \cup D,$$
$$\tilde{Q}_4 = \tilde{Q}_1, \quad \tilde{Q}_3 = \tilde{Q}_- \setminus \tilde{Q}_4,$$

Then

(2.4) $$\tilde{Q}_4 \circ \tilde{Q}_- \subseteq \tilde{Q}_4,$$

(2.5) $$\tilde{Q}_3 \circ \tilde{Q}_d \subseteq \tilde{Q}_3,$$

(2.6) $$\overleftarrow{\tilde{Q}}_4 \circ \tilde{Q}_3 \subseteq \tilde{Q}_+^\circ.$$

In addition, there exists a partition $\{Q_-^\circ, Q_d, Q_+^\circ\}$ *of* $\Lambda \times \Lambda$ *that has the triangular property with* $\overleftarrow{Q}_-^\circ = Q_+^\circ$ *and* $Q_1 \subseteq Q_+^\circ$ *so that*

(2.7) $$Q_1 \circ Q_+ \subseteq Q_1,$$

(2.8) $$Q_2 \circ Q_d \subseteq Q_2,$$

(2.9) $$\overleftarrow{Q}_1 \circ Q_2 \subseteq Q_-^\circ,$$

where $Q_\pm = Q_\pm^\circ \cup Q_d$, $Q_2 = Q_+ \setminus Q_1$, *and also the conditions*

(2.10) $$\tilde{Q}_- \circ Q_2 \subseteq Q_-^\circ \cup Q_2,$$

(2.11) $$Q_2 \circ Q_- \subseteq Q_-^\circ \cup Q_2,$$

(2.12) $$\overleftarrow{\tilde{Q}}_3 \circ Q_- \subseteq Q_-^\circ \cup Q_2,$$

hold

Note that (2.3) and (2.4) are equivalent.

Proof: Let $\leq$ be a linear order on Λ for which (i), (ii) and (iii) in Theorem 2.1 and (iv) in Theorem 2.4 hold. By Theorem 2.1 the partition $\{S, D, \overleftarrow{S}\}$ has the triangular property.

We shall start by defining the partition $\{Q_-^\circ, Q_d, Q_+^\circ\}$. Let $U_j \subseteq \Lambda$ be defined via

$$U_j = \{k \in \Lambda : (k, j) \notin Q_1\}, \quad j \in \Lambda.$$

By (iv) of Theorem 2.4 we get that for all i and j in Λ either $U_i \subseteq U_j$ or $U_j \subseteq U_i$. Indeed, suppose that $\ell \in U_j \setminus U_i$. Since $\ell \notin U_i$, we have that $(\ell, i) \in Q_1$. But then for $p \leq \ell$, we have by (iv) that $(p, i) \in Q_1$. Thus $p \notin U_i$. Consequently,

$$U_i \subseteq \{q \in \Lambda : \ell \leq q, \ell \neq q\}.$$

Next, since $\ell \in U_j$, we have that $(\ell, j) \notin Q_1$. But then for any $p \geq \ell$ we have $(p, j) \notin Q_1$, otherwise (iv) would imply that $(\ell, j) \in Q_1$. Thus any $p \geq \ell$ satisfies $p \in U_j$, giving

$$\{p \in \Lambda : \ell \leq p\} \subseteq U_j.$$

But then $U_i \subseteq U_j$ follows. Now, let

$$
\begin{aligned}
Q_-^\circ &= \{(i, j) \in \Lambda \times \Lambda : U_i \subsetneq U_j\}, \\
Q_d &= \{(i, j) \in \Lambda \times \Lambda : U_i = U_j\}, \\
Q_+^\circ &= \{(i, j) \in \Lambda \times \Lambda : U_j \subsetneq U_i\}.
\end{aligned}
$$

It follows easily from this definition that $\overleftarrow{Q}_-^\circ = Q_+^\circ$, and that $\{Q_-^\circ, Q_d, Q_+^\circ\}$ is a partition of $\Lambda \times \Lambda$ that has the triangular property. When $(i, j) \in Q_1$, then $i \notin U_j$ by definition. Since $(i, i) \in D$ and $D \cap Q_1 = \emptyset$, because $Q_1 \subseteq S$, we have $i \in U_i$. Therefore, $(i, j) \in Q_1$ implies that $i \in U_i \backslash U_j$, and thus $(i, j) \in Q_+^\circ$. It remains to show that (2.4)–(2.12) hold.

For (2.4) let $(i, j) \in \widetilde{Q}_4$ and $(j, k) \in \widetilde{Q}_-$. Thus $(j, i) \in Q_1$ and $(k, j) \in S \cup D$. If $(k, j) \in S$, we have by (i) that $k \leq j$. Thus (iv) implies that $(k, i) \in Q_1$, and thus $(i, k) \in \widetilde{Q}_4$. When $(k, j) \in D$, we obtain from $D \circ Q_1 \subseteq Q_1$ that $(k, i) \in Q_1$ and hence $(i, k) \in \widetilde{Q}_4$.

For (2.5) let $(i, j) \in \widetilde{Q}_3$ and $(j, k) \in \widetilde{Q}_d$. Since $\{\widetilde{Q}_+^\circ, \widetilde{Q}_d, \widetilde{Q}_-^\circ\}$ is a partition of $\Lambda \times \Lambda$ with the triangular property, we have $(i, k) \in \widetilde{Q}_-$. Suppose that $(i, k) \in \widetilde{Q}_4$, in other words $(k, i) \in Q_1$. In this case $(j, k) \in \widetilde{Q}_d = D$ and $D \circ Q_1 \subseteq Q_1$ yield $(j, i) \in Q_1$. Contradiction. Thus $(i, k) \notin \widetilde{Q}_4$, and hence $(i, k) \in \widetilde{Q}_3$.

For (2.6) let $(i, j) \in \widetilde{Q}_4$ and $(j, k) \in \widetilde{Q}_3$. Since $(i, j) \in Q_1$ and $(k, j) \notin Q_1$, we must have that $k \geq i$ and $k \neq i$, otherwise this would contradict (iv). Thus, by (i), we get that $(i, k) \notin \overleftarrow{S}$. If $(i, k) \in D$, then $(k, i) \in D$, and so $(k, j) \in D \circ Q_1 \subseteq Q_1$, which also leads to a contradiction. Thus, by elimination, $(i, k) \in S = \widetilde{Q}_+^\circ$.

For (2.7) let $(i, j) \in Q_1$ and $(j, k) \in Q_+$. Thus $U_k \subseteq U_j$. Since $(i, j) \in Q_1$ we have that $i \notin U_j$. Thus $i \notin U_k$, which gives that $(i, k) \in Q_1$.

For (2.8), let $(i, j) \in Q_2$ and let $(j, k) \in Q_d$. Then $U_k = U_j \subseteq U_i$ so $(i, k) \in Q_+$. Moreover, since $(i, j) \notin Q_1$, we have that $i \in U_j = U_k$, giving that $(i, k) \notin Q_1$.

For (2.9), let $(i, j) \in \overleftarrow{Q}_1$ and $(j, k) \in Q_2$. Then $(j, i) \in Q_1$, thus $j \notin U_i$. As $(j, k) \in Q_2$, we have $j \in U_k$. Thus $U_k \not\subseteq U_i$, and consequently, $U_i \subsetneq U_k$. This yields that $(i, k) \in Q_-^\circ$.

For (2.10), let $(i, j) \in \widetilde{Q}_-$ and $(j, k) \in Q_2$. Thus $(j, i) \in S \cup D$. If $(j, i) \in D$, then we have that $(i, k) \notin Q_1$, since otherwise $(j, i) \in D$ and $(i, k) \in Q_1$ would yield $(j, k) \in D \circ Q_1 \subseteq Q_1$, which contradicts $(j, k) \in Q_2$. Since $Q_1 \subseteq Q_+^\circ$, we have

(2.13) $$Q_-^\circ \cup Q_2 = Q_-^\circ \cup (Q_+ \backslash Q_1) = (\Lambda \times \Lambda) \backslash Q_1,$$

and so we get that $(i, k) \in Q_-^\circ \cup Q_2$. If $(j, i) \in S$, then by (i) we have that $j \leq i$. Consequently, $(i, k) \notin Q_1$, otherwise (iv) would yield that $(j, k) \in Q_1$, giving a contradiction. Thus, also in this case,

$$(i, k) \in (\Lambda \times \Lambda)\backslash Q_1 = Q_-^\circ \cup Q_2.$$

For (2.11), let $(i, j) \in Q_2$ and $(j, k) \in Q_-$. Thus $i \in U_j (\subseteq U_i)$ and $U_j \subseteq U_k$. Thus $i \in U_j \subseteq U_k$, giving that $(i, k) \notin Q_1$. Thus $(i, k) \in Q_-^\circ \cup Q_2$ by (2.13).

Lastly, for (2.12), let $(i, j) \in \overleftarrow{\tilde{Q}_3}$ and $(j, k) \in Q_-$. Then $(j, i) \in \tilde{Q}_3 = \tilde{Q}_-\backslash\tilde{Q}_4$, thus $(i, j) \in (S \cup D)\backslash Q_1$. Thus $i \in U_j$, and since $(j, k) \in Q_-$, we have $U_j \subseteq U_k$. This yields $i \in U_k$, and thus $(i, k) \notin Q_1$. Hence $(i, k) \in Q_-^\circ \cup Q_2$ by (2.13). $\quad\square$

3 Finite Operator Matrices

Let $\mathcal{A}$ be the unital C^*-algebra of $n \times n$ operator matrices $(A_{ij})_{i,j=1}^n$ where $A_{ij} : \mathcal{H}_j \to \mathcal{H}_i$ is a Hilbert space operator. Thus the unit of $\mathcal{A}$ is the identity operator on the Hilbert space direct sum $H_1 \oplus \cdots \oplus H_n$, and the involution on $\mathcal{A}$ is the adjoint of an operator on $H_1 \oplus \cdots \oplus H_n$. We shall consider the following positive extension problem in $\mathcal{A}$.

A subset $J \subseteq \underline{n} \times \underline{n}$, where $\underline{n} = \{1, \ldots, n\}$, is called a $(n \times n)$-*pattern*. We say that J is *reflexive* if $(i, i) \in J$ for all $i \in \underline{n}$, and J is called *symmetric* if $(i, j) \in J$ implies $(j, i) \in J$. Recall from [11], §III.1, that J is called *column-diagonally connected* if $(i, j) \in J$ and $i \leq k \leq j$ imply $(k, j) \in J$.

Given a reflexive, symmetric and column-diagonally connected pattern J, the positive extension problem for J is the following: let $A_{i,j} = A_{j,i}^*$, $(i, j) \in J$ be given operators. When do there exist $A_{ij}(= A_{ji}^*)$, $(i, j) \in (\underline{n} \times \underline{n})\backslash J$, such that

$$(A_{ij})_{i,j=1}^n : \oplus_{i=1}^n \mathcal{H}_i \to \oplus_{i=1}^n \mathcal{H}_1$$

is positive definite? In such case, give a description of all solutions.

The question of when such A_{ij}, $(i, j) \in (\underline{n} \times \underline{n})\backslash J$, exist has been answered already in [11], Theorem III.2.1, and can be traced back further to [13] for the case when $\dim \mathcal{H}_i < \infty$, $i = 1, \ldots, n$. In [15], Section 4.3, a linear fractional representation of all solutions is obtained by reducing the problem in two steps, first to a nonstationary (time-variant) tangential Carathéodory type positive extension problem and next by rewriting the Carathéodory problem as a nonstationary (time-variant) tangential Nevanlinna-Pick interpolation problem. The main contribution in this section is the following linear fractional description for the set of all solutions, the proof of which follows directly from the general abstract scheme developed in [14] and the results of Section 1.

Theorem 3.1 *Let $J \subseteq \underline{n} \times \underline{n}$ be a reflexive, symmetric and column-diagonally connected pattern, and let $A_{ij} = A_{ji}^* : \mathcal{H}_j \to \mathcal{H}_i$ be given Hilbert space operators.*

There exists a positive definite block matrix $B = (B_{ij})_{i,j=1}^n$ with $B_{ij} = A_{ij}$, $(i, j) \in J$, if and only if the operators

$$H_k := (A_{ij})_{i,j \in J_k}, \quad k = 1, \ldots, n,$$

where $J_k = \{p \in \underline{n} : (p, k) \in J \text{ and } p \geq k\}$, are positive definite. In that case, construct $n \times n$ operator matrices U, V and C as follows. If we denote $J_k = \{p_1, \ldots, p_{s_k}\}$ for $k = 1, \ldots, n$ with $k = p_1 < p_2 < \cdots < p_{s_k}$, put

$$\begin{bmatrix} \hat{V}_{p_1,k} \\ \vdots \\ \hat{V}_{p_{s_k},k} \end{bmatrix} := H_k^{-1} \begin{bmatrix} I \\ 0 \\ \vdots \\ 0 \end{bmatrix}, \quad k = 1, \ldots, n,$$

and let $V = (V_{ik})_{i,k=1}^n$ be given by

$$V_{ik} = \begin{cases} 0, & i \notin J_k, \\ \hat{V}_{ik} \hat{V}_{kk}^{-1/2}, & i \in J_k. \end{cases}$$

For $k \in \underline{n}$, let

$$L_k = \{i : ((p, k) \in J \text{ and } p \leq i) \Rightarrow (p, i) \in J\} \cap \{i : (i, k) \in J \text{ or } i > k\}.$$

If we denote $L_k = \{q_1, \ldots, q_{t_k}\}$, put

$$\begin{bmatrix} \hat{U}_{q_1,k} \\ \vdots \\ \hat{U}_{q_{t_k},k} \end{bmatrix} := [(A_{q_i,q_j})_{i,j=1,\ldots,t_k}]^{-1} \begin{bmatrix} \delta_{q_1,k} I \\ \vdots \\ \delta_{q_{t_k},k} I \end{bmatrix},$$

where $\delta_{s,t} = 0$ when $s \neq t$ and $\delta_{ss} = 1$, and let $U = \hat{U} D^{-1/2}$, where

$$D = (D_{ij})_{i,j=1}^n, \quad D_{ij} = \begin{cases} \hat{U}_{ij}, & \text{if } \hat{U}_{ij} \neq 0 \neq \hat{U}_{ji}, \\ 0, & \text{otherwise.} \end{cases}$$

Finally, let $C = (C_{ij})_{i,j=1}^n$ be so that

$$C + C^* = U^{*-1} U^{-1} (= V^{*-1} V^{-1})$$

and $C_{ij} = 0$ for

$$(i, j) \in \{(i, j) : i > j \text{ and } (i, j) \notin J\}.$$

Then every $n \times n$ positive definite operator matrix $B = (B_{ij})_{i,j=1}^n$ with $B_{ij} = A_{ij}$ for $(i, j) \in J$ has the form

(3.1)
$$\begin{aligned} B &= (-C^* V G + CU)(V G + U)^{-1} \\ &\quad + (G^* V^* + U^*)^{-1}(-G^* V^* C + U^* C) \\ &= (V G + U)^{*-1}(I - G^* G)(V G + U)^{-1}, \end{aligned}$$

where

$$G = (G_{ij})_{i,j=1}^n$$

is an operator matrix with $\|G\| < 1$ *and* $G_{ij} = 0$ *for* $(i, j) \in J \cup \{(i, j) : i > j\}$. *Moreover, the correspondence* (3.1) *between such operator matrices* B *and* G *is one-to-one.*

Note that it is not strictly necessary to introduce the matrix C in the statement of the theorem since the last expression of (3.1) does not contain the matrix C. However, when one is interested in extensions with positive real part the representation that includes the matrix C is useful (see [5], [14]).

Before proving this result we consider an example which also appears in [15], Section 4.3. Let A be the partial matrix

$$A = \begin{bmatrix} 5 & -4 & ? & ? & ? \\ -4 & 5 & ? & ? & 2 \\ ? & ? & 4 & ? & -4 \\ ? & ? & ? & 1 & 2 \\ ? & 2 & -4 & 2 & 8 \end{bmatrix},$$

where question marks correspond to unknown entries. The pattern

$$J = \{1, 2\}^2 \cup \{(5, 2), (2, 5), (5, 3), (3, 5), (3, 3)\} \cup \{4, 5\}^2$$

is column-diagonally connected. In this case we get that

$$J_1 = \{1, 2\}, \quad J_2 = \{2, 5\}, \quad J_3 = \{3, 5\}, \quad J_4 = \{4, 5\}, \quad J_5 = \{5\}$$

and

$$L_1 = \{1, 2\}, \quad L_2 = \{1, 2\}, \quad L_3 = \{3, 5\}, \quad L_4 = \{4, 5\}, \quad L_5 = \{2, 5\}.$$

For the matrices U and V we obtain

$$U = \begin{bmatrix} \frac{5}{9} & \frac{4}{9} & 0 & 0 & 0 \\ \frac{4}{9} & \frac{5}{9} & 0 & 0 & -\frac{1}{18} \\ 0 & 0 & \frac{1}{2} & 0 & 0 \\ 0 & 0 & 0 & 2 & 0 \\ 0 & 0 & \frac{1}{4} & -\frac{1}{2} & \frac{5}{36} \end{bmatrix} \begin{bmatrix} \frac{2}{3} & \frac{1}{3} & 0 & 0 & 0 \\ \frac{1}{3} & \frac{2}{3} & 0 & 0 & 0 \\ 0 & 0 & \frac{1}{2}\sqrt{2} & 0 & 0 \\ 0 & 0 & 0 & \sqrt{2} & 0 \\ 0 & 0 & 0 & 0 & \frac{\sqrt{5}}{6} \end{bmatrix}^{-1},$$

$$V = \begin{bmatrix} \frac{5}{9} & 0 & 0 & 0 & 0 \\ \frac{4}{9} & \frac{2}{9} & 0 & 0 & 0 \\ 0 & 0 & \frac{1}{2} & 0 & 0 \\ 0 & 0 & 0 & 2 & 0 \\ 0 & -\frac{1}{18} & \frac{1}{4} & -\frac{1}{2} & \frac{1}{8} \end{bmatrix} \begin{bmatrix} \frac{\sqrt{5}}{3} & 0 & 0 & 0 & 0 \\ 0 & \frac{\sqrt{2}}{3} & 0 & 0 & 0 \\ 0 & 0 & \frac{1}{2}\sqrt{2} & 0 & 0 \\ 0 & 0 & 0 & \sqrt{2} & 0 \\ 0 & 0 & 0 & 0 & \frac{1}{4}\sqrt{2} \end{bmatrix}^{-1}.$$

Then

$$U^{*-1}U^{-1} = V^{*-1}V^{-1} = \begin{bmatrix} 5 & -4 & \frac{4}{5} & -\frac{2}{5} & -\frac{8}{5} \\ -4 & 5 & -1 & \frac{1}{2} & 2 \\ \frac{4}{5} & -1 & 4 & -1 & -4 \\ -\frac{2}{5} & \frac{1}{2} & -1 & 1 & 2 \\ -\frac{8}{5} & 2 & -4 & 2 & 8 \end{bmatrix}.$$

Choose C to be for instance

$$C = \begin{bmatrix} 2\frac{1}{2} & -4 & \frac{4}{5} & -\frac{2}{5} & -\frac{8}{5} \\ 0 & 2\frac{1}{2} & -1 & \frac{1}{2} & 2 \\ 0 & 0 & 2 & -1 & -2 \\ 0 & 0 & 0 & \frac{1}{2} & 2 \\ 0 & 0 & 0 & 0 & 4 \end{bmatrix}.$$

If we let

$$G = \begin{bmatrix} 0 & 0 & g_{13} & g_{14} & g_{15} \\ 0 & 0 & g_{23} & g_{24} & 0 \\ 0 & 0 & 0 & g_{34} & 0 \\ 0 & 0 & 0 & 0 & 0 \\ 0 & 0 & 0 & 0 & 0 \end{bmatrix},$$

then (3.1) gives a one-to-one correspondence between all strict contractions G of the above form and all positive definite completions of the partial matrix A.

Proof of Theorem 3.1: We will apply Theorem 1.6 in [14]. Let $\mathcal{A} = \mathcal{R} = \mathcal{B}(\mathcal{H}, \oplus \cdots \oplus \mathcal{H}_n)$, which is a unital C^*-algebra. Define the following subsets of $\underline{n} \times \underline{n}$:

$$S = \{(i, j) : i < j\}$$
$$D = \{(i, i) : i \in \underline{n}\}$$

and $Q_1 = S \backslash J$. Then, clearly, $D \circ Q_1 \subseteq Q_1$ and properties (i), (ii) and (iii) in Theorem 2.1 and property (iv) in Theorem 2.4 hold with respect to the usual ordering $\leq$ on $\underline{n}$. To check property (iv) one needs that J is column-diagonally connected. We may now apply Theorem 2.5 and introduce $\tilde{Q}_+^\circ, \tilde{Q}_d, \tilde{Q}_-^\circ, \tilde{Q}_+, \tilde{Q}_-,$ $\tilde{Q}_4, \tilde{Q}_3, Q_-^\circ, Q_d, Q_+^\circ, Q_2, Q_-, Q_+$ such that (2.4)–(2.12) hold. With these sets we introduce subsets of $\mathcal{A}$ accordingly as follows:

$$\tilde{\mathcal{A}}_+^0 = \{(A_{ij})_{i,j=1}^n : A_{ij} = 0 \text{ for } (i, j) \notin \tilde{Q}_+^\circ\},$$
$$\tilde{\mathcal{A}}_d = \{(A_{ij})_{i,j=1}^n : A_{ij} = 0 \text{ for } (i, j) \notin \tilde{Q}_d\},$$

etc. Using inclusions (2.4)–(2.12) one sees that, in the terminology of [14], $(\mathcal{A}_-^\circ, \mathcal{A}_d, \mathcal{A}_+^\circ, \mathcal{A}_1, \mathcal{A}_2)$ is a right semi-band structure on $\mathcal{A}$ and the quintet $(\tilde{\mathcal{A}}_-^\circ, \tilde{\mathcal{A}}_d, \tilde{\mathcal{A}}_+^\circ, \tilde{\mathcal{A}}_3, \tilde{\mathcal{A}}_4)$ is a left semi-band structure on $\mathcal{A}$, that the two band structures are coupled and that the conditions $(U1)$ $(U2)$ and $(U3)$ in [14] are satisfied. In addition, if $H \in \mathcal{A}_+$ and $\|H\| < 1$, then $(I - H)^{-1} \in \mathcal{A}_+$, which follows simply from the observation that if H is permutation similar to an upper triangular, then the same holds for $(I - H)^{-1}$ with the same permutation similarity. We may now apply Theorem 1.6 in [14] to $K = (K_{ij})_{i,j=1}^n$ where

$$K_{ij} = \begin{cases} A_{ij}, & i < j \\ 0, & i > j, \\ \frac{1}{2}A_{ii}, & i = j, \end{cases}$$

where $A_{ij} = 0$ for $(i, j) \notin J$. Writing out the equations

$$P_{\mathcal{A}_2}((K + K^*)\hat{U}) = E, \quad P_{\tilde{\mathcal{A}}_3}((K + K^*)\hat{V}) = E,$$

we see that the solutions $\hat{U} \in \mathcal{A}_2$ and $\hat{V} \in \tilde{\mathcal{A}}_3$ are exactly given by $\hat{U}$ and $\hat{V}$ as in the statement of the theorem. Since $\hat{V}$ is lower triangular with invertible (in fact, positive definite) diagonal entries, V^{-1} exists and belongs to $\tilde{\mathcal{A}}_-$. Since after permutation $\hat{U}$ is block upper triangular with positive definite block diagonal entries it is easy to see that U^{-1} exists and belongs to $\mathcal{A}_+$. Thus we obtain that

$$B_0 := U^{*-1}U^{-1} = V^{*-1}V^{-1}$$

satisfies $(B_0)_{ij} = A_{ij}$, $(i, j) \in J$. Apply now Theorem 1.6 in [14] to obtain that (3.1) yields all positive definite B with $B_{ij} = A_{ij}$, $(i, j) \in J$, where $G \in \mathcal{A}_1$ is a strict contraction. The last equality in (3.1) is a simply algebraic exercise that goes back to [7], [18]. The one-to-one correspondence also follows from Theorem 1.6 in [14]. This finishes the proof.

4 Semi-infinite Operator Matrices

Let $\mathcal{L}$ denote the linear space of all semi-infinite operator matrices

(4.1) $$V = (V_{jk})_{j,k=1}^\infty \text{ such that } \sum_{\nu=-\infty}^\infty \sup_{k-j=\nu} \|V_{jk}\| < \infty.$$

The entry V_{jk} of V is assumed to be an operator from the Hilbert space $\mathcal{H}_k$ into the Hilbert space $\mathcal{H}_j$, and the norm in (4.1) is the operator norm. The space $\mathcal{L}$ is an algebra under the usual operations of addition and multiplication for infinite matrices. For $V = (V_{jk})_{j,k=1}^\infty$ we define

$$V^* = (V_{kj}^*)_{j,k=1}^\infty$$

and this operation $*$ is an involution on $\mathcal{L}$. The element $E = \text{diag}(I_{\mathcal{H}_j})_{j=1}^\infty$ is the unit in $\mathcal{L}$.

We write $\mathcal{Z}$ for the Hilbert space direct sum $\oplus_{j=1}^{\infty} \mathcal{H}_j$. Thus $\mathcal{Z}$ consists of all square summable sequences $(\eta_j)_{j=1}^{\infty}$ with $\eta_j \in \mathcal{H}_j$. Note that each $V \in \mathcal{L}$ induces a bounded linear operator on $\mathcal{Z}$.

We are interested in the following extension problem. Given are operators $A_{ij} = A_{ij}^*$ for (i, j) in the symmetric set of indices

$$(4.2) \qquad J = \{(i, j) \in \mathbb{N} \times \mathbb{N} \mid i = 1 \text{ or } j = 1 \text{ or } |j - i| \le m\},$$

where $\mathbb{N} = \{1, 2, 3, \ldots\}$. Thus the given data is centered around the main diagonal in a band of width m and is located in the last row and column. We are looking for $V = (V_{ij})_{i,j=1}^{\infty} \in \mathcal{L}$ with the properties

(a) V induces a positive definite operator on $\mathcal{Z}$;

(b) $V_{ij} = A_{ij}, \quad (i, j) \in J$.

Such an infinite operator matrix $V \in \mathcal{L}$ will be called a *positive* extension of $\{A_{ij} \mid (i, j) \in J\}$. If V satisfies in addition the condition

(c) $(V^{-1})_{ij} = 0, \quad (i, j) \notin J$,

then V is called a *positive band extension* of $\{A_{ij} \mid (i, j) \in J\}$.

Theorem 4.1 *Let J be the index set (4.2), and let $A_{ij} = A_{ji}^*$, $(i, j) \in J$, be an operator from $\mathcal{H}_j$ into $\mathcal{H}_i$. The given data $\{A_{ij} \mid (i, j) \in J\}$ has a positive extension if and only if*

$$(4.3) \qquad\qquad \sum_{\nu=1}^{\infty} \|A_{1\nu}\| < \infty,$$

the operator matrices

$$(4.4) \qquad H_j := \begin{bmatrix} A_{11} & \cdots & A_{1j} \\ \vdots & & \vdots \\ A_{j1} & \cdots & A_{jj} \end{bmatrix}, \qquad j = 1, 2, \ldots, m+1,$$

$$(4.5) \qquad H_j := \begin{bmatrix} A_{11} & A_{1,j-m} & \cdots & A_{1j} \\ A_{j-m,1} & A_{j-m,j-m} & \cdots & A_{j-m,j} \\ \vdots & & & \vdots \\ A_{j-m,1} & A_{j,j-m} & \cdots & A_{jj} \end{bmatrix}, \qquad j \ge m+2,$$

are positive definite, and

$$(4.6) \qquad\qquad \|H_j\|, \|H_j^{-1}\| \le M, \qquad j \ge 1,$$

for a fixed $M > 0$. Let $X = (X_{ij})_{i,j=1}^{\infty}$ and $Y = (Y_{ij})_{i,j=1}^{\infty}$ be given via

$$(4.7) \qquad H_{j+m} \begin{bmatrix} X_{1,j} \\ X_{jj} \\ \vdots \\ X_{m+j,j} \end{bmatrix} = \begin{bmatrix} 0 \\ I \\ 0 \\ \vdots \\ 0 \end{bmatrix}, j \geq 2, \ X_{11} = A_{11}^{-1},$$

$$(4.8) \qquad H_j \begin{bmatrix} Y_{1j} \\ \vdots \\ Y_{jj} \end{bmatrix} = \begin{bmatrix} 0 \\ \vdots \\ 0 \\ I \end{bmatrix}, j = 1, \ldots, m+1,$$

$$(4.9) \qquad H_j \begin{bmatrix} Y_{1j} \\ Y_{j-m,j} \\ \vdots \\ Y_{jj} \end{bmatrix} = \begin{bmatrix} 0 \\ \vdots \\ 0 \\ I \end{bmatrix}, j \geq m+2,$$

and $X_{ij} = 0$ and $Y_{ij} = 0$ otherwise. Put

$$U = (U_{ij})_{i,j=1}^{\infty}, \qquad U_{ij} = X_{ij} X_{jj}^{-1/2},$$
$$V = (V_{ij})_{i,j=1}^{\infty}, \qquad V_{ij} = Y_{ij} Y_{jj}^{-1/2},$$

Then

$$B := U^{*-1} U^{-1} = V^{*-1} V^{-1}$$

is the unique positive band extension of $\{A_{ij} \mid (i, j) \in J\}$. Moreover, the map

$$T(G) = (U + VG)^{*-1}(I - G^*G)(U + VG)^{-1}$$

gives a one-to-one correspondence between the set of all positive extensions of $\{A_{ij} \mid (i, j) \in J\}$ and the parameter set

$$\{G = (G_{ij})_{i,j=1}^{\infty} \mid \|G\| < 1 \text{ and } G_{ij} = 0 \text{ for } (i, j) \in J \cup \{(i, j) : i \leq j\}\}.$$

Proof: The necessity of the conditions follows from Theorem II.1.1 in [11]. For the sufficiency we will apply Theorem 1.6 in [14]. Let $\mathcal{A} = \mathcal{L}$ and $\mathcal{R} = \mathcal{B}(\mathcal{Z})$, the C^*-algebra of bounded linear operators on $\mathcal{Z}$. Let

$$S = \{(i, j) : i > j\} \subset \mathbb{N} \times \mathbb{N},$$
$$D = \{(i, i) : i \in \mathbb{N}\},$$

and $Q_1 = S \backslash J$. Then $D \circ Q_1 \subseteq Q_1$ and properties (i), (ii) and (iii) in Theorem 2.1 and property (iv) in Theorem 2.4 hold with respect to the (reverse) ordering $\geq$

on $\mathbb{N}$. We may now apply Theorem 2.5. We obtain that

$$\tilde{Q}^0_+ = \overleftarrow{\tilde{Q}^0}_- = S, \ \tilde{Q}_d = Q_d = D,$$

$$\tilde{Q}_+ = \overleftarrow{\tilde{Q}}_- = S \cup D, \ \tilde{Q}_4 = \overleftarrow{\tilde{Q}}_1,$$

$$\tilde{Q}_3 = Q_- \backslash \tilde{Q}_4,$$

$$Q^0_+ = \overleftarrow{Q}^0_- = \{(i, j) : i > j > 1 \text{ or } j > i = 1\},$$

$$Q_+ = \overleftarrow{Q}_- = Q^0_+ \cup Q_d, \ Q_2 = Q_+ \backslash Q_1,$$

satisfy (2.4)–(2.12). With these sets we define subspaces of $\mathcal{A}$ accordingly:

$$\tilde{\mathcal{A}}^0_+ = \tilde{\mathcal{A}}^{0*}_- = \{(A_{ij})^\infty_{i,j=1} | A_{ij} = 0 \text{ for } (i, j) \notin \tilde{Q}^0_+\},$$

$$\tilde{\mathcal{A}}_4 = \{(A_{ij})^\infty_{i,j=1} | A_{ij} = 0 \text{ for } (i, j) \notin \tilde{Q}_4\},$$

etc. Using inclusions (2.4)–(2.12) one sees that, in the terminology of [14], $(\mathcal{A}^0_-,$ $\mathcal{A}_d, \mathcal{A}^0_+, \mathcal{A}_1, \mathcal{A}_2)$ is a right semi-band structure on $\mathcal{A}$ and the quintet $(\tilde{\mathcal{A}}^0_-, \tilde{\mathcal{A}}_d, \tilde{\mathcal{A}}^0_+, \tilde{\mathcal{A}}_3, \tilde{\mathcal{A}}_4)$ is a left semi-band structure on $\mathcal{A}$, that the two semi-band structures are coupled and that the conditions (U1), (U2) and (U3) in [14] are satisfied. In addition, if $H \in \mathcal{A}_+$ and $\|H\| < 1$, then

$$(I - H)^{-1} = \sum_{k=0}^\infty H^k \in \mathcal{A}_+.$$

We may now apply Theorem 1.6 in [14] to $K = (K_{ij})^\infty_{i,j=1}$, where

$$K_{ij} = \begin{cases} A_{ij}, & i > j, \\ 0, & i < j, \\ \frac{1}{2} A_{ii}, & i = j, \end{cases}$$

and $A_{ij} = 0$ for $(i, j) \notin J$. Thus we need to solve for $X \in \mathcal{A}_2$ and $Y \in \tilde{\mathcal{A}}_3$ so that

$$P_{\mathcal{A}_2}((K + K^*)X) = E \text{ and } P_{\tilde{\mathcal{A}}_3}((K + K^*)Y) = E.$$

It is straightforward to check that these two equations are equivalent to equations (4.7)–(4.9). Next we need to show that X and Y are invertible and that $X^{-1} \in \mathcal{A}_+$ and $Y^{-1} \in \tilde{\mathcal{A}}_-$. The statement for Y follows directly from Theorem II.1.1 in [11], since the element Y corresponds exactly to $\hat{X}$ in Theorem II.1.1 in [11]. In order to show that $X^{-1} \in \mathcal{A}_+$ we employ a similar argument as in the proof of Theorem II.1.1 in [11]. Write

$$X = \begin{bmatrix} X_{11} & \Psi \\ 0 & \Phi \end{bmatrix}$$

where $\Psi = \text{row } (X_{1j})_{j=2}^\infty$ and $\Phi = (X_{ij})_{i,j=2}^\infty$. Since X_{00} is invertible, one notices that it suffices to show that Φ is invertible and Φ^{-1} is lower triangular. Indeed, then it follows that

$$X^{-1} = \begin{bmatrix} X_{11}^{-1} & -X_{11}^{-1}\Psi\Phi^{-1} \\ 0 & \Phi^{-1} \end{bmatrix} \in \mathcal{A}_+.$$

Observe that

$$\begin{bmatrix} X_{jj} \\ \vdots \\ X_{j+m,j} \end{bmatrix} = L_{j+m} \begin{bmatrix} I \\ 0 \\ \vdots \\ 0 \end{bmatrix},$$

where L_{j+m} is the submatrix of H_{j+m}^{-1} consisting of the last $m+1$ rows and columns. Using the general rule that

$$\begin{bmatrix} P & Q \\ R & S \end{bmatrix}^{-1} = \begin{bmatrix} * & * \\ * & (S - RP^{-1}Q)^{-1} \end{bmatrix}$$

when the indicated inverses exist, we observe that

$$L_{j+m}^{-1} = \begin{bmatrix} A_{jj} - A_{j1}A_{11}^{-1}A_{1j} & \cdots & A_{j,j+m} - A_{j1}A_{11}^{-1}A_{1,j+m} \\ \vdots & & \vdots \\ A_{j+m,j} - A_{j+m,1}A_{11}^{-1}A_{1j} & \cdots & A_{j+m,j+m} - A_{j+m,1}A_{11}^{-1}A_{1,j+m} \end{bmatrix}.$$

Consequently, Φ appears when one applies the band method on the data

$$\{A_{ij} - A_{i1}A_{11}^{-1}A_{1j} \mid i, j \geq 2, |i - j| \leq m\}.$$

For such a band this was done in §II.3 in [8] where Φ corresponds to the solution y in Theorem II.3.3 in [8]. Thus it follows from Section II.3 in [8] that Φ^{-1} exists and is upper triangular.

We may now apply Theorems 1.4 and 1.6 in [14] to obtain that B is the unique positive band extension, and that $T(G)$ provides the stated one-to-one correspondence. $\qquad\square$

It should be observed that, since in this section we considered a bordered band example, the results could also have been obtained by using Theorems 3.1 and 3.3 in [12]. The main purpose of this section is to give another illustration of the theory developed in Section 1 in combination with the results in [12].

Acknowledgements

The research of the second author is partially supported by NSF Grant DMS 9800704.

References

[1]   J.A. Ball, I. Gohberg and M.A. Kaashoek, The Band Method and Grassmanian Approach for Completion and Extension Problems, in: *Recent developments in operator theory and its applications (Winnipeg, MB, 1994)*, OT **87**, Birkhäuser Verlag, Basel, 1996, 17–60.

[2]   J.A. Ball, I. Gohberg and M.A. Kaashoek, Nudelman Interpolation and the Band Method, *Integral Equations and Operator Theory* **27** (1997), 253–284.

[3]   H. Dym and I. Gohberg, Extensions of Matrix Valued Functions with Rational Polynomial Inverse, *Integral Equations and Operator Theory* **2** (1979), 503–528.

[4]   H. Dym and I. Gohberg, Extensions of Kernels of Toeplitz Operators, *J. d'Anal. Math* **42** (1982/83), 51–97.

[5]   R.L. Ellis, I. Gohberg and D.C. Lay, Extensions with Positive Real Part. A New Version of the Abstract Band Method with Applications, *Integral Equations and Operator Theory* **16** (1993), 360–384.

[6]   I. Gohberg, S. Goldberg and M.A. Kaashoek, *Classes of Linear Operators. II*, OT **63**, Birkhäuser Verlag, Basel, 1993.

[7]   I. Gohberg, M.A. Kaashoek and H.J. Woerdeman, The Band Method for Positive and Contractive Extension Problems, *J. Operator Theory* **22** (1989), 109–155.

[8]   I. Gohberg, M.A. Kaashoek and H.J. Woerdeman, The band method for positive and contractive extension problems: An alternative version and new applications, *Integral Equations and Operator Theory* **12** (1989), 343–382.

[9]   I. Gohberg, M.A. Kaashoek and H.J. Woerdeman, A Maximum Entropy Principle in the General Framework of the Band Method, *J. Funct. Anal.* **95** (1991), 231–254.

[10] I. Gohberg, M.A. Kaashoek and H.J. Woerdeman, The Band Method for Extension Problems and Maximum Entropy, in: *Signal Processing; Part I* (L. Auslander, T. Kailath and S. Mitter, Eds.), The IMA Volumes in Mathematics and its Applications **22**, Springer Verlag, New York, 1990, 75–94.

[11] I. Gohberg, M.A. Kaashoek and H.J. Woerdeman, The band method for several positive extension problems of non-band type, *J. Operator Theory* **26** (1991), 191–218.

[12] I. Gohberg, M.A. Kaashoek and H.J. Woerdeman, The band method for bordered algebras, in: *Contributions to Operator Theory and its Applications*, OT **62**, Birkhäuser Verlag 1993, 85–97.

[13] R. Grone, C.R. Johnson, E.M. Sá and H. Wolkowitz, Positive definite completions of partial Hermitian matrices, *Linear Algebra Appl.* **58** (1984), 109–124.

[14] M.A. Kaashoek and C.G. Zeinstra, The band method and generalized Carathéodory Toeplitz interpolation at operator points, *Integral Equations and Operator Theory* **33** (1999), 175–210.

[15] J. Kos, *Time-dependent problems in linear operator theory*, Ph. D. thesis, Department of Mathematics, Vrije Universiteit, Amsterdam, 1995.

[16] L. Rodman, I.M. Spitkovsky and H.J. Woerdeman, Abstract Band Method via Factorization, Positive and Band Extensions of Multivariable Almost Periodic Matrix Functions, and Spectral Estimation, preprint.

[17] S. Willard, *General Topology*, Addison-Wesley, Reading MA, 1970.

[18]   H.J. Woerdeman, *Matrix and Operator Extensions*, CWI Tract **68**. Centre for Mathematics and Computer Science, Amsterdam, The Netherlands, 1989.

[19]   H.J. Woerdeman, Hermitian and normal completions, *Linear Multilinear Algebra* **42** (1997), 239–280.

M.A. Kaashoek
Divisie Wiskunde en Informatica
Faculteit Exacte Wetenschappen
Vrije Universiteit
De Boelelaan 1081a
1081 HV Amsterdam
The Netherlands
e-mail: kaash@cs.vu.nl

H.J. Woerdeman
Department of Mathematics
P. O. Box 8795
The College of William and Mary
Williamsburg
VA 23187-8795
U.S.A.
e-mail: hugo@math.wm.edu

1991 Mathematics Subject Classification. Primary: 47A57; Secondary: 15A30.

Operator Theory:
Advances and Applications, Vol. 124
© 2001 Birkhäuser Verlag Basel/Switzerland

Parametrized Furuta Inequality
and its Application

Eizaburo Kamei

Dedicated to Professor Isreal Gohberg on his seventieth birthday

As generalizations of the Furuta inequality and grand Furuta inequality, we give parametrized forms of them; if $A \geq B > 0$, then

$$A^u \, \natural_{\frac{\delta - u}{p - u}} \, B^p \leq B^\delta$$

for $u \leq 0$, $0 \leq \delta \leq p$ and

$$A^u \, \natural_{\frac{\delta - u}{\beta - u}} \, (A^t \, \natural_{\frac{\beta - t}{p - t}} \, B^p) \leq (A^t \, \natural_{\frac{\beta - t}{p - t}} \, B^p)^{\frac{\delta}{\beta}}$$

for $t \in [0, 1]$, $0 \leq t < p \leq \beta$, $u \leq 0$ and $\delta \in [0, \beta]$.

Applying these results, we can easily show the monotone properties of an operator function for $A \geq B$ defined by

$$H_{p,\delta,t}(A, B, u, \beta) = A^u \, \natural_{\frac{\delta - u}{\beta - u}} \, (A^t \, \natural_{\frac{\beta - t}{p - t}} \, B^p);$$

it is increasing for $u \leq 0$ and decreasing for $\beta \geq p$ where $t \in [0, 1]$, $0 \leq t < p \leq \beta$, $u \leq 0$ and $\delta \in [0, \beta]$.

1 Introduction

About twelve years ago, Furuta established an interesting operator inequality [7] which is an extension of the Löwner-Heinz inequality. This inequality is called now the Furuta inequality and penetrating through the vast area of operator theory. First of all, we recall it for convenience.

Throughout this note, a capital letter means a bounded linear operator on a Hilbert space H. An operator A is said to be positive (in symbol: $A \geq 0$) if $(Ax, x) \geq 0$ for all $x \in H$, and also an operator A is strictly positive (in symbol: $A > 0$) if A is positive and invertible.

The original form of the Furuta inequality [7] given by Furuta himself is the following (cf. [8]).

Furuta Inequality: *If $A \geq B \geq 0$, then for each $r \geq 0$,*

$$(A^{\frac{r}{2}} A^p A^{\frac{r}{2}})^{\frac{1}{q}} \geq (A^{\frac{r}{2}} B^p A^{\frac{r}{2}})^{\frac{1}{q}}$$

and

$$(B^{\frac{r}{2}} A^p B^{\frac{r}{2}})^{\frac{1}{q}} \geq (B^{\frac{r}{2}} B^p B^{\frac{r}{2}})^{\frac{1}{q}}$$

holds for p and q such that $p \geq 0$ and $q \geq 1$ with

$$(1 + r)q \geq p + r.$$

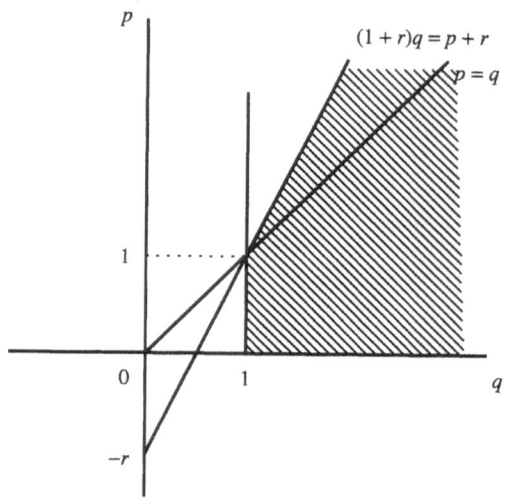

Figure 1

In [21], Tanahashi had shown the domain drawn for p, q and r in the figure is the best possible one for this inequality.

The case of $r = 0$ in this inequality is the Löwner-Heinz inequality:

(LH) $A^\alpha \geq B^\alpha$ for $A \geq B \geq 0$ and $0 \leq \alpha \leq 1$.

Here we use the notations $\natural_\alpha$ and $\flat_\alpha$ defined for positive operators A and B by

$$A \, \flat_\alpha \, B = A^{\frac{1}{2}} (A^{-\frac{1}{2}} B A^{-\frac{1}{2}})^\alpha A^{\frac{1}{2}}, \quad \text{for } \alpha \in \mathbf{R}$$

and $\natural_\alpha = \flat_\alpha$ when $\alpha \in [0, 1]$. Note that $\natural_\alpha$ is an operator mean in the sense of Kubo-Ando [20] which corresponds to the operator monotone function x^α in the Löwner theory.

From the viewpoint of operator means ([2], [3], [15], [16] etc.), the Furuta inequality is rewritten as follows;

$$A^u \, \natural_{\frac{1-u}{p-u}} \, B^p \leq A \quad \text{and} \quad B \leq B^u \, \natural_{\frac{1-u}{p-u}} \, A^p$$

for $p \geq 1$ and $u \leq 0$. As shown in [16], we had arranged these inequalities in one line as follows:

Satellite Theorem of the Furuta Inequality: *If $A \geq B \geq 0$, then*

$$A^u \, \natural_{\frac{1-u}{p-u}} \, B^p \; \leq B \leq A \leq B^u \, \natural_{\frac{1-u}{p-u}} \, A^p$$

for all $p \geq 1$ and $u \leq 0$.

2 Parametrization of the Furuta Inequality

We can generalize the satellite theorem as follows, in which the case of $\delta = 1$ is the satellite theorem [4] (cf. [17], [19]).

Generalization of Satellite Theorem: *If $A \geq B > 0$, then*

(i) *for $0 \leq \delta \leq 1$, $\delta \leq p$ and $u \leq 0$*

$$A^u \, \natural_{\frac{\delta-u}{p-u}} \, B^p \leq B^\delta \leq A^\delta \leq B^u \, \natural_{\frac{\delta-u}{p-u}} \, A^p,$$

and

(ii) *for $-1 \leq \gamma \leq 0$, $u \leq \gamma$ and $p \geq 0$*

$$A^u \, \natural_{\frac{\gamma-u}{p-u}} \, B^p \leq A^\gamma \leq B^\gamma \leq B^u \, \natural_{\frac{\gamma-u}{p-u}} \, A^p.$$

More generally we have the following and call it a parametrization of the Furuta inequality [17], [19] (cf. [11]).

Theorem 2.1 (Parametrized Furuta Inequality). *If $A \geq B > 0$, then*

(i) *for $0 \leq \delta \leq p$ and $u \leq 0$*

$$A^u \, \natural_{\frac{\delta-u}{p-u}} \, B^p \leq B^\delta \text{ and } B^u \, \natural_{\frac{\delta-u}{p-u}} \, A^p \geq A^\delta,$$

and

(ii) *for $u \leq \gamma \leq 0$ and $p \geq 0$*

$$A^u \, \natural_{\frac{\gamma-u}{p-u}} \, B^p \leq A^\gamma \text{ and } B^u \, \natural_{\frac{\gamma-u}{p-u}} \, A^p \geq B^\gamma.$$

Proof: (i) The case of $p \leq 1$ has already shown in [4], we have only to see the case of $p > 1$. We shall prove by induction on $k = 1, 2, \ldots$ that

$$A^u \, \natural_{\frac{\delta-u}{p-u}} \, B^p \leq B^\delta \text{ whenever } -k \leq u < -k+1.$$

When $k = 1$ and $-1 \leq u < 0$, $A^{-u} \geq B^{-u}$ by the Löwner-Heinz, and $A^u \leq B^u$ because inverse formation is order-converting. Then since the mean $\natural_{\frac{\delta-u}{p-u}}$ is monotone in each variable,

$$A^u \; \natural_{\frac{\delta-u}{p-u}} \; B^p \leq B^u \; \natural_{\frac{\delta-u}{p-u}} \; B^p = B^\delta.$$

Now suppose that the assertion is true for k and take u such that $-(k+1) \leq u < -k$. Then

$$
\begin{aligned}
A^u \; \natural_{\frac{\delta-u}{p-u}} \; B^p &= A^{-\frac{k}{2}}(A^{u+k} \; \natural_{\frac{\delta-u}{p-u}} \; A^{\frac{k}{2}}B^p A^{\frac{k}{2}})A^{-\frac{k}{2}} \\
&\leq A^{-\frac{k}{2}}((A^{\frac{k}{2}}B^p A^{\frac{k}{2}})^{\frac{u+k}{p+k}} \; \natural_{\frac{\delta-u}{p-u}} \; A^{\frac{k}{2}}B^p A^{\frac{k}{2}})A^{-\frac{k}{2}} \\
&= A^{-\frac{k}{2}}(A^{\frac{k}{2}}B^p A^{\frac{k}{2}})^{(1-\frac{u+k}{p+k})\frac{\delta-u}{p-u}+\frac{u+k}{p+k}}A^{-\frac{k}{2}} \\
&= A^{-\frac{k}{2}}(A^{\frac{k}{2}}B^p A^{\frac{k}{2}})^{\frac{\delta+k}{p+k}}A^{-\frac{k}{2}} \\
&= A^{-k} \; \natural_{\frac{\delta+k}{p+k}} \; B^p = A^{(-k)} \; \natural_{\frac{\delta-(-k)}{p-(-k)}} \; B^p \leq B^\delta.
\end{aligned}
$$

The first inequality of this calculation is obtained as the following; since

$$(A^{\frac{k}{2}}B^p A^{\frac{k}{2}})^{\frac{1+k}{p+k}} \leq A^{1+k}$$

by the Furuta inequality or satellite theorem, the Löwner-Heinz inequality leads to $(A^{\frac{k}{2}}B^p A^{\frac{k}{2}})^{\frac{1}{p+k}} \leq A$ and $(A^{\frac{k}{2}}B^p A^{\frac{k}{2}})^{\frac{u+k}{p+k}} \geq A^{u+k}$ as before because $-1 \leq u + k < 0$. The second one is the assumption.

The other inequality is easily obtained by the first one since $B^{-1} \geq A^{-1}$.

(ii) If $p \geq 1$, then $0 \leq \gamma - u \leq -u \leq 1 - u$. We can use the Furuta inequality or satellite theorem and have the following;

$$
\begin{aligned}
A^u \; \natural_{\frac{\gamma-u}{p-u}} \; B^p &= A^u \; \natural_{\frac{\gamma-u}{1-u}} \; (A^u \; \natural_{\frac{1-u}{p-u}} \; B^p) \\
&\leq A^u \; \natural_{\frac{\gamma-u}{1-u}} \; A = A^{(1-u)\frac{\gamma-u}{1-u}+u} = A^\gamma.
\end{aligned}
$$

And similarly we have

$$B^u \; \natural_{\frac{\gamma-u}{p-u}} \; A^p = B^u \; \natural_{\frac{\gamma-u}{1-u}} \; (B^u \; \natural_{\frac{1-u}{p-u}} \; A^p) \geq B^u \; \natural_{\frac{\gamma-u}{1-u}} \; B = B^\gamma.$$

Secondly, the case of $1 \geq p \geq 0$ is easily obtained by the property of operator mean as follows;

$$A^u \; \natural_{\frac{\gamma-u}{p-u}} \; B^p \leq A^u \; \natural_{\frac{\gamma-u}{p-u}} \; A^p = A^\gamma$$

and

$$B^u \; \natural_{\frac{\gamma-u}{p-u}} \; A^p \geq B^u \; \natural_{\frac{\gamma-u}{p-u}} \; B^p = B^\gamma.$$

$\square$

We can explain these relations by the following Figure 2. Here we imagine $B^u \, \natural_\alpha \, A^p$ as the path combining B^u and A^p. The Furuta inequality or satellite theorem is only the case of $\delta = 1$.

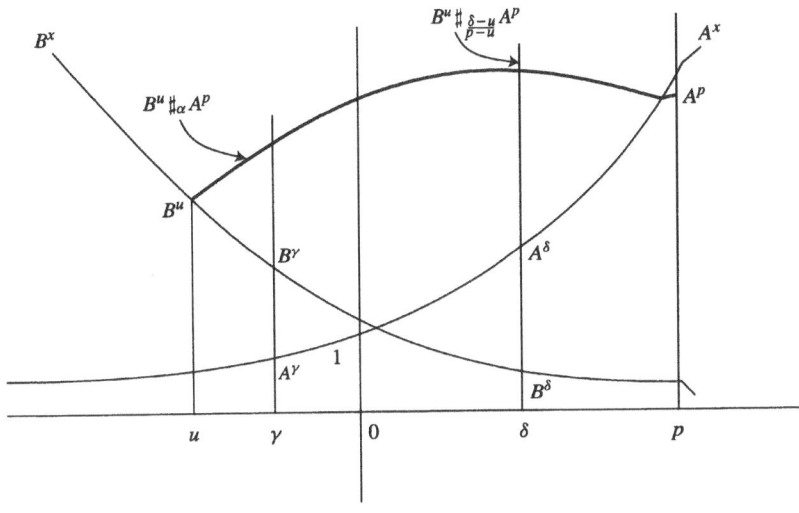

Figure 2

3 Parametrization of the Grand Furuta Inequality

Furuta proposed an inequality as a generalization of the Furuta inequality [10], which we called the grand Furuta inequality in [5]. This inequality interpolates the Furuta inequality and the Ando-Hiai inequality [1] equivalent to the main result of log majorization. We cite it here in terms of operator mean:

The Grand Furuta Inequality: *If $A \geq B \geq 0$ and A is invertible, then for each $t \in [0, 1]$ and $p \geq 1$,*

$$A^{-r+t} \, \natural_{\frac{1-t+r}{(p-t)s+r}} \, (A^t \, \natural_s \, B^p) \leq A$$

and

$$B^{-r+t} \, \natural_{\frac{1-t+r}{(p-t)s+r}} \, (B^t \, \natural_s \, A^p) \geq B$$

holds for $r \geq t$ and $s \geq 1$.

The best possibility of the power $\frac{1-t+r}{(p-t)s+r}$ is shown in [22]. We can combine these inequalities also in one line by the satellite form as follows [17], [18]:

Satellite Form of Grand Furuta Inequality: *If $A \geq B > 0$, then for $t \in [0, 1]$, $0 \leq t < p \leq \beta$, $u \leq 0$, $0 \leq \delta \leq 1$ and $\delta \leq \beta$*

$$A^u \, \natural_{\frac{\delta-u}{\beta-u}} \, (A^t \natural_{\frac{\beta-t}{p-t}} B^p) \leq (A^t \natural_{\frac{\beta-t}{p-t}} B^p)^{\frac{\delta}{\beta}} \leq B^\delta$$

$$\leq A^\delta \leq (B^t \, \natural_{\frac{\beta-t}{p-t}} A^p)^{\frac{\delta}{\beta}} \leq B^u \, \natural_{\frac{\delta-u}{\beta-u}} \, (B^t \natural_{\frac{\beta-t}{p-t}} A^p).$$

We show this result more general form in the following theorem. We prepare the next lemma given first in our approach to grand Furuta inequality by operator mean theory [5]. Here we make a reformulation to fit the context of this note.

Lemma 3.1 *If $A \geq B > 0$, then for $t \in [0, 1]$ and $\beta \geq p > t$,*

$$(A^t \, \natural_{\frac{\beta-t}{p-t}} B^p)^{\frac{q}{\beta}} \leq B^q \leq A^q$$

where $q = \min(1, p)$.

Proof: First of all, we note the following formula obtained by the definition:

$$A \, \natural_\alpha \, B = A(A^{-1} \natural_{-\alpha} B^{-1})A \quad \text{for } A, B > 0.$$

First let us treat the case of $1 \leq \frac{\beta-t}{p-t} \leq 2$;

$$A^t \, \natural_{\frac{\beta-t}{p-t}} B^p = B^p \, \natural_{1-\frac{\beta-t}{p-t}} A^t = B^p(B^{-p} \, \natural_{\frac{\beta-t}{p-t}-1} A^{-t})B^p$$

$$\leq B^p(B^{-p} \, \natural_{\frac{\beta-p}{p-t}} B^{-t})B^p = B^\beta.$$

Hence we have $(A^t \, \natural_{\frac{\beta-t}{p-t}} B^p)^{\frac{1}{\beta}} \leq B \leq A$ for $p \geq 1$. If $p \leq 1$ and $(A^t \, \natural_{\frac{\beta-t}{p-t}} B^p)^{\frac{p}{\beta}} \leq B^p \leq A^p$ for $p \leq 1$ by the Löwner-Heinz theorem.

Next let us consider the case $\frac{\beta-t}{p-t} > 2$. We choose β_1; $1 < \frac{\beta_1-t}{\beta-t} \leq 2$. If $p \geq 1$, then for A and $B_1 = (A^t \, \natural_{\frac{\beta-t}{p-t}} B^p)^{\frac{1}{\beta}}$, we can repeat similar calculations as follows:

$$A^t \, \natural_{\frac{\beta_1-t}{p-t}} B^p = A^t \, \natural_{\frac{\beta_1-t}{\beta-t}} (A^t \, \natural_{\frac{\beta-t}{p-t}} B^p) = A^t \, \natural_{\frac{\beta_1-t}{\beta-t}} B_1^\beta$$

$$= B_1^\beta \, \natural_{1-\frac{\beta_1-t}{\beta-t}} A^t = B_1^\beta(B_1^{-\beta} \, \natural_{\frac{\beta_1-t}{\beta-t}-1} A^{-t})B_1^\beta$$

$$\leq B_1^\beta(B_1^{-\beta} \, \natural_{\frac{\beta_1-\beta}{\beta-t}} B_1^{-t})B_1^\beta = B_1^{\beta_1} = (A^t \, \natural_{\frac{\beta-t}{p-t}} B^p)^{\frac{\beta_1}{\beta}}.$$

So we have $(A^t \, \natural_{\frac{\beta_1-t}{p-t}} B^p)^{\frac{1}{\beta_1}} \leq (A^t \, \natural_{\frac{\beta-t}{p-t}} B^p)^{\frac{1}{\beta}} \leq B \leq A$ by (LH).

In the case of $p \leq 1$, we can calculate similarly for $A_1 = A^p$ and $B_1 = (A^t \,\natural_{\frac{\beta-t}{p-t}}\, B^p)^{\frac{p}{\beta}}$ as follows:

$$
A^t \,\natural_{\frac{\beta_1-t}{p-t}}\, B^p = A^t \,\natural_{\frac{\beta_1-t}{\beta-t}}\, (A^t \,\natural_{\frac{\beta-t}{p-t}}\, B^p) = A_1^{\frac{t}{p}} \,\natural_{\frac{\frac{\beta_1}{p}-\frac{t}{p}}{\frac{\beta}{p}-\frac{t}{p}}}\, (B_1)^{\frac{\beta}{p}}
$$

$$
= B_1^{\frac{\beta}{p}} \,\natural_{1-\frac{\beta_1-t}{\beta-t}}\, A_1^{\frac{\beta}{p}} = B_1^{\frac{\beta}{p}} (B_1^{-\frac{\beta}{p}} \,\sharp_{\frac{\beta_1-t}{\beta-t}-1}\, A_1^{-\frac{t}{p}}) B_1^{\frac{\beta}{p}}
$$

$$
\leq B_1^{\frac{\beta}{p}} (B_1^{-\frac{\beta}{p}} \,\sharp_{\frac{\beta_1-\beta}{\beta-t}}\, B_1^{-\frac{t}{p}}) B_1^{\frac{\beta}{p}} = B_1^{\frac{\beta_1}{p}} = (A^t \,\natural_{\frac{\beta-t}{p-t}}\, B^p)^{\frac{\beta_1}{\beta}}.
$$

(LH) leads us to $(A^t \,\natural_{\frac{\beta_1-t}{p-t}}\, B^p)^{\frac{p}{\beta_1}} \leq (A^t \,\natural_{\frac{\beta-t}{p-t}}\, B^p)^{\frac{p}{\beta}} \leq B^p \leq A^p$.

The third case, we choose β_2; $1 < \frac{\beta_2-t}{\beta_1-t} \leq 2$, and repeating the above method, we can attain the conclusion. $\qquad\square$

Theorem 3.2 (Parametrized Form of the Grand Furuta Inequality). *If $A \geq B > 0$, then for $t \in [0, 1]$, $0 \leq t < p \leq \beta$, $u \leq 0$ and $\delta \in [0, \beta]$*

$$
A^u \,\sharp_{\frac{\delta-u}{\beta-u}}\, (A^t \,\natural_{\frac{\beta-t}{p-t}}\, B^p) \leq (A^t \,\natural_{\frac{\beta-t}{p-t}}\, B^p)^{\frac{\delta}{\beta}}
$$

and

$$
B^u \,\sharp_{\frac{\delta-u}{\beta-u}}\, (B^t \,\natural_{\frac{\beta-t}{p-t}}\, A^p) \geq (B^t \,\natural_{\frac{\beta-t}{p-t}}\, A^p)^{\frac{\delta}{\beta}}.
$$

Proof: Since $(A^t \,\natural_{\frac{\beta-t}{p-t}}\, B^p)^{\frac{q}{\beta}} \leq B^q \leq A^q$ by Lemma 3.1 we have only to apply Theorem 2.1 to $A_1 = A^p$ and $B_1 = (A^t \,\natural_{\frac{\beta-t}{p-t}}\, B^p)^{\frac{q}{\beta}}$. That is, for $u_1 \leq 0$ and $0 \leq \delta_1 \leq p_1$

$$
A_1^{u_1} \,\sharp_{\frac{\delta_1-u_1}{p_1-u_1}}\, B_1^{p_1} \leq B_1^{\delta_1}.
$$

Let $u_1 = \frac{u}{q}$, $\delta_1 = \frac{\delta}{q}$ and $p_1 = \frac{\beta}{q}$, then we have the conclusion. $\qquad\square$

These relations are also shown in the following Figure 3:

4 Application

The Furuta inequality and the grand one have monotone properties as operator functions. Recently Furuta, Yamazaki and Yanagida [13] had shown the following which is an extension of [14].

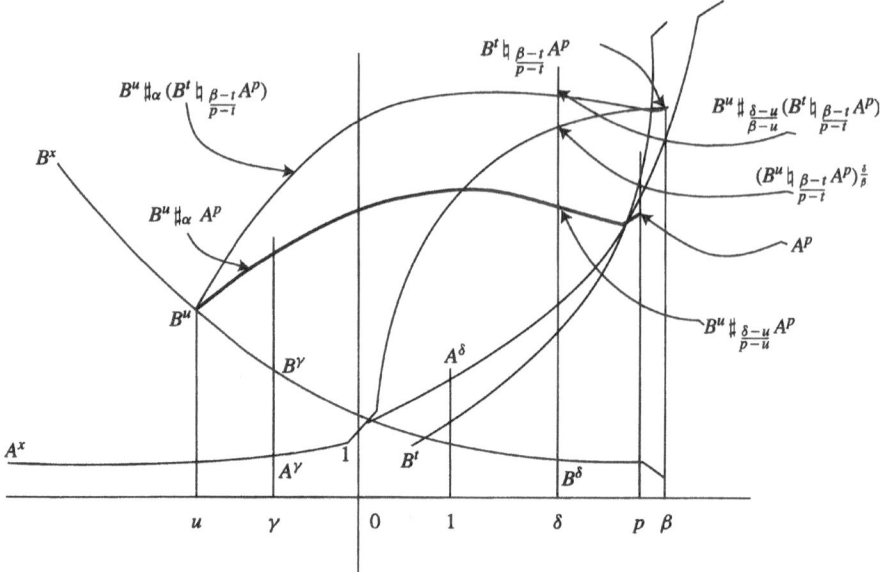

Figure 3

Theorem FYY: *If $A \geq B \geq 0$ with $A > 0$, then for each $\delta \geq 0$, $t \in [0, 1]$ and $p \geq t$,*

$$G_{p,\delta,t}(A, B, r, s) = A^{-\frac{r}{2}}\{A^{\frac{r}{2}}(A^{-\frac{t}{2}}B^p A^{-\frac{t}{2}})^s A^{\frac{r}{2}}\}^{\frac{\delta-t+r}{(p-t)s+r}} A^{-\frac{r}{2}}$$

is decreasing for $r \geq t$ and $s \geq 1$ such that $(p - t)s \geq \delta - t$.

In this section, we show Theorem FYY is easily obtainable from Theorem 2.1 and Theorem 3.2 by reformulating them as operator functions (cf. [6]).

The proof of Lemma 3.1 is also a proof of the operator function $f_q(\beta) = (A^t \natural_{\frac{\beta-t}{p-t}} B^p)^{\frac{q}{\beta}}$ being monotone decreasing for $\beta \geq q$. So we sum up this as the following lemma:

Lemma 4.1 *Let $A \geq B > 0$ and $t \in [0, 1]$, $p > t$ and $q = \min(1, p)$. Then*

$$(f_q(\beta_1))^{\frac{\beta_1}{q}} \leq (f_q(\beta))^{\frac{\beta_1}{q}}$$

holds for $1 \leq \frac{\beta_1-t}{\beta-t} \leq 2$ and

$$f_q(\beta) = (A^t \natural_{\frac{\beta-t}{p-t}} B^p)^{\frac{q}{\beta}}(\leq B^q \leq A^q)$$

is decreasing for $\beta \geq q$.

Now, we prepare a reformulation of Theorem 2.1 as the monotonicity of an operator function, which is already known in [3] and [9].

Lemma 4.2 *If* $A \geq B > 0$, *then for given* $\delta > 0$ *the operator function*

$$F_\delta(A, B; u, p) = A^u \, \natural_{\frac{\delta-u}{p-u}} \, B^p$$

is decreasing in $p \geq \delta$ *and increasing in* $u \leq 0$.

Finally we give an interpretation of Theorem FYY by the form of operator mean and give a proof by the use of Lemma 3.1 and Lemma 4.1.

Theorem 4.3 *If* $A \geq B > 0$, *then for given* $\delta > 0$ *and* p, t *such that* $0 \leq t \leq 1$ *and* $p > t$ *the operator function*

$$H_{p,\delta,t}(A, B; u, \beta) = A^u \, \natural_{\frac{\delta-u}{\beta-u}} \, (A^t \, \natural_{\frac{\beta-t}{p-t}} \, B^p)$$

is increasing in $u \leq 0$ *and decreasing in* $\beta \geq \max(p, \delta)$.

Proof: By using the same notations of the above, we have $H_{p,\delta,t}(A, B; u, \beta) = F_{\frac{\delta}{q}}(A^q, f_q(\beta); \frac{u}{q}, \frac{\beta}{q}) = A^u \, \natural_{\frac{\delta-u}{\beta-u}} \, (f_q(\beta))^{\frac{\beta}{q}}$. By Lemma 4.2 $H_{p,\delta,t}(A, B; u, \beta)$ is increasing for $u \leq 0$. So we have only to show $H_{p,\delta,t}(A, B; u, \beta_1) \leq H_{p,\delta,t}(A, B; u, \beta)$ for $\beta \leq \beta_1$. By Lemma 4.1, $(f_q(\beta_1))^{\frac{\beta_1}{q}} \leq (f_q(\beta))^{\frac{\beta_1}{q}}$ holds for β_1 such that $1 \leq \frac{\beta_1-t}{\beta-t} \leq 2$, that is, $\beta \leq \beta_1 \leq 2\beta - t$. Hence we have

$$
\begin{aligned}
H_{p,\delta,t}(A, B; u, \beta_1) &= A^u \, \natural_{\frac{\delta-u}{\beta_1-u}} \, (f_q(\beta_1))^{\frac{\beta_1}{q}} \\
&\leq A^u \, \natural_{\frac{\delta-u}{\beta_1-u}} \, (f_q(\beta))^{\frac{\beta_1}{q}} = F_{\frac{\delta}{q}}\left(A^q, f_q(\beta); \frac{u}{q}, \frac{\beta_1}{q}\right) \\
&\leq F_{\frac{\delta}{q}}\left(A^q, f_q(\beta); \frac{u}{q}, \frac{\beta}{q}\right) = H_{p,\delta,t}(A, B; u, \beta).
\end{aligned}
$$

The first inequality follows from the monotonicity of a mean in each variable while the second inequality follows from Lemma 4.2. For $\beta' > \beta$ there is a sequence $\beta = \beta_0 < \beta_1 < \beta_2 < \cdots < \beta_n = \beta'$ such that $1 < \frac{\beta_j-t}{\beta_{j-1}-t} \leq 2 (j = 1, 2, \ldots, n)$. Repeating the above arguments, we arrive at the inequality

$$H_{p,\delta,t}(A, B; u, \beta') \leq H_{p,\delta,t}(A, B; u, \beta).$$

$\square$

We remark the relation between operator functions $H_{p,\delta,t}$ and $G_{p,\delta,t}$ in Theorem FYY. By exchanging variables $r = t - u, s = \frac{\beta-t}{p-t}$ for $\beta \geq p$, we have

$$H_{p,\delta,t}(A, B, u, \beta) = A^{\frac{t}{2}} G_{p,\delta,t}\left(A, B, t - u, \frac{\beta - t}{p - t}\right) A^{\frac{t}{2}}.$$

Acknowledgements

The author would like to express his thanks to Professor T. Furuta for his kind information that Furuta, Hashimoto and Itó have developed a similar discussion to ours by their own methods in [12].

References

[1]   T. Ando and F. Hiai, Log majorization and complementary Golden-Thompson type inequality, *Linear Alg. and Its Appl.* **197** (1994), 113–131.

[2]   M. Fujii, Furuta's inequality and its mean theoretic approach, *J. Operator Theory* **23** (1990), 67–72.

[3]   M. Fujii, T. Furuta and E. Kamei, Operator functions associated with Furuta's inequality, *Linear Alg. and its Appl.* **149** (1991), 91–96.

[4]   M. Fujii, J.F. Jiang and E. Kamei, A characterization of orders defined by $A^\delta \geq B^\delta$ via Furuta inequality, *Math. Japon.* **45** (1997), 519–525.

[5]   M. Fujii and E. Kamei, Mean theoretic approach to the grand Furuta inequality, *Proc. Amer. Math. Soc.* **124** (1996), 2751–2756.

[6]   M. Fujii and E. Kamei, Monotone properties of parametrized Furuta inequality, *Sci. Math.* **1** (1998), 277–288.

[7]   T. Furuta, $A \geq B \geq 0$ assures $(B^r A^p B^r)^{1/q} \geq B^{(p+2r)/q}$ for $r \geq 0, p \geq 0, q \geq 1$ with $(1 + 2r)q \geq p + 2r$, *Proc. Amer. Math. Soc.* **101** (1987), 85–88.

[8]   T. Furuta, Elementary proof of an order preserving inequality, *Proc. Japan Acad.* **65** (1989), 126.

[9]   T. Furuta, Two operator functions with monotone property, *Proc. Amer. Math. Soc.* **111** (1991), 511–516.

[10] T. Furuta, Extension of the Furuta inequality and Ando-Hiai log-majorization, *Linear Alg. and Its Appl.* **219** (1995), 139–155.

[11] T. Furuta, Parametric operator function via Furuta inequality, *Sci. Math.* **1** (1998), 1–5.

[12] T. Furuta, M. Hashimoto and M. Itó, Equivalence relation between generalized Furuta inequality and related operator functions, preprint.

[13] T. Furuta, T. Yamazaki and M. Yanagida, Operator functions implying generalized Furuta inequality, *Math. Inequal. Appl.* **1** (1998), 123–130.

[14] T. Furuta and D. Wang, A decreasing operator function associated with the Furuta inequality, *Proc. Amer. Math. Soc.*, to appear.

[15] E. Kamei, Furuta's inequality via operator means, *Math. Japon.* **33** (1988), 737–739.

[16] E. Kamei, A satellite to Furuta's inequality, *Math. Japon.* **33** (1988), 883–886.

[17] E. Kamei, Parametrization of the Furuta inequality, *Math. Japon.* **49** (1999), 65–71.

[18] E. Kamei, Parametrized grand Furuta inequality, *Math. Japon.* **50** (1999), 79–83.

[19] E. Kamei, Parametrization of the Furuta inequality II, *Math. Japon.* **50** (1999), 179–182.

[20] F. Kubo and T. Ando, Means of positive linear operators, *Math. Ann.* **246** (1980), 205–224.

[21] K. Tanahashi, Best possibility of the Furuta inequality, *Proc. Amer. Math. Soc.* **124** (1996), 141–146.

[22] K. Tanahashi, Best possibility of the grand Furuta inequality, *Proc. Amer. Math. Soc.*, to appear.

Eizaburo Kamei
Maebashi Institute of Technology
460-1, Kamisadori-machi
Maebashi, Gunma
371-0816
Japan

1991 Mathematics Subject Classification. Primary: 47A63; Secondary: 47A56.

Operator Theory:
Advances and Applications, Vol. 124
© 2001 Birkhäuser Verlag Basel/Switzerland

J-Symmetric Factorizations and Algebraic Riccati Equations

I. Karelin, L. Lerer and A.C.M. Ran

Dedicated to Israel Gohberg on the occasion of his seventieth birthday with admiration and affection

This paper discusses two interrelated topics: minimal J-symmetric factorizations of rational matrix functions and the algebraic Riccati equation. In particular, necessary and sufficient conditions are presented for the existence of a complete set of minimal J-symmetric factorizations of a selfadjoint rational matrix function with constant signature. For the algebraic Riccati equation the selfadjoint function which is of vital importance is the Popov function. Our first result for the algebraic Riccati equation describes the connection between the hermitian solutions, J-symmetric factorizations of the Popov function and generalized Bezoutians. Then, necessary and sufficient conditions are given for the algebraic Riccati equation to have a complete set of solutions. Both the continuous and discrete algebraic Riccati equation are treated.

0 Introduction

This paper deals mainly with two topics which are known to be strongly interrelated: symmetric factorizations of selfadjoint rational matrix functions and algebraic Riccati equations. Symmetric factorizations of selfadjoint rational matrix functions have been extensively studied (see, e.g., [FP, F1, F2, GLR, GLR1, RR1, RZ]) and have diverse applications in various fields including the theory of Wiener-Hopf operators, linear quadratic optimal control, stochastic filtering, model reduction and H^∞-control (see, e.g., [BR, GGLD, FP, P]). In the present paper we are particularly interested in J-symmetric factorizations of selfadjoint rational matrix functions with constant signature and the specific aspects of such factorizations in the theory of algebraic Riccati equations.

To be more specific, we are concerned with selfadjoint rational matrix functions $G(\lambda)$ that have a constant signature on the imaginary axis and a minimal J-symmetric factorization, i.e., minimal factorizations of the form

$$G(\lambda) = R(-\bar{\lambda})^* J R(\lambda).$$

Here J is a signature matrix, necessarily such that the signatures of J and $G(\lambda)$ coincide. As is well-known (see, e.g., [GLR1]) functions with constant signature always admit a J-symmetric factorization, but minimal J-symmetric factorizations need not always exist. Minimal J-symmetric factorizations of such functions were studied earlier in [RR] and [RZ]. In case $J = I$, i.e., when the function takes positive semidefinite values on the imaginary line it is known that there exists a complete characterization of all minimal symmetric factorizations (see [R]) in terms of the matrices appearing in a minimal realization of G, and involving invariant lagrangian subspaces (see also for other parametrizations [F1, F2, FP, FMP, R1]). In this paper we shall be concerned in particular with completeness of the set of minimal J-symmetric factorizations in the case of a general J. A precise definition of this notion will be given in Section 1; for the moment it suffices to say that completeness means that the same complete parametrization of the minimal J-symmetric factors as in the case $J = I$ holds. We shall establish necessary and sufficient conditions for completeness of the set of J-symmetric factorizations. This is the main result of Section 1 of the present paper.

In Section 2 we study selfadjoint rational matrix functions with constant signature and their J-symmetric factorizations that appear in the theory of the algebraic Riccati equation

$$(0.1) \qquad XBR^{-1}B^*X + XA + A^*X - C = 0,$$

where $R = R^*$ is an invertible $m \times m$ matrix, $C = C^*$ is an $n \times n$ matrix, and the pair (A, B) is mostly assumed to be controllable. We put $W = BR^{-1}B^*$, and decompose W as $W = \Psi J \Psi^*$, where $J = J^* = J^{-1}$ is a signature matrix and Ψ is a matrix of the same order as B. So, if R decomposes as $R = B_0 J B_0^*$, where J is a signature matrix, then $\Psi = BB_0$. Moreover, the signature of R equals the signature of J. Connected to this Riccati equation we introduce several well-known items: the Hamiltonian

$$H = \begin{pmatrix} -A & -W \\ -C & A^* \end{pmatrix},$$

the signature matrix $J_1 = \begin{pmatrix} 0 & I \\ -I & 0 \end{pmatrix}$, and the Popov function

$$(0.2) \qquad G_0(\lambda) = J - \Psi^*(\lambda I - A^*)^{-1}C(\lambda I + A)^{-1}\Psi.$$

Note that $J_1 H = -H^* J_1$, and that $G_0(-\bar\lambda)^* = G_0(\lambda)$. The function $G_0(\lambda)$, as well as related functions $G_K(\lambda)$ to be introduced later on, has constant signature on the imaginary axis in case equation (0.1) has a hermitian solution.

In this part also the notion of generalized Bezoutians as introduced in [KL] plays a central role. In fact, we shall show that the hermitian solutions of (0.1) can be identified as generalized Bezoutians connected to certain J-symmetric factorizations of $G_K(\lambda)$. These results can be viewed as a continuation of our earlier results for the case $C = 0$, obtained in [LRa] on the one hand, and of the results of [L], [KL] for general quadratic equations on the other hand.

Our second theme, which we start to discuss in Section 3, concerns completeness of the set of hermitian solutions of (0.1), as introduced by N.E. Barabanov [B]. Completeness means that every H-invariant J_1-lagrangian subspace is automatically a graph subspace, and hence corresponds to a hermitian solution of (0.1).

We shall give necessary and sufficient conditions for the set of solutions to be complete, thereby sharpening the result in [B]. We show that in a certain sense a complete set of solutions of the algebraic Riccati equation (0.1) corresponds to a complete set of minimal J-symmetric factorizations of $G_K(\lambda)$. Important to note here is that in our result we do not need a-priori the controllability assumption on the pair (A, B). In case $J = I$ and (A, B) is controllable it is well-known that the set of hermitian solutions is complete (see, e.g., [LaR]).

In Sections 4 and 5 we derive analogues for the case of the discrete algebraic Riccati equation of what was obtained for the continuous algebraic Riccati equation in Sections 2 and 3, respectively.

Section 6 contains results concerning minimal factorizations for arbitrary rational matrix functions having hermitian values on the imaginary line, not necessarily having constant signature. For this case we cannot expect J-symmetric factorizations to exist, but minimal factorizations that are very similar to symmetric factorizations do exist in case the function takes a positive value somewhere on the imaginary line (see [R]). Here we consider the analogue of this result for the general case, and again completeness is the central issue here.

Some remarks about notation: in the sequel we shall frequently write $(\lambda - A)^{-1}$ instead of $(\lambda I - A)^{-1}$ for shortness. Also, with $\sigma(A)$ we denote the set of eigenvalues of a matrix A, and with $\rho(A)$ its complement in the complex plane.

1 Completeness of the Set of Minimal J-Symmetric Factors

In this section we shall be concerned with a general rational matrix valued function $G(\lambda)$ taking hermitian values on the imaginary line and having an invertible value at infinity. We shall assume throughout that $G(\lambda)$ has constant signature on the imaginary line. In that case we know that there is a factorization of $G(\lambda)$ as

$$(1.1) \qquad G(\lambda) = R(-\bar{\lambda})^* J R(\lambda),$$

however, it may be the case that none of these factorizations is minimal. Such factorizations have been investigated in [GLR1, RR, RZ], in the latter two papers particular attention was payed to minimal factorizations of this type.

Let $G(\lambda) = G(-\bar{\lambda})^*$ and let $G(\lambda) = J + C(\lambda I_n - A)^{-1} B = G(-\bar{\lambda})^*$ be a minimal realization, where we assume without loss of generality that $G(\infty)$ is a signature matrix. Then it is a well-known consequence of the state space isomorphism theorem that there exists a unique invertible matrix H such that

$$(1.2) \qquad HA = -A^* H, \quad HB = C^*, \quad H^* = -H.$$

From [RR] we know that a necessary condition for a minimal factorization of the type (1.1) to exist is that G has constant pole signature as well as constant zero signature, meaning that both A and $A^\times := A - BJC$ have an invariant H-lagrangian subspace, say M and $M^\times$, respectively. Observe that from (1.2) we also have $HA^\times = -A^{\times *}H$. Recall that a subspace M is called H-*lagrangian* if $(HM)^\perp = M$. (There is a characterization of the notions of constant pole and constant zero signature that avoids the use of the matrices in a realization as well, see [RR], Theorem 2.2.)

For the case where $J = I$, and $G(\lambda)$ takes positive semidefinite values on the imaginary line it was shown in [R] that for any pair of such subspaces M and $M^\times$ we have $M \oplus M^\times = \mathbb{C}^n$, and consequently to any such pair of subspaces there corresponds a minimal factorization (1.1) with $J = I$ (see [BGK]). Conversely, even in the case of a general signature matrix J, any minimal symmetric factorization necessarily corresponds to an A-invariant subspace M and an $A^\times$-invariant subspace $M^\times$ that are H-lagrangian, i.e. $HM = M^\perp$, $HM^\times = M^{\times\perp}$. This was proved in [RR].

We shall say that a hermitian function $G(\lambda)$ with constant signature, constant pole signature and constant zero signature has a *complete set of minimal J-symmetric factorizations* if for any A-invariant lagrangian subspace M and any $A^\times$-invariant lagrangian subspace $M^\times$ we have $M \oplus M^\times = \mathbb{C}^n$. In that case we have for any pair of such subspaces a corresponding minimal J-symmetric factorization (5.1) (see [BGK]). The statement in the paragraph above pertaining to the case of positive semidefinite rational matrix functions can be restated as saying that for such a function there is a complete set of I-symmetric factorizations.

As we see the existence of A-invariant H-lagrangian subspaces plays an important role in the theory. Necessary and sufficient conditions for such a subspace to exist were give in [RR1] in terms of the sign characteristic of the pair (A, H) (see, e.g., [GLR] for the latter notion). We start with a lemma of a geometrical nature that essentially states that if an invariant lagrangian subspace exists, we can always find one that contains a given invariant neutral subspace.

Lemma 1.1 *Let $H = -H^*$, and let A be an $n \times n$ matrix such that $HA = -A^*H$. Let N be an A-invariant H-neutral subspace. Assume there exists an A-invariant H-lagrangian subspace M_0. Then $M = N + (M_0 \cap (HN)^\perp)$ is an A-invariant H-lagrangian subspace such that $N \subset M$.*

Proof: Clearly M is A-invariant, and it is also obviously H-neutral. So it remains to compute its dimension.

$$\dim M = \dim N + \dim (M_0 \cap (HN)^\perp) - \dim (M_0 \cap (HN)^\perp \cap N).$$

Now $M_0 \cap (HN)^\perp \cap N = M_0 \cap N$, as N is H-neutral (so $N \subset (HN)^\perp$). Thus

$$(1.3) \qquad \dim M = \dim N + \dim (M_0 \cap (HN)^\perp) - \dim (M_0 \cap N).$$

Furthermore $(H(M_0 \cap (HN)^\perp))^\perp = M_0 + N$. So

$$\begin{aligned}
\dim M_0 \cap (HN)^\perp &= \operatorname{codim} (M_0 + N) = n - \dim (M_0 + N) \\
&= n - (\dim M_0 + \dim N - \dim (M_0 \cap N))
\end{aligned}$$

$$(1.4) \qquad = n - \frac{n}{2} - \dim N + \dim (M_0 \cap N)$$

$$= \frac{n}{2} - \dim N + \dim (M_0 \cap N).$$

Inserting (1.4) in (1.3) we see that $\dim M = \frac{n}{2}$. So M is H-lagrangian. $\qquad \square$

We have the following result.

Theorem 1.2 *Let $G(\lambda)$ be a rational matrix function whose values on the imaginary line are hermitian, and having constant signature, constant pole signature and constant zero signature. Let $G(\lambda) = J + C(\lambda - A)^{-1}B$ be a minimal realization of $G(\lambda)$, and let H be the unique skew hermitian matrix such that (1.2) holds. Then the following are equivalent:*

(i) *there is a complete set of minimal J-symmetric factorizations (1.1) of $G(\lambda)$,*

(ii) *for every A-invariant H-lagrangian subspace M and for every non-zero vector $x \in M$ we have*

$$\langle H(\lambda - A^{\times})^{-1}x, x \rangle \not\equiv 0, \quad \lambda \in \rho(A^{\times}),$$

(iii) *for every $A^{\times}$-invariant H-lagrangian subspace $M^{\times}$ and for every non-zero vector $x \in M^{\times}$ we have*

$$\langle H(\lambda - A)^{-1}x, x \rangle \not\equiv 0, \quad \lambda \in \rho(A).$$

Proof: Recall that there is a complete set of J-symmetric factorizations if and only if for any A-invariant H-lagrangian subspace M and any $A^{\times}$-invariant H-lagrangian subspace $M^{\times}$ we have $M \oplus M^{\times} = \mathbb{C}^n$.

Assume that (i) holds and (ii) does not hold. Then there is an A-invariant H-lagrangian subspace M and a non-zero vector $x \in M$ such that

$$\langle H(\lambda - A^{\times})^{-1}x, x \rangle \equiv 0, \quad \lambda \in \rho(A^{\times}).$$

Define

$$(1.5) \qquad N = \text{span}\{\{x\} \cup \{(\lambda - A^{\times})^{-1}x \mid \lambda \in \rho(A^{\times})\}\}.$$

Then N is an $A^{\times}$-invariant H-neutral subspace. Indeed, as $x \in M$ clearly $\langle Hx, x \rangle = 0$. Also, $\langle Hx, (\lambda - A^{\times})^{-1}x \rangle = 0$ for $\lambda \in \rho(A^{\times})$ by assumption. For $\lambda \neq -\bar{\mu}, \lambda, \mu \in \rho(A^{\times})$ we have, using (1.2),

$$\Gamma(\lambda, \mu) := \langle H(\lambda - A^{\times})^{-1}x, (\mu - A^{\times})^{-1}x \rangle$$
$$= \langle H(\bar{\mu} + A^{\times})^{-1}(\lambda - A^{\times})^{-1}x, x \rangle.$$

Now using the resolvent identity and the assumption one proves that $\Gamma(\lambda, \mu) = 0$ for all $\lambda, \mu \in \rho(A^\times)$. It remains to see that for $\lambda \in \rho(A^\times)$ we have

$$\langle H(\lambda - A^\times)^{-1}x, (-\bar\lambda - A^\times)^{-1}x \rangle = 0.$$

This can be seen by taking the limit $\mu \to -\bar\lambda$ in $\Gamma(\lambda, \mu)$.

So by Lemma 1.1 there is an $A^\times$-invariant H-lagrangian subspace $M^\times$ with $N \subset M^\times$. But then $x \in M \cap M^\times$, which contradicts (i).

The implication (i)$\Rightarrow$ (iii) is proved in the same way.

Conversely, assume that (ii) holds, but (i) does not. Then there are A-invariant and $A^\times$-invariant H-lagrangian subspaces M and $M^\times$, respectively, with $M \cap M^\times \neq (0)$. Take such M and $M^\times$, and let x be a non-zero vector in their intersection. Then for all $\lambda \in \rho(A^\times)$ the vector $(\lambda - A^\times)^{-1}x$ is in $M^\times$, and hence

$$\langle H(\lambda - A^\times)^{-1}x, x \rangle \equiv 0,$$

which contradicts (ii).

Again, the implication (iii)$\Rightarrow$ (ii) is proved in the same way. $\qquad\square$

Examples: As a first example consider the function $G(\lambda) = \begin{pmatrix} 0 & 1 \\ 1 & \lambda^{-2} \end{pmatrix}$. A minimal realization is given by

$$G(\lambda) = \begin{pmatrix} 0 & 1 \\ 1 & 0 \end{pmatrix} + \begin{pmatrix} 0 & 0 \\ 1 & 0 \end{pmatrix} \left(\lambda - \begin{pmatrix} 0 & 1 \\ 0 & 0 \end{pmatrix} \right)^{-1} \begin{pmatrix} 0 & 0 \\ 0 & 1 \end{pmatrix}.$$

The unique H such that (1.2) holds is the matrix $H = \begin{pmatrix} 0 & 1 \\ -1 & 0 \end{pmatrix}$. Observe that the corresponding $A^\times$ is the same as the matrix A. Also, there is in this case a unique A-invariant H-lagrangian subspace M, namely span $\begin{pmatrix} 1 \\ 0 \end{pmatrix}$. Clearly the conditions of the theorem are violated in this case, and so there is not a complete set of minimal J-symmetric factorizations in this case. As we also see, there really is only one candidate for M and $M^\times$, so in this case there is even no nontrivial minimal factorization of $G(\lambda)$ (symmetric or not). Of course, this is well-known, but the example serves to show that the conditions of our theorem can explain this.

As a second example, consider the function

$$G(\lambda) = \begin{pmatrix} 0 & \frac{-\lambda-2}{\lambda+1} \\ \frac{-\lambda+2}{\lambda-1} & 0. \end{pmatrix}.$$

This function clearly has constant signature on the imaginary line. Put $J = \begin{pmatrix} 0 & -1 \\ -1 & 0 \end{pmatrix}$. A minimal realization is then given by $G(\lambda) = J + C(\lambda - A)^{-1}B$, where $C = \begin{pmatrix} 0 & -1 \\ 1 & 0 \end{pmatrix}$, $A = \begin{pmatrix} 1 & 0 \\ 0 & -1 \end{pmatrix}$, and $B = \begin{pmatrix} 1 & 0 \\ 0 & 1 \end{pmatrix}$. The corresponding

matrix H is given by $H = \begin{pmatrix} 0 & 1 \\ -1 & 0 \end{pmatrix}$. Then $A^\times = A - BJC = \begin{pmatrix} 2 & 0 \\ 0 & -2 \end{pmatrix}$.
There is not a complete set of minimal J-symmetric factorizations. Indeed, take
for M the subspace spanned by the vector $(1 \ 0)^T$, and for $M^\times$ the same subspace. Then these are H-lagrangian and A-invariant, respectively, $A^\times$-invariant.
However, $M \cap M^\times$ is clearly not the zero subspace. Hence, for $x = (1 \ 0)^T$ the
condition in (ii) and (iii) of Theorem 1.2 fails. This is easy to see right away, as
$\langle H(\lambda - A^\times)^{-1}x, x \rangle$ is evidently zero.

Note that this function has all nice properties one could want: it has constant
zero and pole signature, both A and $A^\times$ are diagonal, and moreover, no zero is also
a pole. It follows from [RZ], or by direct checking, that $G(\lambda)$ does admit minimal
J-symmetric factorizations; for instance

$$G(\lambda) = \begin{pmatrix} 0 & \frac{-\lambda+2}{\lambda-1} \\ -1 & 0 \end{pmatrix} \begin{pmatrix} 0 & -1 \\ -1 & 0 \end{pmatrix} \begin{pmatrix} 0 & -1 \\ \frac{-\lambda-2}{\lambda+1} & 0 \end{pmatrix}$$

is such a minimal factorization. However, $G(\lambda)$ does not have a complete set of
minimal J-symmetric factorizations.

Our third example exhibits a function that has a complete set of minimal
J-symmetric factorizations. Consider

$$G(\lambda) = \begin{pmatrix} \frac{\lambda^2-1}{\lambda^2} & 0 \\ 0 & \frac{-\lambda^2+4}{\lambda^2-1}. \end{pmatrix}.$$

This function has constant signature on the imaginary axis, as one readily checks.
No matter what minimal realization we have, it turns out that there are only
two A-invariant H-lagrangian subspaces and that there are four $A^\times$-invariant
H-lagrangian subspaces. In total that leads to eight possible minimal J-symmetric
factorizations, and these can indeed be written down simply by inspection (the factors are diagonal matrix functions).

Next, we consider the case where we start with a rational matrix valued function
$R_0(\lambda)$ having a minimal realization $R_0(\lambda) = I + C(\lambda - A)^{-1}B$ and a signature
matrix J, and we construct $G(\lambda)$ by

$$G(\lambda) = R_0(-\bar{\lambda})^* J R_0(\lambda)$$

(1.6)

$$= J + (-B^* \ JC)\left(\lambda - \begin{pmatrix} -A^* & C^*JC \\ 0 & A \end{pmatrix}\right)^{-1}\begin{pmatrix} C^*J \\ B \end{pmatrix}.$$

This need not be a minimal realization of $G(\lambda)$, however, we shall assume in the
sequel that $\sigma(A) \cap \sigma(-A^*) = \emptyset$, in which case (1.6) is a minimal realization.
This assumption is satisfied, for instance, in case A is stable or antistable. We are
particularly interested in minimal J-symmetric factorizations of the form (1.1),
where $R(\lambda)$ can be written in the form

(1.7) $$R(\lambda) = I + C_R(\lambda - A)^{-1}B,$$

i.e., (A, B) is an extension of the pole pair of $R(\lambda)$. (For the notion of pole pair we refer to e.g. [BGR].) Put

$$\tilde{A} = \begin{pmatrix} -A^* & CJC^* \\ 0 & A \end{pmatrix}.$$

Note that the role of H is played here by the matrix $J_1 = \begin{pmatrix} 0 & I \\ -I & 0 \end{pmatrix}$. Also observe that the main operator in the inverse of the realization (1.6) is given by

$$(1.8) \qquad\qquad \tilde{A}^\times = \begin{pmatrix} -A^{\times *} & 0 \\ BJB^* & A^\times \end{pmatrix},$$

where $A^\times = A - BC$.

We shall say that $G(\lambda)$ as defined in (1.6) has a *complete set of minimal J-symmetric factorizations with factor of the form* (1.7) if every J_1-lagrangian subspace $M^\times$ that is $\tilde{A}^\times$-invariant satisfies $M^\times \cap \operatorname{Im} \begin{pmatrix} I \\ 0 \end{pmatrix} = (0)$.

Observe that $\operatorname{Im} \begin{pmatrix} I \\ 0 \end{pmatrix}$ is $\tilde{A}$-invariant and J_1-lagrangian. Also, if we have a minimal factorization of the form (1.1) where $R(\lambda)$ is given as in (1.7), then the corresponding subspace M must be $\operatorname{Im} \begin{pmatrix} I \\ 0 \end{pmatrix}$ (see [BGK]).

For this problem we have the following result.

Theorem 1.3 *Let $G(\lambda)$ be as in (1.6), where $\sigma(A) \cap \sigma(-A^*) = \emptyset$. Then $G(\lambda)$ has a complete set of minimal J-symmetric factorizations with factor of the form (1.7) if and only if for all non-zero vectors x we have*

$$\langle JB^*(\lambda + A^{\times *})^{-1}x, B^*(\lambda + A^{\times *})^{-1}x \rangle \not\equiv 0, \quad \lambda \in \rho(-A^{\times *}).$$

Proof: Suppose that there is a non-zero vector x such that

$$\langle JB^*(\lambda + A^{\times *})^{-1}x, B^*(\lambda + A^{\times *})^{-1}x \rangle \equiv 0, \quad \lambda \in \rho(-A^{\times *}).$$

Let $N = \operatorname{span}\{\begin{pmatrix} x \\ 0 \end{pmatrix}\} \cup \{(\lambda - \tilde{A}^\times)^{-1} \begin{pmatrix} x \\ 0 \end{pmatrix} \mid \lambda \in \rho(\tilde{A}^\times)\}$. Then N is $\tilde{A}^\times$-invariant and it is J_1-neutral by an argument similar to the one in the proof of the previous theorem. By Lemma 1.1 there is a $\tilde{A}^\times$-invariant J_1-lagrangian subspace $M^\times$ containing N. Then $M^\times \cap \operatorname{Im} \begin{pmatrix} I \\ 0 \end{pmatrix} \neq (0)$, so there is not a complete set of J-symmetric factorizations with factor of the form (1.7).

Conversely, assume there is not a complete set of J-symmetric factorizations with factor of the form (1.7). Then there is a $\tilde{A}^\times$-invariant J_1-lagrangian subspace

$M^\times$ with $M^\times \cap \mathrm{Im} \begin{pmatrix} I \\ 0 \end{pmatrix} \neq (0)$. Take $x \neq 0$ so that $\begin{pmatrix} x \\ 0 \end{pmatrix} \in M^\times \cap \mathrm{Im} \begin{pmatrix} I \\ 0 \end{pmatrix}$.
Then

$$\left\langle J_1(\lambda - \tilde{A}^\times)^{-1} \begin{pmatrix} x \\ 0 \end{pmatrix}, \begin{pmatrix} x \\ 0 \end{pmatrix} \right\rangle \equiv 0,$$

as $M^\times$ is J_1-lagrangian and $\tilde{A}^\times$-invariant. But this is equivalent to

$$\langle J B^*(\lambda + A^{\times *})^{-1} x, B^*(\lambda + A^{\times *})^{-1} x \rangle \equiv 0, \quad \lambda \in \rho(-A^{\times *}).$$

This finishes the proof. □

2 The Popov Function Approach to the Algebraic Riccati Equation

In this section we shall be concerned with the algebraic Riccati equation (0.1), and we shall assume throughout that (A, B) is controllable.

For an arbitrary feedback $m \times n$ matrix K we define $V = A - \Psi K$, and

$$G_K(\lambda) = J + (JK \quad \Psi^*)$$

(2.1)

$$\left(\lambda - \begin{pmatrix} -V & 0 \\ -(C + K^*JK) & V^* \end{pmatrix} \right)^{-1} \begin{pmatrix} \Psi \\ -K^*J \end{pmatrix}.$$

Observe that we have

(2.2) $\quad G_K(\lambda)^{-1} = J - (K \quad J\Psi^*) \left(\lambda - \begin{pmatrix} -A & -W \\ -C & A^* \end{pmatrix} \right)^{-1} \begin{pmatrix} \Psi J \\ -K^* \end{pmatrix}.$

Note that the zeros of any G_K are a subset of the eigenvalues of the Hamiltonian $H = \begin{pmatrix} -A & -W \\ -C & A^* \end{pmatrix}$. As (A, B) is controllable, by taking an appropriate K we can arrange it so that H and $\begin{pmatrix} -V & 0 \\ -(C + K^*JK) & V^* \end{pmatrix}$ have no common eigenvalues, in which case the realization (2.1) is minimal.

Our main aim in this section is to show that there is a bijective correspondence between hermitian solutions of (0.1) and certain J-symmetric factorizations of $G_K(\lambda)$. We shall state the result more explicitly later, but now we start discussing this correspondence in a somewhat informal way. First we mention the following fact.

Fact: We have

(2.3) $\quad G_K(\lambda) = (I - \Psi^*(\lambda - V^*)^{-1}K^*)G_0(\lambda)(I + K(\lambda + V)^{-1}\Psi).$

Proof: Indeed,

$$(I - \Psi^*(\lambda - V^*)^{-1}K^*)G_0(\lambda)(I + K(\lambda + V)^{-1}\Psi)$$
$$= (J - \Psi^*(\lambda - V^*)^{-1}K^*J)(I + K(\lambda + V)^{-1}\Psi)$$
$$+ - (I - \Psi^*(\lambda - V^*)^{-1}K^*)\Psi^*(\lambda - A^*)^{-1}$$
$$C(\lambda + A)^{-1}\Psi(I + K(\lambda + V)^{-1}\Psi).$$

Now we have

$$(\lambda + A)^{-1}\Psi(I + K(\lambda + V)^{-1}\Psi)$$
$$= (\lambda + A)^{-1}\Psi + (\lambda + A)^{-1}(A - V)(\lambda + V)^{-1}\Psi$$
$$= (\lambda + A)^{-1}\Psi + (\lambda + A)^{-1}\{(A + \lambda) - (\lambda + V)\}(\lambda + V)^{-1}\Psi$$
$$= (\lambda + V)^{-1}\Psi.$$

Inserting this in the formula above (together with its symmetric analogue), we obtain that

$$(I - \Psi^*(\lambda - V^*)^{-1}K^*)G_0(\lambda)(I + K(\lambda + V)^{-1}\Psi)$$
$$= (J - \Psi^*(\lambda - V^*)^{-1}K^*J)(I + K(\lambda + V)^{-1}\Psi)$$
$$-\Psi^*(\lambda - V^*)^{-1}C(\lambda + V)^{-1}\Psi.$$

One easily checks that the last expression is equal to the right hand side of (2.1), thereby proving (2.3). □

Now let X be a hermitian solution of (0.1). Then

(2.4) $$G_0(\lambda) = R_0(-\bar{\lambda})^* J R_0(\lambda),$$

where

(2.5) $$R_0(\lambda) = J + \Psi^* X(\lambda + A)^{-1}\Psi.$$

Indeed, if X solves (0.1) then

$$R_0(-\bar{\lambda})^* J R_0(\lambda) = (I - \Psi^*(\lambda - A^*)^{-1}X\Psi J)(J + \Psi^* X(\lambda + A)^{-1}\Psi)$$
$$= J - \Psi^*(\lambda - A^*)^{-1}X\Psi + \Psi^* X(\lambda + A)^{-1}\Psi$$
$$-\Psi^*(\lambda - A^*)^{-1}X\Psi J\Psi^* X(\lambda + A)^{-1}\Psi$$
$$= J + \Psi^*(\lambda - A^*)^{-1}\{-X(\lambda + A)$$
$$+ (\lambda - A^*)X - XWX\}(\lambda + A)^{-1}\Psi$$
$$= J + \Psi^*(\lambda - A^*)^{-1}\{-XA - A^*X - XWX\}(\lambda + A)^{-1}\Psi$$
$$= J - \Psi^*(\lambda - A^*)^{-1}C(\lambda + A)^{-1}\Psi$$

It follows that for any K

(2.6) $$G_K(\lambda) = R_K(-\bar{\lambda})^* J R_K(\lambda),$$

where

$$R_K(\lambda) = R_0(\lambda)(I + K(\lambda + V)^{-1}\Psi)$$

(2.7)

$$= J + (JK + \Psi^*X)(\lambda + V)^{-1}\Psi.$$

To see this we only need to check (2.7):

$$(J + \Psi^*X(\lambda + A)^{-1}\Psi)(I + K(\lambda + V)^{-1}\Psi)$$
$$= J + \Psi^*X(\lambda + A)^{-1}\Psi + JK(\lambda + V)^{-1}\Psi$$
$$+ \Psi^*X(\lambda + A)^{-1}\Psi K(\lambda + V)^{-1}\Psi.$$

Using $V = A - \Psi K$ we obtain that this equals

$$J + \Psi^*X(\lambda + A)^{-1}\Psi + JK(\lambda + V)^{-1}\Psi$$
$$+ \Psi^*X(\lambda + A)^{-1}\{(A + \lambda) - (\lambda + V)\}(\lambda + V)^{-1}\Psi$$
$$= J + (JK + \Psi^*X)(\lambda + V)^{-1}\Psi,$$

as desired.

Conversely, suppose we have a factorization (2.6) for some G_K where $R_K(\lambda) = J + C_R(\lambda + V)^{-1}\Psi$ for some C_R. Then $C_R = JK + \Psi^*X$ for some hermitian solution of (0.1).

To prove this we might use the following line of argument.

If for some K the function G_K admits such a factorization first show that it admits such a factorization for all choices of K. Then take one such choice so that the realization (2.1) is minimal. Then necessarily also the factorization (2.6) is minimal. By the theorem in [BGK] on minimal factorizations such a factorization comes from subspaces M and $M^\times$ such that the first is invariant under the main operator in the realization (2.1), the second is invariant under H, and the matching condition $\mathbb{C}^{2n} = M \dotplus M^\times$ holds. Looking at the formulas for the factors, we see from the particular form that R_K has that we must have $M = \mathrm{Im}\begin{pmatrix} 0 \\ I \end{pmatrix}$. As the matching condition holds we have $M^\times = \mathrm{Im}\begin{pmatrix} I \\ X \end{pmatrix}$ for some matrix X. Since $M^\times$ is H-invariant we get that X solves (0.1). Now the symmetry of the factorization forces $J_1 M = M^\perp$ and $J_1 M^\times = M^{\times\perp}$ (see [RR]). This implies that $X = X^*$.

We prefer, however, to present a different proof in detail, which will also establish the connection with the notion of the generalized Bezoutian for rational matrix functions introduced and studied in [GoL] and [KL]. Let be given rational matrix functions $N(\lambda)$ of size $k \times n$ and $M(\lambda)$ of size $m \times k$, such that

(2.8) $$M(\lambda)N(\lambda) = 0$$

for all points λ that are not poles of $M(\lambda)$ or $N(\lambda)$. Write a controllable realization

(2.9) $$N(\lambda) = D_N + C_N(\lambda I - A)^{-1}B$$

and an observable realization

$$(2.10) \qquad M(\lambda) = D_M + U(\lambda I - V)^{-1} W_M.$$

As proved in [GoL], [KL] there exists a *unique* matrix $\mathbb{B}$ such that

$$(2.11) \qquad (\lambda - \mu)^{-1} M(\lambda) N(\mu) = U(\lambda I - V)^{-1} \mathbb{B}(\mu I - A)^{-1} B.$$

The matrix $\mathbb{B}$ in (2.11) is called *the generalized Bezoutian* based on equation (2.8) with realizations (2.9), (2.10).

One often deals with the situation when $M(\lambda)$ and $N(\lambda)$ are of the form

$$M(\lambda) = (M_1(\lambda) \quad M_2(\lambda) \quad \ldots \quad M_s(\lambda)), \qquad N(\lambda) = \begin{pmatrix} N_1(\lambda) \\ N_2(\lambda) \\ \vdots \\ N_s(\lambda) \end{pmatrix}.$$

The case $s = 2$ and M_i, N_i ($i = 1, 2$) are square size rational matrix functions with invertible values at infinity has been studied in [LR, LR1], where it is shown that Ker $\mathbb{B}$ completely describes the common (right) zero structure of $N_1(\lambda)$ and $N_2(\lambda)$. In the case $s > 2$ this property is not valid anymore, but we still have the property that Ker $\mathbb{B} = (0)$ implies right coprimeness of $N_1(\lambda), N_2(\lambda), \ldots, N_s(\lambda)$.

Decompose $C = PY$ and put

$$\begin{aligned} L_1(\lambda) &= J - \Psi^*(\lambda - V^*)^{-1} K^* J, \\ L_2(\lambda) &= I + K(\lambda + V)^{-1}\Psi, \\ P(\lambda) &= -\Psi^*(\lambda - V^*)^{-1} P, \\ Y(\lambda) &= Y(\lambda + V)^{-1}\Psi. \end{aligned}$$

Then from our computation above we also have

$$(2.12) \qquad G_K(\lambda) = L_1(\lambda) L_2(\lambda) + P(\lambda) Y(\lambda).$$

Now assume that we have a function $R_K(\lambda)$ of the form

$$(2.13) \qquad R_K(\lambda) = J + C_R(\lambda + V)^{-1} \Psi$$

such that

$$(2.14) \qquad G_K(\lambda) = R_K(-\bar{\lambda})^* J R_K(\lambda).$$

Denote

$$M_K(\lambda) = (L_1(\lambda) \quad P(\lambda) \quad - R_K(-\bar{\lambda})^* J), \qquad N_K(\lambda) = \begin{pmatrix} L_2(\lambda) \\ Y(\lambda) \\ R_K(\lambda) \end{pmatrix}.$$

Then, in view of (2.12), one can rewrite (2.14) as

$$(2.15) \qquad M_K(\lambda) N_K(\lambda) = 0.$$

We can write the following realizations for M_K and N_K:

$$(2.16) \quad M_K(\lambda) = \begin{pmatrix} J & 0 & I \end{pmatrix} + \Psi^*(\lambda - V^*)^{-1} \begin{pmatrix} -K^*J & -P & -C_R^*J \end{pmatrix},$$

$$(2.17) \qquad N_K(\lambda) = \begin{pmatrix} I \\ 0 \\ J \end{pmatrix} + \begin{pmatrix} K \\ Y \\ C_R \end{pmatrix} (\lambda + V)^{-1} \Psi.$$

Denoting by $\mathbb{B}_K$ the generalized Bezoutian based on equation (2.15) and associated with realizations (2.16), (2.17), we have the following proposition.

Proposition 2.1 $\mathbb{B}_K$ *is a hermitian matrix.*

Proof: Denote $\Gamma(\lambda, \mu) = M_K(\lambda) N_K(\mu)$. In our case $\mathbb{B}_K$ is defined by the equation

$$(2.18) \qquad (\lambda - \mu)^{-1} \Gamma(\lambda, \mu) = \Psi^*(\lambda - V^*)^{-1} \mathbb{B}_K (\mu + V)^{-1} \Psi.$$

Writing explicitly

$$\Gamma(\lambda, \mu) = L_2(-\bar{\lambda})^* J L_2(\mu) + \Psi^*(\lambda - V^*)^{-1}$$
$$C(\mu + V)^{-1} \Psi - R_K(-\bar{\lambda})^* J R_K(\mu),$$

one easily sees that

$$(2.19) \qquad \Gamma(\lambda, \mu) = \Gamma(-\bar{\lambda}, -\bar{\mu})^*.$$

Form (2.18) we have

$$(-\bar{\mu} + \bar{\lambda})^{-1} \Gamma(-\bar{\lambda}, -\bar{\mu}) = \Psi^*(\bar{\mu} + V^*)^{-1} \mathbb{B}_K (\bar{\lambda} - V)^{-1} \Psi.$$

Taking adjoints to both sides of this equality and using (2.19) we obtain

$$(2.20) \qquad (\lambda - \mu)^{-1} \Gamma(\lambda, \mu) = \Psi^*(\lambda - V^*)^{-1} \mathbb{B}_K^* (\mu + V)^{-1} \Psi.$$

Comparing (2.18) and (2.20) and using the unicity of the generalized Bezoutian we infer that $\mathbb{B}_K = \mathbb{B}_K^*$. $\qquad \square$

Next, we state the main result of this section.

Theorem 2.2 *If X is a hermitian solution of (0.1) then $G_K(\lambda)$ admits a J-symmetric factorization (2.6) with $R_K(\lambda)$ given by (2.7), and in this case $X = \mathbb{B}_K$.*

Conversely, if $G_K(\lambda)$ admits a J-symmetric factorization (2.6) with $R_K(\lambda)$ of the form (2.13), then $\mathbb{B}_K$ is a hermitian solution of (0.1) and the coefficient C_R in (2.13) is given by $C_R = JK + \Psi^* \mathbb{B}_K$.

The above correspondence between hermitian solutions of (0.1) and factorizations (2.6) with $R_K(\lambda)$ of the form (2.13) is one-to-one.

Proof: We need the following formulas

(2.21) $A(\mu + V)^{-1}\Psi = \Psi L_2(\mu) - \mu(\mu + V)^{-1}\Psi,$

(2.22) $\Psi^*(\lambda - V^*)^{-1}A^* = -L_1(\lambda)J\Psi^* + \lambda\Psi^*(\lambda - V^*)^{-1}$

Clearly (2.22) follows form (2.21) by setting $\mu = -\bar{\lambda}$ and taking adjoint to both sides of (2.21).

The equation (2.21) follows from the following computation:

$$\begin{aligned}
A(\mu + V)^{-1}\Psi &= (V + \Psi K + \mu I - \mu I)(\mu + V)^{-1}\Psi \\
&= \Psi + \Psi K(\mu + V)^{-1}\Psi - \mu(\mu + V)^{-1}\Psi \\
&= \Psi L_2(\mu) - \mu(\mu + V)^{-1}\Psi.
\end{aligned}$$

Now assume that X is a hermitian solution of (0.1). Multiply (0.1) by $\Psi^*(\lambda - V^*)^{-1}$ from the left, and by $(\mu + V)^{-1}\Psi$ from the right to obtain

(2.23)
$$\begin{aligned}
\Psi^*(\lambda - V^*)^{-1}A^*X(\mu + V)^{-1}\Psi + \Psi(\lambda - V^*)^{-1}XA(\mu + V)^{-1}\Psi \\
+ \Psi(\lambda - V^*)^{-1}X\Psi J\Psi^*X(\mu + V)^{-1}\Psi + P(\lambda)Y(\mu) = 0.
\end{aligned}$$

Denote

$$S(\mu) = \Psi^*X(\mu + V)^{-1}\Psi, \quad R(\lambda) = \Psi^*(\lambda - V^*)^{-1}X\Psi.$$

Also introduce

$$\mathcal{S}(\lambda, \mu) = \Psi^*(\lambda - V^*)^{-1}X(\mu + V)^{-1}\Psi$$

With these notations and using (2.22) we can rewrite the first term in (2.23) as follows

$$\begin{aligned}
\Psi^*(\lambda - V^*)^{-1}A^*X(\mu + V)^{-1}\Psi \\
= (-L_1(\lambda)J\Psi^* + \lambda\Psi^*(\lambda - V^*)^{-1})X(\mu + V)^{-1}\Psi \\
= -L_1(\lambda)JS(\mu) + \lambda\mathcal{S}(\lambda, \mu).
\end{aligned}$$

Using (2.21) we can rewrite the second term in (2.23) as follows:

$$\begin{aligned}
\Psi^*(\lambda - V^*)^{-1}XA(\mu + V)^{-1}\Psi \\
= \Psi^*(\lambda - V^*)^{-1}X(\Psi L_2(\mu) - \mu(\mu + V)^{-1}\Psi) \\
= R(\lambda)L_2(\mu) - \mu\mathcal{S}(\lambda, \mu).
\end{aligned}$$

The third term in (2.23) clearly equals $R(\lambda)JS(\mu)$. Thus (2.23) can be rewritten as follows

(2.24)
$$\begin{aligned}
(\lambda - \mu)\mathcal{S}(\lambda, \mu) = L_1(\lambda)JS(\mu) - R(\lambda)L_2(\mu) \\
- P(\lambda)Y(\mu) - R(\lambda)JS(\mu).
\end{aligned}$$

Setting $\lambda = \mu$ we obtain

$$P(\lambda)Y(\lambda) = L_1(\lambda)JS(\lambda) - R(\lambda)L_2(\lambda) - R(\lambda)JS(\lambda).$$

Hence

$$
\begin{aligned}
G_K(\lambda) &= L_1(\lambda)L_2(\lambda) + P(\lambda)Y(\lambda) \\
(2.25) \qquad &= L_1(\lambda)L_2(\lambda) + L_1(\lambda)JS(\lambda) - R(\lambda)L_2(\lambda) - R(\lambda)JS(\lambda) \\
&= (L_1(\lambda) - R(\lambda))(L_2(\lambda) + JS(\lambda)).
\end{aligned}
$$

Now

$$
\begin{aligned}
L_2(\lambda) + JS(\lambda) &= I + K(\lambda + V)^{-1}\Psi + J\Psi^*X(\lambda + V)^{-1}\Psi \\
&= I + (K + J\Psi^*X)(\lambda + V)^{-1}\Psi \\
&= J(J + (JK + \Psi^*X)(\lambda + V)^{-1}\Psi)
\end{aligned}
$$

and

$$
\begin{aligned}
L_1(\lambda) - R(\lambda) &= J - \Psi^*(\lambda - V^*)^{-1}K^*J - \Psi^*(\lambda - V^*)^{-1}X\Psi \\
&= J - \Psi^*(\lambda - V^*)^{-1}(K^*J + X\Psi).
\end{aligned}
$$

So, denoting $R_K(\lambda) = J + (JK + \Psi^*X)(\lambda + V)^{-1}\Psi$, we have from (2.25) that

$$G_K(\lambda) = R_K(-\bar{\lambda})^*JR_K(\lambda)$$

as desired. Also, with these notations we can rewrite (2.24) in the form

$$
\begin{aligned}
&L_1(\lambda)L_2(\mu) + P(\lambda)Y(\mu) - R_K(-\bar{\lambda})^*JR(\mu) \\
&= (\lambda - \mu)\Psi^*(\lambda - V^*)^{-1}X(\mu + V)^{-1}\Psi,
\end{aligned}
$$

which shows that $X = \mathbb{B}_K$.

To prove the converse part of the theorem, assume that a J-symmetric factorization of the type (2.6) holds true with some $R_K(\lambda)$ of the form (2.13). Let $\mathbb{B}_K$ be the generalized Bezoutian based on the equation (2.15) and associated with the realizations (2.16), (2.17). Denote $\Omega = A^*\mathbb{B}_K + \mathbb{B}_K A$. Using formulas (2.21), (2.22) we have

$$
\begin{aligned}
&\Psi^*(\lambda - V^*)^{-1}\Omega(\mu + V)^{-1}\Psi \\
&= (-L_1(\lambda)J\Psi^* + \lambda\Psi^*(\lambda - V^*)^{-1})\mathbb{B}_K(\mu + V)^{-1}\Psi \\
&\quad + \Psi^*(\lambda - V^*)^{-1}\mathbb{B}_K(\Psi L_2(\mu) - \mu(\mu + V)^{-1}\Psi) \\
&= (\lambda - \mu)\Psi^*(\lambda - V^*)^{-1}\mathbb{B}_K(\mu + V)^{-1}\Psi \\
&\quad - L_1(\lambda)J\tilde{S}(\mu) + \tilde{R}(\lambda)L_2(\mu),
\end{aligned}
$$

where $\tilde{S}(\mu) = \Psi^* \mathbb{B}_K (\mu + V)^{-1} \Psi$, and $\tilde{R}(\lambda) = \Psi^* (\lambda - V^*)^{-1} \mathbb{B}_K \Psi$. From the definition of $\mathbb{B}_K$ we see that

$$
\begin{aligned}
(2.26) \qquad &-\Psi^* (\lambda - V^*)^{-1} \Omega (\mu + V)^{-1} \Psi \\
&= L_1(\lambda) L_2(\mu) + P(\lambda) Y(\mu) - R_K(-\bar{\lambda})^* J R_K(\mu) \\
&\quad - L_1(\lambda) J \tilde{S}(\mu) + \tilde{R}(\lambda) L_2(\mu),
\end{aligned}
$$

Now let $\mu \to \infty$ in (2.26), then we see that

$$
(2.27) \qquad\qquad R_K(-\bar{\lambda})^* = L_1(\lambda) + \tilde{R}(\lambda).
$$

Similarly, letting $\lambda \to \infty$ we deduce from (2.26) that

$$
(2.28) \qquad\qquad R_K(\lambda) = J L_2(\lambda) + J \tilde{S}(\lambda).
$$

Substituting (2.27) and (2.28) into (2.26) we obtain

$$
\begin{aligned}
-\Psi^* (\lambda - V^*)^{-1} \Omega (\mu + V)^{-1} \Psi &= -\tilde{R}(\lambda) J \tilde{S}(\mu) + P(\lambda) Y(\mu) \\
&= \Psi^* (\lambda - V^*)^{-1} (\mathbb{B}_K \Psi J \Psi^* \mathbb{B}_K - C)(\mu + V)^{-1} \Psi,
\end{aligned}
$$

which shows that $\mathbb{B}_K$ is indeed a solution of (0.1). $\qquad\qquad\square$

The connections between solutions of the algebraic Riccati equation and factorizations of rational matrix functions are in principle well-known. They go back to [P] (compare also [M1, M2], [IW], [IWO], [LaR].) The observation that hermitian solutions of the algebraic Riccati equation can be interpreted as generalized Bezoutians, however, is new (see also our previous works [L], [LRa] where these connections were used for the case $C = 0$). Note that there are explicit formulas for the generalized Bezoutians (see e.g., [LR]). We shall exploit these formulas elsewhere.

3 Complete Set of Solutions of the Algebraic Riccati Equation

In this section, the pair (A, B) is not assumed to be controllable, except when explicitly stated.

Recall that a subspace N is called J_1-*lagrangian* if $J_1 N = N^\perp$.

There is a one-to-one correspondence between the following sets:

(a) solutions $X = X^*$ of

$$
X B R^{-1} B^* X + X A + A^* X - C = 0,
$$

(b) H-invariant J_1-lagrangian subspaces $M^\times$ such that $M^\times \cap \operatorname{Im} \begin{pmatrix} 0 \\ I \end{pmatrix} = (0)$.

This correspondence is given by $M^\times = \operatorname{Im} \begin{pmatrix} I \\ X \end{pmatrix}$.

Moreover, under the condition that the pair (A, B) is controllable, these sets are in one-one correspondence to

(c) symmetric factorizations (2.6) of G_K, where R_K has the pair $(-V, \Psi)$ as an extension of its pole pair, i.e., $R_K(\lambda)$ is of the form $R_K(\lambda) = J + C_R(\lambda + V)^{-1}\Psi$ for some C_R; this correspondence is given by formulas (2.7).

All these connections are well-known, see, e.g. [K1, K2, K3, LaR1, C, S, W] for the connection to invariant subspaces, and [M1, M2, P, IW] for the connection to factorization; see also the books [BLW, LaR, IWO]. Moreover, we have established a connection with generalized Bezoutians: all solutions of the algebraic Riccati equation are generalized Bezoutians corresponding to a J-symmetric factorization of the type (c) above, and vice versa.

For the particular case where $J = I$ and the pair (A, B) is controllable it is known that in (b) above we can omit the condition that $M^\times \cap \mathrm{Im}\begin{pmatrix} 0 \\ I \end{pmatrix} = (0)$, i.e., in that case any H-invariant J_1-lagrangian subspace is automatically a graph subspace. This was first proved independently in [C], [S], [LaR1] (see also the book [LaR]).

Returning to the case of a general J, and without the condition of controllability, we shall say that (0.1) has a *complete set of solutions* if for any H-invariant J_1-lagrangian subspace $M^\times$ we have the matching condition $M^\times \cap \mathrm{Im}\begin{pmatrix} 0 \\ I \end{pmatrix} = (0)$, under the requirement that there exists at least one H-invariant J_1-lagrangian subspace. This concept was introduced in [B], and studied there for the first time.

From [RR1], Theorem 5.2, we see that the existence of an H-invariant J_1-lagrangian subspace is equivalent to saying that for each pure imaginary eigenvalue of H there is an even number of odd partial multiplicities (possibly zero) and that in the sign characteristic of the pair (iH, iJ_1) the number of -1's corresponding to these partial multiplicities is equal to the number of $+1$'s corresponding to these partial multiplicities. (Compare [RR2] for the real case which is more in line with the Hamiltonian approach.) The reader is referred to [GLR] for the notion of sign characteristic of the pair (iH, iJ_1).

Observe that if the realization (2.1) of G_K is minimal and we have a minimal factorization (2.6) then by [RR] the function G_K has constant signature as well as constant zero-signature and constant pole-signature. The condition on the constant zero-signature is equivalent to the statement that there exists an H-invariant J_1-lagrangian subspace. (For the pole-signature the statement is trivial in this particular case, since obviously the main operator in the realization (2.1) has an invariant J_1-lagrangian subspace.)

We start our investigation with the following lemma.

Lemma 3.1 *For any K we have*

$$(3.1) \quad (I \quad 0)(\lambda - H)^{-1}\begin{pmatrix} 0 \\ I \end{pmatrix} = -(\lambda + V)^{-1}\Psi G_K(\lambda)^{-1}\Psi^*(\lambda - V^*)^{-1}.$$

Proof: Using the formulas for multiplication of realizations we have

$$(\lambda + V)^{-1}\Psi G_K(\lambda)^{-1} = (I \quad 0 \quad 0)$$

$$\left(\lambda - \begin{pmatrix} -V & -\Psi K & -\Psi J\Psi^* \\ 0 & -A & -W \\ 0 & -C & A^* \end{pmatrix}\right)^{-1} \begin{pmatrix} \Psi J \\ \Psi J \\ -K^* \end{pmatrix}.$$

Put $S = \begin{pmatrix} I & I & 0 \\ 0 & I & 0 \\ 0 & 0 & I \end{pmatrix}$. Then

$$S^{-1}\begin{pmatrix} \Psi J \\ \Psi J \\ -K^* \end{pmatrix} = \begin{pmatrix} 0 \\ \Psi J \\ -K^* \end{pmatrix}, \quad (I \quad 0 \quad 0)S = (I \quad I \quad 0),$$

and

$$S^{-1}\begin{pmatrix} -V & -\Psi K & -\Psi J\Psi^* \\ 0 & -A & -W \\ 0 & -C & A^* \end{pmatrix}S = \begin{pmatrix} -V & 0 & 0 \\ 0 & -A & -W \\ 0 & -C & A^* \end{pmatrix}.$$

Now the first component is uncontrollable, so we see that

$$(\lambda + V)^{-1}\Psi G_K(\lambda)^{-1} = (I \quad 0)(\lambda - H)^{-1}\begin{pmatrix} \Psi J \\ -K^* \end{pmatrix}.$$

To obtain the final result we repeat the procedure:

$$(\lambda + V)^{-1}\Psi G_K(\lambda)^{-1}\Psi^*(\lambda - V^*)^{-1}$$

$$= (I \quad 0)(\lambda - H)^{-1}\begin{pmatrix} \Psi J \\ -K^* \end{pmatrix}\Psi^*(\lambda - V^*)^{-1}$$

$$= (I \quad 0 \quad 0)\begin{pmatrix} -A & -W & \Psi J\Psi^* \\ -C & A^* & -K^*\Psi^* \\ 0 & 0 & V^* \end{pmatrix}^{-1}\begin{pmatrix} 0 \\ 0 \\ I \end{pmatrix}.$$

Put $T = \begin{pmatrix} I & 0 & 0 \\ 0 & I & I \\ 0 & 0 & I \end{pmatrix}$. Then

$$T^{-1}\begin{pmatrix} -A & -W & \Psi J\Psi^* \\ -C & A^* & -K^*\Psi^* \\ 0 & 0 & V^* \end{pmatrix}T = \begin{pmatrix} -A & -W & 0 \\ -C & A^* & 0 \\ 0 & 0 & V^* \end{pmatrix}.$$

Also,

$$(I \quad 0 \quad 0)T = (I \quad 0 \quad 0), \quad T^{-1}\begin{pmatrix} 0 \\ 0 \\ I \end{pmatrix} = \begin{pmatrix} 0 \\ -I \\ I \end{pmatrix}.$$

As the third component is unobservable we obtain

$$-(\lambda + V)^{-1}\Psi G_K(\lambda)^{-1}\Psi^*(\lambda - V^*)^{-1} = (I \quad 0)(\lambda - H)^{-1}\begin{pmatrix}0\\I\end{pmatrix}.$$

The lemma is proved.   □

Our main result is the following (compare with the main theorem in [B], the statement of which is much more complicated).

Theorem 3.2 *The following are equivalent.*

(i) *There exists a solution to the algebraic Riccati equation (0.1), and the set of solutions of this equation is complete.*

(ii) *The following conditions are satisfied:*

(a) *there exists an H-invariant J_1-lagrangian subspace,*

(b) *the rational matrix*

$$(0 \quad I)(\lambda J_1 - J_1 H)^{-1}\begin{pmatrix}0\\I\end{pmatrix}$$

has constant signature,

(c) *for all non-zero x we have*

$$\left\langle (\lambda - H)^{-1}\begin{pmatrix}0\\x\end{pmatrix}, \begin{pmatrix}x\\0\end{pmatrix}\right\rangle \neq 0, \qquad \lambda \in \rho(H),$$

or equivalently, $\langle J_1(\lambda - H)^{-1}\begin{pmatrix}0\\x\end{pmatrix}, \begin{pmatrix}0\\x\end{pmatrix}\rangle \neq 0, \lambda \in \rho(H)$.

Proof: Suppose X solves the algebraic Riccati equation (0.1), and there is a complete set of solutions. Then condition (a) holds trivially, as for any solution X the subspace $\text{Im}\begin{pmatrix}I\\X\end{pmatrix}$ will be invariant and lagrangian. Condition (b) is satisfied as well, as the (1, 2)-entry of the resolvent of H is connected to $G_0(\lambda)$ by (3.1). So we see that (b) holds if G_0 has constant signature. As there is a solution to (0.1) we have from (2.4) that this is indeed the case. It remains to show that condition (c) holds. Suppose that for some non-zero x we have that

(3.2) $$x^*(I \quad 0)(\lambda - H)^{-1}\begin{pmatrix}0\\I\end{pmatrix}x \equiv 0 \qquad \lambda \in \rho(H).$$

Let

(3.3) $$N = \underset{\lambda \notin \sigma(H)}{\text{span}}\left\{(\lambda - H)^{-1}\begin{pmatrix}0\\x\end{pmatrix} \cup \left\{\begin{pmatrix}0\\x\end{pmatrix}\right\}\right\}.$$

By (3.2) we have that N is J_1-neutral: clearly $\begin{pmatrix} 0 \\ x \end{pmatrix}$ is a J_1-neutral vector, and we have for all λ that are not eigenvalues of H

$$0 = (x^* \quad 0)(\lambda - H)^{-1} \begin{pmatrix} 0 \\ x \end{pmatrix} = \left\langle J_1(\lambda - H)^{-1} \begin{pmatrix} 0 \\ x \end{pmatrix}, \begin{pmatrix} 0 \\ x \end{pmatrix} \right\rangle,$$

as well as for all λ and μ that are not eigenvalues of H with $\lambda \neq \bar{\mu}$

$$\left\langle J_1(\lambda - H)^{-1} \begin{pmatrix} 0 \\ x \end{pmatrix}, (\mu - H)^{-1} \begin{pmatrix} 0 \\ x \end{pmatrix} \right\rangle$$

$$= (x^* \quad 0)(\bar{\mu} - H)^{-1}(\lambda - H)^{-1} \begin{pmatrix} 0 \\ x \end{pmatrix}$$

$$= \frac{1}{\lambda - \bar{\mu}}(x^* \quad 0)(\bar{\mu} - H)^{-1} \begin{pmatrix} 0 \\ x \end{pmatrix} - (x^* \quad 0)$$

$$(\lambda - H)^{-1} \begin{pmatrix} 0 \\ x \end{pmatrix} = 0.$$

Finally, for $\lambda \notin \sigma(H)$ we have

$$\left\langle J_1(\lambda - H)^{-1} \begin{pmatrix} 0 \\ x \end{pmatrix}, (\lambda - H)^{-1} \begin{pmatrix} 0 \\ x \end{pmatrix} \right\rangle$$

$$= \lim_{\mu \to \lambda} \left\langle J_1(\lambda - H)^{-1} \begin{pmatrix} 0 \\ x \end{pmatrix}, (\mu - H)^{-1} \begin{pmatrix} 0 \\ x \end{pmatrix} \right\rangle = 0.$$

By Lemma 1.1 there exists an H-invariant J_1-lagrangian subspace $M^\times$ that contains N. But $\begin{pmatrix} 0 \\ x \end{pmatrix} \in M^\times$, so $M^\times \cap \text{Im} \begin{pmatrix} 0 \\ I \end{pmatrix} \neq (0)$. Thus (0.1) does not have a complete set of solutions.

Conversely, suppose (a), (b) and (c) hold. We shall show that any H-invariant J_1-lagrangian subspace is a graph subspace, thereby showing (i). Let M be such a subspace, and suppose that it is not a graph subspace. Then there is a non-zero x such that $\begin{pmatrix} 0 \\ x \end{pmatrix} \in M$. Construct N as in (3.3). Clearly, by H-invariance of both N and M we have $N \subset M$. But M is J_1-neutral, and so for all $\lambda \notin \sigma(H)$ we have

$$0 = \left\langle J_1(\lambda - H)^{-1} \begin{pmatrix} 0 \\ x \end{pmatrix}, \begin{pmatrix} 0 \\ x \end{pmatrix} \right\rangle = \left\langle (\lambda - H)^{-1} \begin{pmatrix} 0 \\ x \end{pmatrix}, \begin{pmatrix} x \\ 0 \end{pmatrix} \right\rangle$$

which contradicts condition (c).   □

As a corollary we have the following observation.

Corollary 3.3 *Suppose the pair (A, B) is not controllable. Then equation (0.1) does not have a complete set of solutions.*

Proof: First observe that (A, B) is controllable if and only if for any K the pair (V, Ψ) is controllable. So, under our assumption we have that (V, Ψ) is not controllable. Then there is a non-zero vector x such that $\Psi^*(\lambda - V^*)^{-1}x \equiv 0$. By Lemma 3.1 this implies that $\langle(\lambda - H)^{-1}\begin{pmatrix} 0 \\ x \end{pmatrix}, \begin{pmatrix} x \\ 0 \end{pmatrix}\rangle \equiv 0$, i.e., condition (ii) (c) in Theorem 3.2 is not satisfied. Hence there is not a complete set of solutions. $\qquad\square$

Examples: As a first example, consider the case $A = \begin{pmatrix} 0 & 0 \\ 0 & \sqrt{3} \end{pmatrix}$, $B = A$, and $J = I$, while $C = I$ as well. The Hamiltonian now becomes

$$H = \begin{pmatrix} 0 & 0 & 0 & 0 \\ 0 & -\sqrt{3} & 0 & 3 \\ -1 & 0 & 0 & 0 \\ 0 & -1 & 0 & \sqrt{3} \end{pmatrix}.$$

One checks that it has two Jordan chains of length two corresponding to the eigenvalue zero. Also there is a unique H-invariant J_1-lagrangian subspace being given by $\operatorname{Ker} H = \operatorname{span}\{(0\ \ 0\ \ 1\ \ 0)^T, (0\ \ \sqrt{3}\ \ 0\ \ 1)^T\}$. Clearly this is not a graph subspace, so there is not a complete set of solutions in this case, in fact there is no hermitian solution at all. Obviously, the cause of this problem is the fact that the pair (A, B) is not controllable. That this can also be seen from Theorem 3.2 is easily checked as

$$(I\quad 0)(\lambda - H)^{-1}\begin{pmatrix} 0 \\ I \end{pmatrix} = \begin{pmatrix} 0 & 0 \\ 0 & \frac{3\sqrt{3}+3\lambda}{(\sqrt{3}+\lambda)\lambda^2} \end{pmatrix}.$$

As a second example, take $A = 0$, $B = I$, and $C = \begin{pmatrix} 1 & 0 \\ 0 & -1 \end{pmatrix}$, and $J = C$. Then the algebraic Riccati equation becomes

$$X\begin{pmatrix} 1 & 0 \\ 0 & -1 \end{pmatrix}X = \begin{pmatrix} 1 & 0 \\ 0 & -1 \end{pmatrix},$$

which clearly has infinitely many solutions. The corresponding Hamiltonian

$$H = \begin{pmatrix} 0 & 0 & 1 & 0 \\ 0 & 0 & 0 & -1 \\ 1 & 0 & 0 & 0 \\ 0 & -1 & 0 & 0 \end{pmatrix}$$

has two eigenvalues, ± 1, both with geometric multiplicity two. Hence there are, as expected, infinitely many invariant lagrangian subspaces. It turns out that these are not all graph subspaces though. Indeed the subspace $\operatorname{span}\{(1\ \ 1\ \ 0\ \ 0)^T, (0\ \ 0\ \ 1\ \ -1)^T\}$ is invariant and lagrangian, but not a graph subspace. Again, this can also be seen from Theorem 3.2, as in this case

$$(I\quad 0)(\lambda - H)^{-1}\begin{pmatrix} 0 \\ I \end{pmatrix} = \begin{pmatrix} \frac{1}{1-\lambda^2} & 0 \\ 0 & \frac{-1}{1-\lambda^2} \end{pmatrix}.$$

Clearly, this function does have constant signature, but for the vector $\begin{pmatrix} 1 \\ 1 \end{pmatrix}$ the condition (ii) (c) of Theorem 3.2 is violated.

It follows that when considering the question of completeness of the set of hermitian solutions it is natural to assume that the pair (A, B) is controllable.

It is instructive to specialize Theorem 3.2 to the case where $J = I$. In that case, suppose first that condition (c) fails to hold. That means by Lemma 3.1 that for some non-zero vector x one has

$$\langle G_K(\lambda)^{-1}\Psi^*(\lambda - V^*)^{-1}x, \Psi^*(\lambda - V^*)^{-1}x \rangle \equiv 0.$$

Taking λ on the imaginary line, and keeping in mind that $J = I$, we see that for large pure imaginary values of λ the function $G_K(\lambda)^{-1}$ has positive definite values. Hence from the equation above we get that $\Psi^*(\lambda - V^*)^{-1}x = 0$ for large values of λ on the imaginary line. But that implies that $\Psi^*(\lambda - V^*)^{-1}x \equiv 0$, i.e., (V, Ψ) is not controllable. Combining with what was proved already in Corollary 3.3 we see that for the case $J = I$ condition (ii) (c) in Theorem 3.2 is equivalent to the controllability of (A, B).

Now assuming controllability, and still in the case $J = I$, we see by Lemma 3.1 again, that condition (b) in Theorem 3.2 is equivalent to $G_K(\lambda)^{-1}$ having constant signature on the imaginary line. But that is equivalent to $G_K(\lambda)$ having constant signature on the imaginary line, and as $G_K(\infty) = I$, it is equivalent to $G_K(\lambda) \geq 0$ for pure imaginary λ. We conclude that in case $J = I$ there is a complete set of solutions if and only if the following three conditions are satisfied:

(a) there exists an H-invariant J_1-lagrangian subspace,

(b) for any (and hence for all) K we have $G_K(\lambda) \geq 0$ on the imaginary line,

(c) the pair (A, B) is controllable.

However, one can actually show somewhat more. As a matter of fact, if the pair (A, B) is controllable and $J = I$, then (b) is equivalent to (a), see, e.g. [LaR]. Moreover, there is a complete description of the set of invariant lagrangian subspaces in that case, which is based on the fact that partial multiplicities of H at its pure imaginary eigenvalues are all even in this case, and the signs in the sign characteristic of the pair (iH, iJ_1) are all the same. We summarize these considerations in the following proposition.

Proposition 3.4 *The algebraic Riccati equation*

$$XBR^{-1}B^*X + XA + A^*X - C = 0,$$

where $R > 0$ has a complete set of solutions if and only if the pair (A, B) is controllable and there exists an H-invariant J_1-lagrangian subspace.

Another special case deserves special attention as well. Consider the case $C = 0$, i.e., the algebraic Riccati equation now becomes

$$(3.4) \qquad XBR^{-1}B^*X + XA + A^*X = 0.$$

This type of equation was studied in [LRa], along with its relation to J-symmetric factorizations of certain functions. (Compare also Theorem 1.3.) Observe that in this case the Hamiltonian is given by

$$H = \begin{pmatrix} -A & -BR^{-1}B^* \\ 0 & A^* \end{pmatrix},$$

so condition (ii) (a) is trivially fullfilled. Also, we can compute

$$(I \quad 0)(\lambda - H)^{-1} \begin{pmatrix} 0 \\ I \end{pmatrix} = -(\lambda + A)^{-1}BR^{-1}B^*(\lambda - A^*)^{-1},$$

so condition (ii) (b) is also fullfilled. We conclude that the following holds.

Corollary 3.5 *The equation* (3.4) *has a complete set of solutions if and only if for every non-zero vector x we have*

$$\langle R^{-1}B^*(\lambda - A^*)^{-1}x, B^*(\lambda - A^*)^{-1}x \rangle \neq 0, \qquad \lambda \in \rho(A^*).$$

Next we return to the general case.

Theorem 3.6 *Let the pair (A, B) be controllable. Then the following are equivalent.*

(i) *The algebraic Riccati equation* (0.1) *has a complete set of solutions,*

(ii) *there is a K such that the realization* (2.1) *is minimal and the function G_K has the following three properties*

    (a) *$G_K(\lambda)$ has constant signature,*

    (b) *$G_K(\lambda)$ has constant zero-signature,*

    (c) *for all non-zero vectors x we have*

$$\langle G_K(\lambda)^{-1}\Psi^*(\lambda - V^*)^{-1}x, \Psi^*(\lambda - V^*)^{-1}x \rangle \neq 0.$$

If (a), (b) *and* (c) *hold for one choice of K for which the realization* (2.1) *is minimal, then they hold for any such K.*

Proof: (i) $\Rightarrow$ (ii). $G_K(\lambda)$ admits a symmetric factorization as seen in Section 2. So (a) holds. As the realization is minimal, and the factorization is minimal there is an H-invariant J_1-lagrangian subspace. So (b) holds. To prove (c) assume to the contrary that there is a non-zero vector x such that $\langle G_K(\lambda)^{-1}\Psi^*$ $(\lambda - V^*)^{-1}x, \Psi^*(\lambda - V^*)^{-1}x\rangle \equiv 0$. By Lemma 3.1 we have that

$$(3.5) \qquad\qquad x^*\,(I \quad 0)\,(\lambda - H)^{-1}\begin{pmatrix} 0 \\ I \end{pmatrix} x \equiv 0.$$

Now use Theorem 3.2 to see that (0.1) does not have a complete set of solutions.

For the converse, choose K so that the realization of G_K is minimal. Let $M^\times$ be any H-invariant J_1-lagrangian subspace (which exists by condition (b)). Suppose that $M^\times \cap \mathrm{Im}\begin{pmatrix} 0 \\ I \end{pmatrix} \neq (0)$, and choose $\begin{pmatrix} 0 \\ x \end{pmatrix} \in M^\times$. Construct N as in (3.3). Then $N \subset M^\times$ by H-invariance of $M^\times$. Since M is J_1-neutral, in particular,

$$\left\langle J_1(\lambda - H)^{-1}\begin{pmatrix} 0 \\ x \end{pmatrix}, \begin{pmatrix} 0 \\ x \end{pmatrix}\right\rangle \equiv 0.$$

Therefore, by Lemma 3.1,

$$x^*(\lambda + V)^{-1}\Psi G_K(\lambda)^{-1}\Psi^*(\lambda - V^*)^{-1}x \equiv 0.$$

This contradicts condition (c). So there is a complete set of solutions. $\qquad\qquad\square$

We can also give an alternative theorem, where we allow all K's, but replace condition (b) by a condition on H. The proof remains virtually unchanged.

Theorem 3.7 *Let the pair (A, Ψ) be controllable. Then the following are equivalent.*

(i) *The algebraic Riccati equation (0.1) has a complete set of solutions,*

(ii) *the matrix H has an invariant J_1-lagrangian subspace, and there is a K such that the function G_K has the following two properties*

(a) *$G_K(\lambda)$ has constant signature,*

(b) *for all non-zero vectors x we have*

$$\langle G_K(\lambda)^{-1}\Psi^*(\lambda - V^*)^{-1}x, \Psi^*(\lambda - V^*)^{-1}x\rangle \not\equiv 0.$$

Observe that if the properties (a) and (b) hold for one choice of K, then they will hold for all K.

Recall that the first condition in (ii) can also be restated in terms of the eigenvalues of H: for each pure imaginary eigenvalue of H the number of odd partial multiplicities is even, and exactly half of the signs in the sign characteristic of (H, J_1) corresponding to these odd multiplicities is -1, the other half being $+1$ (see [RR1], Theorem 5.2).

4 The Discrete Algebraic Riccati Equation

In this section we shall consider the analogues of what we obtained in Section 2 for the discrete algebraic Riccati equation (DARE). So, we consider

$$(4.1) \qquad X = A^*XA + Q - A^*XB(R + B^*XB)^{-1}B^*XA.$$

We assume that A and R are invertible, $R = R^*$, $Q = Q^*$, and (A, B) is controllable. The analogue of the Hamiltonian in this case is the matrix

$$T = \begin{pmatrix} A + BR^{-1}B^*A^{*-1}Q & -BR^{-1}B^*A^{*-1} \\ -A^{*-1}Q & A^{*-1} \end{pmatrix}.$$

It is easily checked that T is J_1-unitary, i.e., $T^*J_1T = J_1$.

The role of $G_0(\lambda)$ is played by the matrix function

$$G_0(z) = R + B^*(z^{-1} - A^*)^{-1}Q(z - A)^{-1}B,$$

which has hermitian values on the unit circle. Much of the theory for this equation is developed under the assumptions $R > 0$, $Q \geq 0$, which occasionally is replaced by the existence of some η on the unit circle for which $G_0(\eta) > 0$ (see [LaR]).

If X is a solutions of the DARE, then we have the following factorization of G_0:

$$G_0(z) = \Delta(\bar{z}^{-1})^*(R + B^*XB)\Delta(z),$$

where $\Delta(z) = I + (R + B^*XB)^{-1}B^*XA(z - A)^{-1}B$.

We decompose $BR^{-1}B^* = \Psi J\Psi^*$. For an arbitrary K of suitable size we put $V = A - \Psi K$. We assume in addition to the invertibility of A that K is chosen so that V is invertible. We define

$$G_0(z) = J + \Psi^*(z^{-1} - A^*)^{-1}Q(z - A)^{-1}\Psi,$$

and

$$G_K(z) = (\Psi^*(z^{-1} - V^*)^{-1} \quad I) \begin{pmatrix} Q + K^*JK & -K^*J \\ -JK & J \end{pmatrix} \begin{pmatrix} (z - V)^{-1}\Psi \\ I \end{pmatrix}.$$

We claim that

$$(4.2) \quad G_K(z) = (I - \Psi^*(z^{-1} - V^*)^{-1}K^*)G_0(z)(I - K(z - V)^{-1}\Psi),$$

which follows easily from the identity

$$(4.3) \qquad \begin{aligned} &(z - A)^{-1}\Psi(I - K(z - V)^{-1}\Psi) \\ &= (z - A)^{-1}\Psi - (z - A)^{-1}(A - z + z - V) \\ &(z - V)^{-1}\Psi = (z - V)^{-1}\Psi. \end{aligned}$$

From this identity we also see that

$$G_K(z) = (I - \Psi^*(z^{-1} - V^*)^{-1}K^*)J(I - K(z - V)^{-1}\Psi)$$
$$+ \Psi^*(z^{-1} - V^*)^{-1}Q(z - V)^{-1}\Psi.$$

We have the following realization for $G_K(z)$, assuming that it is invertible at infinity:

$$G_K(z) = J + \Psi^*V^{*-1}K^*J$$

(4.4) $$+(-JK - \Psi^*V^{*-1}(Q + K^*JK) \quad \Psi^*V^{*-1})$$

$$\times \left(z - \begin{pmatrix} V & 0 \\ -V^{*-1}(Q + K^*JK) & V^{*-1} \end{pmatrix}\right)^{-1} \begin{pmatrix} \Psi \\ V^{*-1}K^*J \end{pmatrix},$$

(see formula (4.1) in [LRR]).

If X is a hermitian solution of the discrete algebraic Riccati equation then we have the following symmetric factorization

(4.5) $$G_K(z) = R_K(\bar{z}^{-1})^*(J + \Psi^*X\Psi)R_K(z),$$

where

(4.6) $$R_K(z) = I + (J + \Psi^*X\Psi)^{-1}(-JK + \Psi^*XV)(z - V)^{-1}\Psi.$$

(See, e.g., Theorem 1.1 in [LRR] or [M2]. A proof of this fact will also be given in the proof of Theorem 4.3 below.) Consequently, the signature of the matrix $J + \Psi^*X\Psi$ is the same for any hermitian solution X of the discrete algebraic Riccati equation. Conversely, if K is such that the realization (4.4) for $G_K(z)$ is minimal, then a symmetric factorization (4.5), where $R_K(z) = I + C_R(z - V)^{-1}\Psi$ for some C_R is necessarily of the form (4.6) for some hermitian solution X of (4.1). (See Propositions 2.2 and 2.3 in [LRR].)

Next we compute a realization for $G_K(z)^{-1}$. Obviously, this could be done directly from (4.4), but we prefer to use a different approach.

Lemma 4.1 *We have*

$$G_K(z)^{-1} = J - J\Psi^*A^{*-1}K^*$$

(4.7) $$+(J\Psi^*A^{*-1}Q + K \quad -J\Psi^*A^{*-1})(z - T)^{-1}$$

$$\begin{pmatrix} \Psi J(I - \Psi^*A^{*-1}K^*) \\ A^{*-1}K^* \end{pmatrix}.$$

Proof: We have (see [RR3], page 169)

(4.8) $$G_0(z)^{-1} = J - (-J\Psi^*A^{*-1}Q \quad J\Psi^*A^{*-1})(z - T)^{-1}\begin{pmatrix} \Psi J \\ 0 \end{pmatrix}.$$

Also, from (4.2) we have that

$$G_K(z)^{-1} = (I + K(z - A)^{-1}\Psi)G_0(z)^{-1}(I + \Psi^*(z^{-1} - A^*)^{-1}K^*)$$

$$= (I + K(z - A)^{-1}\Psi)G_0(z)^{-1}$$

$$\times \{(I - \Psi^*A^{*-1}K^*) - \Psi^*A^{*-1}(z - A^{*-1})^{-1}A^{*-1}K^*\}$$

(Here we use $(z^{-1} - A^*)^{-1} = -A^{*-1} - A^{*-1}(z - A^{*-1})^{-1}A^{*-1}$.) By the multiplication rule for functions in realized form we have

$$(I + K(z - A)^{-1}\Psi)G_0(z)^{-1}$$
$$= J + (K \quad J\Psi^*A^{*-1}Q \quad -J\Psi^*A^{*-1})$$
$$\times \left(z - \begin{pmatrix} A & \Psi J\Psi^*A^{*-1}Q & -\Psi J\Psi^*A^{*-1} \\ 0 & A + \Psi J\Psi^*A^{*-1}Q & -\Psi J\Psi^*A^{*-1} \\ 0 & -A^{*-1}Q & A^{*-1} \end{pmatrix} \right)^{-1} \begin{pmatrix} \Psi J \\ \Psi J \\ 0 \end{pmatrix}.$$

Take $S_1 = \begin{pmatrix} I & -I & 0 \\ 0 & I & 0 \\ 0 & 0 & I \end{pmatrix}$. Then we have

$$S_1 \begin{pmatrix} \Psi J \\ \Psi J \\ 0 \end{pmatrix} = \begin{pmatrix} 0 \\ \Psi J \\ 0 \end{pmatrix},$$

$$(K \quad J\Psi^*A^{*-1}Q \quad -J\Psi^*A^{*-1}) S_1^{-1} = (K \quad J\Psi^*A^{*-1}Q + K \quad -J\Psi^*A^{*-1}),$$

and one computes that

$$S_1 \begin{pmatrix} A & \Psi J\Psi^*A^{*-1}Q & -\Psi J\Psi^*A^{*-1} \\ 0 & A + \Psi J\Psi^*A^{*-1}Q & -\Psi J\Psi^*A^{*-1} \\ 0 & -A^{*-1}Q & A^{*-1} \end{pmatrix} S_1^{-1} = \begin{pmatrix} A & 0 \\ 0 & T \end{pmatrix}.$$

As the first component is not controllable we arrive at

$$(I + K(z - A)^{-1}\Psi)G_0(z)^{-1}$$
$$= J + (J\Psi^*A^{*-1}Q + K \quad -J\Psi^*A^{*-1}) (z - T)^{-1} \begin{pmatrix} \Psi J \\ 0 \end{pmatrix}.$$

Multiplying this on the right by

$$(I - \Psi^*A^{*-1}K^*) - \Psi^*A^{*-1}(z - A^{*-1})^{-1}A^{*-1}K^*$$

we obtain

$$J - J\Psi^*A^{*-1}K^* + (J\Psi^*A^{*-1}Q + K \quad -J\Psi^*A^{*-1} \quad -J\Psi^*A^{*-1})$$

$$\times \left(z - \begin{pmatrix} A + \Psi J\Psi^*A^{*-1}Q & -\Psi J\Psi^*A^{*-1} & \Psi J\Psi^*A^{*-1} \\ -A^{*-1}Q & A^{*-1} & 0 \\ 0 & 0 & A^{*-1} \end{pmatrix} \right)^{-1}$$

$$\times \begin{pmatrix} \Psi J(I - \Psi^*A^{*-1}K^*) \\ 0 \\ A^{*-1}K^* \end{pmatrix}.$$

Put $S_2 = \begin{pmatrix} I & 0 & 0 \\ 0 & I & -I \\ 0 & 0 & I \end{pmatrix}$. Then we have

$$(J\Psi^*A^{*-1}Q + K \quad -J\Psi^*A^{*-1} \quad -J\Psi^*A^{*-1})\, S_2$$
$$= (J\Psi^*A^{*-1}Q + K \quad -J\Psi^*A^{*-1} \quad 0),$$

$$S_2^{-1} \begin{pmatrix} \Psi J(I - \Psi^*A^{*-1}K^*) \\ 0 \\ A^{*-1}K^* \end{pmatrix} = \begin{pmatrix} \Psi J(I - \Psi^*A^{*-1}K^*) \\ A^{*-1}K^* \\ A^{*-1}K^* \end{pmatrix}$$

and

$$S_2^{-1} \begin{pmatrix} A + \Psi J\Psi^*A^{*-1}Q & -\Psi J\Psi^*A^{*-1} & \Psi J\Psi^*A^{*-1} \\ -A^{*-1}Q & A^{*-1} & 0 \\ 0 & 0 & A^{*-1} \end{pmatrix} S_2 = \begin{pmatrix} T & 0 \\ 0 & A^{*-1} \end{pmatrix}.$$

Obviously, the third component is unobservable, and therefore we have that (4.7) holds.   □

Next, we recall the notion of generalized T-Bezoutian introduced in [KL]. Let $L(\lambda)$ be an $n \times r$ rational matrix function, which is analytic at the origin. Let

$$(4.9) \qquad\qquad L(\lambda) = L(0) + \lambda\Phi(I - \lambda V)^{-1}C_L$$

be its realization, where $L(0)$ is an $n \times r$ matrix, Φ is an $n \times p$ matrix, V is a $p \times p$ matrix, and finally, C_L is a $p \times r$ matrix. We shall assume throughout that the pair (Φ, V) is observable. Likewise, let $M(\lambda)$ be $r \times n$ rational matrix function analytic at ∞ and let

$$(4.10) \qquad\qquad M(\lambda) = M(\infty) + K_M(\lambda - A)^{-1}\Psi$$

be its realization where $M(\infty)$, K_M, A and Ψ are $r \times n, r \times q, q \times q, q \times n$ matrices, respectively. We assume that the pair (A, Ψ) is controllable. Further, we assume that $L(\lambda)$ and $M(\lambda)$ satisfy the equation

$$(4.11) \qquad\qquad L(\lambda)M(\lambda) = 0$$

for $\lambda \in \Omega$, where $\Omega = \{\lambda \mid \lambda^{-1} \in \rho(V) \text{ and } \lambda \in \rho(A)\}$. The following result is established in [KL].

Proposition 4.2 *There exists a unique matrix T_1 such that the following relation holds true:*

$$(4.12) \qquad (L(\lambda)M(\mu))/(\lambda - \mu) = \Phi(I - \lambda V)^{-1} T_1 (\mu - A)^{-1} \Psi$$

The matrix T_1 in (4.12) is called the *generalized T-Bezoutian* of (L, M), associated with equation (4.11) and realizations (4.9), (4.10).

Now consider the DARE (4.1), which we rewrite as

$$(4.13) \qquad X = A^* X A + Q - A^* X \Psi (J + \Psi^* X \Psi)^{-1} \Psi^* X A$$

where we assume that (A, Ψ) is controllable, A is invertible, $Q = Q^*$, and $J = J^*$ is invertible. For any matrix K of appropriate size introduce $V = A - \Psi K$, where K is chosen such that V is invertible. Decompose $Q = PY$. Introducing the functions

$$L_1(\lambda) = J - \Psi^*(\lambda - V^*)^{-1} K^* J, \quad L_2(\lambda) = I - K(\lambda - V)^{-1} \Psi,$$

$$(4.14)$$

$$P(\lambda) = \Psi^*(\lambda - V^*)^{-1} P, \qquad Y(\lambda) = Y(\lambda - V)^{-1} \Psi,$$

we have that

$$G_K(\lambda) = L_1(\lambda) L_2(\lambda) + P(\lambda) Y(\lambda).$$

Assume that $G_K(\lambda)$ can be factored in the following way

$$(4.15) \qquad\qquad G_K(\lambda) = R(\bar{\lambda}^{-1})^* D R(\lambda)$$

where D is some hermitian matrix and where

$$(4.16) \qquad\qquad R(\lambda) = I + K_R(\lambda - V)^{-1} \Psi.$$

Consider the matrix T_K defined by the equation

$$\frac{L_1(\lambda^{-1}) L_2(\mu) + P(\lambda^{-1}) Y(\mu) - R(\bar{\lambda}^{-1})^* D R(\mu)}{\lambda - \mu}$$

$$(4.17)$$

$$= \Psi^*(I - \lambda V^*)^{-1} T_K (\mu - V)^{-1} \Psi$$

Then the matrix T_K in (4.17) is the T-Bezoutian associated with equation (4.11), where $L(\lambda) = (L_1(\lambda) \quad P(\lambda) \quad R(\bar{\lambda}^{-1})^*)$ and $M(\lambda) = \begin{pmatrix} L_2(\lambda) \\ Y(\lambda) \\ -R(\lambda) \end{pmatrix}$ and with the realizations

$$(4.18) \qquad L(\lambda) = (J \quad 0 \quad I) + \lambda \Psi^*(I - \lambda V^*)^{-1}(-K^*J \quad P \quad K_R^*)$$

$$(4.19) \qquad\qquad M(\lambda) = \begin{pmatrix} I \\ 0 \\ I \end{pmatrix} + \begin{pmatrix} -K \\ Y \\ K_R \end{pmatrix} (\lambda - V)^{-1}\Psi$$

Proposition 4.3 *The T-Bezoutian T_K of $(L_1, P, \tilde{R}; L_2, Y, -R)$ defined by (4.17) and associated with the equation (4.15) and the realizations above, where $\tilde{R}(\lambda)$ is defined as $R(\bar{\lambda}^{-1})^* D$, is a Hermitian matrix.*

Proof: Denote

$$G_K(\lambda, \mu) = L_1(\lambda)L_2(\mu) + P(\lambda)Y(\mu) - R(\bar{\lambda}^{-1})^* DR(\mu).$$

We first check that $G_K(\lambda, \mu) = (G_K(\bar{\mu}^{-1}, \bar{\lambda}^{-1}))^*$. Indeed, observe that

$$\begin{aligned} L_1(\bar{\mu}^{-1}) &= J - \Psi^*(\bar{\mu} - V^*)^{-1}K^*J, \\ L_2(\bar{\lambda}^{-1}) &= I - K(\bar{\lambda}^{-1} - V)^{-1}\Psi, \end{aligned}$$

and so $(L_1(\bar{\mu}^{-1})L_2(\bar{\lambda}^{-1}))^* = L_1(\lambda)L_2(\mu)$. In the same way we see that $(P(\bar{\mu}^{-1})Y(\bar{\lambda}^{-1}))^* = P(\lambda)Y(\mu)$, and $(R(\mu)^*R(\bar{\lambda}^{-1}))^* = R(\bar{\lambda}^{-1})^*R(\mu)$ and we have the desired result.

Now we show that T_K is hermitian. Indeed,

$$\begin{aligned} &(G_K(\bar{\mu}^{-1}, \bar{\lambda}^{-1})(\bar{\mu}^{-1} - \bar{\lambda}^{-1})^{-1})^* \\ &= (\Psi^*(I - \bar{\mu}^{-1}V^*)^{-1}T_K^*(\bar{\lambda}^{-1} - V)^{-1}\Psi)^*. \end{aligned}$$

From what we have shown in the previous paragraph we have that

$$\frac{G_K(\lambda, \mu)}{(\lambda - \mu)\lambda^{-1}\mu^{-1}} = \Psi^*(\lambda^{-1} - V^*)^{-1}T_K^*(I - \mu^{-1}V)^{-1}\Psi$$

or

$$\frac{G_K(\lambda, \mu)}{\lambda - \mu} = \Psi(I - \lambda V^*)^{-1}T_K^*(\mu - V)^{-1}\Psi,$$

and so the uniqueness of the Bezoutian (see Proposition 4.2) yields $T_K = T_K^*$.   □

Theorem 4.4 *Let $L_1(\lambda), L_2(\lambda), P(\lambda), Y(\lambda), R(\lambda)$ be rational matrix functions having the realizations as in (4.18) and (4.19) and assume that the equation*

(4.11) *holds true. Then the T-Bezoutian T_K of the functions $(L_1, P, \tilde{R}; L_2, Y, -R)$, defined by (4.17) and associated with realizations (4.18), (4.19) and relation (4.11) is a solution of the algebraic Riccati equation (4.13).*

Conversely, let X be an hermitian solution of (4.13). Then the function $G_K(\lambda)$ defined by (4.15) admits the symmetric factorization

$$G_K(\lambda) = (R_K(\bar{\lambda}^{-1}))^*(J + \Psi^* X \Psi) R_K(\lambda),$$

where

$$R_K(\lambda) = I + (J + \Psi^* X \Psi)^{-1}(-JK + \Psi^* X V)(\lambda - V)^{-1}\Psi.$$

Moreover, this hermitian solution X of the algebraic Riccati equation is the T-Bezoutian of the sextet of functions $(L_1, P, \tilde{R}_K; L_2, Y, -R_K)$ associated with the equation (4.5) and the realizations of L_1, P, Y and L_2 as in (4.14) and the realization (4.6) for $R_K(\lambda)$. Here $\tilde{R}_K(\lambda) = R_K(\lambda^{-1})^* D$, where $D = J + \Psi^* X \Psi$.

This correspondence between symmetric solution of the algebraic Riccati equation and symmetric factorizations of $G_K(\lambda)$ of the form (4.5), (4.6) is one-to-one.

Proof: Let $G_K(\lambda)$ be defined by (4.2) and assume that $G_K(\lambda)$ admits the factorization (4.15) where $R(\lambda)$ satisfies (4.16). Let T_K be the T-Bezoutian of $(L_1, P, \tilde{R}; L_2, Y, -R)$ associated with equation (4.11) and realizations (4.18), (4.19). We have to show that T_K is solution of (4.13). To this end first rewrite (4.17) as

$$\begin{aligned}
&(1 - \lambda\mu)\Psi^*(\lambda - V^*)^{-1}T_K(\mu - V)^{-1}\Psi \\
&= (J - D) - \Psi^*(\lambda - V^*)^{-1}(K^*J + K_R^* D) \\
&\quad - (JK + DK_R)(\mu - V)^{-1}\Psi + \Psi^*(\lambda - V^*)^{-1} \\
&\quad \times (K^*JK + PY - K_R^* DK_R)(\mu - V)^{-1}\Psi.
\end{aligned}$$

Developing both sides of this equation in a neighbourhood of ∞ and comparing the coefficients of $1, \mu^{-j}, \lambda^{-j}$ for $j = 1, 2, \ldots$, on both sides we have $J - D = -\Psi^* T_K \Psi$, i.e.,

$$(4.20) \qquad\qquad D = J + \Psi^* T_K \Psi,$$

and moreover, $\Psi^*(V^*)^{j-1}(K^*J + K_R^* D) = \Psi^*(V^*)^j T_K \Psi$ for $j = 1, 2, \ldots$.

Because of observability of (Ψ^*, V^*) we obtain

$$(4.21) \qquad\qquad K_R^* = (V^* T_K \Psi - K^* J)(J + \Psi^* T_K \Psi)^{-1}.$$

Taking adjoints in both sides of (4.21) we have

$$(4.22) \qquad\qquad K_R = (J + \Psi^* T_K \Psi)^{-1}(\Psi^* T_K V - JK).$$

Then using (4.20)–(4.22) we obtain

$$
\begin{aligned}
(1 - \lambda\mu)\Psi^*(\lambda - V^*)^{-1}T_K(\mu - V)^{-1}\Psi &= -\Psi^*T_K\Psi \\
&\quad -\Psi^*(\lambda - V^*)^{-1}V^*T_K\Psi - \Psi^*T_K V(\mu - V)^{-1}\Psi \\
&\quad +\Psi^*(\lambda - V^*)^{-1}\{K^*JK + PY \\
&\quad -(V^*T_K\Psi - K^*J)(J + \Psi^*T_K\Psi)^{-1} \\
&\quad (\Psi^*T_K V - JK)\}(\lambda - V)^{-1}\Psi.
\end{aligned}
$$

(4.23)

From the easily verified equality

$$
\begin{aligned}
(1 - \lambda\mu)T_K &+ (\lambda - V^*)T_K(\mu - V) + V^*T_K(\mu - V) \\
&+(\lambda - V^*)T_K V = T_K - V^*T_K V
\end{aligned}
$$

we can rewrite (4.23) as follows

$$
\begin{aligned}
\Psi^*(\lambda - V^*)^{-1}(T_K - V^*T_K V)(\mu - V)^{-1}\Psi \\
= \Psi^*(\lambda - V)^{-1}(K^*JK + Q - (V^*T_K\Psi - K^*J)(J + \Psi^*T_K\Psi)^{-1} \\
\times(\Psi^*T_K V - JK)(\mu - V)^{-1}\Psi
\end{aligned}
$$

for any λ, μ such that $(\lambda - V^*)$, $(\mu - V)$ are invertible. Thus T_K satisfies the equation

$$
\begin{aligned}
T_K &= V^*T_K V + K^*JK + Q \\
&\quad -(-K^*J + V^*T_K\Psi)(J + \Psi^*T_K\Psi)^{-1}(\Psi^*T_K V - JK)
\end{aligned}
$$

(4.24)

We can rewrite the last term in (4.24) as follows

$$
\begin{aligned}
(V^*T_K\Psi - K^*J)(J + \Psi^*T_K\Psi)^{-1}(\Psi^*T_K V - JK) \\
= (A^*T_K\Psi - K^*(J + \Psi^*T_K\Psi))(J + \Psi^*T_K\Psi)^{-1} \\
(\Psi^*T_K A - (J + \Psi^*T_K\Psi)K) \\
= A^*T_K\Psi(J + \Psi^*T_K\Psi)^{-1}\Psi^*T_K A \\
-K^*\Psi^*T_K A - A^*T_K\Psi K + K^*(J + \Psi^*T_K\Psi)K
\end{aligned}
$$

Substituting this expression into (4.24) we obtain

$$
\begin{aligned}
T_K &= Q - A^*T_K\Psi(J + \Psi^*T_K\Psi)^{-1}\Psi^*T_K A + V^*T_K V + K^*JK \\
&\quad + K^*\Psi^*T_K A + A^*T_K\Psi K - K^*(J + \Psi^*T_K\Psi)K \\
&= Q - A^*T_K\Psi(J + \Psi^*T_K\Psi)^{-1}\Psi^*T_K A \\
&\quad + (A^* - K^*\Psi^*)T_K(A - \Psi K) \\
&\quad + K^*\Psi^*T_K A + A^*T_K\Psi K - K^*\Psi^*T_K\Psi K \\
&= Q - A^*T_K\Psi(J + \Psi^*T_K\Psi)^{-1}\Psi^*T_K A + A^*T_K A
\end{aligned}
$$

i.e., we have (4.13). Hence the T-Bezoutian T_K is a solution of the algebraic Riccati equation.

To prove the converse part of this theorem we assume that X is a symmetric solution of (4.13). We claim that (4.5) holds true for $R_K(\lambda)$ defined by (4.6). Indeed,

$$
\begin{aligned}
&(I + (J + \Psi^*X\Psi)^{-1}(-JK + \Psi^*XV)(\bar{\lambda}^{-1} - V)^{-1}\Psi)^*(J + \Psi^*X\Psi) \\
&\quad \times (I + (J + \Psi^*X\Psi)^{-1}(-JK + \Psi^*XV)(\lambda - V)^{-1}\Psi) \\
&= (I + \Psi^*(\lambda^{-1} - V^*)^{-1}(-K^*J + V^*X\Psi)(J + \Psi^*X\Psi)^{-1}) \\
&\quad \times (J + \Psi^*X\Psi)(I + (J + \Psi^*X\Psi)^{-1} \\
&\quad (-JK + \Psi^*XV)(\lambda - V)^{-1}\Psi) \\
&= J + \Psi^*X\Psi + \Psi^*(\lambda^{-1} - V^*)^{-1}(-K^*J + V^*X\Psi) \\
&\quad + (-JK + \Psi^*XV)(\lambda - V)^{-1}\Psi \\
&\quad + \Psi^*(\lambda^{-1} - V^*)^{-1}(-K^*J + V^*X\Psi) \\
&\quad \times (J + \Psi^*X\Psi)^{-1}(-JK + \Psi^*XV)(\lambda - V)^{-1}\Psi
\end{aligned}
$$

Compute

$$
\begin{aligned}
&\Psi^*X\Psi + \Psi^*(\lambda^{-1} - V^*)^{-1}V^*X\Psi + \Psi^*XV(\lambda - V)^{-1}\Psi \\
&\quad + \Psi^*(\lambda^{-1} - V^*)^{-1}(-K^*J + V^*X\Psi)(J + \Psi^*X\Psi)^{-1} \\
&\quad \times (-JK + \Psi^*XV)(\lambda - V)^{-1}\Psi \\
&= \Psi^*(\lambda^{-1} - V^*)^{-1}\{(\lambda^{-1} - V^*)X(\lambda - V) \\
&\quad + V^*X(\lambda - V) + (\lambda^{-1} - V^*)XV \\
&\quad + (-K^*J + V^*X\Psi)(J + \Psi^*X\Psi)^{-1} \\
&\quad (-JK + \Psi^*XV)\}(\lambda - V)^{-1}\Psi \\
&= \Psi^*(\lambda^{-1} - V^*)^{-1}\{X - V^*XV \\
&\quad + (-K^*J + V^*X\Psi)(J + \Psi^*X\Psi)^{-1} \\
&\quad (-JK + \Psi^*XV)\}(\lambda - V)^{-1}\Psi
\end{aligned}
$$

Now this equals $\Psi^*(\lambda^{-1} - V^*)^{-1}(Q + K^*JK)(\lambda - V)^{-1}\Psi$, since X is solution of (4.13). So we have

$$
(R_K(\bar{\lambda}^{-1})^*(J + \Psi^*X\Psi)R_K(\lambda) = L_1(\lambda)L_2(\lambda) + P(\lambda)Y(\lambda) = G_K(\lambda)
$$

and this part of the theorem is proved.

Now we show that such X is a T-Bezoutian of $(L_1, P, \tilde{R}_K; L_2, Y, -R_K)$ associated with equation (4.5) and realizations as in (4.6) and (4.14). Denote

$$
G_K(\lambda, \mu) = L_1(\lambda)L_2(\mu) + P(\lambda)Y(\mu) - R_K(\bar{\lambda}^{-1})^*DR_K(\mu)
$$

Then

$$
\begin{aligned}
G_K(\lambda, \mu) = {} & -\Psi^* X \Psi - \Psi^*(\lambda - V^*)^{-1} V^* X \Psi - \Psi^* X V (\mu - V)^{-1} \Psi \\
& + \Psi^*(\lambda - V^*)^{-1}\{(K^* J K + Q) \\
& - (-K^* J + V^* X \Psi)(J + \Psi^* X \Psi)^{-1} \\
& (-J K + \Psi^* X V)\}(\mu - V)^{-1}\Psi.
\end{aligned}
$$

Since X is solution of (4.13) we have

$$
\begin{aligned}
G_K(\lambda, \mu) = {} & -\Psi^* X \Psi - \Psi^*(\lambda - V^*)^{-1} V^* X \Psi - \Psi^* X V (\mu - V)^{-1} \Psi \\
& + \Psi^*(\lambda - V^*)^{-1}(X - V^* X V)(\mu - V)^{-1}\Psi.
\end{aligned}
$$

But

$$
\begin{aligned}
& -\Psi^* X \Psi - \Psi^*(\lambda - V^*)^{-1} V^* X \Psi - \Psi^* X V (\mu - V)^{-1} \\
& = \Psi^*(\lambda - V^*)^{-1}\{-(\lambda - V^*)X(\mu - V) - V^* X(\mu - V) \\
& \quad -(\lambda - V^*)X V + X - V^* X V\}(\mu - V)^{-1}\Psi \\
& = \Psi^*(\lambda - V^*)^{-1}(X - \lambda\mu X)(\lambda - V)^{-1}\Psi
\end{aligned}
$$

and we have that

$$
G_K(\lambda, \mu) = (1 - \lambda\mu)\Psi^*(\lambda - V^*)^{-1} X (\mu - V)^{-1}\Psi.
$$

We have proved that X is the T-Bezoutian of $(L_1, P, \tilde{R}_K; L_2, Y, -R_K)$ associated with equation (4.5) and realizations (4.6) and (4.14).

To prove the one-to-one correspondence suppose we have two factorizations which generate the same solution X of (4.13).

$$
G_K(\lambda) = R_1(\bar{\lambda}^{-1})^* D_1 R_1(\lambda)
$$

$$
G_K(\lambda) = R_2(\bar{\lambda}^{-1})^* D_2 R_2(\lambda)
$$

where $R_1(\lambda)$ and $R_2(\lambda)$ have the form (4.16) for some K_{R_1} and K_{R_2}. Since Bezoutians associated with the above factorizations and with appropriate realizations coincide we conclude

$$
R_1(\bar{\lambda}^{-1})^* D_1 R_1(\mu) = R_2(\bar{\lambda}^{-1})^* D_2 R_2(\mu).
$$

Letting $\lambda \to \infty$, we have

$$
D_1 R_1(\mu) = D_2 R_2(\mu)
$$

and since $R_1(\infty) = R_2(\infty) = I$, and D_1 and D_2 are invertible, it follows that $D_1 = D_2$ and $R_1(\lambda) = R_2(\lambda)$, and the theorem is proved. □

5 Completeness of the Set of Solutions of the Discrete Algebraic Riccati Equation

In this section we shall consider completeness of the set of solutions of the discrete algebraic Riccati equation. To do so, we first establish the following lemma. Here, as in Section 3, controllability of the pair (A, B), or equivalently, controllability of (A, Ψ), is only assumed when explicitly stated.

Lemma 5.1 *The following identity holds*

$$(z - V)^{-1} \Psi G_K(z)^{-1} \Psi^*(z^{-1} - V^*)^{-1} = (I \quad 0)(z - T)^{-1} T \begin{pmatrix} 0 \\ I \end{pmatrix}.$$

Proof: We compute, using the multiplication of realizations,

$$(z - V)^{-1} \Psi G_K(z)^{-1} = (I \quad 0 \quad 0)$$
$$\times \left(z - \begin{pmatrix} V & \Psi J \Psi^* A^{*-1} Q + \Psi K & -\Psi J \Psi^* A^{*-1} \\ 0 & T & \\ 0 & & \end{pmatrix} \right)^{-1}$$
$$\begin{pmatrix} \Psi(J - J\Psi^* A^{*-1} K^*) \\ \Psi(J - J\Psi^* A^{*-1} K^*) \\ -A^{*-1} K^* \end{pmatrix}.$$

With S_1 as in the proof of Lemma 4.1, we get

$$S_1 \begin{pmatrix} \Psi(J - J\Psi^* A^{*-1} K^*) \\ \Psi(J - J\Psi^* A^{*-1} K^*) \\ -A^{*-1} K^* \end{pmatrix} = \begin{pmatrix} 0 \\ \Psi(J - J\Psi^* A^{*-1} K^*) \\ -A^{*-1} K^* \end{pmatrix},$$
$$(I \quad 0 \quad 0) S_1^{-1} = (I \quad I \quad 0),$$

and

$$S_1 \begin{pmatrix} V & \Psi J \Psi^* A^{*-1} Q + \Psi K & -\Psi J \Psi^* A^{*-1} \\ 0 & T & \\ 0 & & \end{pmatrix} S_1^{-1} = \begin{pmatrix} V & 0 \\ 0 & T \end{pmatrix}.$$

As the first component is uncontrollable, we obtain

$$(z - V)^{-1} \Psi G_K(z)^{-1} = (I \quad 0)(z - T)^{-1} \begin{pmatrix} \Psi(J - J\Psi^* A^{*-1} K^*) \\ -A^{*-1} K^* \end{pmatrix}.$$

Now write

$$(z^{-1} - V^*)^{-1} = -V^{*-1} - V^{*-1}(z - V^{*-1})^{-1} V^{*-1}.$$

We shall also make use of the following identity $A^{*-1}K^*\Psi^*V^{*-1} = V^{*-1}-A^{*-1}$. Then we have

$$-(z-V)^{-1}\Psi G_K(z)^{-1}(z^{-1}-V^*)^{-1} = (I \quad 0 \quad 0)$$

$$\times \left(z - \begin{pmatrix} T & \Psi(J-J\Psi^*A^{*-1}K^*)\Psi^*V^{*-1} \\ & A^{*-1}K^*\Psi^*V^{*-1} \\ 0 \quad 0 & V^{*-1} \end{pmatrix}\right)^{-1}$$

$$\times \begin{pmatrix} \Psi(J-J\Psi^*A^{*-1}K^*)\Psi^*V^{*-1} \\ A^{*-1}K^*\Psi^*V^{*-1} \\ V^{*-1} \end{pmatrix}$$

$$= (I \quad 0 \quad 0)\left(z - \begin{pmatrix} T & \Psi J\Psi^*A^{*-1} \\ & V^{*-1}-A^{*-1} \\ 0 \quad 0 & V^{*-1} \end{pmatrix}\right)^{-1}\begin{pmatrix} \Psi J\Psi^*A^{*-1} \\ V^{*-1}-A^{*-1} \\ V^{*-1} \end{pmatrix}.$$

With S_2 as in the proof of Lemma 4.1 we have

$$S_2\begin{pmatrix} \Psi J\Psi^*A^{*-1} \\ V^{*-1}-A^{*-1} \\ V^{*-1} \end{pmatrix} = \begin{pmatrix} \Psi J\Psi^*A^{*-1} \\ -A^{*-1} \\ V^{*-1} \end{pmatrix},$$

$$(I \quad 0 \quad 0)S_2^{-1} = (I \quad 0 \quad 0),$$

and

$$S_2\begin{pmatrix} T & \Psi J\Psi^*A^{*-1} \\ & V^{*-1}-A^{*-1} \\ 0 \quad 0 & V^{*-1} \end{pmatrix}S_2^{-1} = \begin{pmatrix} T & 0 \\ 0 & V^{*-1} \end{pmatrix}.$$

As the third component is not observable, we arrive at

$$-(z-V)^{-1}\Psi G_K(z)^{-1}(z^{-1}-V^*)^{-1}$$

$$= (I \quad 0)(z-T)^{-1}\begin{pmatrix} \Psi J\Psi^*A^{*-1} \\ -A^{*-1} \end{pmatrix}$$

$$= -(I \quad 0)(z-T)^{-1}T\begin{pmatrix} 0 \\ I \end{pmatrix}.$$

This proves the lemma.   □

Theorem 5.2 *The following are equivalent:*

(i) *the set of solutions of the discrete algebraic Riccati equation* (4.1) *is complete and nonempty,*

(ii) *the following three conditions are satisfied:*

(a) *there exists a T-invariant J_1-lagrangian subspace,*

(b) *the rational matrix function*

$$(I \quad 0)(z-T)^{-1}T\begin{pmatrix} 0 \\ I \end{pmatrix}$$

has constant signature on the unit circle,

(c) *for all non-zero vectors x we have*

$$\left\langle (z - T)^{-1} T \begin{pmatrix} 0 \\ x \end{pmatrix}, \begin{pmatrix} x \\ 0 \end{pmatrix} \right\rangle \neq 0,$$

or, equivalently,

$$\left\langle J_1 (z - T)^{-1} T \begin{pmatrix} 0 \\ x \end{pmatrix}, \begin{pmatrix} 0 \\ x \end{pmatrix} \right\rangle \neq 0.$$

Proof: Suppose that (4.1) has a complete set of solutions. Then condition (a) is satisfied, as for any solution X the subspace $\text{Im} \begin{pmatrix} I \\ X \end{pmatrix}$ will be T-invariant and J_1-lagrangian. Also condition (b) is satisfied, as we have seen that $G_K(\lambda)$ has constant signature on the unit circle (by virtue of (4.5)), and we can then use Lemma 5.1 to derive (b). It remains to show that (c) holds. Suppose that there is a non-zero vector x such that

$$\left\langle (z - T)^{-1} T \begin{pmatrix} 0 \\ x \end{pmatrix}, \begin{pmatrix} x \\ 0 \end{pmatrix} \right\rangle \equiv 0.$$

Introduce the subspace N by

(5.1)
$$\begin{aligned} N &= \text{span} \left\{ (z - T)^{-1} T \begin{pmatrix} 0 \\ x \end{pmatrix}, \begin{pmatrix} 0 \\ x \end{pmatrix} \right\} \\ &= \text{span} \left\{ (zT^{-1} - I)^{-1} \begin{pmatrix} 0 \\ x \end{pmatrix}, \begin{pmatrix} 0 \\ x \end{pmatrix} \right\}. \end{aligned}$$

This subspace is obviously T^{-1}-invariant. Since T is invertible (it is J_1-unitary), it must also be T-invariant. We check that N is J_1-neutral in case condition (c) fails to hold. Obviously, the vector $\begin{pmatrix} 0 \\ x \end{pmatrix}$ is J_1-neutral, and we have

$$\left\langle J_1 (z - T)^{-1} T \begin{pmatrix} 0 \\ x \end{pmatrix}, \begin{pmatrix} 0 \\ x \end{pmatrix} \right\rangle = 0.$$

Take $z \neq \bar{\zeta}^{-1}$. Using the fact that T is J_1-unitary, and the resolvent identity, we have that

$$\begin{aligned} T^*(\bar{\zeta} - T^*)^{-1} J_1 (z - T)^{-1} T &= T^* J_1 (\bar{\zeta} - T^{*-1})^{-1} (z - T)^{-1} T \\ &= T^* J T (\bar{\zeta} T - I)^{-1} (z - T)^{-1} T = \frac{1}{\bar{\zeta}} J_1 \left(T - \frac{1}{\bar{\zeta}} \right)^{-1} (z - T)^{-1} T \\ &= \frac{1}{1 - \bar{\zeta} z} \left(J_1 \left(\frac{1}{\bar{\zeta}} - T \right)^{-1} T - (z - T)^{-1} T \right). \end{aligned}$$

From this it follows that for $z \neq \bar{\zeta}^{-1}$

$$\left\langle J_1(z-T)^{-1}T \begin{pmatrix} 0 \\ x \end{pmatrix}, (\zeta - T)^{-1}T \begin{pmatrix} 0 \\ x \end{pmatrix} \right\rangle = 0.$$

Obviously, the above expression is continuous in ζ for $\zeta \notin \sigma(T)$. Taking the limit as $\zeta \to \frac{1}{\bar{z}}$ we see that also

$$\left\langle J_1(z-T)^{-1}T \begin{pmatrix} 0 \\ x \end{pmatrix}, (z-T)^{-1}T \begin{pmatrix} 0 \\ x \end{pmatrix} \right\rangle = 0.$$

Thus, N is J_1-neutral.

So, there exists a T-invariant $M^\times$ containing N such that $J_1 M^\times = M^{\times \perp}$. That contradicts the completeness of the set of solutions, as in that case $M^\times \cap$ Im $\begin{pmatrix} 0 \\ I \end{pmatrix} \neq (0)$.

For the converse, assume that (a)–(c) hold, and that for some T-invariant $M^\times$ such that $J_1 M^\times = M^{\times \perp}$ we have $M^\times \cap$ Im $\begin{pmatrix} 0 \\ I \end{pmatrix} \neq (0)$. Define N by (5.1), where $\begin{pmatrix} 0 \\ x \end{pmatrix} \in M^\times \cap$ Im $\begin{pmatrix} 0 \\ I \end{pmatrix} \neq (0)$. By invariance of $M^\times$ this is contained in $M^\times$. Now $M^\times$ is J_1-neutral which implies that condition (c) is violated. $\qquad \square$

As is Section 3 we observe that completeness of the set of hermitian solutions of the discrete algebraic Riccati equation implies controllability of the pair (A, B). The next theorem rephrases the statement of the previous one in terms of the functions $G_K(\lambda)$ under the additional condition that (A, B) is controllable.

Theorem 5.3 *Assume that the pair (A, B) is controllable. Then there exists a complete set of solutions of the discrete algebraic Riccati equation (4.1) if and only if the following conditions hold for any K such that the realization (4.4) is minimal:*

(a) *$G_K(z)$ has constant signature on the unit circle,*

(b) *$G_K(z)$ has constant zero signature on the unit circle,*

(c) *for all non-zero vectors x we have*

$$\langle G_K(z)^{-1} \Psi^* (z^{-1} - V^*)^{-1} x, \Psi^* (z^{-1} - V^*)^{-1} x \rangle \not\equiv 0.$$

Proof: The proof is essentially the same as for the continuous Riccati equation. $\qquad \square$

6 Completeness of Factorizations for Hermitian Rational Matrix Functions Corresponding to a Sign Definite Pair of Subspaces

Let $G(\lambda)$ be an arbitrary rational matrix function taking hermitian values on the imaginary axis of which the signature need not be constant. Then we cannot expect J-symmetric factorizations. Let $G(\lambda) = D + C(\lambda I_n - A)^{-1}B$ be a minimal realization of $G(\lambda)$, and let H be the unique skew-hermitian matrix for which (1.2) holds, i.e., $HA = -A^*H$, $HB = C^*$. Then there always exists an A-invariant maximal H-nonnegative subspace M and an $A^\times$-invariant maximal H-nonpositive subspace $M^\times$, see, e.g. [GLR]. For the special case where D is positive definite it was proved in [R] that any such pair of subspaces automatically matches, that is, for any such M and $M^\times$ we have $M \oplus M^\times = \mathbb{C}^n$. It follows from [BGK] that corresponding to such a matching pair there is a minimal factorization. We shall say that such a minimal factorization *corresponds to a sign definite pair of subspaces*.

For the general case, i.e., with D invertible, but not necessarily positive definite, we shall say that there is a *complete set of factorizations corresponding to a sign definite pair of subspaces* if for any A-invariant maximal H-nonnegative subspace M and any $A^\times$-invariant maximal H-nonpositive subspace $M^\times$ we have $M \oplus M^\times = \mathbb{C}^n$.

We start with a geometric lemma that is the counterpart of Lemma 1.1.

Lemma 6.1 *Let A and H satisfy $HA = -A^*H$, $H = -H^*$. Let N be an A-invariant H-neutral subspace. Then there is an A-invariant maximal H-nonpositive subspace M_- with $N \subset M_-$ and there is an A-invariant maximal H-nonnegative subspace M_+ with $N \subset M_+$.*

Proof: We shall first prove the existence of M_-. Let M_0 be an arbitrary A-invariant maximal H-nonpositive subspace. Put

$$M_- = N + ((HN)^\perp \cap M_0).$$

Then M_- is A-invariant as one easily checks. It is also obviously H-nonpositive. Let us consider $(HM_-)^\perp$. We shall show that it is H-nonegative, which implies that M_- is maximal H-nonpositive. Indeed,

$$(HM_-)^\perp = (HN)^\perp \cap (N + (HM_0)^\perp).$$

Take $x \in (HM_-)^\perp$. Then it can be written as $x = x_1 + x_2$, where $x_1 \in N$ and $x_2 \in (HM_0)^\perp$. Since N is H-neutral we have $N \subset (HN)^\perp$. As x is also in $(HN)^\perp$, it follows that x_2 is in $(HN)^\perp$. In particular, as x_1 is in N, we get that $\langle Hx_1, x_2 \rangle = 0$, and $\langle Hx_1, x_1 \rangle = 0$. Then we have

$$\langle Hx, x \rangle = \langle H(x_1 + x_2), (x_1 + x_2) \rangle = \langle Hx_2, x_2 \rangle$$

However, $x_2 \in (HM_0)^{\perp}$, and M_0 is maximal H-nonpositive. Then $(HM_0)^{\perp}$ is H-nonnegative. It follows that $\langle Hx, x \rangle \geq 0$.

Observe that $(HM_-)^{\perp}$ is in fact A-invariant and maximal H-nonnegative and that $N \subset (HM_-)^{\perp}$. So it is the subspace desired in the second statement. $\qquad \square$

Our main result of this section is the following.

Theorem 6.2 *Let $G(\lambda) = D + C(\lambda - A)^{-1}B$ be a minimal realization of a rational matrix function having hermitian values on the imaginary line. Let H be the unique skew-hermitian matrix for which (1.2) holds. Then the following are equivalent:*

(i) *there is a complete set of factorizations corresponding to a sign definite pair of subspaces of $G(\lambda)$,*

(ii) *for every A-invariant maximal H-nonnegative subspace M and for every non-zero vector $x \in M$ we have*

$$\langle H(\lambda - A^{\times})^{-1}x, x \rangle \not\equiv 0, \quad \lambda \in \rho(A^{\times}),$$

(iii) *for every $A^{\times}$-invariant maximal H-nonpositive subspace $M^{\times}$ and for every non-zero vector $x \in M^{\times}$ we have*

$$\langle H(\lambda - A)^{-1}x, x \rangle \not\equiv 0, \quad \lambda \in \rho(A).$$

Proof: The proof is analogous to the proof of Theorem 1.2. The role of Lemma 1.1 in the proof of Theorem 1.2 is now taken over by Lemma 6.1. $\qquad \square$

Acknowledgements

The research of L. Lerer is partially supported by the United States-Israel Binational Science Foundation grant 9400271.

References

[BGR]   J.A. Ball, I. Gohberg and L. Rodman, *Interpolation of rational matrix functions*, OT **45**, Birkhäuser, Basel etc., 1990.

[BR]   J.A. Ball and A.C.M. Ran, Optimal Hankel-norm model reduction and Wiener-Hopf factorization I: the canonical case, *SIAM Journal of Control and Optimization* **25** (1987), 362–382.

[BGK]   H. Bart, I. Gohberg and M.A. Kaashoek, *Minimal factorization of matrix and operator functions* OT **1**, Birkhäuser, Basel etc., 1979.

[B]   N.E. Barabanov, On the existence of full set of solutions of Riccati equations, in: *Proceedings of the 1997 European Control Conference* (CD-Rom).

[BLW]   S. Bittanti, A.J. Laub and J.C. Willems, Eds. *The algebraic Riccati Equation*, Springer Verlag, Berlin etc., 1991.

[C]   A.N. Čurilov, On the solvability of certain matrix inequalities, *Vestnik Leningrad Univ. Mat. Mekh. Astronom.* **2** (1980), 51–55, 124. (Russian, English summary).

[FMP]   A. Ferrante, G. Michaletzky and M. Pavon, Parametrization of all minimal square spectral factors, *Systems and & Control Letters* **21** (1993), 249–254.

[FP]   L. Finesso and G. Picci, A characterization of minimal spectral factors, *IEEE Trans. Automatic Control* AC-**27** (1982), 122–127.

[F1]   P.A. Fuhrmann, On symmetric rational transfer functions, *Linear Algebra and its Applications* **50** (1983), 167–250.

[F2]   P.A. Fuhrmann, On the characterization and parametrization of minimal spectral factors, *J. Math. Systems, Estimation and Control* **5** (1995), 383–444.

[GLR]   I. Gohberg, P. Lancaster and L. Rodman, *Matrices and indefinite scalar products*, OT **8**, Birkhäuser, Basel etc., 1983.

[GLR1]   I. Gohberg, P. Lancaster and L. Rodman, Factorization of selfadjoint matrix polynomials with constant signature, *Linear and Multilinear Algebra* **11** (1982), 209–224.

[GoL]   G. Gomez and L. Lerer, Generalized Bezoutians for analytic operator functions and inversion of structured operators. *Systems and Networks: Mathematical Theory and Applications* U. Helmke, R. Mennicken and J. Saurer, Eds., Academie Verlag, 1994, 691–696.

[GGLD]   M. Green, K. Glover, D. Limebeer and J. Doyle, A *J*-spectral approach to H_∞-control, *SIAM Journal of Control and Optimization* **28** (1990), 1350–1371.

[IW]   V. Ionescu and M. Weiss, Continuous and discrete-time algebraic Riccati theory: A Popov function approach, *Linear Algebra and its Applications* **193** (1993) 173–209.

[IWO]   V. Ionescu, M. Weiss and C. Oara, *Generalized Riccati Theory and robust control. A Popov function approach.* John Wiley, Chichester, 1999.

[KL]   I. Karelin and L. Lerer, Generalized Bezoutian, factorization of rational matrix functions and matrix quadratic equations. (submitted for publication).

[K1]   V. Kucera, The discrete Riccati equation of optimal control. *Kybernetika* **8** (1971), 430–477.

[K2]   V. Kucera, Algebraic Riccati Equation: Hermitian and Definite Solutions, in: *The algebraic Riccati Equation*, S. Bittanti, A.J. Laub and J.C. Willems, Eds., Springer Verlag, Berlin etc., 1991, 53–88.

[K3]   V. Kucera, A contribution to matrix quadratic equations, *IEEE Trans. Automat. Control*, AC-**17** (1972), 344–347.

[LaR]   P. Lancaster and L. Rodman, *Algebraic Riccati equations.* Clarendon Press, Oxford, 1995.

[LaR1]   P. Lancaster and L. Rodman, Existence and uniqueness theorems for the algebraic Riccati equation, *Int. J. Control* **32** (1980), 285–309.

[LRR]   P. Lancaster, A.C.M. Ran and L. Rodman, Hermitian solutions of the discrete algebraic Riccati equation, *Int. J. Control* **44** (1986), 777–802.

[L]   L. Lerer, The matrix quadratic equation and factorization of matrix polynomials, *Operator Theory: Advances and Applications* **40** (1989), 279–324.

[LRa]   L. Lerer and A.C.M. Ran, J-pseudo-spectral and J-inner-pseudo-outer factorizations for matrix polynomials, *Integral Equations and Operator Theory* **29** (1997), 23–51.

[LR]   L. Lerer and L. Rodman, Bezoutians of rational matrix functions, *J. Functional Analysis* **141** (1996), 1–36.

[LR1]   L. Lerer and L. Rodman, Common zero structure of rational matrix functions, *J. Functional Analysis* **136** (1996), 1–38.

[M1]   B.P. Molinari, The stabilizing solution of the algebraic Riccati equation, *SIAM J. Control* **11** (1973), 262–271.

[M2]   B.P. Molinari, The stabilizing solution of the discrete algebraic Riccati equation, *IEEE Trans. Automat. Control*, AC-**20** (1975), 396–399.

[P]   V.M. Popov, *Hyperstability of Control Systems*, Die Grundlehren der Mathematischen Wissenschaften in Einzeldarstellungen, vol. **204**, Springer Verlag, Berlin etc., 1973. (Translated and expanded version of the 1966 original in Romanian.)

[R]   A.C.M. Ran, Minimal factorization of selfadjoint rational matrix functions, *Integral Equations and Operator Theory* **5** (1982), 850–869.

[R1]   A.C.M. Ran, Minimal square spectral factors, *Systems & Control Letters* **26** (1994), 621–634.

[RR]   A.C.M. Ran and L. Rodman, On symmetric factorizations of rational matrix functions, *Linear and Multilinear Algebra* **29** (1991), 243–261.

[RR1]   A.C.M. Ran and L. Rodman, Stability of invariant maximal semidefinite subspaces I, *Linear Algebra and its Applications* **62** (1984), 51–86.

[RR2]   A.C.M. Ran and L. Rodman, Stability of invariant lagrangian subspaces I, in: *Topics in Operator Theory, Constantin Apostol Memorial Issue*, OT **32**, Birkhäuser, Basel etc., 1988, 181–218.

[RR3]   A.C.M. Ran and L. Rodman, Stable Hermitian Solutions of Discrete Algebraic Riccati Equations, *Math. Control, Signals and Systems* **5** (1992), 165–193.

[RZ]   A.C.M. Ran and P. Zizler, On selfadjoint matrix polynomials with constant signature, *Linear Algebra and its Applications* **259** (1997), 133–153.

[S]   M.A. Shayman, Geometry of the algebraic Riccati equations, I, II. *SIAM J. Control* **21** (1983), 375–394 and 395–409.

[W]   J.C. Willems, Least squares statinary optimal control and the algebraic Riccati equation, *IEEE Trans. Automatic Control* AC-**16** (1971), 621–634.

I. Karelin and L. Lerer   A.C.M. Ran
Department of Mathematics   Divisie Wiskunde en Informatica
Technion Israel Institute of Technology   Faculteit Exacte Wetenschappen
Haifa 32000   Vrije Universiteit
Israel   de Boelelaan 1081a
e-mail: llerer@techunix.technion.ac.il   1081 HV Amsterdam
e-mail: kirina@techunix.technion.ac.il   The Netherlands
                                                 e-mail: ran@cs.vu.nl

1991 Mathematics Subject Classification. Primary: 47A68, 15A24; Secondary: 47B50, 49N05, 93B36.

Operator Theory:
Advances and Applications, Vol. 124
© 2001 Birkhäuser Verlag Basel/Switzerland

Scattering of Waves by Periodic Gratings and Factorization Problems

A.G. Kostyuchenko and A.A. Shkalikov

Dedicated to Israel Gohberg on the occasion of his 70th birthday

The paper deals with scattering problems of plane and space waves by periodic gratings in $\mathbb{R}^2$ and $\mathbb{R}^3$, respectively. The plane problem caused a vast physical and mathematical literature starting from the papers of Rayleigh. A new approach to treat this problem is proposed in the present paper. It is based on the possibility to reformulate the scattering problem in terms of abstract ordinary differential equation with constant operator coefficients on a Hilbert space. The solvability of such equations with radiation conditions at the infinity is based on the factorization of the operator symbol of the equation. This approach is general and allows, in particular, to solve the scattering problem of a space wave in $\mathbb{R}^3$.

0 Introduction

In this paper we consider two scattering problems for the Helmholtz equation. The study of the first one was originated by Lord Rayleigh [R1, R2]. The problem is to give an analysis of the scattering of a monochromatic plane wave incident on a grating with a periodic curve in $\mathbb{R}^2$ (Rayleigh considered a sinusoidal grating profile). The second problem is a three-dimensional analogue of the first one: to give an analysis of the scattering of a space wave by a periodic surface in $\mathbb{R}^3$.

The scattering of acoustic and electromagnetic waves by periodic gratings plays a significant role in physics and engineering, which caused a vast literature devoted to these problems. The works are mostly connected with the scattering in $\mathbb{R}^2$, because even in this case the problem is quite non-trivial and involves an intricate and lengthy analysis.

A mathematical core of the problem is to prove existence and uniqueness of the solution which describes the scattering. In the plane case and in the case of non-resonant frequencies the proof was given by Badyukov [B1, B2], who reduced the problem to an integral equation of the Fredholm type using the Hankel function expansion for the kernel of the Helmholtz equation. Wilcox and Guilliot [WG] independently obtained similar results using Rayleigh-Bloch wave expansions (which, essentially, coincide with those in [B1, B2]).

Alber [A] and Wilcox [W1] developed an alternative method for solving the scattering problem based on analytic continuations. Further developments, numerical studies, historical remarks and references can be found in the book edited by

Petit [P], in the monographs of Günter [G], Wilcox [W1, W2], Galashnikova and Il'inskii [GI], Nazarov and Plamenevskii [NP], in the papers of Babich [B], Il'inskii and Mikheev [IM], Beljaev, Mikheev and Shamaev [BMS].

For the case of resonant frequencies, we have not found in the literature rigorous results on the solvability of the scattering problems. In this case an additional problem arises: how to select outgoing waves and to pose the radiation conditions? It turns out that the formulation of the radiation condition for the plane scattering problem in the case of resonant frequencies remains the same as in the non-resonant case. However, this circumstance is rather incidental from the mathematical point of view. It is explained by the fact that the Jordan chains of the spectral problem corresponding to the Helmholtz equation have the simplest structure: their lengths equal 2. This is not true if the scattering problem is considered for the system of elasticity (see [KO]).

The aim of this paper is to propose a *new approach* to treat scattering problems. This approach is quite general and allows to consider scattering problems which were not treated before. It is based on the possibility to reformulate these problems in terms of abstract ordinary differential equations with operator coefficients on a Hilbert space. It turns out that the solvability of a scattering problem is equivalent to the solvability of the appropriate differential equation on the semiaxis with the radiation conditions at $+\infty$. To solve the last problem we apply the factorization theorems for the operator symbol of the corresponding equation. This paper can be considered as a continuation of the paper [Sh2], but it can be read independently.

An outstanding role in development of the factorization theory is played by the works of I. Gohberg and his co-authors. A particular mention deserves his pioneering work with M. Krein [GK1]. Further developments and references can be found in the monograph of Gohberg and Feldman [GF], Gohberg and Krupnik [GK], Gohberg, Lancaster and Rodman [GLR], Markus [M], Gohberg, Goldberg and Kaashoek [GGK].

In our case the factorization problem has to be solved for selfadjoint operator pencils with unbounded operator coefficients. This leads to new difficulties. In particular, even if a linear right divisor $\lambda - Z$ of a pencil is found, one has to investigate the properties of Z: does this operator generate a holomorphic C_0-semigroup? The study of factorization of operator pencils with unbounded coefficients was originated in the authors' paper [KS]. Here we give a short review of results on the factorization of elliptic pencils which are essentially used in the sequel.

In contrast with the previous works, an abstract approach of this paper does not use specific properties of the Helmholtz equation and can be applied to scattering problems in electrodynamics described by the Maxwell equation (see [GI]) or the system of elasticity. The Helmholtz equation itself can be modified; the frequency k^2 can be replaced by a periodic function $k^2 c^2(x)$ that corresponds to the scattering in a non-homogeneous medium. All our arguments remain valid; the only change is to replace the exponents by the eigenfunctions of the Sturm-Liouville operator with potential $k^2 c^2(x)$ and with quasi-periodic boundary conditions. We remark also that the resonant case is not an obstacle for the method.

The plan of the paper is the following. In Section 1 we pose the scattering problem for the plane Helmholtz equation and formulate the radiation condition. In Section 2 we give a review of results on factorization of elliptic pencils, properties of divisors and solvability of the corresponding operator equations on the semiaxis with the radiation conditions at $+\infty$. In the subsequent sections we give the detailed analysis of the scattering problems for the two- and three-dimensional cases. To our best knowledge, the space problem has not been considered in the literature before.

1 Scattered Waves and the Radiation Condition for the Two-dimensional Helmholtz Equation

Let (x, y) be the coordinates in $\mathbb{R}^2$ and Γ be a 2π-periodic curve given by a smooth function $y = a(x)$. Let

$$(1.1) \qquad v_\varphi(x, y) = e^{-ik(x \sin \varphi + y \cos \varphi)}$$

be a monochromatic wave incident on the grating Γ. The reflection of this wave generates the scattered waves which are to be found. The number φ coincides with the angle between the axis Oy and the direction of the wave v_φ (see Figure 1). The wave $v_\varphi(x, y)$ satisfies the Helmholtz equation

$$(1.2) \qquad \Delta u + k^2 u = 0, \qquad u = u(x, y),$$

and the quasi-periodic boundary conditions

$$(1.3) \qquad \begin{aligned} u(0, y) &= e^{it - iv} u(2\pi, y), \\ u'_x(0, y) &= e^{-iv} u'_x(2\pi, y), \end{aligned}$$

where $v = 2\pi k \sin \varphi$.

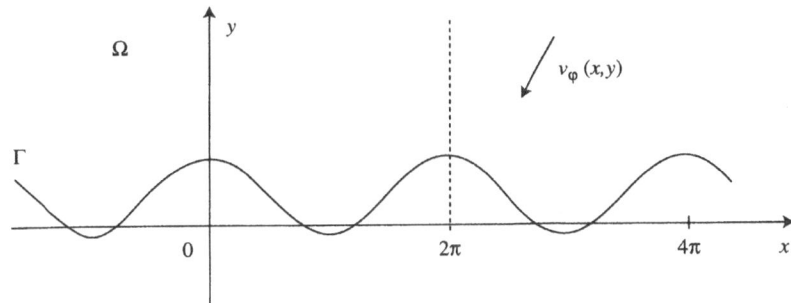

Figure 1

Naturally, the scattered waves also satisfy equation (1.2) and boundary conditions (1.3). We have to declare the law of reflection. Assume that the wave v_φ

reaches all points of the grating. This means that $\cot \varphi > \max a'(x)$. The full reflection means

(1.4) $$u(x, y)|_\Gamma = v_\varphi(x, y)|_\Gamma$$

or

$$u(x, a(x)) = v_\varphi(x, a(x)),$$

where $u(x, y)$ is a solution of the scattering problem in the domain

$$\Omega = \{x, y \mid x \in \mathbb{R}, y > a(x)\},$$

i.e., a solution of equation (1.2) subject to boundary conditions (1.3).

The problem given in the unbounded domain Ω by equation (1.2) and initial condition (1.4) is not well-posed, since the frequency $k^2 \geq 0$ belongs to the continuous spectrum of the Laplace operator in Ω with the Dirichlet boundary condition on $\partial\Omega$. To extract physically reasonable solutions in such cases one claims additional conditions. It is well known that for the Helmholtz equation on the exterior of a bounded domain the condition

$$\frac{\partial u}{\partial r} - iku = o(r) \quad \text{as} \quad r = \sqrt{x^2 + y^2} \to \infty,$$

guarantees existence and uniqueness of a solution. This is the so-called Sommerfeld radiation condition.

For unbounded domains with periodic boundaries the radiation conditions have a more intricate form. To formulate them, we remark that the elementary quasi-periodic solutions of equation (1.2) have the representation

(1.5) $$u_n^\pm(x, y) = e^{\pm i\lambda_n y} e^{i\mu_n x}, \quad \mu_n = \frac{\nu}{2\pi} + n, \quad \lambda_n = \sqrt{k^2 - \mu_n^2}.$$

Here $n \in \mathbb{Z}$, and the main branch of the square root function is chosen, i.e., $\lambda_n > 0$ for $k > |\mu_n|$ and $\mathrm{Im}\lambda_n > 0$ for $|\mu_n| > k$. The solutions $u_n^-(x, y)$ corresponding to the non-real values λ_n grow exponentially in Ω as $y \to \infty$ and have no physical sense in scattering problems. The solutions corresponding to the real numbers λ_n are called propagating waves and play the most important role. The solutions $u_n^+(x, y)$ ($u_n^-(x, y)$) which correspond to $\lambda_n > 0$ ($\lambda_n < 0$) are called outgoing (incoming) waves. The physical sense prompt us that the scattered waves must include only outgoing waves and exponentially decaying waves corresponding to solutions $u_n^+(x, y)$, $\mathrm{Im}\lambda_n > 0$. Actually, it was Rayleigh [R1, R2] who assumed that scattered waves consist only of outgoing and decaying waves. The problem of a choice of physically reasonable propagating waves has been widely discussed in the literature since 30's. In particular, Mandelstam noticed that the Rayleigh hypothesis does not work for some equations of electrodynamics and proposed to choose the waves with positive group velocity (see details in [Sh2]). For the Helmholtz equation the Rayleigh and the Mandelstam hypotheses coincide.

Now we can formulate the scattering problem as follows: *to find a solution of equation* (1.2) *subject to quasi-periodic conditions* (1.3), *initial condition* (1.4) *and the radiation condition*

$$(1.6) \qquad u(x, y) = \sum_{-k \leq \mu_n \leq k} c_n e^{i\mu_n x} e^{i\lambda_n y} + o(1),$$

where the sum contains only outgoing waves corresponding to $\lambda_n > 0$ *and* $o(1)$ *is a decaying function as* $y \to \infty$. Here c_n are the amplitudes of the outgoing waves. They must be determined by initial condition (1.4). Further we will see that $o(1)$ in (1.6) is represented as a convergent series of the exponentially decaying waves (although this is not necessarily true in the three-dimensional problem).

A frequency k^2 is called *resonant* (in some books it is called *cut-off*) if $\lambda_n^2 = k^2 - \mu_n^2 = 0$ for some $n \in \mathbb{Z}$. If $v/\pi \notin \mathbb{Z}$ then the equality $\lambda_n = 0$ may hold for the only value $n_0 \in \mathbb{Z}$. In this case the pair of elementary solutions

$$e^{ikx}, \quad y e^{ikx}, \qquad \mu_0 = v/2\pi + n_0 = k \geq 0,$$

corresponds to the wave number $\lambda_n = 0$. If $v/\pi \in \mathbb{Z}$ and the equality $k^2 - \mu_n^2 = 0$ holds for some $n \in \mathbb{Z}$, then it holds for two values $n_0, n_1 \in \mathbb{Z}$. In this case the resonant elementary solutions have the form

$$e^{ikx}, \quad y e^{ikx}; \qquad e^{-ikx}, \quad y e^{-ikx}.$$

The first functions of these pairs of solutions are degenerated waves (independent of y); the second ones are associated solutions. There is one-to-one correspondence between elementary solutions of problem (1.2), (1.3) and Jordan chains of the pencil (operator symbol) corresponding to this problem. It follows from the subsequent general results that in the presence of a Jordan chain of even length only the first half of the functions of this chain has to be taken into consideration. In our case the lengths of Jordan chains equal 2. Therefore, in this case only the eigenfunctions must be involved in the group of the scattered waves participating in the radiation condition. Hence the radiation condition in the resonant case is given as before by formula (1.6).

2 Factorization of Elliptic Operator Pencils and Solvability of the Corresponding Equations on the Semiaxis

In this section we deal with an operator pencil

$$T(\lambda) = \lambda^2 F + \lambda G + H - V$$

on a Hilbert space $\mathfrak{H}$. It is always assumed that the "main" operator H is selfadjoint and uniformly positive while the other ones are symmetric.

Definition 2.1 $T(\lambda)$ is called *elliptic* if the following conditions are fulfilled:

i) F is bounded and uniformly positive ($0 \ll F \ll \infty$);

ii) $H = H^* \gg 0$ and V is a symmetric H-compact operator, i.e., VH^{-1} is compact in $\mathfrak{H}$;

iii) G is symmetric and $\mathcal{D}(G) \supset \mathcal{D}(H^{1/2})$;

iv) $T(\lambda) > 0$ for all $\lambda \in \mathbb{R}$ with $|\lambda| > r_0$ provided r_0 is large enough.

Definition 2.2 An elliptic pencil $T(\lambda)$ is called *strongly elliptic* if there exist a number $\varepsilon > 0$ and a symmetric H-compact operator V' such that

$$(2.1) \qquad T(\lambda) \geq \varepsilon(\lambda^2 I + H) - V' \qquad \forall \lambda \in \mathbb{R}.$$

Definition 2.3 An elliptic pencil $L(\lambda)$ is called *regular elliptic* if

$$(2.2) \qquad \|HT^{-1}(\lambda)\| + |\lambda|^2 \|T^{-1}(\lambda)\| \leq \text{const} \quad \forall \lambda \in \mathbb{R}, \ |\lambda| > r_0.$$

Let $\mathfrak{H}_\theta$ be the scale of the Hilbert spaces generated by the operator $H^{1/2}$, i.e., $\mathfrak{H}_\theta = \mathcal{D}(H^{\theta/2})$ and $\|x\|_\theta = \|H^{\theta/2}x\|$. Recall that the abstract Sobolev space $W^s(a, b; \mathfrak{H})$ consists of functions $f(t)$ defined on $(a, b) \subset \mathbb{R}$, taking values in $\mathfrak{H}$, and having a finite norm

$$\|\|f\|\|_s^2 = \int_a^b (\|f^{(s)}(t)\|^2 + \|H^{s/2}f(t)\|^2)\, dt.$$

(see details in [LM, Ch. 1]).

Definition 2.4 An elliptic pencil $T(\lambda)$ is called *strongly regular* if for all functions $v(t) \in W^2[0, \infty; \mathfrak{H}]$ subject to the condition $v(0) = 0$, the following estimate holds

$$(2.3) \qquad \left\| T_\varkappa\left(-i\frac{d}{dt}v(t)\right) \right\|_{L_2} \geq \varepsilon \|\|v\|\|_2, \quad \varepsilon > 0,$$

where $T_\varkappa(\lambda) = \lambda^2 F + \varkappa\lambda G + H$ and ε does not depend on v and $\varkappa \in [0, 1]$.

It is known [Sh1, §3] that estimate (2.3) implies (2.2), i.e., a strongly regular elliptic pencil is regular elliptic but not vice versa. We remark that for usual elliptic operators estimate (2.1) is equivalent to the Gårding inequality (see [Sh2]), estimate (2.2) is known as the Agmon-Nirenberg or the Agranovich-Vishik estimate for regular elliptic problems with parameter on smooth bounded domains (in this context $\mathfrak{H} = L_2(\Omega)$, where Ω is a smooth bounded domain in $\mathbb{R}^n$), and estimate (2.3) is known as the Bernstein-Ladyzhenskaya inequality (see [Sh1, §3, 6]). More details on the motivation of the above definitions can be found in the papers [Sh1, Sh2].

The verification of estimate (2.3) is not trivial even for concrete pencils. We shall use the following result.

Proposition 2.5 *Let $T(\lambda)$ be elliptic. Suppose that $F = F_0 + F_1$, $G = G_0 + G_1$, $H = H_0 + H_1$, where F_1, G_1, H_1 are symmetric operators such that $F_1, G_1 H^{-1/2}$, $H_1 H^{-1}$ are compact, and $H_0 \gg 0$. If the estimate*

$$(2.4) \qquad \|F_0^{-\frac{1}{2}} Gy\| \le (2-\varepsilon)\|H_0^{1/2} y\|, \quad y \in \mathcal{D}(H_0^{1/2})$$

holds with some $\varepsilon > 0$, then $T(\lambda)$ is strongly regular.

Proof: It can be found in [Sh1, §9]. □

Let the operator H have discrete spectrum. It follows from the theorem on holomorphic operator functions (see [GK2, Ch. 1] and [Sh1, §1.4] for the version of this theorem for pencils with unbounded coefficients) that the spectrum of an elliptic pencil $T(\lambda)$ in this case is also discrete.

It is known [Ke] that the principal part of the Laurent expansion of the resolvent $T^{-1}(\lambda)$ admits a representation

$$\sum_{k=1}^{N} \sum_{s=0}^{p_k} \frac{(\cdot, z_k^{p_k-s}) x_k^s}{(\lambda - \mu)^{p_k+1-s}},$$

where

$$(2.5) \qquad x_k^0, \ldots, x_k^{p_k}, \qquad k = 1, \ldots, N,$$

is a canonical system of eigen and associated vectors of the pencil $T(\lambda)$, and

$$z_k^0, \ldots, z_k^{p_k}, \qquad k = 1, \ldots, N,$$

is the adjoint canonical system of the eigen and associated vectors of $T(\lambda)$ corresponding to the eigenvalue $\overline{\mu}$. The following result is essential in the sequel.

Proposition 2.6 *Canonical system (2.5) corresponding to a real eigenvalue μ can be chosen so that*

$$x_k^s = \varepsilon_k z_k^s, \qquad k = 1, \ldots, N, \quad \forall s = 0, \ldots, p_k,$$

where $\varepsilon_k = \pm 1$ and $\varepsilon_k = \text{sign}(L'(\mu) x_k^0, x_k^{p_k})$.

Proof: See [KS, Lemma 1.2]. □

Other definitions of the sign characteristics are given in [GLR]. Actually, the sign characteristics are important only for Jordan chains of odd length. Further, we assume for convenience that the *sign characteristics of Jordan chains of even length equal zero.*

Canonical systems that possess the properties formulated in Proposition 2.6 are called *normal*. Normal canonical systems and sign characteristics are important in the analysis of the factorization problem. We shall give here short historical comments concerning the problem of the factorization of operator polynomials with respect to the real axis. Krein and Langer [KL] studied pencils of the form

$$L(\lambda) = I + \lambda B + \lambda^2 C,$$

where B and C are bounded selfadjoint operators and $C > 0$. They proved that $L(\lambda)$ possesses a right divisor of the form $(\lambda Z - I)$, whose spectrum is located in the closed upper half-plane. The real spectrum of this divisor was investigated by Kostyuchenko and Orazov [KO]. The factorization of higher order operator polynomials was out carried by Langer [L]; the detailed analysis of the real spectrum of divisors of polynomials with Hermitian matrix coefficients was done by Gohberg, Lancaster and Rodman [GLR]. We should mention that the problem of factorization of non-negative operator pencils (and operator functions) on the real axis has its own history (see the book of Rosenblum and Rovnjak [RR]). In the paper [KS] the authors proposed a new analytic approach to the factorization of quadratic pencils, investigated the properties of a linear operator Z participating in the factorization and proved the first factorization theorem for pencils with unbounded coefficients. Further developments of the theory was carried out by Shkalikov [Sh1–Sh3].

Now, let us define the half of the eigen and associated vectors of an operator pencil $T(\lambda)$. Let canonical system (2.5) be normal. Its half consists of the vectors

$$(2.6) \qquad\qquad x_k^0, x_k^1, \ldots, x_k^{l_k}, \qquad k = 1, \ldots, N,$$

where $l_k = (p_k - 1 + \varepsilon_k)/2$ and ε_k are the sign characteristics (we assume $\varepsilon_k = 0$ for Jordan chains of even length). We imply that in the case $l_k = -1$ the corresponding set in (2.6) is empty. The set of all canonical systems of $T(\lambda)$ corresponding to the eigenvalues from the open upper half-plane and of the halves of canonical systems corresponding to the real eigenvalues, is called *the half of eigen and associated vectors of* $T(\lambda)$. We point out a particular important case (connected with the scattering problem): if the lengths of Jordan chains corresponding to the real eigenvalues do not exceed 2, then the half contains the canonical systems corresponding to the eigenvalues from the upper half-plane and only the eigenvectors x_k corresponding to $\lambda_k \in \mathbb{R}$ subject to the condition $(T'(\lambda_k)x_k, x_k) \geq 0$.

To formulate the basic results we shall introduce the class of operator pencils whose resolvents are meromorphic functions of finite order having polynomial growth on some rays in the complex plane.

Definition 2.7 We say $T(\lambda)$ belongs to the class K if

i) the eigenvalues of the operator H satisfy the estimate

$$\lambda_j(H) \geq cj^p$$

with some constants c and p.

ii) either $p \geq 2$ or $p < 2$ and there exist rays $z\gamma_j = \{\lambda | \arg \lambda = \theta_j\}$, $j = 1, \ldots, N$, $\theta_{j+1} > \theta_j$, in the upper half-plane such that

$$\max(\theta_1, \theta_{j+1} - \theta_j, \pi - \theta_N) < 2\pi/p$$

and

(2.7) $\qquad \|H^{1/2}T^{-1}(\lambda)H^{1/2}\| \leq c|\lambda|^m \quad \text{for} \quad \lambda \in \gamma_j$

with some constants c, m, provided $|\lambda|$ is large enough.

It is proved in [Sh2] that the inequality

$$|(T(\lambda)x, x)| \geq \varepsilon[r^2(x, x) + (Hx, x)], \qquad \forall \lambda \in \gamma, \; |\lambda| = r > r_0,$$

is sufficient for the validity of estimate (2.7) with $m = 0$ on the ray γ. The last inequality is easier to verify for concrete problems. In particular, (2.7) holds on the real line for strongly elliptic pencils. If $T(\lambda)$ is elliptic and the condition $\lambda_j(H) \geq cj^p$ holds, then the resolvent $T^{-1}(\lambda)$ is a meromorphic function of order $\leq p/2$ (see [Sh1, §2]). Applying the Phragmen-Lindelöf theorem we find: if an elliptic pencil $T(\lambda)$ is strongly regular, belongs to the class K and $F(\lambda) = (T^{-1}(\lambda)f, f)$ is an entire function for some $f \in \mathfrak{H}$, then $F(\lambda) \equiv 0$. Hence conditions i) and ii) in Definition 2.7 are needed to prove the completeness theorems for eigenvectors (see details in [Sh2]).

Now let us specify the understanding of solutions of the equation

(2.8) $\qquad T\left(i\dfrac{d}{dt}\right)u(t) = -Fu''(t) + iGu'(t) + (H + V)u(t) = 0.$

A function $u(t) \in W^1(a, b; \mathfrak{H})$ is said to be *a generalized solution* of equation (2.8) if for all functions $v(t) \in W^1(a, b; \mathfrak{H})$ subject to the conditions $v(a) = v(b) = 0$, the equality

$$(Fu', v') + i(u', Gv) + ((I + V')H^{1/2}u, H^{1/2}v) = 0$$

holds, where $V' = H^{-1/2}VH^{-1/2}$ and the scalar product is taken in $L_2(a, b; \mathfrak{H})$. Details clarifying this definition see in [Sh2].

A function $u(t) \in W^2(a, b; \mathfrak{H})$ is called a *classical solution* of equation (2.8) on (a, b) if (2.8) holds as equality of functions in $L_2(a, b; \mathfrak{H})$.

We say that a classical (generalized) solution of equation (2.8) satisfies the radiation condition at $+\infty$ if

$$(2.9) \qquad u(t) = \sum_{\varepsilon_k \geq 0} c_k e^{i\lambda_k t} x_k + u_0(t),$$

where $u_0(t)$ is a classical (generalized) solution on $\mathbb{R}^+$ satisfying the condition $\|u_0(t)\|_1 \to 0$ as $t \to \infty$. Here the first term in (2.9) is a finite sum of elementary solutions corresponding to the real eigenvalues of nonnegative type, and for simplicity we have assumed that the lengths of Jordan chains corresponding to the real eigenvalues do not exceed 2. In the general case the sum has to contain all elementary solutions corresponding to the halves of Jordan chains (2.6).

Let us formulate the basic results on elliptic pencils.

Theorem 2.8 *The half of eigen and associated vectors of a selfadjoint elliptic pencil $T(\lambda)$ is minimal (i.e., there exists a biorthogonal system) in the spaces $\mathfrak{H}_\theta$, $0 \leq \theta \leq 1$. It is complete in the same spaces if $T(\lambda)$ is either strongly or regular elliptic and belongs to the class K. If $T(\lambda)$ is strongly regular and belongs to the class K, then the half is a complete system in $\mathfrak{H}_\theta$ for $0 \leq \theta \leq 3/2$.*

Theorem 2.9 *Let $T(\lambda)$ be either strongly elliptic or regular elliptic pencil and belong to the class K. Then*

$$(2.10) \qquad T(\lambda) = (\lambda - Z_1) F(\lambda - Z),$$

where the operator Z possesses the properties:

1) *$\mathcal{D}(Z) = \mathfrak{H}_1$ and $Z - \lambda = K H^{1/2}$, where K is bounded and boundedly invertible in $\mathfrak{H}$, provided $\lambda \notin \sigma(Z)$;*

2) *the spectrum of Z lies in the closed upper half-plane and the system of its eigen and associated vectors coincides with the half of those of $T(\lambda)$;*

3) *iZ generates a holomorphic semigroup in the spaces $\mathfrak{H}_\theta$, $0 \leq \theta \leq 1$.*

If in addition $T(\lambda)$ is strongly regular, then property 3) remains valid in the spaces $\mathfrak{H}_\theta$ for $0 \leq \theta \leq 3/2$.

Theorem 2.10 *Let $T(\lambda)$ be either strongly or regular elliptic selfadjoint pencil belonging to the class K. Then for any vector $f \in \mathfrak{H}_\theta$, $0 \leq \theta \leq 1$, there exists a unique function $u(t)$ which is a generalized solution of equation (2.8) on (ε, ∞) for any $\varepsilon > 0$ and satisfies the radiation condition (2.9) and the initial condition*

$$(2.11) \qquad \lim_{t \to 0+} \|u(t) - f\|_\theta = 0.$$

If $T(\lambda)$ is strongly regular, then the same is true for $0 \leq \theta \leq 3/2$. Moreover, $u(t)$ is a classical solution for $t > 0$ and is represented by the formula

$$(2.12) \qquad u(t) = \frac{1}{2\pi i} \int_\gamma e^{it\lambda} (Z - \lambda)^{-1} d\lambda,$$

where a contour γ contains the spectrum of the operator Z and lies asymptotically in the upper half-plane.

Theorem 2.11 *Let T (λ) be a strongly regular selfadjoint pencil. Then there exists a unique classical solution of equation (2.8) on the semiaxis $\mathbb{R}^+$ satisfying radiation condition (2.9) and initial condition (2.11) for $\theta = 3/2$.*

The most important fact of the last theorem is that existence and uniqueness of solutions of the half-range Cauchy problem on the semiaxis $\mathbb{R}^+$ is true for operator pencils not necessarily belonging to the class K.

The proofs of Theorems 2.9–2.11 can be found in [Sh1, Sh2] (see also [KS], where the first results of this kind were obtained).

3 Existence and Uniqueness of the Solution of the Plane Scattering Problem

It suffices to define a solution of the scattering problem in the semistrip

$$\Omega_0 = \{x, y \mid 0 \le x \le 2\pi, \ a(x) \le y < \infty\}.$$

The corresponding solution in the whole half-plane $y > a(x)$ is restored by quasiperiodic conditions (1.3).

The substitution of

$$(3.1) \qquad\qquad \xi = x, \qquad \eta = y - a(x),$$

maps the semistrip Ω_0 onto the standard semistrip Ω_0' (see Figure 2). Taking into account that

$$\xi_x = 1, \quad \xi_y = 0, \quad \xi_{xx} = 0, \quad \xi_{yy} = 0;$$
$$\eta_x = -a'(x), \quad \eta_y = 1, \quad \eta_{xx} = -a''(x), \quad \eta_{yy} = 0,$$

and

$$u_{xx} = u_{\xi\xi}\xi_x^2 + 2u_{\xi\eta}\xi_x\eta_x + u_{\eta\eta}\eta_x^2 + u_\xi\xi_{xx} + u_\eta\eta_{xx},$$
$$u_{yy} = u_{\xi\xi}\xi_y^2 + 2u_{\xi\eta}\xi_y\eta_y + u_{\eta\eta}\eta_y^2 + u_\xi\xi_{yy} + u_\eta\eta_{yy},$$

we find that the Helmholtz equation is transformed into the following one

$$(3.2) \qquad u_{\xi\xi} - 2u_{\xi\eta}a'(\xi) + u_{\eta\eta}(a'(\xi)^2 + 1) - u_\eta a''(\xi) + k^2 u = 0.$$

The form of this equation is more intricate but its coefficients do not depend on η and the advantage is that ξ and η belong to the domain Ω_0', where the separation of

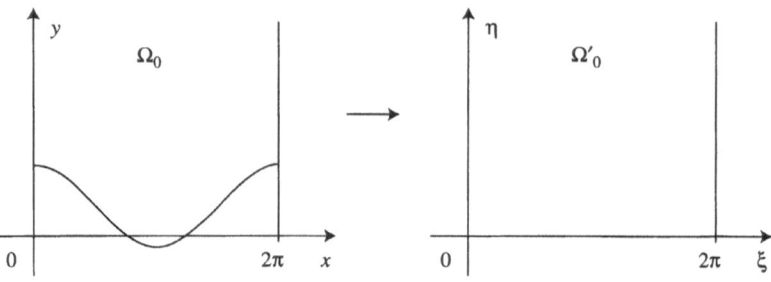

Figure 2

variables can be realized. Looking for solutions of the form $u(\xi, \eta) = e^{i\lambda\eta} f(\xi)$ and taking into account quasi-periodic conditions (1.3) we come to the spectral problem

(3.3) $-f'' - k^2 f + \lambda i (2a'(\xi) f' + a''(\xi) f) + \lambda^2 (a'(\xi)^2 + 1) f = 0$

with the boundary conditions

(3.4) $f(0) = e^{-i\nu} f(2\pi), \qquad f'(0) = e^{-i\nu} f'(2\pi).$

Since functions (1.5) represent a complete set of elementary quasi-periodic solutions of the Helmholtz equation, the functions

(3.5)
$$f_n^\pm(x) = u_n^\pm(x, a(x)) = e^{\pm i\lambda_n a(x)} e^{i\mu_n x},$$
$$\mu_n = \frac{\nu}{2\pi} + n, \quad \lambda_n = \sqrt{k^2 - \mu_n^2},$$

form a complete set of all eigenfunctions of the spectral problem (3.3) and (3.4), which correspond to the eigenvalues λ_n. The set of eigenfunctions of the problem (3.3), (3.4) can also be found by a straightforward calculation. Substituting $f(\xi) = z(\xi) e^{i\lambda a(\xi)}$ in (3.3) and (3.4) we find

$$-z'' + (\lambda^2 - k^2)z = 0,$$
$$z(0) = e^{-i\nu} z(2\pi), \quad z'(0) = e^{-i\nu} z'(2\pi).$$

Since $e^{i\mu_n x}$ form a complete set of eigenfunctions of this problem, we obtain that set (3.5) possesses the same property with respect to problem (3.3), (3.4).

It is easily seen that in the non-resonant case all eigenvalues $\lambda_n = \sqrt{k^2 - \mu_n^2}$ are simple provided $\nu/\pi \notin \mathbb{Z}$. The location of λ_n for values ν close to 0 is shown in Figure 3.

If $\nu/\pi \in \mathbb{Z}$ but $\nu/2\pi \notin \mathbb{Z}$, then two eigenfunctions

$$e^{ia(x)\sqrt{k^2-(n+1/2)^2}} \sin(n+1/2)x, \quad e^{ia(x)\sqrt{k^2-(n+1/2)^2}} \cos(n+1/2)x$$

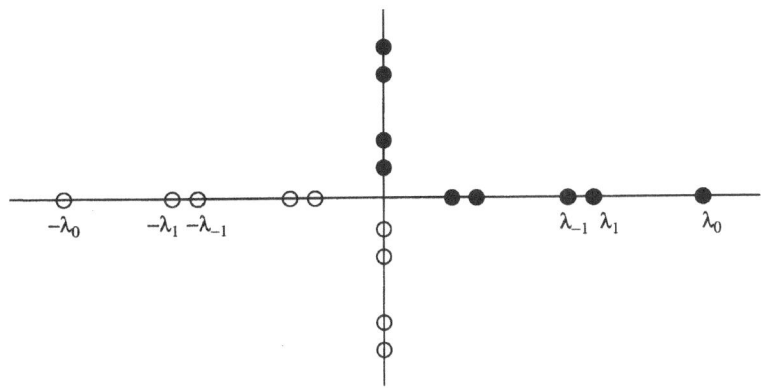

Figure 3

correspond to the eigenvalues $\lambda_n = \sqrt{k^2 - (n + 1/2)^2}$. Finally, if $\nu/2\pi \in \mathbb{Z}$ then the pair of eigenfunctions

$$e^{ia(x)\sqrt{k^2-n^2}} \cos nx, \quad e^{ia(x)\sqrt{k^2-n^2}} \sin nx,$$

correspond to all eigenvalues $\lambda_n = \sqrt{k^2 - n^2} \neq \pm k$. The extremal real eigenvalues $\pm\lambda_0 = \pm k$ are simple; the corresponding eigenfunctions are $e^{\pm ika(x)}$.

In the resonant case zero is the eigenvalue of pencil (3.3), (3.4) of algebraic multiplicity 2 or 4. If $\nu/\pi \notin \mathbb{Z}$, then the only eigenfunction $f_{n_0}(x) = e^{i(n_0+\nu/2\pi)x}$ corresponds to this eigenvalue (here n_0 is defined by the equality $k = n_0 + \nu/2\pi$), and there is an associated function that coincides with $f_{n_0}(x)$ (we omit here elementary calculations). If $\nu/\pi \in \mathbb{Z}$, then two eigenfunctions $e^{\pm i(n_0+\nu/2\pi)x}$ correspond to 0, and there are associate functions coinciding with the previous ones.

According to the general definition (see Section 2) the functions $\{f_n^+(x)\}_{n=-\infty}^{\infty}$ defined in (3.5) form the half of the root functions of pencil (3.3), (3.4). The same is true in the resonant case, since the lengths of Jordan chains do not exceed 2. It is worth mentioning that Rayleigh [R1] calculated these waves in the case of a vertical incident wave ($\nu = 0$) assuming that $a(x)$ is an even function with respect to $x = 0$ and $x = \pi$. In this case a solution $u(x, y)$ of the scattering problem the same property $u'(0, y) = u'(\pi, y) = 0$.

If conditions (3.4) are replaced by $f'(0) = f'(\pi) = 0$, then the functions

$$(3.6) \qquad f_n(x) = e^{ia(x)\sqrt{k^2-n^2}} \cos nx, \qquad n = 0, 1, \ldots.$$

form the half of the root functions of the corresponding pencil. System (3.6) is called the Rayleigh system. Suppose that the Rayleigh system is minimal, say in the space $L_2(0, \pi)$. In this case a solution of the scattering problem can be represented by a formal series

$$u(x, y) = \sum_{n=0}^{\infty} (v_\varphi, f_n^*) f_n(x) e^{i\sqrt{k^2 - n^2}(y - a(x))},$$

where $v_\varphi = e^{ika(x)}$ and $\{f_n^*\}$ is a biorthogonal system with respect to $\{f_n^+\}$. However, we are not aware of papers where the minimality or the completeness of the Rayleigh system (or the generalized Rayleigh system defined by (3.5)) is proved. Moreover, the minimality and the completeness do not guarantee the convergence of the series to the solution $u(x, y)$. Hence a rigorous justification of the Fourier method for the Rayleigh problem seems to be a hard task (see Theorem 3.4 below).

We intend to apply the results of Section 2 to solve the scattering problem. Let us represent pencil (3.3), (3.4) in the abstract form. For $s = 1$ ($s = 2$) denote by $W_U^s[0, 2\pi]$ the subspace of the Sobolev space $W_2^s[0, 2\pi]$ consisting of functions satisfying the first boundary condition (3.4) (both conditions (3.4)). The intermediate spaces

$$W_U^\theta[0, 2\pi] = [W_U^2[0, 2\pi], L_2[0, 2\pi]]_\theta, \quad 0 \le \theta \le 2,$$

are defined by interpolation (see [LM, Ch.1]).

In the space $L_2[0, 2\pi]$, let us define the operators

$$
\begin{aligned}
Hf &= -f'' + f, & \mathcal{D}(H) &= W_U^2[0, 2\pi], \\
Gf &= i(2a'(x)f' + a''(x)f), & \mathcal{D}(G) &= W_U^1[0, 2\pi], \\
Ff &= (a'(x)^2 + 1)f, & \mathcal{D}(F) &= L_2[0, 2\pi].
\end{aligned}
$$

Further it is assumed that $a(x) \in W_2^2[0, 2\pi]$. Now, problem (3.3), (3.4) is represented in the form

$$T(\lambda)f = 0, \qquad T(\lambda) = \lambda^2 F + \lambda G + H - V, \qquad V = (k^2 + 1)I.$$

Proposition 3.1 *The pencil $T(\lambda)$ is selfadjoint and strongly elliptic.*

Proof: It is obvious that $H = H^* \gg 0$. Integrating by parts we find

(3.7) $i(2a'(x)f' + a''(x)f, f) = i(a'(x)f', f) - i(f, a'(x)f').$

Hence the quadratic form (Gf, f) is real and G is symmetric. Since

$$(Hf, f) = (f', f') + (f, f) = \|H^{1/2}f\|^2,$$

we have $\mathcal{D}(H^{1/2}) = W_U^1[0, 2\pi] \subset \mathcal{D}(G)$. Finally, let us prove estimate (2.1). We shall use the inequality

$$(f, f) \geq M^{-2}(a'f, a'f), \quad \text{where} \quad M = \max_{0 \leq x \leq 2\pi} |a'(x)|.$$

Bearing in mind (3.7), we obtain for $\lambda \in \mathbb{R}$

$$
\begin{aligned}
(T(\lambda)f, f) &\geq (f', f') - k^2(f, f) - 2|\lambda \operatorname{Im}(f', a'f)| \\
&\quad + \lambda^2(a'f, a'f) + \lambda^2(f, f) \geq \|f'\|^2 - 2|\lambda| \|f'\| \|a'f\| \\
&\quad + \left(1 + \frac{1}{2}M^{-2}\right) \|a'f\|^2 \lambda^2 + \frac{1}{2} M^{-2} \lambda^2 \|f\|^2 - k^2 \|f\|^2 \\
&\geq \left[1 - \left(1 + \frac{1}{2}M^{-2}\right)^{-1}\right] \|f'\|^2 + \frac{1}{2} M^{-2} \lambda^2 \|f\|^2 - k^2 \|f\|^2 \\
&> (2M^2 + 1)^{-1}[(Hf, f) + \lambda^2(f, f)] - (k^2 + 1)(f, f).
\end{aligned}
$$

This proves the proposition. $\qquad\square$

Proposition 3.2 *The pencil $T(\lambda)$ is regular elliptic and, moreover, strongly regular elliptic.*

Proof: Estimate (2.2) can be obtained by a straightforward calculation of the resolvent kernel of the integral operator $T^{-1}(\lambda)$. First, one has to prove by standard means that there is a pair of solutions of equation (3.3) having the asymptotics

$$f^{\pm}(\lambda, x) = e^{(a(x) \pm ix)\lambda}(1 + O(\lambda^{-1})), \lambda \to \infty,$$

if λ is located in one of the quadrants that are formed by the real and the imaginary axes. Then these solutions have to be substituted in the well-known formulas for the Green function (see [Na, Ch. 1]). A detailed proof of estimates of the type (2.2) for ordinary differential pencils of arbitrary order can be found in the work of Pliev [P].

To prove the strong regularity we recall Proposition 2.5. Since the operator $a''(x)H^{-1/2}$ is compact in the space $\mathfrak{H} = L_2$ (provided $a \in W_2^2$), it suffices to obtain the estimate

(3.8) $$\|F^{-1/2}a'(x)y'\|^2 \leq (1 - \varepsilon)^2 (Hy, y), \quad \varepsilon > 0$$

for functions $y \in \mathcal{D}(H)$. We have

$$|F^{-1/2}a'(x)| = |(1 + a'(x)^2)^{-1/2}a'(x)| < 1 - \varepsilon$$

for some $\varepsilon > 0$. Therefore,

$$\|F^{-1/2}a'(x)y'\|^2 \leq (1 - \varepsilon)^2 \|y'\|^2 = (1 - \varepsilon)^2 (Hy, y).$$

The proposition is proved. $\qquad\square$

Proposition 3.3 *The sign characteristics of the eigenfunctions of $T(\lambda)$ are defined by the relations*

$$\varepsilon_n^\pm = \text{sign}(T'(\pm\lambda_n)f_n^\pm, f_n^\pm) = \text{sign}(\pm\lambda_n), \quad \lambda_n \in \mathbb{R}.$$

Hence the functions $\{f_n^+(x)\}_{n=-\infty}^\infty$ defined in (3.5) form the half of the eigen and associated functions of $T(\lambda)$.

Proof: For $\lambda_n > 0$ we find

$$
\begin{aligned}
(T'(\lambda_n)f_n^+, f_n^+) &= (Gf_n^+, f_n^+) + 2\lambda_n(Ff_n^+, f_n^+) \\
&= -\mu_n(a'f_n^+, f_n^+) + i(a''f_n^+, f_n^+) + 2\lambda_n(f_n^+, f_n^+) \\
&= \mu_n(a', 1) + i(a'', 1) + 4\pi\lambda_n = 4\pi\lambda_n,
\end{aligned}
$$

as the functions a and a' are periodic. $\square$

Theorem 3.4 *The generalized Rayleigh system $\{f_n^+(x)\}_{n=-\infty}^\infty$ defined in (3.5) is minimal and complete in the spaces $W_U^\theta[0, 2\pi]$ if $0 \le \theta \le 3/2$. For any function $g \in W_U^\theta$ the Fourier series*

$$(3.9) \qquad u(x, \eta) = \sum_{n=-\infty}^\infty (g, f_n^*)f_n^+(x)e^{i\lambda_n\eta},$$

converges for $\eta > \eta_0$ in the norm of W_U^θ provided η_0 is large enough (here $\{f_n^*\}$ is the biorthogonal system in $\mathbf{L}_2$ with respect to $\{f_n\}$). Moreover, the function $u(x, \eta)$ admits a holomorphic continuation in a sector $|\arg \eta| < \varepsilon$ for sufficiently small ε, and there exists

$$\text{s}-\lim_{\eta \to +0} u(x, \eta) = g$$

(the limit is understood in the norm of W_2^θ).

Proof: The completeness and the minimality is the consequence of Theorem 2.8. The convergence of series (3.9) follows from representation (2.12) if there is a sequence of semicircles $|\lambda| = r_k \to \infty$ in the upper half-plane such that

$$(3.10) \qquad \|(Z - \lambda)^{-1}\| \le e^{\eta_0|\text{Im}\lambda|} = e^{\eta_0|\sin \varphi|r_k}, \quad |\lambda| = r_k.$$

Let us prove (3.10). Without loss of generality suppose that Z is invertible (equivalently, $T(0)$ is invertible). It follows from Theorem 2.9 that

$$Z = KH^{1/2}, \qquad Z_1 = (H^{1/2} - (k^2 + 1)H^{-1/2})K^{-1}F^{-1},$$

where K and K^{-1} are bounded. Hence

$$(3.11) \qquad (Z - \lambda)^{-1} = T^{-1}(\lambda)(Z_1 - \lambda)F.$$

Recall that the Green function of the integral operator $T^{-1}(\lambda)$ (see Proposition 3.2) is a meromorphic function of order 1 and of finite type. By virtue of the Titchmarsh theorem for any C exceeding the type there is a sequence $r_k \to \infty$ such that $\|T^{-1}(\lambda)\| \leq \exp(C|\lambda|)$ for $|\lambda| = r_k$. Hence estimate (3.10) outside a double sector Λ_{φ_0} containing the real axis follows from (3.11). Since $T(\lambda)$ is strongly elliptic, estimate (3.10) inside a small double sector Λ_{φ_0} follows from [Sh2, Theorem 1.7]. Hence, (3.10) is proved and series (3.9) converges for $\eta > C$. The last assertion of the theorem follows from the fact that the operator Z is a generator of a holomorphic semigroup in the spaces W_U^θ. $\qquad\square$

We remark that series (3.9) does not converge for all $\eta > 0$ and arbitrary functions g, i.e., the system in question does not form a basis for the Abel summability method of order 1. Let us clarify our claim for Rayleigh system (3.6) assuming in addition that $a(x)$ is holomorphic.

Proposition 3.5 *Let* $\lambda_n = \sqrt{k^2 - n^2}$, $f_n(x)$ *be defined by (3.6) and* $a(x)$ *be holomorphic on* $\mathbb{R}$. *If the series*

$$(3.12) \qquad u(x, \eta) = \sum_{n=0}^{\infty} (g, f_n^*) f_n(x) e^{i\lambda_n \eta}$$

converges in L_2 *for all* $\eta > 0$, *then* $g(x)$ *is holomorphic at all points* $x \in (0, \pi)$ *except the points where* $a(x)$ *attains the maximum.*

Proof: Let $M = \max a(x)$. For any $\varepsilon > 0$ we have $\|f_n\| \geq e^{(M-\varepsilon)n}$ for all sufficiently large n. Assuming that series (3.12) converges in L_2 for $\eta = \varepsilon$ we get the estimate $|(g, f_n^*)| < e^{(2\varepsilon - M)n}$ (under our assumption the norms of functions in series (3.12) tend to 0 as $n \to \infty$). If $a(\xi) < M$ and ε is small enough, then there is a neighbourhood U of ξ such that series (3.12) converges uniformly for all $x \in U$ and $0 \leq \eta \leq \varepsilon$. Since the terms of the series are holomorphic functions of x, the sum is also holomorphic at ξ. According to Theorem 3.4 this sum coincides with $g(x)$. Thus $g(x)$ is holomorphic at ξ. $\qquad\square$

Remark 3.6 It follows from the proof of Proposition 3.5 that $g(x)$ admits a holomorphic continuation in the domain

$$\Lambda = \{z \mid Re\, a(z) - M < 0\}$$

if the series (3.12) is summable by the Abel method of order 1 (i.e., converges for all $\eta > 0$). We do not know if the converse assertion is also true.

Let us formulate the basic result of this section.

Theorem 3.7 *There is the only solution* $u(x, y)$ *of scattering problem* (1.2)–(1.4), (1.6). *If* $C_R = \{x, y \mid x^2 + y^2 < R^2\}$ *then* $u(x, y) \in W_{2,loc}^2(\Omega \cap C_R)$ *for any* $R > 0$. *For large* y *the solution* $u(x, y)$ *is represented by the Fourier series with respect to the generalized Rayleigh system* $\{f_n^+(x)\}$.

Proof: It suffices to put $\eta = y - a(x)$ and recall Theorems 2.11 and 3.4.   □

4 Scattering by Two-periodic Surfaces in $\mathbb{R}^3$

Let a smooth function $z = a(x, y)$ be 2π-periodic with respect to both variables x, y. This function defines the surface (the grating) S in $\mathbb{R}^3$. Let

$$v(x, y, z) = e^{-ik(x \cos \alpha + y \cos \beta + z \cos \gamma)}$$

be the wave incident onto this surface with directing vector $\overline{\varphi} = (\cos \alpha, \cos \beta, \cos \gamma)$, $|\overline{\varphi}| = 1$.

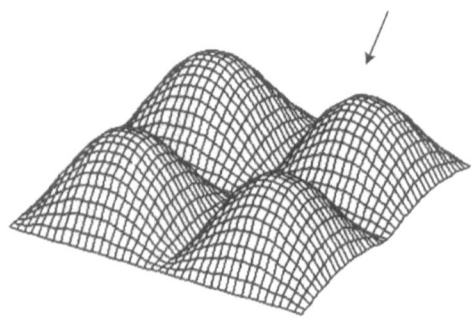

Figure 4

The wave v satisfies the Helmholtz equation

(4.1) $$\Delta u + k^2 u = 0, \qquad u = u(x, y, z)$$

and quasi-periodic conditions

(4.2)
$$u(0, y, z) = e^{-i\nu} u(2\pi, y, z), \quad u'_x(0, y, z) = e^{-i\nu} u'_x(2\pi, y, z),$$
$$u(x, 0, z) = e^{-i\delta} u(x, 2\pi, z), \quad u'_y(x, 0, z) = e^{-i\delta} u'_y(x, 2\pi, z),$$

where $\nu = 2\pi k \cos \alpha$, $\delta = 2\pi k \cos \beta$. Scattered waves do the same. It is assumed that the wave v reaches all the points of the surface (there are no shadows). The problem is to find a quasi-periodic solution of (4.1) that is represented (in some sense) as a superposition of scattered waves and satisfies the "full reflection" condition

(4.3) $$u(x, y, a(x, y)) = v(x, y, a(x, y)).$$

To define the scattered waves let us find all quasi-periodic solutions of equation (4.1) in the cylinder with the base $K_{2\pi} = [0, 2\pi] \times [0, 2\pi]$. Separating the variables x and y, we find that the elementary solutions have the representation

(4.4) $$u^{\pm}_{nj}(x, y, z) = e^{i\mu_n x} e^{i\rho_j y} e^{\pm i\lambda_{nj} z},$$

where

(4.5)
$$\mu_n = \frac{\nu}{2\pi} + n, \quad \rho_j = \frac{\delta}{2\pi} + j,$$
$$\lambda_{nj} = \sqrt{k^2 - \mu_n^2 - \rho_j^2}, \quad n, j \in \mathbb{Z}.$$

The branch of the square root is chosen so that either $\lambda_{nj} \geq 0$ or $\text{Im } \lambda_{nj} > 0$. The set of scattered waves consists of outgoing propagating waves and decaying waves. It coincides with the system $\{u_{nj}^{+}\}$. We can not guarantee that the reflected solution is a finite or infinite superposition of the scattered waves. Therefore, we are looking for solutions satisfying the radiation condition at $z \to \infty$, namely

(4.6)
$$u(x, y, z) = \sum_{\mu_n^2 + \rho_j^2 \leq k^2} c_{nj} e^{i(\mu_n x + \rho_j y + \lambda_{nj} z)} + o(1).$$

Here $o(1) \to 0$ as $z \to \infty$ and c_{nj} are unknown constants to be determined.

The substitution of

$$\xi = x, \quad \eta = y, \quad \zeta = z - a(x, y)$$

maps the half-cylinder

$$\Omega = \{x, y, z \mid 0 \leq x \leq 2\pi, \ 0 \leq y \leq 2\pi, \ a(x, y) \leq z < \infty\}$$

onto a usual half-cylinder whose base is the square $K_{2\pi}$. This substitution transforms equation (4.1) to the form

$$u_{\xi\xi} + u_{\eta\eta} + u_{\zeta\zeta} - 2u_{\xi\zeta} a_{\xi}' - 2u_{\eta\zeta} a_{\eta}'$$
$$+ u_{\zeta\zeta}(a_{\xi}'^{\,2} + a_{\eta}'^{\,2}) - u_{\zeta} a_{\xi}'' - u_{\zeta} a_{\eta}'' + k^2 u = 0.$$

Separating the variable ζ by putting $u = f(\xi, \eta)e^{i\lambda\zeta}$ and taking into account boundary conditions (4.2), we find

(4.7) $$T(\lambda)f = (F\lambda^2 + \lambda G + H - (k^2 + 1)I)f = 0,$$

where the operators H, G, F are defined as follows

$$Hf = -f_{\xi\xi}'' - f_{\eta\eta}'' + f, \qquad\qquad \mathcal{D}(H) = W_U^2[K_{2\pi}],$$
$$Gf = i(2a_{\xi}' f_{\xi}' + 2a_{\eta}' f_{\eta}' + (a_{\xi\xi}'' + a_{\eta\eta}'')f), \qquad \mathcal{D}(G) = W_U^1[K_{2\pi}],$$
$$Ff = (1 + a_{\xi}'^{\,2} + a_{\eta}'^{\,2})f, \qquad\qquad \mathcal{D}(F) = \mathbf{L}_2(K_{2\pi}).$$

Here $W_U^s[K_{2\pi}]$ is the subspace of the Sobolev space $W_2^s[K_{2\pi}]$ consisting of functions subject to the quasiperiodic boundary conditions. It is easily seen that the eigenfunctions of pencil (4.7) coincide with the traces of elementary solutions (4.4) on the surface $z = a(x, y)$, i.e.,

$$(4.8) \qquad f_{nj}^{\pm}(x, y) = e^{\pm i\lambda_{nj} a(x,y)} e^{i\mu_n x} e^{i\rho_j y}.$$

This can be checked independently by a straightforward calculation if one puts in (4.7)

$$f(\xi, \eta) = v(\xi, \eta) e^{i\lambda a(\xi, \eta)}.$$

Then the function v satisfies the equation

$$\Delta v + (k^2 - \lambda^2)v = 0$$

and quasi-periodic boundary conditions. This holds for functions $v_{nj} = e^{i\mu_n x} e^{i\rho_j y}$, where μ_n and ρ_j are defined by (4.5). We remark that the multiplicities of the eigenvalues $\lambda_{nj} = \sqrt{k^2 - \mu_n^2 - \rho_j^2}$ may grow as $k \to \infty$. For example, if $v = \delta = 0$ and $k^2 = 50$, then zero is the eigenvalue of the geometric multiplicity 16 ($\lambda_{nj} = 0$ for $n = \pm 1, \pm 4, \pm 6, \pm 7$ and $j = \pm 7, \pm 6, \pm 4, \pm 1$, respectively) and of algebraic multiplicity 32 (all Jordan chains have length 2 and associated functions coincide with eigenfunctions).

Let us prove that $T(\lambda)$ is an elliptic pencil. The properties

$$H = H^* \gg 0, \quad 0 \ll F \ll \infty, \quad G \subset G^*,$$
$$\mathcal{D}(G) \subset W_U^1(K_{2\pi}) = \mathcal{D}(H^{1/2}),$$

are trivial to check. Further,

$$(T(\lambda)f, f) = \|f_\xi'\|^2 + \|f_\eta'\|^2 - 2(f_\xi', a_\xi' f) - 2\lambda \mathrm{Im} \, (f_\eta', a_\eta' f)$$
$$+ \lambda^2((1 + a_\xi'^2 + a_\eta'^2)f, f) - k^2(f, f) > 0,$$

provided $\lambda^2 > k^2$, $\lambda \in \mathbb{R}$. Since the embedding $I : W_U^2(K_{2\pi}) \to L_2(K_{2\pi})$ is compact, the last estimate implies that $T(\lambda)$ is strongly elliptic. It is important in the sequel to find explicitly a double angle where estimate (2.1) holds.

Proposition 4.1 *Let*

$$M_1 = \max_{x,y \in K_{2\pi}} |a_x'(x, y)|, \qquad M_2 = \max_{x,y \in K_{2\pi}} |a_y'(x, y)|,$$

$$\varphi = \mathrm{arctg} \frac{1}{\sqrt{1 + M_1^2 + M_2^2}}.$$

If $|\arg \pm \lambda| = \theta < \varphi$, *then for* $|\lambda| = r > r_0$ *the following estimate holds*

$$(4.9) \qquad \mathrm{Re} \, (T(\lambda)f, f) \geq \varepsilon(r^2(f, f) + (Hf, f)).$$

Proof: Let $\lambda = re^{i\theta}$ and let α, β be positive numbers such that $\alpha + \beta = 1$. We have

(4.10)
$$Re \; [\| f'_\xi \|^2 - 2re^{i\theta} Im \; (f'_\xi, a'_\xi f) + r^2 e^{2i\theta}(\alpha + a'^2_\xi)(f, f)]$$
$$\geq \varepsilon(\| f'_\xi \|^2 + r^2(f, f))$$

for some $\varepsilon = \varepsilon(\theta)$, provided

(4.11)
$$\cos^2 \theta - \cos 2\theta \left(1 + \frac{\alpha}{M_1^2} \right) < 0.$$

Similarly,

(4.12)
$$Re \; [\| f'_\eta \|^2 - 2re^{i\theta} Im \; (f'_\eta, a'_\eta f) + r^2 e^{2i\theta}(\beta + a'^2_\xi)(f, f)]$$
$$\geq \varepsilon(\| f'_\eta \|^2 + r^2(f, f))$$

provided

(4.13)
$$\cos^2 \theta - \cos 2\theta \left(1 + \frac{\beta}{M_2^2} \right) < 0.$$

The inequalities (4.11) and (4.13) are equivalent to the following ones

$$\mathrm{tg}^2\theta \leq \frac{\alpha}{M_1^2 + \alpha}, \qquad \mathrm{tg}^2\theta < \frac{\beta}{M_2^2 + \beta}.$$

For $\alpha = \frac{M_1^2}{M_1^2 + M_2^2}$, $\beta = \frac{M_2^2}{M_1^2 + M_2^2}$ the last inequalities are equivalent to the condition $|\theta| < \varphi$. Summing inequalities (4.10) and (4.12) with the chosen α and β we obtain estimate (4.9). $\square$

Analyzing the proof of Proposition 4.1 one can understand that the bound for φ is precise, i.e., estimate (4.9) does not hold generally inside the angle $\varphi < \arg \lambda < \pi - \varphi$. Seemingly, $T^{-1}(\lambda)$ has an exponential growth inside this angle. The eigenvalue asymptotics of the Laplace operator on a bounded domain is known, hence, in our problem we have $\lambda_j(H) = cj^p$ with $p = 1$. Therefore, we can guarantee that the pencil $T(\lambda)$ belong to the class K and we can claim (by virtue of Theorem 2.8) the completeness of the traces $\{f^+_{nj}(x, y)\}$ of the scattered waves only in the case $\varphi < \pi/4$. The problem whether the system $\{f^+_{nj}\}$ is complete in the case $\varphi \geq \pi/2$ is open. Nevertheless, the following basic result is true.

Theorem 4.2 *There exists the only solution $u(x, y, z)$ of scattering problem (4.1)–(4.3), (4.6).*

Proof: Elliptic pencil (4.7) is strongly regular. One can prove this fact repeating the arguments of Proposition 2.5. Putting $z = \zeta + a(x, y)$ and recalling Theorem 2.11 we obtain the assertion of the theorem. $\square$

This work is supported by RFBR, grants No. 96-15-96091 and No. 98-01-1000.

References

[A]   H.D. Alber, A quasi-periodic boundary value problem for Laplacian and
           continuation of its resolvent, *Proc. Roy. Soc. Edinburgh. Sect. A. Math.* **82** (1979),
           251–272.

[B]   V.M. Babich, On the existence theorem of solutions of the Dirichlet and the
           Neiman problems for the Helmholtz equation in the quasi-periodical case,
           Siberian Math. Jour. **39** (1988), no. 2, 3–9.

[B1]   V.F. Badyukov, The uniqueness theorem for the scattering problem by a periodical
           grating, *Vestnik Leningrad Univ.* **19** (1977), 88–92 (Russian).

[B2]   V.F. Badyukov, The existence theorem for the scattering problem by a periodical
           grating, *Vestnik Leningrad Univ.* **20** (1978), 81–88 (Russian).

[BMS]   A.G. Belyaev, A.S. Mikheev and A.S. Shamaev, The scattering of a plane wave
           by a rapidly oscillating grating, *Journal of Numer. Math. and Math. Physics* **32**
           (1992), no. 3, 1253–1272 (Russian).

[G]   N.M. Günter, *Potential Theory and its Applications to Basic Problems of Mathe-
           matical Physics*, Ungar, New York, 1967.

[GI]   T.N. Galashnikova and A.S. Il'inskii, *Numerical Methods in Scattering Problems*,
           Nauka, Moscow, 1987 (Russian).

[GF]   I. Gohberg and I.A. Fel'dman, Convolution Equations and Projection Methods
           for Their Solutions, *Transl. Math. Monograph*, vol. **41**, *Amer. Math. Soc.*, R.I.,
           1974.

[GGK]   I. Gohberg, S. Goldberg and M.A. Kaashoek, *Classes of Linear Operators*, vol. **1**.
           Operator theory: Adv. and Appl. **49** (1990).

[GK]   I. Gohberg and N.Ya. Krupnik, *Erfürung in die Theorie der Eindimentionalen
           Singulären Integraloperatoren*, Mathematische Reihe, Band 63, Birkhäuser
           Verlag, Basel, 1979.

[GK1]   I. Gohberg and M.G. Krein, Systems of integral equations on a half line with
           kernels depending on the difference of arguments, *Uspekhi Math. Nauk* 13, **2**
           (80) (1958), 3–72 (Russian); English transl. *Amer. Math. Soc. Transl. (Series 2)*
           14 (1960), 217–287.

[GK2]   I. Gohberg and M.G. Krein, Introduction to the Theory of Linear Nonselfadjoint
           Operators in Hilbert Space, Nauka, Moscow, 1965; English transl. by *Amer.
           Math. Soc.*, Providence, RI, 1969.

[GLR]   I. Gohberg, P. Lancaster and L. Rodman, Matrices and Indefinite Scalar Products,
           Operator theory: Adv. and Appl. **8** (1983).

[IM]   A.S. Il'inskii and A.S. Mikheev, Scattering of waves by periodic grating, *Vest-
           nik Moscow Univ., Ser. 15, Numer. Math. and Cybernetics*, 1990, no. 1, 35–39
           (Russian).

[Ka]   T. Kato, *Perturbation Theory for Linear Operators* (2-nd edition), Springer-
           Verlag, New York, 1976.

[Ke]   M.V. Keldysh, On the completeness of eigenfunction of certain classes of
           nonselfadjoint linear operators, *Russian Math. Surveys* **26** (1971), no. 4,
           295–305.

[KL]   M.G. Krein and G.K. Langer [H. Langer], On some mathematical principles
           in the linear theory of damped oscillations of continua, *Appl. Theory of Functions
           in Continuum Mech.* (Proc. Internat. Sympos. Tbilisi , 1963) vol. II: Fluid and

Gas Mech., Math. Methods, Nauka, Moscow, 1965, 283–322; English transl., Parts I, II: *Integral Equations and Operator Theory* **1** (1978), 364–399, 539–566.

[KO]   A.G. Kostyuchenko and M.B. Orasov, On certain properties of the roots of the selfadjoint quadratic pencil, *J. Funct. Anal. and Appl.* **9** (1975), 28–40.

[KS]   A.G. Kostyuchenko and A.A. Shkalikov, Selfadjoint quadratic operator pencils and elliptic problems, *J. Funct. Anal. and Appl.* **17** (1983), 109–128.

[L]   H. Langer, Factorization of operator pencils, *Acta Scient. Math. Szeged* **38** (1976), 83–96.

[LM]   J.L. Lions and E. Magenes, *Problems aux Limites Nonhomogenes et Applications*, vol. **1**, Dunod Paris, 1968; English transl. in Springer-Verlag.

[M]   A.S. Markus, Introduction to the Spectral Theory of Polynomial Operator Pencils, *Amer. Math. Soc.*, Providence, 1988.

[Na]   M.A. Naimark, *Linear Differential Operators, I, II*, Frederick Ungar Publishing Company, New York, 1967.

[NP]   S.A. Nazarov and B.A. Plamenevskii, *Elliptic Operators in Non-Smooth Domains*, Nauka, Moscow, 1992.

[P]   V.T. Pliev, Problems on Completeness and Basisness in Operator Pencil Theory, *PHD dissertation*, Moscow, MSU, 1990.

[R1]   J.W. Rayleigh, *The Theory of Sound*, vol. 2, Macmillan, London-New York, 1896; 2-nd Ed., Dover, 1945.

[R2]   J.W. Rayleigh, On the dynamical theory of gratings, *Proc. Roy. Soc., Ser A* **79** (1907), 399–416.

[RR]   M. Rosenblum and J. Rovnjak, *Hardy Classes and Operator Theory*, Oxford Univ. Press, New York - Oxford, 1985.

[Sh1]   A.A. Shkalikov, Elliptic equations in Hilbert space and associated spectral problems, *J. Soviet Math.* **51** (1990), no. 4, 2399–2467.

[Sh2]   A.A. Shkalikov, Factorization of elliptic pencils and the Mandelstam hypothesis, *Operator theory: Adv. and Appl.* **106** (1998), Birkhäuser Verlag.

[Sh3]   A.A. Shkalikov, Operator equations in Hilbert space with dissipative symbols. *Selected Rus. Math.* **1** (1999).

[W1]   C.H. Wilcox, *Scattering Theory of the D'Alambert Equation in Exterior Domains*, Springer Verlag, Berlin, 1975. (Lecture Notes in Math., vol. **442**.)

[W2]   C.H. Wilcox, *Scattering Theory for Diffraction Gratings*, Springer Verlag, New York, 1984.

[WG]   C.H. Wilcox and J.C. Guilliot, Scattering theory for acoustic diffraction gratings — preliminary report, *Notices AMS* **25**, A356 (1978).

A.G. Kostyuchenko and A.A. Shkalikov
Department of Mechanics and Mathematics
Moscow Lomonosov University
Moscow
119899 Russia

Operator Theory:
Advances and Applications, Vol. 124
© 2001 Birkhäuser Verlag Basel/Switzerland

Corners of Numerical Ranges

Heinz Langer, Alexander Markus and Christiane Tretter

Dedicated with admiration to our teacher, colleague and friend Israel Gohberg
on the occasion of his 70th birthday

It is well-known that a corner of the numerical range of a bounded linear operator T in a Hilbert
space belongs to the spectrum $\sigma(T)$. In this paper we prove corresponding results for the corners
of the numerical range of an analytic operator function and of the quadratic numerical range of
a 2×2 block operator matrix.

1 Introduction

A boundary point α of a subset W of the complex plane $\mathbb{C}$ is called a *corner* of W
if for some $\varepsilon > 0$ and $\varphi \in [0, 2\pi)$ there exists a $\psi \in [0, \pi)$ such that

$$\varphi \leq \arg(\lambda - \alpha) \leq \varphi + \psi$$

for all $\lambda \in W$, $|\lambda - \alpha| < \varepsilon$, where $\arg(\cdot)$ is suitably defined. The infimum ψ_0 of
all possible ψ (for all ε and ψ) is called the *angle* of the corner α.

For a bounded linear operator T in a Hilbert space it is well-known that a corner
λ_0 of the numerical range $W(T)$ which belongs to $W(T)$ is an eigenvalue of T.
Moreover, any corner of $W(T)$ belongs to the spectrum of T (see [K], [HJ], [GR]).
In this paper we consider the corners of the numerical range $W(\mathcal{L})$ of an analytic
operator function

$$\mathcal{L} : \mathcal{D} \to L(\mathcal{H})$$

defined on a domain $\mathcal{D} \subset \mathbb{C}$ with values in the set $L(\mathcal{H})$ of all bounded linear
operators in a Hilbert space $\mathcal{H}$, and the corners of the quadratic numerical range
$W^2(\mathcal{A})$ of a block operator matrix $\mathcal{A} \in L(\mathcal{H})$,

$$\mathcal{A} = \begin{pmatrix} A & B \\ C & D \end{pmatrix} : \mathcal{H}_1 \times \mathcal{H}_2 \to \mathcal{H}_1 \times \mathcal{H}_2$$

with respect to a given decomposition $\mathcal{H} = \mathcal{H}_1 \times \mathcal{H}_2$.

We recall that a point $\lambda_0 \in \mathcal{D}$ belongs to the *numerical range* $W(\mathcal{L})$ of $\mathcal{L}$ if λ_0
is a root of

$$(1.1) \qquad\qquad (\mathcal{L}(\lambda)f, f) = 0$$

for some $f \in \mathcal{H}$, $\|f\| = 1$ (see [M]). For the particular case $\mathcal{L}(\lambda) = T - \lambda I$, $\lambda \in \mathbb{C}$, with some bounded linear operator $T \in L(\mathcal{H})$, the numerical range $W(\mathcal{L})$ of $\mathcal{L}$ coincides with the usual numerical range $W(T)$ of T. A point $\lambda_0 \in \mathbb{C}$ belongs to the *quadratic numerical range* $W^2(\mathcal{A})$ of $\mathcal{A}$ if λ_0 is a root of

$$\det \begin{pmatrix} (Ax, x) - \lambda & (By, x) \\ (Cx, y) & (Dy, y) - \lambda \end{pmatrix} = 0$$

for some $x \in \mathcal{H}_1$, $y \in \mathcal{H}_2$, $\|x\| = \|y\| = 1$ (see [LT], [LMMT]). We recall that the quadratic numerical range is contained in the numerical range and its closure contains the spectrum.

In Section 2 we show that each point $\lambda_0 \in W(\mathcal{L})$ which is a corner of $W(\mathcal{L})$ and a simple root of the equation (1.1) is an eigenvalue of the operator function $\mathcal{L}$ (for the case of a matrix polynomial this was proved in [MP]). The same is true if $\lambda_0 \in W(\mathcal{L})$ is a corner of $W(\mathcal{L})$ with angle $< \pi/l$ for some $l \in \mathbb{N}$ and a root of multiplicity at most l of (1.1). We also prove that if a corner λ_0 of $W(\mathcal{L})$ belongs only to the closure of $W(\mathcal{L})$ and to $\mathcal{D}$, then, under certain additional assumptions, λ_0 belongs to the spectrum of $\mathcal{L}$.

In Section 3 we prove that if a point $\lambda_0 \in W^2(\mathcal{A})$ is a corner of $W^2(\mathcal{A})$, then λ_0 is either an eigenvalue of the block operator matrix $\mathcal{A}$ or an eigenvalue of one of the diagonal entries A and D. If λ_0 is not attained, i.e., it lies only in the closure of $W^2(\mathcal{A})$, then λ_0 belongs either to the spectrum of $\mathcal{A}$ or to the spectrum of one of the diagonal operators A or D. We mention that a corner of the quadratic numerical range need not be a corner of the numerical range (see Figure 3 and Figures 3, 4 in [LMMT]).

2 Corners of the Numerical Range of an Analytic Operator Function

2.1 In this subsection we consider corners of the numerical range $W(\mathcal{L})$ of an analytic operator function $\mathcal{L} : \mathcal{D} \to L(\mathcal{H})$ which belong to $W(\mathcal{L})$.

Theorem 2.1 *If $\lambda_0 \in W(\mathcal{L})$ is a corner of $W(\mathcal{L})$ and a simple zero of the function $(\mathcal{L}(\cdot) f_0, f_0)$ for some $f_0 \in \mathcal{H}$, $\|f_0\| = 1$, then λ_0 is an eigenvalue of $\mathcal{L}$ with eigenvector f_0.*

Proof: Without loss of generality we assume $\lambda_0 = 0$. For $f \in \mathcal{H}$, $\|f\| = 1$, the function

$$g_f(\lambda, z) := (\mathcal{L}(\lambda)(f_0 + zf), f_0 + \bar{z}f), \quad \lambda \in \mathcal{D}, \ z \in \mathbb{C},$$

is analytic in λ and z with $g_f(0, 0) = 0$, and $g_f(\,\cdot\,, 0)$ has a zero of first order at $\lambda = 0$ due to the assumption. By the Implicit Function Theorem (see e.g. [Ma], Theorem 3.11), the function g_f admits a representation

$$g_f(\lambda, z) = (\lambda - \lambda_f(z)) k(\lambda, z)$$

where λ_f is analytic in a neighbourhood U of 0, $\lambda_f(0) = 0$, and k is analytic and does not vanish in a neighbourhood of $(0, 0)$. Hence for real $t \in U$

$$0 = g_f(\lambda_f(t), t) = (\mathcal{L}(\lambda_f(t))(f_0 + tf), f_0 + tf)$$

and consequently $\lambda_f(t) \in W(\mathcal{L})$. The condition that 0 is a corner of $W(\mathcal{L})$ implies that the curve $\lambda_f(t)$, $t \in U \cap \mathbb{R}$, does not have a tangent in the point 0 which yields

$$\left. \frac{d}{dt}\lambda_f(t) \right|_{t=0} = 0.$$

On the other hand,

$$\begin{aligned} 0 = {} & \frac{d}{dt}g_f(\lambda_f(t), t) = (\mathcal{L}'(\lambda_f(t))(f_0 + tf), f_0 + tf)\frac{d}{dt}\lambda_f(t) \\ & + (\mathcal{L}(\lambda_f(t))f, f_0 + tf) + (\mathcal{L}(\lambda_f(t))(f_0 + tf), f) \end{aligned}$$

and hence for $t = 0$

$$(f, \mathcal{L}(0)^* f_0) + (\mathcal{L}(0) f_0, f) = 0.$$

Since $f \in \mathcal{H}$, $\|f\| = 1$, was arbitrary, the above relation also holds with if instead of f. This implies $\mathcal{L}(0) f_0 = 0$ and $\mathcal{L}(0)^* f_0 = 0$. $\qquad\square$

Remark 2.2 Under the assumptions of Theorem 2.1, we also have $\mathcal{L}(\lambda_0)^* f_0 = 0$.

An analogous result for the special case that $\mathcal{H}$ is finite dimensional and $\mathcal{L}$ is a matrix polynomial has been proved in [MP].

For the particular case that $\mathcal{L}(\lambda) = T - \lambda I$, $\lambda \in \mathbb{C}$, with some bounded linear operator $T \in L(\mathcal{H})$, we obtain the following well-known result (for the finite dimensional case see [K], [HJ], Theorem 1.6.3, and for the general case [GR]).

Corollary 2.3 *If $\lambda_0 \in W(T)$, $\lambda_0 = (Tf_0, f_0)$ for some $f_0 \in \mathcal{H}$, $\|f_0\| = 1$, is a corner of $W(T)$, then λ_0 is an eigenvalue of T with eigenvector f_0 and $\bar{\lambda}_0$ is an eigenvalue of T^* with the same eigenvector f_0.*

The next two lemmata were proved in [MP] for the case $\dim \mathcal{H} < \infty$ by another method. In the following, we denote by $\Re(\,\cdot\,)$ the real part, by $\Im(\,\cdot\,)$ the imaginary part of a complex number or operator, and by $\mathrm{Arg}(\,\cdot\,)$ the argument function with values in $(-\pi, \pi]$.

Lemma 2.4 *Let $A = A^* \in L(\mathcal{H})$ and $(Af_0, f_0) = 0$ for some $f_0 \in \mathcal{H}$. If there exists a $\delta > 0$ such that*

$$(A(f_0 + h), f_0 + h) \geq 0, \qquad h \in \mathcal{H}, \ \|h\| < \delta,$$

then $A \geq 0$.

Proof: By the assumptions,

$$(Af_0, h) + (Ah, f_0) + (Ah, h) \geq 0, \quad h \in \mathcal{H}, \ \|h\| < \delta.$$

Let $g \in \mathcal{H}$, $\|g\| = 1$, be arbitrary and set $h := tg$ with $t \in (-\delta, \delta)$. Then

$$(2.1) \qquad\qquad 2t \, \Re(Af_0, g) + t^2(Ag, g) \geq 0.$$

Hence

$$2\Re(Af_0, g) + t(Ag, g) \begin{cases} \geq 0, & 0 < t < \delta, \\ \leq 0, & -\delta < t < 0. \end{cases}$$

If $t \to 0$, we obtain

$$(2.2) \qquad\qquad \Re(Af_0, g) = 0,$$

and (2.1) yields $(Ag, g) \geq 0$. $\qquad\qquad\qquad\qquad\qquad\qquad\qquad\qquad\square$

Lemma 2.5 *Let $A \in L(\mathcal{H})$ and $(Af_0, f_0) = 0$ for some $f_0 \in \mathcal{H}$, $f_0 \neq 0$. If there exist $\delta > 0$ and α, β such that $\beta - \alpha < \pi$ and*

$$\alpha \leq \operatorname{Arg}(A(f_0 + h), f_0 + h) \leq \beta, \qquad h \in \mathcal{H}, \ \|h\| < \delta,$$

then

$$\alpha \leq \operatorname{Arg}(Ag, g) \leq \beta, \qquad g \in \mathcal{H}.$$

Proof: Apply Lemma 2.4 to the operators $\Im(e^{-i\alpha} A)$ and $\Im(e^{i(\pi - \beta)} A)$. $\qquad\square$

Remark 2.6 Under the assumptions of Lemma 2.4 and Lemma 2.5 we have $Af_0 = 0$. This relation remains true if these assumptions are fulfilled only for elements h which are scalar multiples of Af_0.

Theorem 2.7 *If $\lambda_0 \in W(\mathcal{L})$ is a corner of $W(\mathcal{L})$ with angle $< \pi / l$ for some integer $l \geq 1$ and λ_0 is a zero of multiplicity k with $k \leq l$ of the function $(\mathcal{L}(\cdot) f_0, f_0)$ for some $f_0 \in \mathcal{H}$, $\|f_0\| = 1$, then λ_0 is an eigenvalue of $\mathcal{L}$ with eigenvector f_0.*

Proof: In the following, without loss of generality we suppose that for some $r' > 0$, the points $\lambda \in W(\mathcal{L})$ with $|\lambda - \lambda_0| < r'$ satisfy $0 \leq \mathrm{Arg}(\lambda - \lambda_0) \leq \psi_0 < \pi/l$. By assumption, the function

$$u(\lambda) := \frac{(\mathcal{L}(\lambda) f_0, f_0)}{(\lambda - \lambda_0)^k}$$

is analytic and $u(\lambda) \neq 0$ for $|\lambda - \lambda_0| < r$ with some $r > 0$, $r < r'$. It follows from Rouché's Theorem that for any $\varepsilon > 0$ there exists a $\delta > 0$ such that for all $f \in \mathcal{H}$, $\|f - f_0\| < \delta$, the function $(\mathcal{L}(\cdot)f, f)$ has exactly k roots (counted with multiplicities), say $\{\lambda_j(f)\}_1^k$, in the circle $\{\lambda \in \mathbb{C} : |\lambda - \lambda_0| < r\}$, and the function

$$u_f(\lambda) := \frac{(\mathcal{L}(\lambda) f, f)}{\prod_{j=1}^k (\lambda - \lambda_j(f))}$$

satisfies the condition

(2.3)
$$\max_{|\lambda - \lambda_0| \leq r} \left| \frac{u_f(\lambda)}{u(\lambda)} - 1 \right| < \varepsilon.$$

Note that the enumeration of $\lambda_j(f)$ is not essential, and the functions λ_j may be discontinuous.

In the following we are going to prove that all the values of $(\mathcal{L}(\lambda_0)f, f)$ for $f \in \mathcal{H}$, $\|f - f_0\| < \gamma$ with some $\gamma > 0$, are contained in an angle (cone) with vertex in 0. If $(\mathcal{L}(\lambda_0)f, f) = 0$ (i.e. $\lambda_j(f) = \lambda_0$ for some $j \in \{1, 2, \ldots, k\}$), this is clear. Now suppose $(\mathcal{L}(\lambda_0)f, f) \neq 0$. We have

(2.4)
$$\frac{(\mathcal{L}(\lambda_0)f, f)}{(-1)^k u_f(\lambda_0)} = \prod_{j=1}^k (\lambda_j(f) - \lambda_0).$$

Since $\lambda_j(f) \in W(\mathcal{L})$ and $|\lambda_j(f) - \lambda_0| < r$, we obtain

$$0 \leq \mathrm{Arg}\,(\lambda_j(f) - \lambda_0) \leq \psi_0$$

and hence

(2.5)
$$0 \leq \mathrm{Arg} \prod_{j=1}^k (\lambda_j(f) - \lambda_0) \leq k\psi_0.$$

If ε in (2.3) is small enough, we can assume that

(2.6)
$$\left| \mathrm{Arg}\left(\frac{u_f(\lambda_0)}{u(\lambda_0)} \right) \right| < \frac{\pi - k\psi_0}{3}$$

for all $f \in \mathcal{H}$, $\|f - f_0\| < \delta$. From (2.4)–(2.6) and the relation

$$\mathrm{Arg}\left(\frac{(\mathcal{L}(\lambda_0)f, f)}{(-1)^k u(\lambda_0)}\right) = \mathrm{Arg}\left(\frac{(\mathcal{L}(\lambda_0)f, f)}{(-1)^k u_f(\lambda_0)}\right) + \mathrm{Arg}\left(\frac{u_f(\lambda_0)}{u(\lambda_0)}\right)$$

we find that for all $f \in \mathcal{H}$, $\|f - f_0\| < \delta$, with $\omega := (-1)^k u(\lambda_0)^{-1}(\neq 0)$,

$$(2.7) \quad \alpha := -\frac{\pi - k\psi_0}{3} \leq \mathrm{Arg}((\omega\mathcal{L}(\lambda_0)f, f)) \leq k\psi_0 + \frac{\pi - k\psi_0}{3} =: \beta.$$

Since

$$(2.8) \qquad \beta - \alpha = \frac{2}{3}\pi + \frac{1}{3}k\psi_0 \leq \frac{2}{3}\pi + \frac{1}{3}l\psi_0 < \pi$$

and $(\mathcal{L}(\lambda_0)f_0, f_0) = 0$, the assumptions of Lemma 2.5 are fulfilled for $A = \omega\mathcal{L}(\lambda_0)$, and hence $\mathcal{L}(\lambda_0)f_0 = 0$ according to Remark 2.6. $\square$

The proof of Theorem 2.7 follows ideas from [MP] where an analogous result for the special case that $\mathcal{H}$ is finite dimensional and $\mathcal{L}$ is a matrix polynomial has been proved. We mention that Theorem 2.7 can also be obtained along the lines of the proof of Theorem 2.1 using Puiseux series. Moreover, the previous Theorem 2.1 is a particular case of Theorem 2.7. Nevertheless, we have given another proof of Theorem 2.1 because these ideas will be used again in the next section.

From Lemma 2.5 and from (2.7), (2.8) it follows that under the assumptions of Theorem 2.7 the point $0 \in W(\mathcal{L}(\lambda_0))$ (the usual numerical range of $\mathcal{L}(\lambda_0)$) is a corner of $W(\mathcal{L}(\lambda_0))$. Therefore the statement of Theorem 2.7 also follows using [GR], Theorem 1.5.5.

In the following we illustrate Theorem 2.1 by two examples of quadratic 2×2 matrix polynomials. First we consider the quadratic pencil

$$\mathcal{L}_1(\lambda) = \lambda^2 I_2 + \lambda \begin{pmatrix} -i & -5i \\ 5i & -i \end{pmatrix} + \begin{pmatrix} 2 & 1 \\ 1 & 2 \end{pmatrix}, \quad \lambda \in \mathbb{C},$$

where I_2 denotes the 2×2 unit matrix. The numerical range of $\mathcal{L}_1$ (produced by a C^{++}-code analogous to the one already used in [LMMT] for calculating numerical ranges and quadratic numerical ranges of matrices) is shown in Figure 1 on the left. Its corners are the eigenvalues of $\mathcal{L}_1$ (which are marked by $\circ$).

The second example is the quadratic pencil

$$\mathcal{L}_2(\lambda) = \lambda^2 I_2 + \lambda \begin{pmatrix} 0 & 2.8i \\ -2.8i & 0 \end{pmatrix} + \begin{pmatrix} 2 & 1 \\ 1 & 2 \end{pmatrix}, \quad \lambda \in \mathbb{C},$$

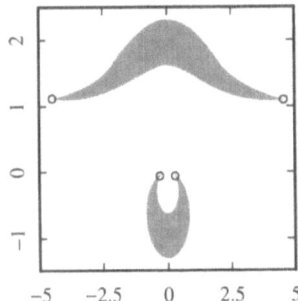 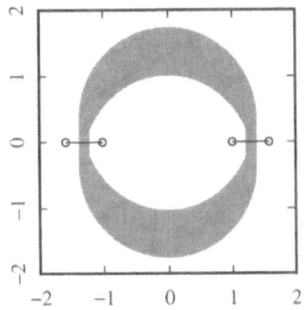

Figure 1 Quadratic numerical ranges of $\mathcal{L}_1$ and $\mathcal{L}_2$

which has been considered also in [LMZ]. The corners of its numerical range displayed in Figure 1 on the right (which are the end points of the segments $W(\mathcal{L}_2) \cap \mathbb{R}$) are the eigenvalues of $\mathcal{L}_2$.

2.2 In this subsection we consider corners of the numerical range $W(\mathcal{L})$ which do not belong to $W(\mathcal{L})$, but only to its closure. We always suppose that there exists a point $\mu_0 \in \mathcal{D}$ and a constant $\delta > 0$ such that

$$(2.9) \qquad |(\mathcal{L}(\mu_0)f, f)| \geq \delta \|f\|^2, \qquad f \in \mathcal{H}.$$

This assumption is equivalent to $0 \notin \overline{W(\mathcal{L}(\mu_0))}$ and implies that $\sigma(\mathcal{L}) \subset \overline{W(\mathcal{L})}$ (see [M], §26.3).

Lemma 2.8 *If the analytic operator function $\mathcal{L}$ on $\mathcal{D}$ satisfies the assumption (2.9), then, for $\lambda \in \mathcal{D}$,*

$$\lambda \in \overline{W(\mathcal{L})} \quad \Longleftrightarrow \quad 0 \in \overline{W(\mathcal{L}(\lambda))}.$$

Proof: If $\lambda \in \overline{W(\mathcal{L})} \cap \mathcal{D}$, there exist sequences $(f_n)_1^\infty \subset \mathcal{H}$, $\|f_n\| = 1$, and $(\lambda_n)_1^\infty \subset \mathbb{C}$ such that $\lambda_n \to \lambda$, $n \to \infty$, and $(\mathcal{L}(\lambda_n)f_n, f_n) = 0$, $n \in \mathbb{N}$. Obviously,

$$|(\mathcal{L}(\lambda)f_n, f_n)| = |(\mathcal{L}(\lambda)f_n, f_n) - (\mathcal{L}(\lambda_n)f_n, f_n)|$$
$$\leq \|\mathcal{L}(\lambda) - \mathcal{L}(\lambda_n)\| \to 0, \quad n \to \infty,$$

and hence $0 \in \overline{W(\mathcal{L}(\lambda))}$.

Conversely, if $0 \in \overline{W(\mathcal{L}(\lambda))}$, there exists a sequence $(f_n)_1^\infty \subset \mathcal{H}$, $\|f_n\| = 1$, with $(\mathcal{L}(\lambda)f_n, f_n) \to 0$, $n \to \infty$. Then the sequence of the functions $\varphi_n := (\mathcal{L}(\cdot)f_n, f_n)$ is bounded on compact subsets of $\mathcal{D}$, and by the Theorem of Montel (see e.g. [T], Chapter V, 5.23) we can assume without loss of generality that

φ_n converges uniformly on compact subsets of $\mathcal{D}$ to some analytic function φ. The condition $(\mathcal{L}(\lambda)f_n, f_n) \to 0$, $n \to \infty$, implies $\varphi(\lambda) = 0$, and from the assumption (2.9), it follows that $\varphi \not\equiv 0$. Now Hurwitz' Theorem (see e.g. [T], Chapter III, 3.45) yields the existence of a sequence $(\lambda_n)_1^\infty \subset \mathbb{C}$ such that $\lambda_n \to \lambda$, $n \to \infty$, and $(\mathcal{L}(\lambda_n)f_n, f_n) = 0$, $n \in \mathbb{N}$. □

Theorem 2.9 *Suppose that the condition* (2.9) *holds. If $\lambda_0 \in \overline{W(\mathcal{L})} \cap \mathcal{D}$ is a corner of $W(\mathcal{L})$ with angle $< \pi/l$ for some integer $l \geq 1$ and if there exist a neighbourhood V of λ_0 and a sequence $(f_n)_1^\infty \subset \mathcal{H}$, $\|f_n\| = 1$, such that the functions $(\mathcal{L}(\cdot)f_n, f_n)$ have k zeros in V, $k \leq l$, and $(\mathcal{L}(\lambda_n)f_n, f_n) = 0$ for a sequence $(\lambda_n)_1^\infty \subset \mathcal{D}$ with $\lambda_n \to \lambda_0$, $n \to \infty$, then $\lambda_0 \in \sigma(\mathcal{L})$.*

Proof: We use a well-known method based on the notion of Banach limits (see [B]). Fix a Banach limit LIM, and let $\mathcal{R}$ be the linear space of all bounded sequences $x = (x_n)_1^\infty$ in $\mathcal{H}$ with the (non-negative, but degenerate) inner product

$$[x, y] := \text{LIM} (x_n, y_n).$$

Let $\mathcal{R}_0$ be the subspace of all $x = (x_n)_1^\infty \in \mathcal{R}$ such that $\text{LIM} (x_n, x_n) = 0$, and let $\widetilde{\mathcal{H}}$ be the completion of the quotient space $\mathcal{R}/\mathcal{R}_0$ with respect to the norm generated by the inner product $[\cdot, \cdot]$. For any operator $T \in L(\mathcal{H})$, we define an operator $\widetilde{T} \in L(\widetilde{\mathcal{H}})$ by $\widetilde{T}\widetilde{x} := (Tx_n)_1^\infty$, $\widetilde{x} = (x_n)_1^\infty \in \widetilde{\mathcal{H}}$. The mapping $T \mapsto \widetilde{T}$ is an isometry from $L(\mathcal{H})$ into $L(\widetilde{\mathcal{H}})$, and for the given analytic operator function $\mathcal{L}$ we consider the corresponding analytic operator function $\widetilde{\mathcal{L}}$ given by $\widetilde{\mathcal{L}}(\lambda) := \widetilde{\mathcal{L}(\lambda)}$, $\lambda \in \mathcal{D}$.

Next we are going to show that $\overline{W(\mathcal{L})} = W(\widetilde{\mathcal{L}})$. Indeed, according to Lemma 2.8 and the relation $\overline{W(\mathcal{L}(\lambda))} = W(\widetilde{\mathcal{L}(\lambda)})$ (see [BO]), we obtain, for $\lambda \in \mathcal{D}$,

$$\lambda \in \overline{W(\mathcal{L})} \iff 0 \in \overline{W(\mathcal{L}(\lambda))} \iff 0 \in W(\widetilde{\mathcal{L}(\lambda)}) \iff \lambda \in W(\widetilde{\mathcal{L}}).$$

Hence λ_0 is also a corner of $W(\widetilde{\mathcal{L}})$ with angle $< \pi/l$.

As in the proof of Lemma 2.8, the sequence $((\mathcal{L}(\cdot)f_n, f_n))_1^\infty$ of analytic functions is bounded on compact subsets of $\mathcal{D}$, and by the Theorem of Montel it contains a subsequence converging uniformly on compact subsets of $\mathcal{D}$ to some analytic function φ. The condition $(\mathcal{L}(\lambda_n)f_n, f_n) = 0$ implies $\varphi(\lambda_0) = 0$, and from the condition (2.9) it follows that $\varphi \not\equiv 0$. By Hurwitz' Theorem, we infer that the multiplicity of the zero λ_0 of φ is k. It is easy to see that $\varphi(\lambda) = [\widetilde{\mathcal{L}}(\lambda)\widetilde{f}, \widetilde{f}]$ where $\widetilde{f} = (f_n)_1^\infty$. Now Theorem 2.7 applies to the operator function $\widetilde{\mathcal{L}}$ which yields that λ_0 is an eigenvalue of $\widetilde{\mathcal{L}}$. The assertion then follows from the equivalences

$$\lambda_0 \in \sigma_p(\widetilde{\mathcal{L}}) \iff 0 \in \sigma_p(\widetilde{\mathcal{L}(\lambda_0)}) \iff 0 \in \sigma_{app}(\mathcal{L}(\lambda_0))$$
$$\iff \lambda_0 \in \sigma_{app}(\mathcal{L}).$$

Here σ_p denotes the point spectrum and σ_{app} the approximate point spectrum. The middle equivalence follows from [B]. □

For the particular case of the numerical range of a bounded linear operator $T \in L(\mathcal{H})$, we obtain the following statement, which has been shown e.g. in [GR].

Corollary 2.10 *If $\lambda_0 \in \overline{W(T)}$ is a corner of $W(T)$, then $\lambda_0 \in \sigma(T)$.*

3 Corners of the Quadratic Numerical Range

3.1 Let $\mathcal{A}$ be the block operator matrix in $\mathcal{H} = \mathcal{H}_1 \times \mathcal{H}_2$ given by

$$\mathcal{A} = \begin{pmatrix} A & B \\ C & D \end{pmatrix} : \mathcal{H}_1 \times \mathcal{H}_2 \to \mathcal{H}_1 \times \mathcal{H}_2.$$

In this subsection we consider corners of the quadratic numerical range $W^2(\mathcal{A})$ of $\mathcal{A}$ which belong to $W^2(\mathcal{A})$.

Theorem 3.1 *Let $\lambda_0 \in W^2(\mathcal{A})$ be a corner of $W^2(\mathcal{A})$. If λ_0 is a zero of the function*

$$(3.1) \qquad \Delta(x_0, y_0; \lambda) := \det \begin{pmatrix} (Ax_0, x_0) - \lambda & (By_0, x_0) \\ (Cx_0, y_0) & (Dy_0, y_0) - \lambda \end{pmatrix}$$

for some $x_0 \in \mathcal{H}_1$, $y_0 \in \mathcal{H}_2$, $\|x_0\| = \|y_0\| = 1$, then at least one of the following statements i), ii) *or* iii) *holds:*

i) *λ_0 is an eigenvalue of A with eigenvector x_0,*

ii) *λ_0 is an eigenvalue of D with eigenvector y_0,*

iii) *λ_0 is an eigenvalue of $\mathcal{A}$ with eigenvector $\begin{pmatrix} x_0 \\ \gamma y_0 \end{pmatrix}$ where*

$$\gamma = -\frac{(Cx_0, y_0)}{((D - \lambda_0)y_0, y_0)} \quad or \quad \gamma = -\frac{((A - \lambda_0)x_0, x_0)}{(By_0, x_0)}.$$

Proof: Without loss of generality we assume $\lambda_0 = 0$. First we consider the case of a simple zero. For $y \in \mathcal{H}_2$, $\|y\| = 1$, and $z \in \mathbb{C}$, the quadratic polynomial in λ

$$g_y(\lambda, z) := ((Ax_0, x_0) - \lambda)((D(y_0 + zy), y_0 + \bar{z}y) - \lambda(y_0 + zy, y_0 + \bar{z}y))$$
$$- (B(y_0 + zy), x_0)(Cx_0, y_0 + \bar{z}y),$$

has a root $\lambda_y(z)$ such that λ_y is analytic in a neighbourhood U of 0 and $\lambda_y(0) = 0$. We can represent $\lambda_y(z)$ in the form

$$\lambda_y(z) = \frac{1}{2}\left((Ax_0, x_0) + \frac{(D(y_0 + zy), y_0 + \bar{z}y)}{(y_0 + zy, y_0 + \bar{z}y)}\right.$$

$$\left. + \sqrt{\frac{1}{4}\left((Ax_0, x_0) - \frac{(D(y_0 + zy), y_0 + \bar{z}y)}{(y_0 + zy, y_0 + \bar{z}y)}\right)^2 + \frac{(B(y_0 + zy), x_0)(Cx_0, y_0 + \bar{z}y)}{(y_0 + zy, y_0 + \bar{z}y)}}\right.$$

(3.2)

where the branch of the square root is chosen such that $\lambda_y(0) = 0$. Obviously, $\lambda_y(t) \in W^2(\mathcal{A})$ for real $t \in U$. We proceed as in the proof of Theorem 2.1 and obtain

$$\frac{d}{dt}\lambda_y(t)\bigg|_{t=0} = 0.$$

Then

$$0 = \frac{d}{dt}g_y(\lambda_f(t), t)$$

$$= -\frac{d}{dt}\lambda_y(t)((D(y_0 + ty), y_0 + ty) - \lambda_y(t)(y_0 + ty, y_0 + ty))$$

$$+ ((Ax_0, x_0) - \lambda_y(t))((Dy, y_0 + ty) + (D(y_0 + ty), y))$$

$$- \frac{d}{dt}\lambda_y(t)(y_0 + ty, y_0 + ty) - \lambda_y(t)((y, y_0 + ty) + (y_0 + ty, y)))$$

$$- (By, x_0)(Cx_0, y_0 + ty) - (B(y_0 + ty), x_0)(Cx_0, y),$$

and hence for $t = 0$

$$0 = (Ax_0, x_0)((Dy, y_0) + (Dy_0, y)) - (By, x_0)$$

$$(Cx_0, y_0) - (By_0, x_0)(Cx_0, y)$$

$$= (y, \overline{(Ax_0, x_0)}D^*y_0 - \overline{(Cx_0, y_0)}B^*x_0)$$

$$+ ((Ax_0, x_0)Dy_0 - (By_0, x_0)Cx_0, y).$$

Since $y \in \mathcal{H}_2$, $\|y\| = 1$, was arbitrary, it follows that

(3.3) $$(Ax_0, x_0)Dy_0 - (By_0, x_0)Cx_0 = 0,$$

(3.4) $$\overline{(Ax_0, x_0)}D^*y_0 - \overline{(Cx_0, y_0)}B^*x_0 = 0.$$

In a similar way, for $x \in \mathcal{H}_1$, $\|x\| = 1$, and $z \in \mathbb{C}$, considering the polynomial

$$h_x(\lambda, z) := ((A(x_0 + zx), x_0 + \bar{z}x) - \lambda(x_0 + zx, x_0 + \bar{z}x))$$

$$((Dy_0, y_0) - \lambda)$$

$$- (By_0, x_0 + \bar{z}x)(C(x_0 + zx), y_0),$$

one arrives at

$$(Dy_0, y_0)Ax_0 - (Cx_0, y_0)By_0 = 0,$$ (3.5)

$$\overline{(Dy_0, y_0)}A^*x_0 - \overline{(By_0, x_0)}C^*y_0 = 0.$$ (3.6)

Due to the fact that the numerical range of A is contained in $W^2(\mathcal{A})$ only if the dimension of $\mathcal{H}_2$ is > 1 (and analogously for D, see [LMMT], Theorem 2.4), we have to distinguish the following cases:

a) $\dim \mathcal{H}_1 = \dim \mathcal{H}_2 = 1$: In this case $W^2(\mathcal{A})$ consists only of the two eigenvalues of $\mathcal{A}$, and the assertion is trivial.

b) $\dim \mathcal{H}_1 > 1$, $\dim \mathcal{H}_2 = 1$: In this case D is the multiplication by a constant, say d. If $Ax_0 = 0$ or $d = 0$, we are in case i) or ii), respectively. If $Ax_0 \neq 0$ or $d \neq 0$, the relation (3.5) yields $(Cx_0, y_0) \neq 0$ and

$$Ax_0 + B\left(-\frac{(Cx_0, y_0)}{d}y_0\right) = 0.$$

Since

$$Cx_0 + D\left(-\frac{(Cx_0, y_0)}{d}y_0\right) = Cx_0 - \frac{Cx_0 \cdot \overline{y_0}}{d}dy_0 = 0,$$

it follows that 0 is an eigenvalue of $\mathcal{A}$ with eigenvector

$$\left(\begin{array}{c} x_0 \\ -\frac{(Cx_0, y_0)}{(Dy_0, y_0)}y_0 \end{array}\right) = \left(\begin{array}{c} x_0 \\ -\frac{1}{d}Cx_0 \end{array}\right).$$

Note that since $\dim \mathcal{H}_2 = 1$, we cannot conclude from $Ax_0 \neq 0$ that $(Ax_0, x_0) \neq 0$ because the numerical range of A need not be contained in $W^2(\mathcal{A})$ (see [LMMT], Theorem 2.4). In this case we can only use the first form of the constant γ in the eigenvectors in iii).

c) $\dim \mathcal{H}_1 = 1$, $\dim \mathcal{H}_2 > 1$: In this case A is the multiplication by a constant, say a. In the same way as in b), we obtain that either $a = 0$, $Dy_0 = 0$ or 0 is an eigenvalue of $\mathcal{A}$ with eigenvector

$$\left(\begin{array}{c} x_0 \\ -\frac{(Ax_0, x_0)}{(By_0, x_0)}y_0 \end{array}\right) = \left(\begin{array}{c} x_0 \\ -\frac{a}{By_0 \cdot \overline{x_0}}y_0 \end{array}\right).$$

d) $\dim \mathcal{H}_1 > 1$, $\dim \mathcal{H}_2 > 1$: Consider first the case $(Ax_0, x_0) = 0$. By [LMMT], Theorem 2.4, it follows that $W(A) \subset W^2(\mathcal{A})$ since $\dim \mathcal{H}_2 > 1$. Hence 0 is also a corner of $W(A)$. Then Corollary 2.3 implies that $Ax_0 = 0$. If $(Dy_0, y_0) = 0$, a similar reasoning yields that $Dy_0 = 0$. If $(Ax_0, x_0) \neq 0$ and $(Dy_0, y_0) \neq 0$, then also $(By_0, x_0) \neq 0$, and we obtain from (3.3) and (3.5)

$$Ax_0 + B\left(-\frac{(Cx_0, y_0)}{(Dy_0, y_0)} y_0\right) = 0,$$

$$Cx_0 + D\left(-\frac{(Ax_0, x_0)}{(By_0, x_0)} y_0\right) = 0.$$

Using the relation $(Ax_0, x_0)(Dy_0, y_0) - (By_0, x_0)(Cx_0, y_0) = 0$, we conclude that 0 is an eigenvalue of $\mathcal{A}$ with an eigenvector of the asserted form.

If λ_0 is a double zero, then, for any $y \in \mathcal{H}_2$, $\|y\| = 1$, there are two root functions $\lambda_y^{(1)}(t)$, $\lambda_y^{(2)}(t)$, $t \in \mathbb{R}$, such that $\lambda_y^{(j)}(0) = 0$ and $g_y(\lambda_y^{(j)}(t), t) \equiv 0$, $j = 1, 2$, near $t = 0$ with Puiseux expansions (see, e.g., [Ka], Chapter II, §1.2)

$$\lambda_y^{(j)}(t) = \alpha_1 e^{\pi i j} t^{1/2} + \alpha_2 e^{2\pi i j} t + \dots, \quad j = 1, 2.$$

If $\alpha_1 \neq 0$, the four one-sided tangents of the functions $\lambda_y^{(1)}(t)$, $\lambda_y^{(2)}(t)$, $\lambda_y^{(1)}(-t)$, and $\lambda_y^{(2)}(-t)$, $t \geq 0$, divide the plane into four sectors of angle $\pi/2$. This contradicts the fact that 0 is a corner of $W^2(\mathcal{A})$. If $\alpha_1 = 0$, then $\lambda_y^{(j)}(t)$ are differentiable at 0 and the claim follows as above. □

Remark 3.2 From the equations (3.4) and (3.6), it follows that in the case iii) of Theorem 3.1 the point $\overline{\lambda_0}$ is an eigenvalue of $\mathcal{A}^*$ with eigenvector $\begin{pmatrix} x_0 \\ \widetilde{\gamma} y_0 \end{pmatrix}$ with

$$\widetilde{\gamma} = -\frac{(B^* x_0, y_0)}{((D^* - \overline{\lambda_0}) y_0, y_0)} \quad \text{or} \quad \widetilde{\gamma} = -\frac{((A^* - \overline{\lambda_0}) x_0, x_0)}{(C^* y_0, x_0)}.$$

Remark 3.3 In order to prove the relation (3.3), it is sufficient to consider roots $\lambda_y(t)$ (for real t) with

$$y = (Ax_0, x_0) Dy_0 - (By_0, x_0) Cx_0 \quad \text{and}$$
$$y = i((Ax_0, x_0) Dy_0 - (By_0, x_0) Cx_0),$$

and for the relation (3.5),

$$y = (Dy_0, y_0) Ax_0 - (Cx_0, y_0) By_0 \quad \text{and}$$
$$y = i((Dy_0, y_0) Ax_0 - (Cx_0, y_0) By_0).$$

In the following we illustrate the statements of Theorem 3.1 by some 4×4 matrix examples. We plot the quadratic numerical range of $\mathcal{A}$ (by means of a C^{++}-code which was already used in [LMMT] where a more detailed description may be found) and calculate the eigenvalues of $\mathcal{A}$ (which are marked by o) and those of the diagonal entries A and D.

The quadratic numerical ranges of the matrices

$$
\mathcal{A}_1 := \left(\begin{array}{cc|cc}
1 & 3+i & 2 & i \\
3+i & 1 & i & 2 \\
\hline
-2 & i & 1 & 3+i \\
i & -2 & 3+i & 1
\end{array}\right), \quad
\mathcal{A}_2 := \left(\begin{array}{cc|cc}
1 & 0 & 2 & 0 \\
0 & -1 & 0 & 2 \\
\hline
-2 & 0 & 1 & 0 \\
0 & -2 & 0 & -1
\end{array}\right).
$$

displayed in Figure 2 both have 6 corners.

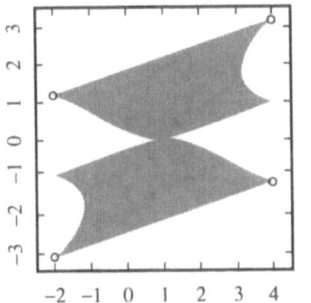 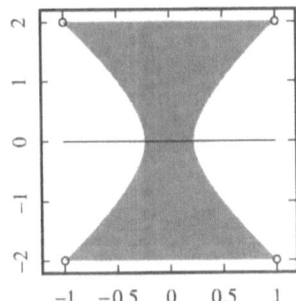

Figure 2 Quadratic numerical ranges of $\mathcal{A}_1$ and $\mathcal{A}_2$

For $\mathcal{A}_1$ the corners $-2-i(1-\sqrt{5})$, $-2-i(1+\sqrt{5})$, $4+i(1+\sqrt{5})$, $4+i(1-\sqrt{5})$ are the eigenvalues of $\mathcal{A}_1$, the corners $4+i$, $-2-i$ are the eigenvalues of the left upper corner A (and also of the right lower corner D). For $\mathcal{A}_2$ the corners $-1+2i$, $-1-2i$, $1+2i$, $1-2i$ are the eigenvalues of $\mathcal{A}_2$, the corners -1, 1 are the eigenvalues of A (and also of D).

The quadratic numerical range of the matrix

$$
\mathcal{A}_3 := \left(\begin{array}{cc|cc}
2 & i & 1 & 3+i \\
i & 2 & 3+i & 1 \\
\hline
1 & 3+i & -2 & i \\
3+i & 1 & i & -2
\end{array}\right)
$$

shown in Figure 3 below consists of two components and has 8 corners.

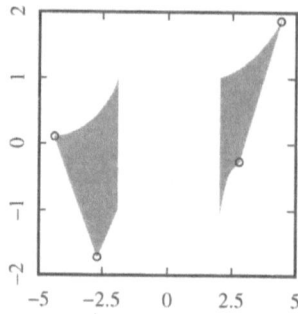

Figure 3 Quadratic numerical range of $\mathcal{A}_3$

The corners $-i + \sqrt{7 + 4i}$, $-i - \sqrt{7 + 4i}$ and $i + \sqrt{19 + 8i}$, $i - \sqrt{19 + 8i}$ are the eigenvalues of $\mathcal{A}_3$, the corners $2 + i$, $2 - i$ are the eigenvalues of A, and the corners $-2 + i$, $-2 - i$ are the eigenvalues of D.

3.2 In this subsection we consider corners of the quadratic numerical range $W^2(\mathcal{A})$ which do not belong to $W^2(\mathcal{A})$, but only to its closure.

Theorem 3.4 *If $\lambda_0 \in \overline{W^2(\mathcal{A})}$ is a corner of $W^2(\mathcal{A})$, then $\lambda_0 \in \sigma(A) \cup \sigma(D) \cup \sigma(\mathcal{A})$.*

Proof: Since λ_0 belongs to $\overline{W^2(\mathcal{A})}$, there exist a sequence $(\lambda_n)_1^\infty$, $\lambda_n \to \lambda_0$ for $n \to \infty$, and sequences $(x_n^0)_1^\infty \subset \mathcal{H}_1$, $(y_n^0)_1^\infty \subset \mathcal{H}_2$, $\|x_n^0\| = \|y_n^0\| = 1$, such that

$$\Delta(x_n^0, y_n^0; \lambda_n) = \det \begin{pmatrix} (Ax_n^0, x_n^0) - \lambda & (By_n^0, x_n^0) \\ (Cx_n^0, y_n^0) & (Dy_n^0, y_n^0) - \lambda \end{pmatrix} = 0.$$

We can assume that there exists a neighbourhood V of λ_0 such that for all $n \in \mathbb{N}$ the quadratic polynomial $\Delta(x_n^0, y_n^0; \cdot)$ has exactly one zero or two zeros in V.

As in the proof of Theorem 2.9, we construct the space $\tilde{\mathcal{H}} = \tilde{\mathcal{H}}_1 \times \tilde{\mathcal{H}}_2$ and the operator

$$\tilde{\mathcal{A}} = \begin{pmatrix} \tilde{A} & \tilde{B} \\ \tilde{C} & \tilde{D} \end{pmatrix} : \tilde{\mathcal{H}}_1 \times \tilde{\mathcal{H}}_2 \to \tilde{\mathcal{H}}_1 \times \tilde{\mathcal{H}}_2$$

by means of a Banach limit. First we consider the case of a simple zero. Since the sequences $(x_n^0)_1^\infty$ and $(y_n^0)_1^\infty$ are bounded, we can assume without loss of generality (by passing to suitable subsequences) that all sequences

(3.7) $$((Fu_n, v_n))_1^\infty$$

converge where F is one of the operators A, B, C, D, A^*, B^*, C^*, D^* or a product of 2 or 3 of them, and u_n, v_n are the elements x_n^0 or y_n^0, whenever the inner products in (3.7) are defined. Now let $\widetilde{x}^0 = (x_n^0)_1^\infty \in \widetilde{\mathcal{H}}_1$, $\widetilde{y}^0 = (y_n^0)_1^\infty \in \widetilde{\mathcal{H}}_2$. It is easy to check (see e.g. the proof of Theorem 2.9) that λ_0 is a simple root of

$$\det \begin{pmatrix} (\widetilde{A}\widetilde{x}^0, \widetilde{x}^0) - \lambda & (\widetilde{B}\widetilde{y}^0, \widetilde{x}^0) \\ (\widetilde{C}\widetilde{x}^0, \widetilde{y}^0) & (\widetilde{D}\widetilde{y}^0, \widetilde{y}^0) - \lambda \end{pmatrix} = 0.$$

Hence $\lambda_0 \in W^2(\widetilde{A})$.

Next we derive the equalities (3.3), (3.5) for the operators $\widetilde{A}$, $\widetilde{B}$, $\widetilde{C}$ and $\widetilde{D}$. To this end we proceed as in the proof of Theorem 3.1 and introduce the quadratic polynomial $g_{\widetilde{y}}(\lambda, z)$ with its root $\lambda_{\widetilde{y}}(z)$ given by a formula analogous to (3.2). By Remark 3.3, for the proof of the analogues of (3.3), (3.5) it is sufficient to consider e.g. elements $\widetilde{y} = (y_n)_1^\infty$ which are certain linear combinations of the four vectors $\widetilde{A}\widetilde{x}^0$, $\widetilde{B}\widetilde{y}^0$, $\widetilde{C}\widetilde{x}^0$ and $\widetilde{D}\widetilde{y}^0$. Since all the sequences of the form (3.7) converge, we can use the ordinary limit instead of the Banach limit in the construction of the quadratic forms occurring in the formula for the root $\lambda_{\widetilde{y}}(z)$. This means that the root $\lambda_{\widetilde{y}}(z)$ is a limit of the corresponding roots $\lambda_{y_n}(z)$. By assumption, all roots $\lambda_y(t)$ for real $t \in U$ lie in a (closed) sector with vertex λ_0 and angle $< \pi$. Hence the roots $\lambda_{\widetilde{y}}(t)$ lie in the same sector, and we obtain the analogues of (3.3) and (3.5) for $\widetilde{A}$, $\widetilde{B}$, $\widetilde{C}$, $\widetilde{D}$, $\widetilde{x}^0$ and $\widetilde{y}^0$. Now we finish the proof for the case of a simple zero as in the proof of Theorem 3.1 (here we only have to consider the case d) since $\dim \widetilde{\mathcal{H}}_1 = \dim \widetilde{\mathcal{H}}_2 = \infty$). This yields that either $\lambda_0 \in \sigma_p(\widetilde{A}) = \sigma_{app}(A) \subset \sigma(A)$, $\lambda_0 \in \sigma_p(\widetilde{D}) = \sigma_{app}(D) \subset \sigma(D)$ or $\lambda_0 \in \sigma_p(\widetilde{\mathcal{A}}) = \sigma_{app}(\mathcal{A}) \subset \sigma(\mathcal{A})$.

If λ_0 is a double zero, the proof is analogous to the corresponding part of Theorem 3.1. $\qquad\square$

Acknowledgements

A. Markus acknowledges the support of the Regensburger Universitätsstiftung Hans Vielberth, C. Tretter has been supported by the German Research Foundation, DFG, under Grant Tr 368/4–1. The authors thank Markus Wagenhofer for his contribution to the numerical calculations and visualizations in this paper.

References

[B]   S.K. Berberian, Approximate proper vectors, *Proc. Amer. Math. Soc.* **13** (1962), 111–114.

[BO]   S.K. Berberian and G.H. Orland, On the closure of the numerical range of an operator, *Proc. Amer. Math. Soc.* **18** (1967), 499–503.

[GR]   K.E. Gustafson and D.K.M. Rao, *Numerical Range*, Springer Verlag, Berlin Heidelberg New York Tokyo 1997.

[HJ]   R. Horn and C. Johnson, *Topics in Matrix Analysis*, Cambridge University Press, New York 1991.

[K]   R. Kippenhahn, Über den Wertevorrat einer Matrix, *Math. Nachr.* **6** (1951), 193–228.

[Ka]   T. Kato, *Perturbation Theory for Linear Operators*, Springer Verlag, Berlin Heidelberg New York 1966.

[LMZ]   P. Lancaster, J. Maroulas and P. Zizler, The numerical range of selfadjoint matrix polynomials, *Operator Theory: Adv. Appl.* **106** (1998), 291–308.

[LMMT]   H. Langer, A. Markus, V. Matsaev and C. Tretter, *A new concept for block operator matrices: The quadratic numerical range*, Linear Algebra Appl., to appear.

[LT]   H. Langer and C. Tretter, Spectral decomposition of some nonselfadjoint block operator matrices, *J. Operator Theory* **39** (1998), 339–359.

[M]   A. Markus, Introduction to the Spectral Theory of Polynomial Operator Pencils, *AMS Transl. Math. Monographs* Providence Rhode Island, **71** (1988).

[Ma]   A.I. Markushevich, *Theory of Functions of a Complex Variable, vol. II*, Prentice Hall, Englewood Cliffs, N. Y. 1965.

[MP]   J. Maroulas and P. Psarrakos, The boundary of the numerical range of matrix polynomials, *Linear Algebra Appl.* **267** (1997), 101–111.

[T]   E.C. Titchmarsh, *The Theory of Functions, 2nd Edition*, Oxford University Press, London 1968.

Heinz Langer   Alexander Markus
Institute of Analysis   Department of Mathematics
  and Technical Mathematics   Ben–Gurion University of the Negev
University of Technology   Beer Sheva 84105
A-1040 Vienna   Israel
Austria   e-mail: markus@cs.bgu.ac.il
e-mail: hlanger@email.tuwien.ac.at

Christiane Tretter
Department of Mathematics
  and Computer Science
University of Leicester
Leicester LE1 7RH
United Kingdom
e-mail: c.tretter@mcs.le.ac.uk

1991 Mathematics Subject Classification. Primary: 47 A 12; Secondary: 47 A 56, 47 A 75.

Operator Theory:
Advances and Applications, Vol. 124
© 2001 Birkhäuser Verlag Basel/Switzerland

On Some Classes of Extensions
of Sectorial Operators and
Dual Pairs of Contractions

M.M. Malamud

Dedicated to Professor Israel Gohberg on the occasion of his seventieth birthday

Some classes of contractive and noncontractive extensions of a dual pair of contractions are investigated. A problem of a description of the set of all m-sectorial extensions of a sectorial operator is solved in terms of its linear-fractional transformation. Some complements to the J. von Neumann inequality is obtained also.

1 Introduction

In the theory of extensions of a nonnegative operator $A(\subset A^*)$ in a Hilbert space $\mathfrak{H}$ to a selfadjoint or m-sectorial [Ka] operator (regularly dissipative in the sense of S.G. Krein [Kr]) there are two well-known approaches in which extensions $\widetilde{A} \supset A$ in various classes are described in diverse forms. One of these, proposed by M.G. Krein in [K] uses the linear fractional transformation $T_1 = (I - A)(I + A)^{-1}$ to reduce the problem to the description of various classes of extensions $T \supset T_1$ of a nondensely defined (on the subspace $\mathfrak{H}_1 = (I + A)\mathfrak{H}$) Hermitian contraction T_1.

The other approach to the description of proper extensions $\widetilde{A}$ of an operator $A > 0$ was proposed by Vishik [V] and Birman [B]. They associate with each extension $\widetilde{A} \supset A$ (not necessarily selfadjoint) a "boundary" operator B acting in an auxiliary space $\mathcal{H}(dim\mathcal{H} = dim(A^* - i)\mathfrak{H})$, and describe the properties of the extension $\widetilde{A} = \widetilde{A}_B$ in terms of the operator B, i.e. essentially in terms of the boundary conditions if A is a differential operator. This approach was subsequently formalized in the concept of a "boundary triplet" and was developed in later papers by many authors (see for instance [DM] and references therein). We observe only that all proper m-sectorial extensions of an operator $A \geq 0$ with zero lower bound were described via boundary triplets and Weyl functions in [KM] and [DMT]. Another description in the framework of first (Krein's) approach has been obtained in [AT1], [AT2].

We remark that the methods used in these approaches are essentially different, as are the descriptions obtained with their help.

In this paper we solve, among others, the following problems in the framework of the first approach:

1) Selfadjoint and nonselfadjoint analogues are obtained for the theorem of Birman-Krein (see [B], [K]) (as well as its generalization to the case of zero lower bound operators [DM]) on the number (counting multiplicity) of negative eigenvalues of the operators $\widetilde{A}_B$ and B;

2) m-sectorial extensions are described for a sectorial operator $A(\not\subset A^*)$.

It is convenient to regard all these problems as problems on the "completion" of a contractive operator matrix $T_1 = \binom{T_{11}}{T_{21}}$ to form a matrix $T = (T_{jk})^2_{j,k=1}$ which is connected in a natural way with the problem of extending a dual pair of contractions (DPC) to operators in various classes. A description is given in terms of operator balls, "holes", and objects close to them.

The starting point of our investigation is a description of the set of all contractive extensions of a DPC $\{T_1 = \binom{T_{11}}{T_{21}}, \ T_2 = \binom{T_{11}^*}{T_{12}^*}\}$ or what is the same a description of all "completions" of a matrix

(1.1) $$T_0 = \begin{pmatrix} T_{11} & T_{12} \\ T_{21} & * \end{pmatrix} = \begin{pmatrix} T_{11} & D_{T_{11}^*} U \\ V D_{T_{11}} & * \end{pmatrix}$$

to form a contractive matrix $T = (T_{ij})^2_{i,j=1}$.

It has been shown in [ArG], [D], [DKW], [ShY] that all missing blocks T_{22} in (1.1) form the operator ball $B(-V T_{11}^* U; D_{V^*}, D_U)$:

(1.2) $$T_{22} = -V T_{11}^* U + D_{V^*} K D_U, \quad \|K\| \leq 1.$$

We denote by $\text{Ext}_{\{T_1,T_2\}}$ the set of all (not necessarily contractive) extensions of a DPC $\{T_1, T_2\}$ of the form (1.2) and describe some classes of such extensions.

The paper is organized as follows.

In Section 2 we present one more proof of the formula (1.2) describing the set of contractive extensions T of a DPC $\{T_1, T_2\}$. Here we also provide a solution to Problem 1) mentioned above. Namely this solution is based on the following formula (see (2.31)–(2.32), (2.57)) for the operator $D^2_{T_K}$ with $T_K := (T_{ij})$ "shorted" to $\mathfrak{H}_2$:

$$(I - T_K^* T_K)_{\mathfrak{H}_2} = D_U \begin{pmatrix} 0 & 0 \\ 0 & I - K^* K \end{pmatrix} D_U$$

Some applications of this result to the boundary value problems are presented also.

In Section 3 we investigate some properties of noncontractive operators T satisfying $(I - T^* T)_- \in \mathfrak{S}_\infty$ and complement the J. von Neumann inequality. Namely, it is proved here that for a bounded operator T in $\mathfrak{H}$ and an arbitrary inner function $f \in H^\infty(\mathbb{D})$ regular on $\sigma(T)$ the following implications hold

$$(I - T^*T)_\pm \in \mathfrak{S}_\infty \implies (I - f(T)^*f(T))_\pm \in \mathfrak{S}_\infty$$

Further in Section 3 we also consider the classes $C_{\mathfrak{H}}(\varphi)(\varphi \in (0, \pi/2])$ of contractions $T \in [\mathfrak{H}]$ satisfying

$$(1.3) \qquad \|T \sin \varphi \pm i \cos \varphi \cdot I\| \le 1, \qquad \varphi \in (0, \pi/2]$$

and $C_{\mathfrak{H}}(0) = \{T = T^* \in [\mathfrak{H}] : \|T\| \le 1\}$ and establish some their properties. These classes naturally arise in the framework of Krein's approach to the class $S_{\mathfrak{H}}(\varphi)$ of m-sectorial operators (and linear relations) in $\mathfrak{H}$: $T = (I - \widetilde{A})$ $(I + \widetilde{A})^{-1} \in C_{\mathfrak{H}}(\varphi) \iff \widetilde{A} \in S_{\mathfrak{H}}(\varphi)$.

We also present here a partial description of the set of extreme points $C_{\mathfrak{H}}^E(\varphi)$ of the lune $C_{\mathfrak{H}}(\varphi)$. We emphasise that even in the finite-dimensional case the set $C_{\mathfrak{H}}^E(\varphi)(\varphi \in (0, \pi/2))$ is much wider then the topological boundary of $C_{\mathfrak{H}}(\varphi)$ in contrast to the well-known result on the extreme points of the unit ball $C_{\mathfrak{H}}(\pi/2)$ in $[\mathfrak{H}]$.

In Section 4 starting with a sectorial operator A we consider the nonhermitian contraction $T_1 := (I - A)(I + A)^{-1}$ from $\mathfrak{H}_1 := (I - A)\mathcal{D}(A)$ to $\mathfrak{H}$ satisfying

$$(1.4) \qquad 2ctg\varphi|Im(T_1 f, f)| \le ((I - T_1^*T_1)f, f).$$

We present a description of the set $Ext_{T_1}(\varphi)$ of all contractive extensions $T(\supset T_1)$ of the class $C_{\mathfrak{H}}(\varphi)$. Block-matrix representations of two extremal extensions T_m and T_M of a nondensely defined contraction T_1 satisfying (1.4), are presented too. Such representations of T_m and T_M allow us to describe some of their properties which are similar to that in the self-adjoint case.

Main results of the paper have been announced in [M2] and partially published (with proofs) in [KM1].

2 Contractive and Non-contractive Extensions of a Dual Pair of Contractions

2.1 In the section we present one more proof of the well known result from [ArG], [DKW], [ShY] and investigate noncontractive extensions of a dual pair of contractions.

Definition 2.1 Let $\mathfrak{H} = \mathfrak{H}_1 \oplus \mathfrak{H}_2 = \mathfrak{H}_1' \oplus \mathfrak{H}_2'$ be orthogonal decompositions of the Hilbert space H. Operators $T_1 \in [\mathfrak{H}_1, \mathfrak{H}]$, $T_2 \in [\mathfrak{H}_1', \mathfrak{H}]$ are said to form a dual pair (DP) if

$$(2.1) \qquad (T_1 f, g) = (f, T_2 g) \qquad \forall f \in \mathfrak{H}_1, \qquad \forall g \ni \mathfrak{H}_1'$$

An operator $T (\in [\mathfrak{H}])$ is termed an extension of the dual pair $\{T_1, T_2\}$ if $T \lceil \mathfrak{H}_1 = T_1$ and $T^* \lceil \mathfrak{H}_2 = T_2$.

When rewritten in the block-matrix representation with respect to the pointed out decompositions of the space $\mathfrak{H}$, the operators T_1 and T_2 form a dual pair if and only if

$$(2.2) \qquad\qquad T_1 = \begin{pmatrix} T_{11} \\ T_{21} \end{pmatrix}, \qquad T_2 = \begin{pmatrix} T_{11}^* \\ T_{21}' \end{pmatrix}$$

with $T_{11} \in [\mathfrak{H}_1, \mathfrak{H}_1']$, $T_{21} \in [\mathfrak{H}_1, \mathfrak{H}_2']$, $T_{21}' \in [\mathfrak{H}_1', \mathfrak{H}_2']$.

Setting $T_{12} = (T_{21}')^*$, an extension T of the DP $\{T_1, T_2\}$ can be rewritten in the form

$$(2.3) \qquad\qquad T = \begin{pmatrix} T_{11} & T_{12} \\ T_{21} & T_{22} \end{pmatrix}.$$

In this case the problem of description of a certain class X of extensions of the DP $\{T_1, T_2\}$ is equivalent to the problem of completing an incomplete block-matrix $\begin{pmatrix} T_{11} & T_{12} \\ T_{21} & * \end{pmatrix}$ with respect to the matrix T of the form (2.3) and such that $T \in X$.

In this section we consider contractive extensions of a DP of contractions (DPC) $\{T_1, T_2\}$. The union of all such extensions will be denoted by $\mathrm{Ext}_{\{T_1, T_2\}}(\pi/2)$.

The set $\mathrm{Ext}_{\{T_1, T_2\}}(\pi/2)$ turns out to be an operator ball in the sence of the following definition.

Definition 2.2 The totality of the operators $Z \in [\mathfrak{H}]$ of the form

$$(2.4) \qquad\qquad Z = Z_0 + R_l K R_r, \qquad \|K\| \leq 1$$

is referred to as an operator ball $B(Z_0; R_l, R_r)$.

Here Z_0 is called the center of the ball, and $R_l = R_l^* \geq 0$ and $R_r = R_r^* \geq 0$ are called left and right radiuses respectively.

For the proof of Theorem 2.7, based on the Sylvester criterion, we need two simple and well-known lemmas.

Lemma 2.3 [Do]. *Let $\mathfrak{H}, \mathfrak{H}_1, \mathfrak{H}_2$ be Hilbert spaces and $A \in [\mathfrak{H}_1, \mathfrak{H}], B \in [\mathfrak{H}_1, \mathfrak{H}]$. The following assertions are equivalent:*

a) $\mathfrak{R}(A) \subset \mathfrak{R}(B)$;

b) *there exists an operator $K \in [\mathfrak{H}_1, \mathfrak{H}_2]$ such that $A = BK$;*

c) *the estimate*

$$(2.5) \qquad\qquad \|A^* f\| \leq c \|B^* f\| \qquad \forall f \in \mathfrak{H}$$

holds true with some constant $c > 0$.

Proof: a)$\Leftarrow$ b). Since $\Re(|T^*|) = \Re(T)$ for each operator $T \in [\mathfrak{H}_1, \mathfrak{H}]$ (see [Ka]) then $\Re(|A^*|) = \Re(A) \subset \Re(B) = \Re(|B^*|)$. But then $\ker |A^*| \supset \ker |B^*|$ and therefore the operator[1] $T = |B^*|^{-1}|A^*|$ is well defined. Let also $A = U_1|A|$ and $B^* = U_2|B^*|$ be polar decompositions of the operators A and B^* respectively. Then (see [Ka]) $A = |A^*|U_1$ and $BU_2 = |B^*|$. Therefore

$$A = |A^*|U_1 = |B^*|TU_1 = BU_2TU_1 = BK$$

with $K = U_2TU_1$.

b)$\Longrightarrow$c) This implication is obvious.

c)$\Longrightarrow$a) The equality $|A^*| = \sqrt{c}K_1|B^*|$ with some contraction $K_1 \in [\mathfrak{H}]$ directly follows from the inequality $|A^*|^2 = AA^* < cB^*B = c|B^*|^2$ which is equivalent to the inequality (2.5). But then $\Re(A) = \Re(|A^*|) \subset \Re(|B^*|) = \Re(B)$. $\square$

Lemma 2.4 [GL]. *Let* $Q_i \in [\mathfrak{H}]$ $(i = 1, 2, 3)$, $Q_3 = Q_3^*$, $Q_1 > 0$ *and* $0 \in \rho(Q_1)$. *Then the equality*

$$(2.6) \qquad\qquad Z^*Q_1Z + Z^*Q_2 + Q_2^*Z + Q_3 \le 0$$

has a solution if and only if

$$(2.7) \qquad\qquad Q_2^*Q_1^{-1}Q_2 - Q_3 \ge 0.$$

Under this condition the set of the solutions of the inequality (2.6) *makes up an operator ball* $B(Z_0; R_l; R_r)$ *of the form* (2.4) *with*

$$(2.8) \qquad \begin{aligned} Z_0 &= -Q_1^{-1}Q_2, \qquad R_l = Q_1^{-1/2}, \\ R_r &= (Q_2^*Q_1^{-1}Q_2 - Q_3)^{1/2} \end{aligned}$$

Proof: It is clear that $Z_0 = -Q_1^{-1}Q_2$ is a solution of (2.6) if the condition (2.7) is satisfied.

Conversely, if Z is a solution of (2.6) then

$$Q_2^*Q_1^{-1}Q_2 - Q_3 \ge (Z^*Q_1^{1/2} + Q_2^*Q_1^{-1/2})$$
$$(Q_1^{1/2}Z + Q_1^{-1/2}Q_2) \ge 0,$$

and (2.7) is fulfilled, whence we arrive at the equality

$$(2.9) \quad Q_1^{1/2}Z + Q_1^{-1/2}Q_2 = K(Q_2^*Q_1^{-1}Q_2 - Q_3)^{1/2}, \qquad \|K\| \le 1.$$

with some contraction K. The relations (2.8) and (2.9) together mean that $Z \in B(Z_0; R_l, R_r)$. On the other hand it is obvious that each Z belonging to $B(Z_0; R_l, R_r)$ is a solution of (2.6). $\square$

---

[1] Here and in the sequel we set $A^{-1}f = 0$ $\quad \forall f \in \ker A$.

Lemma 2.5 *Let* $A_n, B_n, C_n \in [\mathfrak{H}]$ *and* $s - \lim_{n\to\infty} A_n^* = A^*$, $w - \lim_{n\to\infty} B_n = B$, $s - \lim_{n\to\infty} C_n = C$. *Then there exists*

$$(2.10) \qquad\qquad w - \lim_{n\to\infty} A_n B_n C_n = ABC.$$

Proof: Since the sequence $\{\|B_n\|\}_1^\infty$ is bounded the existence of the weak limits $w - \lim_{n\to\infty} A_n B_n = AB$ and $w - \lim_{n\to\infty} B_n^* A_n^* = B^* A^*$ follows from the identity

$$((A_n B_n - AB)f, g) = (B_n f, (A_n^* - A^*)g) + ((B_n - B)f, A^* g).$$

Now one derives the required assertion from the identity

$$((A_n B_n C_n - ABC)f, g) = ((C_n - C)f, B_n^* A_n^* g)$$
$$+ (Cf, (B_n^* A_n^* - B^* A^*)g).$$

$\qquad\qquad\qquad\qquad\qquad\qquad\qquad\qquad\qquad\qquad\qquad\qquad\qquad\qquad\qquad\square$

Remark 2.6 Note that the equality (2.10) may be violated if the condition $s - \lim_{n\to\infty} A_n^* = A^*$ is replaced by the one $s - \lim_{n\to\infty} A_n = A$. Let for example U be the unilaterial shift, $A_n = (U^*)^n$, $B_n = U^n$, $C_n = I$, then $s - \lim_{n\to\infty} A_n = 0$, $w - \lim_{n\to\infty} B_n = 0$ but $w - \lim_{n\to\infty} A_n B_n C_n = I$.

2.2 The operators T_1 and T_2 of the form (2.2) are contractive if

$$(2.11) \qquad T_{11}^* T_{11} + T_{21}^* T_{21} \le I \iff T_{21}^* T_{21} \le D_{T_{11}} := I - T_{11}^* T_{11}$$

$$(2.12) \qquad T_{11} T_{11}^* + T_{12} T_{12}^* \le I \iff T_{12} T_{12}^* \le D_{T_{11}^*} := I - T_{11} T_{11}^*$$

By Lemma 2.3 these relations are equivalent to the following ones

$$(2.13) \qquad\qquad T_{21} = V D_{T_{11}}, \qquad T_{12} = D_{T_{11}^*} U$$

with contractions V and U ($V \in [\mathfrak{H}_1, \mathfrak{H}_2']$, $U \in [\mathfrak{H}_2, \mathfrak{H}_1']$) which are uniquely determined provided that $\ker V \supset \ker D_{T_{11}}$ and $\ker U^* \supset \ker D_{T_{11}^*}$, that is $V^* = D_{T_{11}}^{-1} T_{21}^*$ and $U = D_{T_{11}^*}^{-1} T_{12}$.

Theorem 2.7 ([ArG], [D], [DKW], [ShY]). *Let* $\{T_1, T_2\}$ *be a dual pair of contractions,*

$$(2.14) \qquad T_1 = \begin{pmatrix} T_{11} \\ T_{21} \end{pmatrix} = \begin{pmatrix} T_{11} \\ V D_{T_{11}} \end{pmatrix}, \qquad T_2 = \begin{pmatrix} T_{11}^* \\ T_{12} \end{pmatrix} = \begin{pmatrix} T_{11}^* \\ U^* D_{T_{11}^*} \end{pmatrix},$$

$\mathcal{H}_1 = \overline{\mathfrak{R}(D_U)}$ *and* $\mathcal{H}_2 = \overline{\mathfrak{R}(D_{V^*})}$. *Then the formula*

$$T := T_K = \begin{pmatrix} T_{11} & D_{T_{11}^*} U \\ V D_{T_{11}} & T_{22} \end{pmatrix},$$

$$(2.15)$$

$$T_{22} = -V T_{11}^* U + D_{V^*} K D_U,$$

establishes a bijective correspondence between all contractive extensions $T :=$ $T_K = (T_{ij}) \in \mathrm{Ext}_{\{T_1,T_2\}}(\pi/2)$ and all contractions $K \in [\mathcal{H}_1, \mathcal{H}_2]$.
   Thus the set $\mathrm{Ext}_{\{T_1,T_2\}}(\pi/2)$ forms an operator ball.

Proof: Let

$$
G = \begin{pmatrix} G_{11} & G_{12} \\ G_{21} & G_{22} \end{pmatrix}
$$

(2.16)

$$
:= \begin{pmatrix} I - T_{11}^* T_{11} - T_{21}^* T_{21} & -T_{11}^* T_{12} - T_{21}^* T_{22} \\ -T_{12}^* T_{11} - T_{22}^* T_{21} & I - T_{12}^* T_{12} - T_{22}^* T_{22} \end{pmatrix} \geq 0
$$

At the beginning we assume in addition that $0 \in \rho(D_{T_{11}}) \cap \rho(D_V)$, that is $\|T_{11}\| < 1$ and $\|V\| < 1$. Then $0 \in \rho(G_{11})$ and in accordance with the Silvester criterion

$$
G \geq 0 \Longleftrightarrow G_{22} - G_{21} G_{11}^{-1} G_{12} \geq 0.
$$

Letting $Z := T_{22}$ one rewrites the last inequality in the form

$$
Z^*(I + T_{21} G_{11}^{-1} T_{21}^*) Z + Z^* T_{21} G_{11}^{-1} T_{11}^* T_{12} + T_{12}^* T_{11} G_{11}^{-1} T_{21}^* Z
$$

(2.17)

$$
\leq I - T_{12}^* T_{12} - T_{12}^* T_{11} G_{11}^{-1} T_{11}^* T_{12}
$$

Since $0 \in \rho(I + T_{12} G_{11}^{-1} T_{21}^*)$ one can apply Lemma 2.4 to the inequality (2.17) and conclude that all its solutions form an operator ball $B(Z_0; R_l, R_r)$. Let us transform expressions for Z_0, R_l, R_r given by the formulas (2.8). Since

(2.18) $G_{11} = D_T^2 = I - T_{11}^* T_{11} - D_{T_{11}} V^* V D_{T_{11}} = D_{T_{11}} D_V^2 D_{T_{11}}$

we have

$$
\begin{aligned}
R_l^2 &= (I + T_{21} G_{11}^{-1} T_{21}^*)^{-1} \\
&= (I + V D_{T_{11}} D_{T_{11}}^{-1} D_V^{-2} D_{T_{11}}^{-1} D_{T_{11}} V^*)^{-1}
\end{aligned}
$$

(2.19)

$$
(I + V D_V^{-2} V^*)^{-1} = (I + D_{V^*}^{-2} V V^*)^{-1} = D_{V^*}^2.
$$

It follows from (2.18) and (2.19) that

(2.20)
$$
\begin{aligned}
Z_0 &= -R_l^2(T_{21} G_{11}^{-1} T_{11}^* T_{12}) \\
&= -D_{V^*}^2 V D_{T_{11}} D_{T_{11}}^{-1} D_V^{-2} D_{T_{11}}^{-1} T_{11}^* D_{T_{11}^*} U \\
&= -V D_V^2 D_V^{-2} D_{T_{11}}^{-1} D_{T_{11}} T_{11}^* U = -V T_{11}^* U
\end{aligned}
$$

And finally

$$
\begin{aligned}
R_r^2 &= U^* D_{T_{11}^*} T_{11} D_{T_{11}}^{-1} D_V^{-2} D_{T_{11}}^{-1} D_{T_{11}} V^* D_{V^*}^2 V D_{T_{11}} \\
&\quad D_{T_{11}}^{-1} D_V^{-2} D_{T_{11}}^{-1} T_{11}^* D_{T_{11}^*} U \\
&\quad + I - U^* D_{T_{11}}^2 U - U^* D_{T_{11}^*} T_{11} D_{T_{11}}^{-1} D_V^{-2} D_{T_{11}}^{-1} T_{11}^* D_{T_{11}^*} U \\
&= U^* T_{11} (I - D_V^2) D_V^{-2} T_{11}^* U + I - U^* U + U^* T_{11} T_{11}^* U \\
&\quad - U^* T_{11} D_V^{-2} T_{11}^* U = D_U^2 + U^* T_{11} (D_V^{-2} - I + I - D_V^2) = D_U^2
\end{aligned}
$$

Note that in view of (2.21) the solvability criterion of the inequality (2.17) is fulfilled. Therefore on account of Lemma 2.4 and the relations (2.19)–(2.21) we have

(2.22)
$$
Z = T_{22} = -V T_{11}^* U + D_{V^*} K D_U, \quad \|K\| \le 1
$$

Thus the formula (2.15) is proved under the additional condition $0 \in \rho(D_{T_{11}}) \cap \rho(D_V)$.

The general case may be obtained from the previous one by passage to the limit. Actually, if $\{T_1, T_2\}$ is an arbitrary DPC of the form (2.14), then a DPC $\{rT_1, rT_2\}$ with $r \in (0, 1)$ satisfies the condition $0 \in \rho(r D_{T_{11}}) \cap \rho(D_{rV})$. Therefore an arbitrary contraction $K \in [\mathfrak{H}_2, \mathfrak{H}_2']$ determines a contractive extension

(2.23)
$$
T_K(r) := \begin{pmatrix} r T_{11} & r T_{12} \\ r T_{21} & T_{22}(r) \end{pmatrix},
$$
$$
T_{22}(r) = -r V(r) T_{11}^* U(r) + D_{V^*(r)} K D_{U(r)}
$$

with

(2.24)
$$
V(r) = r V D_{T_{11}} D_{rT_{11}}^{-1}, \qquad U(r) = r D_{T_{11}^*} D_{rT_{11}^*}^{-1} U
$$

The relations (2.24) yield the existence of strong limits

(2.25)
$$
s - \lim_{r \uparrow 1} D_{V^*(r)} = D_{V^*}, \quad s - \lim_{r \uparrow 1} D_{U(r)} = D_U,
$$
$$
s - \lim_{r \uparrow 1} r V(r) T_{11}^* U(r) = V T_{11}^* U
$$

Therefore passage to the limit in (2.23) as $r \to 1$ leads us to the relation (2.22) proving the implication $\|K\| \le 1 \Longrightarrow T_K = (T_{ij}) \in \mathrm{Ext}_{\{T_1, T_2\}}(\pi/2)$.

Conversely, if $T = (T_{ij}) \in \text{Ext}_{\{T_1, T_2\}}(\pi/2)$ then $rT \in \text{Ext}_{\{rT_1, rT_2\}}(\pi/2)$ for all $r \in (0, 1)$ and $0 \in \rho(D_{rT_{11}}) \cap \rho(D_{V(r)})$. Therefore $rT = T_{K(r)}$ with some contraction $K(r)$ that is

$$rT_{22} = -rV(r)T_{11}^* U(r) + D_{V^*(r)} K(r) D_{U(r)},$$

(2.26)
$$\|K(r)\| \leq 1,$$

where $U(r)$ and $V(r)$ are determined by the equalities (2.24).

In view of compactness of the unite operator ball $B_1 := \{K \in [\mathcal{H}_1, \mathcal{H}_2] : \|K\| \leq 1\}$ in the weak operator topology there exists a sequence $r_n \uparrow 1$ such that the operators $K(r_n)$ converge weakly to an operator $K_0 (\in B_1)$ as $n \to \infty$. On account of Lemma 2.5 and equalities (2.25) we have

(2.27)
$$w - \lim_{r \uparrow 1} D_{V^*}(r) K(r) D_{U(r)} = D_{V^*} K_0 D_U$$

Besides, it follows from (2.24) that $s - \lim_{r \uparrow 1} V(r) = V$, $s - \lim_{r \uparrow 1} U(r) = U$.

Passing to the limit in (2.26) as $r \uparrow 1$ and taking arrangement of (2.27) one arrives at the converse implication: $T_K \in \text{Ext}_{\{T_1, T_2\}}(\pi/2) \implies T_{22} = Z_0 + R_l K R_r$. □

Corollary 2.8 *The set* $\text{Ext}_{\{T_1, T_2\}}(\pi/2)$ *consists of only one element iff at least one of the two following conditions*

$$a) \ D_V^* = 0; \qquad b) \ D_U = 0$$

is satisfied.

Remark 2.9 Let us make some historical remarks concerning Theorem 2.7. The case $T_{11} = T_{11}^*$, $T_{12} = T_{21}^*$ was considered by M.G. Krein [K] in connection with selfadjoint extensions of positive unbounded operators. The existence of contractive extensions of a DPC $\{T_1, T_2\}$ (that is the fact $\text{Ext}_{\{T_1, T_2\}}(\pi/2) \neq \varnothing$) was first established by B.S. Nagy and C. Foias [SNF1, p. 190] by means of a corresponding generalization of the Krein's method [K]. Note also that the claim $\text{Ext}_{\{T_1, T_2\}}(\pi/2) \neq \varnothing$ is implicitly contained in [Ph1], [Ph2]. Another proof of the existence part of Theorem 2.7 has also been obtained by S. Parrot [P].

The complete description of the set $\text{Ext}_{\{T_1, T_2\}}(\pi/2)$, i.e., Theorem 2.7, was obtained in [ArG], [Da], [DKW], [ShY]. In the special case $(T_{21} = 0)$ Theorem 2.7 has been obtained by B. Sz.-Nagy and C. Foias [SNF1] much earlier. The given proof of Theorem 2.7 is due to the author and is borrowed from [KM1]. Note also that an elegant deduction of Theorem 2.7 directly from M. Krein's formula [K], [KO] due to V. Kolmanovich is also contained therein (see [KM1, pp. 32–34]). One more proof of Theorem 2.7 based on Redheffer's products has been obtained in [FF].

Next we apply Theorem 2.7 to generalize some Vishik's results [V]. For this purpose we consider a dual pair $\{A_1, A_2\}$ of closed (not necessary bounded) operators in $\mathfrak{H}$ with domains $\mathcal{D}(A_1)$ and $\mathcal{D}(A_2)$ not necessary dense in $\mathfrak{H}$. This means that $(A_1 f, g) = (f, A_2 g)$, for $f \in \mathcal{D}(A_1)$, $g \in \mathcal{D}(A_2)$.

Proposition 2.10 *Let $\{A_1, A_2\}$ be a dual pair of closed operators. Suppose that for some $\lambda_0 \in \mathbb{C}$ and $\varepsilon > 0$*

$$\|A_1 f - \lambda_0 f\| > \varepsilon \|f\|, \quad f \in \mathcal{D}(A_1), \quad \|A_2 g - \bar{\lambda}_0 g\| > \varepsilon \|g\|, \quad g \in \mathcal{D}(A_2).$$

Then: 1) *There exists a linear relation $\tilde{A} \in Ext_{\{A_1, A_2\}}$ such that $\mathbb{D}(\lambda_0; \varepsilon) := \{\lambda \in \mathbb{C} : |\lambda - \lambda_0| < \varepsilon\} \subset \rho(\tilde{A})$;*

2) *If additionally $\mathcal{D}(A_2)$ is dense in $\mathfrak{H}$ $(\overline{\mathcal{D}(A_2)} = \mathfrak{H})$ then each extension $\tilde{A} \in Ext_{\{A_1, A_2\}}$ is an operator;*

3) *If $B_i := (A_i - \lambda_0)^{-1} \in \mathfrak{S}_p$ (that is $(B_i B_i^*)^{1/2} \in \mathfrak{S}_p$) with $p \in (0, \infty]$ then there exists $\tilde{A} \in Ext_{\{A_1, A_2\}}$ with $(\tilde{A} - \lambda_0)^{-1} \in \mathfrak{S}_p$ and satisfying 1).*

Proof: 1) Setting $T_i = \varepsilon(A_i - \lambda_0)^{-1}$ $(i = 1, 2)$ one arrives at DPC $\{T_1, T_2\}$ with contractions $T_i \in [\mathfrak{H}_i, \mathfrak{H}]$ where $\mathfrak{H}_i := \mathcal{R}(A_i - \lambda_0)$, $(i = 1, 2)$. By Theorem 2.7 $Ext_{\{T_1, T_2\}}(\pi/2) \neq \emptyset$. For an arbitrary $T := T_K \in Ext_{\{T_1, T_2\}}$ $(\pi/2)$ one puts $\tilde{A} := \varepsilon T^{-1} + \lambda_0 I$. It is clear that $\tilde{A}$ is a linear relation, $\tilde{A} \in Ext_{\{A_1, A_2\}}$ and $\mathbb{D}(\lambda_0; \varepsilon) \subset \rho(\tilde{A})$.

2) Since $\overline{\mathcal{D}(A_2)} = \mathfrak{H}$ then A_2^* is a densely defined operator in $\mathfrak{H}$. The inclusion $\tilde{A} \in Ext_{\{A_1, A_2\}}$ means that $A_1 \subset \tilde{A} \subset A_2^*$ and hence $\tilde{A}$ is an operator.

3) If $(A_i - \lambda_0)^{-1} \in \mathfrak{S}_p$ then $T_i \in \mathfrak{S}_p$ $(i = 1, 2)$. By formula (2.3) the extension $T := T_K \in \mathfrak{S}_p$ iff $T_{22} \in \mathfrak{S}_p$. Since $T_{11}^* \in \mathfrak{S}_p$ along with T_1 formula (2.15) yields: $K \in \mathfrak{S}_p \Longrightarrow T_{22} \in \mathfrak{S}_p$ which is what had to be proved. $\qquad\square$

Remark 2.11 a) M. Vishik [V] has proved that a DP $\{A_1, A_2\}$ possessing the conditions of Proposition 2.10 (with $\lambda_0 = \bar{\lambda}_0$) admits an extension $\tilde{A} \in Ext_{\{A_1, A_2\}}$ with $\lambda_0 \in \rho(\tilde{A})$, and moreover $(\tilde{A} - \lambda_0)^{-1} \in \mathfrak{S}_\infty$ if $(A_i - \lambda_0)^{-1} \in \mathfrak{S}_\infty$ $(i = 1, 2)$. One obtains both results by choosing an arbitrary $T_{22} \in \mathfrak{S}_\infty$ and setting $T = (T_{ij})_{i,j=1}^2 \in Ext_{\{T_1, T_2\}}$ and $\tilde{A} = \varepsilon T^{-1} + \lambda_0 I$. Thus, Proposition 2.10 improves and complements mentioned results from [V].

b) Let now $A_1 = A_2 =: A$, i.e., A be a symmetric operator. In this case the extension $\tilde{A} = \tilde{A}^*$ with $\lambda_0 \in \rho(\tilde{A})$ may be choosen as $\tilde{A} = A_{\lambda_1}$ with $\mathcal{D}(A_{\lambda_1}) = \mathcal{D}(A) \dotplus \mathfrak{N}_{\lambda_1}$, where $\lambda_1 \in \mathbb{R}$, $|\lambda_1 - \lambda_0| < \varepsilon$ and ε is small enough.

c) Let A_λ be an extension of A defined as a restriction of A^* to $\mathcal{D}(A_\lambda) = \mathcal{D}(A) \dotplus \mathfrak{N}_\lambda, \lambda \in \mathbb{C}_+ \cup \mathbb{C}_-$. These extensions play an essential role in extension theory of symmetric operators. In particular they allow us to obtain the following generalization of J. von Neumann formula

$$\mathcal{D}(A^*) = \mathcal{D}(A_{\lambda_1}) + \mathcal{D}(A_{\lambda_2})$$

(2.28)
$$= \mathcal{D}(A) \dotplus \mathfrak{N}_{\lambda_1} \dotplus \mathfrak{N}_{\lambda_2} \qquad (\lambda_1 \in \mathbb{C}_+, \ \lambda_2 \in \mathbb{C}_-).$$

If $\lambda_1 \in \mathbb{C}_+$ then the extension $A_{\lambda_1} (\in Ext_A)$ is a maximal dissipative operator and thus $\mathbb{C}_- \subset \rho(A_{\lambda_1})$. It is shown in [M5] that for each proper extension $\tilde{A}$ $(A \subset \tilde{A} \subset A^*)$ the inclusion $\lambda_0 \in \rho(\tilde{A})$ holds iff the extensions $\tilde{A}$ and $\tilde{A}_{\lambda_0}$ are transversal, that is $\mathcal{D}(\tilde{A}) \cap \mathcal{D}(A_{\lambda_0}) = \mathcal{D}(A)$ and $\mathcal{D}(A^*) = \mathcal{D}(\tilde{A}) + \mathcal{D}(A_{\lambda_0})$. Thus for each $\lambda_2 \in \mathbb{C}_-$ the extensions A_{λ_1} and $A_{\lambda_2} \in Ext_A$ are transversal, what amounts to saying that the formula (2.28) holds.

In turn this formula makes it possible to present a very simple proof of the well known result [AG] on the constancy in each of the half-planes $\mathbb{C}_+$ and $\mathbb{C}_-$ of the deficiency indices $dim\mathfrak{N}_\lambda(A) = dimker(A^* - \lambda)$. Indeed, setting in (2.28) $\lambda_2 = -i$ and $\lambda := \lambda_1$ one gets

$$dim\mathfrak{N}_\lambda(A) = dim\mathcal{D}(A^*)/(\mathcal{D}(A) \dotplus \mathfrak{N}_{-i}), \quad \lambda \in \mathbb{C}_+.$$

Thus $dim\mathfrak{N}_\lambda(A) = dim\mathfrak{N}_i(A) = n_+(A) \quad for \ \lambda \in \mathbb{C}_+$.

2.3 Here we investigate some classes of noncontractive extensions of a DPC.

Definition 2.12 A dual pair of contractions $\{T_1, T_2\}$ of the form (2.14) will be called nonsingular if ker $D_U = $ ker $D_{V^*} = \{0\}$ and will be called transversal if $0 \in \rho(D_U) \cap \rho(D_{V^*})$.

Definition 2.13 Let $\{T_1, T_2\}$ be a DPC of the form (2.14). We denote by $\mathrm{Ext}_{\{T_1,T_2\}}$ the set of all its (not necessary contractive) bounded extensions $T_K := (T_{ij})_{i,j=1}^2$ with

(2.29)
$$T_{22} = -VT_{11}^*U + D_{V^*}KD_U, \ K \in [\mathcal{H}_1, \mathcal{H}_2]$$

Observe that only if a DPC $\{T_1, T_2\}$ is transversal, the totality of all its bounded extensions coincides with $\mathrm{Ext}_{\{T_1,T_2\}}$, though Theorem 2.7 yields $\mathrm{Ext}_{\{T_1,T_2\}} \supset \mathrm{Ext}_{\{T_1,T_2\}}(\pi/2)$. Note also that in the nonsingular case $\mathcal{H}_1 = \mathfrak{H}_2$ and $\mathcal{H}_2 = \mathfrak{H}_2'$.

From now on we as a rule consider nonsingular DPCs, since the general case may be considered without additional difficulties.

Theorem 2.14 *Let* $\mathfrak{H} = \mathfrak{H}_1 \oplus \mathfrak{H}_2 = \mathfrak{H}'_1 \oplus \mathfrak{H}'_2$, $\{T_1, T_2\}$ *be a dual pair of contractions of the form* (2.14), $T_K \in \mathrm{Ext}_{\{T_1, T_2\}}$ *and*

$$(2.30) \qquad (G_{ij})_{i,j=1}^2 := I - T_K^* T_K, \qquad (S_{ij})_{i,j=1}^2 := I - T_K T_K^*$$

where $G_{ij} \in [\mathfrak{H}_j, \mathfrak{H}_i]$, $S_{ij} \in [\mathfrak{H}'_j, \mathfrak{H}'_i]$. *Then*

1. $\mathfrak{R}(G_{11}^{1/2}) \supset \mathfrak{R}(G_{12})$, $\mathfrak{R}(S_{11}^{1/2}) \supset \mathfrak{R}(S_{12})$ *and consequently the operators* $G_{11}^{-1/2} G_{12}$ *and* $S_{11}^{-1/2} S_{12}$ *are well-defined and bounded;*
2. *the identities*

   $$(2.31) \quad G_{22} - (G_{11}^{-1/2} G_{12})^* (G_{11}^{-1/2} G_{12}) = D_U (I - K^* K) D_U$$

   $$(2.32) \qquad S_{22} - (S_{11}^{-1/2} S_{12})^* (S_{11}^{-1/2} S_{12}) = D_{V^*} (I - K K^*) D_{V^*}$$

   are valid.

Proof: 1) It is easily seen that

$$(2.33) \qquad G_{11} = I - T_{11}^* T_{11} - D_{T_{11}} V^* V D_{T_{11}} = D_{T_{11}} D_V^2 D_{T_{11}}$$

$$
\begin{aligned}
-G_{21} &= -G_{12}^* = T_{12}^* T_{11} + T_{22}^* T_{21} = U^* D_{T_{11}^*} T_{11} \\
&\quad -(U^* T_{11} V^* - D_U K^* D_{V^*}) V D_{T_{11}}
\end{aligned}
$$

$$
\begin{aligned}
(2.34) \qquad &= U^* T_{11} D_V^2 D_{T_{11}} + D_U K^* V D_V D_{T_{11}} \\
&= (U^* T_{11} D_V + D_U K^* V) D_V D_{T_{11}}
\end{aligned}
$$

The relations (2.33) and (2.34) together yield the inequality $\|G_{12}^* f\|^2 \le c^2 (G_{11} f, f) \; \forall f \in \mathfrak{H}_1$ with $c = \|U^* T_{11} D_V + D_U K^* V\|$. Combining this inequality with Lemma 2.3 one derives the inclusion $\mathfrak{R}(G_{12}) \subset \mathfrak{R}(G_{11}^{1/2})$. Thus the operator $G_{11}^{-1/2} G_{12}$ is well defined and bounded.

The inclusion $\mathfrak{R}(S_{12}) \subset \mathfrak{R}(S_{11}^{1/2})$ also follows from Lemma 2.3 and the equalities

$$S_{11} = D_{T_{11}^*} D_{U^*}^2 D_{T_{11}^*},$$

$$(2.35)$$

$$-S_{21} = -S_{12}^* = (V T_{11}^* D_{U^*} + D_{V^*} K U^*) D_{U^*} D_{T_{11}^*},$$

which may be established similarly.

2a) At first we prove the equality (2.31) under additional assumption $0 \in \rho(D_{T_{11}}) \cap \rho(D_V)$. In this case $0 \in \rho(G_{11})$ and the left-hand side of (2.31) takes the form

$$G_{22} - G_{12}^* G_{11}^{-1} G_{12} = I - T_{12}^* T_{12} - Z^* Z$$

$$(2.36)$$

$$-(T_{12}^* T_{11} + Z^* T_{21})(I - T_{11}^* T_{11} - T_{21}^* T_{21})^{-1} (T_{11}^* T_{12} + T_{21}^* Z)$$

where as before $Z = T_{22}$. Let us transform (2.36) with regard to (2.14) and (2.33) and the obvious equalities $T^*D_{T^*} = D_T T^*$, $TD_T = D_{T^*}T$. We have

(2.37) $\qquad T_{21}G_{11}^{-1}T_{11}^*T_{12} = VD_V^{-2}T_{11}^*U = D_{V^*}^{-2}VT_{11}^*U,$

(2.38) $\qquad T_{12}^*T_{11}G_{11}^{-1}T_{21}^* = U^*T_{11}V^*D_{V^*}^{-2}.$

And finally

$$T_{12}^*[I + T_{11}G_{11}^{-1}T_{11}^*]T_{12}$$
$$= U^*[D_{T_{11}^*}^2 + D_{T_{11}^*}^*T_{11}(D_{T_{11}}^{-1}D_V^{-2}D_{T_{11}}^{-1})T_{11}^*D_{T_{11}}]U$$
$$= U^*[D_{T_{11}^*}^2 + T_{11}D_V^{-2}T_{11}^*]U$$

$$T_{21}G_{11}^{-1}T_{21}^* = VD_{T_{11}}(D_{T_{11}}^{-1}D_V^{-2}D_{T_{11}}^{-1})D_{T_{11}}V^*$$
(2.39)
$$= VD_V^{-2}V^*$$

Combining the relations (2.37)–(2.39) we obtain

$$G_{22} - G_{12}^*G_{11}^{-1}G_{12} = I - U^*[D_{T_{11}^*}^2 + T_{11}D_V^{-2}T_{11}^*]U$$
(2.40)
$$-Z^*D_{V^*}^{-2}VT_{11}^*U - U^*T_{11}V^*D_{V^*}^{-2}Z - Z^*(I + VD_V^{-2}V^*)Z$$

Let us now transform the right-hand side of the equality (2.31). It follows from (2.29) that $KD_U = D_{V^*}^{-1}(Z + VT_{11}^*U)$ and, consequently

$$D_U(I - K^*K)D_U$$
$$= D_U^2 - (Z^* + U^*T_{11}V^*)D_{V^*}^{-2}(Z + VT_{11}^*U)$$
(2.41)
$$= I - U^*[I + T_{11}V^*D_{V^*}^{-2}VT_{11}^*]U$$
$$-Z^*D_{V^*}^{-2}VT_{11}^*U - U^*T_{11}V^*D_{V^*}^{-2}Z - Z^*D_{V^*}^{-2}Z.$$

Now one derives (2.31) from (2.30) and (2.41) and the following relations:

$$I + T_{11}V^*D_{V^*}^{-2}VT_{11}^* = I + T_{11}D_V^{-2}V^*VT_{11}^* = D_{T_{11}^*}^2 + T_{11}D_V^{-2}T_{11}^*,$$
$$I + VD_V^{-2}V^* = I + D_{V^*}^{-2}VV^* = I + D_{V^*}^{-2}(I - D_{V^*}^2) = D_{V^*}^{-2}.$$

Thus under the condition $0 \in \rho(D_{T_{11}}) \cap \rho(D_V)$ the equality (2.31) is proved.

2b) Let us now prove (2.31) not assuming that $0 \in \rho(D_{T_{11}}) \cap \rho(D_V)$. In order to do that we consider the DPC $\{rT_1, rT_2\}$ with $r \in (0, 1)$ along with the DPC $\{T_1, T_2\}$. Then $rT_{21} = V(r)D_{rT_{11}}$, $rT_{12} = D_{rT_{11}^*}U(r)$ where $U(r)$ and $V(r)$ are defined by the equalities (2.24). Since

$$(2.42) \qquad \|U(r)f - Uf\|^2 = \int_0^1 \left|\frac{t(1-r^2)}{1-r^2t}\right|^2 d(E_t Uf, Uf),$$

where E_t is the resolution of the identity of the operator $T_{11}T_{11}^*$, and the integrand in (2.42) is bounded by the number 1 one obtains from the Lebesgue dominated convergence theorem that $s - \lim_{r \uparrow 1} U(r) = U$. The relations

$$s - \lim_{r \uparrow 1} V(r) = V, \qquad s - \lim_{r \uparrow 1} D_{U(r)} = D_U,$$

$$(2.43)$$
$$s - \lim_{r \uparrow 1} D_{V^*(r)} = D_{V^*}$$

are established similarly.

Further let $T_K(r) \in Ext_{\{rT_1, rT_2\}}$ be an extension of the DPC $\{rT_1, rT_2\}$ defined by the operator $K \in [\mathcal{H}_1, \mathcal{H}_2]$, that is

$$T_K(r) = \begin{pmatrix} rT_{11} & rT_{12} \\ rT_{21} & T_{22}(r) \end{pmatrix},$$

$$(2.44)$$
$$T_{22}(r) = rV(r)T_{11}^*U(r) + D_{V^*(r)}KD_{U(r)}$$

Since $0 \in \rho(D_{rT_{11}}) \cap \rho(D_{V(r)})$ then for the operator $I - T_K^*(r)T_K(r) =: (G_{ij}(r))_{i,j=1}^2$ the equality (2.31) is already proved, that is

$$(2.45) \quad G_{22}(r) - G_{12}^*(r)G_{11}^{-1}(r)G_{12}(r) = D_{U(r)}(I - K^*K)D_{U(r)}$$

Assume without loss of generality that $\ker G_{11} = \{0\}$. Then by virtue of (2.33) one derives the equality $G_{11}^{1/2} = U_1 D_V D_{T_{11}} = D_{T_{11}}D_V U_1^*$ with operator U_1 being unitary in $\mathfrak{H}_1$. Combining the last equality with (2.34) one obtains

$$(2.46) \quad -G_{11}^{-1/2}G_{12} = U_1 X^*, \qquad X = U^*T_{11}D_V + D_V K^*V$$

On the other hand similarly to (2.33) one obtains $G_{11}(r) = D_{rT_{11}}D_{V(r)}^2 D_{rT_{11}}$ and consequently $G_{11}^{1/2}(r) = U_1(r)D_{V(r)}D_{rT_{11}}$ with unitary operator $U_1(r)$.

Further in complete analogy with (2.34) we derive

$$-G_{21}^*(r) = X(r)D_{V(r)}D_{rT_{11}},$$
$$X(r) := rU^*(r)T_{11}D_{V(r)} + D_{U(r)}K^*V(r)$$

It follows that

$$-G_{12}^*(r)G_{11}^{-1/2}(r) = X(r)U_1^*(r),$$

(2.47)

$$-G_{11}^{-1/2}(r)G_{12}(r) = U_1(r)X^*(r)$$

Combining (2.46) and (2.47) with (2.43) we arrive at the relation

$$s - \lim_{r \uparrow 1} G_{12}^*(r)G_{11}^{-1}(r)G_{12}(r) = s - \lim_{r \uparrow 1} X(r)X^*(r)$$

(2.48)

$$= (G_{11}^{-1/2}G_{12})^*(G_{11}^{-1/2}G_{12})$$

To complete the proof of the equality (2.31) it remains to pass to the limit in (2.45) as $r \uparrow 1$ and take advantage of (2.44) and (2.48).

One can prove the relation (2.32) in just the same way making use of the equalities (2.35). □

Remark 2.15 We refer the paper [ACG] in connection with Theorem 2.14. In this paper the authors started with a bounded operator $T \in [\mathcal{K}_1, \mathcal{K}_2]$, where $\mathcal{K}_i$ is a Pontryagin space with a fundamental symmetry J_i, $\kappa_-(\mathcal{K}_i) := \kappa_-(J_i) < \infty$ ($i = 1, 2$). Assuming that $\kappa_-(J_1 - T^*J_2T) = \kappa$ (κ is a given cardinal) they have presented a description of all extensions $\widetilde{T} = \begin{pmatrix} T & A \\ B & X \end{pmatrix} \in [\widetilde{\mathcal{K}}_1, \widetilde{\mathcal{K}}_2]$ ($\widetilde{\mathcal{K}}_i$ is a Pontryagin space $\widetilde{\mathcal{K}}_i = \mathcal{K}_i \oplus \mathcal{K}_i'$), satisfying the inequality

$$\widetilde{\kappa} := \kappa_-(\widetilde{T}) = \kappa - \kappa_-(\widetilde{\mathcal{K}}_2) + \kappa_-(\mathcal{K}_2) \geq \kappa_-(\widetilde{\mathcal{K}}_1) - \kappa_-(\widetilde{\mathcal{K}}_2).$$

In the Hilbert spaces case $\kappa_-(\mathcal{K}_i) = \kappa_-(\widetilde{\mathcal{K}}_i) = 0$ ($i = 1, 2$) and the above relation takes a form $\kappa_-(\widetilde{T}) = \kappa_-(T) \geq 0$. If T is a Hilbert space contraction, that is $\kappa_-(T) = 0$ then so is $\widetilde{T}$ and the result from [ACG] is reduced to Theorem 2.7. Emphasize that in Theorem 2.14 as distinguished from [ACG] the extensions of class $\text{Ext}_{\{T_1, T_2\}}$ satisfying $0 = \kappa_-(T_1) = \kappa_-(T_2) < \kappa_-(\widetilde{T})$ are described.

It is natural to complete Theorem 2.14 by the following

Proposition 2.16 *Let* $T = (T_{ij})_{i,j=1}^2$ *be a bounded operator in* $\mathfrak{H} = \mathfrak{H}_1 \oplus \mathfrak{H}_2 = \mathfrak{H}_1' \oplus \mathfrak{H}_2'$ *and* $G = (G_{ij})$ *and* $S = (S_{ij})$ *be as above. Then the following equivalences hold*

(2.49)   1) $\mathfrak{R}(G_{11}^{1/2}) \supset \mathfrak{R}(G_{12}) \Longleftrightarrow T_{22} = -VT_{11}^*U + D_{V^*}K_1$

(2.50)   2) $\mathfrak{R}(S_{11}^{1/2}) \supset \mathfrak{R}(S_{12}) \Longleftrightarrow T_{22} = -VT_{11}^*U + K_2D_U$

Proof: Let us establish the first equivalence. The implication $\Longleftarrow$ was established in the course of the proof of Theorem 2.14.

Conversely writing T_{22} as $T_{22} = -VT_{11}^* U + X$ one obtains the relation

$$(2.51) \qquad G_{12}^* = U^* T_{11} D_V^2 D_{T_{11}} + X^* V D_{T_{11}}$$

just like it was done in the proof of (2.31). By Lemma 2.3 the inclusion $\Re(G_{11}^{1/2}) \supset \Re(G_{12})$ is equivalent to the inequality $\|G_{12}^* f\| \le c_1 \|G_{11}^{1/2} f\|^2 \, \forall f \in \mathfrak{H}_1$ with some $c_1 > 0$. Combining this inequality with (2.51) and (2.33) one derives the estimate $\|X^* V f\|^2 \le c_2 \|D_V f\|^2 \, \forall f \in \overline{\Re(D_V)}$ with some $c_2 > 0$. Hence by virtue of Lemma 2.3 $V^* X = D_V K_3$ with some bounded operator K_3 and consequently

$$
\begin{aligned}
(I - D_{V^*}^2)X &= V D_V K_3 = D_{V^*} V K_3 \Longrightarrow X \\
&= D_{V^*}(D_{V^*} X + V K_3) =: D_{V^*} K_1
\end{aligned}
$$

which is what had to be proved.

The equivalence (2.50) can be proved similarly.   □

Remark 2.17 Note that the equalities (2.49) and (2.50) are necessary but not sufficient for the inclusion $T \in \mathrm{Ext}_{\{T_1, T_2\}}$ which may be characterized by means of the inequality

$$|((T_{22} + VT_{11}^* U)f, g)| \le c(\|D_U f\|^2 + \|D_{V^*} g\|^2) \qquad \forall f, g \in \mathfrak{H}_2.$$

2.4 Here we present some corollaries from Theorem 2.14. To formulate them we need some definitions and two elementary lemmas.

Let $\varkappa(t)$ be the number of negative squares of the symmetric quadratic form t, that is the maximum dimensions of the "negative" lineals

$$L_- = \{f \in \mathcal{D}(t) \setminus \{0\} : t[f] < 0\} \cup \{0\}$$

For a selfadjoint operator $T = T^* \in \mathcal{C}(\mathfrak{H})$ with the resolution of the identity $E_T(\lambda)$ we define $\varkappa_-(T)$ by the equation $\varkappa_-(T) = \dim E_T(-\infty, 0)\mathfrak{H}$. If the form t is closed and the operator T is associated with it, $(t = t_T)$ then by virtue of the minimax principle $\varkappa_-(t) = \varkappa_-(T)$.

Lemma 2.18 *Let a monotonically decreasing sequence of symmetric upper semibounded forms t_n converge to a form t, i.e.*

$$\lim_{n \to \infty} t_n[f] = t[f] \qquad \forall f \in \mathcal{D}(t) = \cup_{n=1}^{\infty} \mathcal{D}(t_n)$$

Then $\varkappa_-(t) = \varkappa_-(t_n)$ for sufficiently large $n \ge N$. If the forms t_n are closed and $T_n = T_n^*$ are operators associated with them then $\varkappa_-(t) = \varkappa_-(T_n) \, \forall n \ge N$.

Lemma 2.19 *Let $T_0 = \binom{T_{11}}{T_{21}}(\in [\mathfrak{H}_1, \mathfrak{H}])$ be a Hermitian operator admitting a bounded nonnegative selfadjoint extension. Then:*

1) $\Re(T_{11}^{1/2}) \supset \Re(T_{12})$ *and the operator* $S := T_{11}^{-1/2}T_{12}$ *is well-defined and bounded;*

2) *for each selfadjoint extension* $T (\supset T_0)$ *with the block-matrix representation* $T = T^* = (T_{ij})_{i,j=1}^2$ *with respect to the orthogonal decomposition* $\mathfrak{H} = \mathfrak{H}_1 \oplus \mathfrak{H}_2$, *the equality* $\varkappa_-(T) = \varkappa_-(T_{22} - S^*S)$ *holds true.*

Definition 2.20 Let $\varkappa \in \mathbb{Z}_+$ and $\mathfrak{S}$ be a two-sided ideal in $[\mathfrak{H}]$. We shall write

a) $T \in C(\pi/2; \varkappa)$ if $T \in [\mathfrak{H}, \mathfrak{H}']$ and $\varkappa_-(I - T^*T) = \varkappa$;

b) $T \in C(\pi/2; \mathfrak{S})$ if $T \in [\mathfrak{H}, \mathfrak{H}']$ and $(I - T^*T)_- \in \mathfrak{S}$.

Now we are ready to present the corollaries.

Corollary 2.21 *Let* $\{T_1, T_2\}$ *be a nonsingular DPC and* $T_K \in \mathrm{Ext}_{\{T_1,T_2\}}$. *Then the equivalence*

$$(2.52) \qquad T_K \in C(\pi/2; \varkappa) \iff K \in C(\pi/2; \varkappa)$$

is valid.

Proof: Let as before $(G_{ij})_{i,j=1}^2 := I - T_K^*T_K$. The operator $G_0 := \binom{G_{11}}{G_{21}}$ is nonnegative $(G_{11} \geq 0)$ and admits a bounded nonnegative selfadjoint extension. For example the operator

$$\begin{pmatrix} G_{11}^{1/2} & 0 \\ S^* & 0 \end{pmatrix} \begin{pmatrix} G_{11}^{1/2} & S \\ 0 & 0 \end{pmatrix} = \begin{pmatrix} G_{11} & G_{12} \\ G_{21} & S^*S \end{pmatrix} \geq 0$$

with bounded $S = G_{11}^{-1/2}G_{12}$ is such a one. Therefore from Lemma 2.19 and the equality (2.31) we deduce $\varkappa_-(I - T_K^*T_K) = \varkappa_-(G_{22} - (G_{11}^{-1/2}G_{12})^*(G_{11}^{-1/2})) = \varkappa_-(I - K^*K)$. $\qquad\square$

Corollary 2.22 *Let* $\{T_1, T_2\}$ *be a DPC such that* $0 \in \rho(D_{T_{11}}) \cap \rho(D_V)$ *and let* $T_K \in \mathrm{Ext}_{\{T_1,T_2\}}$. *Then the implication*

$$(2.53) \qquad (I - K^*K)_- \in \mathfrak{S} \Longrightarrow (I - T_{K^*}T_K)_- \in \mathfrak{S}$$

holds. If the DPC is transversal (that is in addition $0 \in \rho(D_U)$*), then the implication* (2.53) *is replaced by equivalence.*

Proof: The required assertion immediately follows from (2.31) and the identity

$$\begin{pmatrix} I & 0 \\ -G_{21}G_{11}^{-1} & I \end{pmatrix} \begin{pmatrix} G_{11} & G_{12} \\ G_{21} & G_{22} \end{pmatrix} \begin{pmatrix} I & -G_{11}^{-1}G_{12} \\ 0 & I \end{pmatrix}$$
$$= \begin{pmatrix} G_{11} & 0 \\ 0 & G_{22} - G_{21}G_{11}^{-1}G_{12} \end{pmatrix}.$$

$\qquad\square$

2.5 In this subsection we complement Theorem 2.14 by two additional corollaries. For this purpose we recall some well-known results and the definition of shorted operator.

Proposition 2.23 ([Sh]). *Let $\mathfrak{H} = \mathfrak{H}_1 \oplus \mathfrak{H}_2$ be an orthogonal decomposition of a Hilbert space $\mathfrak{H}$, $A^0 = \begin{pmatrix} A_{11} & A_{12} \\ A_{21} & \end{pmatrix}$ be an incomplete block matrix with entries $A_{ij} \in [\mathfrak{H}_j, \mathfrak{H}_i]$. If $A_{11} \geq 0$ then*

1) *An operator A_{22} complementary to the matrix A^0 and such that*

$$(2.54) \qquad\qquad A := \begin{pmatrix} A_{11} & A_{12} \\ A_{21} & A_{22} \end{pmatrix} \geq 0$$

exists iff $\mathfrak{R}(A_{11}^{1/2}) \supset \mathfrak{R}(A_{12})$;

2) *In this case the operator $S := A_{11}^{-1/2} A_{12}$ is well defined, $S \in [\mathfrak{H}_2, \mathfrak{H}_1]$ and the operator $A_{22} = S^* S$ is the smallest in the set*

$$M = \{A_{22} \in [\mathfrak{H}_2] : A = (A_{ij})_{i,j=1}^2 \geq 0\}$$

Definition 2.24 ([K]). For each nonnegative operator $A(\in [\mathfrak{H}])$ and subspace $\mathfrak{N}(\subset \mathfrak{H})$ there exists the largest element in the set of all bounded operators not exceeding A and annihilating $\mathfrak{N}^\perp = \mathfrak{H} \ominus \mathfrak{N}$. This element, is denoted by $A_\mathfrak{N}$ and is called the shorted to $\mathfrak{N}$ operator. The transformation $A \to A_\mathfrak{N}$ is called the Krein transformation.

Corollary 2.25 ([K], [Sh]). *Let $A = (A_{ij})_{i,j=1}^2$ be a block-matrix representation of an operator $A \geq 0$ $(A \in [\mathfrak{H}])$ with respect to the decomposition $\mathfrak{H} = \mathfrak{H}_1 \oplus \mathfrak{N}$. Then the shorted to $\mathfrak{N}$ operator $A_\mathfrak{N}$ is of the form*

$$(2.55) \qquad A_\mathfrak{N} = \begin{pmatrix} 0 & 0 \\ 0 & A_{22} - S^* S \end{pmatrix}, \qquad S = A_{11}^{-1/2} A_{12}$$

If in addition $0 \in \rho(A_{11})$ then $S^* S = A_{21} A_{11}^{-1} A_{12}$.

Corollary 2.26 ([K], [Sh]). *Let $\mathfrak{H} = \mathfrak{H}_1 \oplus \mathfrak{A}$, $A \in [\mathfrak{H}]$ and $A \geq 0$. Then*

$$(2.56) \qquad \inf_{g \in \mathfrak{H}_1} (A(f - g), f - g) = (A_\mathfrak{N} f, f) \qquad \forall f \in \mathfrak{H}$$

Now we are ready to present two complementary corollaries from Theorem 2.14.

Corollary 2.27 *Let under the conditions of Theorem 2.7 $T_K \in \text{Ext}_{\{T_1,T_2\}}(\pi/2)$. Then the operators D_T^2 and D_{T*}^2, shorted to $\mathfrak{H}_1$ and $\mathfrak{H}_2'$ respectively, have the form*

(2.57)

$$(D_{T_K}^2)_{\mathfrak{H}_2} = \begin{pmatrix} 0 & 0 \\ 0 & D_U D_K^2 D_U \end{pmatrix},$$

$$(D_{T_K^*}^2)_{\mathfrak{H}_2'} = \begin{pmatrix} 0 & 0 \\ 0 & D_{V*} D_K^2 D_{V*} \end{pmatrix}$$

Proof: One deduces the proof combining Theorem 2.14 and Corollary 2.25. □

The equalities (2.57) allow us to describe the set $\text{Ext}_{\{T_1,T_2\}}^E(\pi/2)$ of extreme points of the set $\text{Ext}_{\{T_1,T_2\}}(\pi/2)$ in the following manner.

Corollary 2.28 *Let $T_K \in \text{Ext}_{\{T_1,T_2\}}(\pi/2)$. Then $T_K \in \text{Ext}_{\{T_1,T_2\}}^E(\pi/2)$ if and only if at least one of the following equalities is valid:*

$$(2.58) \qquad (I - T_K^* T_K)_{\mathfrak{H}_2} = 0 \ (\Longleftrightarrow \Re(I - T_K^* T_K)^{1/2} \cap \mathfrak{H}_2 = \{0\})$$

$$(2.59) \qquad (I - T_K T_K^*)_{\mathfrak{H}_2'} = 0 \ (\Longleftrightarrow \Re(I - T_K T_K^*)^{1/2} \cap \mathfrak{H}_2' = \{0\})$$

Proof: As it is known, the extreme points of the unite ball $B_1 = \{X \in [\mathfrak{H}_2, \mathfrak{H}_2'] : \|X\| \leq 1\}$ are the isometries $(D_X^2 = I - X^*X = 0)$ and co-isometries $(D_{X*}^2 = I - XX^* = 0)$. It is clear that the extreme points of $\text{Ext}_{\{T_1,T_2\}}(\pi/2)$ are exactly the images of the extreme points of B_1 under the mapping $K \to T_K$ of the form (2.29). Due to (2.57) these points are described by the equalities (2.58), (2.59). □

Remark 2.29 We also note that by virtue of Corollary 2.26 the relation (2.58) (2.59) is equivalent to the equality

$$\inf_{g \in \mathfrak{H}_1} \|D_{T_K}(f - g)\| = 0 \quad \forall f \in \mathfrak{H}$$

$$\left(\inf_{g \in \mathfrak{H}_1} \|D_{T_K^*}(f - g)\| = 0 \quad \forall f \in \mathfrak{H} \right).$$

2.6 Here we present some applications of Theorem 2.14 to the boundary value problems.

For this we recall the following definitions.

Definition 2.30 ([GG]). Let A be a nondensely defined symmetric operator in a separable Hilbert space $\mathfrak{H}$ with equal defect numbers $n_+(A) = n_-(A)$. A collection $\{\mathcal{H}, \Gamma_0, \Gamma_1\}$ in which $\mathcal{H}$ is a Hilbert space and $\Gamma_i (i = 0, 1)$ linear mappings from $A^* \subset \mathfrak{H} \oplus \mathfrak{H}$ to $\mathcal{H}$ is said to be a boundary triplet for A^* if:

a) $(f', g) - (f, g') = (\Gamma_1 \hat{f}_1 \Gamma_0 \hat{g}) - (\Gamma_0 \hat{f}, \Gamma_1 \hat{g}) \quad \forall \hat{f} = \{f, f'\}, \hat{g} = \{g, g'\}$

b) the mapping $\Gamma : \hat{f} \to (\Gamma_0 \hat{f}, \Gamma_1 \hat{f})$ from A^* to $\mathcal{H} \oplus \mathcal{H}$ is surjective.

The mapping Γ establishes a bijective correspondence between the proper extensions $\widetilde{A}$ $(A \subset \widetilde{A} \subset A^*)$ of the operator A and the closed linear relations $\theta := \Gamma\widetilde{A}$ in $\mathcal{H}$:

$$\widetilde{A} = \widetilde{A}_\theta \Longleftrightarrow \theta := \Gamma\widetilde{A} = \{\{\Gamma_0 \hat{f}, \Gamma_1 \hat{f}\} : \hat{f} \in \widetilde{A}\}$$

It is clear that the extensions $A_0 := ker\Gamma_0$ and $A_1 := ker\Gamma_1$ correspond to the relations $\theta_0 := \{\{0, h\} : h \in \mathcal{H}\}$ and $\theta_1 := \{\{h, 0\} : h \in \mathcal{H}\}$ respectively. Let $\mathfrak{N}_\lambda = \mathfrak{H} \ominus (A - \bar{\lambda})\mathcal{D}(A)$ be the defect subspace of A, $\hat{\mathfrak{N}}_\lambda = \{\{f_\lambda, \lambda f_\lambda\} : f_\lambda \in \mathfrak{N}_\lambda\}$.

Definition 2.31 ([DM], [M5]). The operator-valued function $M(\lambda)$ defined for $\lambda \in \rho(A_0)$ by the equality

$$M(\lambda)\Gamma_0 \hat{f}_\lambda = \Gamma_1 \hat{f}_\lambda , \qquad \hat{f}_\lambda = \{f_\lambda, \lambda f_\lambda\} \in \hat{\mathfrak{N}}_\lambda$$

is called the Weyl function of the operator A corresponding to the boundary triplet $\{\mathcal{H}, \Gamma_0, \Gamma_1\}$.

It is shown (see [DM] , [M5]) that the well-known formula of M.G. Krein for resolvents can in a fixed boundary triplet $\{\mathcal{H}, \Gamma_0, \Gamma_1\}$ be written in the form

$$(2.60) \qquad (\widetilde{A}_\theta - \lambda)^{-1} = (A_0 - \lambda)^{-1} + \gamma(\lambda)(\theta - M(\lambda))^{-1}\gamma^*(\bar{\lambda}),$$

where $\gamma(\lambda) := (\Gamma_0 \lceil \mathfrak{N}_\lambda)^{-1} = (A_0 - i)(A_0 - \lambda)^{-1}\gamma(i)$ is the γ-field of the operator A generated by A_0.

As usual we denote by A_F the Friedrich's extension of a nonnegative operator $A \geq 0$. If A is nondensely defined then A_F is always a linear relation.

Combining Proposition 2.23 and Corollary 2.21 one arrives at the following lemma.

Lemma 2.32 *Let* $T_0 = \binom{T_{11}}{T_{21}} \in [\mathfrak{H}_1, \mathfrak{H}]$ *be a symmetric operator from* $\mathfrak{H}_1$ *to* $\mathfrak{H}$ *and* $\mathcal{R}(T_{11}^{1/2}) \supset \mathcal{R}(T_{21}^*)$. *If* $T = T^* = (T_{ij})_{i,j=1}^2$ *is a bounded extension of* T_0 $(T \in [\mathfrak{H}])$ *and* $S = T_{11}^{-1/2}T_{21}^*$, *then*

$$(2.61) \qquad \kappa_-(T) = \kappa_-(T_{22} - S^*S).$$

Theorem 2.33 *Let* $A \geq 0$ *be a nondensely defined nonnegative operator in* $\mathfrak{H}$, $\{\mathcal{H}, \Gamma_0, \Gamma_1\}$ *be a boundary triplet for which* $A_0 = A_F$ *and* $M(\lambda)$ *be the corresponding Weyl function. If* $\theta = \theta^*$ *is semibounded below linear relation and* $\mathcal{D}[\theta] \subset \mathcal{D}[M(0)]$, *then:*

1)
$$(2.62) \qquad \kappa_-(\widetilde{A}_\theta) = \kappa_-(\mathfrak{t}_\theta - \mathfrak{t}_{M(0)}).$$

2) *If in addition* $\mathcal{D}(\theta) \subset \mathcal{D}(M(0))$, *(say, for example* $M(0) \in [\mathcal{H}])$ *then the equality* (2.62) *takes the form* $\kappa_-(\widetilde{A}_\theta) = \kappa_-(\theta - M(0))$.

3) *With $\beta < 0$ the following equivalence holds:*

$$(\tilde{A}_\theta - \beta)E_{\tilde{A}_\theta}(\beta) \in \mathfrak{S}_p \Longleftrightarrow (\theta - M(\beta))$$

(2.63)

$$E_{\theta - M(\beta)}(0) \in \mathfrak{S}_p$$

4) *If in addition $M(0) \in [\mathcal{H}]$ then the equivalence (2.63) is replaced by the implication:*

$$\theta - M(0) \in \mathfrak{S}_p \Longrightarrow \tilde{A}_\theta E_{\tilde{A}_\theta}(0) \in \mathfrak{S}_p.$$

Proof: Firstly we assume that there exists $\varepsilon > 0$ such that $(-\varepsilon_0, 0) \subset \rho(\tilde{A}_\theta)$. The operator $(A + \varepsilon)^{-1}$ is a bounded nonnegative operator from $\mathfrak{M}_{-\varepsilon} := (A + \varepsilon)\mathcal{D}(A)$ to $\mathfrak{H}$. Consider its block-matrix representation $(A + \varepsilon)^{-1} = \binom{T_{11}}{T_{21}}$ with respect to the orthogonal decomposition $\mathfrak{H} = \mathfrak{M}_{-\varepsilon} \oplus \mathfrak{N}_{-\varepsilon}$. Since $(A + \varepsilon)^{-1}$ admits a bounded nonnegative selfadjoint extension, say $(A_F + \varepsilon)^{-1}$ then by Proposition 2.23 $\mathcal{R}(T_{11}^{1/2}) \supset \mathcal{R}(T_{21}^*)$. According to the extremal property of the Friedrichs extension (see [K]) the operator $(A_F + \varepsilon)^{-1}$ is the smallest element in the set $\{(\tilde{A}_\theta + \varepsilon)^{-1} : \tilde{A}_\theta \supset A, \ \tilde{A}_\theta \geq 0\}$. By Proposition 2.23 one has

$$(A_F + \varepsilon)^{-1} = \begin{pmatrix} T_{11} & T_{21}^* \\ T_{21} & S^*S \end{pmatrix},$$

$$(\tilde{A}_\theta + \varepsilon)^{-1} = \begin{pmatrix} T_{11} & T_{21}^* \\ T_{21} & T_{22} \end{pmatrix},$$

where $S = T_{11}^{-1/2}T_{21}^*$. Now Lemma 2.32 yields

(2.64) $$\kappa_-(\tilde{A}_\theta + \varepsilon)^{-1} = \kappa_-((\tilde{A}_\theta + \varepsilon)^{-1} - (A_F + \varepsilon)^{-1})$$

On the other hand it follows from the resolvent formula (2.60) that $\kappa_-((\tilde{A}_e + \varepsilon) - (A_F + \varepsilon)^{-1}) = \kappa_-(\theta - M(-\varepsilon))^{-1}$. Combining this equality with (2.64) one arrives at the equality

$$\kappa_-(A_\theta + \varepsilon) = \kappa_-(A_\theta + \varepsilon)^{-1} = \kappa_-(\theta - M(-\varepsilon))^{-1}$$

(2.65)

$$= \kappa_-(\theta - M(-\varepsilon)), \quad \varepsilon \in (0, \varepsilon_0)$$

We obtain the required relation (2.62) by applying Lemma 2.32. For this purpose we set $t_\varepsilon := t_\theta - t_{M(-\varepsilon)}, t := t_{\theta - M(0)}$ and note that $\mathcal{D}(t_\varepsilon) = \mathcal{D}[\theta - M(-\varepsilon)] = \mathcal{D}[\theta]$. It follows from the condition $\mathcal{D}[\theta] \subset \mathcal{D}[M(0)]$, that $\mathcal{D}(t) = \mathcal{D}[\theta - M(0)] = \mathcal{D}[\theta]$. By Lemma 2.32 $\kappa_-(t_\theta - t_{M(0)}) = \kappa_-(\tilde{A}_\theta + \varepsilon)$ for $\varepsilon \in (0, \varepsilon_0)$ and thus $\kappa_-(t_\theta - t_{M(0)}) = \kappa_-\tilde{A}_\theta$).

If now $(-\varepsilon, 0) \cap \sigma(\tilde{A}_\theta) \neq \emptyset$ for each $\varepsilon > 0$ then $\kappa_-(\tilde{A}_\theta) = \infty$. To complete the proof it remains to note that $\lambda_0 \in \sigma(\tilde{A}_\theta) \cap (-\infty, 0)$ iff $0 \in \sigma(\theta - M(\lambda_0))$ and hence $\kappa_-(t_{\theta - M(0)}) = \infty$.

2) This assertion is a simple corollary of the previous one.

3) $+$ 4 The proof of these assertions based on Corollary 2.22 are similar to that contained in [M4]. □

Remark 2.34 Another proof of Lemma 2.32 is contained in [M4]. The proof of Theorem 2.33 is similar to that contained in [M4] for the case $\overline{\mathcal{D}(A)} = \mathfrak{H}$. For this case Theorem 2.33 has been proved in [DM] by a more complicated method. For $M(0) = 0$ the relation (2.62) takes the form $\kappa_-(A_\theta) = \kappa_-(\theta)$. This is the case when $m(A) := \inf\{(Af, f) : \|f\| = 1\} > 0$ and a boundary triplet $\{\mathcal{H}, \Gamma_0, \Gamma_1\}$ is such that $A_0 = A_F$, $A_1 = A_K$. In the last case the equality has been established in [B] by another method.

3 Some Properties of Operators of Classes $C_{\mathfrak{H}}(\varphi; \varkappa)$ and $C_{\mathfrak{H}}(\varphi; \mathfrak{S})$

In the first part of the section the classes of operators pointed out in the heading are defined and their simplest properties are established. Analogs of von Neumann inequality (partiallly with proofs) are presented also. In the second part of the section the extensions on nondensely defined Hermitian contraction are described.

3.1

Definition 3.1 [Ka]. A closed linear relation A in a Hilbert space $\mathfrak{H}$ is called sectorial with vertex zero and half-angle $\varphi \in (0; \pi/2)$ if its numerical range is contained in sector $G_\varphi = \{z \in \mathbb{C} : |\arg z| \le \varphi < \pi/2\}$, that is

$$(3.1) \qquad \operatorname{ctg}\varphi|\operatorname{Im}(f', f)| \le \operatorname{Re}(f', f) \qquad \forall\{f, f'\} \in A$$

If in addition A has no sectorial extensions ($\Longleftrightarrow \rho(A) \ne \varnothing$) it is called m-sectorial[2] and is put in the class $S_{\mathfrak{H}}(\varphi)$.

Further by $S_{\mathfrak{H}}(\pi/2)$ we denote the class of m-accretive operators in $\mathfrak{H}$, i.e. $A \in S_{\mathfrak{H}}(\pi/2)$ if $\operatorname{Re}(f', f) \ge 0$ for all $\{f, f'\} \in A$ and $\rho(A) \ne \varnothing$.

And finally, $S_{\mathfrak{H}}(0)$ means the set of all nonnegative selfadjoint linear relations in $\mathfrak{H}$.

Let A be a sectorial closed linear relation in $\mathfrak{H}$. In the framework of the approach accepted in this paper with each A it is connected a linear transformation

$$(3.2) \qquad\qquad T_1 = X(A) := -I + 2(I + A)^{-1},$$

being a contraction with a nondense in $\mathfrak{H}$ domain of the definition $\mathfrak{H}_1 = \mathcal{D}(T_1) = (I + A)\mathcal{D}(A)$. In so doing the condition (3.1) is transformed to the following one

$$(3.3) \qquad\qquad 2\operatorname{ctg}\varphi|\operatorname{Im}(T_1 f, f)| \le ((I - T_1^* T_1)f, f)$$

---

[2] In [Kr] an operator A with $\overline{\mathcal{D}(A)} = \mathfrak{H}$ is called regularly dissipative if $-A$ is m-sectorial.

Under such transformation the class $Ext_A((0, \infty); \varphi)$ of all m-sectorial extensions $\tilde{A}$ of A is transformed into the class $Ext_{T_1}(\varphi)$ of contractions $T \in [\mathfrak{H}]$ characterized by condition (3.3).

The following definition naturally arises from what has been said.

Definition 3.2 We put an operator $T \in [\mathfrak{H}]$ in the class $C_{\mathfrak{H}}(\varphi)$ if

(3.4) $$\|T \sin \varphi \pm i \cos \varphi \cdot I\| \leq 1, \qquad (\varphi \in (0, \pi/2])$$

and in the class $C_{\mathfrak{H}}(0)$ if $T = T^*$ and $\|T\| \leq 1$.

Lemma 3.3 *Let* $T \in [\mathfrak{H}]$, $\varphi \in (0, \pi/2]$, $\mathcal{H}_T = \overline{\mathfrak{R}(D_T)}$, $\mathcal{H}_{T^*} = \overline{\mathfrak{R}(D_{T^*})}$. *Then the following properties of an operator T are equivalent:*

(3.5)   1) $T \in C_{\mathfrak{H}}(\varphi)$
          2) $2 \operatorname{ctg} \varphi |\operatorname{Im}(Tf, f)| \leq \|D_T f\|^2 \qquad \forall f \in \mathfrak{H}$
(3.6)   3) $2 \operatorname{ctg} \varphi |\operatorname{Im}(Tf, f)| \leq \|D_{T^*} f\|^2 \qquad \forall f \in \mathfrak{H}$
(3.7)   4) $2 \operatorname{ctg} \varphi |(T_I f, g)| \leq \|D_T f\| \cdot \|D_T g\| \qquad \forall f, g \in \mathfrak{H}$
(3.8)   5) $2 \operatorname{ctg} \varphi |(T_I f, g)| \leq \|D_{T^*} f\| \cdot \|D_{T^*} g\| \qquad \forall f, g \in \mathfrak{H}$
(3.9)   6) $T \in C(\pi/2)$ *and* $2T_I = \operatorname{tg} \varphi D_T Q_1 D_T \quad (Q_1 \in C_{\mathcal{H}_T}(0))$
(3.10)   7) $T \in C(\pi/2)$ *and* $2T_I = \operatorname{tg} \varphi D_{T^*} Q_2 D_{T^*} \quad (Q_2 \in C_{\mathcal{H}_{T^*}}(0))$

Proof: The condition (3.4) rewritten as

$$(T^* \sin \varphi \mp i \cos \varphi \cdot I)(T \sin \varphi \pm i \cos \varphi \cdot I) \leq I$$

is equivalent to the condition (3.3), that is to the condition (3.5). The equivalences 1) $\Longleftrightarrow$ i) for $i = 3, 4, 5$ may be proved similarly. And finally the equivalences 4) $\Longleftrightarrow$ 6) and 5) $\Longleftrightarrow$ 7) are rather elementary. $\square$

Proposition 3.4 *The following relations are valid:*

1) $T \in C_{\mathfrak{H}}(\varphi) \Longleftrightarrow T^* \in C_{\mathfrak{H}}(\varphi)$;

2) *The class $C_{\mathfrak{H}}(\varphi)$ is closed in the weak operator topology;*

3) $T \in C_{\mathfrak{H}}(\varphi) \Longrightarrow \sigma(T) \subset L_\varphi = \{z \in \mathbb{C} : |z \sin \varphi \pm i \cos \varphi| \leq 1\}$;

4) $T \in C_{\mathfrak{H}}(\varphi)$ *and* $\{\pm 1\} \notin \sigma_p(T) \Longrightarrow T \in C_{00}$;

5) *the minimal unitary dilation of the operator $T \in C_{\mathfrak{H}}(\varphi)$ is the bilaterial shift;*

6) $\ker D_T = \ker D_{T^*}$;

7) $\mathfrak{R}(D_T) = \mathfrak{R}(D_{T^*})$;

8) $D_T = X_1 D_{T^*}$, $D_{T^*} = X_2 D_T$ $(X_1 \in [\mathcal{H}_{T^*}, \mathcal{H}_T], \; X_2 \in [\mathcal{H}_T, \mathcal{H}_{T^*}])$.

Proof: The properties 1)–3) and 6) obviously follow from Definition 3.2.

The properties 4) and 5) follow from [SNF2, p. 102] in view of the inclusion $\sigma(T) \cap \mathbb{T} \subset L_\varphi \cap \mathbb{T} \subset \{\pm 1\}$.

The properties 7) and 8) follow from Lemma 2.3 and the inequalities

$$(3.11) \qquad C_\varphi^{-1} \|D_{T^*} h\| \leq \|D_T h\| \leq C_\varphi \|D_{T^*} h\| \qquad \forall h \in \mathfrak{H}$$

which are for example valid with constant $C_\varphi = \cos^{-1} \varphi (\sin \varphi + \sqrt{1 + \sin^2 \varphi})$, does not depend on $T \in C_\mathfrak{H}(\varphi)$.

In fact, in accordance with (3.9) and (3.10)

$$
\begin{aligned}
(3.12) \qquad D_T^2 &= I - (T - i\operatorname{tg}\varphi D_T Q_1 D_T)(T^* + i\operatorname{tg}\varphi D_T Q_1 D_T) \\
&= D_{T^*}^2 - 2\operatorname{tg}\varphi \operatorname{Im}(D_T Q_1 T^* D_{T^*}) - \operatorname{tg}^2\varphi D_T Q_1 D_T^2 Q_1 D_T.
\end{aligned}
$$

Hence for all $h \in \mathfrak{H}$ and $\forall \varepsilon \in (0, 1)$ we have

$$
\begin{aligned}
(3.13) \qquad (1 + \operatorname{tg}^2\varphi)\|D_T h\|^2 &\geq \|D_{T^*} h\|^2 - \varepsilon\|D_{T^*} h\|^2 \\
&\quad - \varepsilon^{-1}\operatorname{tg}^2\varphi\|D_T h\|^2.
\end{aligned}
$$

Rewriting this inequality in the form

$$(3.14) \qquad \|D_{T^*} h\|^2 \leq (1 - \varepsilon)^{-1}[\varepsilon^{-1}\operatorname{tg}^2\varphi + \cos^{-2}\varphi] \cdot \|D_T h\|^2$$

and minimizing (3.14) with respect to ε we arrive at the first of the inequalities (3.11). The second one can be proved by the same way. Finally we note that another proof of relations 7) and 8) have been obtained in [Ar1]. $\qquad \square$

3.2 Some interesting properties of the classes $S_\mathfrak{H}(\varphi)$ and $C_\mathfrak{H}(\varphi)$ are contained in the following two results.

Theorem 3.5 *Suppose that $f(z)$ is a function holomorphic in the sector $G_\varphi = \{z \in \mathbb{C} : |\arg z| \leq \varphi \leq \pi/2\}$ and maps it into itself $(f : G_\varphi \to G_\varphi)$. Then the following implication holds:*

$$(3.15) \qquad A \in S_\mathfrak{H}(\varphi) \Longrightarrow f(A) \in S_\mathfrak{H}(\varphi)$$

Theorem 3.6 *Suppose that $\varphi \in (0, \pi/2]$, and $f(z)$ is holomorphic in the lune $L_\varphi = \{z \in \mathbb{C} : |z \sin \varphi \pm i \cos \varphi| \leq 1\}$ and maps it into itself $(f : L_\varphi \to L_\varphi)$. Then the following implication is valid:*

$$(3.16) \qquad T \in C_\mathfrak{H}(\varphi) \Longrightarrow f(T) \in C_\mathfrak{H}(\varphi)$$

For $\varphi = \pi/2$ Theorem 3.6 coincides with the well-known result of von Neumann (see [SNF2]).

It is clear that Theorem 3.5 and Theorem 3.6 are equivalent and moreover this equivalence may be easily established by means of the transformation (3.2). The proofs of Theorem 3.5 and Theorem 3.6 together with some applications will be published elsewhere. Here we mention only the following corollary proved in [KM1], [KM2] (see also [M1]).

Corollary 3.7 *Suppose that $f(z)$ is a function holomorphic in the right half-plane $\mathbb{C}_r := G_{\pi/2}$, and maps into itself both $\mathbb{C}_r$ and the real semiaxis $\mathbb{R}_+$ (i.e. $f(x) = \overline{f(x)}\ \forall x \in \mathbb{R}_+$). Then for each $\varphi \in (0, \pi/2]$ the implication (3.15) holds.*

Proof immediately follows from Theorem 3.5 and the following analog of the Shwarz lemma.

Lemma 3.8 ([M1]). *If $f(z)$ satisfies the conditions of Corollary 3.7 then*

$$(3.17) \qquad f(G_\varphi) \subset G_\varphi \qquad \forall \varphi \in [0, \pi/2]$$

It should be pointed out that Theorem 3.5 and Corollary 3.7 are completely different. Indeed under the hypothesis of Corollary 3.7 $f(z)$ is holomorphic in the half-plane $G_{\pi/2}$ and by Lemma 3.8 the condition (3.17) is satisfied $\forall \varphi \in [0, \pi/2]$. On the other hand the function $f(z)$ satisfying the conditions of Theorem 3.5 is defined and holomorphic only in the angle G_{φ_0} with fixed $\varphi_0 \in (0, \pi/2)$ and generally speaking does not satisfy the condition (3.17) for other $\varphi \in [0, \varphi_0]$. Thus Theorem 3.5 is very special in character and the holomorphy of $f(z)$ only in the angle G_φ makes its proof much more complicated.

3.3 Here some complements to von Neumann inequality are presented.

Definition 3.9 Let $\varphi \in [0, \pi/2]$, $\varkappa^\pm \in \mathbb{Z}_+$, $\mathfrak{S}^\pm$ be two-sided ideals in $[\mathfrak{H}]$, B be a closed linear operator in $[\mathfrak{H}]$. Let further the quadratic forms

$$\mathfrak{t}_\pm[f] = \operatorname{Re}(Bf, f) \pm \operatorname{ctg}\varphi\,\operatorname{Im}(Bf, f), \qquad (f \in \mathcal{D}(B))$$

be semibounded below and closable[3], and $B^\pm$ be the linear operators associated with their closures. We write

a) $B \in S_\mathfrak{H}(\varphi; \varkappa^\pm)$, if $\rho(B) \neq \varnothing$ and the operators $(B_{op}^\pm)_-$ are $\varkappa^\pm$-dimensional;

b) $B \in S_\mathfrak{H}(\varphi; \mathfrak{S}^\pm)$, if $(B^\pm)_- \in \mathfrak{S}^\pm$ and $\rho(B) \neq \varnothing$.

A closed linear relation θ in $\mathfrak{H}$ is put also in the class $S_\mathfrak{H}(\varphi; \varkappa^\pm)$ $(S_\mathfrak{H}(\varphi; \mathfrak{S}^\pm))$ if $\operatorname{Re}(f', f) \geq \beta\|f\|^2$ for all $\{f, f'\} \in \theta$ (with some $\beta \in \mathbb{R}$) and its operator part is in $S_\mathfrak{H}(\varphi; \varkappa^\pm)$ $(S_\mathfrak{H}(\varphi; \mathfrak{S}^\pm))$.

It is clear that $S_\mathfrak{H}(\varphi; 0)$ coinsides with $S_\mathfrak{H}(\varphi)$.

---

[3]For $\varphi < \pi/2$ the closability of the form $\mathfrak{t}_\pm$ is a consequence of their semiboundness (see [Ka]).

Definition 3.10 Let $\varphi \in (0, \pi/2]$, $\varkappa^{\pm} \in \mathbb{Z}^+$, $\mathfrak{S}^{\pm}$ be two-sided ideals in $[\mathfrak{H}]$. An operator T is put

a) in the class $C_{\mathfrak{H}}(\varphi; \varkappa^{\pm})$, if

$$(3.18) \qquad\qquad T \sin \varphi \pm i \cos \varphi \cdot I \in C_{\mathfrak{H}}(\pi/2; \varkappa^{\pm});$$

b) in the class $C_{\mathfrak{H}}(\varphi; \mathfrak{S}^{\pm})$, if $T \sin \varphi \pm i \cos \varphi \cdot I \in C_{\mathfrak{H}}(\pi/2; \mathfrak{S}^{\pm})$.

We write $C_{\mathfrak{H}}(\varphi; \varkappa)$ and $C_{\mathfrak{H}}(\varphi; \mathfrak{S})$ instead of $C_{\mathfrak{H}}(\varphi; \varkappa^{\pm})$ and $C_{\mathfrak{H}}(\varphi; \mathfrak{S}^{\pm})$ respectively if $\varkappa := \varkappa^+ = \varkappa^-$ and $\mathfrak{S} := \mathfrak{S}^+ = \mathfrak{S}^-$.

Observe that the class $C_{\mathfrak{H}}(\pi/2; \varkappa^{\pm})$ is not empty only if $\varkappa^+ = \varkappa^-$.

For $\varphi = 0$ the inclusion $T \in C_{\mathfrak{H}}(0; \varkappa)$ ($T \in C_{\mathfrak{H}}(0; \mathfrak{S})$) means that $T = T^*$ and $\varkappa_-(T) = \varkappa(T_- \in \mathfrak{S})$.

It is clear that the classes $C_{\mathfrak{H}}(\varphi; \varkappa^{\pm})$ and $S_{\mathfrak{H}}(\varphi; \varkappa^{\pm})$ are connected by means of the linear fractional transformation (3.2). The same is also true for the classes $C_{\mathfrak{H}}(\varphi; \mathfrak{S}^{\pm})$ and $S_{\mathfrak{H}}(\varphi; \mathfrak{S}^{\pm})$.

Now we present a property of the class $C(\pi/2; \mathfrak{S}_{\infty})$ similar to that of the class $C(\varphi)$, described in Corollary 3.7.

In the sequel we as usual denote by A_f the set of holomorphy of a function $f(z)$.

Theorem 3.11 ([M2]). *Let $f(z) \in H^{\infty}(\mathbb{D})$ be an inner function on the unit disk $\mathbb{D}$ (i.e. $|f(e^{it})| = 1$ for a.e. $t \in [0, 2\pi]$) and $T \in [\mathfrak{H}]$. If $\sigma(T) \subset A_f$, then the following implications hold:*

$$(3.19) \qquad a) \qquad (I - T^*T)_- \in \mathfrak{S}_{\infty} \Longrightarrow (I - f(T)^*f(T))_- \in \mathfrak{S}_{\infty};$$
$$(3.20) \qquad b) \qquad (I - T^*T)_+ \in \mathfrak{S}_{\infty} \Longrightarrow (I - f(T)^*f(T))_+ \in \mathfrak{S}_{\infty}.$$

Proof: 1. Let us first prove the implications

$$(3.21) \quad (I - T_i^*T_i)_- \in \mathfrak{S}_{\infty} \ (i = 1, 2) \Longrightarrow (I - T_1^*T_2^*T_2T_1)_- \in \mathfrak{S}_{\infty}$$

$$(3.22) \quad (I - T_i^*T_i)_+ \in \mathfrak{S}_{\infty} \ (i = 1, 2) \Longrightarrow (I - T_1^*T_2^*T_2T_1)_- \in \mathfrak{S}_{\infty},$$

Setting $A := I - T_1^*T_2^*T_2T_1$, $B = I - T_1^*T_1$, $C = T_1^*(I - T_2^*T_2)T_1$, one rewrites the obvious euqality $A = B + C$ in the form

$$(3.23) \qquad\qquad A = (B_+ + C_+) - (B_- + C_-)$$

The inequalities

$$(3.24) \qquad A \le R_+ := B_+ + C_+, \quad -A \le R_- := B_- + C_-$$

immediately follow from (3.23). Hence we conclude that for each $\varepsilon > 0$

$$L_A(\varepsilon) := \{f \in \mathfrak{H} : ((A - \varepsilon)f, f) \geq 0\} \subset L_{R_+}(\varepsilon)$$
$$:= \{f \in \mathfrak{H} : (R_+ f, f) \geq \varepsilon \|f\|^2\}$$

and in accordance with the minimax principle

(3.25) $\qquad dim E_A(\varepsilon, \infty) \leq dim E_{R_+}(\varepsilon, \infty) \qquad \forall \varepsilon > 0$

Now the implication (3.22) follows from (3.25). Using the second of the inequalities (3.24) one similarly obtains the implication (3.21).

2. Let $f(z) = b_a(z) = \frac{\bar{a}}{|a|} \frac{a-z}{1-\bar{a}z}$ - be a primary Blaschke factor ($|a| < 1$). In this case the implications (3.19) and (3.20) (together with the inverse) follow from the identity

$$I - b_a(T)^* b_a(T) = X^*(I - T^*T)X$$

where

$$X := (1 - |a|^2)^{1/2}(I - aT)^{-1} \in [\mathfrak{H}], \qquad X^{-1} \in [\mathfrak{H}]$$

3. One obtaines the implications (3.19) and (3.20) for a finite Blaschke product $f(z) = B_n(z) = \prod_{k=1}^{n} b_{a_k}(z)$ ($|a_k| < 1 \ \forall k \leq n$) by comparing the assertions 1 and 2 proved above.

4. Let $f(z) = B(z) = \prod_{k=1}^{\infty} b_{a_k}(z)$ - be an infinite Blaschke product. Since $B_n(z)$ uniformly converges to $B(z)$ on each compact $G \subset A_{B(z)} = \mathbb{C} \backslash K_1$, where $K_1 = clos \ (\cup_{k=1}^{\infty} (\bar{a}_k)^{-1})$, and $\sigma(T) \subset A_{B(z)}$, then

$$\lim_{n \to \infty} \|B_n(T) - B(T)\| = 0$$

It follows that $\lim_{n \to \infty} \|D^2_{B_n(T)} - D^2_{B(T)}\| = 0$. Let further $E_n(a, b)$ and $E(a, b)$ be the spectral projections of the operators $D^2_{B_n(T)}$ and $D^2_{B(T)}$ respectively. Then for each $\varepsilon > 0$ and such that $\pm\varepsilon \in \rho(D^2_{B(T)})$ we have

$$\lim_{n \to \infty} \|E_n(-\infty, -\varepsilon) - E(-\infty, -\varepsilon)\|$$
$$= \lim_{n \to \infty} \|E_n(\varepsilon, \infty) - E_n(\varepsilon, \infty)\| = 0$$

Thus $dim(E(-\infty, -\varepsilon)\mathfrak{H}) < \infty$ and $dim(E(\varepsilon, \infty)\mathfrak{H}) < \infty$ for each $\varepsilon > 0$ and the implications (3.19) and (3.21) are proved.

5. Let finally $f(z)$ be an arbitrary inner function on the unit disk $\mathbb{D}$. According to Frostman Theorem [Ga] an inner function

$$f_\lambda(z) := b_\lambda(f(z)) = (f(z) - \lambda)(1 - \bar{\lambda}f(z))^{-1}$$

is an infinite Blaschke product for almost all $\lambda \in \mathbb{D}$. But for each compact K (not necessary lying in the disk $\mathbb{D}$) the estimate

$$\|f(z) - f_\lambda(z)\|_{L_\infty(K)} = \sup_{z \in K} \left| \frac{\lambda - \bar{\lambda} f(z)^2}{1 - \bar{\lambda} f(z)} \right|$$

$$\leq \frac{1 + M^2}{|\lambda^{-1}| - M} \leq 2|\lambda|(1 + M^2).$$

holds with $M = \|f\|_{L_\infty(K)}$ and $|\lambda| < (2M)^{-1}$. This implies that $\sigma(T) \subset A_{f_\lambda}$ for $|\lambda|$ small enough and $\lim_{\lambda \to 0} \|f_\lambda(T) - f(T)\| = 0$. Since for the operators $f_\lambda(T)$ the implications (3.19) and (3.20) are already proved (for almost all $\lambda \in \mathbb{D}$), we may complete the proof just like it was done on the previous step. $\qquad \square$

Making use of linear fractional transformation (3.2) one derives from Theorem 3.11 the following

Theorem 3.12 ([M3]). *Suppose that $f(z)$ is holomorphic in the right half-plane $\mathbb{C}_r = \{z \in \mathbb{C} : \operatorname{Re} z > 0\}$, and maps into itself both[4] $\mathbb{C}_r$ and the imaginary axis (i.e. $\overline{f(it)} = -f(it)$). If $\sigma(A) \subset A_f$ then the following implication holds true*

$$A \in S_{\mathfrak{H}}(\pi/2; \mathfrak{S}_\infty) \Longrightarrow f(A) \in S_{\mathfrak{H}}(\pi/2; \mathfrak{S}_\infty).$$

Remark 3.13   a) The classes $C_{\mathfrak{H}}(\pi/2; \mathfrak{S})$ are multiplicative:

$$T_1, T_2 \in C_{\mathfrak{H}}(\pi/2; \mathfrak{S}) \Longrightarrow T_1 T_2 \in C_{\mathfrak{H}}(\pi/2; \mathfrak{S}).$$

For $\mathfrak{S} = \mathfrak{S}_\infty$ this fact is contained in the implication (3.21). In general it is a consequence of the inequalities $\lambda_j(A_+) \leq \lambda_j(B_+ + C_+)$, $\lambda_j(A_-) \leq \lambda_j(B_- + C_-)$ for the eigenvalues of such operators implied by (3.24).

b) For a contraction T ($\Longleftrightarrow (I - T^*T)_- = 0$) the implication (3.19) after replacement of the ideal $\mathfrak{S}_\infty$ be the zero ideal $\mathfrak{S} = \{0\}$, coincides with a known inequality (see [SNF2]) of von Neumann and the implication (3.20) adds to it: $D_T \in \mathfrak{S}_\infty \Longrightarrow D_{f(T)} \in \mathfrak{S}_\infty$. However whereas the von Neumann's inequality remains true when the equality $|f(e^{it})| = 1$ is replaced by the inequality $|f(e^{it})| \leq 1$, it does not hold true for the implication (3.20).

3.4 Here we complement the results of Section 2 for the case of a dual pair of contractions $\{T_1, T_1\}$. In this case $T_{11} = T_{11}^*$, $T_{12} = T_{21}^*$, and the set Ext_{T_1}

---

[4] If $f : \mathbb{C}_r \to \mathbb{C}_r$, then for almost all $t \in \mathbb{R}$ there exist nontangential limit values: $f(it) := \lim_{x \to 0} f(x + it)$.

coincides with the set of quasihermitian extensions[5] $T = T_K$ of the form (2.29) of an Hermitian operator T_1.

One easily derives a description of extensions $T_K (\in \text{Ext}_{T_1})$, of classes $C_{\mathfrak{H}}(\varphi; \varkappa^\pm)$ and $C_{\mathfrak{H}}(\varphi; \mathfrak{S}^\pm)$ from Corollary 2.21 and Corollary 2.22 respectively.

Proposition 3.14 *Let* $\mathfrak{H} = \mathfrak{H}_1 \oplus \mathfrak{H}_2$ *and* $T_1 = \binom{T_{11}}{T_{21}} = \binom{T_{11}}{V D_{T_{11}}}$ *be a Hermitian contraction* $(T_1 \in [\mathfrak{H}_1, \mathfrak{H}])$ *and* $T_K = (T_{ij})_{i,j=1}^2 \in \text{Ext}_{T_1}$. *If*

$$(G_{ij}^\pm)_{i,j=1}^2 := I - T_K^* T_K \pm i \operatorname{ctg} \varphi (T_K - T_K^*),$$
$$(S_{ij})_{i,j=1}^2 := I - T_K T_K^* \pm i \operatorname{ctg} \varphi (T_K - T_K^*)$$

then the following identities hold

(3.26)
$$G_{22}^\pm - (G_{11}^{-1/2} G_{12})^* (G_{11}^{-1/2} G_{12})$$
$$= D_U (I - K^* K \pm i \operatorname{ctg} \varphi (K - K^*)) D_U,$$

$$S_{22}^\pm - (S_{11}^{-1/2} S_{12})^* (S_{11}^{-1/2} S_{12})$$
$$= D_U (I - K K^* \pm i \operatorname{ctg} \varphi (K - K^*)) D_U,$$

where G_{ij} *and* S_{ij} *are the same as in Theorem* 2.14 *and* $U = V^*$.

Proof: Since $T_K \in \text{Ext}_{T_1}$, then

$$T_K = \begin{pmatrix} T_{11} & T_{21}^* \\ T_{21} & T_{22} \end{pmatrix} = \begin{pmatrix} T_{11} & D_{T_{11}} U \\ U^* D_{T_{11}} & T_{22} \end{pmatrix},$$

$$T_{22} = -U^* T_{11} U + D_U K D_U$$

Hence one derives

(3.27) $\qquad (G_{ij}^\pm)_{i,j=1}^2 = \begin{pmatrix} G_{11} & G_{12} \\ G_{21} & G_{22} \end{pmatrix} \pm i \operatorname{ctg} \varphi \begin{pmatrix} 0 & 0 \\ 0 & T_{22} - T_{22}^* \end{pmatrix}$

Theorem 2.14 and the relations (3.27) together yield the equality

$$G_{22}^\pm - (G_{11}^{-1/2} G_{12})^* (G_{11}^{-1/2} G_{12})$$
$$= G_{22} - (G_{11}^{-1/2} G_{12})^* (G_{11}^{-1/2} G_{12}) \pm i \operatorname{ctg} \varphi (T_{22} - T_{22}^*)$$
$$= D_U (I - K^* K) D_U \pm i \operatorname{ctg} \varphi D_U (K - K^*) D_U,$$

coinciding with the first of the equalities (3.26). The second one can be proved similarly. □

---

[5] An extension $T \supset T_1$ is called quasihermitian if in addition $T^* \supset T_1$.

Corollary 3.15 *If* ker $D_U = \{0\}$ *and* $T_K \in \text{Ext}_{T_1}$, *then the following equivalence holds*

$$T_K \in C_{\mathfrak{H}}(\varphi; \varkappa^{\pm}) \iff K \in C_{\mathfrak{H}_2}(\varphi; \varkappa^{\pm})$$

Corollary 3.16 *Let* $0 \in \rho(D_U)$ *and the operator* $G_{11}^{-1} G_{12}$ *be bounded. If* $T_K \in \text{Ext}_{T_1}$, *then the following equivalence holds true*

$$T_K \in C_{\mathfrak{H}}(\varphi; \mathfrak{S}^{\pm}) \iff K \in C_{\mathfrak{H}_2}(\varphi; \mathfrak{S}^{\pm}).$$

Corollary 3.17 *Let* $T_K \in \text{Ext}_{T_1}(\pi/2) \, (= \text{Ext}_{T_1} \cap C_{\mathfrak{H}}(\pi/2))$. *Then the operators*

$$G^{\pm} := I - T_K^* T_K \pm i \text{ ctg } \varphi (T_K - T_K^*),$$
$$S^{\pm} := I - T_K T_K^* \pm i \text{ ctg } \varphi (T_K - T_K^*)$$

shorted to $\mathfrak{N} := \mathfrak{H}_2$, *are of the form*

$$(G^{\pm})_{\mathfrak{N}} = \begin{pmatrix} 0 & 0 \\ 0 & D_U(D_K^2 \pm i \text{ ctg } \varphi(K - K^*))D_U \end{pmatrix},$$

$$(S^{\pm})_{\mathfrak{N}} = \begin{pmatrix} 0 & 0 \\ 0 & D_U(D_{K^*}^2 \pm i \text{ ctg } \varphi(K - K^*))D_U \end{pmatrix}$$

One proves Corollary 3.15, Corollary 3.16 and Corollary 3.17 in complete analogy with the proofs of Corollary 2.21, Corollary 2.22 and Corollary 2.27.

3.5 Let us denote by $C_{\mathfrak{H}}^E(\varphi)$ and $\text{Ext}_{T_1}^E(\varphi)$ the sets of extreme points of the closed convex sets $C_{\mathfrak{H}}(\varphi)$ and $\text{Ext}_{T_1}(\varphi) := C_{\mathfrak{H}}(\varphi) \cap \text{Ext}_{T_1}(\pi/2)$ respectively.

Proposition 3.18 *Let* $T \in C_{\mathfrak{H}}(\varphi)$ *and* $\mathfrak{R}(D_T)$ *be closed. Then:*

1) $2T_I = \text{tg}\varphi \, D_T Q_1 D_T$ *with* $Q_1 = Q_1^* \in C_{\mathcal{H}_T}(0)$;

2) *the following implication holds*

$$(3.28) \qquad\qquad \sigma(Q_1) = \{\pm 1\} \implies T \in C_{\mathfrak{H}}^E(\varphi)$$

Proof: 1) This assertion was proved in Lemma 3.5 (see (3.9)).

2) Let $2T = B_1 + B_2$, where $B_i \in C_{\mathfrak{H}}(\varphi)$ $(i = 1, 2)$. If $f \in \ker D_T$, $\|f\| = 1$, then $\|Tf\| = \|f\| = 1 \implies \|B_1 f\| + \|B_2 f\| = 2\|Tf\| = 2$. Hence in view of strict convexity of the unit ball in $\mathfrak{H}$ we obtain the equalities $B_1 f = B_2 f = Tf$. Further letting $T_{\pm} = T \sin \varphi \pm i \cos \varphi$, one obtains

$$D_{T_{\pm}}^2 = I - (T^* \sin \varphi \mp i \cos \varphi)(T \sin \varphi \pm i \cos \varphi)$$
$$(3.29) \qquad\qquad = (I - T^*T)\sin^2 \varphi \pm i \sin \varphi \cos \varphi (T - T^*)$$
$$= \sin^2 \varphi D_T (I \mp Q_1) D_T$$

If $f \in \mathcal{H}_T = (\ker D_T)^\perp$ and $D_T f \in \ker(I - Q_1)$, then according to (3.29) $D^2_{T_+} f = 0 \Longleftrightarrow \|T_+ f\| = \|f\| = 1$. Since B_1, $B_2 \in C_{\mathfrak{H}}(\varphi)$, it follows from the equality

$$2T_+ f = (B_1 \sin \varphi + i \cos \varphi) f + (B_2 \sin \varphi + i \cos \varphi) f$$

and strict convexity of the unit ball in $\mathfrak{H}$, that $Tf = B_1 f = B_2 f$. Similarly we arrive at the identity $Tf = B_1 f = B_2 f$ for $f \in \mathcal{H}_T$ and satisfying the condition $D_T f \in \ker(I + Q_1)$. But according to the hypothesis of proposition $D_T \mathcal{H}_T = \mathcal{H}_T = \ker(I + Q_1) \oplus \ker(I - Q_1)$ and therefore $B_i f = Tf$ for $f \in \mathcal{H}_T$. It remains to note that $\mathfrak{H} = \ker D_T \oplus \mathcal{H}_T$, and the identities $B_i f = Tf$ $(i = 1, 2)$ $\forall f \in \ker D_T$ are obvious. $\square$

Remark 3.19 a) Closability of the lineal $\mathfrak{R}(D_T)$ may be replaced by the conditions $\mathfrak{R}(D_T) \cap \mathcal{H}_\pm = \mathcal{H}_\pm$, with $\mathcal{H}_\pm = \ker(I \pm Q_1)$, which are, for example, valid if $dim \mathcal{H}_+ < \infty$ or $dim \mathcal{H}_- < \infty$.

b) If T is a normal operator $(T^*T = TT^*)$, $T \in [\mathfrak{H}]$ and $\sigma(T) \subset \partial L_\varphi = \{z \in \bar{\mathbb{D}} : |z \sin \varphi \pm i \cos \varphi| = 1\}$, then, as it follows from Proposition 3.14, $T \in C^E_{\mathfrak{H}}(\varphi)$.

Conversely, if $T \in C_{\mathfrak{H}}(\varphi), \sigma(T) \subset \partial L_\varphi$ and the spectrum $\sigma(T)$ is purely point, then T is normal. Actually if $\lambda_j \in L_\varphi \cap \sigma_p(T)$, and $\mathcal{H}_j := \ker(T - \lambda_j)$, then either $\mu^+_j = \lambda_j \sin \varphi + i \cos \varphi \in \sigma(T_+) \cap \mathbb{T}$, or $\mu^-_j = \lambda_j \sin \varphi - i \cos \varphi \in \sigma(T_-) \cap \mathbb{T}$ where $T_\pm = T \sin \varphi \pm i \cos \varphi$, and $\mathbb{T} = \{z \in \mathbb{C} : |z| = 1\}$. Therefore the subspace $\mathcal{H}_j = \ker(T - \lambda_j)$ reduces the operator T for each $j \in \mathbb{Z}_+$ since either $\mathcal{H}_j = \ker(T_+ - \mu^+_j)$, or $\mathcal{H}_j = \ker(T - \mu^-_j)$. Consequently $T = \oplus^\infty_{j=1} \lambda_j I_{\mathcal{H}_j}$ that is T is normal.

c) It is not clear whether the inverse to (3.28) implication holds true even in the case $dim \mathfrak{H} < \infty$. In accordance with (3.29) $T \in C^E_{\mathfrak{H}}(\varphi)$, if $T \in C_{\mathfrak{H}}(\varphi)$ and satisfies at least one of the four equations

$$(3.30) \qquad D^2_T \pm 2 \operatorname{ctg} \varphi(T_I) = 0, \qquad D^2_{T^*} \pm 2 \operatorname{ctg} \varphi(T_I) = 0.$$

However even in the case $dim \mathfrak{H} < \infty$ the set $C^E_{\mathfrak{H}}(\varphi)$ is much wider than the set defined by the equations (3.30).

Corollary 3.20 Let $\mathfrak{H} = \mathfrak{H}_1 \oplus \mathfrak{N}$, $T_K \in C_{\mathfrak{H}}(\varphi) \cap Ext_{T_I}$. Then:

1) $T_K \in C^E_{\mathfrak{H}}(\varphi) \Longleftrightarrow K \in C^E_{\mathcal{H}_U}(\varphi)$, $(\mathcal{H}_U := \overline{\mathfrak{R}(D_U)})$;

2) $2K_I = \operatorname{tg} \varphi D_K Q D_K$ with $Q \in C_{\mathcal{H}_K}(0)$, and the implication

$$\sigma(Q) = \{\pm 1\}, \qquad \mathfrak{R}(D_K) = \overline{\mathfrak{R}(D_K)} \Longrightarrow T_K \in Ext_{T_1}(\varphi)$$

is valid.

3) $T_K \in Ext^E_{T_1}(\varphi)$, if at least one of the identities (3.30) holds.

A proof may be derived if one compares Corollary 3.17 with Proposition 3.18. Moreover each of the equalities (3.30) is equivalent to one of the following four equalities:

$$(D^2_{T_K} \pm 2 \operatorname{ctg} \varphi(\operatorname{Im}T_K))_{\mathfrak{N}} = 0,$$
$$(D^2_{T^*_K} \pm 2 \operatorname{ctg} \varphi(\operatorname{Im}T_K))_{\mathfrak{N}} = 0.$$

3.6 Here we consider in a brief form a connection between the results of the previous two subsections with the M.G. Krein's formula [K] and its generalizations from [AT1].

Let $\{T_1, T_2\}$ be a dual pair with $T_1 = T_2$, i.e. $T_{11} = T^*_{11}$, $T_{12} = T^*_{21} (\Longleftrightarrow V = U^*)$. Letting in (2.15) $K = \mp I$ we arrive at the equalities

$$(T_{-I})_{22} = -VT_{11}V^* - D^2_{V*} = -I + V(I - T_{11})V^*,$$
$$(T_{+I})_{22} = -VT_{11}V^* + D^2_{V*} = I - V(I + T_{11})V^*,$$

which present a description of missing blocks T_{22} of the extremal extensions $T_{\min}$, $T_{\max} \in \operatorname{Ext}_{T_1}(\pi/2)$ satisfying the inequality

$$T_{\min} \le T = T^* \le T_{\max} \quad for \ all \ \ T = T^* \in \operatorname{Ext}_{T_1}(\pi/2)$$

Thus, $T_m = T_{\min} = T_{-I}$, $T_M = T_{\max} = T_{+I}$, that is

(3.31)

$$T_m = \begin{pmatrix} T_{11} & D_{T_{11}}V^* \\ VD_{T_{11}} & -I + V(I - T_{11})V^* \end{pmatrix},$$

$$T_M = \begin{pmatrix} T_{11} & D_{T_{11}}V^* \\ VD_{T_{11}} & I - V(I + T_{11})V^* \end{pmatrix}$$

These equalities yield

(3.32)

$$2^{-1}(T_M - T_m) = \begin{pmatrix} 0 & 0 \\ 0 & D^2_{V*} \end{pmatrix},$$

$$T_0 := (T_m + T_M)/2 = \begin{pmatrix} 0 & 0 \\ 0 & -VT_{11}V^* \end{pmatrix}.$$

Consequently the formula (2.15) (in the case $T_1 = T_2$) takes the form

(3.33)

$$2T_K = (T_M + T_m) + (T_M - T_m)^{1/2}K(T_M - T_m)^{1/2},$$

$$K \in C_{\mathcal{H}_{V*}}(\pi/2)$$

in which the sc-extensions were described by M.G. Krein as far back as [K] and the qsc-extensions (quasihermitian contractive extensions) in [AT1], [AT2].

By means of the formula (3.33) we may also describe some different classes of quasihermitian extensions of a contraction T_1. For example according to Corollary 3.15 the equality (3.33) establishes a bijective correspondence

$$(3.34) \qquad T_K \in Ext_{T_1} \cap C_{\mathfrak{H}}(\varphi; \varkappa^\pm) \iff K \in C_{\mathcal{H}_{V^*}}(\varphi; \varkappa^\pm)$$

In particular for $\varkappa^+ = \varkappa^- = 0$ we arrive at the following description of the class $C_{T_1}(\varphi)$ obtained in [AT1], [AT2].

Proposition 3.21 [AT1], [AT2]. *Let T_1 be a Hermitian contraction from $\mathfrak{H}_1$ to $\mathfrak{H} = \mathfrak{H}_1 \oplus \mathfrak{H}_2$, $\mathcal{H} = \overline{\mathfrak{R}(T_M - T_m)}$. Then the equality (3.33) establishes a bijective correspondence*

$$T_K \in C_{\mathfrak{H}}(\varphi) \cap Ext_{T_1}(\pi/2) \iff K \in C_{\mathcal{H}}(\varphi),$$

(3.35)

$$\varphi \in (0, \pi/2]$$

between the quasihermitian extensions $T = T_K$ of the operator T_1, belonging to the class $C_{\mathfrak{H}}(\varphi)$ and the operators K of the class $C_{\mathcal{H}}(\varphi)$.

Other proofs of the equivalences (3.34), (3.35) are contained in [KM1]. We complement Proposition 3.21 by the following simple

Proposition 3.22 [KM1]. *Let A, $B \in [\mathfrak{H}]$, $A = A^*$, $B \geq 0$, $\|A \pm B\| \leq 1$, $\mathcal{H} = \overline{\mathfrak{R}(B)}$ and $\varphi \in [0, \pi/2]$. Then the following implication holds true:*

$$K \in C_{\mathcal{H}}(\varphi) \implies A + B^{1/2} K B^{1/2} \in C_{\mathfrak{H}}(\varphi)$$

Proof: Since $\|A \pm B\| \leq 1$, then $-I \leq A \pm B \leq I$ and therefore $|(Af, f)| \leq ((I - B)f, f)$ for all $f \in \mathfrak{H}$. The last inequality by virtue of Lemma 2.3 yields the relation $A = (I - B)^{1/2} M (I - B^{1/2}$, with M being a Hermitian contraction. Since $K \in C_{\mathcal{H}}(\varphi)$, and $M \in C_{\mathcal{H}}(0) \subset C_{\mathfrak{H}}(\varphi)$ then the operators

$$\begin{aligned} T_1^\pm &= M \sin \varphi \pm i \cos \varphi \cdot I, \\ T_2^\pm &= K \sin \varphi \pm i \cos \varphi \cdot I \end{aligned}$$

are contractive. Therefore $\forall f, g \in \mathfrak{H}$:

$$\begin{aligned} &2 \operatorname{Re} ((\sin \varphi(A + B^{1/2}KB^{1/2}) \pm i \cos \varphi \cdot I)f, g) \\ &= 2 \operatorname{Re} (((I - B)^{1/2} T_1^\pm (I - B^{1/2}) + B^{1/2} T_2^\pm B^{1/2})f, g) \\ &\leq ((I - B)f, f) + ((I - B)g, g) + (Bf, f) + (Bg, g) = \|f\|^2 + \|g\|^2. \end{aligned}$$

The last inequality proves the inclusion $A + B^{1/2} K B^{1/2} \in C_{\mathfrak{H}}(\varphi)$. $\qquad \square$

Remark 3.23 Observe that Parlett [PaB, p. 252] pointed out that W.M. Kahan had obtained the following formula

$$T_0 = \lim_{r \downarrow 1} \begin{pmatrix} T_{11} & T_{21}^* \\ T_{21} & -T_{21}(rI - T_{11}^2)^{-1}T_{11}T_{21}^* \end{pmatrix}$$

for one of the selfadjoint matrix of the class $\text{Ext}_T(\pi/2)$, $\dim \mathfrak{H} < \infty$. It follows from (3.31) that $T_0 = (T_m + T_M)/2$.

The formulas (3.31) have been obtained by V. Kolmanovich and the author (see [Kol, p. 5]) without using Theorem 2.7. Next, they were reproved (by algebraic method) and used in [KM1]. Besides, as it is mentioned in this subsection, they easily follow from Theorem 2.7. Thus, with regard to what has been said about the matrix case ($\dim \mathfrak{H} < \infty$), they should be apparently considered as a mathematical folklore.

In the next section starting with a contraction T_1 satisfying (3.3), and following the method proposed by the author in [KM1] we construct extremal extensions $T_m, T_M \in \text{Ext}_{T_1}(\varphi)$ with the properties similar to that of the extensions T_m and T_M of the form (3.31) and coinsiding with them if $\varphi = 0$.

4 Extensions of the Class $C(\varphi)$ of Nonhermitian Contractions and m-Sectorial Extensions of Sectorial Operators

This section is devoted to the study of the family $\text{Ext}_{T_1}(\varphi)$ of extensions of the class $C_{\mathfrak{H}}(\varphi)$ of a nondensely defined contraction $T_1 = \begin{pmatrix} T_{11} \\ T_{21} \end{pmatrix} \in [\mathfrak{H}_1, \mathfrak{H}]$ on a Hilbert space $\mathfrak{H} = \mathfrak{H}_1 \oplus \mathfrak{H}_2$ and satisfying the condition

(4.1) $2 \, \text{ctg} \, \varphi |\text{Im}(T_1 f, f)| \leq ((I - T_1^* T_1)f, f)$

with $\varphi \in (0, \pi/2)$.

4.1 Here it is shown that the class $\text{Ext}_{T_1}(\varphi)$ contains the extremal extensions T_m and T_M and in particular is nonempty. These extensions, are the analogs of the extensions T_m, T_M of the form (3.31) of a Hermitian contraction T_1 and coincide with them for $\varphi = 0 (\Longleftrightarrow T_{11} = T_{11}^*)$.

Proposition 4.1 *Let $\mathfrak{H} = \mathfrak{H}_1 \oplus \mathfrak{H}_2$, $T_1 = \begin{pmatrix} T_{11} \\ T_{21} \end{pmatrix}$ be a contraction from $\mathfrak{H}_1$ to $\mathfrak{H}$. Then for each $\lambda \in \mathbb{C}$ with $|\lambda| > 1$ there exists the sole extension $T(\lambda)$ of the operator T_1 with:*

                              a) $\|T(\lambda)\| = |\lambda|$;

(4.2)

                              b) $\mathfrak{H} = \mathfrak{H}_1 \dotplus \ker(T(\lambda) - \lambda)$.

This extension is of the form

(4.3) $T(\lambda) = \begin{pmatrix} T_{11} & (T_{11} - \lambda I)(T_{11}^* - \bar{\lambda}I)^{-1}T_{21}^* \\ T_{21} & \lambda I + T_{21}(T_{11}^* - \bar{\lambda}I)^{-1}T_{21}^* \end{pmatrix}.$

Proof: Let $T(\lambda) = \begin{pmatrix} T_{11} & X \\ T_{21} & Y \end{pmatrix}$ be the required extension. Since $\|T(\lambda)\| = |\lambda|$, the subspace $\ker(T - \lambda)$ reduces the operator $T(\lambda)$ and therefore $\ker(T - \lambda) = \ker(T^* - \bar{\lambda})$. Hence for $f = \begin{pmatrix} f_1 \\ f_2 \end{pmatrix} \in \ker(T - \lambda)$ we have two systems of equations

$$(4.4) \qquad \begin{cases} T_{11} f_1 + X f_2 = \lambda f_1 \\ T_{21} f_1 + Y f_2 = \lambda f_2 \end{cases}, \qquad \begin{cases} T_{11}^* f_1 + T_{21}^* f_2 = \bar{\lambda} f_1 \\ X^* f_1 + Y^* f_2 = \bar{\lambda} f_2 \end{cases}$$

We consequently derive from the equations (4.4)

$$(4.5) \qquad f_1 = (\bar{\lambda} \cdot I - T_{11}^*)^{-1} T_{21}^* f_2$$

$$(4.6) \qquad Y f_2 = \lambda f_2 - T_{21} f_1 = \lambda f_2 - T_{21}(\bar{\lambda} \cdot I - T_{11}^*)^{-1} T_{21}^* f_2$$

For the decomposition (4.2) to be valid it is necessary that the identity (4.6) hold true for each $f_2 \in \mathfrak{H}_2$ and therefore $Y = \lambda \cdot I + T_{21}(T_{11}^* - \bar{\lambda} \cdot I)^{-1} T_{21}^*$. Further (4.4) and (4.5) yield

$$X f_2 = (\lambda I - T_{11}) f_1 = (\lambda I - T_{11})(\bar{\lambda} I - T_{11}^*)^{-1} T_{21}^* f_2.$$

So, $X = (T_{11} - \lambda I)(T_{11}^* - \bar{\lambda} I)^{-1} T_{21}^*$ and the operator $T(\lambda)$ is of the form (4.3). Conversely it is not difficult to see that the operator $T = T(\lambda)$ of the form (4.3) obeys the equivalences

$$(4.7) \qquad f = \begin{pmatrix} f_1 \\ f_2 \end{pmatrix} \in \ker(T - \lambda I) \iff f \in \ker(T^* - \bar{\lambda} I) \iff$$

$$f = \begin{pmatrix} (\bar{\lambda} I - T_{11}^*)^{-1} T_{21}^* f_2 \\ f_2 \end{pmatrix}$$

yielding the relation (4.2).

The proof of the equality $\|T(\lambda)\| = |\lambda|$ is omitted. $\qquad \square$

Corollary 4.2 *If* $1 \in \hat{\rho}(T_1)$ $(-1 \in \hat{\rho}(T_1))$*, then the extension* $T(1)$ $(T(-1))$ *of the form* (4.3) *with* $\lambda = 1 (\lambda = -1)$ *possesses the properties* a) *and* b).

In what follows we assume without loss of generality that $\pm 1 \notin \sigma_p(T_{11})$, that is $\ker(T_{11} \pm I) = \{0\}$.

Proposition 4.3 *Let* $T_1 = \begin{pmatrix} T_{11} \\ T_{21} \end{pmatrix}$ *be a contraction on* $\mathfrak{H} = \mathfrak{H}_1 \oplus \mathfrak{H}_2$*, satisfying the condition* (4.1). *Then:*

1) *There exist the limits*

$$s - \lim_{r \uparrow 1}(T_{11} - T_{11}^*)(I \pm r T_{11}^*)^{-1} T_{21}^*$$

$$(4.8) \qquad =: (T_{11} - T_{11}^*)(I \pm T_{11}^*)^{-1} T_{21}^*$$

$$(4.9) \qquad s - \lim_{r \uparrow 1} T_{21}(I \pm r T_{11}^*)^{-1} T_{21}^* =: T_{21}(I \pm T_{11}^*)^{-1} T_{21}^*$$

2) *the class* $\text{Ext}_{T_1}(\varphi)$ *is not empty: it contains the operators*

$$T_m = s - \lim_{\lambda \downarrow 1} T(-\lambda)$$

(4.10)
$$= \begin{pmatrix} T_{11} & T_{21}^* + (T_{11} - T_{11}^*)(T_{11}^* + I)^{-1} T_{21}^* \\ T_{21} & -I + T_{21}(T_{11}^* + I)^{-1} T_{21}^* \end{pmatrix},$$

$$T_M = s - \lim_{\lambda \downarrow 1} T(\lambda)$$

(4.11)
$$= \begin{pmatrix} T_{11} & T_{21}^* + (T_{11} - T_{11}^*)(T_{11}^* - I)^{-1} T_{21}^* \\ T_{21} & I + T_{21}(T_{11}^* - I)^{-1} T_{21}^* \end{pmatrix}.$$

Proof:

1) Since T_1 satisfies the inequality (4.1) and $D_{T_1}^2 \le D_{T_{11}}^2$, then $T_{11} \in C_{\mathfrak{H}_1}(\varphi)$ and by Lemma 3.5

(4.12) $T_{11} - T_{11}^* = i \tan \varphi D_{T_{11}} Q_1 D_{T_{11}} = i \tan \varphi D_{T_{11}^*} Q_2 D_{T_{11}^*}$

with Q_1 and Q_2 being Hermitian contractions. Moreover, $T_{21} = V D_{T_{11}}$ ($\|V\| \le 1$) and according to (3.12), $D_{T_{11}} = X_1 D_{T_{11}^*}$, $D_{T_{11}^*} = X_2 D_{T_{11}}$. Therefore

$$T_{21}(I + rT_{11}^*)^{-1} T_{21}^* = V D_{T_{11}}(I + rT_{11}^*)^{-1} D_{T_{11}} V^*$$
$$= V X_1 D_{T_{11}^*}(I + rT_{11}^*)^{-1} D_{T_{11}} V^*$$
$$(T_{11} - T_{11}^*)(I + rT_{11}^*)^{-1} T_{21}^* = i \tan \varphi D_{T_{11}^*} Q_2 D_{T_{11}^*}(I + rT_{11}^*)^{-1} D_{T_{11}} V^*$$

It is clear from these equalities, that, in order to prove the relations (4.10), (4.11) it suffices to prove the existence of the strong limits

(4.13) $s - \lim_{r \uparrow 1} D_{T_{11}^*}(I \pm rT_{11}^*)^{-1} D_{T_{11}} =: D_{T_{11}^*}(I \pm T_{11}^*)^{-1} D_{T_{11}}$

Let at first $f = (I \pm T_{11}^* X_1^*)h$, $h \in \mathfrak{H}_1$. Then

$$D_{T_{11}} f = D_{T_{11}^*} X_1^* h \pm D_{T_{11}} T_{11}^* X_1^* h = (I \pm T_{11}^*) D_{T_{11}^*} X_1^* h$$

and consequently (with regard to the inequality $(1 - r)\|(I + rT)^{-1}\| \le 1$ for $r \in (0, 1)$) we have

$$\lim_{r \uparrow 1} D_{T_{11}^*}(I \pm rT_{11}^*)^{-1} D_{T_{11}}$$

(4.14)
$$= \lim_{r \uparrow 1} D_{T_{11}^*}(I \pm rT_{11}^*)^{-1}(I \pm T_{11}^*) D_{T_{11}^*} X_1^* h$$

$$= D_{T_{11}^*}^2 X_1^* h.$$

It is clear that the sets $\mathcal{H}^{\pm} := \{f = (I \pm T_{11}^* X_1^*)h \;:\; h \in \mathfrak{H}_1\}$ are dense $\mathfrak{H}_1$.

Thus to complete the proof it remains to note that $\|D_{T_{11}^*}(I \pm rT_{11}^*)^{-1}D_{T_{11}}\|$ ≤ 4 for $r \in [1/2, 1]$. Indeed this estimate is implied by the well known fact (see [SNF2]) that the characteristic operator – function

$$\theta_{T_{11}}(\lambda) := [-T_{11} + \lambda D_{T_{11}^*}(I - \lambda T_{11}^*)^{-1}D_{T_{11}}]$$

(4.15)

$$\lceil \overline{D_{T_{11}}\mathfrak{H}_1}, \; \lambda \in \mathbb{D}.$$

of the contraction T_{11} is contractive in the unit disk: $\|\theta_{T_{11}}(\lambda)\| \leq 1$ for $\lambda \in \mathbb{D}$.

2) Let $R := D_{T_{11}}(I + T_{11})^{-1}D_{T_{11}}$. Since $D_{T_{11}} = X_1 D_{T_{11}^*}$, then by virtue of (5.13) the operator $R^* = s - \lim_{r \uparrow 1} X_1 D_{T_{11}^*}(I + rT_{11}^*)^{-1}D_{T_{11}}$ is bounded and $R^* = D_{T_{11}}(I + T_{11}^*)^{-1}D_{T_{11}}$.

Let us now transform the entries of the operator – matrix $(B_{ij})_{i,j=1}^2 := D_{T_m}^2$. We have

$$
\begin{aligned}
B_{11} &= D_{T_1}^2 \\
&= I - T_{11}^* T_{11} - D_{T_{11}}VV^*D_{T_{11}} \\
&= D_{T_{11}}D_V^2 D_{T_{11}}
\end{aligned}
$$

(4.16)

$$
\begin{aligned}
B_{12} &= B_{21}^* = -[T_{11}^*(I + T_{11}^*) - (I - T_{11}^*) \\
&\quad + T_{11}^*(T_{11} - T_{11}^*) + T_{21}^* T_{21}](I + T_{11}^*)^{-1}T_{21}^* \\
&= D_{T_1}^2(I + T_{11}^*)^{-1}T_{21}^* \\
&= D_{T_{11}}D_V^2 D_{T_{11}}(I + T_{11}^*)^{-1}D_{T_{11}}V^* \\
&= D_{T_{11}}D_V^2 R^*V^*
\end{aligned}
$$

(4.17)

$$
\begin{aligned}
B_{22} &= -T_{21}(I + T_{11})^{-1}\{(I + T_{11})(I + T_{11}^*) \\
&\quad + (I + T_{11})(T_{11} - T_{11}^*) \\
&\quad + (T_{11}^* - T_{11})(I + T_{11}^*) \\
&\quad + (T_{11}^* - T_{11})(T_{11} - T_{11}^*) \\
&\quad - (I + T_{11}) - (I + T_{11}^*) + T_{21}^* T_{21}\}(I + T_{11}^*)^{-1}T_{21}^* \\
&= T_{21}(I + T_{11})^{-1}D_{T_1}^2(I + T_{11}^*)^{-1}T_{21}^* = VRD_V^2 R^*V^*
\end{aligned}
$$

(4.18)

Introducing the operators $C^{\pm} := D_{T_m}^2 \mp i \operatorname{ctg} \varphi(T_m - T_m^*)$ we rewrite them in the block form $C^{\pm} =: (C_{ij})_{i,j=1}^2$ with respect to the orthogonal decomposition $\mathfrak{H} = \mathfrak{H}_1 \oplus \mathfrak{H}_2$. Taking (4.16) and (4.17) into account, we obtain the following expressions for the entries $C_{ij}^{\pm}$:

$$
\begin{aligned}
C_{11}^{\pm} &= B_{11} \mp i \operatorname{ctg} \varphi (T_{11} - T_{11}^*) \\
&= D_{T_{11}} D_V^2 D_{T_{11}} \pm D_{T_{11}} Q_1 D_{T_{11}} \\
&= D_{T_{11}} (D_V^2 \pm Q_1) D_{T_{11}} \geq 0, \\
C_{12}^{\pm} &= B_{12} \mp i \operatorname{ctg} \varphi (T_{12} - T_{12}^*) \\
&= D_{T_{11}} D_V^2 R^* V^* \mp i \operatorname{ctg} \varphi (T_{11} - T_{11}^*)(I + T_{11}^*)^{-1} T_{21}^* \\
&= D_{T_{11}} D_V^2 R^* V^* \pm D_{T_{11}} Q_1 R^* V^* = D_{T_{11}} (D_V^2 \pm Q_1) R^* V^*, \\
C_{22}^{\pm} &= V R D_V^2 R^* V^* \pm V D_{T_{11}} (I + T_{11})^{-1} \\
&\quad D_{T_{11}} Q_1 D_{T_{11}} (I + T_{11}^*)^{-1} D_{T_{11}} V^* \\
&= V R D_V^2 R^* V^* \pm V R Q_1 R^* V^* = V R (D_V^2 \pm Q_1) R^* V^*.
\end{aligned}
$$

It is clear that $D_V^2 \pm Q_1 \geq 0$. Therefore

$$
\begin{aligned}
C^{\pm} &= \begin{pmatrix} D_{T_{11}} (D_V^2 \pm Q_1) D_{T_{11}} & D_{T_{11}} (D_V^2 \pm Q_1) R^* V^* \\ V R (D_V^2 \pm Q_1) D_{T_{11}} & V R (D_V^2 \pm Q_1) R^* V^* \end{pmatrix} \\
&= M_{\pm}^* M \pm \geq 0
\end{aligned}
$$

(4.19)

with

(4.20) $M_{\pm} = ((D_V^2 \pm Q_1)^{1/2} D_{T_{11}} \ (D_V^2 \pm Q_1)^{1/2} R^* V^*).$

But the inequalities (4.19) amount to the inclusion $T_m \in C_{\mathfrak{H}}(\varphi)$. The inclusion $T_M \in C_{\mathfrak{H}}(\varphi)$ may be proved similarly.   □

Remark 4.4 a) By Corollary 4.2 the extension $(rT)_m ((rT)_M)$, where $r \in (0, 1)$, is of the form (4.10) (4.11) with T_{11} and T_{21} replaced by $r T_{11}$ and $r T_{21}$ respectively. Therefore the relations $s - \lim_{r \uparrow 1} (rT)_m = T_m$, $s - \lim_{r \uparrow 1} (rT)_M = T_M$ are implied by Proposition 4.3.

b) In the case $T_{11} = T_{11}^*$ (that is for $\varphi = 0$), the extensions T_m and T_M coinside with $T_m = T_m^*$ and $T_M = T_M^*$ of the form (3.31). The construction of extensions T_m and T_M presented in Proposition 4.1 and Proposition 4.6 is similar to that contained in [KM1] for the case $\varphi = 0$. In the class $\operatorname{Ext}_{T_1} (\varphi)$ the extensions T_m and T_M play the role similar to their role in the class $\operatorname{Ext}_{T_1} (0)$.

c) The nonemptiness of the class $\text{Ext}_{T_1}(\varphi)$ established in passing in Proposition 4.3 has been first discovered in [Mir] in 1969. However it should be pointed out that this fact can be easily derived (via linear fractional transformation (4.2)) from the Shecter's result (see [Ka, p. 399]) proved earlier. In his paper Shecter has constructed an analog of Friedrich's extension involving in particular the nonemptiness of the set $\text{Ext}_A((0, \infty); \varphi)$ for a sectorial operator A.

d) Existence of the strong limits $\theta_T(\pm 1) := s - \lim_{\lambda \uparrow 1} \theta_T(\pm \lambda)$ for the characteristic function $\theta_T(\lambda)$ of the operator $T \in C_{\mathfrak{H}}(\varphi)$ is proved in [Ar1]. This immediately implies the relation (4.13). Its proof is presented for the sake of completeness. However, it should be pointed out that the existence of the week limits $w - \lim_{\lambda \uparrow 1} \theta_T(\pm \lambda) = \theta_T(\pm 1)$ (in the case $T \in C_{\mathfrak{H}}(\varphi)$) and thus the existence of the week limits in (4.8) and (4.9) easily follow from the contractivity of $\theta_T(\lambda)$ in the unit disk $\mathbb{D}$. If $-1 \in \rho(T_{11})$ $(1 \in \rho(T_{11}))$, (it is always the case when $\dim \mathfrak{H} < \infty$) then formula (4.10) (4.11) as well as its proof is purely algebraic in character.

In the following proposition we describe some extremal properties of the extensions T_m and $T_M \in \text{Ext}_{T_1}(\varphi)$. For the case $\varphi = 0$ these properties are well-known (see [AG], [K], [Sh]) and easily follow from the formulas (3.31) for $T_m = T_m^*$ and $T_M = T_M^*$.

At the beginning we present a simple lemma.

Lemma 4.5 *Let* $\mathfrak{H} = \mathfrak{H}_1 \oplus \mathfrak{N}$, $X \in [\mathfrak{H}_1, \mathfrak{H}_3]$, $Y, P \in [\mathfrak{N}, \mathfrak{H}_3]$ *and* $\mathfrak{R}(X) \supseteq \mathfrak{R}(Y)$. *Then for the operator*

$$
\begin{aligned}
A &= \begin{pmatrix} X^*X & X^*Y \\ Y^*X & Y^*Y \end{pmatrix} + \begin{pmatrix} 0 & 0 \\ 0 & P^*P \end{pmatrix} \\
&= \begin{pmatrix} X^* \\ Y^* \end{pmatrix} (X, Y) + \begin{pmatrix} 0 & 0 \\ 0 & P^*P \end{pmatrix} \geq 0
\end{aligned}
$$

the following relation

$$
A_{\mathfrak{N}} = P^*P.
$$

holds. In particular $A_{\mathfrak{N}} = 0$ *if* $P = 0$. *The converse is also true.*

Proof: The required relation $A_{\mathfrak{N}} = P^*P$ is implied by Corollary 2.26 and the identity

$$
(A(f - g), f - g) = \|X(f_1 - g) + Yf_2\|^2 + \|Pf_2\|^2
$$

with $f = \begin{pmatrix} f_1 \\ f_2 \end{pmatrix}$ and $g \in \mathfrak{H}_1$. $\qquad \square$

Proposition 4.6 *Let $T_1 = \binom{T_{11}}{T_{21}}$ be a contraction on $\mathfrak{H} = \mathfrak{H}_1 \oplus \mathfrak{N}$, obeying the condition (4.1). Then its extensions T_m and T_M of the form (4.10) and (4.11) satisfy the equalities*

1) $(D_{T_m}^2)_{\mathfrak{N}} = 0;$ 1') $(D_{T_M}^2)_{\mathfrak{N}} = 0;$

2) $(D_{T_m}^2 \pm 2 \operatorname{ctg} \varphi \cdot (T_m)_I)_{\mathfrak{N}} = 0;$ 2') $(D_{T_M}^2 \pm 2 \operatorname{ctg} \varphi \cdot (T_M)_I)_{\mathfrak{N}} = 0;$

3) $(I + \operatorname{Re} T_m)_{\mathfrak{N}} = 0;$ 3') $(I - \operatorname{Re} T_M)_{\mathfrak{N}} = 0;$

4) $\inf_{g \in \mathfrak{H}_1} ((I + \operatorname{Re} T_m)$ 4') $\inf_{g \in \mathfrak{H}_1} ((I - \operatorname{Re} T_M)$

  $(f - g), f - g) = 0;$ $(f - g), f - g) = 0;$

5) $(-T)_m = -T_M;$ 5') $(-T)_M = -T_m.$

Proof: 1) The relations (4.16)–(4.18) yield

$$D_{T_m}^2 = \begin{pmatrix} D_{T_{11}} D_V^2 D_{T_{11}} & D_{T_{11}} D_V^2 R^* V^* \\ V R D_V^2 D_{T_{11}} & V R D_V^2 R^* V^* \end{pmatrix}$$

(4.21)

$$= \begin{pmatrix} D_{T_{11}} D_V \\ V R D_V \end{pmatrix} (D_V D_{T_{11}} \quad D_V R^* V^*)$$

Since $\mathfrak{R}(D_V D_{T_{11}}) \supseteq \mathfrak{R}(D_V R^* V^*)$, then (4.21) and Lemma 4.5 imply the equality $(D_{T_m}^2)_{\mathfrak{N}} = 0$.

2) By virtue of (4.19) $C^\pm = M_\mp^* M_\pm$, with $M_\pm$ defined by the equalities (4.20). Since

$$\mathfrak{R}((D_V^2 \pm Q_1)^{1/2} D_{T_{11}}) \supseteq \mathfrak{R}(D_V^2 \pm Q_1)^{1/2} R^* V^*,$$

then the required relations again follow from Lemma 4.5.

Besides, these relations immediately follow from the implication: $A \geq B \geq 0 \implies A_{\mathfrak{N}} \geq B_{\mathfrak{N}} \geq 0$ applied to the nonnegative operators $C^\pm = D_{T_m}^2 \mp i \operatorname{ctg} \varphi(T_m - T_m^*)$, and the equality

$$(C^+ + C^-)_{\mathfrak{N}} = (D_{T_m}^2)_{\mathfrak{N}} = 0.$$

has already been proved.

3) Observe that the operators $Y = (I + \operatorname{Re} T_{11})^{1/2}(I + T_{11}^*)^{-1} T_{21}^*$ and Y^* are bounded since so is the operator $Y^* Y = T_{21}(I + T_{11}^*)^{-1} T_{21}^* + T_{21}(I + T_{11})^{-1} T_{21}^*$. Therefore

$I + \mathrm{Re} T_m =$

$$\begin{pmatrix} I + \mathrm{Re} T_{11} & (I + \mathrm{Re} T_{11})(I + T_{11}^*)^{-1} T_{21}^* \\ T_{21}(I + T_{11})^{-1}(I + \mathrm{Re} T_{11}) & T_{21}(I + T_{11})^{-1}(I + \mathrm{Re} T_{11})(I + T_{11}^*)^{-1} T_{21}^* \end{pmatrix}$$

$$= \begin{pmatrix} (I + \mathrm{Re} T_{11})^{1/2} \\ T_{21}(I + T_{11})^{-1}(I + \mathrm{Re} T_{11})^{1/2} \end{pmatrix}$$

$$\cdot \left((I + \mathrm{Re} T_{11})^{1/2} \quad (I + \mathrm{Re} T_{11})^{1/2}(I + T_{11}^*)^{-1} T_{21}^* \right).$$

The identity $(I + \mathrm{Re} T_m)_{\mathfrak{N}} = 0$ follows now from Lemma 4.5 and the relation $\mathfrak{R}(X) \supseteq \mathfrak{R}(Y)$, where $X = (I + \mathrm{Re} T_{11})^{1/2}$.

4) In accordance with Corollary 2.26 the required relation is equivalent to the previous one.

5) The required relations are implied by the formulas (4.10) and (4.11).

To complete the proof it remains to note that the relations 1')–5') may be derived from the representation (4.11) for T_M.   □

Remark 4.7 a) Another approach to the investigation of the extremal extensions T_m, $T_M \in \mathrm{Ext}_{T_1}(\varphi)$ of an operator T_1 and their properties, described in Proposition 4.6 has been proposed in [Ar2].

b) It is not difficult to show (compare with Proposition 4.13) that the properties 3) and 3'), as well as in the case $\varphi = 0$ (see [AG], [K]), characterise the extentions T_m and T_M of an operator T_1 among all the extensions of the class $\mathrm{Ext}_{T_1}(\varphi)$. Namely, for $T \in \mathrm{Ext}_{T_1}(\varphi)$ the following equivalences hold true:

$$(I + \mathrm{Re} T)_{\mathfrak{N}} = 0 \Longleftrightarrow T = T_m, \quad (I - \mathrm{Re} T)_{\mathfrak{N}} = 0 \Longleftrightarrow T = T_M.$$

4.2 Here we present a description of the class $\mathrm{Ext}_{T_1}(\varphi)$. Our description is based on the following simple lemma on intersection of two operator balls.

Lemma 4.8 (On a Parametrization of an Operator Hole). *Let $B_1 = B(C_1; R_{l1}, R_{r1})$ and $B_2 = B(C_2; R_{l2}, R_{r2})$ be two operator balls in $\mathfrak{H}$. Suppose also that their intersection is nonempty, i.e. $B_1 \cap B_2 \neq \emptyset$. Then:*

1) *For each fixed $T_0 \in L := B_1 \cap B_2$ the operators*

$$(4.22) \qquad Q_i^0 = R_{ie}^{-1}(T_0 - C_i) R_{ir}^{-1} \qquad (i = 1, 2)$$

are bounded and their closures $Q_i := \overline{Q_i^0}$ are contractions.

2) *The operator hole L admits a parametrization*

$$(4.23) \qquad T \in L := B_1 \cap B_2 \Longleftrightarrow T = T_0 + K,$$

with K satisfying the relations

(4.24) $||R_{il}^{-1} K R_{ir}^{-1} + Q_i|| \leq 1, \qquad i \in \{1, 2\}.$

Proof: 1) Let $T_0 \in B_1 \cap B_2$. Then with some contractions $K_i (i = 1, 2)$ one has

(4.25) $T_0 = C_1 + R_{1l} K_{10} R_{1r} = C_2 + R_{2l} K_{20} R_{2r}, \qquad K_i \in C(\pi/2)$

It follows that for each $i \in \{1, 2\}$ the operator Q_i^0 defined by (4.22) is

bounded and its closure is K_{i0}, $\overline{Q_i^0} = K_{i0}$.

2) Let now $T \in L = B_1 \cap B_2$. Then

(4.26) $T = C_1 + R_{1l} K_1 R_{1r} = C_2 + R_{2l} K_2 R_{2r}$

with some $K_i \in C(\pi/2)$, $(i = 1, 2)$. Setting $K_i = K_{i0} + K_i'$ $(i = 1, 2)$ and
taking (4.25) and (4.26) into account one gets

$$K := R_{1l} K_1' R_{1r} = R_{2l} K_2' R_{2r}.$$

Thus $K_i' = R_{il}^{-1} K R_{ir}^{-1}$ $(i = 1, 2)$ and the relations (4.24) are fullfiled. $\square$

Remark 4.9 a) If $R_l := R_{1l} = R_{2l}$ and $R_r := R_{ir} = R_{2r}$ then as it is shown in
[KM1] the operator hole $L = B_1 \cap B_2$ is nonempty $(L \neq \emptyset)$ iff

$$|((C_1 - C_2)f, g)| \leq ||R_l f|| \cdot ||R_r g|| \quad \text{for} \quad f, g \in \mathfrak{H}.$$

In this case $2^{-1}(C_1 + C_2) \in L$ and the parametrization (4.23)–(4.24) may
be rewritten as

$$T = 2^{-1}(C_1 + C_2) + R_l K R_r, \qquad K \pm Q \in C(\pi/2)$$

where Q is the closure of $Q^0 := 2^{-1} R_l^{-1}(C_1 - C_2)R_r^{-1}$.

b) If at least one of the conditions $R_{1l} = R_{2l}$ and $R_{1r} = R_{2r}$ is violated, a
criterion of nonemptiness of the intersection $B_1 \cap B_2$ is unknown.

The following lemma is well-known.

Lemma 4.10 *Let $\mathfrak{H}_1$ be a subspace of a Hilbert space $\mathfrak{H} = \mathfrak{H}_1 \oplus \mathfrak{H}_2$ and $T_1 \in$
$[\mathfrak{H}_1, \mathfrak{H}]$ be a contraction. Then the formula*

$$T = T_1 P_1 + D_{T_1^*} K P_2, \qquad K \in [\mathfrak{H}_2, \mathfrak{H}], \quad ||K|| \leq 1$$

*where P_i is the orthoprojection from $\mathfrak{H}$ onto $\mathfrak{H}_i (i = 1, 2)$, establishes a bijective
correspondence between the set of all contractive extensions $T \in [\mathfrak{H}]$ and the set
of contractions $K \in [\mathfrak{H}_2, \mathfrak{H}]$.*

Now we are ready to present a description of the class $Ext_{T_1}(\varphi)$ for a contraction T_1 satisfying (4.1).

Theorem 4.11 [KM1]. *Suppose that* $\mathfrak{H} = \mathfrak{H}_1 \oplus \mathfrak{H}_2$ *and* $T_1 = \binom{T_{11}}{T_{21}}$ *is a contraction, satisfying*

$$2ctg\varphi|\text{Im}(T_1 f, f)| \le \|D_{T_1} f\|^2, \qquad f \in \mathfrak{H}_1.$$

Let also $T_m (\in Ext_{T_1}(\varphi))$ *be a contraction defined by* (4.10), *$S := (T_1\ 0)(\in [\mathfrak{H}])$ and* $Q_\pm := (T_m \pm i\ \text{ctg}\ \varphi I)P_2$. *Then the following equivalence holds*

$$T \in Ext_{T_1}(\varphi) \Longleftrightarrow T = T_m + K P_2, \qquad R_\pm^{-1}(K P_2 + Q_\pm) \in C(\pi/2)$$

where $K \in [\mathfrak{H}_2, \mathfrak{H}]$ *and*

(4.27) $$R_\pm^2 = D_{S^*}^2 \pm i\ \text{ctg}\ \varphi(S - S^*) + \text{ctg}^2\varphi P_2.$$

Proof: Let $T \in Ext_{T_1}(\varphi)$. This means that $T_+ := T\sin\varphi + i\cos\varphi I$ and $T_- := T\sin\varphi - i\cos\varphi I$ are contractive extensions of

$$S_+ := \begin{pmatrix} T_{11}\sin\varphi + i\cos\varphi \cdot I_1 \\ T_{21}\sin\varphi \end{pmatrix} \quad \text{and}$$

(4.28)
$$S_- := \begin{pmatrix} T_{11}\sin\varphi - i\cos\varphi \cdot I_1 \\ T_{21}\sin\varphi \end{pmatrix}$$

respectively. It is clear that $S_\pm = T_\pm P_1 \lceil \mathfrak{H}_1$ where P_j is the orthoprojection from $\mathfrak{H}$ onto $\mathfrak{H}_j (j = 1, 2)$. Therefore by Lemma 4.10

(4.29) $$T_\pm = T_\pm P_1 + D_{S_\pm^*} K_\pm P_2, \qquad K_\pm \in [\mathfrak{H}_2, \mathfrak{H}], \qquad \|K_\pm\| \le 1.$$

One rewrites (4.29) as

$$(T\sin\varphi \pm i\cos\varphi)P_2 = D_{S_\pm^*} K_\pm P_2,$$

(4.30)
$$K_\pm \in [\mathfrak{H}_2, \mathfrak{H}], \qquad \|K_\pm\| \le 1$$

Hence

(4.31) $$T P_2 = \mp i\ \text{ctg}\ \varphi P_2 + \frac{1}{\sin\varphi} D_{S_\pm^*} K_\pm P_2, \qquad \|K_\pm\| \le 1,$$

that is $T P_2 \in L := B_+ \cap B_-$, where $B_\pm := B(C_\pm; D_{S_\pm^*}, P_2)$.

By Proposition 4.6 the extensions T_m and T_M defined by (4.10) and (4.11) belong to $Ext_{T_1}(\varphi)$ and consequently $L := B_+ \cap B_- \ne \emptyset$.

Therefore to complete the proof it suffices to note that

$$D_{S_\pm^*}^2 = I - S_\pm S_\pm^* = D_{S^*}^2 \pm i\ \text{ctg}\ \varphi(S - S^*) + \text{ctg}^2\varphi P_2$$

where $S = (T_1\ 0) = \begin{pmatrix} T_{11} & 0 \\ T_{21} & 0 \end{pmatrix} \in C_{\mathfrak{H}}(\pi/2)$, and apply Lemma 4.8. $\qquad\square$

Remark 4.12 The block-matrix representations of $R^2_\pm$ are:

$$R^2_\pm = \begin{pmatrix} D^2_{T^*_{11}} \pm i(T_{11} - T^*_{11})\mathrm{ctg}\,\varphi & -T_{11}T^*_{21} \mp iT^*_{21}\mathrm{ctg}\,\varphi \\ -T_{21}T^*_{11} \pm iT_{21}\mathrm{ctg}\,\varphi & D^2_{T^*_{21}} + \mathrm{ctg}^2\varphi \cdot I_2 \end{pmatrix}.$$

4.3 Let A be a closed sectorial operator in $\mathfrak{H}$ with vertex at zero and a half-angle $\varphi \in [0, \pi/2)$, $\overline{D(A)} \subseteq \mathfrak{H}$, that is

$$(4.32) \qquad |\mathrm{Im}(Af, f)| \le \mathrm{tg}\,\varphi \mathrm{Re}(Af, f) \qquad \forall f \in D(A).$$

As in Section 3 we denote by $\mathrm{Ext}_A((0, \infty); \varphi)$ the set of all m-sectorial extensions $\widetilde{A}$ of A in $\mathfrak{H}$. Recall that $\rho(\widetilde{A}) \ne 0$ for each $\widetilde{A} \in \mathrm{Ext}_A((0, \infty); \varphi)$ and moreover $\sigma(\widetilde{A}) \subset G_\varphi = \{z \in \mathbb{C}: |\arg z| \le \varphi\}$.

Following M.G. Krein [K], we use, as in Section 3 the linear fractional transformation

$$(4.33) \qquad T_1 = X(A) := -I + 2(I + A)^{-1}$$

to investigate the properties of m-sectorial extensions $\widetilde{A} \in \mathrm{Ext}_A((0, \infty); \varphi)$.

It is easily seen that T_1 is a contraction from $\mathfrak{H}_1 := (I + A)\mathcal{D}(A)$ to $\mathfrak{H}$ satisfying the conditions (4.1). Moreover, the formulas

$$(4.34) \qquad \begin{aligned} \widetilde{A} &= X^{-1}(T) = -I + 2(I + T)^{-1}, \\ T &= X(\widetilde{A}) = -I + 2(I + \widetilde{A})^{-1} \end{aligned}$$

establish a bijective correspondence between $\mathrm{Ext}_A((0, \infty); \varphi)$ and $\mathrm{Ext}_{T_1}(\varphi)$.

As in the case of a symmetric operator A we set

$$(4.35) \qquad \begin{aligned} A_F &:= X^{-1}(T_m) = -I + 2(I + T_m)^{-1}, \\ A_K &:= X^{-1}(T_M) = -I + 2(I + T_M)^{-1}. \end{aligned}$$

We call the extensions A_F and $A_K \in \mathrm{Ext}_A((0, \infty); \varphi)$ by Friedrichs and Krein's extentions of A respectively.

In what follows we assume without loss of generality that $Ker\,A = \{0\}$ ($\Longleftrightarrow$ $Ker(I - T_1) = \{0\}$).

For each sectorial operator A satisfying (4.32) it is naturally connected (see [Ka]) a Hilbert space $D[A]$ being a completion of $\mathcal{D}(A)$ with respect to the norm

$$(4.36) \qquad \|f\|^2_{ReA} = \|f\|^2 + \mathrm{Re}\,t_A[f] := \|f\|^2 + \mathrm{Re}(Af, f).$$

Proposition 4.13 *Let A be a sectorial operator in $\mathfrak{H}$ and A_F and A_K be its Friedrichs and Krein's extensions, defined by (4.35). Then:*

1) $(A^{-1})_F = (A_K)^{-1}, (A^{-1})_K = (A_F)^{-1};$

2) $\mathcal{D}[A] = \mathcal{D}[A_F] \subset \mathcal{D}[\widetilde{A}] \subseteq \mathcal{D}[A_K]$;

3) *If* $\widetilde{A} \in \mathrm{Ext}_A((0, \infty); \varphi)$ *and* $\mathcal{D}[A] = \mathcal{D}[\widetilde{A}]$, *then* $\widetilde{A} = A_F$;

4) *For each* $f \in \mathcal{D}(A_F)$ *the following relation holds*

$$\inf_{f_A \in \mathcal{D}(A)} \{\|f - f_A\|^2 + Re(A_F(f - f_A), f - f_A)\} = 0, \ f \in \mathcal{D}(A_F).$$

Moreover, if this identity holds with A_F *replaced by* $\widetilde{A} \in \mathrm{Ext}_A((0, \infty); \varphi)$, *then* $\widetilde{A} = A_F$;

5) *For each* $f \in \mathcal{D}(A_K)$

$$\inf_{f_A \in \mathcal{D}(A)} \{\|A_K(f - f_A)\|^2 + Re(A_K(f - f_A), f - f_A)\} = 0, \quad f \in \mathcal{D}(A_K).$$

Moreover, if this identity holds with A_K *replaced by* $\widetilde{A} \in \mathrm{Ext}_A((0, \infty); \varphi)$, *then* $\widetilde{A} = A_K$.

6) *The relations*

$$\inf_{f_A \in \mathcal{D}(A)} \{Re(1 \pm i \ ctg \ \varphi)t_{\widetilde{A}}(f - f_A)\} = 0 \quad \text{for all } f \in \mathcal{D}(\widetilde{A})$$

hold true with $\widetilde{A} = A_F$ *and* $\widetilde{A} = A_K$.

Proof:

1) The required relations are implied by (4.25) and relations 5) and 5') of Proposition 4.6.

2) Let $\widetilde{A} \in Ext A$ and $\widetilde{T} = X(\widetilde{A})$. Since $A \subset \widetilde{A}$ it is clear that $\mathcal{D}[A] \subset \mathcal{D}[\widetilde{A}]$. On the other hand, setting $g = (I + \widetilde{A})f$, $g_T = (I + A)f_A$ where $f \in \mathcal{D}(\widetilde{A})$, $f_A \in \mathcal{D}(A)$, one derives from (4.34) and (4.36)

$$2\|f - f_A\|^2_{Re\widetilde{A}}$$

(4.37)

$$= \|g - g_T\|^2 + Re(\widetilde{T}(g - g_T), g - g_T).$$

It follows from (4.37) and Proposition 4.6 4) that $\mathcal{D}(A)$ is dense in $\mathcal{D}[A_F]$, that is $\mathcal{D}[A] = \mathcal{D}[A_F]$. Other assertions follow now from (4.37), Proposition 4.6 and Remark 4.7 b).

3) This assertion is well-known (see [Ka, Sect. 6.2]).

4) In view of Corollary 2.26 this relation is implied by the previous one.

5) With previous notations one deduces from (4.35) and (4.36) that

$(A_K f, (I + A_K)f) = ((I - T_M)g, g)$. Hence

$$2Ret_{A_K}(f - f_A) + 2\|A_K(f - f_A)\|^2$$

(4.38)

$$= Re((I - T_M)(g - g_T), g - g_T).$$

Combining this relation with assertion 4') of Proposition 4.6 one obtains the desired one.

Let now for some $\widetilde{A} \in \mathrm{Ext}_A((0, \infty); \varphi)$ the relation

$$\inf_{f_A \in \mathcal{D}(A)} \{\|\widetilde{A}(f - f_A)\|^2 + Re(\widetilde{A}(f - f_A), f - f_A)\} = 0,$$

(4.39)

$$f \in \mathcal{D}(\widetilde{A})$$

holds for each $f \in \mathcal{D}(\widetilde{A})$. Assume without loss of generality that $ker A = \{0\}$. Setting $g_A := Af_A \in \mathcal{D}(A^{-1})$, $g := \widetilde{A}f \in \mathcal{D}(\widetilde{A}^{-1})$ one rewrites (4.39) as

$$\inf_{g_A \in \mathcal{D}(A^{-1})} \{\|g - g_A\|^2 + Re(\widetilde{A}^{-1}(g - g_A), g - g_A)\} = 0,$$

(4.40)

$$g \in \mathcal{D}(\widetilde{A}^{-1}).$$

In view of assertion 4) relation (4.40) yields $(\widetilde{A})^{-1} = (A^{-1})_F$. By assertion 1) $(A_K)^{-1} = (A^{-1})_F = (\widetilde{A})^{-1}$ and hence $\widetilde{A} = A_K$.

6) Keeping previous notations one easily derives from (4.34)

$$Re((1 \pm i \operatorname{ctg} \varphi)(\widetilde{A}f, f)) = (D_{\overline{T}}^2 g, g) \mp 2 \operatorname{ctg} \varphi(T_I g, g).$$

The required relations are implied now by assertions 2) and 2') of Proposition 4.6.   □

Remark 4.14 Another approach to the investigation of the properties of the extensions A_F and A_K, described in Proposition 4.13 has been proposed in [Ar3].

The most part of the results on the class $Ext_A((0, \infty))$ may be extended to the class $\mathrm{Ext}_A((0, \infty); \varphi)$ with slight changes in the proofs. We present below only a generalization of the result of T. Ando and K. Nishio.

If A is nondensely defined, the class $\mathrm{Ext}_A((0, \infty); \varphi)$ may no contain an operator. To formulate a condition for the existence of m-sectorial operator in the class $\mathrm{Ext}_A((0, \infty); \varphi)$ we introduce a notation.

Following T. Ando and K. Nishio [AN] an operator A satisfying (4.32) is called sectorialy closable if $\lim_{n \to \infty} Re(Af_n, f_n) = 0$ and $\lim_{n \to \infty} Af_n = g$ implies $g = 0$. If $\overline{\mathcal{D}(A)} = \mathfrak{H}$ then A is sectorially closable.

Proposition 4.15 *Let A be a closed sectorial operator in $\mathfrak{H}$ satisfying (4.32). Then:*

1. *The class $\mathrm{Ext}_A((0, \infty); \varphi)$ contains an operator iff A is sectorially closable;*

2. *If this requirement is fulfilled, A_K is the operator such that*

$$\mathcal{D}[A_K] = \left\{ h \in \mathfrak{H} : C_h^2 := \sup_{f \in \mathcal{D}(A)} \frac{|(Af, h)|^2}{Re(Af, f)} < \infty \right\}.$$

Proof: 1) This assertion may be proved in just the same way as the corresponding assertion for a nonnegative operator A in [AN].

2) Let $B = A^{-1}$. Then B_F is m-sectorial and therefore it admits a representation $B_F = T^{1/2}(I + iK)T^{1/2}$, with $T = T^* = \operatorname{Re}T \geq 0$ and $K = K^* \in [\mathfrak{H}]$ (see [Ka]). It is clear that $(B_F)^{-1} \subset T^{-1/2}(I + iK)^{-1}T^{-1/2}$. Since $(B_F)^{-1}$ is m-sectorial and $T^{-1/2}(I + iK)^{-1}T^{-1/2}$ is a sectorial operator one has $(B_F)^{-1} = T^{-1/2}(I + iK)^{-1}T^{-1/2}$. By Proposition 4.13 $A_K = (B_F)^{-1} = T^{-1/2}(I+iK)^{-1}T^{-1/2}$ and hence $\mathcal{D}[A_K] = \mathcal{D}(T^{-1/2})$.

Setting $g = Af$ one derives

$$
(4.41) \qquad
\begin{aligned}
C_h^2 &= \sup_{g \in \mathcal{D}(B)} \frac{|(g,h)|^2}{\operatorname{Re}(Bg,g)} = \sup_{g \in \mathcal{D}[B]} \frac{|(g,h)|^2}{t_{\operatorname{Re}B}(g)} \\
&= \sup_{g \in \mathcal{D}(T^{1/2})} \frac{|(g,h)|^2}{\|T^{1/2}g\|^2}
\end{aligned}
$$

The relation (4.41) yields the equivalence $h \in \mathfrak{R}(T^{1/2}) \iff C_h < \infty$. To complete the proof it remains to note that $\mathfrak{R}(T^{1/2}) = \mathcal{D}(T^{-1/2}) = \mathcal{D}[A_K]$. $\qquad\square$

References

[AG]   N.I. Akhiezer and I.M. Glazman, *Theory of Linear Operators in Hilbert Spase*, Moscow "Nauka" 1966.

[AN]   T. Ando and K. Nishio, Positive self-adjoint extensions of positive symmetric operators, *Tohoku Math. J.* **22** (1970), 65–75.

[ArG]   Gr. Arsene and A. Gheondea, Completing matrix contractions, *J. Operator Theory* **7** (1982), 179–189.

[ACG]   Gr. Arsene, T. Constantinescu and A. Gheondea, Lifting of operators and prescribed numbers of negative squares, *Michigan Math. J.* **34** (1987), 201–216.

[Ar1]   Yu.M. Arlinskii, On a class of contractions on a Hilbert space, *Ukrain. Math. Journ.* **39** no. 6 (1987), 691–696.

[Ar2]   Yu.M. Arlinskii, On a class of nondensely defined contractions and their extensions, *Journ. Math. Sci.* **97**, no. 5 (1999), 4390– 4419.

[Ar3]   Yu.M. Arlinskii, Maximal sectorial extensions of sectorial operators, *Dokl. Acad. Nauk Ukraine* no. 6 (1995), 22–27.

[AT1]   Yu.M. Arlinskii and E.R. Tsekanovskii, Maximal sectorial extensions of positive Hermitian operators and their resolvents, *Dokl Acad Nauk Armyan SSR* **79**, no. 5 (1984), 199–202.

[AT2]   Yu.M. Arlinskii and E.R. Tsekanovskii, Quasiselfadjoint contractive extensions of Hermitian contractions, *Teor. Funkts., Funksional. Anal. i Prilozen.* **50** (1988), 9–16.

[B]   M.S. Birman, On self-adjoint extensions of positive definite operators, *Mat. Sb.* **38**, no. 4 (1956), 431–450.

[BN1]   J.F. Brasche and H. Neidhardt, Has every symmetric operator a closed symmetric restriction whose square has a trivial domain?, *Acta Sci. Math. (Szeged)* **58** (1993), 425–430.

[BN2]   J.F. Brasche and H. Neidhardt, Some remarks on Krein's extension Theory, *Math. Nachr.* **165** (1994), 159–181.

[CS]   E.A. Coddington and H.S.V. de Snoo, Positive self-adjoint extensions of positive symmetric subspaces, *Math. Z.* **159** (1978), 203–214.

[Da]   Ch. Davis, Some dilation representation Theorems, *in the book Proc. of the Second Intern. Symp. in West Africa on funct. anal. and its appl.-Kunasi* (1979), 159–182.

[DKW]   Ch. Davis, W.M. Kahan and H.F. Weinberger, Norm-preserveng dilations and their applications to optimal error bounds, *Siam J. Numerical Anal.* **19**, no. 3 (1982), 445–469.

[DM]   V.A. Derkach and M.M. Malamud, Generalized Resolvents and the boundary value problems for Hermitian operators with gaps, *J. Func. Anal.* **95**, no. 1 (1991), 1– 95.

[Do]   R.G. Douglas, On majorization, factorization and range inclusion of operators in Hilbert spase, *Proc. Amer. Math. Soc.* **17** (1966), 413–415.

[FF]   C. Foias and A.E. Frazho, Redheffer products and the lifting of contractions on hilbert space, *J. Operator theory* **11** (1984), 193–196.

[Ga]   J.B. Garnett, *Bounded analytic functions*, Academic Press 1981.

[GL]   I.M. Glazman and Yu.I. Lyubitch, *Finite dimensional Linear Analysis*, Nauka, Moscow 1969.

[GG]   V.I. Gorbachuk and M.L. Gorbachuk, *Boundary value problems for operator-differential equations*, Naukova Dumka, Kiev 1984.

[Ka]   T. Kato, Perturbation *Theory for Linear Operators*, Springer Verlag 1966.

[Kol]   V.U. Kolmanovich, Some properties of self-adjoint extensions of Hermitian contractions (Russian), *Manuscript No 3192-83 Deposited at Vses. Nauchn.-Issled. Inst. Nauchno-Tekhn. Informatsii 01 06 83 Moscow*, 1983, 1–10.

[KM1]   V.U. Kolmanovich and M.M. Malamud, Extensions of Sectorial operators and dual pair of contractions (Russian), *Manuscript No 4428-85 Deposited at Vses Nauchn-Issled Inst. Nauchno-Techn. Informatsii VINITI 19 04 85 Moscow RJ. Mat 1985 10B1144*, 1985, 1–57.

[KM2]   V.U. Kolmanovich and M.M. Malamud, An Operator Analog of Schwarz-Levner's Lemma (Russian), *Teor. Funkts., Funksional Anal.i Prilozen.* **45** (1987), 71–75.

[K]   M.G. Krein, The theory of self-adjoint extensions of semi-bounded Hermitian operators and its applications 1, *Math. sb.* **20**, no. 3 (1947), 431–495.

[KO]   M.G. Krein and I.E. Ovcharenko, On the Q-functions and sc-resolvents of a nondensely defined Hermitian contraction, *Sib. Math. J.* **18**, no. 5 (1977), 1032–1056.

[Kr]   S.G. Krein, Linear Differential Equations in a Banach Space, *Amer. Math. Soc.*, Providence. Rhode Island 1971.

[M1]   M.M. Malamud, On some analogs of J. von Neuman inequality for J-contraction, *Zap. Nauchn. Sem. Leningrad. Otdel. Mat. Inst. Steklov. LOMI* **157** (1987), 165–172.

[M2]   M.M. Malamud, On extensions of Hermitian and sectorial operators and dual pairs of contractions, *Sov. Math. Dokl.* **39**, no. 2 (1989), 253–259.

[M3]   M.M. Malamud, Boundary value problems for Hermitian operators with gaps, *Soviet. Math. Dokl.* **42**, no. 1 (1991), 190–196.

[M4]   M.M. Malamud, Certain classes of extensions of a lacunary Hermitian operator, *Ukranian Math. Journal* **44**, no. 2 (1992), 190–204.

[M5]   M.M. Malamud, On a formula of the generalized resolvents of a nondensely defined Hermitian operator, *Ukrainian Math. Journal* **44**, no. 12 (1992), 1658–1688.

[Mir]   D.A. Mirman, On maximal extension of – bounded operator, *Theory Functions, Funct. Anal. and Applications.* **8** (1969), 52–56.

[PaB]   B.N. Parlett, *The symmetric eigenvalue problem*, Prentice Hall Inc. Englewood Cliffs NJ **1980.**

[Parr]   S. Parrot, On a quotient norm and the Sz.-Nagy-Foias Lifting Theorem, *J. Funct. Anal.* **30** (1978), 311–328.

[P1]   R.S. Phillips, Dissipative operators and hyperbolic systems of partial differential eguations, *Trans. Amer. Math. Soc.* **90** (1959), 192–254.

[P2]   R.S. Phillips, The extension of dual subspaces invariant under an algebra, *in the book Proc. Inter. Symp. Linear Algebra, Israel, 1960, Academic Press*, 1961, 366–398.

[SNF1]   B. Sz.-Nagy and C. Foias, Forme triangulaire d'un contraction et factorization de la fonction caracteristigue, *Acta Sci. Math. (Szeged)* **28** (1967), 201–212.

[SNF2]   B. Sz.-Nagy and C. Foias, *Harmonic analysis of operators on Hilbert space*, Amsterdam – Budapest 1970.

[Sh]   Yu.L. Shmul'yan, A Hellinger operator integral, *Mat. Sb.* **49(91)** (1959), 381–430.

[ShY]   Yu.L. Shmul'yan and R.N. Yanovskaya, On matrices whose entries are contractions, *Izv. Vissh. Ucheb. Zaved. Matematica 7,* **230** (1981), 72–75.

[V]   M.I. Vishik, On general boundary problems for elliptic differential equations, *Trans. Moscow Math. Soc.* **1** (1952), 186–246.

M.M. Malamud
Department of Mathematics
Donetsk State University
Universitetskaya st. 24
340055 Donetsk
Ukraine
e-mail: mmm@univ.donetsk.ua

Operator Theory:
Advances and Applications, Vol. 124
© 2001 Birkhäuser Verlag Basel/Switzerland

Liftings of Intertwining Operators

S.A.M. Marcantognini and M.D. Morán

To Professor Israel Gohberg with affection and admiration

Let $\mathcal{E}_1$ and $\mathcal{E}_2$ be two Hilbert spaces and let T_1 and T_2 be two contractions on $\mathcal{E}_1$ and $\mathcal{E}_2$, respectively. Consider a closed subspace $\mathcal{E}_0$ of $\mathcal{E}_2$ invariant to T_2^* and set $T_0 := P_{\mathcal{E}_0}^{\mathcal{E}_2} T_2|_{\mathcal{E}_0}$, where $P_{\mathcal{E}_0}^{\mathcal{E}_2}$ is the orthogonal projection from $\mathcal{E}_2$ onto $\mathcal{E}_0$. Given a continuous linear operator A from $\mathcal{E}_1$ into $\mathcal{E}_0$ intertwining T_1 and T_0, i.e., such that $AT_1 = T_0 A$, the problem we deal with is to find an intertwining lifting of A, i.e., a continuous linear operator B from $\mathcal{E}_1$ into $\mathcal{E}_2$ such that $BT_1 = T_2 B$ and $P_{\mathcal{E}_0}^{\mathcal{E}_2} B = A$. The problem was formulated by C. Foiaş who also asked for the computation of min $\|B\|$ where B runs over all intertwining liftings of A. Foiaş Lifting Problem is not always solvable. So necessary and sufficient conditions for the existence of such an intertwining lifting and then the minimal norm of all such intertwining liftings are to be found. We obtain all the intertwining liftings with minimal norm in some particular interesting cases. We also give a necessary and sufficient condition for the solvability of the problem that can be easily checked in the particular cases discussed in the paper and in some other settings as well.

1 Introduction

The Sz.-Nagy-Foiaş Commutant Lifting Theorem ([NF]) provides a complete answer to a lifting problem that is of substantial interest in both pure and applied operator theory. It states that every continuous linear operator intertwining an isometry and a contraction can be lifted, by preserving its norm, to a continuous linear operator intertwining the given isometry and the minimal isometric dilation of the given contraction.

If $T_0 : \mathcal{E}_0 \to \mathcal{E}_0$ is the contraction and $T_2 : \mathcal{E}_2 \to \mathcal{E}_2$ is its minimal isometric dilation, it is worth recalling that $\mathcal{E}_0$ is a closed subspace of $\mathcal{E}_2$ such that $T_2^* \mathcal{E}_0 \subseteq \mathcal{E}_0$ and that T_0 is the $\mathcal{E}_0$-compression of T_2, in the sense that T_0 can be recovered from T_2 as the orthogonal projection from $\mathcal{E}_2$ onto $\mathcal{E}_0$ of the restriction of T_2 to $\mathcal{E}_0$. A more general situation than that in the Commutant Lifting Theorem arises from considering a continuous linear operator that intertwines a contractive (not necessarily isometric) operator T_1 on $\mathcal{E}_1$ and the $\mathcal{E}_0$-compression T_0 of a contractive (not necessarily isometric) operator T_2 on $\mathcal{E}_2$, where $\mathcal{E}_0$ is a closed subspace of $\mathcal{E}_2$ invariant to T_2^*.

The lifting problem in this more general framework is not always solvable. So one has to find necessary and sufficient conditions for the existence of intertwining liftings and then the minimal norm of such intertwining liftings. The problem was formulated by C. Foiaş with the aim of developing for each solvable case the full analogue of the theory of the Commutant Lifting Theorem. A concise presentation of some particular interesting cases is [F]. Therein a definite characterization of the solvability of the problem is given. However, since that characterization can not be used in a simple way to deduce the cases for which a positive answer to the problem is known to exist, to quote from C. Foiaş ([F], p. 234), the "problem...must be considered still open."

In this note we give all the intertwining liftings with minimal norm in three cases, namely, the cases when: (a) T_1 and T_2 are isometries, (b) only T_1 is supposed to be an isometry, and (c) T_1 and T_2 are coisometries. Case (a) includes the Commutant Lifting Theorem while case (b) corresponds with a generalization of the Commutant Lifting Theorem ([G].) Case (c) is referred to as the Commutant Extension Theorem ([CS], [C].)

In those cases Foiaş Lifting Problem is known to be solvable. In particular, the existence of intertwining liftings in the cases (a) and (b) is covered, amongst others, by a recent result of A. Biswas, C. Foiaş and A.E. Frazho ([BFF]) that includes the Commutant Lifting Theorem and the Treil-Volberg Generalization of the Commutant Lifting Theorem ([TV].) Our approach makes use of a parametrization by Schur functions of the γ-intertwining liftings in the Commutant Lifting Theorem as presented by the second author in [M] and employs Douglas Factorization Lemma as stated in [GGK] (for its original version as a relationship between the notions of majorization, factorization, and range inclusion for Hilbert space operators see [D].)

We also give a necessary and sufficient condition for the existence of solutions of the problem. The condition, though abstract, has the advantage that can be easily checked in the three cases discussed in the paper and in some other examples.

The paper is organized as follows. In Section 2 we fix the notation, set out the problem and present an example that shows that the problem is not always solvable. Also, for the sake of a self-contained treatment, we state Douglas Factorization Lemma and the result that describes, by means of Schur functions, all the γ-intertwining liftings in the Commutant Lifting Theorem. Sections 3, 4 and 5 are devoted to the discussions of the aforementioned particular cases. In Section 6 we treat of the solvability of the general problem.

2 Preliminaries

Throughout this note, all Hilbert spaces are assumed to be complex.

As usual, $\mathcal{L}(\mathcal{E}, \mathcal{F})$ denotes the space of all continuous linear operators from a Hilbert space $\mathcal{E}$ to a Hilbert space $\mathcal{F}$, and $\mathcal{L}(\mathcal{E})$ is used instead of $\mathcal{L}(\mathcal{E}, \mathcal{E})$.

If $\mathcal{G}$ is a closed subspace of a Hilbert space $\mathcal{F}$, then $P_{\mathcal{G}}^{\mathcal{F}}$ is the orthogonal projection from $\mathcal{F}$ onto $\mathcal{G}$. If $\{\mathcal{F}_i\}_{i \in I}$ is a collection of closed subspaces of $\mathcal{F}$ then $\vee_{i \in I} \mathcal{F}_i$ is the least closed subspace of $\mathcal{F}$ containing all the subspaces $\mathcal{F}_i$.

For a linear operator $T : \mathcal{E} \to \mathcal{F}$, the symbol $\ker(T)$ denotes the kernel of T, namely, $\ker(T) := \{e \in \mathcal{E} : Te = 0\}$. If $T \in \mathcal{L}(\mathcal{E}, \mathcal{F})$ and $\|T\| \le \gamma$, then $D_T^{\gamma} := (\gamma^2 - T^*T)^{\frac{1}{2}}$ and $\mathcal{D}_T^{\gamma} := \overline{D_T^{\gamma} \mathcal{E}}$. When $\gamma = 1$, we use the standard notation D_T and $\mathcal{D}_T$.

Foiaş Lifting Problem

We are given as a data set for the problem a collection $(A; T_1, T_2; \mathcal{E}_0)$, where:

- $T_1 \in \mathcal{L}(\mathcal{E}_1)$ and $T_2 \in \mathcal{L}(\mathcal{E}_2)$ are two Hilbert space contractions,
- $\mathcal{E}_0$ is a closed subspace of $\mathcal{E}_2$ such that $T_2^* \mathcal{E}_0 \subseteq \mathcal{E}_0$, and
- $A \in \mathcal{L}(\mathcal{E}_1, \mathcal{E}_0)$ intertwines T_1 and the contraction $T_0 := P_{\mathcal{E}_0}^{\mathcal{E}_2} T_2|_{\mathcal{E}_0}$, in the sense that $AT_1 = T_0 A$.

The problem we deal with may be split into two parts:

Problem 1 Find $B \in \mathcal{L}(\mathcal{E}_1, \mathcal{E}_2)$ such that:

(P1.1) $BT_1 = T_2 B$, and

(P1.2) $P_{\mathcal{E}_0}^{\mathcal{E}_2} B = A$,

and, in case a positive answer to the previous problem exists,

Problem 2 Obtain all solutions B of Problem 1 satisfying $\|B\| = \alpha$, with α the minimum amongst the norms of the solutions.

We write $\mathrm{LIF}(A; T_1, T_2; \mathcal{E}_0)$ to indicate the set of all operators $B \in \mathcal{L}(\mathcal{E}_1, \mathcal{E}_2)$ that verifies (P1.1) and (P1.2), so that $\mathrm{LIF}(A; T_1, T_2; \mathcal{E}_0)$ is either empty or the set of all solutions of Problem 1.

For each $B \in \mathrm{LIF}(A; T_1, T_2; \mathcal{E}_0)$ and a fixed positive number γ, we denote by $\mathrm{LIF}_{\gamma}(B)$ the set of all γ-intertwining liftings of B. More precisely, if $V_i \in \mathcal{L}(\mathcal{K}_i)$ is the minimal isometric dilation of T_i ($i = 1, 2$), then $\tilde{B} \in \mathrm{LIF}_{\gamma}(B)$ if, and only if, $\tilde{B} \in \mathcal{L}(\mathcal{K}_1, \mathcal{K}_2)$ satisfies the relations:

(i) $\tilde{B} V_1 = V_2 \tilde{B}$,

(ii) $P_{\mathcal{E}_2}^{\mathcal{K}_2} \tilde{B} = B P_{\mathcal{E}_1}^{\mathcal{K}_1}$, and

(iii) $\|\tilde{B}\| \le \gamma$.

In a similar fashion, if γ is a positive number and $V_0 \in \mathcal{L}(\mathcal{K}_0)$ is the minimal isometric dilation of T_0, then $\mathrm{LIF}_{\gamma}(A)$ denotes the set of all $\tilde{A} \in \mathcal{L}(\mathcal{K}_1, \mathcal{K}_0)$ such that:

(i) $\tilde{A} V_1 = V_0 \tilde{A}$,

(ii) $P_{\mathcal{E}_0}^{\mathcal{K}_0} \tilde{A} = A P_{\mathcal{E}_1}^{\mathcal{K}_1}$, and

(iii) $\|\tilde{A}\| \le \gamma$.

We may assume that $\mathcal{K}_0 = \vee_{n \geq 0} V_2^n \mathcal{E}_0$ and $V_0 = V_2|_{\mathcal{K}_0}$. Since $\mathcal{K}_0$ turns out to reduce V_2, we get that V_2 admits a matrix partitioning of the form

$$
V_2 = \begin{bmatrix} V_0 & 0 \\ 0 & V_2|_{\mathcal{K}_2 \ominus \mathcal{K}_0} \end{bmatrix} : \left(\begin{array}{c} \mathcal{K}_0 \\ \mathcal{K}_2 \ominus \mathcal{K}_0 \end{array} \right) \rightarrow \left(\begin{array}{c} \mathcal{K}_0 \\ \mathcal{K}_2 \ominus \mathcal{K}_0 \end{array} \right).
$$

In tackling Problems 1 and 2, it could be helpful to have a mind to consider the following two auxiliary problems:

Problem 3 Find $Y \in \mathcal{L}(\mathcal{K}_1, \mathcal{K}_2)$ such that:

(P3.1) $Y V_1 = V_2 Y$,

(P3.2) $P_{\mathcal{E}_0}^{\mathcal{K}_2} Y = A P_{\mathcal{E}_1}^{\mathcal{K}_1}$, and

(P3.3) $P_{\mathcal{E}_2}^{\mathcal{K}_2} Y (V_1 - T_1)|_{\mathcal{E}_1} = 0$,

and

Problem 4 Obtain all solutions Y of Problem 3 (if there is any) satisfying $\|Y\| = \beta$, with β the minimum amongst the norms of the solutions.

In fact, Problems 1 and 3, as well as Problems 2 and 4, are closely connected. We mean the following:

Proposition 2.1 *There is a solution B of Problem 1 with $\|B\| \leq \gamma$ if, and only if, there is a solution Y of Problem 3 with $\|Y\| \leq \gamma$.*

Proof: If there is $B \in \mathrm{LIF}(A; T_1, T_2; \mathcal{E}_0)$ with $\|B\| \leq \gamma$, then, by a direct application of the Commutant Lifting Theorem, we find $\tilde{B} \in \mathrm{LIF}_{\|B\|}(B)$. We set $Y := \tilde{B}$, so that

$$
Y V_1 = \tilde{B} V_1 = V_2 \tilde{B} = V_2 Y.
$$

Since $P_{\mathcal{E}_0}^{\mathcal{E}_2} B = A$ and $P_{\mathcal{E}_2}^{\mathcal{K}_2} \tilde{B} = B P_{\mathcal{E}_1}^{\mathcal{K}_1}$, we have that

$$
P_{\mathcal{E}_0}^{\mathcal{K}_2} Y = P_{\mathcal{E}_0}^{\mathcal{K}_2} \tilde{B} = P_{\mathcal{E}_0}^{\mathcal{E}_2} P_{\mathcal{E}_2}^{\mathcal{K}_2} \tilde{B} = P_{\mathcal{E}_0}^{\mathcal{E}_2} B P_{\mathcal{E}_1}^{\mathcal{K}_1} = A P_{\mathcal{E}_1}^{\mathcal{K}_1},
$$

and

$$
P_{\mathcal{E}_2}^{\mathcal{K}_2} Y (V_1 - T_1)|_{\mathcal{E}_1} = P_{\mathcal{E}_2}^{\mathcal{K}_2} \tilde{B}(V_1 - T_1)|_{\mathcal{E}_1} = B P_{\mathcal{E}_1}^{\mathcal{K}_1}(V_1 - T_1)|_{\mathcal{E}_1} = 0.
$$

Finally, as $\|\tilde{B}\| = \|B\|$ and $\|B\| \leq \gamma$, we also obtain that $\|Y\| \leq \gamma$. So Y is shown to be a solution of Problem 3 with the norm constraint $\|Y\| \leq \gamma$.

Conversely, if Y is a solution of Problem 3 and $\|Y\| \leq \gamma$, set $B := P_{\mathcal{E}_2}^{\mathcal{K}_2} Y|_{\mathcal{E}_1}$. Since $Y V_1 = V_2 Y$ and $P_{\mathcal{E}_2}^{\mathcal{K}_2} Y (V_1 - T_1)|_{\mathcal{E}_1} = 0$, we get that

$$
B T_1 = P_{\mathcal{E}_2}^{\mathcal{K}_2} Y T_1 = P_{\mathcal{E}_2}^{\mathcal{K}_2} Y V_1|_{\mathcal{E}_1} = P_{\mathcal{E}_2}^{\mathcal{K}_2} V_2 Y|_{\mathcal{E}_1} = T_2 P_{\mathcal{E}_2}^{\mathcal{K}_2} Y|_{\mathcal{E}_1} = T_2 B.
$$

Also, from the relation $P_{\mathcal{E}_0}^{\mathcal{K}_2} Y = A P_{\mathcal{E}_1}^{\mathcal{K}_1}$, we conclude that

$$P_{\mathcal{E}_0}^{\mathcal{E}_2} B = P_{\mathcal{E}_0}^{\mathcal{E}_2} P_{\mathcal{E}_2}^{\mathcal{K}_2} Y|_{\mathcal{E}_1} = P_{\mathcal{E}_0}^{\mathcal{K}_2} Y|_{\mathcal{E}_1} = A P_{\mathcal{E}_1}^{\mathcal{K}_1}|_{\mathcal{E}_1} = A.$$

Hence B is a solution of Problem 1. To finish the proof it suffices to notice that $\|B\| \leq \|Y\| \leq \gamma$. $\qquad\square$

In some particular cases the set of all solutions of Problem 1 with a certain norm constraint and the set of all solutions of Problem 3 with the same norm constraint are furthermore in bijective correspondence.

An Example

In some concrete situations, LIF$(A; T_1, T_2; \mathcal{E}_0)$ happens to be empty. For instance, in the following example:

In $L^2(\mathbb{T})$ (the space of the square integrable functions on the unit circle $\mathbb{T}$ of the complex plane) consider the functions $e_n(\zeta) := \zeta^n$, with $n \in \mathbb{Z}$ and $\zeta \in \mathbb{T}$. Define $\mathcal{E}_1$ as the subspace spanned by e_0 and e_1, and $\mathcal{E}_2$ as the whole space $L^2(\mathbb{T})$. Consider $T_1 := P_{\text{span}\{e_1\}}^{\mathcal{E}_1} : \mathcal{E}_1 \to \mathcal{E}_1$, and $T_2 := S^* : \mathcal{E}_2 \to \mathcal{E}_2$, where S is the forward shift on $L^2(\mathbb{T})$. Set $\mathcal{E}_0 := H^2(\mathbb{D})$ (the Hardy space of the functions f defined and holomorphic on the unit disk $\mathbb{D}$ of the complex plane such that $f(z) = \sum_{n=0}^{\infty} a_n z^n$ with $\sum_{n=0}^{\infty} |a_n|^2 < \infty$) and $A := P_{\text{span}\{e_0\}}^{\mathcal{E}_1} : \mathcal{E}_1 \to \mathcal{E}_0$.

Though $T_2^* \mathcal{E}_0 \subseteq \mathcal{E}_0$ and $A T_1 = T_0 A = 0$ $(T_0 = P_{\mathcal{E}_0}^{\mathcal{E}_2} T_2|_{\mathcal{E}_0})$, it turns out that $B = 0$ is the only operator that intertwines T_1 and T_2. Since $A \neq 0$, the relation $P_{\mathcal{E}_0}^{\mathcal{E}_2} B = A$ can hardly be true, and, consequently, LIF$(A; T_1, T_2; \mathcal{E}_0)$ comes out to be empty.

Two Auxiliary Results

In some other situations, relevant to pure and applied operator theory, LIF$(A; T_1, T_2; \mathcal{E}_0) \neq \emptyset$. As we already mentioned it in the introduction, our purpose is to discuss some of those situations. Our approach is based on two results.

The first result is known as Douglas Factorization Lemma. We formulate it as in [GGK] and we refer the reader to [D] as the original source.

The other result can be found in [M]. It gives a description of LIF$_\gamma(A)$ in terms of $\mathcal{S}(\mathcal{N}, \mathcal{M})$ functions, where $\mathcal{S}(\mathcal{N}, \mathcal{M})$ denotes the Schur class of all $\mathcal{L}(\mathcal{N}, \mathcal{M})$-valued functions f defined and holomorphic on $\mathbb{D}$ such that $\|f(z)\| \leq 1$ for each $z \in \mathbb{D}$, where $\mathcal{N}$ and $\mathcal{M}$ are certain Hilbert spaces.

Proposition 2.2 (Douglas Factorization Lemma.) *Let* $X : \mathcal{H} \to \mathcal{F}_1$ *and* $Z : \mathcal{H} \to \mathcal{F}_2$ *be given Hilbert space operators. Then* $Z^*Z \leq X^*X$ *if, and only if,* $Z = \Gamma X$ *with* $\Gamma : \mathcal{F}_1 \to \mathcal{F}_2$ *a contraction. Furthermore, in this case the contraction* Γ *can be chosen such that* $\ker \Gamma \supseteq \ker X^*$, *and, with this additional condition,* Γ *is uniquely determined.*

Proposition 2.3 *Let* $(A; T_1, T_2; \mathcal{E}_0)$ *be the data set of Foiaş Lifting Problem, and let* γ *be a real number such that* $\gamma \geq \|A\|$. *Then there is a bijective correspondence between* $\mathrm{LIF}_\gamma(A)$ *and* $\mathcal{S}(\mathcal{N}, \mathcal{M})$, *where*

$$\mathcal{N} := \mathcal{D}_A^\gamma \oplus \mathcal{D}_{T_0} \ominus \overline{\{D_A^\gamma e_1 \oplus D_{T_0} A e_1 : e_1 \in \mathcal{E}_1\}},$$

and

$$\mathcal{M} := \mathcal{D}_A^\gamma \oplus \mathcal{D}_{T_1} \ominus \overline{\{D_A^\gamma T_1 e_1 \oplus \gamma^2 D_{T_1} e_1 : e_1 \in \mathcal{E}_1\}}.$$

3 First Case

We recall that the data set for the problem is the collection $(A; T_1, T_2; \mathcal{E}_0)$, where $T_1 \in \mathcal{L}(\mathcal{E}_1)$ and $T_2 \in \mathcal{L}(\mathcal{E}_2)$ are contractions, $\mathcal{E}_0 \subseteq \mathcal{E}_2$ is invariant to T_2^*, and $A \in \mathcal{L}(\mathcal{E}_1, \mathcal{E}_0)$ satisfies $AT_1 = T_0 A$, with $T_0 := P_{\mathcal{E}_0}^{\mathcal{E}_2} T_2|_{\mathcal{E}_0}$. We also remind that we look for $B \in \mathrm{LIF}(A; T_1, T_2; \mathcal{E}_0)$, i.e., for an operator $B \in \mathcal{L}(\mathcal{E}_1, \mathcal{E}_2)$ such that: (P1.1) $BT_1 = T_2 B$, and (P1.2) $P_{\mathcal{E}_0}^{\mathcal{E}_2} B = A$.

In this section T_1 and T_2 are supposed to be isometries.

As before, let $V_0 \in \mathcal{L}(\mathcal{K}_0)$ denote the minimal isometric dilation of T_0. Then we have that $\mathcal{K}_0 = \vee_{n \geq 0} T_2^n \mathcal{E}_0$ and $V_0 = T_2|_{\mathcal{K}_0}$, with $\mathcal{K}_0$ a T_2-reducing closed subspace of $\mathcal{E}_2$.

Let $\tilde{A} \in \mathrm{LIF}_\alpha(A)$, with $\alpha = \|A\|$, so that $\tilde{A} \in \mathcal{L}(\mathcal{E}_1, \mathcal{K}_0)$ satisfies: (i) $\tilde{A}T_1 = T_2\tilde{A}$, (ii) $P_{\mathcal{E}_0}^{\mathcal{K}_0} \tilde{A} = A$, and (iii) $\|\tilde{A}\| = \|A\| = \alpha$. Define B from $\mathcal{E}_1$ into $\mathcal{E}_2$ by

$$B := \begin{bmatrix} \tilde{A} \\ 0 \end{bmatrix} : \mathcal{E}_1 \rightarrow \begin{pmatrix} \mathcal{K}_0 \\ \mathcal{E}_2 \ominus \mathcal{K}_0 \end{pmatrix}.$$

Then B belongs to $\mathcal{L}(\mathcal{E}_1, \mathcal{E}_2)$ and satisfies (P1.1) and (P1.2). Therefore, $B \in \mathrm{LIF}(A; T_1, T_2; \mathcal{E}_0)$. Moreover, $\|B\| = \|\tilde{A}\| = \alpha$.

This shows that $\mathrm{LIF}(A; T_1, T_2; \mathcal{E}_0) \neq \emptyset$ and, furthermore, that there are intertwining liftings $B \in \mathrm{LIF}(A; T_1, T_2; \mathcal{E}_0)$ with $\|B\| = \alpha$.

In general, if $B \in \mathrm{LIF}(A; T_1, T_2; \mathcal{E}_0)$ and $\|B\| = \alpha$, then $P_{\mathcal{K}_0}^{\mathcal{E}_2} B$ turns out to belong to $\mathrm{LIF}_\alpha(A)$. So every $B \in \mathrm{LIF}(A; T_1, T_2; \mathcal{E}_0)$ with $\|B\| = \alpha$ admits a 2×1 block matrix representation of the form

$$B = \begin{bmatrix} \tilde{A} \\ X \end{bmatrix} : \mathcal{E}_1 \rightarrow \begin{pmatrix} \mathcal{K}_0 \\ \mathcal{E}_2 \ominus \mathcal{K}_0 \end{pmatrix},$$

where $\tilde{A} \in \mathrm{LIF}_\alpha(A)$, $\|\tilde{A}\| = \|A\| = \alpha$, and $X \in \mathcal{L}(\mathcal{E}_1, \mathcal{E}_2 \ominus \mathcal{K}_0)$. Since $\|B\| = \alpha$, then $\tilde{A}^*\tilde{A} + X^*X = B^*B \leq \alpha^2$ and, thus, $X^*X \leq (D_{\tilde{A}}^\alpha)^2$. We apply Douglas Lemma (Proposition 2.2) and we obtain the following result:

Proposition 3.1 *Let* $(A; T_1, T_2; \mathcal{E}_0)$ *be the data set of Foiaş Lifting Problem, with* T_1 *and* T_2 *isometries, and let* $\alpha = \|A\|$. *Then the map* ϕ *from*

$$\{(\tilde{A}, \Gamma) \mid \tilde{A} \in \mathrm{LIF}_\alpha(A), \Gamma : \mathcal{D}_{\tilde{A}}^\alpha \rightarrow \mathcal{E}_2 \ominus \mathcal{K}_0, \|\Gamma\| \leq 1, \Gamma D_{\tilde{A}}^\alpha T_1 = T_2 \Gamma D_{\tilde{A}}^\alpha\}$$

into

$$\{B \in \text{LIF}(A; T_1, T_2; \mathcal{E}_0) \mid \|B\| = \alpha\},$$

given by

$$\phi(\tilde{A}, \Gamma) := \tilde{A} + \Gamma D_{\tilde{A}}^{\alpha},$$

is a bijection.

According with Proposition 3.1, the choice $\Gamma = 0$ may happen to provide only a subset of solutions of Problem 1 with minimal norm.

In the case we are discussing in this section there is no need of considering Problems 3 and 4, since T_1 and T_2 are isometries and, hence, coincide with their minimal isometric dilations V_1 and V_2, respectively.

4 Second Case

In this section we consider the case when only T_1 is supposed to be an isometry.

We recall that $V_2 \in \mathcal{L}(\mathcal{K}_2)$ and $V_0 \in \mathcal{L}(\mathcal{K}_0)$ denote the minimal isometric dilations of T_2 and T_0, respectively.

To begin with, let us assume that Problem 3 is solvable and let us see how a solution Y looks like.

Since T_1 on $\mathcal{E}_1$ is isometric and, hence, coincides with its minimal isometric dilation V_1 on $\mathcal{K}_1$, we have that $Y \in \mathcal{L}(\mathcal{E}_1, \mathcal{K}_2)$ and satisfies: (P3.1) $YT_1 = V_2Y$, and (P3.2) $P_{\mathcal{E}_0}^{\mathcal{K}_2} Y = A$.

Let $\alpha = \|A\|$ and assume that $\|Y\| = \alpha$.

The relations (P3.1) and (P3.2), together with the fact that $\mathcal{K}_0$ reduces V_2, assure that $P_{\mathcal{K}_0}^{\mathcal{K}_2} Y$ belongs to $\text{LIF}_\alpha(A)$. Therefore, Y can be represented in the form

$$Y = \begin{bmatrix} \tilde{A} \\ X \end{bmatrix} : \mathcal{E}_1 \to \begin{pmatrix} \mathcal{K}_0 \\ \mathcal{K}_2 \ominus \mathcal{K}_0 \end{pmatrix},$$

where $\tilde{A} \in \text{LIF}_\alpha(A)$, $\|\tilde{A}\| = \|A\| = \alpha$, and $X \in \mathcal{L}(\mathcal{E}_1, \mathcal{K}_2 \ominus \mathcal{K}_0)$.

A direct application of Douglas Lemma provides a uniquely determined contraction $\Gamma : \mathcal{D}_{\tilde{A}}^{\alpha} \to \mathcal{K}_2 \ominus \mathcal{K}_0$ such that $X = \Gamma D_{\tilde{A}}^{\alpha}$. As (P3.1) holds and $\tilde{A} \in \text{LIF}_\alpha(A)$, we have that $\Gamma D_{\tilde{A}}^{\alpha} T_1 = V_2 \Gamma D_{\tilde{A}}^{\alpha}$.

The above discussion allows us to say that Problem 3 has always a solution Y with $\|Y\| = \alpha$, since $\Gamma = 0$ is a contraction from $\mathcal{D}_{\tilde{A}}^{\alpha}$ into $\mathcal{K}_2 \ominus \mathcal{K}_0$ that verifies $\Gamma D_{\tilde{A}}^{\alpha} T_1 = V_2 \Gamma D_{\tilde{A}}^{\alpha} = 0$. Also, from the above reasoning, we get the following result concerning Problem 4:

Proposition 4.1 *Let $(A; T_1, T_2; \mathcal{E}_0)$ be the data set of Foiaş Lifting Problem, with T_1 an isometry, and let $\alpha = \|A\|$. Then the map ϕ from*

$$\{(\tilde{A}, \Gamma) \mid \tilde{A} \in \text{LIF}_\alpha(A), \Gamma : \mathcal{D}_{\tilde{A}}^{\alpha} \to \mathcal{K}_2 \ominus \mathcal{K}_0,$$
$$\|\Gamma\| \le 1, \Gamma D_{\tilde{A}}^{\alpha} T_1 = V_2 \Gamma D_{\tilde{A}}^{\alpha}\}$$

into

$$\{Y \text{ solution of Problem 3} \mid \|Y\| = \alpha\},$$

given by

$$\phi(\tilde{A}, \Gamma) := \tilde{A} + \Gamma D_{\tilde{A}}^\alpha,$$

is a bijection.

Now we deal with Problems 1 and 2. According with Proposition 2.1, there exists $B \in \text{LIF}(A; T_1, T_2; \mathcal{E}_0)$ such that $\|B\| = \alpha$, with again $\alpha = \|A\|$.

We have that $B \in \mathcal{L}(\mathcal{E}_1, \mathcal{E}_2)$ satisfies: (P1.1) $BT_1 = T_2 B$, and (P1.2) $P_{\mathcal{E}_0}^{\mathcal{E}_2} B = A$.

Relation (P1.2) and Douglas Lemma imply that there exists a uniquely determined contraction $\Gamma : \mathcal{D}_A^\alpha \to \mathcal{E}_2 \ominus \mathcal{E}_0$ such that $P_{\mathcal{E}_2 \ominus \mathcal{E}_0}^{\mathcal{E}_2} B = \Gamma D_A^\alpha$. On the other hand, from condition (P1.1), it follows that Γ must verify

$$\Gamma D_A^\alpha T_1 - T_2 \Gamma D_A^\alpha = P_{\mathcal{E}_2 \ominus \mathcal{E}_0}^{\mathcal{E}_2} T_2 A.$$

Proposition 4.2 *Let $(A; T_1, T_2; \mathcal{E}_0)$ be the data set of Foiaş Lifting Problem, with T_1 an isometry, and let $\alpha = \|A\|$. Then the map ϕ from*

$$\{\Gamma \mid \Gamma : \mathcal{D}_A^\alpha \to \mathcal{E}_2 \ominus \mathcal{E}_0, \|\Gamma\| \le 1, \Gamma D_A^\alpha T_1 - T_2 \Gamma D_A^\alpha = P_{\mathcal{E}_2 \ominus \mathcal{E}_0}^{\mathcal{E}_2} T_2 A\}$$

into

$$\{B \in \text{LIF}(A; T_1, T_2; \mathcal{E}_0) \mid \|B\| = \alpha\},$$

given by

$$\phi(\Gamma) := A + \Gamma D_A^\alpha,$$

is a bijection.

If $\tilde{A} \in \text{LIF}_\alpha(A)$, $\|\tilde{A}\| = \|A\| = \alpha$, then the operator $\Gamma : \mathcal{D}_A^\alpha \to \mathcal{E}_2 \ominus \mathcal{E}_0$, defined as

(4.1) $\Gamma D_A^\alpha e_1 := P_{\mathcal{E}_2 \ominus \mathcal{E}_0}^{\mathcal{K}_2} \tilde{A} e_1,$ $e_1 \in \mathcal{E}_1,$

is a contraction that satisfies $\Gamma D_A^\alpha T_1 - T_2 \Gamma D_A^\alpha = P_{\mathcal{E}_2 \ominus \mathcal{E}_0}^{\mathcal{E}_2} T_2 A$. This shows that each $\tilde{A} \in \text{LIF}_\alpha(A)$, with $\|\tilde{A}\| = \|A\| = \alpha$, gives rise to a $B \in \text{LIF}(A; T_1, T_2; \mathcal{E}_0)$, with $\|B\| = \alpha$. However, the contractions Γ associated with α-intertwining liftings $\tilde{A}$ by means of (4.1) may happen to provide only a subset of solutions of Problem 1 with minimal norm.

5 Third Case

In this section we assume that both T_1 and T_2 are coisometries, we mean that T_1^* and T_2^* are isometries.

First of all, we notice that, since $T_0^* = T_2^*|_{\mathcal{E}_0}$, also T_0 is a coisometry.

Besides, the minimal isometric dilations of T_1, T_2 and T_0 are all unitary operators. Recall that we denote them by $V_1 \in \mathcal{L}(\mathcal{K}_1)$, $V_2 \in \mathcal{L}(\mathcal{K}_2)$ and $V_0 \in \mathcal{L}(\mathcal{K}_0)$, respectively, and that $V_0 = V_2|_{\mathcal{K}_0}$, with $\mathcal{K}_0$ a reducing subspace for V_2.

It can be seen that there is only one γ-intertwining lifting in $\mathrm{LIF}_\gamma(A)$, the same one for all $\gamma \geq \|A\|$. This follows from the fact that, when T_1 is a coisometry, $\mathcal{D}_{T_1} = \ker(T_1)$, $T_1\mathcal{E}_1 = \mathcal{E}_1$, and, hence, $\mathcal{M}$ as in Proposition 2.3 turns out to be $\{0\}$. Furthermore, the only $\tilde{A} \in \mathrm{LIF}_\gamma(A)$ is explicitly given by the strong limit

$$\tilde{A}k_1 := \lim_{n\to\infty} V_2^n A P_{\mathcal{E}_1}^{\mathcal{K}_1} V_1^{*n} k_1, \qquad k_1 \in \mathcal{K}_1,$$

and satisfies $\|\tilde{A}\| = \|A\|$.

As in the previous case, let us assume that Problem 3 is solvable and let us consider a solution Y of Problem 3.

Then $Y \in \mathcal{L}(\mathcal{K}_1, \mathcal{K}_2)$ satisfies: (P3.1) $YV_1 = V_2Y$, (P3.2) $P_{\mathcal{E}_0}^{\mathcal{K}_2} Y = A P_{\mathcal{E}_1}^{\mathcal{K}_1}$, and (P3.3) $P_{\mathcal{E}_2}^{\mathcal{K}_2} Y(V_1 - T_1)|_{\mathcal{E}_1} = 0$.

Let $\gamma = \|Y\|$, so that

$$(5.1) \qquad\qquad\qquad \|A\| \leq \gamma.$$

The relations (P3.1) and (P3.2), together with the fact that $\mathcal{K}_0$ reduces V_2, imply that $P_{\mathcal{K}_0}^{\mathcal{K}_2} Y \in \mathrm{LIF}_\gamma(A)$. Hence $P_{\mathcal{K}_0}^{\mathcal{K}_2} Y = \tilde{A}$. On the other hand, from (P3.1) and (P3.3), it follows that $Y(\mathcal{K}_1 \ominus \mathcal{E}_1) \subseteq \mathcal{K}_2 \ominus \mathcal{E}_2$. Thus, if $E := Y|_{\mathcal{K}_1 \ominus \mathcal{E}_1}$, then $E \in \mathcal{L}(\mathcal{K}_1 \ominus \mathcal{E}_1, \mathcal{K}_2 \ominus \mathcal{E}_2)$ and $P_{\mathcal{K}_0}^{\mathcal{K}_2} E = \tilde{A}|_{\mathcal{K}_1 \ominus \mathcal{E}_1}$. Therefore

$$(5.2) \qquad\qquad E^* Q_2 k_0 = Q_1 \tilde{A}^* k_0, \qquad k_0 \in \mathcal{K}_0,$$

where $Q_i = 1 - P_{\mathcal{E}_i}^{\mathcal{K}_i}$ ($i = 1, 2$). It follows that

$$(5.3) \qquad\qquad \|Q_1 \tilde{A}^* k_0\| \leq \gamma \|Q_2 k_0\|, \qquad k_0 \in \mathcal{K}_0.$$

Since T_1 and T_2 are coisometries, it can be seen that (5.3) is equivalent to

$$(5.4) \qquad \|D_{T_1^n} A^* e_0\| \leq \gamma \|D_{T_2^n} e_0\|, \qquad n \in \mathbb{N}, \ e_0 \in \mathcal{E}_0,$$

where $D_{T_i^n} = P_{\ker(T_i^n)}^{\mathcal{E}_i}$, for all $n \in \mathbb{N}$ and $i = 1, 2$.

As far as (5.4) is granted, we can define

$$E_0 : \mathcal{K}_1 \ominus \mathcal{E}_1 \to \overline{Q_2 \mathcal{K}_0}$$

by its adjoint, setting

$$E_0^* Q_2 k_0 := Q_1 \tilde{A}^* k_0, \qquad k_0 \in \mathcal{K}_0.$$

It readily follows that

(5.5) $\|E_0\| \le \gamma,$

and

$$P_{\mathcal{K}_0}^{\mathcal{K}_2} E_0 = \tilde{A}|_{\mathcal{K}_1 \ominus \mathcal{E}_1}.$$

Also, for all $k_1 \in \mathcal{K}_1$ and $k_0 \in \mathcal{K}_0$,

$$
\begin{aligned}
\langle E_0 V_1 Q_1 k_1, Q_2 k_0 \rangle_{\mathcal{K}_2} &= \langle V_1 Q_1 k_1, E_0^* Q_2 k_0 \rangle_{\mathcal{K}_1} = \langle V_1 Q_1 k_1, \tilde{A}^* k_0 \rangle_{\mathcal{K}_1} \\
&= \langle Q_1 k_1, V_1^* \tilde{A}^* k_0 \rangle_{\mathcal{K}_1} = \langle Q_1 k_1, \tilde{A}^* V_2^* k_0 \rangle_{\mathcal{K}_1} \\
&= \langle Q_1 k_1, E_0^* Q_2 V_2^* k_0 \rangle_{\mathcal{K}_1} = \langle V_2 E_0 Q_1 k_1, k_0 \rangle_{\mathcal{K}_2} \\
&= \langle V_2 E_0 Q_1 k_1, Q_2 k_0 \rangle_{\mathcal{K}_2},
\end{aligned}
$$

since $V_i Q_i = Q_i V_i Q_i$ $(i = 1, 2)$. If $k_0 = e_0 + \sum_{n \ge 0} V_2^n (V_2 - T_0) e_0(n) \in \mathcal{K}_0$, then

$$Q_2 k_0 = \sum_{n \ge 0} V_2^n (V_2 - T_2) \left[e_0(n) + \sum_{k \ge n+1} T_2^{k-(n+1)} (T_2 - T_0) e_0(k) \right].$$

From this it can be shown that $V_2 Q_2 \mathcal{K}_0 \subseteq Q_2 \mathcal{K}_0$. Then we get that

(5.6) $E_0 V_1|_{\mathcal{K}_1 \ominus \mathcal{E}_1} = V_2 E_0.$

Also, from (5.2), it follows that

(5.7) $P_{Q_2 \mathcal{K}_0}^{\mathcal{K}_2 \ominus \mathcal{E}_2} Y|_{\mathcal{K}_1 \ominus \mathcal{E}_1} = P_{Q_2 \mathcal{K}_0}^{\mathcal{K}_2 \ominus \mathcal{E}_2} E = E_0.$

So far, under the assumption that Problem 3 is solvable, we have shown that if Y is any solution of Problem 3, with $\|Y\| = \gamma$, on one hand, A satisfies the norm constraints (5.1), (5.4), and, on the other hand, the operator E_0, which is well defined from (5.4), partakes of Y.

Next, under the hypothesis that there exists $\gamma > 0$ such that (5.1), (5.4) hold, we are going to construct a particular solution Y_0 of Problem 3 from E_0.

The relations (5.5) and (5.6) say that E_0 is a continuous linear operator from $\mathcal{K}_1 \ominus \mathcal{E}_1$ into $\overline{Q_2 \mathcal{K}_0}$ that intertwines $V_1|_{\mathcal{K}_1 \ominus \mathcal{E}_1}$ and $V_2|_{\overline{Q_2 \mathcal{K}_0}}$. Set

$$\mathcal{F}_1 := \vee_{n \geq 0} V_1^{*n} (\mathcal{K}_1 \ominus \mathcal{E}_1).$$

There is a uniquely determined continuous linear operator $\tilde{E}_0 : \mathcal{F}_1 \to \mathcal{K}_2$ such that:

(i) $\|\tilde{E}_0\| \leq \gamma$,

(ii) $\tilde{E}_0 V_1|_{\mathcal{F}_1} = V_2 \tilde{E}_0$,

(iii) $\tilde{E}_0|_{\mathcal{K}_1 \ominus \mathcal{E}_1} = E_0$, and

(iv) $P_{\mathcal{K}_0}^{\mathcal{K}_2} \tilde{E}_0 = \tilde{A}|_{\mathcal{F}_1}$.

Furthermore, $\tilde{E}_0$ is given by

$$\tilde{E}_0 V_1^{*n} Q_1 k_1 := V_2^{*n} E_0 Q_1 k_1, \qquad n \geq 0, k_1 \in \mathcal{K}_1.$$

We claim that there exists $X_0 : \mathcal{K}_1 \ominus \mathcal{F}_1 \to \mathcal{K}_2$ such that $Y_0 = \tilde{E}_0 P_{\mathcal{F}_1}^{\mathcal{K}_1} + X_0 P_{\mathcal{K}_1 \ominus \mathcal{F}_1}^{\mathcal{K}_1}$ is a solution of Problem 3.

The operator Y_0 must belong to $\mathcal{L}(\mathcal{K}_1, \mathcal{K}_2)$ and verify: (P3.1) $Y_0 V_1 = V_2 Y_0$, (P3.2) $P_{\mathcal{E}_0}^{\mathcal{K}_2} Y_0 = A P_{\mathcal{E}_1}^{\mathcal{K}_1}$, and (P3.3) $P_{\mathcal{E}_2}^{\mathcal{K}_2} Y_0 (V_1 - T_1)|_{\mathcal{E}_1} = 0$.

In order to have $\|Y_0\| \leq \gamma$ it is necessary and sufficient that $\|X_0^* k_2\| \leq \|D_{\tilde{E}_0^*}^\gamma k_2\|$, for all $k_2 \in \mathcal{K}_2$. Hence, according with Douglas Lemma, $\|Y_0\| \leq \gamma$ if, and only if, there exists a contraction $\Gamma^* : \mathcal{D}_{\tilde{E}_0^*}^\gamma \to \mathcal{K}_1 \ominus \mathcal{F}_1$ that verifies $X_0^* = \Gamma^* D_{\tilde{E}_0^*}^\gamma$, so that $X_0 = D_{\tilde{E}_0^*}^\gamma \Gamma$. As $\mathcal{K}_1 \ominus \mathcal{E}_1 \subseteq \mathcal{F}_1$ and $Y_0|_{\mathcal{K}_1 \ominus \mathcal{E}_1} = \tilde{E}_0|_{\mathcal{K}_1 \ominus \mathcal{E}_1}$, (P3.3) is yielded by (iii) above and the fact that E_0 maps $\mathcal{K}_1 \ominus \mathcal{E}_1$ into a subspace of $\mathcal{K}_2 \ominus \mathcal{E}_2$. From (iv) above, it turns out that $P_{\mathcal{K}_0}^{\mathcal{K}_2} Y_0 = \tilde{A}$, and so (P3.2) holds, if, and only if, $P_{\mathcal{K}_0}^{\mathcal{K}_2} D_{\tilde{E}_0^*}^\gamma \Gamma = \tilde{A}|_{\mathcal{K}_1 \ominus \mathcal{F}_1}$. Finally, the identity (P3.1) is equivalent to $\Gamma V_1|_{\mathcal{K}_1 \ominus \mathcal{F}_1} = V_2 \Gamma$, since condition (ii) is satisfied, $V_2 D_{\tilde{E}_0^*}^\gamma = D_{\tilde{E}_0^*}^\gamma V_2$ and $\mathcal{K}_1 \ominus \mathcal{F}_1$ reduces V_1. Therefore, we must look for a contraction $\Gamma : \mathcal{K}_1 \ominus \mathcal{F}_1 \to \mathcal{D}_{\tilde{E}_0^*}^\gamma$ such that:

(5.8)
$$P_{\mathcal{K}_0}^{\mathcal{K}_2} D_{\tilde{E}_0^*}^\gamma \Gamma = \tilde{A}|_{\mathcal{K}_1 \ominus \mathcal{F}_1},$$

and

(5.9) $\Gamma V_1|_{\mathcal{K}_1 \ominus \mathcal{F}_1} = V_2 \Gamma.$

Define $\sigma_1 : \mathcal{K}_1 \ominus \mathcal{F}_1 \to \overline{D^\gamma_{\tilde{E}^*_0} \mathcal{K}_0}$ by means of

$$\sigma_1^* D^\gamma_{\tilde{E}^*_0} k_0 := P^{\mathcal{K}_1}_{\mathcal{K}_1 \ominus \mathcal{F}_1} \tilde{A}^* k_0, \qquad k_0 \in \mathcal{K}_0.$$

Then σ_1 is a contraction, since, for all $k_0 \in \mathcal{K}_0$,

$$\|\sigma_1^* D^\gamma_{\tilde{E}^*_0} k_0\|^2 = \|\tilde{A}^* k_0\|^2 - \|P^{\mathcal{K}_1}_{\mathcal{F}_1} \tilde{A}^* k_0\|^2 = \|\tilde{A}^* k_0\|^2 - \|\tilde{E}^*_0 k_0\|^2$$

$$\leq \gamma^2 \|k_0\|^2 - \|\tilde{E}^*_0 k_0\|^2 = \|D^\gamma_{\tilde{E}^*_0} k_0\|^2.$$

Besides, the contraction σ_1 satisfies (5.8), by definition, and (5.9), as

$$\sigma_1^* V_2^* D^\gamma_{\tilde{E}^*_0} k_0 = \sigma_1^* D^\gamma_{\tilde{E}^*_0} V_2^* k_0 = P^{\mathcal{K}_1}_{\mathcal{K}_1 \ominus \mathcal{F}_1} \tilde{A}^* V_2^* k_0$$

$$= P^{\mathcal{K}_1}_{\mathcal{K}_1 \ominus \mathcal{F}_1} V_1^* \tilde{A}^* k_0 = V_1^* P^{\mathcal{K}_1}_{\mathcal{K}_1 \ominus \mathcal{F}_1} \tilde{A}^* k_0 = V_1^* \sigma_1^* D^\gamma_{\tilde{E}^*_0} k_0,$$

for all $k_0 \in \mathcal{K}_0$. Set $\sigma_0 : \mathcal{K}_1 \ominus \mathcal{F}_1 \to \mathcal{D}^\gamma_{\tilde{E}^*_0}$ by means of

$$\sigma_0^* x := \sigma_1^* P^{\mathcal{D}^\gamma_{\tilde{E}^*_0}}_{\overline{D^\gamma_{\tilde{E}^*_0} \mathcal{K}_0}} x, \qquad x \in \mathcal{D}^\gamma_{\tilde{E}^*_0}.$$

Then $Y_0 = \tilde{E}_0 P^{\mathcal{K}_1}_{\mathcal{F}_1} + D^\gamma_{\tilde{E}^*_0} \sigma_0 P^{\mathcal{K}_1}_{\mathcal{K}_1 \ominus \mathcal{F}_1}$ is the solution of Problem 3 we were looking for.

From the above discussion we conclude that Problem 3 is solvable if, and only if, there exists $\gamma > 0$ such that (5.1) and (5.4) hold. We can also say that $\beta := \min\{\|Y\| \mid Y \text{ is a solution of Problem 3}\}$ equals $\min \gamma$, where γ runs over all positive numbers satisfying

$$\|A\| \leq \gamma,$$

and

$$\|D_{T_1^n} A^* e_0\| \leq \gamma \|D_{T_2^n} e_0\|, \qquad n \in \mathbb{N}, \ e_0 \in \mathcal{E}_0$$

(which are conditions (5.1) and (5.4), respectively.)

Now, if Y is a solution of Problem 3 with $\|Y\| = \beta$, $E := Y|_{\mathcal{K}_1 \ominus \mathcal{E}_1}$, and $\tilde{E} \in \mathcal{L}(\mathcal{F}_1, \mathcal{K}_2)$ is defined by

$$\tilde{E} V_1^{*n} Q_1 k_1 := V_2^{*n} E Q_1 k_1, \qquad n \geq 0, k_1 \in \mathcal{K}_1,$$

then there exists a uniquely determined contraction $\Gamma : \mathcal{K}_1 \ominus \mathcal{F}_1 \to \mathcal{D}_{\tilde{E}}^\beta$ (given by Douglas Lemma) such that

$$Y = \tilde{E} P_{\mathcal{F}_1}^{\mathcal{K}_1} + D_{\tilde{E}^*}^\beta \Gamma P_{\mathcal{K}_1 \ominus \mathcal{F}_1}^{\mathcal{K}_1},$$

where

$$P_{\mathcal{K}_0}^{\mathcal{K}_2} D_{\tilde{E}^*}^\beta \Gamma = \tilde{A}|_{\mathcal{K}_1 \ominus \mathcal{F}_1},$$

and

$$\Gamma V_1|_{\mathcal{K}_1 \ominus \mathcal{F}_1} = V_2 \Gamma,$$

(respectively, conditions (5.8) and (5.9) as before.) In a similar way as in the construction of the particular solution Y_0, the first relation yields two contractions $\Gamma_1 : \mathcal{K}_1 \ominus \mathcal{F}_1 \to \overline{D_{\tilde{E}^*}^\beta \mathcal{K}_0}$ and $\Gamma_0 : \mathcal{K}_1 \ominus \mathcal{F}_1 \to \mathcal{D}_{\tilde{E}^*}^\beta$ defined by their adjoints by means of $\Gamma_1^* D_{\tilde{E}^*}^\beta k_0 := P_{\mathcal{K}_1 \ominus \mathcal{F}_1}^{\mathcal{K}_1} \tilde{A}^* k_0$ ($k_0 \in \mathcal{K}_0$) and $\Gamma_0^* x := \Gamma_1^* P_{\overline{D_{\tilde{E}^*}^\beta \mathcal{K}_0}}^{\mathcal{D}_{\tilde{E}^*}^\beta} x$

($x \in \mathcal{D}_{\tilde{E}^*}^\beta$), respectively. In view of Douglas Lemma and the second relation, we have that Γ can be represented as $\Gamma = \Gamma_0 + \delta D_{\Gamma_0}$, where $\delta : \mathcal{D}_{\Gamma_0} \to \overline{\mathcal{D}_{\tilde{E}^*}^\beta \ominus D_{\tilde{E}^*}^\beta \mathcal{K}_0}$ is a contraction such that $\delta D_{\Gamma_0} V_1 = V_2 \delta D_{\Gamma_0}$.

Furthermore, $E \in \mathcal{L}(\mathcal{K}_1 \ominus \mathcal{E}_1, \mathcal{K}_2 \ominus \mathcal{E}_2)$, $\|E\| \leq \beta$, $E V_1|_{\mathcal{K}_1 \ominus \mathcal{E}_1} = V_2 E$, and $P_{Q_2 \mathcal{K}_0}^{\mathcal{K}_2 \ominus \mathcal{E}_2} E = E_0$ (cf. (5.7).) Once again, a direct application of Douglas Lemma yields a uniquely determined contraction $\rho : \mathcal{D}_{E_0}^\beta \to (\mathcal{K}_2 \ominus \mathcal{E}_2) \ominus Q_2 \mathcal{K}_0$ such that

$$E = E_0 + \rho D_{E_0}^\beta,$$

and

$$\rho D_{E_0}^\beta V_1 = V_2 \rho D_{E_0}^\beta.$$

Proposition 5.1 *Let $(A; T_1, T_2; \mathcal{E}_0)$ be the data set of Foiaş Lifting Problem, with T_1 and T_2 coisometries, and let $\beta = \min \gamma$, where γ runs over all positive numbers satisfying*

$$\|A\| \leq \gamma,$$

and

$$\|D_{T_1^n} A^* e_0\| \leq \gamma \|D_{T_2^n} e_0\|, \qquad n \in \mathbb{N}, \ e_0 \in \mathcal{E}_0.$$

Let $V_1 \in \mathcal{L}(\mathcal{K}_1)$ and $V_2 \in \mathcal{L}(\mathcal{K}_2)$ denote the minimal isometric dilations of T_1 and T_2, respectively. Put $\mathcal{K}_0 = \vee_{n \geq 0} V_2^n \mathcal{E}_0$. Let $\tilde{A}$ be the unique β-intertwining lifting in $LIF_\beta(A)$. Set

$$\mathcal{F}_1 := \vee_{n \geq 0} V_1^{*n}(\mathcal{K}_1 \ominus \mathcal{E}_1),$$

and

$$E_0 : \mathcal{K}_1 \ominus \mathcal{E}_1 \to \overline{Q_2 \mathcal{K}_0}, \qquad E_0^* Q_2 k_0 := Q_1 \tilde{A}^* k_0, \qquad k_0 \in \mathcal{K}_0,$$

where $Q_i = 1 - P_{\mathcal{E}_i}^{\mathcal{K}_i}$ ($i = 1, 2$). Define

$$\mathcal{R} := \{ \rho \mid \rho : \mathcal{D}_{E_0}^\beta \to (\mathcal{K}_2 \ominus \mathcal{E}_2) \ominus Q_2 \mathcal{K}_0,$$
$$\|\rho\| \leq 1, \rho D_{E_0}^\beta V_1 = V_2 \rho D_{E_0}^\beta \}.$$

For $\rho \in \mathcal{R}$, put $\tilde{E}(\rho) = \tilde{E}$, $\Gamma_1(\rho) = \Gamma_1$, and $\Gamma_0(\rho) = \Gamma_0$, where

$$\tilde{E} : \mathcal{F}_1 \to \mathcal{K}_2, \qquad \tilde{E} V_1^{*n} Q_1 k_1 := V_2^{*n}(E_0 + \rho D_{E_0}^\beta) Q_1 k_1,$$
$$n \in \mathbb{N}, k_1 \in \mathcal{K}_1,$$
$$\Gamma_1 : \mathcal{K}_1 \ominus \mathcal{F}_1 \to \overline{D_{\tilde{E}*}^\beta \mathcal{K}_0}, \qquad \Gamma_1^* D_{\tilde{E}*}^\beta k_0 := P_{\mathcal{K}_1 \ominus \mathcal{F}_1}^{\mathcal{K}_1} \tilde{A}^* k_0,$$
$$k_0 \in \mathcal{K}_0,$$

and

$$\Gamma_0 : \mathcal{K}_1 \ominus \mathcal{F}_1 \to \mathcal{D}_{\tilde{E}*}^\beta, \qquad \Gamma_0^* x := \Gamma_1^* P_{\overline{D_{\tilde{E}*}^\beta \mathcal{K}_0}}^{\mathcal{D}_{\tilde{E}*}^\beta} x, \qquad x \in \mathcal{D}_{\tilde{E}*}^\beta.$$

Then the map ϕ from

$$\{(\rho, \delta) \mid \rho \in \mathcal{R}, \delta : \mathcal{D}_{\Gamma_0} \to \mathcal{D}_{\tilde{E}*}^\beta \ominus \overline{D_{\tilde{E}*}^\beta \mathcal{K}_0},$$
$$\|\delta\| \leq 1, \delta D_{\Gamma_0} V_1 = V_2 \delta D_{\Gamma_0} \}$$

into

$$\{Y \text{ solution of Problem 3} \mid \|Y\| = \beta\},$$

given by

$$\phi(\rho, \delta) := \tilde{E} P_{\mathcal{F}_1}^{\mathcal{K}_1} + D_{\tilde{E}*}^\beta (\Gamma_0 + \delta D_{\Gamma_0}) P_{\mathcal{K}_1 \ominus \mathcal{F}_1}^{\mathcal{K}_1},$$

is a bijection.

We point out that the existence of a number $\gamma > 0$ such that

$$\|A\| \leq \gamma,$$

and

$$\|D_{T_1^n} A^* e_0\| \leq \gamma \|D_{T_2^n} e_0\|, \qquad n \in \mathbb{N}, \quad e_0 \in \mathcal{E}_0,$$

is a necessary and sufficient condition for the solvability of Problem 3, and, hence, according with Proposition 2.1, for the solvability of Problem 1.

Proof: We already proved that ϕ is surjective. In order to see that ϕ is injective, let us consider two pairs (ρ, δ), (ρ', δ'), with the required properties, such that

$$\phi(\rho, \delta) = \tilde{E} P_{\mathcal{F}_1}^{\mathcal{K}_1} + (\Gamma_0 + \delta D_{\Gamma_0}) D_{\tilde{E}_*}^{\beta} P_{\mathcal{K}_1 \ominus \mathcal{F}_1}^{\mathcal{K}_1}$$

$$= \tilde{E}' P_{\mathcal{F}_1}^{\mathcal{K}_1} + (\Gamma_0' + \delta' D_{\Gamma_0'}) D_{\tilde{E}'_*}^{\beta} P_{\mathcal{K}_1 \ominus \mathcal{F}_1}^{\mathcal{K}_1} = \phi(\rho', \delta').$$

For all $k_1 \in \mathcal{K}_1$,

$$\phi(\rho, \delta) Q_1 k_1 = (E_0 + \rho D_{E_0}^{\beta}) Q_1 k_1$$

$$= (E_0 + \rho' D_{E_0}^{\beta}) Q_1 k_1 = \phi(\rho', \delta') Q_1 k_1.$$

From this we can conclude that $\rho = \rho'$. Consequently, $\tilde{E} = \tilde{E}'$, $D_{\tilde{E}_*}^{\beta} = D_{\tilde{E}'_*}^{\beta}$, $\Gamma_1 = \Gamma_1'$, and $\Gamma_0 = \Gamma_0'$.
If $k_1 \in \mathcal{K}_1 \ominus \mathcal{F}_1$ then

$$\phi(\rho, \delta) k_1 = (\Gamma_0 + \delta D_{\Gamma_0}) D_{\tilde{E}_*}^{\beta} k_1 = (\Gamma_0 + \delta' D_{\Gamma_0}) D_{\tilde{E}_*}^{\beta} k_1 = \phi(\rho', \delta') k_1.$$

Thus, $\delta = \delta'$. The proof is complete. $\qquad\qquad\qquad\square$

We notice that the choice $\rho = 0$, $\delta = 0$ yields the particular solution Y_0 constructed above.

Now we consider Problems 1 and 2. Since T_1 and T_2 are coisometries, we have that the relation given in Proposition 2.1

$$Y \mapsto B := P_{\mathcal{E}_2}^{\mathcal{K}_2} Y |_{\mathcal{E}_1}$$

establishes a bijection between the set of solutions Y of Problem 3 with $\|Y\| \leq \beta$ and the set of solutions B of Problem 1 with $\|B\| \leq \beta$. Thus LIF$(A; T_1, T_2; \mathcal{E}_0) \neq \emptyset$. Furthermore:

Proposition 5.2 *Under the same hypotheses as in the above proposition and with the notation introduced therein. The map ϕ from*

$$\{(\rho, \delta) \mid \rho \in \mathcal{R}, \delta : \mathcal{D}_{\Gamma_0} \to \overline{D_{\tilde{E}_*}^{\beta} \ominus D_{\tilde{E}_*}^{\beta} \mathcal{K}_0}, \|\delta\| \leq 1, \delta D_{\Gamma_0} V_1 = V_2 \delta D_{\Gamma_0}\}$$

into

$$\{B \in \text{LIF}(A; T_1, T_2; \mathcal{E}_0) \mid \|B\| = \beta\},$$

given by

$$\phi(\rho, \delta) := P_{\mathcal{E}_2}^{\mathcal{K}_2}[\tilde{E}P_{\mathcal{F}_1}^{\mathcal{K}_1} + D_{\tilde{E}^*}^{\beta}(\Gamma_0 + \delta D_{\Gamma_0})P_{\mathcal{K}_1 \ominus \mathcal{F}_1}^{\mathcal{K}_1}]|_{\mathcal{E}_1},$$

is a bijection.

6 The General Problem

We recall that the data set for the general problem is the collection $(A; T_1, T_2; \mathcal{E}_0)$, where:

- $T_1 \in \mathcal{L}(\mathcal{E}_1)$ and $T_2 \in \mathcal{L}(\mathcal{E}_2)$ are two Hilbert space contractions,

- $\mathcal{E}_0$ is a closed subspace of $\mathcal{E}_2$ such that $T_2^* \mathcal{E}_0 \subseteq \mathcal{E}_0$, and

- $A \in \mathcal{L}(\mathcal{E}_1, \mathcal{E}_0)$ intertwines T_1 and $T_0 := P_{\mathcal{E}_0}^{\mathcal{E}_2} T_2|_{\mathcal{E}_0}$, so that $AT_1 = T_0 A$.

Reduction

It is known that we may consider only the case when T_2 is a coisometry. For this, let T_2 be an arbitrary contraction and let $T_2' \in \mathcal{L}(\mathcal{F}_2)$ be the adjoint operator of the minimal isometric dilation of T_2^*. Define $\mathcal{E}_0' := \mathcal{E}_0 \oplus (\mathcal{F}_2 \ominus \mathcal{E}_2)$ and $A' := A$ (viewed as a continuous linear operator from $\mathcal{E}_1$ into $\mathcal{E}_0'$.)

It can be seen that $T_2'^* \mathcal{E}_0' \subseteq \mathcal{E}_0'$ and $A'T_1 = T_0'A'$, where $T_0' := P_{\mathcal{E}_0'}^{\mathcal{F}_2} T_2'$.

If $B' \in \text{LIF}(A'; T_1, T_2'; \mathcal{E}_0')$ then $P_{\mathcal{F}_2 \ominus \mathcal{E}_2}^{\mathcal{F}_2} B' = 0$, so that $B' \in \mathcal{L}(\mathcal{E}_1, \mathcal{E}_2)$; also, $B'T_1 = T_2 B'$ and $P_{\mathcal{E}_0}^{\mathcal{E}_2} B' = A$. Therefore, $B' \in \text{LIF}(A; T_1, T_2; \mathcal{E}_0)$.

Conversely, if $B \in \text{LIF}(A; T_1, T_2; \mathcal{E}_0)$ then, in a similar way, it can be proved that $B \in \text{LIF}(A'; T_1, T_2'; \mathcal{E}_0')$.

This shows that $\text{LIF}(A; T_1, T_2; \mathcal{E}_0) = \text{LIF}(A'; T_1, T_2'; \mathcal{E}_0')$.

In what follows, unless otherwise stated, T_2 is a coisometry.

Example

A useful example is the following:

Assume that $\ker(T_1^*) = \{0\}$, $\mathcal{E}_2 = \vee_{n=0}^{\infty} \mathcal{E}_{-n}$, where

$$\mathcal{E}_{-n} := \{e_2 \in \mathcal{E}_2 \mid T_2^{*n} e_2 \in \mathcal{E}_0\}, \qquad n \geq 0,$$

and there exists $\alpha > 0$ such that

(6.1) $$\|P_{\mathcal{E}_{-n}}^{\mathcal{E}_2} T_2^n A e_1\| \leq \alpha \|T_1^n e_1\|, \qquad n \geq 0, e_1 \in \mathcal{E}_1.$$

Notice that $\mathcal{E}_0 \subseteq \mathcal{E}_{-1} \subseteq \cdots \subseteq \mathcal{E}_{-n} \subseteq \mathcal{E}_{-(n+1)} \subseteq \cdots \subseteq \mathcal{E}_2$.

The above conditions imply that, for each $n \in \mathbb{N}$, the operator $A_{-n} : \mathcal{E}_1 \to \mathcal{E}_{-n}$, given as

$$A_{-n} T_1^n e_1 = P_{\mathcal{E}_{-n}}^{\mathcal{E}_2} T_2^n A e_1, \quad e_1 \in \mathcal{E}_1,$$

is a well defined operator with norm at most α.

Moreover, for $n \in \mathbb{N}$,

$$A_{-n} T_1 = P_{\mathcal{E}_{-n}}^{\mathcal{E}_2} T_2 A_{-n},$$

$$P_{\mathcal{E}_0}^{\mathcal{E}_{-n}} A_{-n} = A,$$

and, for every $k \leq n$,

$$P_{\mathcal{E}_{-k}}^{\mathcal{E}_{-n}} A_{-n} = A_{-k}.$$

From these relations it can be shown that there exists an operator $B \in \mathrm{LIF}(A; T_1, T_2; \mathcal{E}_0)$ such that

$$B = \lim_{n \to \infty} A_{-n},$$

where the limit is taken in the norm operator topology.

Since $P_{\mathcal{E}_{-n}}^{\mathcal{E}_2} T_2^n = P_{\mathcal{E}_{-n}}^{\mathcal{E}_2} T_2^n P_{\mathcal{E}_0}^{\mathcal{E}_2}$, it also can be seen that B is the unique element in $\mathrm{LIF}(A; T_1, T_2; \mathcal{E}_0)$.

Necessary and Sufficient Condition

Let $R \in \mathcal{L}(\mathcal{E})$ and $R' \in \mathcal{L}(\mathcal{E}')$ be two contractions with minimal isometric dilations $V \in \mathcal{L}(\mathcal{K})$ and $V' \in \mathcal{L}(\mathcal{K}')$, respectively. If $F \in \mathcal{L}(\mathcal{E}, \mathcal{E}')$ intertwines R and R', i.e., $FR = R'F$, and α is any positive number, the symbol $\mathrm{LIF}_\alpha(F)$ is used to denote the set of all operators $\tilde{F} \in \mathcal{L}(\mathcal{K}, \mathcal{K}')$ such that $\tilde{F} V = V' \tilde{F}$, $P_{\mathcal{E}'}^{\mathcal{K}'} \tilde{F} = F P_{\mathcal{E}}^{\mathcal{K}}$, and $\|\tilde{F}\| \leq \alpha$. As in Proposition 2.3, $\mathrm{LIF}_\alpha(F)$ can be labeled by Schur functions. For each $\alpha \geq \|F\|$, the α-intertwining lifting corresponding with the Schur function $f \equiv 0$ is called the central lifting in $\mathrm{LIF}_\alpha(F)$.

Proposition 6.1 *Let $(A; T_1, T_2; \mathcal{E}_0)$ be the data set of Foiaş Lifting Problem, with T_2 coisometry. Let $W_1 \in \mathcal{L}(\mathcal{F}_1)$ be the minimal isometric dilation of T_1^*. Then $\mathrm{LIF}(A; T_1, T_2; \mathcal{E}_0) \neq \emptyset$ if, and only if, there exist $\alpha \geq \|A\|$ and $C \in \mathrm{LIF}_\alpha(A^*)$ such that*

$$(6.2) \qquad \|D_{W_1^{*n}} C e_0\| \leq \alpha \|D_{T_2^n} e_0\|, \quad n \in \mathbb{N}, e_0 \in \mathcal{E}_0.$$

Proof: Assume that LIF$(A; T_1, T_2; \mathcal{E}_0) \neq \emptyset$ and let B belong to LIF$(A; T_1, T_2; \mathcal{E}_0)$. Then $B \in \mathcal{L}(\mathcal{E}_1, \mathcal{E}_2)$ satisfies: (P1.1) $BT_1 = T_2B$, and (P1.2) $P_{\mathcal{E}_0}^{\mathcal{E}_2}B = A$. A direct application of the Commutant Lifting Theorem yields an operator $Z \in \mathcal{L}(\mathcal{E}_2, \mathcal{F}_1)$ such that: (i) $ZT_2^* = W_1Z$, (ii) $P_{\mathcal{E}_1}^{\mathcal{F}_1}Z = B^*$, and (iii) $\|Z\| = \|B^*\|(= \|B\|)$. If $\alpha := \|B\|$, then $\alpha \geq \|A\|$. If $C := Z|_{\mathcal{E}_0}$ then, clearly, $C \in \mathrm{LIF}_\alpha(A^*)$. Since, for $n \in \mathbb{N}$, $D_{W_1^{*n}}ZT_2^{*n}T_2^n = D_{W_1^{*n}}W_1^nZT_2^n = 0$, it also holds true that

$$\|D_{W_1^{*n}}Ce_0\| = \|D_{W_1^{*n}}ZD_{T_2^n}e_0\| \leq \alpha\|D_{T_2^n}e_0\|, \quad n \in \mathbb{N}, e_0 \in \mathcal{E}_0.$$

Conversely, assume there exist $\alpha \geq \|A\|$ and $C \in \mathrm{LIF}_\alpha(A^*)$. Then $C^*W_1^* = T_0C^*$ and $C^*|_{\mathcal{E}_1} = A$. If also (6.2) is satisfied, then, by Proposition 5.1, we obtain an operator $D \in \mathrm{LIF}(C^*; W_1^*, T_2; \mathcal{E}_0)$. Clearly, the operator $B := D^*|_{\mathcal{E}_1}$ belongs to LIF$(A; T_1, T_2; \mathcal{E}_0)$ and, hence, LIF$(A; T_1, T_2; \mathcal{E}_0)$ comes out to be nonempty. $\qquad\square$

We remark that if Y is a solution of Problem 3 (if there is any), $\alpha = \|Y\|$, $U_1 \in \mathcal{L}(\mathcal{G}_1)$ is the minimal unitary dilation of T_1, and $\mathcal{L} := \overline{(1 - U_1T_1^*)(\mathcal{E}_1)}$, then

$$C := \left(A^* + \sum_{s \geq 0} U_1^{-(s+1)}P_{\mathcal{L}}^{\mathcal{G}_1}Y^*V_2^{s+1}\right)\Bigg|_{\mathcal{E}_0}$$

belongs to $\mathrm{LIF}_\alpha(A^*)$ and verifies (6.2).

From Proposition 6.1 and the reduction discussed in the beginning of this section, we readily obtain the following:

Proposition 6.2 *Let $(A; T_1, T_2; \mathcal{E}_0)$ be the data set of Foiaş Lifting Problem. Let $W_i \in \mathcal{L}(\mathcal{F}_i)$ be the minimal isometric dilation of T_i^* ($i = 1, 2$). Then* LIF$(A; T_1, T_2; \mathcal{E}_0) \neq \emptyset$ *if, and only if, there exist $\alpha \geq \|A\|$ and $C \in$* $\mathrm{LIF}_\alpha(A^*P_{\mathcal{E}_0}^{\mathcal{E}_0\oplus(\mathcal{F}_2\ominus\mathcal{E}_2)})$ *such that*

(6.3) $\|D_{W_1^{*n}}Ce_0'\| \leq \alpha\|D_{W_2^{*n}}e_0'\|, \quad n \in \mathbb{N}, e_0' \in \mathcal{E}_0 \oplus (\mathcal{F}_2 \ominus \mathcal{E}_2).$

In Cases 1 and 2, T_1 is assumed to be an isometry, so that W_1 is unitary. Hence, $D_{W_1^{*n}} = 0$, for all $n \in \mathbb{N}$, and (6.3) is satisfied. In Case 3, T_2 is a coisometry, hence (6.3) reduces to (6.2); and (6.2) is just the necessary and sufficient condition given in Proposition 5.1 for the existence of solutions of Problem 1.

In the example, (6.3) reduces again to (6.2), since T_2 is a coisometry. There is a certain resemblance between conditions (6.2) and (6.1). C. Foiaş, in [F], p. 232, asks if there is any deep reason for that resemblance. As we will see next, the answer is that (6.2) and (6.1) are in fact equivalent.

On one hand, assume that there exist $\alpha \geq \|A\|$ and $C \in \text{LIF}_\alpha(A^*)$ such that (6.2) holds. Then, for all $e_1 \in \mathcal{E}_1$, $e_2 \in \mathcal{E}_2$ and $n \in \mathbb{N}$,

$$\langle P_{\mathcal{E}_{-n}}^{\mathcal{E}_2} T_2^n A e_1, e_2 \rangle_{\mathcal{E}_2} = \langle P_{\mathcal{E}_{-n}}^{\mathcal{E}_2} T_2^n C^* e_1, e_2 \rangle_{\mathcal{E}_2} = \langle e_1, C T_2^{*n} P_{\mathcal{E}_{-n}}^{\mathcal{E}_2} e_2 \rangle_{\mathcal{E}_1}$$

$$= \langle e_1, D_{W_1^{*n}} C T_2^{*n} P_{\mathcal{E}_{-n}}^{\mathcal{E}_2} e_2 \rangle_{\mathcal{E}_1} + \langle e_1, W_1^n W_1^{*n} C T_2^{*n} P_{\mathcal{E}_{-n}}^{\mathcal{E}_2} e_2 \rangle_{\mathcal{E}_1},$$

hence

$$|\langle P_{\mathcal{E}_{-n}}^{\mathcal{E}_2} T_2^n A e_1, e_2 \rangle_{\mathcal{E}_2}| \leq \alpha \|e_1\| \|D_{T_2^n} T_2^{*n} P_{\mathcal{E}_{-n}}^{\mathcal{E}_2} e_2\|$$

$$+ |\langle e_1, W_1^n W_1^{*n} C T_2^{*n} P_{\mathcal{E}_{-n}}^{\mathcal{E}_2} e_2 \rangle_{\mathcal{E}_1}|$$

$$= |\langle T_1^n e_1, W_1^{*n} C T_2^{*n} P_{\mathcal{E}_{-n}}^{\mathcal{E}_2} e_2 \rangle_{\mathcal{E}_1}| \leq \alpha \|T_1^n e_1\| \|e_2\|.$$

Therefore, for all $e_1 \in \mathcal{E}_1$ and $n \in \mathbb{N}$,

$$\|P_{\mathcal{E}_{-n}}^{\mathcal{E}_2} T_2^n A e_1\| \leq \alpha \|T_1^n e_1\|.$$

That is, (6.1) holds.

Conversely, assume that (6.1) is satisfied. As we showed above, from (6.1) we get for each $n \in \mathbb{N}$ a continuous linear operator $A_{-n} : \mathcal{E}_1 \to \mathcal{E}_{-n}$ such that: $\|A_{-n}\| \leq \alpha$, $A_{-n} T_1 = P_{\mathcal{E}_{-n}}^{\mathcal{E}_2} T_2 A_{-n}$, $P_{\mathcal{E}_0}^{\mathcal{E}_{-n}} A_{-n} = A$, and, for every $k \leq n$, $P_{\mathcal{E}_{-k}}^{\mathcal{E}_{-n}} A_{-n} = A_{-k}$. So, $A_{-n}^* T_2^* |_{\mathcal{E}_{-n}} = T_1^* A_{-n}^*$. Let $B_{-n} : \mathcal{F}_1 \to \mathcal{E}_{-n}$ be the central lifting in $\text{LIF}_\alpha(A_{-n}^*)$ ($n \in \mathbb{N}$). The limit

$$C^* := \lim_{n \to \infty} P_{\mathcal{E}_0}^{\mathcal{E}_2} B_{-n}^*,$$

taken in the norm operator topology, can be shown to exist. Moreover, it can be easily proved that C belongs to $\text{LIF}_\alpha(A^*)$ and satisfies (6.2).

To conclude we verify (6.3), in fact (6.2), in other two situations.

Case 4

Assume that T_2 is a coisometry and T_1 is similar to an isometry, i.e., there exists an invertible operator $X \in \mathcal{L}(\mathcal{E}_1, \mathcal{E})$, with $\mathcal{E}$ a Hilbert space, such that $T := X T_1 X^{-1}$ is an isometry.

Then $(A X^{-1}) T = T_0 (A X^{-1})$. Let $\alpha := \|A X^{-1}\|$. As we saw in Case 2, there exists $B_1 \in \text{LIF}(A X^{-1}; T, T_2; \mathcal{E}_0)$ with $\|B_1\| = \alpha$. Therefore, $B := B_1 X \in \text{LIF}(A; T_1, T_2; \mathcal{E}_0)$ and $\|B\| \leq \alpha \|X\|$. Let $W \in \mathcal{L}(\mathcal{F})$ be the minimal isometric dilation of T^*. Then W is a unitary operator on $\mathcal{F}$. If $Z \in \text{LIF}_{\|X\|}(X^*)$ and $D \in \text{LIF}_\alpha(X^{*-1} A^*)$, set $C := Z D$. Then C belongs to $\text{LIF}_{\alpha \|X\|}(A^*)$ and, for all $e_0 \in \mathcal{E}_0$ and $n \in \mathbb{N}$, $\|D_{W_1^{*n}} C e_0\| = \|D_{W_1^{*n}} Z D e_0\| = \|D_{W_1^{*n}} Z W^n W^{*n} D e_0\| = \|D_{W_1^{*n}} W_1^n Z W^{*n} D e_0\| = 0$.

Case 5

Assume that T_2 is a coisometry and T_1 is similar to a coisometry, i.e., there exists an invertible operator $X \in \mathcal{L}(\mathcal{E}_1, \mathcal{E})$, with $\mathcal{E}$ a Hilbert space, such that $T := XT_1X^{-1}$ is a coisometry.

Then $(AX^{-1})T = T_0(AX^{-1})$. As we showed in Case 3, there exists $B_1 \in$ LIF$(AX^{-1}; T, T_2; \mathcal{E}_0)$ if, and only if, there exists $\alpha > 0$ such that $\|AX^{-1}\| \le \alpha$ and $\|D_{T^n}X^{*-1}A^*e_0\| \le \alpha\|D_{T_2^n}e_0\|$, for all $e_0 \in \mathcal{E}_0$ and $n \in \mathbb{N}$. Under these conditions, let $B := B_1X$. Then $B \in$ LIF$(A; T_1, T_2; \mathcal{E}_0)$ and $\|B\| \le \alpha\|X\|$. If $Z \in$ LIF$_{\|X\|}(X^*)$, then $C := ZX^{*-1}A^*$ belongs to LIF$_{\alpha\|X\|}(A^*)$ and, for all $e_0 \in \mathcal{E}_0$ and $n \in \mathbb{N}$,

$$
\begin{aligned}
\|D_{W_1^{*n}}Ce_0\| &= \|D_{W_1^{*n}}ZX^{*-1}A^*e_0\| \\[2mm]
&= \|D_{W_1^{*n}}ZD_{T^n}X^{*-1}A^*e_0 + D_{W_1^{*n}}ZT^{*n}T^nX^{*-1}A^*e_0\| \\[2mm]
&= \|D_{W_1^{*n}}ZD_{T^n}X^{*-1}A^*e_0 + D_{W_1^{*n}}W_1^nZT^nX^{*-1}A^*e_0\| \\[2mm]
&= \|D_{W_1^{*n}}ZD_{T^n}X^{*-1}A^*e_0\| \le \|Z\|\|D_{T^n}X^{*-1}A^*e_0\| \\[2mm]
&\le \alpha\|X\|\|D_{T_2^n}e_0\|.
\end{aligned}
$$

Due to the reduction and somehow inspired by the example discussed in this section, we conjecture that LIF$(A; T_1, T_2; \mathcal{E}_0) \ne \emptyset$ if, and only if, for some $\beta \ge \|A\|$, the central lifting in LIF$_\beta(A^*)$, say C_0^β, satisfies the relation $\|D_{W_1^{*n}}C_0^\beta e_0\| \le \beta\|D_{T_2^n}e_0\|$, for all $e_0 \in \mathcal{E}_0$ and $n \in \mathbb{N}$.

Acknowledgements

We thank Professor Ciprian Foiaş for his encouragement.

References

[BFF]   A. Biswas, C. Foiaş and A.E. Frazho, *Weighted Commutant Lifting*, preprint.

[CS]   J.G.W. Carswell and C.F. Schubert, Lifting of operators that commute with shifts, *Mich. Math. J.* **22** (1975), 65–69.

[C]   Z. Ceausescu, Lifting of a contraction intertwining two isometries, *Mich. Math. J.* **26** (1979), 231–241.

[D]   R.G. Douglas, On majorization, factorization, and range inclusion of operators on Hilbert space, *Proc. Amer. Math. Soc.* **17** (1966), 413–415.

[F]   C. Foiaş, On the extension of Intertwining Operators, in: *Harmonic Analysis and Operator Theory* (Marcantognini, S.A.M. et al, eds.),Contemporary Mathematics, vol. 189, American Mathematical Society, Providence, Rhode Island, 1995, 227–234.

[GGK]   I. Gohberg, S. Goldberg and M.A. Kaashoek, *Classes of Linear Operators vol. II*, Operator Theory: Advances and Applications, vol. 63, Birkhäuser Verlag, Basel, 1993.

[G]   C. Gu, A generalization of Cowen's characterization of hyponormal Toeplitz operators, *J. Funct. Anal.* **124** (1994), no. 1, 135–148.

[M]   M.D. Morán, On the Commutant Lifting Theorem *Series de Investigación y Desarrollo, Postgrado en Matemáticas, Facultad de Ciencias, Universidad Central de Venezuela*, preprint.

[NF]   B. Sz.-Nagy and C. Foiaş, Dilation des commutants d'opérateurs, *C.R. Acad. Sci. Paris, série A* **266** (1968), 493–495.

[TV]   S. Treil and A. Volberg, A fixed point approach to Nehari's problem and its applications, *Operator Theory: Advances and Applications*, vol. 71, Birkhäuser Verlag, Basel, 1994, 165–186.

S.A.M. Marcantognini
Instituto Venezolano de
    Investigaciones Científicas
Dept. of Mathematics
Apartado 21827
Caracas 1020A
Venezuela &
    Universidad Simón Bolívar
Dept. of Mathematics
Apartado 89000
Caracas 1086A
Venezuela

M.D. Morán
Universidad Central de Venezuela
Facultad de Ciencias
Escuela de Matemáticas
Caracas
Venezuela

1991 Mathematics Subject Classification. Primary: 47A20; Secondary: 47A99.

Operator Theory:
Advances and Applications, Vol. 124
© 2001 Birkhäuser Verlag Basel/Switzerland

Some Estimates for the Resolvent and for the Lengths of Jordan Chains of an Analytic Operator Function

Alexander Markus and Vladimir Matsaev

To Israel Gohberg with gratitude and admiration.

We consider a bounded component of the numerical range of an analytic operator function $A(\lambda)$ and prove an estimate for the inverse operator function in a neighborhood of the component. This gives us a possibility to estimate the lengths of Jordan chains of $A(\lambda)$ corresponding to a boundary point of the component.

1 Introduction

Let H be a Hilbert space and $\mathcal{L}(H)$ be the set of all linear bounded operators in H. Denote by $A(\lambda)$ an operator function with values in $\mathcal{L}(H)$ which is defined and analytic in a domain $U(\subset \mathbf{C})$. The *spectrum* of $A(\lambda)$ is defined to be the set $\sigma(A)$ of all numbers $\lambda \in U$ such that $A(\lambda)$ does not have an inverse in $\mathcal{L}(H)$. The set

$$W(A) = \{\lambda \in U : (A(\lambda)f, f) = 0 \text{ for some } f \neq 0\}$$

is known as the *numerical range* of $A(\lambda)$.

The following lemma explains the connection between the spectrum and the numerical range.

Lemma 1 *If there exists a number $\lambda \in U$ such that*

$$(1) \qquad \inf\{|(A(\lambda)f, f)| : \|f\| = 1\} > 0,$$

then

(a) $\sigma(A) \subset \overline{W(A)}$;

(b) *condition* (1) *holds for any number* $\lambda \in U \setminus \overline{W(A)}$.

The part (a) of this assertion is well-known (see, e.g., [5], Theorem 26.6). A proof of the part (b) in essence is contained in the proof of the mentioned theorem.

In this paper we are interested in *components* of $W(A)$, i.e., maximal connected subsets of $W(A)$, and especially in bounded components. They play an important role in factorization problems (see, e.g., [2]; [4]; [5], §§26, 27, 31; [7]; [8]), and some properties of these components were studied in [4], [6], [8].

We introduce some numerical characteristic of a bounded component of $W(A)$.

Lemma 2 *Suppose that condition* (1) *holds for some number* $\lambda \in U$, *and let* $F (\neq \emptyset)$ *be a bounded component of* $W(A)$ *such that*

(2) $\overline{F} \subset U, \ \overline{F} \cap \overline{W(A) \backslash F} = \emptyset.$

Then there exists a natural number $m(F)$ *such that the function* $(A(\lambda)f, f)$ *for arbitrary* $f \neq 0$ *has exactly* $m(F)$ *roots in the set* F *(counting multiplicities).*

Proof: It follows from condition (2) that there exists a bounded domain U_0 with boundary Γ consisting of a finite number of piecewise smooth closed non-intersecting Jordan curves such that

$$\overline{F} \subset U_0, \ \overline{U}_0 \subset U, \ \overline{W(A) \backslash F} \cap \overline{U}_0 = \emptyset$$

(see, e.g., [1], p. 6). Now the assertion of Lemma 2 follows from [5], Lemmas 26.8, 26.9.   □

The main result of this paper is contained in Theorem 1. We prove that in a neighborhood of a component F the inverse operator function $A^{-1}(\lambda)$ admits an estimate

$$\|A^{-1}(\lambda)\| \leq \frac{K}{[\text{dist}(\lambda, F)]^{m(F)}}$$

where the constant K depends on the chosen neighborhood, and $\text{dist}(\lambda, F)$ denotes the distance between the point λ and the set F. This result allows us to prove (Theorem 2) that for any eigenvalue of $A(\lambda)$ which is a boundary point of F the lengths of corresponding Jordan chains do not exceed $m(F)$ (under some additional assumption of geometrical nature). Analogous results for the quadratic numerical range of block operator matrix were proved in [3].

We complete this introduction with a remark that both mentioned results have the well-known predecessor for the simplest case $A(\lambda) = \lambda I - A \ (A \in \mathcal{L}(H))$. In this case $F = W(A)$ (a convex set), $m(F) = 1$,

$$\|(\lambda I - A)^{-1}\| \leq \frac{1}{\text{dist}(\lambda, W(A))},$$

and the lengths of Jordan chains for a boundary point of $W(A)$ which is an eigenvalue of A are all equal 1.

2 Estimate of the Resolvent

Theorem 1 *Let $A(\lambda)$ be an operator function analytic in a domain U, let condition (1) hold for some $\lambda \in U$, and let F be a bounded component of the numerical range $W(A)$ such that*

(3) $$\overline{F} \subset U, \ \overline{F} \cap \overline{W(A)\backslash F} = \emptyset.$$

If a bounded domain U_1 satisfies the conditions

(4) $$\overline{F} \subset U_1, \ \overline{U}_1 \subset U, \ \overline{U}_1 \cap \overline{W(A)\backslash F} = \emptyset,$$

then there exists a positive number K such that

(5) $$\|A^{-1}(\lambda)\| \leq \frac{K}{[\mathrm{dist}(\lambda, F)]^{m(F)}} \quad (\lambda \in U_1\backslash\overline{F}).$$

Proof: First we note that a domain U_1 with the properties (4) exists according to the conditions (3). Choose an additional bounded domain U_0 such that

$$\overline{U}_1 \subset U_0, \ \overline{U}_0 \subset U, \ \overline{U}_0 \cap \overline{W(A)\backslash F} = \emptyset,$$

and the boundary Γ of U_0 consists of a finite number of piecewise smooth closed non-intersecting Jordan curves. By Lemma 2, for any f, $\|f\| = 1$, the analytic function $(A(\lambda)f, f)$ has exactly $m = m(F)$ roots in the set $\overline{U}_0$, and all of them belong to F. Let us denote these roots by $\lambda_1(f), \ldots, \lambda_m(f)$. The order of the enumeration is not essential here, and the functionals $\lambda_j(f)$ can be discontinuous.

The function

(6) $$b_f(\lambda) = (A(\lambda)f, f)/\prod_{r=1}^{m}(\lambda - \lambda_r(f))$$

is analytic and non-vanishing on the set U_0. Denote

(7) $$M = \max\{\|A(\lambda)\| : \lambda \in \overline{U}_0\}$$

If $\lambda \in \Gamma$ and $\|f\| = 1$ then (6) and (7) imply

(8) $$|b_f(\lambda)| \leq \frac{M}{[\mathrm{dist}(\Gamma, F)]^m}.$$

By the Maximum Principle this inequality remains valid for all $\lambda \in U_0$. Let us prove now that there exists a number $\gamma > 0$ such that

(9) $$|b_f(\lambda)| \geq \gamma \quad (\lambda \in U_1, \ \|f\| = 1).$$

Assuming the opposite we obtain sequences $\lambda_n \in U_1$ and $f_n \in H$ ($\|f_n\| = 1$) such that

$$(10) \qquad b_{f_n}(\lambda_n) \to 0 \quad (n \to \infty).$$

We can suppose that the sequence $\{\lambda_n\}$ converges to some λ_0 ($\in \overline{U}_1$).

It follows from (8) that the sequence $b_{f_n}(\lambda)$ contains a subsequence which converges uniformly on compact subsets of U_0 to some analytic function $b(\lambda)$ (see, e.g., [9], p. 169). We can suppose that this subsequence coincides with the initial sequence $b_{f_n}(\lambda)$. It follows from (10) that $b(\lambda_0) = 0$. On the other hand, it follows from Lemma 1(b) and from equality (6) that for any $\lambda \in U_0 \backslash \overline{F}$ and for any f, $\|f\| = 1$,

$$|b_f(\lambda)| \geq \delta(\lambda)[\operatorname{dist}(\lambda, F)]^{-m} \quad (\delta(\lambda) > 0).$$

Hence, for any $\lambda \in U_0 \backslash \overline{F}$

$$|b(\lambda)| \geq \delta(\lambda)[\operatorname{dist}(\lambda, F)]^{-m}.$$

This implies that $b(\lambda) \not\equiv 0$. According to the Hurwitz theorem ([9], p. 119) it follows from the equality $b(\lambda_0) = 0$ that there exists a sequence $\mu_n \to \lambda_0$ such that $b_{f_n}(\mu_n) = 0$. This contradicts the condition $b_f(\lambda) \neq 0$ ($\lambda \in U_0$, $\|f\| = 1$), proving the inequality (9). Now (6) and (9) imply that for any f ($\|f\| = 1$) and $\lambda \in U_1 \backslash \overline{F}$

$$|(A(\lambda)f, f)| \geq \gamma[\operatorname{dist}(\lambda, F)]^m.$$

Hence

$$\|A(\lambda)f\| \geq \gamma[\operatorname{dist}(\lambda, F)]^m \quad (\|f\| = 1, \ \lambda \in U_1 \backslash \overline{F}),$$

or

$$(11) \qquad \|A(\lambda)g\| \geq \gamma[\operatorname{dist}(\lambda, F)]^m \|g\| \quad (g \in H, \ \lambda \in U_1 \backslash \overline{F}).$$

By Lemma 1, the inverse operator $A^{-1}(\lambda)$ exists for all $\lambda \in U_1 \backslash \overline{F}$. If we substitute $g = A^{-1}(\lambda)h$ in (11), we obtain inequality (5) (with the constant $K = \gamma^{-1}$). $\quad \square$

3 Estimate for the Lengths of Jordan Chains

A number $\lambda_0 \in U$ is called an *eigenvalue* of $A(\lambda)$ if the equation $A(\lambda_0)g_0 = 0$ has a nonzero solution g_0. This solution is called an *eigenvector* of $A(\lambda)$ corresponding to λ_0. The collection of vectors $g_0, g_1, \ldots, g_{k-1}$ is called a *Jordan chain of length* k for $A(\lambda)$ corresponding to the eigenvalue λ_0 if g_0 is a corresponding eigenvector and the equations

$$\sum_{j=0}^{m} \frac{1}{j!} A^{(j)}(\lambda_0) g_{m-j} = 0 \quad (m = 1, \ldots, k-1)$$

hold.

A vector polynomial $g(\lambda)$ such that $g(\lambda_0) \neq 0$ and $A(\lambda_0)g(\lambda_0) = 0$ is called a *root polynomial* of $A(\lambda)$ corresponding to λ_0. The order of λ_0 as a zero of $A(\lambda)g(\lambda)$ is called the *order* of the root polynomial $g(\lambda)$. It easy to check (see also [5], Lemma 11.3) that a collection of vectors $g_0, g_1, \ldots, g_{k-1}$ is a Jordan chain for $A(\lambda)$ corresponding to an eigenvalue λ_0 if and only if the vector polynomial of the form

$$g(\lambda) = \sum_{j=0}^{k-1} (\lambda - \lambda_0)^j g_j + (\lambda - \lambda_0)^k h(\lambda),$$

where $h(\lambda)$ is a vector polynomial, is a root polynomial of $A(\lambda)$ of order k corresponding to λ_0. Hence the lengths of Jordan chains corresponding to λ_0 coincide with the orders of root polynomials corresponding to λ_0.

Lemma 3 *Let $\lambda_0 \in U$ be an eigenvalue of $A(\lambda)$. If for a natural number m there exists a sequence $\lambda_n \to \lambda_0$ such that*

$$(12) \qquad \|A^{-1}(\lambda_n)\| = O(|\lambda_n - \lambda_0|^{-m}) \quad (n \to \infty)$$

then the lengths of all Jordan chains corresponding to λ_0 do not exceed m.

Proof: Let $g(\lambda)$ be a root polynomial of $A(\lambda)$ of the order k corresponding to λ_0. By definition,

$$(13) \qquad \|A(\lambda)g(\lambda)\| = O(|\lambda - \lambda_0|^k) \quad (\lambda \to \lambda_0).$$

Obviously,

$$\|g(\lambda)\| \leq \|A^{-1}(\lambda)\| \|A(\lambda)g(\lambda)\|,$$

and it follows from (12), (13) that

$$\|g(\lambda_n)\| = O(|\lambda_n - \lambda_0|^{k-m}) \quad (n \to \infty).$$

According to the definition of root polynomial, $g(\lambda_0) \neq 0$, and hence $k \leq m$. $\square$

A boundary point λ_0 of the component F is said to have the *exterior cone property* if there exist a closed cone (angle) with a vertex λ_0

$$\Lambda = \{\lambda \in \mathbf{C} : \alpha \leq \arg(\lambda - \lambda_0) \leq \beta\} \quad (\beta - \alpha > 0)$$

and a number $r > 0$ such that

$$\Lambda \cap \{\lambda \in \mathbf{C} : |\lambda - \lambda_0| \leq r\} \cap \overline{F} = \{\lambda_0\}.$$

Theorem 2 *Let the conditions of Theorem 1 be satisfied. If a boundary point λ_0 of F is an eigenvalue of $A(\lambda)$ and has the exterior cone property, then the lengths of all corresponding Jordan chains do not exceed $m(F)$.*

Proof: If λ lies on the axis of the cone Λ (i.e., $\arg(\lambda - \lambda_0) = (\alpha + \beta)/2$) and $|\lambda - \lambda_0| \leq r$, then

$$|\lambda - \lambda_0| \leq C \operatorname{dist}(\lambda, F)$$

with some constant $C > 0$. It follows from this inequality and Theorem 1 that

$$\|A^{-1}(\lambda)\| \leq K(C|\lambda - \lambda_0|^{-1})^{m(F)} \quad (\lambda \in U_1, \ \arg(\lambda - \lambda_0)$$
$$= (\alpha + \beta)/2, \ |\lambda - \lambda_0| \leq r).$$

Hence the statement of Theorem 2 follows from Lemma 3.   □

Corollary: *Let $A(\lambda)$ be an operator polynomial of degree m:*

$$A(\lambda) = \sum_{k=0}^{m} \lambda^k A_k,$$

and let

(14) $\inf\{|(A_m f, f)| : \|f\| = 1\} > 0.$

If a boundary point λ_0 of $W(A)$ is an eigenvalue and has the exterior cone property, then the lengths of all corresponding Jordan chains do not exceed m.

Proof: It is clear that under condition (14) the numerical range $W(A)$ of the polynomial $A(\lambda)$ is bounded. If we use Theorem 2 for the set $W(A)$ (instead of F), we obtain the statement of Corollary.   □

Remark: A trivial example $A(\lambda) = (\lambda - \lambda_0)^m I$ shows that the estimate of the corollary can be reached.

Acknowledgements

The second named author is supported by the Israel Science Foundation founded by the Israel Academy of Sciences and Humanities under Grant No 93/97-1.

References

[1]   I. Gohberg, S. Goldberg and M.A. Kaashoek,, *Classes of Linear Operators, vol. 1*, Birkhäuser Verlag, Basel 1990.

[2]   P. Lancaster and A. Markus, A note on factorization of analytic matrix functions, *Operator Theory: Adv. Appl.*, 1999, accepted.

[3]   H. Langer, A. Markus, V. Matsaev and C. Tretter, A new concept for block operator matrices: The quarqtic numerical range, *Math. Ann.*, 1999, submitted.

[4]   C.-K. Li and L. Rodman, Numerical range of matrix polynomials, *SIAM J. Matrix Anal. Appl.* **15** (1994), 1256–1265.

[5]   A. Markus, Introduction to the Spectral Theory of Polynomial Operator Pencils, *Amer. Math. Soc.*, Providence 1988.

[6]   A. Markus, J. Maroulas and P. Psarrakos, Spectral properties of a matrix polynomial connected with a component of its numerical range, *Operator Theory: Adv. Appl.* **106** (1998), 305–308.

[7]   A. Markus and V. Matsaev, On the spectral theory of holomorphic operator-valued functions in Hilbert space, *Funct. Anal. Appl.* **9** (1975), 76–77.

[8]   A. Markus and L. Rodman, Some results on numerical ranges and factorization of matrix polynomials, *Linear and Multilinear Algebra* **42** (1997), 169–185.

[9]   E.C. Titchmarsh, *The Theory of Functions*, Second Edition, Oxford University Press, London 1968.

Alexander Markus
Department of Math. and Comp. Sci
Ben-Gurion University of the Negev
Beer-Sheva 84 105
Israel

Vladimir Matsaev
School of Mathematical Sciences
Tel-Aviv University
Ramat-Aviv 69 978
Israel

1991 Mathematics Subject Classification. Primary: 47A56.

Operator Theory:
Advances and Applications, Vol. 124
© 2001 Birkhäuser Verlag Basel/Switzerland

The Parrott Problem for Singular Values

David Ogle and Nicholas Young

Dedicated to Israel Gohberg with affection and esteem

We discuss the problem of completing a partially specified matrix $\begin{bmatrix} A & B \\ C & ? \end{bmatrix}$ subject to a bound on the mth singular value. A complete solution was given by Arsene, Constantinescu, and Gheondea [ACG]. We present a more elementary approach appropriate to the matrix case based on the use of Möbius transformations.

1 Introduction

A well known theorem of Parrott [P] asserts that, for the partially specified block matrix,

$$\begin{bmatrix} A & B \\ C & ? \end{bmatrix},$$

if the norms of the first block row and the first block column are bounded by 1, then the (2, 2) block may be completed so that the whole matrix has norm at most 1. It is natural to ask whether an analogous assertion holds for other singular values, and indeed the answer to this question can be used to study interpolation problems for multipliers of Hilbert function spaces [AY]. The problem has been completely solved by Arsene, Constantinescu and Gheondea [AG, ACG, CG1, CG2, CG3, G], who not only give necessary and sufficient conditions for the existence of the desired completion, but also parametrize the set of completions. They do this in a general setting, allowing A, B etc. to be operators on infinite-dimensional Krein spaces. Their solution is complicated, and some effort is needed to interpret it in the matrix context. An earlier complete treatment, in an even more abstract setting, can be found in [Ph], although some effort is required to compare the results of that work with those presented here. A more elementary approach has been taken by Gohberg, Rodman, Shalom and Woerdeman [GRSW]. These authors' reasoning is easier to follow, but gives less: they show that the matrix can be completed with approximate satisfaction of the singular value constraint, but they do not consider necessary and sufficient conditions for exact satisfaction, nor do they obtain a formula for a solution.

In the present paper we study this matrix completion problem within the framework of finite matrices and without recourse to Krein space theory. It will become clear that the problems of the existence and determination of completions with the desired properties are very delicate. The abstract methods of Arsene, Constantinescu and Gheondea are ideally suited to a full resolution of these problems. We nevertheless wish to put forward an alternative approach, which (we believe) provides different insights into the subtleties of the problem. Although our results are less far reaching than those of [CG2], they are somewhat more concrete as far as they go.

Here is a formal statement of the questions we wish to answer.

Problem 1.1 Let m be a non-negative integer and let A, B, C be matrices of suitable dimensions such that

$$s_m[A \ B] \le 1, \qquad s_m\begin{bmatrix} A \\ C \end{bmatrix} \le 1.$$

Under what conditions does there exist a matrix D such that

$$s_m\begin{bmatrix} A & B \\ C & D \end{bmatrix} \le 1?$$

What is a formula for D when it exists?

We shall answer these questions in various special cases, in particular whenever $s_{m-1}(A) > 1$ (see Theorem 3.4) and when $m = 1$ (Sections 3.1 and 4). We do not, however, address the delicate issue of the parametrization of the set of solutions. Our results are incomplete, and for some cases the only successful treatment is that in [CG1, CG2, Ph]. For others, though, we believe we offer an accessible alternative, and we were encouraged by participants in the conference to provide this account even though it does not give a full solution.

We adopt the convention that the singular values of a matrix X are denoted by $s_0(X) \ge s_1(X) \ge \cdots$. Thus $s_j(X)$ is the distance (in the operator norm for matrices acting on Euclidean space) of X from the set of matrices of rank at most j. Parrott's theorem asserts that in the case $m = 0$ the desired D always exists. Furthermore there is a fairly straightforward parametrization (see, for example [AG, ACG]) of the set of possible solutions. The results of [CG2] show that the issue is much more subtle when $m \ge 1$. Let us begin with a simple example to show that the desired D need not exist. We shall denote the $r \times r$ identity matrix by I_r.

Example 1.2 Let $A = I_2$, $B = [1 \ 0]^T$, $C = [0 \ 1]$. For $d \in \mathbb{C}$ consider

$$T_d \stackrel{\text{def}}{=} \begin{bmatrix} A & B \\ C & d \end{bmatrix} = \begin{bmatrix} 1 & 0 & 1 \\ 0 & 1 & 0 \\ 0 & 1 & d \end{bmatrix}.$$

We have

$$s_1[A \ B] = s_1 \begin{bmatrix} 1 & 0 & 1 \\ 0 & 1 & 0 \end{bmatrix} = 1$$

and

$$s_1 \begin{bmatrix} A \\ C \end{bmatrix} = s_1 \begin{bmatrix} 1 & 0 \\ 0 & 1 \\ 0 & 1 \end{bmatrix} = 1.$$

It follows that $s_1(T_d) \geq 1$ for every choice of d in $\mathbb{C}$. However, there is no d for which $s_1(T_d) = 1$. This follows from the fact that

$$\det(T_d T_d^* - I) = \begin{vmatrix} 1 & 0 & \bar{d} \\ 0 & 0 & 1 \\ d & 1 & |d|^2 \end{vmatrix} = -1,$$

which shows that none of the singular values of T_d can equal 1 for any choice of d.

The result of [GRSW] mentioned above is that, in the notation of the Main Problem,

$$\inf_D s_m \begin{bmatrix} A & B \\ C & D \end{bmatrix} = 1.$$

The method of [GRSW] does not lend itself to the ascertaining of conditions for the attainment of the infimum, nor to a simple way of finding a suitable D. In this article we use a combination of unitary and Möbius transformations to reduce the general problem to special cases which are easier to analyse.

2 Möbius Transformations

The material contained in this section will be used to reduce the main problem to one of a few special cases. We wish to manipulate the partially specified block matrix until it is of a certain special form. These manipulations take two forms: permutations and Möbius transformations. Most of this section will address the latter.

From this point onwards we shall assume that the matrices A, B, C satisfy the conditions of the main problem. That is, for some non-negative integer m we have

(2.1) $$s_m[A \ B] \leq 1, \qquad s_m \begin{bmatrix} A \\ C \end{bmatrix} \leq 1.$$

We begin with a definition.

Definition 2.1 If T and X are matrices with $\|T\| < 1$ and $I + T^*X$ nonsingular then we define the Möbius transformation $\Phi_T(X)$ of X to be

$$\Phi_T(X) = (I - TT^*)^{-\frac{1}{2}}(T + X)(I + T^*X)^{-1}(I - T^*T)^{\frac{1}{2}}.$$

The following result shows how Möbius transformations play a key role in the study of singular values.

Lemma 2.2 *Let X and T be as in the definition above. Then*

$$s_m(X) \le 1$$

if and only if

$$s_m(\Phi_T(X)) \le 1.$$

Proof: Let $Y = \Phi_T(X)$. Then it is well known (e.g. [S]) that

$$I - Y^*Y = (I - TT^*)^{\frac{1}{2}}(I + X^*T)^{-1}$$
$$(I - X^*X)(I + T^*X)^{-1}(I - T^*T)^{\frac{1}{2}}.$$

Hence $I - Y^*Y$ and $I - X^*X$ are congruent, which implies that they have the same number of negative eigenvalues. Thus X and Y have the same number of singular values bounded by one. The result follows.   □

We can utilise the above result to simplify the form of the matrices in the Main Problem. Möbius transformations allow us to assume that the singular values of A take only three values; 0, 1, or α (where $\alpha > 1$). Assume that A is square, of type $n + 1 \times n + 1$. We begin by finding two unitary matrices U_1 and U_2 such that

$$U_1 A U_2^* = \text{diag}\{s_0(A), \ldots s_n(A)\} = \begin{bmatrix} A_1 & 0 & 0 \\ 0 & I_r & 0 \\ 0 & 0 & A_3 \end{bmatrix}.$$

Here A_1 is the diagonal matrix composed of those singular values of A which are greater than one and A_3 contains those which are less than one. The $r \times r$ identity matrix contains the singular values which are equal to one. If none of the singular values fall into one of these sets, then the corresponding block row and column are deleted. When A is not square, we may add the required number of rows or columns of zeros to form a square matrix. This operation will simply increase the number of zero singular values by the number of rows or columns added. Although it will not be explicitly stated, we shall employ this method whenever we require a singular value decomposition of a non-square matrix. Since multiplication by unitary matrices does not affect the singular values of a matrix, and since

$$(2.2) \quad \begin{bmatrix} U_1 & 0 \\ 0 & I \end{bmatrix} \begin{bmatrix} A & B \\ C & D \end{bmatrix} \begin{bmatrix} U_2^* & 0 \\ 0 & I \end{bmatrix} = \begin{bmatrix} A_1 & 0 & 0 & B_1 \\ 0 & I_r & 0 & B_2 \\ 0 & 0 & A_3 & B_3 \\ C_1 & C_2 & C_3 & D \end{bmatrix},$$

where

$$U_1 B = \begin{bmatrix} B_1 \\ B_2 \\ B_3 \end{bmatrix} \quad \text{and} \quad CU_2^* = [C_1 \quad C_2 \quad C_3],$$

it follows that the Main Problem is equivalent to finding a D such that $s_m(W) \le 1$ where

$$W = \begin{bmatrix} A_1 & 0 & 0 & B_1 \\ 0 & I_r & 0 & B_2 \\ 0 & 0 & A_3 & B_3 \\ C_1 & C_2 & C_3 & D \end{bmatrix}.$$

When A_1, A_3, B_i, C_i $(i = 1, 2, 3)$ are as above, and given $\alpha > 1$ we define the following matrices:

$$M = \text{diag}\{0, 0, A_3, 0\}, \quad T = \text{diag}\left\{ \frac{\alpha - s_j(A_1)}{1 - \alpha s_j(A_1)} \right\},$$

$$N = \text{diag}\{T, 0, 0, 0\},$$

$$\tilde{B}_1 = (I - T^2)^{-\frac{1}{2}}(I - \alpha T)B_1,$$

$$\tilde{C}_1 = C_1(I + TA_1)^{-1}(I - T^2)^{\frac{1}{2}},$$

$$\overline{B}_3 = (I - A_3^2)^{-\frac{1}{2}}B_3, \quad \overline{C}_3 = C_3(I - A_3^2)^{-\frac{1}{2}},$$

$$\overline{W} = \begin{bmatrix} A_1 & 0 & 0 & B_1 \\ 0 & I & 0 & B_2 \\ 0 & 0 & 0 & \overline{B}_3 \\ C_1 & C_2 & \overline{C}_3 & D_1 \end{bmatrix}, \quad \tilde{W} = \begin{bmatrix} \alpha I & 0 & 0 & \tilde{B}_1 \\ 0 & I & 0 & B_2 \\ 0 & 0 & 0 & \overline{B}_3 \\ \tilde{C}_1 & C_2 & \overline{C}_3 & D_2 \end{bmatrix}.$$

For any matrix X we shall denote by X_T and X_S the matrices formed by removing the bottom row and right hand column of X respectively.

Lemma 2.3 *Given A, B, C as above, the following four statements are equivalent:*

(a) *The conditions given in (2.1) are satisfied. That is*

$$s_m[A \quad B] \le 1, \quad s_m\begin{bmatrix} A \\ C \end{bmatrix} \le 1.$$

(b) $s_m(W_T) \leq 1$ and $s_m(W_S) \leq 1$;

(c) $s_m(\overline{W_T}) \leq 1$ and $s_m(\overline{W_S}) \leq 1$;

(d) $s_m(\tilde{W}_T) \leq 1$ and $s_m(\tilde{W}_S) \leq 1$.

Furthermore the following three statements are also equivalent:

(A) *There exists a matrix D such that* $s_m(W) \leq 1$;

(B) *There exists a matrix* D_1 *such that* $s_m(\overline{W}) \leq 1$;

(C) *There exists a matrix* D_2 *such that* $s_m(\tilde{W}) \leq 1$.

Moreover, when any of these three matrices exist, we can construct the other two using the following equality:

$$\begin{aligned} D &= D_2 + C_1(I + TA_1)^{-1}TB_1 - C_3A_3(I - A_3^2)^{-1}B_3 \\ &= D_1 - C_3A_3(I - A_3^2)^{-1}B_3. \end{aligned}$$

Proof: (a) $\Leftrightarrow$ (b) This follows from (2.2).

To prove each of the remaining equivalences we will simply make repeated use of Lemma 2.2. We shall prove one equivalence in full, but omit the details of the other proofs as the method is identical.

(A) $\Leftrightarrow$ (B) Consider $\Phi_M(\overline{W})$.

$$\Phi_M(\overline{W}) = (I - MM^*)^{-\frac{1}{2}}(M + \overline{W})(I + M^*\overline{W})^{-1}(I - M^*M)^{\frac{1}{2}}$$

$$= \begin{bmatrix} 1 & 0 & 0 & 0 \\ 0 & 1 & 0 & 0 \\ 0 & 0 & (I - A_3^2)^{-\frac{1}{2}} & 0 \\ 0 & 0 & 0 & 1 \end{bmatrix} \begin{bmatrix} A_1 & 0 & 0 & B_1 \\ 0 & I & 0 & B_2 \\ 0 & 0 & A_3 & B_3 \\ C_1 & C_2 & C_3 & D_1 \end{bmatrix}$$

$$\times \begin{bmatrix} 1 & 0 & 0 & 0 \\ 0 & 1 & 0 & 0 \\ 0 & 0 & 1 & -A_3\overline{B_3} \\ 0 & 0 & 0 & 1 \end{bmatrix} \begin{bmatrix} 1 & 0 & 0 & 0 \\ 0 & 1 & 0 & 0 \\ 0 & 0 & (I - A_3^2)^{\frac{1}{2}} & 0 \\ 0 & 0 & 0 & 1 \end{bmatrix}$$

$$= \begin{bmatrix} 1 & 0 & 0 & 0 \\ 0 & 1 & 0 & 0 \\ 0 & 0 & (I - A_3^2)^{-\frac{1}{2}} & 0 \\ 0 & 0 & 0 & 1 \end{bmatrix} \begin{bmatrix} A_1 & 0 & 0 & B_1 \\ 0 & I & 0 & B_2 \\ 0 & 0 & A_3 & (I - A_3^2)\overline{B_3} \\ C_1 & C_2 & C_3 & D_1 - \overline{C_3A_3B_3} \end{bmatrix}$$

$$\times \begin{bmatrix} 1 & 0 & 0 & 0 \\ 0 & 1 & 0 & 0 \\ 0 & 0 & (I - A_3^2)^{\frac{1}{2}} & 0 \\ 0 & 0 & 0 & 1 \end{bmatrix}$$

$$= \begin{bmatrix} A_1 & 0 & 0 & B_1 \\ 0 & I & 0 & B_2 \\ 0 & 0 & A_3 & B_3 \\ C_1 & C_2 & C_3 & D_1 - \overline{C_3 A_3 B_3} \end{bmatrix}.$$

Finally, since this is equal to W with D replaced by $D_1 - \overline{C_3 A_3 B_3}$, the statement holds by Lemma 2.2.

(B) $\Leftrightarrow$ (C) $\Phi_N(\overline{W}) = \tilde{W}$ with D_2 replaced by $D_1 - \overline{C_1}(I + TA_1)^{-1} T\overline{B_1}$

(b) $\Leftrightarrow$ (c) $\Phi_{M_T}(\overline{W_T}) = W_T$ and $\Phi_{M_S}(\overline{W_S}) = W_S$

(c) $\Leftrightarrow$ (d) $\Phi_{N_T}(\overline{W_T}) = \tilde{W}_T$ and $\Phi_{N_S}(\overline{W_S}) = \tilde{W}_S$ □

Lemma 2.3 allows us to assume that A takes the special form $\alpha I \oplus I \oplus 0 \, (\alpha > 1)$.

3 Some Special Cases

We shall now study cases of the Main Problem in which A takes various simple forms. This will not only give us an insight into the problem but will also lead to a full solution in the case $m = 1$ by way of the reductions described above. Our goal here is to solve the problem in the case where A has the special form

$$A = \begin{bmatrix} \alpha I_r & 0 & 0 \\ 0 & I_s & 0 \\ 0 & 0 & 0 \end{bmatrix}$$

for suitable integers r, s (which may be zero, in which case the corresponding row and column are deleted), and for any $\alpha > 1$. We shall begin by considering the case $A = 0$.

Lemma 3.1 *If B and C are matrices such that, for some non-negative integer m,*

$$s_m [0 \ B] \leq 1 \qquad \text{and} \qquad s_m \begin{bmatrix} 0 \\ C \end{bmatrix} \leq 1$$

then there exists a matrix D such that

$$s_m \begin{bmatrix} 0 & B \\ C & D \end{bmatrix} \leq 1.$$

Furthermore D can be taken to be

$$U_2(\kappa I_m \oplus 0) V_2^*$$

where

$$U_1^* B V_2 \quad and \quad U_2^* C V_1$$

are the singular value decompositions of B and C respectively. The number κ is taken as any complex number which satisfies

$$|\kappa| \geq s_0(C) s_0(B).$$

Proof: First assume B and C are scalars. Parrott's Theorem [P] shows that the result holds when $m = 0$. Consider the case $m = 1$. We have

$$|BC| = \left| \det \begin{bmatrix} 0 & B \\ C & D \end{bmatrix} \right| = s_0 \begin{bmatrix} 0 & B \\ C & D \end{bmatrix} s_1 \begin{bmatrix} 0 & B \\ C & D \end{bmatrix}$$

$$= \left\| \begin{bmatrix} 0 & B \\ C & D \end{bmatrix} \right\| s_1 \begin{bmatrix} 0 & B \\ C & D \end{bmatrix}.$$

It is clear that

$$\left\| \begin{bmatrix} 0 & B \\ C & D \end{bmatrix} \right\| \geq |D|$$

and hence

$$0 \leq s_1 \begin{bmatrix} 0 & B \\ C & D \end{bmatrix} \leq \left| \frac{BC}{D} \right|.$$

It follows that, for sufficiently large $D \in \mathbb{R}$, we have

$$s_1 \begin{bmatrix} 0 & B \\ C & D \end{bmatrix} \leq 1.$$

Clearly, any complex D with $|D| > |BC|$ will also suffice. We now return to the case where B and C are matrices. Let $\beta_j = s_j(B)$ and $\gamma_j = s_j(C)$ for $j \geq 0$. Choose unitary matrices U_1, U_2, V_1, V_2 such that

$$U_1^* B V_2 = \text{diag} \{\beta_j\}, \qquad\qquad U_2^* C V_1 = \text{diag} \{\gamma_j\}.$$

Since

$$\begin{bmatrix} U_1^* & 0 \\ 0 & U_2^* \end{bmatrix} \begin{bmatrix} 0 & B \\ C & D \end{bmatrix} \begin{bmatrix} V_1 & 0 \\ 0 & V_2 \end{bmatrix} = \begin{bmatrix} 0 & U_1^* B V_2 \\ U_2^* C V_1 & U_2^* D V_2 \end{bmatrix}$$

it will suffice to find $\{\delta_j : j \geq 0\}$ such that

$$
s_m \left[
\begin{array}{ccc|cc}
0 & \cdots & 0 & \beta_0 & 0 \\
\vdots & \ddots & \vdots & & \ddots \\
0 & \cdots & 0 & 0 & \beta_n \\
\hline
\gamma_0 & & 0 & \delta_0 & 0 \\
& \ddots & & & \ddots \\
0 & & \gamma_n & 0 & \delta_n
\end{array}
\right] \leq 1.
$$

Multiplication by suitable permutation matrices shows that this condition is equivalent to the δ_j satisfying

$$
(3.1) \qquad s_m \left[
\begin{array}{cccc}
\begin{bmatrix} 0 & \beta_0 \\ \gamma_0 & \delta_0 \end{bmatrix} & 0 & \cdots & 0 \\
0 & \ddots & & \vdots \\
\vdots & & \begin{bmatrix} 0 & \beta_j \\ \gamma_j & \delta_j \end{bmatrix} & \vdots \\
\vdots & & & 0 \\
0 & \cdots & 0 & \begin{bmatrix} 0 & \beta_n \\ \gamma_n & \delta_n \end{bmatrix}
\end{array}
\right] \leq 1.
$$

We can now apply the scalar result above to the first m summands of this direct sum by choosing $\delta_j \in \mathbb{C}$ such that $|\delta_j| \geq \beta_j \gamma_j$ for $j < m$. This will ensure that each of the first m block diagonal entries will have at most one singular value greater than one. For $j \geq m$ we have $\beta_j \leq 1$ and $\gamma_j \leq 1$, therefore we can take $\delta_j = 0$ to ensure that none of the block diagonal entries of this matrix, after the mth, has any singular value greater than one. With the δ_j chosen in this way, there can be a maximum of m singular values greater than one. Finally, since the set of singular values of a direct sum is the union of the sets of singular values of its components, (3.1) holds. The given formula for D can be found by working backwards through the proof. $\qquad\square$

Theorem 3.2 *Let A, B, C be as in (2.1) with the added condition that $\|A\| < 1$. Then there exists a matrix D such that*

$$
(3.2) \qquad s_m \begin{bmatrix} A & B \\ C & D \end{bmatrix} \leq 1.
$$

Moreover, D can be taken to be

$$
U_2(\kappa I_m \oplus 0)V_2^* - CA^*(I - A^*A)^{-1}B
$$

where

$$U_1^*(I - AA^*)^{-\frac{1}{2}} B V_2 \qquad and \qquad U_2^* C(I - A^*A)^{-\frac{1}{2}} V_1$$

are the singular value decompositions of $(I - AA^*)^{-\frac{1}{2}} B$ *and* $C(I - A^*A)^{-\frac{1}{2}}$
respectively. The number κ *can be taken as any complex number which satisfies*

$$|\kappa| \geq s_0[C(I - A^*A)^{-\frac{1}{2}}] s_0[(I - AA^*)^{-\frac{1}{2}} B].$$

Proof: Since $\|A\| < 1$ we may apply a Möbius transformation to reduce A to the zero matrix (Lemma 2.2). We can now apply Lemma 3.1, to find a suitable matrix D_1 which solves the reduced problem. A final application of Lemma 2.2 then provides a D for which (4) holds. The given formula for D can be found by combining Lemmas 2.2 and 3.1. $\qquad\qquad\qquad\qquad\qquad\qquad\qquad\qquad\qquad\qquad\quad\square$

Recall from Section 2 that any matrix A can be reduced to the following diagonal form

$$\begin{bmatrix} A_1 & 0 & 0 \\ 0 & I_r & 0 \\ 0 & 0 & A_3 \end{bmatrix}.$$

The next result shows that, in the search for a solution to the main problem, r can be taken to be zero if the matrix A has m singular values strictly greater than one.

Lemma 3.3 *Suppose* $\alpha > 1$ *and*

$$s_m \begin{bmatrix} \alpha I_m & 0 & 0 & B_1 \\ 0 & I_r & 0 & B_2 \\ 0 & 0 & 0 & B_3 \end{bmatrix} \leq 1.$$

Then $B_2 = 0$.

Proof: We have

$$s_m \begin{bmatrix} \alpha I_m & 0 & 0 & B_1 \\ 0 & I_r & 0 & B_2 \end{bmatrix} \leq 1,$$

which implies that the matrix

$$\begin{bmatrix} \alpha I_m & 0 & B_1 \\ 0 & I_r & B_2 \end{bmatrix} \begin{bmatrix} \alpha I_m & 0 \\ 0 & I_r \\ B_1^* & B_2^* \end{bmatrix} - I = \begin{bmatrix} (\alpha^2 - 1)I_m + B_1 B_1^* & B_1 B_2^* \\ B_2 B_1^* & B_2 B_2^* \end{bmatrix}$$

has at most m positive eigenvalues. By taking Schur complements we see that

$$B_2 B_2^* - B_2 B_1^* ((\alpha^2 - 1) I_m + B_1 B_1^*)^{-1} B_1 B_2^* \leq 0,$$

or alternatively

$$B_2 [I - B_1^* ((\alpha^2 - 1) I_m + B_1 B_1^*)^{-1} B_1] B_2^* \leq 0.$$

However,

$$I - B_1^* ((\alpha^2 - 1) I_m + B_1 B_1^*)^{-1} B_1 = (\alpha^2 - 1)$$
$$((\alpha^2 - 1) I_m + B_1^* B_1)^{-1} > 0.$$

It follows that $B_2 = 0$ as required. $\qquad\qquad\qquad\qquad\square$

Theorem 3.4 *Let A, B, C be matrices with $s_{m-1}(A) > 1$ and*

(3.3) $$s_m [A \ \ B] \leq 1, \qquad\qquad s_m \begin{bmatrix} A \\ C \end{bmatrix} \leq 1.$$

Then there exists a matrix D such that

(3.4) $$s_m \begin{bmatrix} A & B \\ C & D \end{bmatrix} \leq 1.$$

The matrix D can be chosen as

$$C_1 A_1 (A_1^2 - I)^{-1} B_1 + C_3 A_3^* (A_3^2 - I)^{-1} B_3$$

where

$$U A V = \begin{bmatrix} A_1 & 0 & 0 \\ 0 & I & 0 \\ 0 & 0 & A_3 \end{bmatrix}$$

is the singular value decomposition of A (with entries in decreasing order and $\|A_3\| \leq 1$) with

$$CV = [C_1 \ \ 0 \ \ C_3] \qquad and \qquad UB = \begin{bmatrix} B_1 \\ 0 \\ B_3 \end{bmatrix}.$$

Proof: We may assume that $A = \alpha I_m \oplus I \oplus 0$ with $\alpha > 1$. By Lemma 3.3, in the notation of (2.2), $B_2 = 0$ and $C_2 = 0$ so it will suffice to find a matrix D_1 such that

$$s_m \begin{bmatrix} \alpha I_m & 0 & \tilde{B}_1 \\ 0 & 0 & B_3 \\ \tilde{C}_1 & C_3 & D_1 \end{bmatrix} \leq 1.$$

We have

$$s_m \begin{bmatrix} \alpha I_m & 0 & \tilde{B}_1 \\ 0 & 0 & B_3 \end{bmatrix} \leq 1,$$

which implies that

$$\begin{bmatrix} \alpha I_m & 0 \\ \tilde{B}_1^{*} & B_3^{*} \end{bmatrix} \begin{bmatrix} \alpha I_m & \tilde{B}_1 \\ 0 & B_3 \end{bmatrix} - I = \begin{bmatrix} (\alpha^2 - 1)I_m & \alpha \tilde{B}_1 \\ \alpha \tilde{B}_1^{*} & \tilde{B}_1^{*}\tilde{B}_1 + B_3^{*}B_3 - I \end{bmatrix}$$

has at most m positive eigenvalues. We can now take Schur complements to show that

(3.5) $$\qquad B_3^{*}B_3 - I - (\alpha^2 - 1)^{-1}\tilde{B}_1^{*}\tilde{B}_1 \leq 0.$$

A similar argument shows that

(3.6) $$\qquad \overrightarrow{C_3^{*}}(I - \tilde{C}_1((\alpha^2 - 1)I_m + \tilde{C}_1^{*}\tilde{C}_1)^{-1}\tilde{C}_1^{*})\overrightarrow{C_3} - I \leq 0.$$

Next we define D_1 and T_{D_1} by:

$$D_1 = \frac{\alpha}{\alpha^2 - 1}\tilde{C}_1\tilde{B}_1, \qquad T_{D_1} = \begin{bmatrix} \alpha I_m & 0 & \tilde{B}_1 \\ 0 & 0 & B_3 \\ \tilde{C}_1 & C_3 & D_1 \end{bmatrix}.$$

We claim that $s_m(T_{D_1}) \leq 1$.

To simplify the following calculations define

$$\zeta = \frac{\alpha}{\alpha^2 - 1}, \qquad \xi = \alpha^2 - 1.$$

Consider $T_{D_1}^{*}T_{D_1} - I$ which expands to give

$$\begin{bmatrix} \xi I_m + \tilde{C}_1^{*}\tilde{C}_1 & \tilde{C}_1^{*}\overrightarrow{C_3} & (\zeta\tilde{C}_1^{*}\tilde{C}_1 + \alpha I_m)\tilde{B}_1 \\ \overrightarrow{C_3^{*}}\tilde{C}_1 & \overrightarrow{C_3^{*}}\overrightarrow{C_3} - I & \zeta\overrightarrow{C_3^{*}}\tilde{C}_1\tilde{B}_1 \\ \tilde{B}_1^{*}(\alpha I_m + \zeta\tilde{C}_1^{*}\tilde{C}_1) & \zeta\tilde{B}_1^{*}\tilde{C}_1^{*}\overrightarrow{C_3} & \tilde{B}_1^{*}(I + \zeta^2\tilde{C}_1^{*}\tilde{C}_1)\tilde{B}_1 + \overrightarrow{B_3^{*}B_3} - I \end{bmatrix}.$$

It will suffice to show that this matrix has at most m positive eigenvalues. Since the $(1, 1)$ block entry of this matrix is positive, we may take a Schur complement. We require

$$\begin{bmatrix} \overline{C_3}^* \overline{C_3} - I & \varsigma \overline{C_3}^* \tilde{C}_1 \tilde{B}_1 \\ \varsigma \tilde{B}_1^* \tilde{C}_1^* \overline{C_3} & \tilde{B}_1^*(I + \varsigma^2 \tilde{C}_1^* \tilde{C}_1)\tilde{B}_1 + \overline{B_3}^* \overline{B_3} - I \end{bmatrix}$$
$$- \begin{bmatrix} \overline{C_3}^* \tilde{C}_1 \\ \tilde{B}_1^*(\alpha I_m + \varsigma \tilde{C}_1^* \tilde{C}_1) \end{bmatrix} [\xi I_m + \tilde{C}_1^* \tilde{C}_1]^{-1}$$
$$[\tilde{C}_1^* \overline{C_3} \quad (\varsigma \tilde{C}_1^* \tilde{C}_1 + \alpha I_m)\tilde{B}_1]$$

to be negative semi-definite. We must therefore show

$$\begin{bmatrix} \overline{C_3}^*(I - \tilde{C}_1(\xi I_m + \tilde{C}_1^* \tilde{C}_1)^{-1}\tilde{C}_1^*)C_3 - I & 0 \\ 0 & \overline{B_3}^* \overline{B_3} - I - \xi^{-1}\tilde{B}_1^* \tilde{B}_1 \end{bmatrix} \leq 0.$$

This statement follows directly from (3.5) and (3.6) and so $s_m(T_{D_1}) \leq 1$. The formula for D_1 given in the proof above gives rise to the proposed formula for D by application of an inverse Möbius transformation. $\square$

The above results appear to be leading to the conclusion that a solution to the Main Problem always exists. However, we know this is not so, and the following result shows where the obstruction lies.

Theorem 3.5 *If B and C are matrices such that*

(3.7) $$s_m[I \ B] \leq 1 \qquad and \qquad s_m \begin{bmatrix} I \\ C \end{bmatrix} \leq 1$$

then there exists a matrix D such that

(3.8) $$s_m \begin{bmatrix} I & B \\ C & D \end{bmatrix} \leq 1$$

if and only if

(3.9) $$\dim(\text{Range } B + \text{Range } C^*) \leq m.$$

Furthermore, when this condition is satisfied, D can be taken to be CB.

Proof: First note that it follows from condition (3.7) that B and C have rank no greater than m.
($\Rightarrow$) Assume there exists a D which satisfies (3.8) and define T by

$$T = \begin{bmatrix} I & B \\ C & D \end{bmatrix}.$$

Consider

$$R \stackrel{\text{def}}{=} [I \quad B](T^*T - I)\begin{bmatrix} I \\ B^* \end{bmatrix} = (C + DB^*)^*$$
$$(C + DB^*) + BB^* + (BB^*)^2.$$

Clearly R is positive semi-definite, and has at most m positive eigenvalues. It follows that $\dim(\text{Range } R) \leq m$. However,

$$R = [(C + DB^*)^* \quad B(I + B^*B)^{\frac{1}{2}}]\begin{bmatrix} C + DB^* \\ (I + B^*B)^{\frac{1}{2}}B^* \end{bmatrix},$$

so Range R is equal to

$$\text{Range } [(C + DB^*)^* \quad B(I + B^*B)^{\frac{1}{2}}]$$
$$= \text{Range } [B \quad C^*]\begin{bmatrix} D^* & (I + B^*B)^{\frac{1}{2}} \\ I & 0 \end{bmatrix}.$$

Finally, since the square matrix on the RHS of the above equation is invertible, it follows that

$$\text{Range R} = \text{Range } [B \quad C^*] = \text{Range B} + \text{Range C}^*$$

and hence,

$$\dim(\text{Range B} + \text{Range C}^*) \leq m.$$

($\Leftarrow$) Suppose $\dim(\text{Range C}^* + \text{Range B}) \leq m$. Without loss of generality we may assume that B and C^* have at most m rows. This follows since condition (3.9) implies the existence of a unitary matrix U such that UB and UC^* have all rows but the first m equal to zero. We may further assume that the identity is also of dimension m. Let $D = CB$. Then the matrix

$$\begin{bmatrix} I & B \\ C & D \end{bmatrix} = \begin{bmatrix} I \\ C \end{bmatrix}[I \quad B]$$

has rank at most m and so (3.8) holds. $\square$

3.1 The Case $m = 1$

The results of the previous section can be specialised to give a full solution in the case $m = 1$.

Corollary 3.6 *Suppose*

$$s_1[A \quad B] \leq 1, \qquad s_1\begin{bmatrix} A \\ C \end{bmatrix} \leq 1.$$

and $\|A\| \neq 1$. *Then there exists a matrix D such that*

$$s_1\begin{bmatrix} A & B \\ C & D \end{bmatrix} \leq 1.$$

Proof: Theorem 3.2 deals with the case when A is a strict contraction, and Theorem 3.4 completes the proof. □

A full solution to the problem when $m = 1$ requires one further result.

Lemma 3.7 *Let β, γ be non-negative real numbers, at least one of which is different from zero, and assume that*

$$s_1 \begin{bmatrix} 1 & \beta \\ 0 & B \end{bmatrix} \leq 1, \qquad s_1 \begin{bmatrix} 1 & 0 \\ \gamma & C \end{bmatrix} \leq 1$$

for some vectors B and C. Then there exists $\delta \in \mathbb{C}$ such that

(3.10)
$$s_1 \begin{bmatrix} 1 & 0 & \beta \\ 0 & 0 & B \\ \gamma & C & \delta \end{bmatrix} \leq 1$$

if and only if

    (a) *β and γ are both non-zero, or*

    (b) *$\beta = 0$ and $\|B\| \leq 1$, or*

    (c) *$\gamma = 0$ and $\|C\| \leq 1$.*

When condition (b) or (c) is satisfied $\delta = 0$ will suffice, and when (a) holds, we may take δ to be any complex number satisfying

$$\operatorname{Re}\delta \geq \frac{\beta^2(\|C\|^2 - 1) + \gamma^2(\|B\|^2 - 1) + \beta^2\gamma^2}{2\beta\gamma}.$$

Proof: Let U and V be unitary matrices such that

$$UB = \begin{bmatrix} \|B\| \\ 0 \\ 0 \\ \vdots \end{bmatrix}, \qquad V^*C^* = \begin{bmatrix} \|C\| \\ 0 \\ 0 \\ \vdots \end{bmatrix}.$$

Then, since

$$\begin{bmatrix} 1 & 0 & 0 \\ 0 & U & 0 \\ 0 & 0 & 1 \end{bmatrix} \begin{bmatrix} 1 & 0 & \beta \\ 0 & 0 & B \\ \gamma & C & \delta \end{bmatrix} \begin{bmatrix} 1 & 0 & 0 \\ 0 & V & 0 \\ 0 & 0 & 1 \end{bmatrix} = \begin{bmatrix} 1 & 0 & \beta \\ 0 & 0 & \|B\| \\ \gamma & \|C\| & \delta \end{bmatrix} \oplus 0,$$

we may assume that B and C are non-negative scalars. Suppose $\gamma \neq 0$. Now $\delta \in \mathbb{C}$ satisfies (3.10) if and only if

$$\begin{bmatrix} 1 & 0 & \beta \\ 0 & 0 & B \\ \gamma & C & \delta \end{bmatrix}^* \begin{bmatrix} 1 & 0 & \beta \\ 0 & 0 & B \\ \gamma & C & \delta \end{bmatrix} - I$$

$$= \begin{bmatrix} \gamma^2 & \gamma C & \beta + \gamma\delta \\ \gamma C & C^2 - 1 & C\delta \\ \beta + \gamma\bar{\delta} & C\bar{\delta} & \beta^2 + B^2 + |\delta|^2 - 1 \end{bmatrix}$$

has at most one positive eigenvalue. This statement is equivalent (by virtue of Schur complements) to

$$\begin{bmatrix} C^2 - 1 & C\delta \\ C\overline{\delta} & \beta^2 + B^2 + |\delta|^2 - 1 \end{bmatrix} - \gamma^{-2} \begin{bmatrix} \gamma C \\ \beta + \gamma\overline{\delta} \end{bmatrix} [\gamma C \quad \beta + \gamma\delta] \leq 0.$$

That is,

$$\begin{bmatrix} -1 & -\dfrac{\beta C}{\gamma} \\ -\dfrac{\beta C}{\gamma} & \dfrac{\beta^2\gamma^2 + B^2\gamma^2 - \gamma^2 - \beta^2 - \beta\gamma\delta - \beta\gamma\overline{\delta}}{\gamma^2} \end{bmatrix} \leq 0,$$

which is so if and only if

$$2\mathrm{Re}\,\{\beta\gamma\delta\} + \beta^2 + \gamma^2 - \beta^2\gamma^2 - \beta^2 C^2 - \gamma^2 B^2 \geq 0.$$

Clearly this is equivalent to one of the following.

(a) β and γ are non-zero and Re δ is sufficiently large, or

(b) $\beta = 0$ and $B \leq 1$, in which case $\delta = 0$ will suffice.
An identical argument in which we assume $\beta \neq 0$ instead of $\gamma \neq 0$ gives rise to the final condition:

(c) $\gamma = 0$ and $C \leq 1$, in which case $\delta = 0$ will suffice.   □

We are now in a position to prove our Main Theorem for the case $m = 1$. Let P_K represent the orthogonal projection operator onto the space K.

Theorem 3.8 *Suppose A, B, C are matrices of suitable dimensions with*

$$(3.11) \qquad\qquad s_1[A \ B] \leq 1 \qquad \text{and} \qquad s_1 \begin{bmatrix} A \\ C \end{bmatrix} \leq 1.$$

Let $\overline{B} = P_{\mathrm{Ker}(I-AA^*)} B$ and $\overline{C} = C|_{\mathrm{Ker}(I-A^*A)}$. Then there exists a matrix D with

$$(3.12) \qquad\qquad s_1 \begin{bmatrix} A & B \\ C & D \end{bmatrix} \leq 1$$

if and only if $\overline{B}$ and $\overline{C}$ have rank no greater than one, and one of the following conditions is satisfied:

(a) *Range $\overline{B} = $ Range $\overline{C}^*$;*

(b) *$\overline{B} = 0$, $\overline{C}^* \neq 0$ and $s_0(B) \leq 1$;*

(c) *$\overline{C}^* = 0$, $\overline{B} \neq 0$ and $s_0(C) \leq 1$.*

Proof: Theorem 3.4 shows that we can always find a D satisfying (3.12) whenever $\|A\| \neq 1$. To avoid a contradiction we must therefore show that this condition on A is equivalent to one of the three conditions of the Theorem. First assume $\|A\| < 1$; then $\mathrm{Ker}(I - A^*A) = \mathrm{Ker}(I - AA^*) = \{0\}$, so that $\overline{B} = 0$, $\overline{C} = 0$ and condition (a) holds. If $\|A\| > 1$ then Lemma 3.3 shows that condition (a) holds once more. It remains to prove the result in the case $\|A\| = 1$.

($\Rightarrow$) Assume $\|A\| = 1$ and that there exists a D satisfying (3.12). Then by Lemma 2.2, there exists a D' satisfying

$$(3.13) \qquad s_1 \begin{bmatrix} 1 & 0 & \overline{B} \\ 0 & 0 & B_1 \\ \overline{C} & C_1 & D' \end{bmatrix} \leq 1.$$

By Lemma 3.5 we have

$$\overline{B} = 0 \text{ or } \overline{C}^* = 0 \text{ or Range } \overline{C}^* = \text{Range } \overline{B}$$

where $\overline{B}$ and $\overline{C}$ have rank at most one. Let us assume that $\overline{B} = 0$ and $\overline{C} \neq 0$; then by Lemma 3.5 there exist unitaries U_1, U_2, V_1, V_2 such that

$$U_1 \overline{C} V_1 = \mathrm{diag}\{\gamma, 0, \ldots, 0\}$$

and

$$U_2 B_1 V_2 = \mathrm{diag}\{\beta_0, \beta_1 \ldots\}.$$

Furthermore,

$$\begin{bmatrix} V_1^* & 0 & 0 \\ 0 & U_2 & 0 \\ 0 & 0 & U_1 \end{bmatrix} \begin{bmatrix} I & 0 & 0 \\ 0 & 0 & B_1 \\ \overline{C} & C_1 & D \end{bmatrix} \begin{bmatrix} V_1 & 0 & 0 \\ 0 & U_1^* & 0 \\ 0 & 0 & V_2 \end{bmatrix}$$

$$= \left[\begin{array}{ccccc|cccc} 1 & 0 & \cdots & & 0 & 0 & \cdots & & 0 \\ \hline 0 & 0 & \cdots & \cdots & 0 & \beta_0 & 0 & \cdots & 0 \\ \vdots & \vdots & & & \vdots & 0 & \beta_1 & \ddots & \vdots \\ \vdots & \vdots & & & \vdots & \vdots & \ddots & \ddots & 0 \\ 0 & 0 & \cdots & \cdots & 0 & 0 & \cdots & 0 & \beta_t \\ \hline \gamma & & & & & & & & \\ 0 & & X & & & & Y & & \\ \vdots & & & & & & & & \\ 0 & & & & & & & & \end{array} \right] \oplus I$$

for some matrices X and Y. We can now take a sub-matrix of the RHS to see that

$$s_1 \begin{bmatrix} 1 & 0 & 0 \\ 0 & 0 & \beta_0 \\ \gamma & X' & \delta \end{bmatrix} \leq 1,$$

where X' is the top row of X. By Lemma 3.6 we must have $\beta_0 \le 1$. A similar result holds when we assume $\overline{C} = 0$ which shows that the given conditions are necessary for the existence of a D satisfying (3.12).

($\Leftarrow$) First note that if conditions (b) or (c) are satisfied then $D = 0$ satisfies (3.12). Assume Range $\overline{B} = $ Range $\overline{C}^* \ne 0$ and $\overline{B}$ and $\overline{C}^*$ are rank 1. By Lemma 3.5 there exists unitaries U_1, U_2, V_1 such that

$$V_1 \overline{B} U_1 = \mathrm{diag}\,\{\beta, 0, \ldots, 0\}$$

and

$$U_2 \overline{C} V_1^* = \mathrm{diag}\,\{\gamma, 0, \ldots, 0\}$$

where β and γ are strictly positive real numbers. Now

$$
\begin{bmatrix} V_1 & 0 & 0 \\ 0 & I & 0 \\ 0 & 0 & U_2 \end{bmatrix}
\begin{bmatrix} I & 0 & \overline{B} \\ 0 & 0 & B_1 \\ \overline{C} & C_1 & D \end{bmatrix}
\begin{bmatrix} V_1^* & 0 & 0 \\ 0 & I & 0 \\ 0 & 0 & U_1 \end{bmatrix}
$$

$$
= \left[
\begin{array}{ccccc|ccc}
1 & 0 & \cdots & \cdots & 0 & \beta & 0 \cdots & 0 \\
\hline
0 & 0 & \cdots & \cdots & 0 & & & \\
\vdots & \vdots & & & \vdots & & B_1 U_1 & \\
\vdots & \vdots & & & \vdots & & & \\
0 & 0 & \cdots & \cdots & 0 & & & \\
\hline
\gamma & & & & & & & \\
0 & & U_2 C_1 & & & & U_2 D U_1 & \\
\vdots & & & & & & & \\
0 & & & & & & & \\
\end{array}
\right] \oplus I,
$$

and so it suffices to find a D such that the matrix on the RHS has $s_1 \le 1$. Let

$$B_1 U_1 = [B_{11} \ \ B_{12}], \qquad U_2 C_1 = \begin{bmatrix} C_{11} \\ C_{12} \end{bmatrix}.$$

Relabel $U_2 D U_1$ as

$$\begin{bmatrix} \delta & D_4 \\ D_5 & D_6 \end{bmatrix}.$$

It will suffice to find D_4, D_6 and δ such that

$$
s_1 \begin{bmatrix} 1 & 0 & \beta & 0 \\ 0 & 0 & B_{11} & B_{12} \\ \gamma & C_{11} & \delta & D_4 \\ 0 & C_{12} & & \boxed{D_6} \end{bmatrix} \le 1.
$$

By Lemma 3.7 there exists a δ such that

(3.14)
$$
s_1 \begin{bmatrix} 1 & 0 & \beta \\ 0 & 0 & B_{11} \\ \gamma & C_{11} & \delta \end{bmatrix} \le 1.
$$

Now let

$$M = \begin{bmatrix} 1 & 0 & \beta \\ 0 & 0 & B_{11} \end{bmatrix}, \qquad N = \begin{bmatrix} 0 \\ B_{12} \end{bmatrix}, \qquad P = [\gamma \quad C_{11} \quad \delta].$$

Then $\|M\| > 1$ and

$$s_1[M \quad N] \le 1, \qquad s_1 \begin{bmatrix} M \\ P \end{bmatrix} \le 1$$

by (3.11) and (3.14). Therefore, by Theorem 3.4, there exists a D_4 such that

$$(3.15) \qquad s_1 \begin{bmatrix} M & N \\ P & D_4 \end{bmatrix} = s_1 \begin{bmatrix} 1 & 0 & \beta & 0 \\ 0 & 0 & B_{11} & B_{12} \\ \gamma & C_{11} & \delta & D_4 \end{bmatrix} \le 1.$$

Define

$$X = \begin{bmatrix} 1 & 0 \\ 0 & 0 \\ \gamma & C_{11} \end{bmatrix}, \qquad Y = [0 \quad C_{12}], \qquad Z = \begin{bmatrix} \beta & 0 \\ B_{11} & B_{12} \\ \delta & D_4 \end{bmatrix}.$$

Then $\|X\| > 1$ and

$$s_1[X \quad Z] \le 1, \qquad s_1 \begin{bmatrix} X \\ Y \end{bmatrix} \le 1$$

by (3.15) and (3.11). Finally, Theorem 3.4 states that there exists a D_6 such that

$$s_1 \begin{bmatrix} X & Z \\ Y & D_6 \end{bmatrix} = s_1 \left[\begin{array}{cc|cc} 1 & 0 & \beta & 0 \\ 0 & 0 & B_{11} & B_{12} \\ \gamma & C_{11} & \delta & D_4 \\ \hline 0 & C_{12} & & D_6 \end{array} \right] \le 1.$$

This settles the case of Range B = Range $C^* \ne 0$. If Range B = Range $C^* = 0$ then a suitable D always exists by Lemma 3.1. $\qquad\qquad\qquad\square$

The results of this section can be combined to give a full solution to the problem when $m1$. Because of the complexity of the problem, no single formula will deal with every case, and we therefore need a number of solutions.

4 Construction of D in the Case m = 1

Let A, B, C be matrices of suitable dimensions such that

$$s_1[A \quad B] \le 1, \qquad s_1 \begin{bmatrix} A \\ C \end{bmatrix} \le 1.$$

Define the following matrices:

$$\overline{B} = P_{Ker(I - AA^*)} B \qquad\qquad \overline{C} = C|_{Ker(I - A^* A)}.$$

Then a matrix D such that

$$s_1 \begin{bmatrix} A & B \\ C & D \end{bmatrix} \leq 1,$$

can be constructed (provided one exists) by the following algorithm:

If $\|A\| < 1$
  Choose unitary matrices U_i and V_i ($i = 1, 2$) such that

$$U_1^*(I - AA^*)^{-\frac{1}{2}} B V_2$$

and

$$U_2^* C (I - A^* A)^{-\frac{1}{2}} V_1$$

are the singular value decompositions of $(I - AA^*)^{-\frac{1}{2}} B$ and $C(I - A^* A)^{-\frac{1}{2}}$.
  Choose a real number

$$\kappa \geq s_0 [C(I - A^* A)^{-\frac{1}{2}}] s_0 [(I - AA^*)^{-\frac{1}{2}} B].$$

A solution D is then given by

$$U_2(\kappa \oplus 0) V_2^* - C A^* (I - A^* A)^{-1} B.$$

If $\|A\| > 1$
  Choose unitary matrices U and V such that

$$UAV = \begin{bmatrix} A_1 & 0 & 0 \\ 0 & I & 0 \\ 0 & 0 & A_3 \end{bmatrix}$$

is the singular value decomposition of A. Next define matrices B_1, B_3, C_1, C_3 such that

$$CV = [C_1 \quad 0 \quad C_3] \qquad \text{and} \qquad UB = \begin{bmatrix} B_1 \\ 0 \\ B_3 \end{bmatrix}.$$

Take D to be

$$C_1 A_1 (A_1^2 - I)^{-1} B_1 + C_3 A_3 (A_3^2 - I)^{-1} B_3.$$

If $\|A\| = 1, \overline{B} = 0, \overline{C} \neq 0$ and $s_0(B) \leq 1$
  Take $D = 0$.

If $\|A\| = 1, \overline{C} = 0, \overline{B} \neq 0$ and $s_0(C) \leq 1$

Take $D = 0$.

If $\|A\| = 1, \overline{B} = 0$ and $\overline{C} = 0$

Define $B_1 = P_{\text{Range}(I-AA^*)}B$ and $C_1 = C|_{\text{Range}(I-A^*A)}$. Let

$$\begin{bmatrix} I & 0 \\ 0 & A_3 \end{bmatrix}$$

be the singular value decomposition of A. Choose unitary matrices U_i and V_i ($i = 1, 2$) such that $U_1^*(I - A_3^2)^{-\frac{1}{2}} B_1 V_2$ and $U_2^* C_1(I - A_3^2)^{-\frac{1}{2}} V_1$ are the singular value decompositions of $(I - A_3^2)^{-\frac{1}{2}} B_1$ and $C_1(I - A_3^2)^{-\frac{1}{2}}$.

Choose a real number

$$\kappa \geq s_0[C_1(I - A_3^2)^{-\frac{1}{2}}]s_0[(I - A_3^2)^{-\frac{1}{2}} B_1].$$

A solution D is given by

$$U_2(\kappa \oplus 0)V_2^* - C_1 A_3(I - A_3^2)^{-1} B_1.$$

If $\|A\| = 1$ and Range $\overline{C}^* = \text{Range}\overline{B}$ has dimension one.

Define $B_1 = P_{\text{Range}(I-AA^*)}B$ and $C_1 = C|_{\text{Range}(I-A^*A)}$.

Next choose a complex number δ such that:

$$\text{Re } \delta \geq \frac{\|\overline{B}\|^2(\|C_1\|^2 - 1) + \|\overline{C}\|^2(\|B_1\|^2 - 1) + \|\overline{B}\|^2\|\overline{C}\|^2}{2\|\overline{B}\|\|\overline{C}\|}.$$

Continue by choosing unitary matrices U_i and V_i ($i = 1, 2$) such that

$$V_1\overline{B}U_1 = \text{diag}\{\beta, 0, \ldots, 0\} \qquad \text{and} \qquad U_2\overline{C}V_1^* = \text{diag}\{\gamma, 0, \ldots, 0\}.$$

Let B_{11} and B_{12} represent the first and remaining columns of B_1 respectively. Similarly, let C_{11} and C_{12} represent the first and remaining rows of C_1 respectively.

Set

$$M = \begin{bmatrix} 1 & 0 & \|\overline{B}\| \\ 0 & 0 & B_{11} \end{bmatrix} \qquad \text{and} \qquad X = \begin{bmatrix} 1 & 0 \\ 0 & 0 \\ \|\overline{C}\| & C_{11} \end{bmatrix}.$$

Define unitaries U_3, U_4, , V_3, V_4 and matrices G_1, G_2, H_1, H_2 such that

$$U_3 M V_3 = \begin{bmatrix} M_1 & 0 & 0 \\ 0 & I & 0 \\ 0 & 0 & M_3 \end{bmatrix} \qquad \text{and} \qquad U_4 X V_4 = \begin{bmatrix} X_1 & 0 & 0 \\ 0 & I & 0 \\ 0 & 0 & X_3 \end{bmatrix}$$

are the singular value decompositions of M and X, and

$$[[\|\overline{C}\|\ C_{11}\ \delta]V_3 = [H_1\ 0\ H_3], \qquad U_3 \begin{bmatrix} 0 \\ B_{12} \end{bmatrix} = \begin{bmatrix} G_1 \\ 0 \\ G_2 \end{bmatrix}.$$

Set

$$D_4 = H_1 M_1 (M_1^2 - I)^{-1} G_1 + H_2 M_3 (M_3^2 - I)^{-1} G_2.$$

Next choose matrices K_1, K_2, L_1, L_2 such that

$$[0\ C_{12}]V_4 = [L_1\ 0\ L_3], \qquad U_4 \begin{bmatrix} \|\overline{B}\| & 0 \\ B_{12} & B_{22} \\ \delta & D_4 \end{bmatrix} = \begin{bmatrix} K_1 \\ 0 \\ K_2 \end{bmatrix}.$$

Then define

$$D_5 = L_1 X_1 (X_1^2 - I)^{-1} K_1 + L_2 X_3 (X_3^2 - I)^{-1} K_2$$

and

$$D_3 = \begin{bmatrix} \delta & D_4 \\ \hline & D_5 \end{bmatrix}.$$

Finally, a completion D is given by

$$D = D_3 - C_3 A_3 (I - A_3^2) B_3.$$

The list of solutions given here is exhaustive when $m = 1$. That is, if none of the above conditions are satisfied, then no suitable completion D exists.

Acknowledgements

This work was supported by a grant from the Engineering and Physical Sciences Research Council and also partially by NATO Collaborative grant CRG 971129.

References

[AY]   J. Agler and N.J. Young, A converse to a theorem of Adamyan, Arov and Krein, *J. Amer. Math. Soc.* **12** (1999), 305–333.

[AG]   Gr. Arsene and A. Gheondea, Completing matrix contractions, *J. Operator Theory* **7** (1982), 179–189.

[ACG]   Gr. Arsene, T. Constantinescu and A. Gheondea, Lifting of operators and prescribed numbers of negative squares, *Michigan Math. J.* **34** (1987), 201–216.

[CG]   T. Constantinescu and A. Gheondea, Completing matrix contractions, *J. Operator Theory* **7** (1982), 179–189.

[CG1]   T. Constantinescu and A. Gheondea, Minimal signature in lifting of operators I, *J. Operator Theory* **22** (1989), 345–367.

[CG2]   T. Constantinescu and A. Gheondea, Minimal signature in lifting of operators II, *J. Functional Analysis* **103** (1992), 317–351.

[CG3]   T. Constantinescu and A. Gheondea, The negative signature of some hermitian matrices, *Lin. Alg. Appl.* **178** (1993), 17–42.

[G]   A. Gheondea, One-step completions of hermitian partial matrices with minimal negative signature, *Lin. Alg. Appl.* **173** (1992), 99–114.

[Ph]   R.S. Phillips, The extension of dual subspaces invariant under and algebra, In: *Proc. Internat. Sympos. Linear Spaces*, 366–398, Jerusalem Academic Press and Peragamon Jerusalem and Oxford 1961.

[GRSW]   I. Gohberg, L. Rodman, T. Shalom and H. Woerdeman, Bounds for eigenvalues and singular values of matrix completions, *Linear and Multilinear Analysis* **33** (1992), 233–249.

[P]   S. Parrott, On a quotient norm and the Sz-Nagy-Foiaş lifting theorem, *J. Functional Analysis* **30** (1978), 311–328.

[S]   C.L. Siegel, *Symplectic geometry*, Academic Press New York 1964.

David Ogle and Nicholas Young
Department of Mathematics
University of Newcastle
Newcastle upon Tyne
NE1 7RU
England

1991 Mathematics Subject Classification. Primary: 15A18.

Operator Theory:
Advances and Applications, Vol. 124
© 2001 Birkhäuser Verlag Basel/Switzerland

Unconditional Decompositions
and Schur-type Multipliers

B. de Pagter, F.A. Sukochev and H. Witvliet

Dedicated to Professor I.C. Gohberg on the occasion of his 70th birthday

The purpose of this paper is to present a general framework for the treatment of Schur-type multipliers. Several applications are given, in particular, multiplier results for non-commutative L_p-spaces are obtained, and we indicate a general approach to the theory of double operator integrals.

1 Introduction

In this paper we continue our study of the interplay between unconditional decompositions, related multiplier theorems and the R-boundedness of collections of operators, as was started in a previous paper of the authors together with Ph. Clément ([CPSW]). In the present paper we concentrate mainly on multipliers of Schur-type and on generalizations of so-called double operator integrals in the sense of M.Sh. Birman and M.Z. Solomyak (see e.g. [BS66], [BS89]). The results are applicable to symmetrically normed ideals of compact operators (in the sense of [GK69]), as well as in the more general setting of symmetric operator spaces associated with von Neumann algebras.

The question of boundedness of Schur-type multipliers is related to the following general problem in the theory of Boolean algebras of projections in Banach spaces. Let $\mathcal{A}$ and $\mathcal{B}$ be two commuting bounded Boolean algebras of projections in the Banach space X. Under which conditions on X is the Boolean algebra $\mathcal{C}$, generated by $\mathcal{A}$ and $\mathcal{B}$, again bounded, i.e., does there exist a constant $C > 0$ such that

$$(1.1) \qquad \left\| \sum_{k,l} \alpha_{k,l} P_k Q_l x \right\| \leq C \|x\|$$

for all $\alpha_{k,l} = \pm 1$, all mutually disjoint $P_k \in \mathcal{A}$, mutually disjoint $Q_l \in \mathcal{B}$ and all $x \in X$? For example, $\mathcal{C}$ is bounded whenever X is any Lebesgue space L_p (see [McC64] and [DS3], pp. 2099–2100). Recently this result of C.A. McCarthy was strengthened and improved by T.A. Gillespie ([Gil97]).

However, in general $\mathcal{C}$ may be unbounded (see e.g. [McC61]). This is in particular the case if X is a Schatten class $\mathcal{C}_p$, $p \neq 2$ (see [McC67]), or more generally, if X is any ideal of compact operators different from the Hilbert-Schmidt class (see

[GL74], [GL75], [Lew75], [Pis78]). In the present paper we are concerned with situations where $\mathcal{C}$ is unbounded and we study the class of sequences $\mu = \{\mu_{k,l}\}_{k,l=1}^{\infty}$ such that for some constant $C > 0$ we have

(1.2)
$$\left\| \sum_{k,l} \mu_{k,l} P_k Q_l x \right\| \leq C \|x\|,$$

for all mutually disjoint $P_k \in \mathcal{A}$, mutually disjoint $Q_l \in \mathcal{B}$ and all $x \in X$. In this form the question has turned out to be relevant in the study of multiplier operators with respect to a given Schauder decomposition $\{D_n\}_{n=1}^{\infty}$ of a Banach space X, i.e., operators of the form

$$T_\lambda(x) = \sum_{k=1}^{\infty} \lambda_k D_k x, \quad x \in X,$$

where $\lambda = \{\lambda_k\}_{k=1}^{\infty}$ is a sequence of complex numbers. Characterizations of sequences λ for which T_λ is a bounded operator in X are treated in [CPSW], based on the notion of R-bounded collections of operators, recently introduced in [BG94] (in connection with the study of similar problems). For the convenience of the reader we have collected the necessary information from [CPSW] in the next section.

Our main result (see Theorem 3.3) presents a reasonably large class of sequences μ for which (1.2) holds in all UMD-spaces and it turns out that this has a variety of applications, in particular in the setting of symmetric operator spaces (see Examples 3.5, 3.7 and 3.8). In Examples 3.7 and 3.8 an alternative approach is presented to the proof of the key Schur-type estimates which are needed in the study of the Lipschitz continuity of the absolute value mapping in the p-Schatten classes (see [Dav88]) and non-commutative L_p-spaces (see [DDPS97]).

Using the results of Section 3, the last section of the paper is a first attempt to treat the double operator integrals of Birman and Solomyak in the general setting of UMD-Banach spaces. In this case the corresponding multiplier operators can be formally written as

$$T_\phi = \int_{\mathbb{R}^2} \phi(\lambda, \mu) dP \otimes Q,$$

for some appropriate Borel function ϕ on $\mathbb{R}^2$, where $P \otimes Q$ is the product of two spectral measures P and Q in X. The main obstruction to be overcome is the fact that $P \otimes Q$ is in general not countably additive, reflecting the unboundedness of the Boolean algebra generated by the ranges of P and Q. The results of the last section are of special interest, since this opens an avenue for the extension of the theory of double operator integrals to the class of operators affiliated with general semifinite von Neumann algebras.

2 Unconditional Decompositions

In the following X will always denote a complex Banach space. The range of a linear operator T on X is denoted by $\mathrm{Ran}(T)$. By $\mathcal{L}(X)$ we denote the space of all bounded linear operators in X.

Definition 2.1 (cf. [LT77], Section 1.g). A sequence $\Delta = \{\Delta_k\}_{k=1}^{\infty}$ of bounded linear projections in X is called a *Schauder decomposition* of X if

1. $\Delta_k \Delta_l = 0$ whenever $k \neq l$,
2. $x = \sum_{k=1}^{\infty} \Delta_k x$ for all $x \in X$.

The corresponding *partial sum projections* $\{P_n\}_{n=1}^{\infty}$ are defined by

$$(2.1) \qquad P_n = \sum_{k=1}^{n} \Delta_k.$$

If the series $\sum_{k=1}^{\infty} \Delta_k x$ is unconditionally convergent for all $x \in X$, then Δ is called an *unconditional decomposition*.

Remark 2.2 Let $\{\Delta_k\}_{k=1}^{\infty}$ be an unconditional decomposition of the Banach space X. For any subset $F \subseteq \mathbb{N}$ the series

$$P_F x = \sum_{k \in F} \Delta_k x, \quad x \in X$$

is convergent and P_F is a bounded linear projection in X, satisfying $\|P_F\| \leq C_\Delta$. The collection $\{P_F : F \subseteq \mathbb{N}\}$ is a bounded Boolean algebra of projections in X (in the sense of [DS3], XV.2). Conversely, if $\mathcal{B}$ is a bounded Boolean algebra of projections in X, and if $\{\Delta_k\}_{k=1}^{\infty} \subseteq \mathcal{B}$ such that $\Delta_k \Delta_l = 0$ whenever $k \neq l$ and $\sum_{k=1}^{\infty} \Delta_k x = x$ for all $x \in X$, then $\{\Delta_k\}_{k=1}^{\infty}$ is an unconditional decomposition of X.

By $\{\varepsilon_k\}_{k=1}^{\infty}$ we shall denote a sequence of independent symmetric $\{-1, 1\}$-valued random variables on some probability space $(\Omega, \mathcal{F}, P)$. Furthermore, $L_p(\Omega; X)$ denotes the space of all X-valued p-Bochner integrable functions on $(\Omega, \mathcal{F}, P)$.

If $\Delta = \{\Delta_n\}_{n=1}^{\infty}$ is an unconditional decomposition of X, then the smallest constant C_Δ such that

$$\left\| \sum_{k=1}^{N} \alpha_k \Delta_k x \right\|_X \leq C_\Delta \left\| \sum_{k=1}^{N} \Delta_k x \right\|_X$$

holds for all $\alpha_k = \pm 1$, $k = 1, 2, \ldots, N$, for all $N \in \mathbb{N}$ and all $x \in X$, is called the *unconditional constant* of the decomposition.

We will need the following property of unconditional decompositions, which is a well known consequence of unconditionality (see e.g. [DJT95], Lemma 1.4).

Lemma 2.3 *Let* $\Delta = \{\Delta_n\}_{n=1}^{\infty}$ *be an unconditional decomposition of the Banach space* X. *Then for all* $1 \le p < \infty$ *we have*

$$(2.2) \qquad C_\Delta^{-1} \left\| \sum_{k=1}^N \Delta_k x \right\|_X \le \left\| \sum_{k=1}^N \varepsilon_k \Delta_k x \right\|_{L_p(\Omega; X)} \le C_\Delta \left\| \sum_{k=1}^N \Delta_k x \right\|_X$$

for all $x \in X$ *and all* $N \in \mathbb{N}$.

Definition 2.4 A collection $\mathcal{T} \subset \mathcal{L}(X)$ is said to be *R-bounded* (Randomized bounded) if there exists a constant $M > 0$ such that

$$(2.3) \qquad \left\| \sum_{k=1}^N \varepsilon_k T_k x_k \right\|_{L_2(\Omega; X)} \le M \left\| \sum_{k=1}^N \varepsilon_k x_k \right\|_{L_2(\Omega; X)}$$

for all $\{T_k\}_{k=1}^N \subset \mathcal{T}$, all $\{x_k\}_{k=1}^N \subset X$ and all $N \in \mathbb{N}$. A constant M such that (2.3) holds is called an R-bound of $\mathcal{T}$.

Remark 2.5 The notion of R-boundedness was introduced by E. Berkson and T.A. Gillespie in [BG94], where it is called the R-property. We emphasize that the operators in the collections $\{T_k\}_{k=1}^N$ in the above definition need not be mutually distinct. Note that by Kahane's inequality (see e.g. [DJT95] Theorem 11.1) we can replace in the above definition $L_2(\Omega; X)$ by $L_p(\Omega; X), 1 \le p < \infty$, adjusting the constant M appropriately.

If (2.3) holds for all collections $\{T_k\}_{k=1}^N \subset \mathcal{T}$ with all T_k mutually different, then the collection $\mathcal{T}$ is R-bounded (see [CPSW]). Also, if $\mathcal{T} = \{T_k\}_{k=1}^\infty$, then it is enough that (2.3) holds for $\{T_k\}_{k=1}^N$ for all $N \in \mathbb{N}$.

The proof of the following two lemmas is straightforward (for details see [CPSW]).

Lemma 2.6 *Suppose that* $\mathcal{T} \subset \mathcal{L}(X)$ *is R-bounded (with R-bound M), then the absolute convex hull* $\mathrm{abco}(\mathcal{T})$ *is also R-bounded (with R-bound $2M$).*

Lemma 2.7 *If* $\mathcal{T} \subset \mathcal{L}(X)$ *is R-bounded, then the strong closure* $\overline{\mathcal{T}}^s$ *is also R-bounded (with the same R-bound).*

The following theorem (see [CPSW], Theorem 3.4) illustrates the interplay between R-boundedness and unconditional decompositions. For the convenience of the reader we include the proof.

Theorem 2.8 *Let* $\{\Delta_k\}_{k=1}^\infty$ *be an unconditional decomposition of the Banach space* X, *with unconditional constant* C_Δ. *Suppose that the collection* $\mathcal{T} \subset \mathcal{L}(X)$

is R-bounded with R-bound M. If $\{T_k\}_{k=1}^{\infty} \subset \mathcal{T}$ such that $\Delta_k T_k = T_k \Delta_k$ for all $k \in \mathbb{N}$, then the series

$$Sx := \sum_{k=1}^{\infty} T_k \Delta_k x$$

is convergent in X for all $x \in X$, and S defines a bounded linear operator $S : X \to X$ with $\|S\| \le C_\Delta^2 M$.

Proof: Let $x \in X$ and $1 \le m \le n$ in $\mathbb{N}$ be given. Then

$$\left\| \sum_{k=m}^{n} T_k \Delta_k x \right\|_X = \left\| \sum_{k=m}^{n} \Delta_k T_k \Delta_k x \right\|_X$$

$$= \left\| \sum_{k=1}^{n} \Delta_k \left(\sum_{j=m}^{n} \Delta_j T_j \Delta_j x \right) \right\|_X$$

$$\le C_\Delta \left\| \sum_{k=1}^{n} \varepsilon_k \Delta_k \left(\sum_{j=m}^{n} \Delta_j T_j \Delta_j x \right) \right\|_{L_2(\Omega; X)}$$

$$= C_\Delta \left\| \sum_{k=m}^{n} \varepsilon_k T_k \Delta_k x \right\|_{L_2(\Omega; X)}$$

$$\le C_\Delta M \left\| \sum_{k=m}^{n} \varepsilon_k \Delta_k x \right\|_{L_2(\Omega; X)}$$

$$\le C_\Delta^2 M \left\| \sum_{k=m}^{n} \Delta_k x \right\|_X.$$

Since $x = \sum_{k=1}^{\infty} \Delta_k x$, the result now follows immediately. $\qquad\square$

Corollary 2.9 *Let $\mathcal{B}$ be a Bade σ-complete Boolean algebra of projections in the Banach space X. Suppose that $\mathcal{T} \subset \mathcal{L}(X)$ is R-bounded with R-bound M, which commutes with $\mathcal{B}$. For any disjoint sequence $\{\Delta_k\}_{k=1}^{\infty}$ in $\mathcal{B}$ and every choice of $T_k \in \mathcal{T}$, the series*

$$Sx = \sum_{k=1}^{\infty} T_k \Delta_k x, \quad x \in X$$

is convergent in X, and $S : X \to X$ is a bounded linear operator with $\|S\| \le 4B^2 M$ (where $B = \sup\{\|\Delta\| : \Delta \in \mathcal{B}\}$).

Proof: First note that $\mathcal{B}$ is bounded (see [DS3], XVII.3.3). Define $\Delta_0 = I - \bigvee_{k=1}^{\infty} \Delta_k$. Since $\mathcal{B}$ is σ-complete, a moment's reflection shows that $\{\Delta_k\}_{k=0}^{\infty}$ is an unconditional decomposition of X with $C_\Delta \le 2B$. Now the result of the corollary follows from the previous theorem. $\qquad\square$

Next we recall some facts concerning UMD-spaces. Given a probability space $(S, \mathcal{A}, \mu)$ and an increasing sequence

$$\mathcal{A}_1 \subset \mathcal{A}_2 \subset \cdots$$

of sub-σ-algebras of $\mathcal{A}$, we denote by $\mathbb{E}(\cdot|\mathcal{A}_j)$ and $\mathbb{E}^X(\cdot|\mathcal{A}_j)$ the conditional expectation operators with respect to $\mathcal{A}_j$ in $L_p(S)$ and $L_p(S; X)$ respectively $(1 < p < \infty)$, where X is a Banach space (for information concerning conditional expectations in spaces of vector valued functions we refer the reader to [DU77]). The Banach space X is called a UMD-space if there exists a constant $C_2(X)$ (the UMD-constant of X) such that

$$\left\| \alpha_1 \mathbb{E}^X(f|\mathcal{A}_1) + \sum_{j=2}^{n} \alpha_j \{ \mathbb{E}^X(f|\mathcal{A}_j) - \mathbb{E}^X(f|\mathcal{A}_{j-1}) \} \right\|_{L_2(S;X)}$$
$$\leq C_2(X) \| f \|_{L_2(S;X)}$$

for all choices of $\alpha_j = \pm 1$, for all $f \in L_2(S; X)$, for all $n = 1, 2, \ldots$ and for all $(S, \mathcal{A}, \mu)$ and $\{\mathcal{A}_j\}_{j=1}^{\infty}$ as above. We note that in this definition of the UMD-property, the space $L_2(S; X)$ may be replaced by any $L_p(S; X)$ with $1 < p < \infty$ (replacing $C_2(X)$ by $C_p(X)$; we refer to e.g. [Bur83], [Bou83] and [RdF86] for more information on UMD-spaces). The following theorem will play an important role in the next sections.

Theorem 2.10 *Let X be a UMD space and let $\{\Delta_k\}_{k=1}^{\infty}$ be an unconditional decomposition with unconditional constant C_Δ. Let $P_n = \sum_{k=1}^{n} \Delta_k$. Then*

$$\left\| \sum_{k=1}^{n} \varepsilon_k P_k x_k \right\|_{L_2(\Omega;X)} \leq C_2(X) C_\Delta^2 \left\| \sum_{k=1}^{n} \varepsilon_k x_k \right\|_{L_2(\Omega;X)},$$

for all $x_1, x_2, \ldots, x_n \in X$ and all $n \in \mathbb{N}$. Consequently, the collection $\{P_n\}_{n \in \mathbb{N}}$ is R-bounded.

Remark 2.11 This theorem is an extension of results of E.M. Stein (see [Ste70]) and J. Bourgain (see [Bou83]), who treat the scalar and UMD-valued cases respectively for martingale decompositions. The complete proof of the above theorem can be found in [CPSW], where this general version is reduced to the martingale case.

3 Schur-type Multipliers

Let X be a Banach space and let $\{\Delta_k\}_{k=1}^{\infty}$ and $\{\Delta_l'\}_{l=1}^{\infty}$ be two commuting unconditional decompositions of X, i.e., $\Delta_k \Delta_l' = \Delta_l' \Delta_k$ for all $k, l \in \mathbb{N}$. The collection $\{\Delta_k \Delta_l'\}_{k,l=1}^{\infty}$ of projections in X is in general not an unconditional decomposition of X. However, choosing a suitable enumeration, we can obtain a Schauder decomposition of X.

Lemma 3.1 *Let X be a Banach space and let $\{\Delta_k\}_{k=1}^{\infty}$ and $\{\Delta_l'\}_{l=1}^{\infty}$ be two commuting Schauder decompositions of X. If we take the ordering "by squares", $I = \{(1, 1), (1, 2), (2, 2), (2, 1), (1, 3), (2, 3), \ldots\}$, then $\{\Delta_k \Delta_l'\}_{(k,l) \in I}$ is a Schauder decomposition of X.*

Proof: Since both $\{\Delta_k\}_{k=1}^{\infty}$ and $\{\Delta_l'\}_{l=1}^{\infty}$ are Schauder decompositions, it is easy to see that $\bigcup\{\mathrm{Ran}(\Delta_k \Delta_l') : k, l \in \mathbb{N}\}$ is dense in X. It remains to show that the partial sum operators P_n of $\{\Delta_k \Delta_l'\}_{(k,l) \in I}$ are uniformly bounded. However, it follows easily from the enumeration of the index set that

$$\| P_n \| \le 3 \left(\sup_N \left\| \sum_{k=1}^{N} \Delta_k \right\| \right) \left(\sup_N \left\| \sum_{l=1}^{N} \Delta_l' \right\| \right)$$

for all $n \in \mathbb{N}$. $\qquad\qquad\square$

Example 3.2 Let H be a separable Hilbert space with an orthonormal basis $\{e_n\}_{n=1}^{\infty}$ and take $X = C_p(H)$, $1 < p < \infty$, the p-Schatten class of compact operators on H (see e.g. [GK70]). For $m, n \in \mathbb{N}$ we define the matrix units $E_{m,n}$ by $E_{m,n}(x) := \langle x, e_n \rangle e_m$ for all $x \in H$. We define the row projections $\{R_k\}_{k=1}^{\infty}$ and column projections $\{C_l\}_{l=1}^{\infty}$ by $R_k(A) := E_{k,k} A$ and $C_l(A) := A E_{l,l}$ respectively for all $A \in X$. It is easy to see that $\{R_k\}_{k=1}^{\infty}$ and $\{C_l\}_{l=1}^{\infty}$ are unconditional decompositions of X (with unconditional constant equal to 1) which clearly commute. Now $R_k C_l$ is a projection onto $\mathrm{span}[E_{k,l}]$. It is well-known that the matrix units do not form an unconditional basis in $X = C_p(H)$ whenever $p \ne 2$ (see [KP70]) and hence $\{R_k C_l : k, l = 1, 2, \ldots\}$ is not an unconditional decomposition for $p \ne 2$. The Boolean algebra generated by $\{R_k C_l : k, l = 1, 2, \ldots\}$ is not bounded.

Let $\Delta = \{\Delta_k\}_{k=1}^{\infty}$ and $\Delta' = \{\Delta_l'\}_{l=1}^{\infty}$ be two commuting unconditional decompositions in the Banach space X. An infinite scalar matrix $\mu = (\mu_{k,l})_{k,l=1}^{\infty}$ is called a (Δ, Δ')-multiplier if there exists a bounded linear operator $M_\mu : X \mapsto X$ such that

$$M_\mu(\Delta_k \Delta_l' x) = \mu_{k,l} \Delta_k \Delta_l' x, \quad \forall x \in X, \forall k, l = 1, 2, \ldots.$$

In this case M_μ is called a Schur-type multiplier operator for (Δ, Δ'). Without loss of generality we may assume μ to be bounded (i.e., $\|\mu_{k,l}\| \le K$ for all $k, l = 1, 2, \ldots$).

We also write

$$(3.1) \qquad M_\mu \left(\sum_{k,l=1}^{\infty} \Delta_k \Delta_l' x \right) = \sum_{k,l=1}^{\infty} \mu_{k,l} \Delta_k \Delta_l' x, \quad \forall x \in X,$$

where the double series are interpreted with respect to an appropriate ordering.

It was pointed out to us by one of the referees, that if for a given x the series on the left converges for a certain ordering, then the boundedness of the operator M_μ ensures that the series on the right also converges for the same ordering.

In the case that $\{\Delta_k \Delta'_l\}_{k,l=1}^\infty$ is an unconditional decomposition, every entry-wise bounded matrix μ is such a multiplier.

In the theorem which follows we will give a general sufficient condition on matrices μ to be a multiplier. For this purpose we introduce the class $\mathfrak{N}$ of bounded infinite matrices $\mu = (\mu_{k,l})_{k,l=1}^\infty$ for which

$$\sup_l \sum_{k=1}^\infty |\mu_{k+1,l} - \mu_{k,l}| < \infty,$$

equipped with the norm given by

(3.2) $$\|\mu\|_{\mathfrak{N}} := \sup_l \sum_{k=1}^\infty |\mu_{k+1,l} - \mu_{k,l}| + \sup_l |\mu_{\infty,l}|,$$

where $\mu_{\infty,l} = \lim_{k \to \infty} \mu_{k,l}$.

Theorem 3.3 *Let $\{\Delta_k\}_{k=1}^\infty$ and $\{\Delta'_k\}_{k=1}^\infty$ be two commuting unconditional decompositions of the UMD-space X. Then every $\mu \in \mathfrak{N}$ is a (Δ, Δ')-multiplier and there exists a constant $C > 0$, only depending on the UMD-constant of X and the unconditional constants of Δ and Δ', such that*

(3.3) $$\|M_\mu\| \leq C\|\mu\|_{\mathfrak{N}}, \quad \forall \mu \in \mathfrak{N}.$$

Proof: For $l = 1, 2, \ldots$ define the bounded linear operators T_l in X by

$$T_l x := \sum_{k=1}^\infty \mu_{k,l} \Delta_k x, \quad \forall x \in X.$$

Via summation by parts we can write $T_l = T_l^{(1)} + \mu_{\infty,l} I$ with

$$T_l^{(1)} = \sum_{k=1}^\infty (\mu_{k,l} - \mu_{k+1,l}) P_k,$$

where $P_k = \sum_{j=1}^k \Delta_j$. It follows from Theorem 2.10 that the collection $\{P_k\}_{k=1}^\infty$ is R-bounded (with R-bound $C' = C_2(X)C_\Delta^2$).

Now it follows from Lemmas 2.6 and 2.7 that $\{T_l^{(1)}\}_{l=1}^\infty$ is R-bounded (with R-bound $2C'\|\mu\|_{\mathfrak{N}}$). It is clear that $T_l^{(1)} \Delta'_l = \Delta'_l T_l^{(1)}$ for all $l = 1, 2, \ldots$ and so we can apply Theorem 2.8. It follows that

$$T^{(1)} x = \sum_{l=1}^\infty T_l^{(1)} \Delta'_l x$$

converges for all $x \in X$ and $\|T^{(1)}\| \leq 2C'C_{\Delta'}^2\|\mu\|_{\mathfrak{N}}$. Now it is clear that

$$M_\mu = T^{(1)} + \sum_{l=1}^{\infty} \mu_{\infty,l}\Delta_l',$$

and so $\|M_\mu\| \leq (2C'C_{\Delta'}^2 + 2C_{\Delta'})\|\mu\|_{\mathfrak{N}}$. $\qquad\qquad\square$

Remark 3.4 The condition in the above theorem on the matrix $\mu = (\mu_{k,l})$ is not symmetric in k and l. A similar result can be obtained by interchanging the two indices k and l. So instead of being of bounded variation in index k, it is also sufficient to be of bounded variation in index l. Combining the two results we see that if $\mu = \{\mu_{k,l}\}_{k,l=1}^{\infty}$ can be written as $\mu_{k,l} = a_{k,l}b_{k,l}$ for all $k, l \in \mathbb{N}$ such that $\{a_{k,l}\}_{k,l=1}^{\infty}$ is bounded and of uniform bounded variation in index k and $\{b_{k,l}\}_{k,l=1}^{\infty}$ is bounded and of uniform bounded variation in index l, then μ is also a (Δ, Δ')-multiplier.

An example of this situation is given by $\mu_{k,l} = (-1)^{k+l}k^{-1}$. Neither the columns nor the rows are of bounded variation. However, if we take $a_{k,l} = (-1)^l$ and $b_{k,l} = (-1)^k k^{-1}$, then these matrices are of uniform bounded variation in k and l respectively.

Example 3.5 As in Example 3.2, take $X = C_p(H)$, $1 \leq p < \infty$, with the unconditional decompositions $\{R_k\}_{k=1}^{\infty}$ and $\{C_l\}_{l=1}^{\infty}$ (with respect to some orthonormal basis of H). For $1 < p < \infty$ the p-Schatten class is a UMD-space (see [Bou85]). Hence it follows from Theorem 3.3 that every $\mu \in \mathfrak{N}$ defines a bounded Schur multiplier operator in $C_p(H)$. In particular, taking $\mu = (\mu_{k,l})$ with $\mu_{k,l} = 1$ if $l \geq k$ and $\mu_{k,l} = 0$ otherwise, it follows that the triangular truncation is a bounded operator in $C_p(H)$ for $1 < p < \infty$, which is a classical result of V.I. Matsaev (see [Mac64]). It is well known that (assuming H to be infinite dimensional) triangular truncation is not a bounded operator in $C_1(H)$ (see e.g. [GK70]), which shows that the condition that X is a UMD-space cannot be omitted in Theorem 3.3.

Furthermore, defining for $0 < \theta < \pi/2$ the matrix $\mu_\theta = \{\mu_{k,l}\}_{k,l=1}^{\infty}$ by $\mu_{k,l} = 1$ if $k/l \leq \tan\theta$ and $\mu_{k,l} = 0$ otherwise, it follows from Theorem 3.3 that each μ_θ is a Schur multiplier for $C_p(H)$, $1 < p < \infty$, and there exists a constant $K_p > 0$ (only depending on p) such that $\|M_{\mu_\theta}\| \leq K_p$ for all $0 < \theta < \pi/2$. This last result was proved by E.B. Davies ([Dav88]) by different methods.

Corollary 3.6 *Let the Banach space X and the two unconditional decompositions be as in Theorem 3.3. Let $D \subset \mathbb{R}^2$, $J \subset \mathbb{R}$ an interval, and let $\phi : D \to J$ be such that $\phi(x_1, y) \leq \phi(x_2, y)$ whenever $(x_1, y), (x_2, y) \in D$ with $x_1 \leq x_2$. Furthermore, let $g : J \to \mathbb{C}$ be a function of bounded variation.*

For every choice of $\{a_k\}_{k=1}^{\infty}$ and $\{b_l\}_{l=1}^{\infty}$ in $\mathbb{R}$ such that $(a_k, b_l) \in D$ for all $k, l \in \mathbb{N}$, the matrix $\mu = (\mu_{k,l})$, given by $\mu_{k,l} = g(\phi(a_k, b_l))$ for all $k, l \in \mathbb{N}$, *is a* (Δ, Δ')-*multiplier and there exists a constant* $C > 0$ *(only depending on the space X and the unconditional constants of* Δ *and* Δ' *) such that*

$$\|M_\mu\| \leq C(\mathrm{Var}(g) + \|g\|_\infty)$$

(where $\mathrm{Var}(g)$ *denotes the total variation of g on J).*

Proof: Let $\{a_k\}_{k=1}^\infty$ and $\{b_l\}_{l=1}^\infty$ be given as above and $\mu_{k,l} = g(\phi(a_k, b_l))$. It is enough to show that there exists a constant $C > 0$ such that

$$(3.4) \qquad \left\| \sum_{k,l=1}^N \mu_{k,l} \Delta_k \Delta'_l x \right\| \leq C(\mathrm{Var}(g) + \|g\|_\infty)\|x\|,$$

for all $N \in \mathbb{N}$ and all $x \in X$. Let N be fixed. Define μ^N by $\mu_{k,l}^N = \mu_{k,l}$ if $k, l \leq N$ and $\mu_{k,l}^N = 0$ otherwise. We can find a permutation π of $\{1, 2, \ldots, N\}$ such that $\{a_{\pi(k)}\}_{k=1}^N$ is increasing. Replacing in $\{\Delta_k\}_{k=1}^\infty$ the first N projections by $\{\Delta_{\pi(k)}\}_{k=1}^N$, we may assume, without loss of generality, that π is the identity.
For μ^N we now have

$$\sum_{k=1}^\infty |\mu_{k+1,l}^N - \mu_{k,l}^N| = \sum_{k=1}^{N-1} |\mu_{k+1,l} - \mu_{k,l}| + |\mu_{N,l}|$$

$$= \sum_{k=1}^{N-1} |g(\phi(a_{k+1}, b_l)) - g(\phi(a_k, b_l))| + |g(\phi(a_N, b_l))|$$

$$\leq \mathrm{Var}(g) + \|g\|_\infty,$$

where the last inequality holds because $\{\phi(a_k, b_l)\}_{k=1}^{N-1}$ is increasing for every fixed $l \in \mathbb{N}$.
Applying Theorem 3.3 to μ^N, inequality (3.4) follows immediately. $\square$

Example 3.7 Let $X = \mathcal{C}_p(H)$, $1 < p < \infty$, as in Example 3.2. Then, for any two strictly positive sequences $\{a_k\}_{k=1}^\infty$ and $\{b_k\}_{k=1}^\infty$ the matrix $\mu = \{\mu_{k,l}\}_{k,l=1}^\infty$, defined by

$$(3.5) \qquad\qquad\qquad \mu_{k,l} = \frac{a_k - b_l}{a_k + b_l},$$

is, with respect to any orthonormal basis, a Schur multiplier for $\mathcal{C}_p(H)$ with $\|M_\mu\| \leq K_p$, where K_p only depends on p (see [Dav88]). Indeed, apply the previous corollary to $g(x) = (x - 1)(x + 1)^{-1}$ on $J = (0, \infty)$, $\phi(a, b) = ab^{-1}$ on $D = (0, \infty) \times (0, \infty)$ and the unconditional row and column decompositions.

Example 3.8 The above results are strong enough to be applied in non-commutative spaces more general than the p-Schatten classes. Let $\mathcal{M}$ be a semi-finite von Neumann algebra on the Hilbert space H equipped with a faithful and normal semifinite trace τ. The identity in $\mathcal{M}$ is denoted by $\mathbb{I}$. The corresponding non-commutative spaces $L_p(\mathcal{M}, \tau)$, $1 < p < \infty$ are UMD-spaces (see e.g. [BGM87]).

Now take sequences $\{p_n\}_{n=1}^\infty$ and $\{q_n\}_{n=1}^\infty$ of orthogonal projections in $\mathcal{M}$ such that $\sum_{n=1}^\infty p_n = \sum_{n=1}^\infty q_n = \mathbb{I}$. Define the projections Δ_k and Δ_l' in $L_p(\mathcal{M}, \tau)$ by $\Delta_k(x) = p_k x$ and $\Delta_l'(x) = x q_l$ respectively for all $x \in L_p(\mathcal{M}, \tau)$. It is easy to see that $\Delta = \{\Delta_k\}_{k=1}^\infty$ and $\Delta' = \{\Delta_l'\}_{l=1}^\infty$ are unconditional decompositions of $L_p(\mathcal{M}, \tau)$ which clearly commute.

Now it follows from Theorem 3.3 that every $\mu \in \mathfrak{N}$ is a (Δ, Δ')-multiplier in $L_p(\mathcal{M}, \tau)$. In particular, "triangular truncation" with respect to $\{p_n\}_{n=1}^\infty$ and $\{q_n\}_{n=1}^\infty$ is a bounded operator in $L_p(\mathcal{M}, \tau)$ (cf. [DDPS97], Theorem 3.3).

Moreover, it follows from Corollary 3.6 that for any two strictly positive sequences $\{a_n\}_{n=1}^\infty$ and $\{b_n\}_{n=1}^\infty$ the matrix $\mu = (\mu_{k,l})$ given by (3.5) is a (Δ, Δ')-multiplier. The latter result was obtained in [DDPS97], Lemma 3.2, by completely different methods.

4 Continuous Schur-type Multipliers

The purpose of this section is to discuss continuous Schur-type multipliers, where we replace the discrete unconditional decompositions Δ and Δ' of the previous section by continuous families $P = \{P_\lambda\}_{\lambda \in \mathbb{R}}$ and $Q = \{Q_\mu\}_{\mu \in \mathbb{R}}$ of projections in the Banach space X, which are unconditional in a sense to be made precise in Definition 4.2. Formally, the corresponding multiplier operators can be written as

$$(4.1) \qquad T_\phi = \int_{\mathbb{R}^2} \phi(\lambda, \mu) d(P \otimes Q),$$

where ϕ is a scalar function on $\mathbb{R}^2$. The main obstruction in this definition is the fact that the product measure $P \otimes Q$ occurring in (4.1) is in general not σ-additive. Consequently, the class of functions ϕ in (4.1) has to be restricted appropriately (see Definition 4.10) and the "integral" in (4.1) has to be defined with some care (see Definition 4.8).

As mentioned in the Introduction, in the case that $X = \mathcal{C}_p(H)$, the operators T_ϕ considered in the present section extend the notion of double operator integrals in the sense of M.Sh. Birman and M.Z. Solomyak (see e.g. [BS66], [BS89]).

As before X will denote a complex Banach space. Since our main interest in this paper is in the case that X is a UMD-space, we will not make an attempt to discuss the results below in the most general form possible. We start this section by recalling the definition of a spectral family (see e.g. [Dow78] Ch.17).

Definition 4.1 A *spectral family* in X is a collection $\{P_\lambda\}_{\lambda \in \mathbb{R}}$ of bounded linear projections in X such that

1. $P_\lambda P_\mu = P_\mu P_\lambda = P_\lambda$ for all $\lambda < \mu$;
2. $\lim_{\mu \downarrow \lambda} P_\mu x = P_\lambda x$ for all $x \in X$ and all $\lambda \in \mathbb{R}$;
3. $P_{\lambda-} x := \lim_{\mu \uparrow \lambda} P_\mu x$ exists for all $x \in X$ and all $\lambda \in \mathbb{R}$;
4. $P_\lambda x \to x$ as $\lambda \to \infty$ and $P_\lambda x \to 0$ as $\lambda \to -\infty$ for all $x \in X$.

Observe that a spectral family is bounded by the uniform boundedness principle.

Definition 4.2 The spectral family $\{P_\lambda\}_{\lambda \in \mathbb{R}}$ is called *unconditional* if there exists a constant $C \geq 1$ such that for all $N \in \mathbb{N}$ and for all $\lambda_0 < \lambda_1 < \cdots < \lambda_N$ in $\mathbb{R}$ we have

$$(4.2) \qquad \left\| \sum_{i=1}^{N} \alpha_i (P_{\lambda_i} - P_{\lambda_{i-1}}) x \right\| \leq C \|x\|, \qquad \forall x \in X, \forall \alpha_i = \pm 1.$$

Let $\{P_\lambda\}_{\lambda \in \mathbb{R}}$ be a spectral family in the Banach space X. For any cell $\delta = (a, b]$, $-\infty < a < b < \infty$, we define

$$(4.3) \qquad\qquad\qquad P(\delta) := P_b - P_a.$$

It is clear that $P(\cdot)$ is finitely additive on the semi-ring of these cells, and hence $P(\cdot)$ extends uniquely to an additive vector measure $P : \mathcal{R} \to \mathcal{L}(X)$ on the ring $\mathcal{R}$ generated by these cells. Observe that $P(\delta_1 \cap \delta_2) = P(\delta_1) P(\delta_2)$ for all $\delta_1, \delta_2 \in \mathcal{R}$, so in particular $P(\cdot)$ is projection valued.

For a general spectral family this vector measure P is not bounded, i.e., $\{P(\delta) : \delta \in \mathcal{R}\}$ is not bounded in $\mathcal{L}(X)$.

Lemma 4.3 *Let $\{P_\lambda\}_{\lambda \in \mathbb{R}}$ be a spectral family in X. Then $\{P(\delta) : \delta \in \mathcal{R}\}$ is bounded in $\mathcal{L}(X)$ if and only if $\{P_\lambda\}_{\lambda \in \mathbb{R}}$ is unconditional.*

Proof: Suppose that $\{P_\lambda\}_{\lambda \in \mathbb{R}}$ is unconditional and take $\delta \in \mathcal{R}$. Then δ can be written as a disjoint union

$$\delta = (a_1, b_1] \cup (a_2, b_2] \cup \cdots \cup (a_n, b_n],$$

where $b_k < a_{k+1}$ for $k = 1, 2, \ldots, n - 1$. Define $\{\lambda_i\}_{i=1}^{2n}$ by $\lambda_{2k-1} = a_k$ and $\lambda_{2k} = b_k$ for $k = 1, 2, \ldots, n$. It follows from the unconditionality that

$$\|P(\delta)x\| = \left\| \sum_{k=1}^{n} (P_{\lambda_{2k}} - P_{\lambda_{2k-1}})x \right\|$$

$$= \left\| \frac{1}{2} \left\{ \sum_{i=2}^{2n} (-1)^i (P_{\lambda_i} - P_{\lambda_{i-1}})x + \sum_{i=2}^{2n} (P_{\lambda_i} - P_{\lambda_{i-1}})x \right\} \right\| \leq C \|x\|,$$

where $C > 0$ is the constant in (4.2) above. Hence $\|P(\delta)\| \leq C$ for all $\delta \in \mathcal{R}$.

Now suppose that $\{P(\delta) : \delta \in \mathcal{R}\}$ is bounded in $\mathcal{L}(X)$. Let $\lambda_0 < \lambda_1 < \cdots < \lambda_N$ in $\mathbb{R}$ and $\{\alpha_i\}_{i=1}^N \in \{-1, 1\}^N$ be given. Put $I_1 = \{i : \alpha_i = 1\}$ and $I_2 = \{i : \alpha_i = -1\}$, and let $\delta_k = \bigcup_{i \in I_k} (\lambda_{i-1}, \lambda_i]$ for $k = 1, 2$. Then

$$\left\| \sum_{i=1}^N \alpha_i (P_{\lambda_i} - P_{\lambda_{i-1}}) x \right\| = \left\| \sum_{i \in I_1} (P_{\lambda_i} - P_{\lambda_{i-1}}) x - \sum_{i \in I_2} (P_{\lambda_i} - P_{\lambda_{i-1}}) x \right\|$$

$$= \|P(\delta_1)x - P(\delta_2)x\| \leq 2 \sup_{\delta \in \mathcal{R}} \|P(\delta)\| \|x\|,$$

which shows that $\{P_\lambda\}_{\lambda \in \mathbb{R}}$ is unconditional. $\qquad\square$

Before formulating the next lemma we recall that a vector measure $E : \mathcal{B}(\mathbb{R}) \to \mathcal{L}(X)$ is called a *spectral measure* if E is strongly σ-additive, $E(B_1 \cap B_2) = E(B_1)E(B_2)$ for all Borel sets $B_1, B_2 \in \mathcal{B}(\mathbb{R})$ (in particular, $E(B)$ is a projection for all $B \in \mathcal{B}(\mathbb{R})$) and $E(\mathbb{R}) = I$. Note that $E(\cdot)$ is uniformly bounded in $\mathcal{L}(X)$.

Lemma 4.4 *If $\{P_\lambda\}_{\lambda \in \mathbb{R}}$ is an unconditional spectral family in X, then $P(\cdot)x : \mathcal{R} \to X$ is a bounded σ-additive vector measure for all $x \in X$.*

Furthermore, if the Banach space X is reflexive, then there exists a unique spectral measure $P(\cdot) : \mathcal{B}(\mathbb{R}) \to \mathcal{L}(X)$ such that $P(-\infty, \lambda] = P_\lambda$ for all $\lambda \in \mathbb{R}$.

Proof: Let $x \in X$ be given. We claim that for every $a \in \mathbb{R}$ and $\varepsilon > 0$ there exists $a' > a$ such that $\|P(R)x\| < \varepsilon$ for all $R \in \mathcal{R}$, $R \subseteq (a, a']$. Indeed, since P_λ is right-continuous, there exists an $a' > a$ such $\|(P_{a'} - P_a)x\| < \varepsilon C^{-1}$ (where C is the constant in (4.2)). Any $R \in \mathcal{R}$ such that $R \subseteq (a, a']$ can be written as a disjoint union $R = \bigcup_{k=1}^n (a_k, b_k]$. Since $P_{b_k} - P_{a_k} = (P_{b_k} - P_{a_k})(P_{a'} - P_a)$, we get

$$\|P(R)x\| = \left\| \sum_{k=1}^n (P_{b_k} - P_{a_k})(P_{a'} - P_a)x \right\|$$

$$\leq C \|(P_{a'} - P_a)x\| < \varepsilon,$$

by the unconditionality of the spectral family. This proves the claim.

A standard argument (cf. [DS1], III.5.13) now shows that $P(\cdot)x$ is σ-additive on $\mathcal{R}$. It has been observed already in the lemma above that $\|P(R)x\| \leq C\|x\|$ for all $R \in \mathcal{R}$.

Now assume in addition that X is reflexive. It follows e.g. from Kluvanek's extension theorem (see [Klu72] or [DU77], Theorem I.5.2) that $P(\cdot)x$ extends to a σ-additive vector measure $P_x : \mathcal{B}(\mathbb{R}) \to X$ for all $x \in X$. For $B \in \mathcal{B}(\mathbb{R})$ and $x \in X$ we now define $P(B)x = P_x(B)$. It is readily verified that $P : \mathcal{B}(\mathbb{R}) \to \mathcal{L}(X)$ is a spectral measure with the desired properties. $\qquad\square$

Proposition 4.5 *If $\{P_\lambda\}_{\lambda \in \mathbb{R}}$ is an unconditional spectral family in the UMD-space X, then $\{P_\lambda : \lambda \in \mathbb{R}\}$ is R-bounded.*

Proof: If $\lambda_0 < \lambda_1 < \cdots < \lambda_n$ in $\mathbb{R}$, then $\{P_{\lambda_0}, P_{\lambda_1} - P_{\lambda_0}, \ldots, P_{\lambda_n} - P_{\lambda_{n-1}}, I - P_{\lambda_n}\}$ is an unconditional decomposition of X with unconditional constant C, as given by Definition 4.2. It follows from Theorem 2.10 that the partial sums, i.e., $\{P_{\lambda_k}\}_{k=1}^n$, are R-bounded with an R-bound independent of n and λ_k.   □

Assume that $\{P_\lambda\}_{\lambda \in \mathbb{R}}$ and $\{Q_\lambda\}_{\lambda \in \mathbb{R}}$ are two commuting unconditional spectral families in the UMD space X. Recall that any UMD-space is reflexive (see e.g. [RdF86]). Let $P(\cdot)$ and $Q(\cdot)$ denote the corresponding spectral measures on the Borel sets $\mathcal{B}(\mathbb{R})$, as given by Lemma 4.4. Observe that $P(\cdot)$ and $Q(\cdot)$ commute.

For $A, B \in \mathcal{B}(\mathbb{R})$ we define

$$(4.4) \qquad\qquad F(A \times B) = P(A)Q(B).$$

It is easy to see that F extends to a finitely additive projection-valued vector measure $F : \Lambda \to \mathcal{L}(X)$, where Λ denotes the algebra generated by all Borel rectangles $A \times B$ in $\mathbb{R}^2$. Note that $F(R \cap S) = F(R)F(S)$ for all $R, S \in \Lambda$. This measure F will also be denoted by $P \otimes Q$.

Remark 4.6 If the vector measure $F : \Lambda \to \mathcal{L}(X)$ is bounded and if the space X does not contain a copy of c_0 (in particular, if the space X is reflexive), then F extends to a spectral measure on $\mathcal{B}(\mathbb{R}^2)$ (see [Gil97], Proposition 3.4). If X is a Hilbert space, then F is always bounded, as a consequence of a theorem of J. Wermer ([Wer54]; see also [DS3], Section XV.6). Other situations in which this occurs are discussed by T.A. Gillespie in [Gil97], and we refer the interested reader to this paper for further details. However, if $X = \mathcal{C}_p(H)$, or a more general non-commutative L_p-space ($p \neq 2$), then the measure F is in general not bounded, as was already mentioned in the Introduction (see also Example 3.2).

For $x \in X$, $x^* \in X^*$ we define the finitely additive measure $F_{x,x^*} : \Lambda \to \mathbb{C}$ by

$$(4.5) \qquad\qquad F_{x,x^*}(R) := \langle F(R)x, x^* \rangle, \quad \forall R \in \Lambda.$$

In what follows we will assume that the following condition is fulfilled.

Condition 4.7 There exist a dense subspace $V \subset X$ and a dense subspace $W \subset X^*$ such that for all $x \in V$ and $x^* \in W$ the measure F_{x,x^*} is σ-additive and of bounded variation on Λ, and hence extends to a (complex) σ-additive Borel measure F_{x,x^*} on $\mathcal{B}(\mathbb{R}^2)$.

If $\phi : \mathbb{R}^2 \to \mathbb{C}$ is a bounded Borel measurable function, we define

$$(4.6) \qquad b_\phi(x, x^*) := \int_{\mathbb{R}^2} \phi(\lambda, \mu) dF_{x,x^*}, \quad \forall x \in V, \forall x^* \in W.$$

Definition 4.8 We say that ϕ is a (P, Q)-*multiplier*, if there exists a (necessarily unique) operator $T_\phi \in \mathcal{L}(X)$ such that

$$(4.7) \qquad\qquad \langle T_\phi x, x^* \rangle = b_\phi(x, x^*), \quad \forall x \in V, \forall x^* \in W.$$

We denote this by $\phi \in \mathfrak{M}(P, Q)$. In this case we will say that T_ϕ is a Schur-type multiplier operator for (P, Q). The operator T_ϕ can be written formally as

$$T_\phi = \int_{\mathbb{R}^2} \phi(\lambda, \mu) dP \otimes Q.$$

Remark 4.9 (1) Let the Banach space X be reflexive. If ϕ is a bounded Borel function on $\mathbb{R}^2$, then $\phi \in \mathfrak{M}(P, Q)$ if and only if there exists a constant $C > 0$ such that

$$(4.8) \qquad \left| \int_{\mathbb{R}^2} \phi(\lambda, \mu) dF_{x,x^*} \right| \leq C \|x\| \|x^*\|, \quad \forall x \in V, \forall x^* \in W.$$

In this case $\|T_\phi\| \leq C$. Indeed, if (4.8) is satisfied, then the bilinear form b_ϕ, defined on $V \times W$ by (4.6), extends to a bilinear form b_ϕ on $X \times X^*$. Consequently, there exists $T_\phi \in \mathcal{L}(X, X^{**})$ such that $\langle T_\phi x, x^* \rangle = b_\phi(x, x^*)$ for all $x \in X, x^* \in X^*$ and $\|T_\phi\| \leq C$. Since X is reflexive, it follows that $T_\phi \in \mathcal{L}(X)$ and $\phi \in \mathfrak{M}(P, Q)$.

(2) Let $\{\phi_n\} \subset \mathfrak{M}(P, Q)$, and let $\phi : \mathbb{R}^2 \to \mathbb{C}$ be a bounded Borel measurable function such that $\phi_n \to \phi$ pointwise on $\mathbb{R}^2$. If $|\phi_n| \leq M$ for all $n \in \mathbb{N}$ and

$$(4.9) \qquad \left| \int_{\mathbb{R}^2} \phi_n(\lambda, \mu) dF_{x,x^*} \right| \leq C \|x\| \|x^*\|, \forall x \in V, \forall x^* \in W, \forall n \in \mathbb{N},$$

then $\phi \in \mathfrak{M}(P, Q)$.

By $\text{Var}(f)$ we denote the variation of a complex valued function $f : \mathbb{R} \to \mathbb{C}$, i.e.,

$$(4.10) \qquad \text{Var}(f) := \sup \sum_{i=1}^{n} |f(t_i) - f(t_{i-1})|,$$

where the supremum is taken over all partitions $t_0 < t_1 < \cdots < t_n$ of $\mathbb{R}$.

Definition 4.10 We denote by $\mathfrak{M}$ the collection of all bounded Borel measurable functions $\phi : \mathbb{R}^2 \to \mathbb{C}$ such that $\phi(\lambda, \cdot)$ is a right continuous function of bounded variation on $\mathbb{R}$ for all $\lambda \in \mathbb{R}$ and $\sup_{\lambda \in \mathbb{R}} \text{Var}(\phi(\lambda, \cdot)) < \infty$.
For $\phi \in \mathfrak{M}$ we define

$$(4.11) \qquad \|\phi\|_\mathfrak{M} := \|\phi(\cdot, -\infty)\|_\infty + \sup_{\lambda \in \mathbb{R}} \text{Var}(\phi(\lambda, \cdot)),$$

where $\phi(\lambda, -\infty) = \lim_{\mu \to -\infty} \phi(\lambda, \mu)$ for all $\lambda \in \mathbb{R}$. By $\mathfrak{M}_\mathbb{R}$ we denote the subspace of all real valued functions in $\mathfrak{M}$.

It is easy to see that $(\mathfrak{M}, \| \cdot \|_\mathfrak{M})$ is a Banach space.

Theorem 4.11 *Let* $\{P_\lambda\}_{\lambda \in \mathbb{R}}$ *and* $\{Q_\mu\}_{\mu \in \mathbb{R}}$ *be commuting unconditional spectral families in the UMD-space* X, *such that Condition 4.7 is satisfied.*
Then $\mathfrak{M} \subset \mathfrak{M}(P, Q)$ *and there exists a constant* $K = K(P, Q, X)$ *such that*

$$(4.12) \qquad \|T_\phi\| \le K \|\phi\|_{\mathfrak{M}}, \quad \forall \phi \in \mathfrak{M}.$$

Before we start with the proof of this theorem, we first give a preliminary result.

Lemma 4.12 *If* $\phi \in \mathfrak{M}_{\mathbb{R}}$, *then* ϕ *can be written as* $\phi = \phi_1 - \phi_2$, *with* $\phi_i \in \mathfrak{M}_{\mathbb{R}}$, $i = 1, 2$, *such that* $\phi_i(\lambda, \cdot)$ *is an increasing function for all* $\lambda \in \mathbb{R}$ *and*

$$(4.13) \qquad \|\phi_i\|_{\mathfrak{M}} \le \|\phi\|_{\mathfrak{M}}.$$

Proof: For $(\lambda, \mu) \in \mathbb{R}^2$ we define $V_\phi : \mathbb{R}^2 \to \mathbb{R}$ by

$$(V_\phi)(\lambda, \mu) = \sup \sum_{i=1}^{n} |\phi(\lambda, \mu_i) - \phi(\lambda, \mu_{i-1})|,$$

where the supremum is taken over all partitions $\mu_0 < \mu_1 < \cdots < \mu_n = \mu, n \in \mathbb{N}$, of $(-\infty, \mu]$. We will show first that V_ϕ is Borel measurable. To this end, for any $v \in C_c^\infty(\mathbb{R})$, define the function $f_v : \mathbb{R}^2 \to \mathbb{R}$ by

$$f_v(\lambda, \mu) = \int_{(-\infty, \mu]} v(t) d\phi(\lambda, t).$$

Note that, via integration by parts, it follows that

$$f_v(\lambda, \mu) = v(\mu)\phi(\lambda, \mu) - \int_{(-\infty, \mu]} v'(t)\phi(\lambda, t)dt,$$

and so f_v is a Borel function. Now let $\{v_n\}_{n=1}^\infty$ be a sequence in $C_c^\infty(\mathbb{R})$ which is dense in $\{w \in C_0(\mathbb{R}) : \|w\|_\infty \le 1\}$. Then

$$\begin{aligned}
(V_\phi)(\lambda, \mu) &= |d\phi(\lambda, \cdot)|(-\infty, \mu] \\
&= \sup \left\{ \left| \int_{(-\infty, \mu]} w d\phi(\lambda, \cdot) \right| : w \in C_0(\mathbb{R}), \|w\|_\infty \le 1 \right\} \\
&= \sup_n |f_{v_n}(\lambda, \mu)|,
\end{aligned}$$

for all $(\lambda, \mu) \in \mathbb{R}^2$. This shows that V_ϕ is a Borel function. It is clear that

$$\|V_\phi\|_{\mathfrak{M}} = \sup_{\lambda \in \mathbb{R}} \text{Var}(\phi(\lambda, \cdot)) \le \|\phi\|_{\mathfrak{M}}.$$

Now define $\phi_1 = \frac{1}{2}(V_\phi + \phi)$ and $\phi_2 = \frac{1}{2}(V_\phi - \phi)$, then $\phi = \phi_1 - \phi_2$ is the desired decomposition. $\square$

Having obtained the preliminary result, we can now give the proof of Theorem 4.11.

Proof: By considering the real and imaginary part separately and by Lemma 4.12, we may restrict ourselves to $\phi \in \mathfrak{M}_{\mathbb{R}}$ for which $\phi(\lambda, \cdot)$ is increasing for all $\lambda \in \mathbb{R}$.

First we consider the special case that $\phi = \mathbb{1}_B$ for some Borel set $B \subset \mathbb{R}^2$ for which $\mathbb{1}_B(\lambda, \cdot)$ is increasing and right continuous.

For $n \in \mathbb{N}$ and $j \in \mathbb{Z}$ define

$$A_j^{(n)} := \{\lambda \in \mathbb{R} : (\lambda, j2^{-n}) \notin B, (\lambda, (j+1)2^{-n}) \in B\},$$
$$A := \{\lambda \in \mathbb{R} : (\lambda, \mu) \in B, \forall \mu \in \mathbb{R}\}.$$

Let

$$(4.14) \qquad B_n := (A \times \mathbb{R}) \cup \left(\bigcup_{j=-\infty}^{\infty} A_j^{(n)} \times (j2^{-n}, \infty) \right).$$

Note that $B_1 \supset B_2 \supset \cdots \supset B$. Since $\mathbb{1}_B(\lambda, \cdot)$ is increasing and right-continuous, it follows easily that $B = \bigcap_{n=1}^{\infty} B_n$, and so $B_n \downarrow B$.

Since $\{I - Q_\mu\}_{\mu \in \mathbb{R}}$ is R-bounded by Proposition 4.5, it follows from Corollary 2.9, applied to the range $\mathcal{B}$ of the spectral measure $P(\cdot)$ and the collection $\mathcal{T} = \{I - Q_\mu\}_{\mu \in \mathbb{R}}$, that the operators T_n, defined by

$$(4.15) \qquad T_n x := P(A)x + \sum_{j=-\infty}^{\infty} P(A_j^{(n)})(I - Q_{j2^{-n}})x,$$

are bounded and satisfy

$$(4.16) \qquad \qquad \|T_n\| \leq K, \quad \forall n \in \mathbb{N}$$

for some constant K which is independent of n.

Now observe that for all $x \in V$ and $x^* \in W$ we have

$$(4.17) \qquad \qquad F_{x,x^*}(B_n) = \langle T_n x, x^* \rangle.$$

Indeed,

$$\langle T_n x, x^* \rangle = \langle P(A)x, x^* \rangle + \sum_{j=-\infty}^{\infty} \langle P(A_j^{(n)})(I - Q_{j2^{-n}})x, x^* \rangle$$

$$= F_{x,x^*}(A \times \mathbb{R}) + \sum_{j=-\infty}^{\infty} F_{x,x^*}(A_j \times (j2^{-n}, \infty))$$

$$= F_{x,x^*}(B_n),$$

using the assumption that F_{x,x^*} is countably additive on $\mathcal{B}(\mathbb{R}^2)$.

Consequently

$$
\begin{aligned}
|F_{x,x^*}(B_n)| &= |\langle T_n x, x^* \rangle| \leq \|T_n\| \|x\| \|x^*\| \\
&\leq K\|x\|\|x^*\|.
\end{aligned}
$$

Since $B_n \downarrow B$, it follows that

(4.18) $|F_{x,x^*}(B)| \leq K\|x\|\|x^*\| \quad \forall x \in V, \forall x^* \in W,$

hence $\mathbb{I}_B \in \mathfrak{M}$ by Remark 4.9.

Now assume that $\phi \in \mathfrak{M}_{\mathbb{R}}$ is such that $\phi(\lambda, \cdot)$ is increasing for all $\lambda \in \mathbb{R}$. For $t \in \mathbb{R}$ we define

$$
B_t := \{(\lambda, \mu) : \phi(\lambda, \mu) \geq t\}.
$$

Then $\mathbb{I}_{B_t} \in \mathfrak{M}$ for all $t \in \mathbb{R}$.

Put $m = \inf\{\phi(\lambda, \mu) : (\lambda, \mu) \in \mathbb{R}^2\}$ and $M = \sup\{\phi(\lambda, \mu) : (\lambda, \mu) \in \mathbb{R}^2\}$. For $n = 1, 2, \ldots$ let $m = t_0^{(n)} < t_1^{(n)} < \cdots < t_{k_n}^{(n)} = M$ be a partition of $[m, M]$ with $0 < t_j^{(n)} - t_{j-1}^{(n)} < \frac{1}{n}$ for all $j = 1, \ldots, k_n$. Define

$$
\phi_n := m\,\mathbb{I}_{B_{t_0^{(n)}}} + \sum_{j=1}^{k_n} (t_j^{(n)} - t_{j-1}^{(n)})\,\mathbb{I}_{B_{t_j^{(n)}}}.
$$

Then

$$
\begin{aligned}
\left| \int_{\mathbb{R}^2} \phi_n d F_{x,x^*} \right| &= \left| m F_{x,x^*}(B_{t_0^{(n)}}) + \sum_{j=1}^{k_n} (t_j^{(n)} - t_{j-1}^{(n)}) F_{x,x^*}(B_{t_j^{(n)}}) \right| \\
&\leq |m| |F_{x,x^*}(B_{t_0^{(n)}})| + (M - m) \sup_j |F_{x,x^*}(B_{t_j^{(n)}})| \\
&\leq |m| K + (M - m) K \leq 3K\|\phi\|_{\mathfrak{M}},
\end{aligned}
$$

using the previously obtained estimates and the triangle inequality. Since $\phi_n \to \phi$ pointwise on $\mathbb{R}^2$, the result follows from Remark 4.9(2).   □

Example 4.13 We can now apply Theorem 4.11 to the case of the Schatten classes. Let $\{E_\lambda\}_{\lambda \in \mathbb{R}}$ and $\{F_\mu\}_{\mu \in \mathbb{R}}$ be two spectral families of orthogonal projections in the Hilbert space H. Let $X = C_p(H)$ for some $1 < p < \infty$. As mentioned before X is a UMD-space. For $\lambda \in \mathbb{R}$ we define projections P_λ and Q_λ in X by $P_\lambda(x) = E_\lambda x$ and $Q_\lambda(x) = x F_\lambda$ for all $x \in X$. Then $P = \{P_\lambda\}_{\lambda \in \mathbb{R}}$ and $Q = \{Q_\mu\}_{\mu \in \mathbb{R}}$ are two commuting unconditional spectral families in X. If $p = 2$, then P and Q are commuting spectral families of orthogonal projections in the Hilbert space C_2 and hence, in this case, the product measure $P \otimes Q$ extends to a spectral measure on $\mathcal{B}(\mathbb{R}^2)$. This shows that for a general $1 < p < \infty$, Condition 4.7 is fulfilled with $V = C_p \cap C_2$ and $W = C_q \cap C_2$, $p^{-1} + q^{-1} = 1$. Note that V and W are dense in C_p and C_q respectively, as the finite rank operators are dense in these spaces. Consequently, Theorem 4.11 can be applied in this situation. The operators T_ϕ correspond in this case with the double operator integrals in the sense of Birman and Solomyak, and Theorem 4.11 now corresponds to a result of M.Z. Solomyak [Sol67] (see also [BS89], Theorem 2.9).

Example 4.14 The above ideas can also be applied to more general non-commutative spaces. Indeed, let $(\mathcal{M}, \tau)$ be a semifinite von Neumann algebra on the Hilbert space H, as in Example 3.8. Let $\{e_\lambda\}_{\lambda \in \mathbb{R}}$ and $\{f_\mu\}_{\mu \in \mathbb{R}}$ be two spectral families of orthogonal projections in $\mathcal{M}$. Take $X = L_p(\mathcal{M}, \tau)$ for some $1 < p < \infty$ and define for $\lambda \in \mathbb{R}$ the projections P_λ and Q_λ in X by $P_\lambda(x) = e_\lambda x$ and $Q_\lambda(x) = x f_\lambda$ for all $x \in X$. Then $P = \{P_\lambda\}_{\lambda \in \mathbb{R}}$ and $Q = \{Q_\mu\}_{\mu \in \mathbb{R}}$ are commuting unconditional spectral families in X. As in Example 4.13, Condition 4.7 is fulfilled with $V = (L_p \cap L_2)(\mathcal{M}, \tau)$ and $W = (L_q \cap L_2)(\mathcal{M}, \tau)$, $p^{-1} + q^{-1} = 1$. Therefore, the result of Theorem 4.11 implies that the Schur-type multiplier operators T_ϕ are bounded on $L_p(\mathcal{M}, \tau)$ for all $\phi \in \mathfrak{M}$ with $\|T_\phi\| \leq K_p \|\phi\|_{\mathfrak{M}}$, where the constant K_p only depends on p.

Of course, the boundedness of the operators T_ϕ can now also be obtained for certain non-commutative interpolation spaces in the L_p-scale.

In a subsequent paper we will discuss the applications of the above results to the study of operator functions in non-commutative spaces.

Acknowledgements

The authors would like to thank W.A.J. Luxemburg for indicating the present proof of Lemma 4.12, replacing a more involved argument in a preliminary version of the paper.

References

[BG94]   E. Berkson and T.A. Gillespie, Spectral decompositions and harmonic analysis on UMD spaces, *Studia Math.* **112** (1994), 13–49.

[BGM87]   E. Berkson, T.A. Gillespie and P.S. Muhly, A generalization of Macaev's theorem to noncommutative L^p-spaces, *Integral Equations Operator Theory* **10** (1987), 164–186.

[BS66]   M.Sh. Birman and M.Z. Solomyak, Double Stieltjes operator integrals, *Problemy Mat. Fiz.* **1** (1966), 33–67.

[BS89]   M.Sh. Birman and M.Z. Solomyak, Operator integration, perturbations, and commutators, *Zap. Nauchn. Sem. Leningrad. Otdel. Mat. Inst. Steklov. (LOMI) Issled. Linein. Oper. Teorii Funktsii. 17,* **170** (1989), 34–66.

[Bou83]   J. Bourgain, Some remarks on Banach spaces in which martingale differences are unconditional, *Arkiv Math.* **21** (1983), 163–168.

[Bou85]   J. Bourgain, Vector-valued singular integrals and the H^1-BMO duality, _Probability Theory and Harmonic Analysis, Dekker_, 1985, 1–19.

[Bur83]   D. Burkholder, A geometric condition that implies the existence of certain singular integrals of Banach-space-valued functions, _Proceedings of Conference on Harmonic Analysis in Honor of Antoni Zygmund (Chicago 1981)_, 1983, 270–286.

[CPSW]   Ph. Clément, B. de Pagter, F.A. Sukochev and H. Witvliet, Schauder decompositions and multiplier theorems, _Studia Math. (to appear)_, 2000.

[Dav88]   E.B. Davies, Lipschitz continuity of functions of operators in the Schatten classes, _J. London Math. Soc. (2)_ **37** (1988), 148–157.

[DJT95]   J. Diestel, H. Jarchow and A. Tonge, _Absolutely summing operators_, Cambridge University Press 1995.

[DU77]   J. Diestel and J.J. Uhl, _Vector measures_, AMS. Surveys 15, Providence 1977.

[DDPS97]   P.G. Dodds, T.K. Dodds, B. de Pagter and F.A. Sukochev, Lipschitz continuity of the absolute value and Riesz projections in symmetric operator spaces, _Journal of Functional Analysis_ **148** (1997), 28–69.

[Dow78]   H.R. Dowson, _Spectral theory of linear operators_, Academic Press 1978.

[DS1]   N. Dunford and J.T. Schwartz, _Linear operators, part I_, Wiley-Interscience 1958.

[DS3]   N. Dunford and J.T. Schwartz, _Linear operators, part III_, Wiley-Interscience 1971.

[Gil97]   T.A. Gillespie, Boundedness criteria for Boolean algebras of projections, _J. Funct. Anal._ **148** (1997), 70–85.

[GK69]   I.C. Gohberg and M.G. Kreĭn, _Introduction to the theory of linear nonselfadjoint operators_, American Mathematical Society 1969.

[GK70]   I.C. Gohberg and M.G. Kreĭn, _Theory of Volterra operators in Hilbert space and its applications_, American Mathematical Society 1970.

[GL74]   Y. Gordon and D. Lewis, Absolutely summing operators and local unconditional structures, _Acta Math._ **133** (1974), 27–48.

[GL75]   Y. Gordon and D. Lewis, Banach ideals on Hilbert spaces, _Studia Math._ **54** (1975), 161–175.

[Klu72]   I. Kluvánek, The extension and closure of vector measure, _Vector and operator valued measures and applications (Proc. Sympos., Alta, Utah, 1972)_, Academic Press, 1973, 175–190.

[KP70]   A. Kwapień and A. Pełczyński, The main triangle projection in matrix spaces and its applications, _Studia Math._ **34** (1970), 43–68.

[Lew75]   D. Lewis, An isomorphic characterization of the Schmidt class, _Compositio Math._ **30** (1975), 293–297.

[LT77]   J. Lindenstrauss and L. Tzafriri, _Classical Banach spaces. I_, Springer-Verlag, Berlin 1977.

[Mac64]   V.I. Macaev, A method of estimation for resolvents of non-selfadjoint operators, _Dokl. Akad. Nauk SSSR_ **154** (1964), 1034–1037.

[McC61]   C.A. McCarthy, Commuting Boolean algebras of projections, _Pacific J. Math._ **11** (1961), 295–307.

[McC64]   C.A. McCarthy, Commuting Boolean algebras of projections. II Boundedness in L_p, _Proc. Amer. Math. Soc._ **15** (1964), 781–787.

[McC67]   C.A. McCarthy, c_p _Israel J. Math._ **5** (1967), 249–271.

[Pis78]   G. Pisier, Some results on Banach spaces without local unconditional structure, *Compositio Math.* **37** (1978), 3–19.

[RdF86]   J.L. Rubio de Francia, Martingale and integral transforms of Banach space valued functions, *Probability and Banach Spaces (Zaragoza 1985), Lecture Notes in maths.* **1221** (1986), 195–222.

[Sol67]   M.Z. Solomyak, On transformers generated by Stieltjes double operator integrals, *Zap. Nauchn. Sem. Leningr. Otd. Mat. Inst.* **5** (1967), 201–231.

[Ste70]   E.M. Stein, *Topics in harmonic analysis related to the Littlewood-Paley theory*, Princeton University Press 1970.

[Wer54]   J. Wermer, Commuting spectral operators on Hilbert space, *Pacific J. Math.* **4** (1954), 355–361.

B. de Pagter and F.A. Sukochev
Department of Mathematics
Faculty ITS
Delft University of Technology
P.O.Box 5031
2600 Ga Delft
The Netherlands

H. Witvliet
Department of Mathematics
and Statistics
The Flinders University
of South Australia
G.P.O.Box 2100
Adelaide
South Australia 5001
Australia

1991 Mathematics Subject Classification. Primary: 47B49; Secondary: 47B10, 46L50.

Operator Theory:
Advances and Applications, Vol. 124
© 2001 Birkhäuser Verlag Basel/Switzerland

Scattering in a Loop-shaped Waveguide

Vyacheslav Pivovarchik

Dedicated to Israel Gohberg on the occasion of his 70-th birthday

Wave scattering is considered in a loop-shaped one-dimensional waveguide. For finite potential the problem can be reduced to a Regge type one. The conditions are obtained sufficient for a function to be the S-matrix of a problem of the class considered. The algorithm of recovering the potential from given Jost function is presented. It is shown that the solution of this inverse problem is not unique.

1 Introduction

The scattering problems on graphs were considered in many publications [ES], [AP], [GP] in connection with their importance in the theory of electronic microschemes [A]. The corresponding inverse problems were solved in [G]. However, in [G] sufficient conditions were not obtained for a function to be the "scattering matrix" of a Sturm-Liouville problem on a graph. Here we give such conditions for the case of a simple loop-shaped graph. The following spectral problem on a semiaxis describes one-dimensional scattering of a quantum particle when the way of propagation has the form of a loop.

$$(1.1) \qquad y'' + (\lambda^2 - q(x))y = 0,$$

$$(1.2) \qquad y(\lambda, 0) = y(\lambda, a - 0) = y(\lambda, a + 0),$$

$$(1.3) \qquad y'(\lambda, 0) + y'(\lambda, a + 0) - y'(\lambda, a - 0) = 0.$$

For the sake of simplicity let us suppose $q(x)$ to be finite, i.e.

$$(1.4) \qquad q(x) = \begin{cases} q_1(x) \in L_2(0, x), & \text{if } x \in (0, a), \\ 0, & \text{if } x \in [a, \infty) \end{cases} .$$

Evidently, there exist the following Jost-type solutions of (1.1):

$$e(\pm\lambda, x)$$

$$(1.5)$$
$$= \begin{cases} s(\lambda, a)e^{\mp i\lambda x}, & \text{if } x \in (a, \infty), \\ (1 - c(\lambda, a))e^{\mp i\lambda a}s(\lambda, x) + s(\lambda, a)e^{\mp i\lambda a} \\ c(\lambda, x), & \text{if } x \in [0, a], \end{cases}$$

which satisfies condition (1.2). Here $s(\lambda, x)$ $(c(\lambda, x))$ is the solution of (1.1) which satisfies the conditions $s(\lambda, 0) = s'(\lambda, 0) - 1 = 0$ $(c(\lambda, 0) - 1 = c'(\lambda, 0) = 0)$. Let us introduce the operator A which acts in $L_2(0, \infty)$ according to the formulae $Ay = -y'' + q(x)y$, $D(A) = \{y \in L_2(0, \infty), -y'' + q(x)y \in L_2(0, \infty), y(0) = y(a - 0) = y(a + 0), y'(0) + y'(a + 0) - y'(a - 0) = 0\}$. This operator is selfadjoint and bounded below. The spectrum of this operator consists of normal eigenvalues (see [GK] for definition) which can occur on the half-axis $(-\infty, 0)$ and of essential spectrum which covers the half-axis $[0, \infty)$. This statement follows from the existence of the Jost solutions of the form (1.5). There can exist finite or infinite set of eigenvalues on essential spectrum (in physical terms "bound states embedded into continuous spectrum"). The solution of (1.1) which satisfies conditions (1.2) and (1.3) behaves asymptotically as follows

$$\phi(\lambda, x) \underset{x \to \infty}{=} C(\lambda)(e^{-i\lambda x} - S(\lambda)e^{i\lambda x}),$$

where so-called S-matrix [N, M] is of the form

(1.6) $$S(\lambda) = e^{2i\lambda a}\frac{e(\lambda)}{e(-\lambda)}$$

and

(1.7) $$e(\lambda) = 2 - c(\lambda, a) - s'(\lambda, a) - i\lambda s(\lambda, a).$$

It is known [M] that the set of zeroes $\{\mu_k\}_{-\infty, k \neq 0}^{\infty}$ of the function $\chi(\lambda) \overset{def}{=} 2 - c(\lambda, a) - s'(\lambda, a)$ coincides with the spectrum of the periodic problem, i.e. the problem generated be equation (1.1) and the following boundary conditions

(1.8) $$y(0) = y(a),$$

(1.9) $$y'(0) = y'(a).$$

The set of zeroes $\{\nu_k\}_{-\infty, k \neq 0}^{\infty}$ of the function $s(\lambda, a)$ coincides with the spectrum of the Dirichlet problem, i.e. the problem generated by equation (1.1) and the boundary conditions $y(0) = y(a) = 0$. Let us introduce the operator A_1 which acts in $L_2(0, a)$ according to the formulae $A_1 y = -y'' + q_1(x)y$, $D(A_1) = \{y : y \in W_2^2(0, a), y(0) = y(a), y'(0) = y'(a)\}$. The set of zeroes of the function $e(\lambda) = \chi(\lambda) - i\lambda s(\lambda, a)$ is the spectrum of the following Regge type [R] problem

$$y'' + (\lambda^2 - q(x))y = 0,$$

$$y(0) = y(a),$$

(1.10) $$y'(0) = y'(a) + i\lambda y(a).$$

Problem (1.1), (1.8), (1.10) describes propagation of a quantum particle (or a wave) in a ring-shaped one dimensional medium with an absorptive membrane.

2 Direct Problem

We identify the spectrum of problem (1.1), (1.8), (1.10) as the spectrum of the following nonmonic quadratic operator pencil

$$L(\lambda) = \lambda^2 P - i\lambda K - A_2,$$

where the operators involved act in the direct product $L_2(0, a) \oplus \mathbb{C}$ according to the formulae

$$A_2 \begin{pmatrix} y(x) \\ y(a) \end{pmatrix} = \begin{pmatrix} -y''(x) + q_1(x)y(x) \\ y'(a) - y'(0) \end{pmatrix},$$

$$D(A_2) = \left\{ \begin{pmatrix} y(x) \\ y(a) \end{pmatrix}, y(x) \in W_2^2(0, a), y(a) = y(0) \right\},$$

$$P = \begin{pmatrix} I & 0 \\ 0 & 0 \end{pmatrix}, \quad K = \begin{pmatrix} 0 & 0 \\ 0 & I \end{pmatrix}.$$

The domain $D(L(\lambda))$ is supposed to be equal to $D(A_2)$ by definition, i.e. $D(L(\lambda))$ is independent of λ. It is easy to check up that A_2 is selfadjoint operator and the operators P and K are symmetric and moreover, $P \geq 0$, $K \geq 0$ and A_2 is bounded below. The spectrum of problem (1.1), (1.8), (1.10) which we identify as the spectrum of $L(\lambda)$ consists of normal eigenvalues because it coincides with the set of zeroes of the entire function $e(\lambda)$. This spectrum is symmetric with respect to the imaginary axis and symmetrically located eigenvalues possess equal multiplicities. The part of the spectrum of problem (1.1), (1.2), (1.3) located in the open lower half-plane coincides with that of problem (1.1), (1.8), (1.10).

Theorem 2.1 1. *All the eigenvalues of problem* (1.1), (1.8), (1.10) *located in the open lower half-plane are pure imaginary.*

  2. *All of them are semisimple (i.e. they do not possess associated eigenvectors).*

  3. *Their total algebraic (or geometric what is the same here) multiplicity is equal to that of the negative part of the spectrum of problem* (1.1), (1.8), (1.9).

Proof: Assertion 1 follows from Lemma 2.3 of [P] and Assertion 2 follows from Lemma 2.4 of [P]. These lemmas were proved in [P] under the additional assumption $P = I$, but the proof remains true under the following less restrictive assumption $ker P \cap ker K = \{0\}$, which is satisfied in our case. Now we make use of Theorem 2.1 of [P]. That theorem was proved in [P] assuming $P = I$ but the proof remains true under the following less restrictive conditions $P + I >> 0$, $P \geq 0$, $K \geq 0$ which are valid in our case. Hence, applying that theorem we obtain that the parts of spectra of the pencils $L(\lambda)$ and $L_0(\lambda) = \lambda^2 P - A_2$ located in the open lower half-plane possess equal total algebraic multiplicities. If $\begin{pmatrix} y(x) \\ y(a) \end{pmatrix}$ is an eigenvector of $L_0(\lambda)$ which corresponds to a pure imaginary nonzero eigenvalue λ_0 then

$$\lambda_0^2 y(x) + y''(x) - q(x)y(x) = 0$$

and (1.8) and (1.9) valid and $y(x)$ is not equal to 0 identically. It means that $y(x)$ is an eigenvector of A_1 which corresponds to the negative eigenvalue λ_0^2 and vice versa every negative eigenvalue of A_1 corresponds to a pair of pure imaginary nonzero eigenvalues of $L_0(\lambda)$ of opposite signs.   □

We make use of the following integral representations due to [M] (p. 32):

$$
\begin{aligned}
s(\lambda, x) &= \frac{\sin \lambda x}{\lambda} + \int_0^x K(x, t) \frac{\sin \lambda t}{\lambda} dt
\end{aligned}
$$

(2.1)

$$
= \frac{\sin \lambda x}{\lambda} - \frac{K(x, x) \cos \lambda x}{\lambda^2} + \int_0^x K_t(x, t) \frac{\cos \lambda t}{\lambda^2} dt,
$$

(2.2) $\quad s'(\lambda, x) = \cos \lambda x + K(x, x) \dfrac{\sin \lambda x}{\lambda} + \displaystyle\int_0^x K_x(x, t) \dfrac{\sin \lambda t}{\lambda} dt,$

$$
c(\lambda, x) = \cos \lambda x + \int_0^x G(x, t) \cos \lambda t \, dt
$$

(2.3)

$$
= \cos \lambda x + G(x, x) \frac{\sin \lambda x}{\lambda} - \int_0^x G_t(x, t) \frac{\sin \lambda x}{\lambda} dt,
$$

where

(2.4) $$K(x, t) = R(x, t) - R(x, -t),$$

(2.5) $$G(x, t) = R(x, t) + R(x, -t)$$

and $R(x, t)$ is the unique solution of the following integral equation

$$
R(x, t) = \frac{1}{2} \int_0^{\frac{x+t}{2}} q_1(\alpha) d\alpha
$$

(2.6)

$$
+ \int_0^{\frac{x+t}{2}} d\alpha \int_0^{\frac{x-t}{2}} q_1(\alpha + \beta) R(\alpha + \beta, \alpha - \beta) d\beta.
$$

Substituting (2.1)–(2.3) into (1.7) we obtain

$$e(\lambda) = 2 - 2\cos\lambda a - 2K(a,a)\frac{\sin\lambda a}{\lambda}$$

(2.7)
$$+ \int_0^a (G_t(a,t) - K_x(a,t))\frac{\sin\lambda t}{\lambda}dt - i$$

$$\left(\sin\lambda a - K(a,a)\frac{\cos\lambda a}{\lambda} + \int_0^a K_t(a,t)\frac{\cos\lambda t}{\lambda}dt\right).$$

Theorem 2.2 *The spectrum of problem (1.1), (1.8), (1.10) can be arranged into two subsequences $\{\lambda_k^{(1)}\}_{-\infty}^\infty$ and $\{\lambda_k^{(2)}\}_{-\infty}^\infty$ in such a way that $\lambda_{-k}^{(j)} = -\overline{\lambda_k^{(j)}}$ for all λ_k not pure imaginary and*

(2.8)
$$\lambda_k^{(1)} \underset{k\to\infty}{=} \frac{2\pi k}{a} + \frac{\int_0^a q_1(x)dx}{2\pi k} + \frac{b_k^{(1)}}{k},$$

(2.9)
$$\lambda_k^{(2)} \underset{k\to\infty}{=} \frac{2\pi k}{a} + \frac{i\log 3}{a} + \frac{\int_0^a q_1(x)dx}{2\pi k} + \frac{b_k^{(2)}}{k},$$

where $\{b_k^{(j)}\}_{-\infty}^\infty \in l_2$, $(j = 1,2)$.

Proof: Let us consider problem (1.1), (1.8), (1.10) with $q(x) = q_1(x) = 0$ and denote the set of its eigenvalues by $\{\lambda_k^{(01)}\}_{-\infty}^\infty \bigcup \{\lambda_k^{(02)}\}_{-\infty}^\infty$. It is clear this spectrum coincides with the set of zeroes of the function

$$e_0(\lambda) \overset{def}{=} 2 - 2\cos\lambda a - i\sin\lambda a$$

(2.10)
$$= 2\sin\frac{\lambda a}{2}\left(2\sin\frac{\lambda a}{2} - i\cos\frac{\lambda a}{2}\right).$$

It is clear that under an appropriate enumeration

$$\lambda_k^{(01)} = \frac{2\pi k}{a},$$

$$\lambda_k^{(02)} = \frac{2\pi k}{a} + \frac{i\log 3}{a}.$$

We need now to prove an auxiliary lemma

Lemma 2.3 *The following formulae are true*

(2.11)
$$\lambda_k^{(1)} = \frac{2\pi k}{a} + o(1),$$

(2.12)
$$\lambda_k^{(2)} = \frac{2\pi k}{a} + \frac{i\log 3}{a} + o(1).$$

Proof: Suppose there exists a subsequence $\{\lambda_{k_m}^{(1)}\}$ of $\{\lambda_k^{(1)}\}_{-\infty}^{\infty}$ such that $\text{Im } \lambda_{k_m}^{(1)} \underset{m \to \infty}{\to} \infty$. Then from (2.7) we obtain

$$e(\lambda_{k_m}^{(1)}) - e^{-i\lambda_{k_m}^{(1)}a}\left(1 + \frac{i}{2}\right) = o(e^{|\text{Im } \lambda_{k_m}^{(1)}|a}),$$

what contradicts the identity $e(\lambda_{k_m}^{(1)}) = 0$. It means that $\text{Im } \lambda_k^{(1)}$ is bounded above. In the same way we can prove that $\sup_k \max\{|\text{Im } \lambda_k^{(1)}|, |\text{Im } \lambda_k^{(2)}|\} < M < \infty$. Then we compare (2.7) with (2.10) and obtain that there exist numbers $C > 0$ and $\epsilon > 0$ such that

$$|e(\lambda) - e_0(\lambda)| < \frac{C}{|\lambda|}$$

for all $\lambda \in \Theta$ where by Θ we denote the strip $\Theta = \{\lambda : |\text{Im } \lambda| \le M + \epsilon\}$. The function $e_0(\lambda)$ is periodic and hence for every $r \in (0, \epsilon)$ it is possible to find $d > 0$ such that the inequality

$$|e_0(\lambda)| > d$$

is valid for all $\lambda \in \Theta \backslash (\bigcup_k C_k^1 \bigcup_k C_k^2)$, where C_k^j are the circles of radii r with the centres at the points $\lambda = \lambda_k^{(0j)}$. Consequently, for all $\lambda = \{\lambda : \lambda \in \Theta \backslash (\bigcup_k C_k^1 \bigcup_k C_k^2), |\lambda| > \frac{C}{d}\}$ the following inequalities are valid

$$|e_0(\lambda)| > d > \frac{C}{|\lambda|} > |e(\lambda) - e_0(\lambda)|.$$

As $r > 0$ can be chosen arbitrary small we apply the Rouché theorem and obtain

$$\lambda_k^{(j)} = \lambda_k^{(0j)} + \Delta_k^{(j)} \quad (j = 1, 2),$$

where $\Delta_k^{(j)} \underset{k \to \infty}{=} o(1)$. $\square$

To finish the proof of Theorem 2.2 it is enough to substitute (2.11) and (2.12) into the equation $e(\lambda_k^{(j)}) = 0$ where $e(\lambda)$ is defined by (2.7) and to expand into power series. $\square$

Definition 2.4 [L]. An entire function $\omega(\lambda)$ which has no zeroes in the closed lower half-plane and satisfies the condition

$$\left|\frac{\omega(\lambda)}{\bar{\omega}(\lambda)}\right| < 1$$

for $\text{Im } \lambda > 0$ is said to be a function of the Hermite-Biehler class. Here by $\bar{\omega}(\lambda)$ we denote the function obtained from $\omega(\lambda)$ by replacing the coefficients in its Tailor series by their complex-conjugates.

Definition 2.5 [L]. An entire function $\omega(\lambda)$ which has no zeroes in the open lower half-plane and satisfies the condition

$$\left| \frac{\omega(\lambda)}{\overline{\omega}(\lambda)} \right| \leq 1$$

for $\operatorname{Im} \lambda > 0$ is said to be a function of the generalized Hermite-Biehler class.

Corollary 2.6 *If problem* (1.1), (1.8), (1.10) *has no eigenvalues in the open lower half-plane, i.e. if the operator A_1 is nonnegative, then the function $e(\lambda)$ belongs to the generalized Hermite-Biehler class.*

Proof: Due to (2.7) the function $e(\lambda)$ can be expressed via its zeroes in the following way

$$e(\lambda) = C \prod_{-\infty}^{\infty} \left(1 - \frac{\lambda}{\lambda_k^{(1)}} \right) \left(1 - \frac{\lambda}{\lambda_k^{(2)}} \right),$$

where C is a real constant. The function $e(\lambda)$ satisfies the condition

$$\left| \frac{e(\lambda)}{\overline{e}(\lambda)} \right| \leq 1$$

because every multiplier in the product satisfies it. $\square$

Consider the following auxiliary function $g(\lambda) \overset{def}{=} 4 - e(\lambda) = \chi_1(\lambda) + i\lambda s(\lambda, a)$, where $\chi_1(\lambda) \overset{def}{=} 4 - \chi(\lambda)$. The set of zeroes of $\chi_1(\lambda)$ coincides with the spectrum of the antiperiodic problem generated by equation (1.1) and the following boundary conditions

$$y(0) = -y(a),$$
$$y'(0) = -y'(a).$$

The set of zeroes of the function $g(\lambda)$ coincides with the spectrum of the following boundary problem

$$y'' + (\lambda^2 - q_1(x))y = 0,$$

(2.13) $$y(0) = -y(a),$$

(2.14) $$y'(0) = -y'(a) - i\lambda y(a).$$

Theorem 2.7 1. *All the eigenvalues of problem* (1.1), (2.13), (2.14) *located in the open lower half-plane are pure imaginary.*

   2. *All of them are semisimple.*

3. *Their total algebraic (or geometric) multiplicity is equal to the total algebraic (or geometric) multiplicity of the negative spectrum of the following operator acting in $L_2(0, a)$: $A_3y = -y'' + q_1(x)y$, $D(A_3) = \{y : y \in W_2^2(0, a), y(0) = -y(a), y'(0) = -y'(a)\}$.*

The proof of this theorem is quite the same that of Theorem 2.1.

Theorem 2.8 *The spectrum of problem (1.1), (2.13), (2.14), i.e. the set of zeroes of $g(\lambda) = 4 - e(\lambda)$ can be arranged into two subsequences $\{\zeta_k^{(1)}\}_{-\infty, k \neq 0}^{\infty}$ and $\{\zeta_k^{(2)}\}_{-\infty, k \neq 0}^{\infty}$ which behave asymptotically as follows*

$$\zeta_k^{(1)} \underset{k \to \infty}{=} \frac{\pi(2k-1)}{a} + \frac{\int_0^a q_1(x)dx}{2\pi k} + \frac{\kappa_k^{(1)}}{k},$$

$$\zeta_k^{(2)} \underset{k \to \infty}{=} \frac{\pi(2k-1)}{a} + i\frac{\log 3}{a} + \frac{\int_0^a q_1(x)dx}{2\pi k} + \frac{\kappa_k^{(2)}}{k},$$

and $\zeta_{-k}^{(j)} = -\overline{\zeta_k^{(j)}}$ for all not pure imaginary $\zeta_k^{(j)}$ and $\{\kappa_k^{(j)}\} \in l_2$, $(j = 1, 2)$

The proof of this theorem is the same as that of Theorem 2.2.

Corollary 2.9 *If $A_3 \geq 0$, then the function $g(\lambda)$ belongs to the generalized Hermite-Biehler class.*

This statement follows from Theorems 2.7 and 2.8.

Theorem 2.10 *Let $q_1(x) \in W_2^3(0, a)$. Then*

$$
\begin{aligned}
e(\lambda) = {} & 2 - 2\cos\lambda a - i\sin\lambda a - 2K\frac{\sin\lambda a}{\lambda} + iK\frac{\cos\lambda a}{\lambda} \\
& + K^2\frac{\cos\lambda a}{\lambda^2} + iB\frac{\sin\lambda a}{\lambda^2} + F\frac{\sin\lambda a}{\lambda^3} + iE\frac{\cos\lambda a}{\lambda^3} \\
& + J\frac{\cos\lambda a}{\lambda^4} + iT\frac{\sin\lambda a}{\lambda^4} + \frac{\psi(\lambda)}{\lambda^4},
\end{aligned}
$$

(2.15)

where $K = K(a, a) = G(a, a) = \frac{1}{2}\int_0^a q_1(x)dx$, $B \in \mathbb{R}$, $F \in \mathbb{R}$, $E \in \mathbb{R}$, $J \in \mathbb{R}$, $T \in \mathbb{R}$, $\psi(\lambda) \in \mathcal{L}_a$, ($\mathcal{L}_a$ is the class of entire functions of exponential type $\leq a$ which belong to $L_2(-\infty, \infty)$ when $\lambda \in \mathbb{R}$) and

(2.16) $K^4 - FK - J \geq 0.$

Proof: If $q_1(x) \in W_2^m(0, a)$ then [M] (Theorem 1.2.2) the function $R(x, t)$ and, consequently, the kernels $K(x, t)$ and $G(x, t)$ possess all the partial derivatives up to $m + 1$ order inclusive which belong to $L_2(0, a)$. Hence, if $q_1(x) \in W_2^3(0, a)$ then we can integrate three times per parts in (2.1)–(2.3) and (2.7) and obtain

$$
\begin{aligned}
s(\lambda, a) \;=\;& \frac{\sin \lambda a}{\lambda} - K \frac{\cos \lambda a}{\lambda^2} + K_t(a, a) \frac{\sin \lambda a}{\lambda^3} \\
(2.17) \qquad & + K_{tt}(a, a) \frac{\cos \lambda a}{\lambda^4} - K_{ttt}(a, a) \frac{\sin \lambda a}{\lambda^5} \\
& + \int_0^a K_{tttt}(a, t) \frac{\sin \lambda t}{\lambda^5},
\end{aligned}
$$

$$
\begin{aligned}
s'(\lambda, a) \;=\;& \cos \lambda a + K \frac{\sin \lambda a}{\lambda} - K_x(a, a) \frac{\cos \lambda a}{\lambda^2} \\
(2.18) \qquad & + K_{xt}(a, a) \frac{\sin \lambda a}{\lambda^3} + K_{xtt}(a, a) \frac{\cos \lambda a}{\lambda^4} \\
& - \int_0^a K_{xttt}(a, t) \frac{\cos \lambda t}{\lambda^4} dt,
\end{aligned}
$$

$$
\begin{aligned}
c(\lambda, a) \;=\;& \cos \lambda a + K \frac{\sin \lambda a}{\lambda} + G_t(a, a) \frac{\cos \lambda a}{\lambda^2} \\
(2.19) \qquad & - G_{tt}(a, a) \frac{\sin \lambda a}{\lambda^3} - G_{ttt}(a, a) \frac{\cos \lambda a}{\lambda^4} \\
& + \int_0^a G_{tttt}(a, t) \frac{\cos \lambda t}{\lambda^4} dt,
\end{aligned}
$$

$$
\begin{aligned}
e(\lambda) \;=\;& 2 - 2\cos \lambda a - i \sin \lambda a - 2K \frac{\sin \lambda a}{\lambda} \\
& + (K_x(a, a) - G_t(a, a)) \frac{\cos \lambda a}{\lambda^2} + iK \frac{\cos \lambda a}{\lambda} \\
& - iK_t(a, a) \frac{\sin \lambda a}{\lambda^2} + (G_{tt}(a, a) - K_{xt}(a, a)) \frac{\sin \lambda a}{\lambda^3} \\
(2.20) \qquad & + (G_{ttt}(a, a) - K_{xtt}(a, a)) \frac{\cos \lambda a}{\lambda^4} \\
& - iK_{tt}(a, a) \frac{\cos \lambda a}{\lambda^3} - iK_{ttt}(a, a) \frac{\sin \lambda a}{\lambda^4} \\
& - \int_0^a (G_{tttt}(a, t) - K_{xttt}(a, t)) \frac{\cos \lambda t}{\lambda^4} dt \\
& - i \int_0^a K_{tttt}(a, t) \frac{\sin \lambda t}{\lambda^4} dt.
\end{aligned}
$$

Here following identities are taken into account which can be deduced from (2.4)–(2.6): $K(x, 0) = K_{tt}(x, 0) = K_x(x, 0) = K_{xtt}(x, 0) = G_t(x, 0) = G_{ttt}(x, 0) = 0$.

Using (2.4)–(2.6) we obtain

$$K_x(x,t) - G_t(x,t) = \int_0^{\frac{x+t}{2}} d\alpha q_1\left(\alpha + \frac{x-t}{2}\right)$$

$$R\left(\alpha + \frac{x-t}{2}, \alpha - \frac{x-t}{2}\right) - \int_0^{\frac{x-t}{2}} d\alpha q_1\left(\alpha + \frac{x+t}{2}\right)$$

$$R\left(\alpha + \frac{x+t}{2}, \alpha - \frac{x+t}{2}\right)$$

and, consequently,

$$K_x(a,a) - G_t(a,a) = \int_0^a d\alpha q_1(\alpha) R(\alpha,\alpha)$$

(2.21)

$$= \frac{1}{2} \int_0^a d\alpha \int_0^\alpha d\beta q_1(\beta) = \frac{1}{4}\left(\int_0^a d\alpha q_1(\alpha)\right)^2 = K^2.$$

It is clear that

$$\int_0^a (-G_{ttt}(a,t) + K_{xttt}(a,t)) \sin\lambda t\, dt - i$$

(2.22)

$$\int_0^a K_{tttt}(a,t)\sin\lambda t\, dt \in \mathcal{L}_a.$$

Formulae (2.20)–(2.22) imply (2.15) where

$$B = -K_t(a,a), \quad F = -K_{xt}(a,a) + G_{tt}(a,a), \quad E = -K_{tt}(a,a),$$
$$J = G_{ttt}(a,a) - K_{xtt}(a,a), \quad T = -K_{ttt}(a,a).$$

Now we prove inequality (2.16). Let us set $\lambda = \nu_k$ into the identity

$$s'(\lambda,a)c(\lambda,a) - s(\lambda,a)c'(\lambda,a) = 1.$$

Then we obtain

(2.23) $$s'(\nu_k,a) = c^{-1}(\nu_k,a)$$

for all $\pm k \in \mathbb{N}$. Setting $\lambda = \nu_k$ into the identity

$$s'(\lambda,a) + c(\lambda,a) = 2 - \frac{e(\lambda) + e(-\lambda)}{2}$$

and using (2.15) and (2.23) we obtain

$$c(\nu_k,a) + c^{-1}(\nu_k,a) = 2\cos\nu_k a + 2K\frac{\sin\nu_k a}{\nu_k} - K^2\frac{\cos\nu_k a}{\nu_k^2}$$

(2.24)

$$- F\frac{\sin\nu_k a}{\nu_k^3} - J\frac{\cos\nu_k a}{\nu_k^4} + \frac{\psi_1(\nu_k)}{\nu_k^4},$$

where $\psi_1(\lambda) = \frac{-\psi(\lambda) - \psi(-\lambda)}{2} \in \mathcal{L}_a$. Now we make use of the following formula (see [M], Sec.3.4)

$$(2.25) \qquad v_k = \frac{\pi k}{a} + \frac{K}{\pi k} + \frac{Q}{k^3} + o(k^{-3}),$$

where $Q \in \mathbb{R}$. Substituting (2.19) and (2.25) into (2.24) and expanding into power series in k^{-1} up to the fourth order inclusive we obtain

$$(2.26) \qquad (G_t(a, a) + K)^2 - K^4 + FK + J = o(1)$$

and (2.16) follows. $\qquad\qquad\qquad\qquad\qquad\qquad\qquad\qquad\qquad\qquad\qquad\qquad\square$

3 Inverse Problem

Now we deal with the problem of recovering $q_1(x)$ from the scattering data.

Theorem 3.1 *Let the following conditions be satisfied:*

1) $e(\lambda) = \overline{e(-\bar{\lambda})}$ *is an entire function of exponential type a of the form* (2.15), *where* $K \in \mathbb{R}$, $B \in \mathbb{R}$, $F \in \mathbb{R}$, $J \in \mathbb{R}$, $T \in \mathbb{R}$, $\psi(\lambda) \in \mathcal{L}_a$;

2) *the functions* $e(\lambda)$ *and* $4 - e(\lambda)$ *belong to the generalized Hermite-Biehler class. Then there exists a real-valued* $q_1(x) \in W_2^1(0, a)$ *which generates problem* (1.1)–(1.4) *with the S-matrix*

$$(3.1) \qquad S(\lambda) = e^{2i\lambda a} \frac{e(\lambda)}{e(-\lambda)}.$$

Proof: Due to (2.15)

$$(3.2) \qquad \begin{aligned} \tilde{\chi}(\lambda) &\overset{def}{=} \frac{e(\lambda) + e(-\lambda)}{2} = 2 - 2\cos\lambda a - 2K\frac{\sin\lambda a}{\lambda} \\ &+ K^2 \frac{\cos\lambda a}{\lambda^2} + F\frac{\sin\lambda a}{\lambda^3} + J\frac{\cos\lambda a}{\lambda^4} + \frac{\psi_1(\lambda)}{\lambda^4}, \end{aligned}$$

where $\psi_1(\lambda) \overset{def}{=} \frac{\psi(\lambda) + \psi(\lambda)}{2} \in \mathcal{L}_a$, and $\psi_1(\lambda)$ is real-valued for $\lambda \in \mathbb{R}$. Let us set

$$(3.3) \qquad \begin{aligned} \tilde{\chi}_2(\lambda) &\overset{def}{=} \frac{e(\lambda) - e(-\lambda)}{2i} = -\sin\lambda a + K\frac{\cos\lambda a}{\lambda} + B\frac{\sin\lambda a}{\lambda^2} \\ &+ E\frac{\cos\lambda a}{\lambda^3} + T\frac{\sin\lambda a}{\lambda^4} + \frac{\psi_2(\lambda)}{\lambda^4}, \end{aligned}$$

where $\psi_2(\lambda) \overset{def}{=} \frac{\psi(\lambda) - \psi(-\lambda)}{2i} \in \mathcal{L}_a$ is real valued for $\lambda \in \mathbb{R}$.

Let us denote by $\{\nu_k\}_{-\infty}^{\infty}$ the set of zeroes of $\tilde{\chi}_2(\lambda)$ and by $\{\mu_k\}_{-\infty, k\neq 0}^{\infty}$ the set of zeroes of $\tilde{\chi}(\lambda)$. It is clear that $\{\nu_k\}_{-\infty}^{\infty}$ behave asymptotically according to (2.25). Let us denote by

(3.4)
$$\tilde{\chi}_1(\lambda) \overset{def}{=} 4 - \tilde{\chi}(\lambda) = 2 + 2\cos\lambda a + 2K\frac{\sin\lambda a}{\lambda} - K^2\frac{\cos\lambda a}{\lambda^2}$$
$$-F\frac{\sin\lambda a}{\lambda^3} - J\frac{\cos\lambda a}{\lambda^4} - \frac{\psi_1(\lambda)}{\lambda^4}.$$

To continue the proof we need the following auxiliary propositions

Proposition 3.2 *The following inequalities are valid*

(3.5)
$$\tilde{\chi}(\nu_{2k}) \leq 0, \quad \tilde{\chi}_1(\nu_{2k-1}) \leq 0$$

for all $\pm k \in \mathbb{N}\bigcup\{\nVdash\}$.

Proof: As $e(\lambda) = \tilde{\chi}(\lambda) + i\tilde{\chi}_2(\lambda)$ belongs to the generalized Hermite-Biehler class $\{\nu_k\}_{-\infty}^{\infty}$ and $\{\mu_k\}$ interlace in the following meaning

(3.6)
$$\cdots \leq \nu_{-1} \leq \mu_{-1} \leq \nu_0 = 0 \leq \mu_1 \leq \nu_1 \leq \cdots \leq \mu_k \leq \nu_k \leq \cdots$$

This statement follows from [L] (Sec.7.2, Theorem 3'). The sequence $\{\nu_k\}_{-\infty}^{\infty}$ satisfies (2.25) (see [M], Lemma 3.4.2). If $K^4 - FK - J > 0$, then substituting (2.25) into (3.2) and (3.4) we obtain (3.5) for k large enough. Then taking into account (3.6) we conclude that (3.5) is valid for all k. If $K^4 - FK - J = 0$ then Proposition 3.2 can be proved as follows. The zeroes ν_k and μ_k are piecewise analytic functions of J (or any of the other parameters) and if $\tilde{\chi}(\nu_{2k}) > 0$ for some k and some J, then μ_{2k-1} and μ_{2k+1} are complex what contradicts (3.6). Because of the same reasons $K^4 - FK - J$ cannot be less then 0.   □

This proposition implies the following inequalities

(3.7)
$$\Theta_{2k} \overset{def}{=} 2\cos\nu_{2k}a + 2K\frac{\sin\nu_{2k}a}{\nu_{2k}} - K^2\frac{\cos\nu_{2k}a}{\nu_{2k}^2}$$
$$-F\frac{\sin\nu_{2k}a}{\nu_{2k}^3} - J\frac{\cos\nu_{2k}a}{\nu_{2k}^4} - \frac{\psi_1(\nu_{2k})}{\nu_{2k}^4} \geq 2$$

and

(3.8)
$$\Theta_{2k-1} \overset{def}{=} 2\cos\nu_{2k-1}a + 2K\frac{\sin\nu_{2k-1}a}{\nu_{2k-1}}$$
$$-K^2\frac{\cos\nu_{2k-1}a}{\nu_{2k-1}^2} - F\frac{\sin\nu_{2k-1}a}{\nu_{2k-1}^3} - J\frac{\cos\nu_{2k-1}a}{\nu_{2k-1}^4}$$
$$-\frac{\psi_1(\nu_{2k-1})}{\nu_{2k-1}^4} \leq -2.$$

This means that for every k the equation

$$(3.9) \qquad\qquad X_k + X_k^{-1} = \Theta_k$$

possesses two real solutions $X_k^{(1)}$ and $X_k^{(2)}$ which can coincide. It follows from (3.7) and (3.8) that

$$(3.10) \qquad\qquad X_{2k}^{(j)} > 0,$$

$$(3.11) \qquad\qquad X_{2k-1}^{(j)} < 0$$

for all k and $j = 1, 2$.

Proposition 3.3 *The set of solutions* $\{X_k^{(1)}\}_{-\infty}^{\infty}$ *of the equations* (3.9) *can be chosen in such a way that*

$$(3.12) \qquad X_k^{(1)} = \cos v_k a + K \frac{\sin v_k a}{v_k} + \frac{Y^{(1)} \cos v_k a}{v_k^2} + \frac{b_k}{v_k^2},$$

where $Y^{(1)}$ *is any of the two solutions (both real and maybe equal) of the equation*

$$(3.13) \qquad\qquad (Y^{(1)} + K)^2 - K^4 + FK + J = 0$$

and $\{b_k\}_{-\infty}^{\infty} \in l_2$.

Proof: Let us substitute (3.12) into (3.9) where Θ_k is defined by (3.7) and (3.8) and expand into power series. Then we obtain

$$(3.14) \qquad (Y^{(1)} + K + b_k)^2 - K^4 + FK + J + \psi_1(v_k) = O(k^{-1}).$$

As $\psi_1(v_k) \in l_2$ (see Lemma 1.4.3 of [M]) the statement of Proposition 3.3 follows. $\qquad\square$

Let us construct the following function

$$(3.15) \qquad \phi(\lambda) \overset{def}{=} \tilde{\chi}_2(\lambda) \sum_{-\infty, k \neq 0}^{\infty} \frac{b_k}{\frac{d\tilde{\chi}_2(\lambda)}{d\lambda}\big|_{\lambda = v_k} (\lambda - v_k)}.$$

The interpolation series in the right hand side of (3.15) converges uniformly on any compact of $\mathbb{C}$ and in the norm of $L_2(-\infty, \infty)$ for $\lambda \in \mathbb{R}$ and $\phi(\lambda) \in \mathcal{L}_a$ [LL]. Let us introduce the following function

$$(3.16) \qquad \tilde{\chi}_3(\lambda) \overset{def}{=} \cos \lambda a + K \frac{\sin \lambda a}{\lambda} + \frac{Y^{(1)} \cos \lambda a}{\lambda^2} + \frac{\phi(\lambda)}{\lambda^2}.$$

Proposition 3.4 *The set* $\{\tau_k\}_{-\infty,k\neq0}^{\infty}$ *of zeroes of* $\tilde{\chi}_3(\lambda)$ *and the set* $\{v_k\}_{-\infty}^{\infty}$
interlace in the following meaning

$$\cdots < v_{-1} < \tau_{-1} < v_0 = 0 < \tau_1 < v_1 < \tau_2 < v_2 < \cdots$$

Proof: We compare (3.16) with (3.12) and obtain taking into account (3.15)

(3.17) $$\tilde{\chi}_3(v_k) = X_k^{(1)}.$$

Then due to (3.10) and (3.11) we obtain

(3.18) $$\tilde{\chi}_3(v_{2k}) > 0,$$

(3.19) $$\tilde{\chi}_3(v_{2k-1}) < 0,$$

for all $k \in \{0, \pm1, \pm2 \ldots\}$. The statement of Proposition 3.4 follows. $\qquad\square$

Proposition 3.5 [M] (Sec.3.4).

(3.20) $$\tau_k = \frac{\pi(k - \frac{1}{2})}{a} + \frac{K}{\pi k} + \frac{\gamma_k}{k^2},$$

where $\{\gamma_k\}_{-\infty,k\neq0}^{\infty} \in l_2$.

Propositions 3.4 and 3.5 mean that the two sequences $\{v_k\}_{-\infty,k\neq0}^{\infty}$ and $\{\tau_k\}_{-\infty,k\neq0}^{\infty}$ satisfy the conditions of Theorem 3.4.1 of [M] and consequently there exists a unique $q_1(x) \in W_2^1(0, a)$ such that: 1) the sequence $\{v_k\}_{-\infty,k\neq0}^{\infty}$ coincides with the spectrum of the problem generated by (1.1) and the Dirichlet boundary conditions $y(\lambda, 0) = y(\lambda, a) = 0$; 2) the sequence $\{\tau_k\}_{-\infty,k\neq0}^{\infty}$ coincides with the spectrum of the problem generated by equation (1.1) and the Dirichlet-Neumann boundary conditions $y(\lambda, 0) = y'(\lambda, a) = 0$. Because of the condition $|\tilde{\chi}_3(\lambda) - \cos \lambda a| \underset{\lambda\to\infty}{=} o(1)$ the set $\{\tau_k\}_{-\infty,k\neq0}^{\infty}$ determine uniquely the function $\tilde{\chi}_3(\lambda)$ via the formula

$$\tilde{\chi}_3(\lambda) = \prod_{k=1}^{\infty} \left(k - \frac{1}{2}\right)^{-2} \left(\frac{a^2(\tau_k^2 - \lambda^2)}{\pi^2}\right).$$

Then the function

$$\tilde{e}_1(\lambda) \overset{def}{=} (\tilde{\chi}_3(\lambda) - i\tilde{\chi}_2(\lambda)) e^{-i\lambda a}$$

is the Jost-function of the following boundary problem on the half-axis

$$y'' + (\lambda^2 - q(x))y = 0,$$
$$y(\lambda, 0) = 0$$

with

$$q(x) = \begin{cases} q_1(x) \in W_2^1(0, x), & \text{if } x \in (0, a), \\ 0, & \text{if } x \in [0, \infty) \end{cases}.$$

Proposition 3.6 *The function $\tilde{e}_1(\lambda)$ belongs to the Hermite-Biehler class.*

Proof: Due to (3.3), (3.16) and Proposition 3.4 the function $\tilde{e}_1(\lambda)$ satisfies the conditions of N.N. Meiman's theorem (see Theorem 3 of Sec.6.2 of [L]). $\qquad\square$

The above mentioned potential $q_1(x)$ can be recovered in the following way. Let us denote by

$$S_1(\lambda) = \frac{\tilde{e}_1(\lambda)}{\tilde{e}_1(-\lambda)} = \frac{\tilde{\chi}_3(\lambda) - i\tilde{\chi}_2(\lambda)}{\tilde{\chi}_3(\lambda) + i\tilde{\chi}_2(\lambda)} e^{-2i\lambda a}.$$

This function is the S-matrix of the problem generated be equation (1.1) and the boundary condition $y(\lambda, 0) = 0$ on the semi-axis $[0, \infty)$ with the finite potential of the form (1.4). Now following [M] we construct the function

$$F(x) = \frac{1}{2\pi} \int_{-\infty}^{\infty} (1 - S_1(\lambda)) e^{-i\lambda x} d\lambda.$$

As $S_1(\lambda)$ is meromorphic and $\tilde{e}_1(\lambda)$ belongs to the generalized Hermite-Biehler class the kernel $F(x)$ does not include the part corresponding to the discrete spectrum. The Marchenko integral equation

$$M(x, t) + F(x + t) + \int_{x}^{\infty} M(x, v) F(v + t) dv = 0.$$

possesses a unique solution. The potential can be recovered from this solution via the formula

$$q_1(x) = -2 \frac{dM(x, x)}{dx}.$$

Now we prove that the constructed potential generates problem (1.1)–(1.4) which possesses the initial Jost function $e(\lambda)$. Let us construct the following function

$$(3.21) \quad \tilde{e}(\lambda) \overset{def}{=} 2 - 2\cos\lambda a - 2K\frac{\sin\lambda a}{\lambda} + K^2\frac{\cos\lambda a}{\lambda^2} - i\tilde{\chi}_2(\lambda) - \frac{\varphi_1(\lambda)}{\lambda^2},$$

where

$$(3.22) \quad \varphi_1(\lambda) \overset{def}{=} \tilde{\chi}_2(\lambda) \sum_{-\infty, k\neq 0}^{\infty} \frac{h_k}{\frac{d\tilde{\chi}_2(\lambda)}{d\lambda}|_{\lambda=v_k} (\lambda - v_k)}$$

and

$$(3.23) \quad h_k = v_k^2\left(\tilde{\chi}_3^{-1}(v_k) + \tilde{\chi}_3(v_k) - 2\cos v_k a - 2K\frac{\sin v_k a}{v_k} + K^2\frac{\cos v_k a}{v_k^2}\right),$$

where

$$K = \lim_{k \to \infty} 2\pi k \tilde{\chi}_2(2\pi k).$$

Proposition 3.7

$$\{h_k\}_{-\infty}^{\infty} \in l_2$$

and consequently

$$\varphi_1(\lambda) \in \mathcal{L}_a.$$

Proof: To prove the first statement it is sufficient to substitute (3.16) into (3.23) and use (2.25) and to expand in power series. The second statement follows from the first one (see [LL]).   □

Definitions (3.2) and (3.3) imply

$$e(\nu_k) = \tilde{\chi}(\nu_k).$$

Using formulae (3.7)–(3.9) and (3.17) we obtain

(3.24)
$$\begin{aligned}
e(\nu_k) &= 2 - \Theta_k = 2 - X_k^{(1)} - (X_k^{(1)})^{-1} \\
&= 2 - \tilde{\chi}_3(\nu_k) - (\tilde{\chi}_3(\nu_k))^{-1}.
\end{aligned}$$

Due to (3.24) and (3.23) we obtain

(3.25)
$$\begin{aligned}
\nu_k^2 &\left(e(\nu_k) - 2 + 2\cos \nu_k a + 2K \frac{\sin \nu_k a}{\nu_k} - K^2 \frac{\cos \nu_k a}{\nu_k^2} \right) \\
&= \varphi_1(\nu_k) = h_k.
\end{aligned}$$

for all k. As $\varphi_1(\lambda) \in \mathcal{L}_a$ and the formula (3.22) establishes one-to-one correspondence between l_2 and $\mathcal{L}_a$ (see [LL]) we conclude from (3.25) that

$$e(\lambda) = \tilde{e}(\lambda).$$

Theorem 3.1 is proved.   □

Remark 3.8 In general the constructed potential is not unique because the ambiguity of the choice of the solutions $X_k^{(1)}$ of equations (3.9) and of the solutions $Y^{(1)}$ of (3.13) (if $Y^{(1)} \neq Y^{(2)}$).

Remark 3.9 It was the unknown referee who directed our attention to the fact that the Assertions 1 and 2 of 2.1 as well as the Assertions 1 and 2 of 2.7 immediately follow from the Lax-Phillips approach.

Acknowledgements

The research described in this publication was made possible in part by Award No.UM1-298 of the Government of Ukraine and U.S. Civilian Research and Development Foundation for the Independent States of the Former Soviet Union.

References

[A]   V. Adamyan, Scattering matrices for Microschemes, *Operator Theory: Advances and Appl.* **59** (1992), 1–10.

[AP]   V.M. Adamyan and B.S. Pavlov, Null-Range Potentials and M.G. Krein's Formula of Generalized Resolvents (Russian), *Studies on Linear Operators of Functions. IV. Research Notes of Scientific Seminar of the LBMI* **149** (1986), 7–23.

[ES]   P. Exner and P. Seba, A New Type of Quantum Interference Transistor, *Phys. Lett.* **A 129:8, 9** (1988), 477–480.

[G]   N.I. Gerasimenko, Inverse Scattering Problem on Noncompact Graph (Russian), *Teoreticheskaya i matematicheskaya fisika* **75:2** (1988), 187–200.

[GP]   N.I. Gerasimenko and B.S. Pavlov, Scattering Problem on Noncompact Graphs (Russian), *Teoreticheskaya i matematicheskaya fisika* **74:3** (1988), 345–359.

[GK]   I.C. Gohberg and M.G. Krein, Introduction to the Theory of Linear Nonselfadjoint Operators, *Amer. Math. Soc.*, Providence, 1969.

[L]   B.Ya. Levin, *Lectures on Entire Functions*, Translations of Mathematical Monographs, AMS **150** (1996).

[LL]   B.Ya. Levin and Yu.I. Lyubarskii, Interpolation by Entire Functions of Special Classes and Related Expansions in Series of Exponents (Russian), *Izv. Akad. Nauk USSR, Sec. Mat.* **43** (1979), 87–110.

[M]   V.A. Marchenko, *Sturm-Liouville Operators and Applications*, Birkhauser, OT **22** (1986).

[N]   R.G. Newton, *Scattering Theory of Waves and Particles*, McGraw-Hill, New York 1966.

[P]   V. Pivovarchik, On Positive Spectra of One Class of Polynomial Operator Pencils, *Integral Equations and Operator Theory* **19** (1994), 314–326.

[R]   T. Regge, Construction of Potential from Resonances, *Nuovo Cimento* **9:3, 5** (1958), 491–503, 671–679.

Vyacheslav Pivovarchik
Preobrajenskaya str. 59/61
a.17, 270045
Odessa
Ukraine

1991 Mathematics Subject Classification. Primary: 34A55, 34B24; Secondary: 34L20, 34C25.

Operator Theory:
Advances and Applications, Vol. 124
© 2001 Birkhäuser Verlag Basel/Switzerland

Pseudospectra of Operator Polynomials

Steffen Roch

Dedicated to Professor Israel C. Gohberg on the occasion of his 70th birthday

The recent interest in ε-pseudospectra of operators results from their (in comparison with usual spectra) excellent continuity properties. The goal of the present paper is to introduce and to examine ε-pseudospectra of operator polynomials with main emphasis on the continuity aspect.

1 Introduction

Let H be a separable Hilbert space, $L(H)$ the C^*-algebra of all linear and bounded operators on H, and $A \in L(H)$ an operator on H with matrix representation $(a_{ij})_{i,j=1}^{\infty}$ with respect to some fixed basis of H. The nth finite section $(a_{ij})_{i,j=1}^{n}$ of A will be denoted by A_n. What can be said about the relation between the spectrum $\sigma(A)$ of the operator A and the eigenvalues of the matrix A_n as n tends to infinity?

In case the operator A is self-adjoint and has a connected spectrum, the answer is affirmative: In this case the spectra of A_n converge to the spectrum of A in the sense that

$$\lim \sigma_{\mathbb{C}^{n \times n}}(A_n) = \sigma_{L(H)}(A)$$

where here and hereafter the following notation is used: Given a sequence (M_n) of subsets of the complex plane, let $\lim M_n$ denote the set of all *partial* limits of sequences (m_n) of points $m_n \in M_n$. On the other hand, this convergence can fail drastically in case the operator A is not normal. To have at least one (trivial) example, consider the shift operator $A = (a_{ij})_{i,j=1}^{\infty}$ where $a_{ij} = 1$ if $i - j = 1$ and $a_{ij} = 0$ otherwise. The spectrum of A is the closed unit disk $\{z \in \mathbb{C} : |z| \leq 1\}$, whereas the spectrum of every of its finite sections A_n consists of the point 0 only.

Reichel and Trefethen [8, 12, 13] observed that, for several classes of operators, the convergence behaviour becomes much better if the usual spectrum is replaced by a certain approximate spectrum which was introduced by Landau [6, 7], and which is known as the ε-pseudospectrum. By definition, given $\varepsilon > 0$, the ε-pseudospectrum $\sigma^{(\varepsilon)}(A)$ of the operator $A \in L(H)$ consists of all $\lambda \in \mathbb{C}$ such that either $A - \lambda I$ is not invertible, or it is invertible, but $\|(A - \lambda I)^{-1}\| \geq 1/\varepsilon$. Analogously one defines the ε-pseudospectrum of an element of a C^*-algebra with identity.

A computer working with finite accuracy cannot distinguish between a non-invertible matrix and an invertible matrix the inverse of which has a very large norm. Thus, the definition of pseudospectra reflects in some sense that finite accuracy.

The usual spectrum follows from its ε-counterpart by letting ε formally go to zero. More precise:

$$\sigma(A) = \cap_{\varepsilon > 0} \sigma^{(\varepsilon)}(A).$$

As usual spectra, ε-pseudospectra are compact and non-empty subsets of the complex plane. Also the upper semi-continuity of spectra has its analogue for pseudospectra: If $(A_n) \subseteq L(H)$ is a norm-convergent sequence of operators, then

$$\lim \sigma^{(\varepsilon)}(A_n) \subseteq \sigma^{(\varepsilon)}(\lim A_n),$$

which can be easily checked.

The results of Reichel and Trefethen, which had been extended later by Böttcher [1], concern pseudospectra of Toeplitz operators and of their finite sections. Given a function $a \in L^\infty(\mathbb{T})$ with Fourier coefficients a_k, $k \in \mathbb{Z}$, the Toeplitz operator $T(a)$ generated by a is given by the matrix $(a_{i-j})_{i,j=1}^\infty$ thought of as acting on the Hilbert space l^2 provided with its standard basis. In case the function a belongs to the Wiener algebra, i.e. if $\sum |a_k| < \infty$, Reichel and Trefethen verified that

(1.1)
$$\lim_{n \to \infty} \sigma^{(\varepsilon)}(T_n(a)) = \sigma^{(\varepsilon)}(T(a))$$

for every $\varepsilon > 0$. Böttcher extended this result to the case of arbitrary piece-wise continuous functions a. A quite general approach to the study of the asymptotic behaviour of pseudospectra, which bases essentially on Böttcher's ideas, was proposed in [9].

The goal of the present paper is to extend results such as (1.1) to the case of operator polynomials. Thereby, the general approach of [9] will serve as a guide.

2 Pseudospectra of Operator Polynomials

The following considerations are not restricted to operators on a Hilbert space; so let B be an arbitrary C^*-algebra with identity e. Given elements $b_0, b_1, \ldots, b_m$ of B, the expression

$$L(\lambda) := b_0 + b_1\lambda + \cdots + b_m\lambda^m$$

is well-defined for every complex number λ, and L is (not quite correctly) called an *operator polynomial* with coefficients in B. The *spectrum* of L is the set $\sigma(L)$ of all $\lambda \in \mathbb{C}$ for which $L(\lambda)$ is not invertible; similarly, for $\varepsilon > 0$, the *ε-pseudospectrum* of L is the set $\sigma^{(\varepsilon)}(L)$ of all $\lambda \in \mathbb{C}$ such that either $L(\lambda)$ is not invertible, or $\|L(\lambda)^{-1}\| \geq 1/\varepsilon$. Evidently, if $L(\lambda) = b - \lambda e$, then the spectrum and the ε-pseudospectrum of the polynomial L coincide with the common spectrum and ε-pseudospectrum of the element $b \in \mathcal{B}$, respectively.

The following result establishes another characterization of pseudospectra of operator polynomials. Its proof follows that for the special case $L(\lambda) = b - \lambda e$ which belongs to Ehrhardt and Finck and is published in [9].

Theorem 2.1 *Let $\mathcal{B}$ be a unital C^*-algebra and $\varepsilon > 0$. Then, for every operator polynomial L with coefficients in $\mathcal{B}$,*

$$\sigma^{(\varepsilon)}(L) = \{\lambda \in \mathbb{C} : \text{there is a } p \in \mathcal{B}$$

(2.1)

$$\text{with } \|p\| \leq \varepsilon \text{ and } t \in \sigma(L + p)\}.$$

Clearly, $L + p$ refers to the polynomial $(L + p)(\lambda) = (b_0 + p) + b_1\lambda + \cdots + b_m\lambda^m$.

Proof: Abbreviate the sets on the left and right hand side of (2.1) by S_1 and S_2, respectively. The first claim is the inclusion $S_2 \subseteq S_1$.

Given $t \in S_2$, choose p in $\mathcal{B}$ with $\|p\| \leq \varepsilon$ such that $t \in \sigma(L + p)$. If t already lies in $\sigma(L)$, then it also belongs to S_1, and we are done. So suppose $t \notin \sigma(L)$. Then $L(t)$ is invertible, and the identity

$$L(t) + p = L(t)(e + L(t)^{-1}p)$$

yields that $e + L(t)^{-1}p$ cannot be invertible. Hence, $\|L(t)^{-1}p\| \geq 1$ (otherwise invertibility would follow via Neumann series). Because of

$$1 \leq \|L(t)^{-1}p\| \leq \|L(t)^{-1}\| \|p\|$$

one has

$$\|L(t)^{-1}\| \geq 1/\|p\| \geq 1/\varepsilon,$$

i.e. $t \in S_1$ in this case, too.

For the reverse inclusion $S_1 \subseteq S_2$, assume there is a $t \in S_1$ such that $L(t) + p$ is invertible for all $p \in \mathcal{B}$ with $\|p\| \leq \varepsilon$. Choosing $p = 0$, one gets the invertibility of $L(t)$ and, hence, of $L(t)^* := b_0^* + b_1^*\bar{t} + \cdots + b_n^*\bar{t}^n$ and, choosing $p = \lambda(L(t)^*)^{-1}$ with a certain complex number λ satisfying

(2.2)

$$0 < |\lambda| \leq \varepsilon/\|(L(t)^*)^{-1}\|,$$

one obtains invertibility of

$$L(t) + \lambda(L(t)^*)^{-1} = \lambda L(t) \left(\frac{1}{\lambda} e + L(t)^{-1}(L(t)^*)^{-1} \right)$$

(observe that $\|p\| \leq \varepsilon$ in both cases). Hence, $\frac{1}{\lambda} e + L(t)^{-1}(L(t)^*)^{-1}$ is invertible for all λ satisfying (2.2). This gives

$$\rho(L(t)^{-1}(L(t)^*)^{-1}) < \|(L(t)^*)^{-1}\| / \varepsilon,$$

whence via the self-adjointness of $L(t)^{-1}(L(t)^*)^{-1}$ (implying the coincidence of spectral radius ρ and norm) follows

$$\|L(t)^{-1}\|^2 = \|L(t)^{-1}(L(t)^*)^{-1}\| < \|(L(t)^*)^{-1}\| / \varepsilon.$$

Finally one has $\|L(t)^{-1}\| = \|(L(t)^*)^{-1}\|$ which yields $\|L(t)^{-1}\| < 1/\varepsilon$ in contrast to $t \in S_1$. This contradiction verifies the inclusion $S_1 \subseteq S_2$.   □

In case $\mathcal{B} = \mathbb{C}^{n \times n}$, Theorem 2.1 has been widely used for numerical computations of ε-pseudospectra by randomly choosing $n \times n$ matrices p with $\|p\| \leq \varepsilon$ and then plotting the eigenvalues of $a + p$ for $a \in \mathbb{C}^{n \times n}$. Beautiful plots of pseudospectra of Toeplitz matrices can be found in [1, 3, 8, 12, 13].

Observe furthermore that if $\mathcal{B}$ is a unital C^*-algebra and $\mathcal{A}$ a C^*-subalgebra of $\mathcal{B}$ which contains the identity, then

$$\sigma_{\mathcal{A}}^{(\varepsilon)}(L) = \sigma_{\mathcal{B}}^{(\varepsilon)}(L) \quad \text{for every polynomial } L \text{ with coefficients in } \mathcal{A},$$

i.e. C^*-algebras are inverse closed with respect to ε-pseudoinvertibility.

3 Limiting Sets of Pseudospectra

The goal of this section is to derive a description of the limiting set $\lim \sigma^{(\varepsilon)}(L_n)$ of the pseudospectra of a given sequence of operator polynomials L_n. Again, it is not essential that the coefficients of the L_n are operators; so assume that $\mathcal{C}_1, \mathcal{C}_2, \ldots$ is a sequence of unital C^*-algebras, and let $\mathcal{F}$ stand for the set of all bounded sequences (c_n) where $c_n \in \mathcal{C}_n$. Further let $m > 0$, and let $(a_n^{(0)}), (a_n^{(1)}), \ldots, (a_n^{(m)})$ be given sequences in $\mathcal{F}$. For every n, consider the polynomial with coefficients in $\mathcal{C}_n$,

(3.1) $L_n(\lambda) := a_n^{(0)} + a_n^{(1)}\lambda + \cdots + a_n^{(m)}\lambda^m.$

The desired description consists in identifying the limiting set $\lim \sigma^{(\varepsilon)}(L_n)$ with the ε-pseudospectrum of a certain element of an accordingly chosen C^*-algebra. A natural choice for that algebra (which also plays a prominent role for other problems in numerical analysis) is as follows. First observe that the set $\mathcal{F}$ forms a C^*-algebra itself when this set is provided with elementwise operations, an elementwise involution, and the supremum norm. Then note that the set of all sequences $(c_n) \in \mathcal{F}$ with $\|c_n\| \to 0$ as $n \to \infty$ is a closed ideal of $\mathcal{F}$ which will be denoted by $\mathcal{G}$. It turns out that the quotient $\mathcal{F}/\mathcal{G}$ is the algebra we looked for. Thus, it should be possible to identify the limiting set of the pseudospectra of the polynomials L_n with the pseudospectrum of some element in $\mathcal{F}/\mathcal{G}$. Indeed, as we will see in a moment, the wanted element is actually the polynomial

$$(3.2) \qquad L(\lambda) := ((a_n^{(0)}) + \mathcal{G}) + ((a_n^{(1)}) + \mathcal{G})\lambda + \cdots + ((a_n^{(m)}) + \mathcal{G})\lambda^m$$

with coefficients in $\mathcal{F}/\mathcal{G}$. The precise statement is in the following theorem. Recall that a sequence $(c_n) \in \mathcal{F}$ is said to be *stable* if there is an n_0 such that all elements c_n with $n \geq n_0$ are invertible and $\sup_{n \geq n_0} \|c_n^{-1}\| < \infty$ or, equivalently, if the coset $(c_n) + \mathcal{G}$ is invertible in the quotient algebra $\mathcal{F}/\mathcal{G}$.

Theorem 3.1 *Let* $m > 0$, *let* $(a_n^{(0)}), (a_n^{(1)}), \ldots, (a_n^{(m)})$ *be sequences in* $\mathcal{F}$ *with* $(a_n^{(m)})$ *being stable, and let* L_n *and* L *be the polynomials* (3.1) *and* (3.2). *Then, for every* $\varepsilon > 0$,

$$\lim \sigma_{C_n}^{(\varepsilon)}(L_n) = \sigma_{\mathcal{F}/\mathcal{G}}^{(\varepsilon)}(L).$$

The proof is based on the following generalization of an observation by Daniluk (see [1] or [9]), concerning the special case $L(\lambda) = b_0 - \lambda e$, which is of its own interest.

Theorem 3.2 *Let* $\mathcal{B}$ *be a* C^*-*algebra with identity* e, *let* $m > 0$, *and let* L *be the polynomial* $L(\lambda) = b_0 + b_1\lambda + \cdots + b_m\lambda^m$ *with coefficients in* $\mathcal{B}$ *where the coefficient* b_m *is supposed to be invertible. Assume further there is an open subset* U *of the complex plane such that* $L(\lambda)$ *is invertible and* $\|L(\lambda)^{-1}\| \leq M$ *for all* $\lambda \in U$. *Then* $\|L(\lambda)^{-1}\| < M$ *for all* $\lambda \in U$.

In other words: the analytic function $U \to \mathcal{B}$, $\lambda \mapsto L(\lambda)^{-1}$ satisfies the maximum principle. This result is quite surprising since – in contrast to complex-valued analytic functions – the maximum principle for operator-valued analytic functions does *not* hold in general as already the simple example

$$\{z \in \mathbb{C} : |z| < 1\} \to \mathbb{C}^{2 \times 2}, \quad \lambda \mapsto \begin{pmatrix} \lambda & 0 \\ 0 & 1 \end{pmatrix}$$

indicates.

Proof: The Proof of Theorem 3.2 will be subdivided into several steps; the third one being identical with Daniluk's original proof.

Step 1 Suppose there exists a $\lambda_0 \in U$ such that $L(\lambda_0) = M$. For the shifted polynomial

$$Q(\lambda) := L(\lambda + \lambda_0) = b_0 + b_1(\lambda + \lambda_0) + \cdots + b_m(\lambda + \lambda_0)^m$$
$$=: a_0 + a_1\lambda + \cdots + a_m\lambda^m,$$

one easily checks that $Q(\lambda)$ is invertible and $\|Q(\lambda)^{-1}\| \leq M$ for all $\lambda \in (U - \lambda_0)$ (= the algebraic difference), and that $\|Q(0)^{-1}\| = M$. Moreover, both $a_m = b_m$ and $a_0 = Q(0)$ are invertible. Further, let $\hat{Q}$ refer to the monic polynomial defined by $Q(\lambda) = \hat{Q}(\lambda)a_m$. Clearly, $\hat{Q}(\lambda)$ is invertible if and only if $Q(\lambda)$ is so.

Step 2 Here some facts concerning a special representation called *linearization* of the inverse of a monic operator polynomial are recalled. For details see [10], Chapter 2, Theorem 2.5.2.

For the desired representation, introduce the following vectors of length m and matrices of order $m \times m$ with entries in the algebra $\mathcal{B}$:

$$X_0 := (e \quad 0 \quad 0 \quad \cdots \quad 0), \qquad I := \text{diag}(e \quad e \quad e \quad \cdots \quad e),$$

$$Y := \begin{pmatrix} 0 \\ 0 \\ \vdots \\ 0 \\ e \end{pmatrix}, \quad T := \begin{pmatrix} 0 & e & 0 & \cdots & 0 \\ 0 & 0 & e & \cdots & 0 \\ 0 & 0 & 0 & \cdots & 0 \\ \vdots & \vdots & \vdots & & \vdots \\ -a_0a_m^{-1} & -a_1a_m^{-1} & -a_2a_m^{-1} & \cdots & -a_{m-1}a_m^{-1} \end{pmatrix},$$

i.e. T is the companion matrix of the monic polynomial $\hat{Q}$. Then, for all λ which are not in the spectrum of $\hat{Q}$,

$$\hat{Q}(\lambda)^{-1} = X_0(\lambda I - T)^{-1}Y.$$

Since $Q(\lambda)^{-1} = a_m^{-1}\hat{Q}(\lambda)^{-1}$, this yields the representation

$$(3.3) \qquad\qquad Q(\lambda)^{-1} = X(\lambda I - T)^{-1}Y$$

with $X = a_m^{-1}X_0 = (a_m^{-1} \quad 0 \quad 0 \quad \cdots \quad 0)$, which holds for all λ such that $Q(\lambda)$ is invertible.

Step 3 Beginning with this step, it will be advantageous to think of $\mathcal{B}$ as an algebra of linear bounded operators acting on some Hilbert space H with scalar product $\langle .,. \rangle$ which is possible due to the GNS construction. Thus, T is actually an operator acting on the orthogonal sum of m exemplaries of H.

Since $Q(0)$ is invertible, the operator T is invertible; hence, for all λ in the disk $|\lambda| \leq r$ with sufficiently small radius r,

$$(T - \lambda I)^{-1} = \sum_{j=0}^{\infty} \lambda^j T^{-j-1},$$

whence

$$X(T - \lambda I)^{-1}Y = \sum_{j=0}^{\infty} \lambda^j X T^{-j-1} Y$$

follows. Thus, for all $f \in H$ and $|\lambda| \leq r$,

$$\|Q(\lambda)^{-1}f\|^2 = \|X(T - \lambda I)^{-1}Yf\|^2$$
$$= \sum_{j,k \geq 0} \lambda^j \overline{\lambda}^k \langle XT^{-j-1}Yf, XT^{-k-1}Yf \rangle.$$

Integrating this identity with respect to λ against the circle $|\lambda| = r$ yields

$$\frac{1}{2\pi r} \int_{|\lambda|=r} \|Q(\lambda)^{-1}f\|^2 \, |d\lambda| = \sum_{j \geq 0} r^{2j} \|XT^{-j-1}Yf\|^2.$$

Because of $\|Q(\lambda)^{-1}f\| \leq M\|f\|$ by hypothesis, this yields for all $j \geq 1$ the estimate

(3.4)
$$\|XT^{-1}Yf\|^2 + r^{2j}\|XT^{-j-1}Yf\|^2$$
$$\leq \sum_{j \geq 0} r^{2j} \|XT^{-j-1}Yf\|^2 \leq M^2\|f\|^2.$$

Let $\varepsilon > 0$ be arbitrarily given. Since, by assumption, $\|Q(0)^{-1}\| = \|XT^{-1}Y\| = M$, there is an f_ε in H with norm 1 such that

$$\|XT^{-1}Yf_\varepsilon\|^2 \geq M^2 - \varepsilon^2,$$

which together with estimate (3.4) shows that $M^2 - \varepsilon^2 + r^{2j}\|XT^{-j-1}Yf_\varepsilon\|^2 \leq M^2$ or

(3.5) $$\|XT^{-j-1}Yf_\varepsilon\| \leq \varepsilon r^{-j} \qquad \text{for all} \quad j \geq 1.$$

Step 4 Set for brevity $c_j = -a_i a_m^{-1}$. The operator c_0 is invertible, and it is easy to check that the inverse of the companion matrix T is a companion matrix again:

$$T^{-1} = \begin{pmatrix} -c_0^{-1}c_1 & -c_0^{-1}c_2 & \cdots & -c_0^{-1}c_{n-1} & c_0^{-1} \\ e & 0 & \cdots & 0 & 0 \\ 0 & e & \cdots & 0 & 0 \\ \vdots & \vdots & & \vdots & \vdots \\ 0 & 0 & \cdots & e & 0 \end{pmatrix}.$$

Computing step by step the last columns of the operator matrices

$$T^{-2},$$
$$T^{-2} + c_0^{-1}c_1 T^{-1},$$
$$T^{-1}(T^{-2} + c_0^{-1}c_1 T^{-1}) = T^{-3} + c_0^{-1}c_1 T^{-2},$$
$$T^{-3} + c_0^{-1}c_1 T^{-2} + c_0^{-1}c_2 T^{-1},$$

one obtains

$$\begin{pmatrix} -c_0^{-1}c_1 c_0^{-1} \\ c_0^{-1} \\ 0 \\ \vdots \\ 0 \end{pmatrix}, \quad \begin{pmatrix} 0 \\ c_0^{-1} \\ 0 \\ \vdots \\ 0 \end{pmatrix}, \quad \begin{pmatrix} -c_0^{-1}c_2 c_0^{-1} \\ 0 \\ c_0^{-1} \\ \vdots \\ 0 \end{pmatrix}, \quad \begin{pmatrix} 0 \\ 0 \\ c_0^{-1} \\ \vdots \\ 0 \end{pmatrix},$$

respectively, what finally shows that the last columns of the matrices

$$T^{-m} + c_0^{-1}c_1 T^{-m+1} + \cdots + c_0^{-1}c_{m-1} T^{-1},$$
$$T^{-m-1} + c_0^{-1}c_1 T^{-m} + \cdots + c_0^{-1}c_{m-1} T^{-2}$$

are as follows:

$$\begin{pmatrix} 0 \\ 0 \\ \vdots \\ 0 \\ c_0^{-1} \end{pmatrix}, \quad \begin{pmatrix} c_0^{-2} \\ 0 \\ \vdots \\ 0 \\ 0 \end{pmatrix}.$$

Consequently,

(3.6)
$$X(T^{-m-1} + c_0^{-1}c_1 T^{-m} + \cdots + c_0^{-1}c_{m-1} T^{-2})$$
$$Y = a_m^{-1} c_0^{-2} = a_0^{-1} a_m a_0^{-1}.$$

Step 5 Identity (3.6) implies that

$$\|a_0^{-1} a_m a_0^{-1} f_\varepsilon\| = \|X(T^{-m-1} + c_0^{-1}c_1 T^{-m} + \cdots + c_0^{-1}c_{m-1} T^{-2})Y f_\varepsilon\|,$$

and since $Xc_0^{-1}c_j = a_m^{-1}c_0^{-1}c_j a_m X$ for all j, one further concludes that

$$\|a_0^{-1} a_m a_0^{-1} f_\varepsilon\|$$
$$= \|(XT^{-m-1}Y + a_m^{-1}c_0^{-1}c_1 a_m X T^{-m} Y$$
$$+ \cdots + a_m^{-1}c_0^{-1}c_{m-1} a_m X T^{-2} Y) f_\varepsilon\|$$
$$\leq \|XT^{-m-1}Y f_\varepsilon\| + \|a_m^{-1}c_0^{-1}c_1\| \cdot \|XT^{-m} Y f_\varepsilon\|$$
$$+ \cdots + \|a_m^{-1}c_0^{-1}c_{m-1} a_m\| \cdot \|XT^{-2}Y f_\varepsilon\|.$$

Together with (3.5) this yields

$$\|a_0^{-1} a_m a_0^{-1} f_\varepsilon\| \leq \left(\frac{1}{r^m} + \frac{a_m^{-1} c_0^{-1} c_1 a_m}{r^{m-1}} + \cdots + \frac{a_m^{-1} c_0^{-1} c_{m-1} a_m}{r} \right) \varepsilon =: C\varepsilon$$

with a constant C depending on the polynomial Q and the radius r only. Letting ε go to zero in

$$1 = \|f_\varepsilon\| \leq \|a_0 a_m^{-1} a_0\| \, \|a_0^{-1} a_m a_0^{-1} f_\varepsilon\| \leq C \|a_0 a_m^{-1} a_0\| \, \varepsilon$$

one arrives at a contradiction. $\square$

Having this result at disposal, one can prove Theorem 3.1 essentially by repeating arguments from Böttcher's paper [1] (although he did not mention the corresponding special case of Theorem 3.1).

Proof: (Theorem 3.1). Let $\lambda \in \sigma_{\mathcal{F}/\mathcal{G}}^{(\varepsilon)}(L)$. Then either $L(\lambda)$ is not invertible in $\mathcal{F}/\mathcal{G}$, or the inverse of this coset exists, but $\|L(\lambda)^{-1}\| \geq 1/\varepsilon$.

In the first case, the sequence

$$(a_n^{(0)} - a_n^{(1)} \lambda + \cdots + a_n^{(m)} \lambda^m)_{n \geq 1} = (L_n(\lambda))_{n \geq 1}$$

fails to be stable whence easily follows that there is an infinite subsequence (n_k) of $\mathbb{N}$ such that $\lambda \in \sigma_{C_{n_k}}^{(\varepsilon)}(L_{n_k})$ for every k. Clearly, $\lambda \in \lim \sigma^{(\varepsilon)}(L_n)$ in this case.

In the second case, Theorem 3.2 implies that, in every open neighbourhood U of λ, there is an λ_0 such that $\|L(\lambda_0)^{-1}\| > 1/\varepsilon$ (otherwise one would have $\|L(\lambda_0)-1\| \leq 1/\varepsilon$ for all $\lambda_0 \in U$ which would imply via Theorem 3.2 that $\|L(\lambda_0)^{-1}\|$ is strictly less than $1/\varepsilon$ for all $\lambda_0 \in U$ including $\lambda_0 = \lambda$.)

Thus, for all sufficiently large k, there are numbers λ_k with $\lambda_k \to \lambda$ as $k \to \infty$ such that

$$\|L(\lambda_k)^{-1}\| \geq 1/(\varepsilon - 1/k).$$

It is an easy exercise (see also [1]) to check that this is equivalent to

$$\limsup_{n \to \infty} \|L_n(\lambda_k)^{-1}\| \geq 1/(\varepsilon - 1/k)$$

(with the invertibility of $L_n(\lambda_k)$ for all sufficiently large n being a consequence of a Neumann series argument). Since $1/\varepsilon < 1/(\varepsilon - 1/k)$, there are numbers n_k tending to infinity as $k \to \infty$ such that

$$\|L_{n_k}(\lambda_k)^{-1}\| \geq 1/\varepsilon \quad \text{for all } k;$$

in other words, $\lambda_k \in \sigma^{(\varepsilon)}(L_{n_k})$ for every sufficiently large k, which implies that $\lambda = \lim \lambda_k$ belongs $\lim \sigma^{(\varepsilon)}(L_n)$.

For the reverse inclusion, let $\lambda \in \lim \sigma^{(\varepsilon)}(L_n)$, but assume for contrary that $\lambda \notin \sigma_{\mathcal{F}/\mathcal{G}}^{(\varepsilon)}(L)$. Thus, $L(\lambda)$ is invertible in $\mathcal{F}/\mathcal{G}$ and $\|L(\lambda)^{-1}\| = 1/\varepsilon - 2\delta < 1/\varepsilon$ with some $\delta > 0$. As already mentioned, then the $L_n(\lambda)$ are invertible for sufficiently large n and, by the lim sup formula for the norm in $\mathcal{F}/\mathcal{G}$ again,

$$\limsup_{n \to \infty} \| L_n(\lambda)^{-1} \| = 1/\varepsilon - 2\delta$$

whence follows that $\| L_n(\lambda)^{-1} \| < 1/\varepsilon - \delta$ for n sufficiently large, say $n \geq n_0$.

The uniform boundedness of the coefficient sequences $(a_n^{(1)})$, $(a_n^{(2)})$, ... $(a_n^{(m)})$ implies the existence of a constant C independent of n such that

$$\| L_n(\lambda) - L_n(\mu) \| \leq C |\lambda - \mu| \quad \text{for all } \mu \in U_1(\lambda).$$

Hence, there exists an open neighbourhood $U \subseteq U_1(\lambda)$ which is independent of n such that

$$\| L_n(\lambda) - L_n(\mu) \| \leq C |\lambda - \mu| \leq \frac{\varepsilon \delta}{1/\varepsilon - \delta} \quad \text{for all } \mu \in U.$$

A Neumann series argument shows that then, for all $\mu \in U$,

$$
\begin{aligned}
\| L_n(\mu)^{-1} \| &= \| L_n(\lambda)^{-1} (e - (L_n(\lambda) - L_n(\mu)) L_n(\lambda)^{-1})^{-1} \| \\
&\leq \frac{\| L_n(\lambda)^{-1} \|}{1 - \| L_n(\lambda) - L_n(\mu) \| \, \| L_n(\lambda)^{-1} \|} \\
&< \frac{1/\varepsilon - \delta}{1 - \varepsilon\delta(1/\varepsilon - \delta)^{-1}(1/\varepsilon - \delta)} = \frac{1}{\varepsilon},
\end{aligned}
$$

hence $\mu \notin \sigma^{(\varepsilon)}(L_n)$ for all μ belonging to a certain open neighbourhood of λ and for all sufficiently large n. But then λ cannot belong to the limiting set of the ε-pseudospectra $\sigma^{(\varepsilon)}(L_n)$ which is a contradiction. $\qquad\square$

4 Finite Sections of Operator Polynomials

In this section, the usefulness of Theorem 3.1 for establishing generalizations of (1.1) (as well as of similar results) to operator polynomials will be illustrated. The crucial point is that there are a lot of C^*-subalgebras of $\mathcal{F}/\mathcal{G}$ the structure of which is sufficiently well understood (see, e.g., [?] for algebras arising from spline approximation sequences for singular integral and Mellin operators). To mention at least one non-trivial (but well known) example, consider again the finite section method for Toeplitz operators, say with piecewise continuous coefficients (where piecewise continuity of a function f on the unit circle means that both one-sided limits of f exist and are finite at every point of $\mathbb{T}$). That is, the algebra $\mathcal{F}$ is specified accordingly by choosing $C_n = \mathbb{C}^{n \times n}$, and one considers the smallest closed subalgebra $\mathcal{S}(PC)$ of $\mathcal{F}$ which contains all finite section sequences $(T_n(f))_{n \geq 1}$

with f being piecewise continuous. The algebra $S(PC)$ contains the ideal G (see [2]), thus it makes sense to consider the quotient algebra $S(PC)/G$.

One property which belongs to every sequence (A_n) in $S(PC)$ is the existence of two associated strong limits: If P_n and R_n stand for the projection and reflection operators

$$P_n : l^2 \to l^2, \quad (x_1, x_2, \ldots) \mapsto (x_1, x_2, \ldots, x_n, 0, 0, \ldots),$$

$$R_n : l^2 \to l^2, \quad (x_1, x_2, \ldots) \mapsto (x_n, x_{n-1}, \ldots, x_1, 0, 0, \ldots),$$

respectively, then the strong limits

$$W(A_n) := \text{s-} \lim_{n \to \infty} A_n P_n \quad \text{and} \quad \tilde{W}(A_n) := \text{s-} \lim_{n \to \infty} R_n A_n R_n$$

exist for every sequence $(A_n) \in S(PC)$. This is evident for the generating sequences $(T_n(f)) = (P_n T(f) P_n)$ of $S(PC)$ since $P_n \to I$ strongly as $n \to \infty$ and since $R_n T_n(f) R_n = P_n T(\tilde{f}) P_n$ with $\tilde{f}(t) = f(1/t)$, and it follows easily for arbitrary sequences by taking into account that the mappings W and $\tilde{W}$ are continuous algebra homomorphisms. Clearly, the operators $W(A_n)$ and $\tilde{W}(A_n)$ depend on the coset $(A_n) + G$ only. The following result due to Silbermann [11] describes the quotient algebra $S(PC)/G$ completely by means of the homomorphisms $W, \tilde{W} : S(PC)/G \to L(l^2)$.

Theorem 4.1 *The mapping*

$$\text{smb} : (A_n) + G \mapsto (W(A_n), \tilde{W}(A_n))$$

*is a *-isomorphism from the C^*-algebra $S(PC)/G$ into the C^*-algebra $L(l^2) \times L(l^2)$.*

Let now $f_0, f_1, \ldots, f_m$ be piecewise continuous functions and consider the matrix resp. operator polynomials

$$L_n(\lambda) := T_n(f_0) + T_n(f_1)\lambda + \cdots + T_n(f_m)\lambda^m,$$
$$P(\lambda) := T(f_0) + T(f_1)\lambda + \cdots + T(f_m)\lambda^m.$$

Theorem 4.2 *Let $m > 0$ and the functions f_j as well as the polynomials L_n and P be as above, and suppose the Toeplitz operator $T(f_m)$ to be invertible. Then, for every $\varepsilon > 0$,*

$$\lim \sigma^{(\varepsilon)}(L_n) = \sigma^{(\varepsilon)}(P).$$

Proof: Let L stand for the polynomial

$$L(\lambda) := ((T_n(f_0)) + \mathcal{G}) + ((T_n(f_1)) + \mathcal{G})\lambda + \cdots + ((T_n(f_m)) + \mathcal{G})\lambda^m.$$

Due to Theorem 3.1,

$$\lim \sigma^{(\varepsilon)}(L_n) = \sigma^{(\varepsilon)}_{\mathcal{F}/\mathcal{G}}(L)$$

and, since C^*-algebras are inverse closed with respect to ε-pseudo-invertibility, one even has

$$\lim \sigma^{(\varepsilon)}(L_n) = \sigma^{(\varepsilon)}_{S(PC)/\mathcal{G}}(L).$$

Now apply the *-isomorphism smb from Theorem 4.1 to obtain

$$\lim \sigma^{(\varepsilon)}(L_n) = \sigma^{(\varepsilon)}_{L(l^2) \times L(l^2)}(\text{smb } L).$$

For the image of L under the homomorphisms W and $\tilde{W}$ one easily finds

$$W(L(\lambda)) = T(f_0) + T(f_1)\lambda + \cdots + T(f_m)\lambda^m = P(\lambda),$$

$$\tilde{W}(L(\lambda)) = T(\tilde{f}_0) + T(\tilde{f}_1)\lambda + \cdots + T(\tilde{f}_m)\lambda^m =: \tilde{P}(\lambda)$$

where again $\tilde{f}(t) := f(1/t)$, which yields

$$\lim \sigma^{(\varepsilon)}(L_n) = \sigma^{(\varepsilon)}_{L(l^2)}(P) \cup \sigma^{(\varepsilon)}_{L(l^2)}(\tilde{P}).$$

Finally, if C refers to the operator of conjugation,

$$C : l^2 \to l^2, \quad (x_1, x_2, \ldots) \mapsto (\overline{x_1}, \overline{x_2}, \ldots)$$

then it is elementary to check that the identity

$$T(\tilde{f}) = CT(f)^*C$$

holds even for arbitrary $f \in L^\infty(\mathbb{T})$. Thus,

$$\begin{aligned}
\tilde{P}(\lambda) &= CT(f_0)^*C + CT(f_1)^*C\lambda + \cdots + CT(f_m)^*C\lambda^m \\
&= C(T(f_0) + T(f_1)\lambda + \cdots + T(f_m)\lambda^m)^*C = CP(\lambda)^*C
\end{aligned}$$

and, hence, $\sigma^{(\varepsilon)}(P) = \sigma^{(\varepsilon)}(\tilde{P})$. $\square$

Observe that a similar assertion (with obvious modifications in the proof) holds for operator polynomials whose coefficients are arbitrary sequences in $\mathcal{S}(PC)$. Moreover, one can derive analogous results in any case where a subalgebra of $\mathcal{F}/\mathcal{G}$ is sufficiently well understood. Finally, the result of Theorem 4.2 can be refined by considering besides the partial limiting set $\lim \sigma^{(\varepsilon)}(L_n)$ the corresponding uniform limiting set $\operatorname{Lim} \sigma^{(\varepsilon)}(L_n)$. By definition, the *uniform* limiting set $\operatorname{Lim} M_n$ of a sequence of subsets $M_n \subseteq \mathbb{C}$ consists of all (uniform) limits of sequences (m_n) of points $m_n \in M_n$.

Theorem 4.3 *Let the hypothesis be as in the preceding theorem. Then*

$$\lim \sigma^{(\varepsilon)}(L_n) = \operatorname{Lim} \sigma^{(\varepsilon)}(L_n).$$

The point is that the algebra $\mathcal{S}(PC)$ is *fractal* in the sense of [9] which roughly speaking means that, given a subsequence $(A_{n_k})_{k \geq 1}$ of a sequence $(A_n) \in \mathcal{S}(PC)$, it is possible to reconstruct the complete sequence (A_n) up to sequences in the ideal $\mathcal{G}$. In the present situation, this fractality is a simple consequence of Theorem 4.1: If a subsequence (A_{n_k}) is known, then it is already possible to compute the strong limits $W(A_n)$ and $\tilde{W}(A_n)$, which then determine the coset $(A_n) + \mathcal{G}$ uniquely. It is easy to check that it is this fractality which makes several limiting processes (such as that one considered above) *uniform*. For more details see [9] and also the textbook [5] which will be finished soon.

References

[1]   A. Böttcher, Pseudospectra and singular values of large convolution operators, *J. Integral Equations and Appl.* **6** (1994), 267–301.

[2]   A. Böttcher and B. Silbermann, The finite section method for Toeplitz operators on the quarter-plane with piecewise continuous symbols, *Math. Nachr.* **110** (1983), 279–291.

[3]   A. Böttcher and B. Silbermann, *Introduction to Large Truncated Toeplitz Matrices*, Springer-Verlag, New York, Berlin, Heidelberg 1999.

[4]   R. Hagen, S. Roch and B. Silbermann, *Spectral Theory of Approximation Methods for Convolution Equations*, Birkhäuser Verlag, Basel, Boston, Berlin 1995.

[5]   R. Hagen, S. Roch and B. Silbermann, *C^*-algebras and Numerical Analysis*, Textbook in preparation.

[6]   H. Landau, On Szegö's eigenvalue distribution theorem and non-Hermitian kernels, *J. Anal. Math.* **28** (1975), 335–357.

[7]   H. Landau, The notion of approximate eigenvalues applied to an integral equation of laser theory, *Q. Appl. Math.*, 1977, 165–171.

[8]   L. Reichel and L.N. Trefethen, Eigenvalues and pseudo-eigenvalues of Toeplitz matrices, *Linear Algebra Appl.* **162** (1992), 153–185.

[9]   S. Roch and B. Silbermann, C^*-algebra techniques in numerical analysis, *J. Oper. Theory* **35** (1996), 241–280.

[10]   L. Rodman, *An Introduction to Operator Polynomials*, Birkhäuser Verlag, Basel, Boston, Berlin 1989.

[11]   B. Silbermann, Symbol constructions in numerical analysis, In: *Petkov, V., Lazarov, R. (Eds.): Integral equations and inverse problems, Pitman Research Notes in Mathematics Series* **235** (1991), 241–252.

[12]   L.N. Trefethen, Pseudospectra of matrices, In: *D.F. Griffiths, G.A. Watson (Eds.), Numerical Analysis 1991, Longman*, 1992, 234–266.

[13]   L.N. Trefethen, *Non-normal matrices and pseudospectra*, Monograph in preparation.

Steffen Roch
Technische Universität Darmstadt
Fachbereich Mathematik
Schlossgartenstrasse 7
D - 64289 Darmstadt
FRG

1991 Mathematics Subject Classification. Primary: 47A10; Secondary: 47B35, 65R20.